THE NORTON HISTORY OF
THE HUMAN SCIENCES

Roger Smith teaches the history of science and intellectual history at Lancaster University. His research and publications concern the history of scientific belief about human nature – especially the relations between mind and brain. He is also interested in literature and in the similarities and contrasts between different European cultures. He was President of the European Society for the History of the Behavioral and Social Sciences, 1987–92.

His books include *Trial by Medicine: Insanity and Responsibility in Victorian Trials* (1981) and *Inhibition: History and Meaning in the Sciences of Mind and Brain* (1992).

NORTON HISTORY OF SCIENCE
GENERAL EDITOR: ROY PORTER

already published

Environmental Sciences PETER J. BOWLER
Chemistry W. H. BROCK
Technology DONALD CARDWELL
Astronomy and Cosmology JOHN NORTH

forthcoming

Physics RUSSELL MACCORMMACH
Biology ROBERT OLBY
Medicine ROY PORTER
Science in Society LEWIS PYENSON &
SUSAN SHEETS-PYENSON

NORTON HISTORY OF SCIENCE
(General Editor: Roy Porter)

THE NORTON HISTORY OF
THE HUMAN SCIENCES

Roger Smith

W. W. NORTON & COMPANY

New York London

The author and publishers are grateful to the following for permission to reproduce illustrative material: Plate 1, Wellcome Institute Library, London; Plate 2, J. B. de C. M. Saunders and C. D. O'Malley, *The Illustrations from the Works of Andreas Vesalius of Brussels* (Cleveland: World Publishing Company, 1950), plate 68:2.; Plate 3, R. Descartes, *Treatise of Man*, trans. and commentary by T. S. Hall (Cambridge, MA: Harvard University Press, 1972), figure 50; Plate 4, Wellcome Institute Library, London; Plate 5, Wellcome Institute Library, London; Plate 6, C. F. A. Marmoy, 'The Auto-icon of Jeremy Bentham at University College, London', *Medical History*, **2** *(1958): 77–86, figure 2, Wellcome Institute Library, London; Plate 7, J. C. Lavater, Physiognomik. Zur Beförderung der Menschenkenntniß und Menschenliebe* (new edn, Vienna: J. P. Sollinger, 1829), Vol. 2, Sect. XVI, figs. 16–18, Lancaster University Library; Plate 8, Wellcome Institute Library, London; Plate 9, I. M. Sechenov, *Selected Works* (Leningrad-Moscow: XVth International Physiological Congress, 1935); Plate 10, F. Galton, *Inquiries into Human Faculty and Its Development* (2nd edn, London: J. M. Dent & Co., n.d); Plate 11, Wellcome Institute Library, London; Plate 12, Wellcome Institute Library, London; Plate 13, Wellcome Institute Library, London; Plate 14, Wellcome Institute Library, London; Plate 15, Wellcome Institute Library, London; Plate 16, Wellcome Institute Library, London.

First American edition 1997
Published by arrangement with HarperCollins Publishers Ltd

Originally published in England under the title *The Fontana History of the Human Sciences*

Printed in the United States of America

The text of this book is composed in Meridien
Composition by Rowland Phototypesetting Ltd.
Manufacturing by The Haddon Craftsmen, Inc.

Library of Congress Cataloging-in-Publication Data

Smith, Roger, 1945–
The Norton history of the human sciences / Roger Smith. — 1st American ed.
p. cm. — (Norton history of science)
Originally published: The Fontana history of the human sciences. London : Fontana, 1997
Includes bibliographical references and index.
ISBN 0-393-04543-9.—ISBN 0-393-31733-1 (pbk.)
1. Social sciences—History. I. Title. II. Series.
H51.S56 1997
300'.9—dc21 97-22287
CIP

W. W. Norton & Company, Inc., 500 Fifth Avenue, New York, N.Y. 10110
http://www.wwnorton.com

W. W. Norton & Company Ltd., 10 Coptic Street, London WC1A 1PU

1 2 3 4 5 6 7 8 9 0

CONTENTS

LIST OF PLATES

PREFACE

The human sciences matter. They speak about our interests, community, sense of self-identity and purpose. They seek to tell us who and what we are. They provide knowledge about how to live better, and sometimes they even tell us directly how we should live. At the same time, the sciences of our own nature are a labyrinth; they cover a vastly complex subject matter that is divided up in ill-defined ways; they exhibit disagreement about concepts, theories and ways of research. And, in many quarters, these sciences are thought to be 'soft' – to lack the status and authority of the natural sciences – though they are an ever-present part of the modern world, indeed of the modern conception of what it is to be human.

The history of the human sciences is the part of the history of science whose subject is 'Man'. There is a deeply intriguing history of how scholars, intellectuals and scientists have created knowledge about men and women and the human world. The way we picture ourselves affects the way we live, and the way we live enters into the knowledge we form about ourselves. To write this history is to write about ideas that shape the life of people in the past and the present. Historical knowledge is part of what we seek when we desire to know 'the nature' of being human.

The field of the human sciences is huge – as large as human life. I focus on what we call psychological knowledge, as it has existed in the West since the fifteenth century. But this history is linked to studies of the body, the social world, philosophy, language, political thought, economics and, indeed, the human spirit. Historians and other scholars have published much on the human sciences, but it is scattered or the preserve of specialists. This book casts the work that has been done into a connected narrative (though individual chapters may be read separately), and it seeks to be an accessible introduction to the field for the general reader, students and non-specialists. I take a broad approach, inclusive of the variety of views characteristic of the human sciences and the great range of settings in which knowledge has existed; for my subject straddles the modern divisions between the natural sciences, the social sciences and the humanities. The book gives much historical information in order to achieve an overview of ideas of human nature, and the ideas are set in the contexts in which they were expressed and debated.

To write on a large scale requires selection and ordering. This process necessarily involves the adoption of a position in relation to current debates in the human sciences. I try to make my position clear and not to hide personal judgements behind academic conventions. The introductory chapter describes this position, outlines the book's content and explains my approach to the history of the human sciences. There is great scope for discussion and difference of view. I often venture my own interpretations, in points of detail as well as on a larger scale. The work represents a dialogue between my reading and the present state of scholarship. But two purposes intrinsic to the book and the series of which it is part should be borne in mind: to integrate information and interpretation across a very wide area; and to contribute to the *history* of science.

The latter point requires emphasis. Many different kinds of scientists and academics lay claim to privileged status, perhaps a position of leadership, in the domain of the human sciences. Some are not interested in, even devalue, historical knowledge. More deeply, current opinion questions the view, which historians are too often thought to hold, that historical knowledge is simply knowledge of 'the past' unrelated to the present. This book is written in the conviction that historical knowledge is a valid form of knowledge, that, in part, it enters into what we accept as knowledge in the present and that it is essential to the full development of the human sciences themselves. I write history as an exciting and indispensable way to make our human world intelligible.

The book aims to be international and comparative. It goes

further in this direction than has been common in the English-speaking world. Nevertheless, much more needs to be done. In particular, though the bibliographic essay is wide-ranging – and may prove a helpful resource in its own right – it includes only English-language sources.

A book that covers so broad a field is especially liable to error, though I have tried to read widely in the primary sources, arrive at my own interpretations, draw accurately on the expertise of others and check facts and judgements. I take responsibility for the errors that occur and will be glad to be put right. Where specialists find factual errors, or see gaps in what I have said about their favourite areas, I hope they will also appreciate the purposes of a larger picture.

It has taken time to write a book that crosses modern disciplinary boundaries and presumptions about what is science and not science, or what is history, or what is philosophy. I would not have written the book without the privilege of continuous academic employment in an environment that supports scholarship, over twenty-five years. My first great debt is therefore to my colleagues and students and to Lancaster University. My second debt is equally large: to all those scholars whose work I have drawn on to write this history. It has been a large part of my business to use what others have done; what I say and the bibliographic essay make clear where my debts lie. I feel special affection for work which opened up for me new theoretical perspectives or previously unknown areas of knowledge.

Many scholars have helped in particular ways. It is also a pleasure to acknowledge some special debts. Stephen Pumfrey was a much valued commentator on my efforts to deal with the difficult Renaissance and early-modern period; John Soyland and other scholars made helpful comments on parts of the twentieth-century chapters. Robert M. Young will recognize the flowering of seeds of ideas planted a quarter of a century ago, however much these ideas now appear in a different light. Roy Porter patiently stood by me and sustained enthusiasm and suggestions during the writing. At different times, Sonja Bradshaw and Ghil O'Neill contributed to the word-processing. My copy-editor, Betty Palmer, gave invaluable and expert help. The completion of the book was made possible by periods of leave from Lancaster University; by a Wellcome Trust Research Leave Fellowship (1991–2); by the Waern Guest Professorship at the university of Göteborg (1993–4); and by a Royal Society Grant in the History of Science (to visit libraries for work on the bibliographic essay). The book simply would not exist without these opportunities, and I am deeply grateful. I would especially like to acknowledge the support in writing a book of this nature given to me by Ingemar Nilsson and by students and staff of the Institutionen för idé- och lärdomshistoria (Department for History of Science and Ideas), Göteborgs Universitet. Friendship and information from participants in the meetings of Cheiron: The European Society for the History of the Behavioural and Social Sciences, now The European Society for the History of the Human Sciences, have greatly enlarged my perspective on the history of science. To these people and institutions, and to all those other scholars who have given help or stimulus, I extend my thanks.

Roger Smith
September 1996

I

INTRODUCTION

1

The History of the Human Sciences

> The mind nobly expands, when it is able to emerge from the narrow circle, which climate and education have drawn round it . . . Numberless ideas, which we have often admitted as the most general principles of the human understanding, disappear, in this place and that, with the climate, as the land vanishes like a mist from the eye of the navigator . . . Thus we wander over the Earth in a labyrinth of human fancies: but the question is; where is the central point of the labyrinth, to which all our wanderings may be traced, as refracted rays to the Sun?
>
> Johann Gottfried Herder, *Ideen zur Philosophie der Geschichte der Menschheit* (*Reflections on the Philosophy of the History of Mankind*, 1784–91)[1]

i *Introduction*

How could anyone avoid having views about what it is to be human? We are all curious about human nature and, even if unselfconsciously, use a language full of assumptions and speculation about it. People of different generations, women and men, different classes and occupations, and people of different ethnic identity, religion and nationality think and act about human nature in both comparable and contrasting ways. The human sciences attempt to make sense of all this, aiming to establish objective knowledge of what human nature is and to explain why people, including scientists, believe what they do about it. Since ordinary people as well as scientists have strong views about our nature, about being human, the human sciences connect directly with the everyday world we all inhabit.

What are the human sciences? While they all in some sense have men and women as their subject, no one definition is in fact possible. There is disagreement both about whether to describe different kinds of study as science, and about what to be human really is. The area of the human sciences certainly includes psychology, sociology, anthropology, linguistics, economics and political science. Disciplines like history, geography, jurisprudence, business management, literary criticism and art history are also possible, but more controversial, contenders. In this book, the term is firstly a label of convenience. It is an anachronistic label, one not used at the time, when applied to any area of scholarship before the twentieth century. It is used here in relation to earlier centuries in order not to prejudge what knowledge in the past did or did not contribute to the sciences of human nature. We need to be open to the possible contributions legal thought, rhetoric or theology, for example, made to science. If we look in the past for early psychology or early sociology we would miss the contribution made by other subjects. When it comes to the twentieth century, the book's reference to the human sciences and not simply to psychology and the other modern disciplines indicates that one particular view about what scientific knowledge is – that it is natural science knowledge like physics or biology – will not be taken for granted. The term 'the human sciences' is therefore one of convenience; but it also implies that questions are on the agenda about what *is* a science of being human. There is no agreement, this book suggests, either in the past or in the twentieth century, about what a human science should be like.

The book runs in a broadly chronological sequence from the Renaissance of the fifteenth and sixteenth centuries to the present, though the arrangement is primarily thematic, with the consequence that any one chapter may move back and forth over a considerable time span. The chapters discuss separable topics, though they also interrelate; while chapters can be read independently, there are also themes that develop throughout the book.

Chapters 2, 3 and 4 cover the sixteenth and seventeenth centuries. They discuss a mixture of ancient inheritance and novel thought about the soul, the self, the personal body and the political body, and the idea of natural law in human affairs. Chapter 4 also relates these matters to other developments in the sciences, that is, to what is often referred to as the scientific revolution. Chapters 5 to 10 show how these ideas changed in the long eighteenth century – the century and a half that began about 1660 and included the period historians call the Enlightenment. These chapters contain different levels of detail – there is specific argument about John Locke but general argument about notions of the social world underlying political economy. Each chapter helps build up a picture of belief about the core concept of human nature.

Chapters 11 to 15 carry the discussion through the nineteenth century. This period was dominated by the historical and evolutionary imagination, an imagination that sought to understand ourselves by writing the history of how men and women have become what they are. The preceding chapter 10 traces sources for this historical approach back into the eighteenth century and to the reaction against its beliefs. At the end of the nineteenth century, as discussed in chapters 14 and 15, the modern human science disciplines like psychology and sociology acquired an institutional identity. Before this time, there were psychological, sociological or other human science ways of thought – in the limited senses which this book describes – but not the human science subjects found in the twentieth century.

Chapters 16 to 20 take the history through the twentieth century. In these chapters it is no longer possible to describe the full range of the human sciences. Instead, I focus on one discipline, psychology, the science central to modern theories of human nature. Even this one discipline, however, is exceedingly diverse and its history is marked by contests about how to explain human action. These chapters show how scientific knowledge correlates in revealing ways with the cultural and political upheavals that scored the century. In addition, even though psychology became a very large academic discipline and a very large expert occupation – in 1992 there were 118,200 fellows, members, affiliates and students of the American Psychological Association – there remained deep disagreement about whether and in what sense it was a science. The final chapter brings the history into the late twentieth century and explores the continuing debate about scientific views of human nature.

The potential scope of a history of the human sciences is very wide indeed, and it merges without clear boundaries into the history of ideas in general. I therefore keep the concept of human nature in mind. This leads to a concentration on topics that, in modern terms, are psychological rather than sociological, on the capacities and actions attributed to men and women rather than the properties of language, culture, economy, society and the state. Chapter 4 on Descartes, chapter 5 on Locke, chapter 7 on the Enlightenment and chapter 14 on the psychology discipline, as well as the chapters on the twentieth century, are concretely about what we recognize as psychological theories. At many points, however, it is imperative also to describe how belief about human nature incorporated ideas and values that concern language, culture, economy, society and the state. Chapter 3 describes early modern natural law theories of the state, chapter 8 the social thought of the eighteenth century, chapter 9 the science of political economy and chapter 12 the historical worldviews of Auguste Comte and Karl Marx. In each case there were significant connections between social thought and what was claimed to be an attribute of individual human beings. Chapter 15 specifically describes the nineteenth-century origins of sociology as an academic discipline. Even in the chapters on the twentieth century, which focus only on psychology, the social dimension is present; in particular, chapter 19 discusses social psychology and the relations between the science of the individual and the science of society.

The history of scientific belief about human nature is surpassingly rich. It includes many authors of unquestionable standing in Western intellectual life, like Descartes, Locke, Marx, Darwin and Freud. It encompasses themes of great moral or spiritual significance: the nature of the soul, the fate of the Enlightenment project to found a science of man, the meaning of evolutionary thought and the exploration of the unconscious and the irrational in human existence. The book provides a systematic guide to what late twentieth-century historians conclude about these mainstream authors and themes. At the same time, it touches on many lesser-known topics, or topics which earlier historians of science considered marginal, and it thus puts famous names and themes in a wider context and enables them to be understood in more historical ways.

A summary might look like this. When scholarly and cultural life flourished in sixteenth-century Europe, Christian ideas of man's nature – ideas in turmoil following the Protestant Reformation – were enriched by a reworking of classical culture. Concern about the political, theological and intellectual division of Europe, conditions which fostered scepticism, led after 1600 to a search to give knowledge and the practical life new foundations, to provide rational authority for an understanding of man's nature and calling. In this context, men like the political and legal theorist Hugo Grotius discussed man as subject to natural law; the philosopher Descartes concluded that man is uniquely of a dual nature, reflective soul and mechanical body; intensely independent thinkers like Thomas Hobbes and Benedict Spinoza unified all nature, man included, under a common causal determinism; and Locke analysed the sources of knowledge in experience. These authors created a substantial and profound literature on the subject, 'man', as a subject that was later studied in ways driven primarily by a secular rather than a theological interest. Their work interacted with many other preoccupations in the early modern world, like medicine's longstanding interest in the body and the Europeans' wide-eyed confrontation with what they called the New World. All this formed the resources which the élite of eighteenth-century intellectual culture, figures like the Scotsman David Hume and the French nobleman the baron de Montesquieu, utilized in a search for what they called a science of man. Deeply stimulated by achievements in natural philosophy, personified in Isaac Newton, they sought to transform moral philosophy, the study of man, in as impressive a manner. One consequence was what critics branded materialism, knowledge that implies that human beings exist only as part of material nature. Concepts characteristic of modern human self-understanding, which related bodily individuals to their position as social actors, became common. A new language for economic activity or for morality as the pursuit of happiness explained man's social life and laid the basis for detailed studies of the different aspects of human activity in the world. Alongside such new and, to conservative opinion, disturbing developments, there were also arguments – elaborated mainly in the German-speaking world – which viewed man's nature as the history of the rational spirit, a linguistic and cultural rather than physical phenomenon. This view of man's spirit was strongly reinforced and made much more personal in the reaction against Enlightenment values and what many people believed to be their consequence, the French Revolution of 1789. Western authors constructed the subject matter of modern psychology when they fashioned the modern experience of the subjective self, the feeling and autonomous individual. There were elements of this change in Descartes' thinking soul, in Adam Smith's assumptions about economic man, and in Maine de Biran's writings on the reflective *moi*, the 'I'.

Learning in general was transformed in the nineteenth century in the reformed German universities, which created models of academic life later followed elsewhere. There was an increasingly specialized pursuit of systematic knowledge and a concern with the objectivity of scholarly methods. Newly rigorous and distinct disciplines like history, philology, ethnology, experimental physiology and historical economics contributed formal thought about human nature and human action. Historical thought, in both Marx's economic version and Charles Darwin's biological version, became central to the search to explain what man is. Science itself, in which the physical sciences appeared to take the lead, acquired cultural authority in the West. To its most ardent advocates, science was a unique source of truth about the world, the vanguard and the essence of human progress. Many arguments, with the theory of evolution the most prominent, pointed towards the continuity of man and nature, raising hopes or fears – depending on point of view – that science was now about to encompass the human sphere. Those who saw this as a hope, and they were supported institutionally by the expansion of education, pushed at the end of the century for the development of the modern human science disciplines like psychology, sociology, economics and political science.

Yet the human sciences did not develop simply as bodies of knowledge. With industrialization, evident in Britain from about 1780, the shift before 1914 of the majority of the population in Western Europe and North America from a rural to an urban existence, and new forms of government, social administration, education and management of daily life, there was a reconstitution of what it is to be an individual and a social agent. Much of the new activity in the human sciences developed in conjunction with these changes, and it thus served the new ways of life through the provision of techniques to sustain order, purpose and identity. This interdependence was very evident in the twentieth century, when the human sciences flourished on a large scale, first in the United States and then, after 1945, in Europe and the wider world. On the one hand, these sciences existed as the occupations of specialized academics, like experimental psychologists or linguists, and expert practitioners, like educational psychologists or political analysts, with all the professional organization of institutions, journals, technical language and esoteric training generally associated with science. On the other hand, since ordinary people provided these sciences with their subject matter, the human sciences existed in a circle of interactions between science and ordinary life, a circle in which they influenced and were influenced by popular culture. The story of the modern human sciences is therefore bound up with the story of modern Western life itself. Scientists and lay people alike refer to the mind and not to the soul (though 'soul' has its own cultural resonance), to personality rather than to virtue; and they discuss society, the economic sphere, differences between individual abilities, between ethnic groups or between men and women or between childhood and adolescence, the power of language, the brain as a computer. In all these topics it is very unclear where to locate the boundaries of science. They all raise questions about our nature. What are we and how are we to be explained? As Victor Frankenstein's creature, the monster to whom he gave life but no name, exclaimed: 'What was I?'[2] The human sciences exist as our means to find answers in systematic, objective and rigorous ways.

ii *Ambivalence and Diversity in the Human Sciences*

One theme of this book is that people are special, in that they reflect and comment on what they do as well as doing it. As Hume observed, 'we ourselves are not only the beings, that reason, but also one of the objects, concerning which we reason'.[3] The collective reflection and commentary of people through language and symbols is called culture. Physical events do not create a culture, and, indeed, nature and culture are often paired as opposite terms. Much of the history of the human sciences pivots around this opposition and how it has been understood. At one extreme of opinion, as the twentieth-century Spanish philosopher, José Ortega y Gasset, expressed it: 'Man has no nature; what he has is history . . .'[4] By contrast, many modern scientists believe that the exciting moments are those when it was shown how culture is continuous with and not in opposition to nature. It is contentious to claim that knowledge about human beings differs from knowledge about physical events, and therefore to claim that the history of knowledge in the human sciences differs from the history of knowledge in the physical sciences. This book traces the argument about how the human and the natural sciences have related; it does not presuppose that the human sciences are natural sciences, though many scientists have thought they are.

The underlying questions are as simple as they are basic. Is human nature given to us by physical nature, or does the human capacity for reflection and language enable human beings to create their own nature? Are the human sciences comparable with the natural sciences or does their subject matter, including reflection and language, require a different conception of science?

I shall explain my underlying theme with a myth, the story of Tristan and Isolde. The tale exists in many versions and has been given many interpretations. It is a romance: a quest for love and a quest to live and die by ideals. This retelling stresses a fundamental ambivalence at the turning-point of the story, at the moment when Tristan and Isolde declare their love and events begin to unfold towards their shared death. Their declaration of love is a discovery, a revelation to themselves, an ecstasy that comes to them; at the same time, they declare in love what they have always known within themselves, what they have created and what they have given life through their history together.

The scene opens with a ship sailing towards the cliffs of Cornwall. It carries the hero Tristan who is bringing the captured Isolde from Ireland as a bride for his sovereign. In this loyal deed he pledges to King Mark the woman who could with justice but less honour be his own. The wise, noble and proud Isolde rages at her position. She is bitterly conscious that she had earlier lost the chance to avenge Ireland by killing Tristan. When he was wounded and in her power she chose instead the gentle arts of healing; but she did not anticipate his return to make her the subject bride of another man and of hated Cornwall.

Enraged at her humiliation, she challenges Tristan to share poison; he agrees, but her anguished maid disobeys and substitutes a love potion. As the coast of Cornwall comes into sight, Tristan and Isolde declare their mutual love, a rapture with no earthly consummation.

When Tristan and Isolde drink from the same cup, they share what they believe to be their death in order to end her humiliation and his abrogation of his debt to her. Secretly, however, they also share knowledge of mutual love, an inadmissable love because of the conflict between Ireland and Cornwall. When they drink a love potion instead of poison, the potion releases what already lies within, their love for each other, which they already but only partly know. Their love is their fate, their nature. Declaring their love breaks the bonds of loyalty, obligation and subjugation binding them to King Mark. Yet, as they drink, each is noble; in each, inward character and outward action match, exemplifying what their culture most values. Their fate is to be victims of circumstance.

The ambivalence of the story lies in the love potion. Are we to understand that the potion creates a passion that would otherwise never have existed and is an intervention by fate in human history? Or are we to understand that the love potion only releases the passions that Tristan and Isolde have already created for themselves? Who or what is the active agent, and who or what is passive, in this drama – the love potion, the maid, the consciously virtuous Tristan and Isolde, the undeclared passionate lovers, the social mores of the age, love itself? The truth of Tristan and Isolde emerges out of the human spirit but unfolds within the circumstances of history. Tristan and Isolde create love, though love creates their history.

There is no one-to-one relation between myth and history. The story of the human sciences, however, is also a story with an ambivalence about human nature. In one version of this story there is an emphasis on human nature as fate, as a condition given in human life to which people have to accommodate. In another version, what it is to be human is described as continuously recreated through relationships and language – through declaration. If, to return to the myth, the love potion is the immediate means by which knowledge of what is really true of Tristan and Isolde, their love, becomes shared knowledge, are we to understand the potion as a true cause of events or merely as a symbol for the events the actors themselves have set in motion? Long before the time of Tristan and Isolde, the ancient Greeks built at Delphi a temple to Apollo, the home of an oracle to whom people came before they took decisions in their lives. According to tradition, the oracle was associated with a series of principles about how to live, the best remembered of which is 'Know thyself'. This is as ambiguous as it is deep. Is it an injunction to recognize what one is, to come to terms with one's own nature? Or is it a longing for a life of self-conscious creation in which one gains knowledge of what one makes of oneself? The modern philosopher Friedrich Nietzsche played on the same ambiguity in one of his best-known moral experiments, 'How one becomes what one is'.[5]

The story of Tristan and Isolde is a tragedy, a myth and an unconventional though not arbitrary starting-point for a history of science. The usual picture of science illuminates knowledge and progress, not literature and tragedy, humanity's overcoming of its conditions of life; and the language of science is factual and not imaginative. The human sciences, however, are not well served by the usual imagery. Can it make sense to ask the human sciences to overcome the conditions of our own nature? There is something disturbingly paradoxical about a science that has for its subject the agent that creates the science. How are we to stand back from being human in order to observe what it is to be human? Even to attempt this standing back – and there are many ways in which it has been undertaken in pursuit of scientific truth – is a way of being human that, in turn, some other person will be able to study. Are we then condemned to travel in self-reflecting circles, to create knowledge of human beings only to find that what has been done is to create another mode of life rather than a lasting truth?

The Delphic 'Know thyself' directed supplicants to become what they could be while accepting what they are. The sybil, the oracle of the temple, often seemed to urge her audiences towards decisive action while she also told them to accept their fate. There are parallel pairs of opposites here: activity and passivity, self-creation and self-discovery, and the conception of a person as an agent and belief in historical circumstances as causes. Human life appears to exist somehow as a tension between these opposites, and the sciences of human life have had to struggle to achieve coherence while subject to the same tension. The result has been recurrent disagreement over what sort of knowledge the human sciences can in fact achieve.

This tension, though couched in a very different idiom, was expressed in Marx's opposition to the philosophical idealism dominant in his youth. Marx was angry at claims about the spirit (or, in German, *Geist*) in human affairs, since he believed such claims were used to rationalize political oppression, and he therefore argued instead that modes of thought result from the way people satisfy material needs. In the 1840s, Marx and Engels famously stated that people 'developing their material production and their material intercourse, alter, along with this their real existence, their thinking and the products of their thinking'.[6] His words make it appear that he regarded human nature as a material nature and thought as its product, the historical means by which people satisfy material needs. Yet Marx's views were more complex and ambivalent than this. While he described human nature as a product of material history, at the same time he described human thought as making history. His whole life, like his revolutionary politics, was a testament to belief in the power of thought to change the course of the world. Marx, too, in his way, declared, 'Know thyself', in the sense that he both wanted to describe man's nature and to change it.

There is another tension in modern Western culture, a culture that makes the self supreme. However celebrated, our conception and experience of self is a social activity: we cannot but know, even ourselves, except in terms that we acquire through living in a particular time and place. To conceive otherwise would be to conceive of having knowledge without language or other symbolism, which is nonsense. Even Tristan and Isolde, who transcend their selves in the ecstasy of love, find a voice for their experience in socially conventional metaphors and use the language of darkness and light to represent the qualities of love. The history of the sciences that characterize the self is therefore also a history of the social worlds in which the self has existed.

This book argues that to know human nature is to know what has been thought and said about human nature. The history of the human sciences is not so much about uncovering the truths of human nature as about adventures of human expression of such power that they have acquired the status of truths. Some readers may find this frustrating, since they may want to know what is 'really' there in human nature. In this account of the history of science, what is 'really' there is the quest itself. Similarly, readers who are natural scientists may wait impatiently for the 'real' science to be described, for the historical story to reveal when and how objective natural science assimilated the human sphere. But what is 'really' science in this account of its history is contested and unsettled.

I write, then, in time-honoured style, about a quest. This book describes the search, in the modern Western manner, for systematic, coherent, objective and empirical knowledge about the nature of being human. What disasters have befallen? What marvels have been discovered? What meaning has there been for those on the journey? What glimpses have there been of treasure?

The quest has been particularly difficult in the human sciences. The pursuit of objectivity has raised special problems. In the twentieth century, disciplines such as psychology and sociology have been preoccupied, sometimes obsessed, with objectivity, a preoccupation which has most commonly taken the form of a search for objective research *methods*. The core problem is simple enough: how, objectively, can we observe ourselves? By the process of observation, do we not make ourselves into something different? This is the problem of the human capacity for reflection; it is the nature of being human to reflect and through reflection to alter experience and action. The subject does not stay still. The result, as suggested by the myth of Tristan and Isolde, is an ambivalence about whether humans are passive or active, creatures of nature or creatures of culture, lovers by fate or lovers by design.

The historical record indicates that none of the methods adopted by the human sciences nor any of the knowledge generated succeeds in commanding universal assent. What counts as objective knowledge in the human sciences is always in dispute. From the historical perspective adopted here, the reasons for this dispute are fascinating, and the dispute itself is not regarded as a failure to achieve something better but as the record of an unending quest.

Philosophers have a reputation for playing with words. The history of this 'play', however, is one of the most fruitful ways to study the history of intellectual culture. The choice of a word may be very significant in the human sciences, as it may reinforce, legitimate or even bring into existence one rather than another view of life. Late twentieth-century readers are familiar with this point through debates about the gendered content of language – and readers may already have questioned how this book will deal with 'man'. Different beliefs and actions are supported by describing the Tupinambá people as Indians, Amerindians, Brazilians, savages, cannibals, native people – or Tupinambá. The German philosopher G. W. Leibniz commented: 'I really believe that languages are the best mirror of the human mind, and that a precise analysis of the significations of words would tell us more than anything else about the operations of the understanding.'[7] One word of crucial importance for the history of the human sciences is the word 'science' itself.

The word '*scientia*' was commonplace in the Latin learned culture of late medieval and Renaissance Europe, when it had various connotations. The principal and most precise usage was to describe systematic knowledge of the true causes of particular things. Scholars at the universities therefore studied deductive knowledge of such topics as optics, music, metaphysics, theology, astronomy and arithmetic, the branches of learning or sciences. Such knowledge was clearly differentiated from revealed knowledge, the articles of Christian faith, the basis of which was taken to be the grace of God alone. Above all, 'science' did not denote 'natural science', as it did in twentieth-century English-language usage. Aspects of what later became the natural sciences were spread across a wide range of subjects in the early universities. A way to drive home the changed meaning of the word 'science' is to quote from Locke, whom some historians have thought to be the founder of a secular science of psychology. Long after the medieval period, Locke wrote: 'There is one science . . . incomparably above all the rest . . . I mean theology . . . This is that noble study which is every man's duty, and every one that can be called a rational creature is capable of.'[8] For Locke, as for his medieval predecessors, the premises of theology are primarily revealed, but, since these premises are true, deductions based on them are also true and these deductions constitute science. Locke was perpetuating an early usage of the word 'science', which makes it difficult to claim he is the founder of any 'science' in the modern English-language sense.

Some medieval scholars drew a distinction between *certum* and *verum* rather than between science and non-science. '*Certum*' is a term applied to what is known with certainty about mundane facts in this world, as in physics, and it applied especially to confirmed experience; '*verum*' is applied to what must be true because it is deducible from the spiritual reality known by faith, and it was thus used in theology. In the late medieval and early modern period the word 'science', strictly speaking, applied to the systematic study of demonstrable truths, that is, truths that convey something of the necessary structure of reality, such as the spherical shape of the earth; but it sometimes referred to certain as opposed to true knowledge. To make matters even more complicated, the distinction between 'the arts' and 'the sciences' was also not clear-cut, though in practice the former was used primarily to denote learning relevant to active, civil life, while the latter usually denoted learning as a theoretical discipline.

Very importantly for the history of the human sciences, early connotations of the word 'science' continued to inform continental European usage even into the late twentieth century. In German- and Scandinavian-language cultures, the academic study of literature, languages, history, theology, politics, the history of art and archaeology, continued to be described as '*Wissenschaft*', the German word conventionally translated as 'science'. In the English-speaking world, it was often questioned whether psychology and sociology *are* sciences, and the question meant: do they use the methods and develop the law-like explanations characteristic of the natural sciences? The same issues about the methods and forms of explanation in the human and natural sciences were central in non-English-language cultures, but there the question was framed as one that concerned the relations *between* the sciences. Thus, the subject area of the theory of science, in the European setting, addressed itself to the humanities as well as to the social sciences and the natural sciences.

Different meanings of the word 'science' in the history of the human sciences cut across debate about what sort of knowledge of human beings is possible and about what is required to explain human action. The following chapters return to the question of relations between the human sciences and the natural sciences but, following the continental European example, they discuss what sort of science each has been, not whether the human sciences are sciences. This makes possible openness to different views. There are further incentives, such as the collapse of belief that the natural sciences themselves possess any specific practice that could be called the scientific method, and recognition that what constitutes an explanation differs even in the natural sciences between, for example, an evolutionary biologist and a particle physicist. It all adds up to a need not to prejudge the boundaries or nature of science.

There is a further dimension to these debates. Historians of science and ideas, especially when referring to the eighteenth century, often generalize about knowledge in terms of rationalist and empiricist traditions. Rationalism, linked to the name of Descartes, claimed that knowledge originates in reason; empiricism, linked to the name of Locke, claimed that it springs from experience. It is possible, however, to avoid reference to these – and to many other – 'isms': they are not needed, and they invent traditions in the place of historical explanation. Nevertheless, there have of course been very different conceptions of what it is to establish a science and to have knowledge about human beings, differences related to the ways reason and experience have been understood. The philosophical issue is to decide whether knowledge is grounded in experience or in reason's analysis of its own activity or in some relationship between reason and experience. In continental Europe, there was a deep-seated commitment to the primacy of reason, encouraging belief that the terms in which reasoning conceptualizes what it is to reason are fundamental to scientific knowledge. This perspective, in turn, supported the description of all rigorous fields of knowledge as sciences, since all fields exercise reason. Conversely, when English-speakers committed themselves to experience as primary, they encouraged the view that the fields of knowledge based on experience – the natural sciences – are models of what science should be and with which other fields should be compared.

The philosophy of the human sciences raises large and intractable problems. Fortunately it is still possible to write history. The search for a science of human beings has been a spectacular undertaking. It has also been deeply significant: there is a very great difference between the contemplative life of the soul and the statistical analysis of voting patterns. To find out how such differences come about and what they have meant in human experience is the stuff of this history.

A glance around a modern university reveals a bewildering variety of disciplines. Departments of management science, information systems, political science, economics, linguistics, sociology, social anthropology and psychology vie with each other to explain human activity. These subjects claim to be sciences. Within that vast field, this book concentrates on theories of human nature and, for the twentieth century, on psychology. It is not possible to deal systematically with such areas as political thought, economics, management science and jurisprudence, though the claim has been made that each of these is a human science. Similarly, linguistics and anthropology remain in the background, though work in these areas has sometimes been crucial for beliefs about human nature.

This book, however, is not a conventional history of psychology. The chapters covering the period from about 1500 to 1850 discuss many topics that we would think of as psychological, though they did not constitute psychology as a subject in the modern sense, however much they may be assessed for the part they may have had in the background of modern psychology. Many other topics will be taken into account in order to do justice to the richness of learning about man. Psychology was not a separate subject until it in fact acquired a distinct identity as a discipline and expert occupation, and this happened largely in the twentieth century. Even then, it was unclear where the boundaries of psychology actually lay.

This question of boundaries is exemplified by the historical relations of the human sciences and medicine. It is possible to imagine a medical historian claiming that medicine is the original human science, the area of practical and theoretical engagement with human life which did most to foster systematic knowledge about people. At many historical junctures it was a medical context that gave purpose and content to human self-knowledge: Renaissance notions of woman related to the way procreation was understood; J. O. de La Mettrie's scandalous views in the eighteenth century on what he called the man machine grew out of medical disputes; Sigmund Freud's claims about the unconscious mind originated with clinical experience. Through the existential realities of health and illness, well-being and suffering, birth and death, human nature necessarily confronted itself – and its limits. As Descartes observed in the mid-seventeenth century: 'the mind depends so much on the temperament and disposition of the bodily organs that if it is possible to find some means of making men in general wiser and more skilful than they have been up till now, I believe we must look for it in medicine'.[9] Symbols of mortality – from the grim reaper portrayed in sixteenth-century woodblock prints to the patient with AIDS in the 1980s – marked out a baseline for any claim to wisdom in the human sciences. All the same, the label 'medicine', like the label 'the human sciences', did not define a unified occupation and area of knowledge until a century ago. It is therefore not possible to talk about medicine's influence on the human sciences since what existed was a common context rather than separate spheres.

What is true for medicine is equally true for religion and theology. For much of the time and in many of the settings described in this book, it was a matter of faith that the most profound knowledge depended on the grace of God, a grace given, as it was differently understood by different faiths, through the church itself, through scripture, or through the soul and conscience. Many modern people, impressed by natural science or man's material existence, presuppose that there is opposition between religious faith and scientific knowledge. This is a modern judgement neither universally shared nor of value to historians. Christian faith and knowledge has been integral to thought about human nature, the thought here anachronistically called the human sciences. How this integration has come undone in many modern settings, particularly in the twentieth-century human sciences, is a topic for subsequent chapters. In concrete terms, the history of the human sciences includes, not excludes, the history of the soul, earlier the subject of a science.

What does a history of the human sciences show? Succinctly, it is *diversity*. Common proverbs and common language, it is true, treat human nature as at all times and everywhere much the same. Voltaire once stated simply that 'Man, generally speaking, was always what he is now'.[10] Why there should be such a belief, whether it is thought true or untrue, is an intriguing cultural and historical question. Voltaire, it may be noted, also remarked: 'The empire of custom . . . sheds variety on the scene of the universe', posing, even in the writings of one man, the question of the balance between nature and culture in what is claimed to be knowledge of man.[11] The record of the human sciences is a record of a bewildering variety of claims about human nature – including claims which deny the value of the concept as well as the universality of what it supposedly describes. Many people have looked to science to evaluate such competing claims. This expectation has been disappointed in the case of the human sciences; their history is a record of division and debate between sometimes fundamentally opposed positions. A glance at some contrasts in twentieth-century psychology makes the point: behaviourism (Watson); the theory of higher nervous activity (Pavlov); operationalism (Stevens); genetic epistemology (Piaget); psychoanalysis (Freud); humanistic psychology (Maslow); differential psychology (Spearman); phenomenological psychology (Buytendijk); cultural-historical theory (Vygotsky); personalism (Stern); gestalt psychology (Wertheimer); hormic psychology (McDougall); field theory (Lewin); sociobiology (Wilson); analytical psychology (Jung); typology (Teplov); social behaviourism (Mead); and so on. Each of these labels describes an approach to what at least the researchers mentioned thought to be psychology. It compounds the point to recognize that these psychologies represent just one facet of the modern human sciences.

There are two main ways of thinking about this diversity. The first concludes that it is a sign of immaturity in a scientific domain, the inevitable consequence of the attempt to pioneer scientific methods and concepts in a vastly complex new field. Proponents of this point of view anticipate that at some stage, however distant it may still be, unification around an agreed theoretical core and agreed research practice will become possible. Writers sometimes follow the language of the historian of science Thomas S. Kuhn, and describe psychology and other human sciences as in a pre-paradigmatic stage. By this they mean that they accept that, *as yet*, there is no single unified theory, model of research practice and social identity for a field. In the 1970s some psychologists thought that the cluster of activities called cognitive psychology established a unifying paradigm, but they did not persuade all their colleagues that this was so. No one could intelligibly claim that there is unified activity in, say, sociology or economics. But of course an existing lack of unity is no proof whatsoever that such a unity will not be forthcoming. It follows, however, in an actual state of disunity, that to write history cannot be neutral in relation to the divergent positions in the current field. This point was well made by the historian John Burrow: 'To write the history of a discipline is to state what the discipline is, and this, in the social sciences, is often highly contentious.'[12]

The sceptic will note that unity has been forthcoming for rather a long time. It is therefore possible to turn to a second way of viewing diversity – as a positive rather than a negative characteristic of the human sciences. This second point of view argues that diversity is intrinsic to the field's subject matter, to the reflexive character of human beings, and that it is not the contingent result of lack of progress in the field. The argument is that human beings, who are reflective and active agents, create themselves anew as they create knowledge of themselves. If this is accepted, then it follows that the variety of approaches to knowledge in the human sciences corresponds to the variety of modes of life in history. We may suspect that unity in the human sciences would become possible only if people were to lead a uniform life. Such a uniformity in a way of life does exist within specialist occupations, and we know that within occupations experts can and do agree. Speech therapists, for example, agree about much of what they do in responding to a child in difficulty. In the wider world of even one culture, let alone of different cultures, however, there is no uniformity in how people live and, according to the present argument, there is therefore also diversity in forms of knowledge about the human subject.

The chapters that follow view diversity in the second way and stress the range of thought about human nature rather than the content of a particular theory. All the same, it is the achievement of the sciences, including the human sciences, to obtain agreement within particular communities as to the truth or correctness of claims about the world. This achievement must be explained and must be central to any history of science; and hence this book certainly deals with agreement as well as difference.

Two major restrictions on the book's range require brief comment: it considers only Western science and only post-medieval science. It is probably the case that all cultures have ways to represent in language or other symbols beliefs about what it is to be human. This appears to be true, for example, for those early people who painted the lovely animal and human figures in the caves at Lascaux and elsewhere. It is certainly true for modern non-Western cultures, and Western writers commonly refer, for example, to Tibetan and to Buddhist psychology. Yet some cultures do not have or use categories such as 'individual' and 'society' in the way that Western cultures do. It is therefore questionable whether cultures around the world possess a psychology or an anthropology in the way that these subjects exist in the West. Be all that as it may, this book is about Western thought – European culture and its derivatives elsewhere. That is not to denigrate other cultures; indeed, rather the opposite. The limitation respects what is other, as opposed to making a condescending gesture towards the other as if it were merely an alternative version of Western thought. Were this book to include non-Western thought – and the word 'non-Western' already indicates bias – that thought would inevitably appear a diluted version of Western ideas. A constructive way forward will become possible when psychology and the social sciences are fully understood as characteristic aspects of modernity, and when the relationship of Western science to other belief systems can be seen as part of the material and political interaction of the West and the wider world.

The restriction to the post-medieval period – that period since about the fifteenth century still known at the end of the twentieth century in the university of Oxford as Modern History – is more difficult to defend. It is reasonable to a degree to argue the point which has just been made about non-Western cultures: we should respect earlier European peoples for the otherness of their beliefs. Yet of course there is also much continuity between ancient and modern Europe and, given the Greek and Latin roots of many European languages, we must accept that this continuity is profound in just the way that is important to knowledge. Historians of science and medicine have indeed conventionally traced their subjects back to Greece and to the language that represents both physical nature and human nature as a knowable order. The word '*psyche*', built into so much twentieth-century language, and Freud's term 'the Oedipus complex', are emblematic of this ancient heritage.

Nevertheless, this history of the human sciences begins with what we know as the Renaissance. The first reason is obviously the practical one that there is sufficient material in the modern period for one volume. More interestingly, there is a question mark over the whole area of the relation between the ancient and the modern intellectual worlds. It would be very difficult and perhaps impossible to provide a reasoned synthesis to cover the soul, self, natural law, human nature and conceptions of the social without a great deal of scholarly discussion about problems of evidence and interpretation. This is not attempted here. It is of course necessary, at the beginning of the book, to refer back to the ancient world and to medieval Christian culture; but the vexed question of the continuity between the ancient and modern worlds is not directly addressed.

Historians and philosophers at the end of the twentieth century paid particular attention to 'the self', perhaps because they lived in a culture where self-identity was both so valued and so questioned. To write the history of the self is to pose a question about continuity and discontinuity in its most difficult form. Authors variously located the origins or invention of the self in the Christian reflection exemplified by Saint Augustine in the fifth century of the Christian era, in the late medieval development of the Catholic confessional, in Descartes' arguments for knowledge based on the reflective *moi* or 'I' in the 1630s, and in the Romantic ideal of inner truth about 1800. There is no doubt that how the history of the self is understood is of considerable importance for how the history of the human sciences is written; but much of this argument must be left for future historians to resolve. On the one hand, for example, there is evidence that a modern subjective self and a modern conception of the social individual became the subject matter for knowledge in the seventeenth century. When we gaze at Rembrandt's self-portraits we also gaze inwardly at what we feel ourselves to be. On the other hand, we must bear in mind that the concept of the individual person was ancient, embedded in the legal concept of a person able to act freely.

iii *Writing History*

New historical writing about the human sciences reacts against earlier conventions in the history of science and ideas, conventions which appear intellectually biased in favour of a restricted range of academic norms. In particular, there is a reaction against textbook histories of psychology (or other disciplines), written primarily to satisfy the huge North American market in the subject. At times, psychology has had the largest intake of undergraduate majors of any academic discipline in the United States, and many students have had to take a course in the history of the subject. Writers of textbook histories in the human sciences drew on a relatively small number of well-known academic histories, all of which covered individual disciplinary fields: for example, E. G. Boring's *A History of Experimental Psychology* (1929; 2nd edition 1950), Raymond Aron's lectures on *Les étapes de la pensée sociologique*, published in English as *Main Currents of Sociological Thought* (1967) and Joseph A. Schumpeter's *History of Economic Analysis* (1954). These studies discussed and confirmed an existing canon of scientific texts by the great men in their respective fields. The end result was that each modern discipline in the human sciences possessed a relatively well-defined notion of its historical identity, an identity constructed out of classic studies, experiments or texts that, it was claimed, established the organizing concepts and objective methods of the discipline as a science. Thus each field had its founding classics: in psychology there was Locke's *An Essay Concerning Human Understanding* (1690); in sociology, Montesquieu's *L'esprit des lois* (*The Spirit of the Laws*, 1748); and in economics, Smith's *The Wealth of Nations* (1776).

To conceive and to write history in this way continues to appear natural to many psychologists and social scientists. This convention follows from a belief that there can be only one story worth telling in the history of science, its plot the achievement of a body of scientific methods and the progressive accumulation of truth. Depending on the storyteller's breadth and education, the history that results may be crudely triumphalist, when reason is portrayed vanquishing ignorance and superstition, or a sensitive and complex exploration of the paths, by-ways and dead ends of intellectual endeavour. When the stories reappear in student textbooks, which have their own pedagogical agendas, they tend to become accounts of step-by-step progress in the core topics of the relevant modern discipline. The end product is what the English historian Herbert Butterfield called Whig history: history written as if the present is the inevitable and desirable outcome of the past.

Historical scholarship in the history of science has changed all this, largely through many detailed studies of restricted scope. These studies have become increasingly historical in the academic sense, which means that they have concentrated on the past in its own context and on its own terms, and have not given priority to what modern psychologists and social scientists might think important. The new scholarship is not easily accessible to the non-specialist. It has also not been obvious or well publicized that historical scholarship might add up to a new way of thinking about the human sciences. This book argues that it does.

Questions of objectivity in writing histories of science are one thing, questions of objectivity in the practice of science are another. The fields legitimately have different interests. Yet many modern scientists have assumed that it is their interests that should direct the writing of the history of their fields. Behind this assumption is the belief that the twentieth century is a secular age and that the history of science is therefore a history of the best truths we have – a history of 'the edge of objectivity' in one historian's memorable phrase.[13] From such a stance, the history of science must have as its core subject matter the conditions that have made possible objective knowledge, doubtless with allowance for asides as to why at times the enterprise of science in some measure failed. Nevertheless, this point of view has long been debated, and – while we may leave the philosophical issues to one side – many practising historians now reject it as a basis for their work. Instead, they seek to understand in a way that is appropriate to the time and place of historical actors and not to modern scientists. This approach can be illustrated by looking at the history of modern academic disciplines.

The disciplines grouped together as the human sciences – psychology, sociology, linguistics, and so forth – are modern social entities; none of them existed before 1800 and most were created in the late nineteenth or early twentieth century. It is unacceptable to project the existence of such modern academic disciplines as psychology or sociology back into the past, as was done in the single discipline histories that once dominated the human sciences. The study of when, how and why these disciplines, with their institutional structures – things like academic departments, technical training in the subjects, occupational organizations, conferences and journals – came into existence is an important area of historical research; and it is also important to study the effect that discipline formation had on the production, authority and content of knowledge about human action. It is not possible to begin the history of psychology with Aristotle in the fourth century BC and continue to the present as if there were a continuous subject of psychology. Historians do not now accept, as the Columbia University psychologist Robert S. Woodworth wrote at the beginning of a popular student text, that: 'It is safe to say that psychology is as old as the inquiring, self-conscious mind of man.'[14] History like this colonized areas of human thought in order to claim all the best bits of high culture for modern science disciplines. We now wish to decolonize the territory of the past and understand the past in its own terms. The history of the human sciences in this book is therefore not a history 'of psychology'; rather, it seeks to escape disciplinary blinkers and to describe knowledge and its organization in the context of human history.

A further consequence of the historical study of learning is that it draws attention to the contingent nature of the division of intellectual labour within the modern academy. Much effort and many words have gone into the construction and maintenance of the boundaries between different subjects – for example, between sociology and history, even though both have a systematic interest in society. Yet the way knowledge and learning are divided up changes over time. The divisions characteristic of the twentieth century are social constructions largely of the twentieth century. The complexities of the human sphere make possible many different sciences, but the boundaries between them are neither essential nor immutable.

A subtle difficulty exists in relation to references to psychological thought before the nineteenth century. It follows from what has just been argued that such references are unacceptable if they imply that earlier writers possessed ideas and language strictly comparable with the ideas and language of modern disciplinary subjects. Yet it would be a very laborious and pedantic matter to excise all such references where we find earlier authors discussing what are to us psychological activities such as memory, vision and passion. This problem prompts a practical question and a philosophical question. The practical question is faced by all historians: they must find a balance between making what they say accessible by using ordinary modern language and misleadingly using that language to impose on the past something that was not there. To give an illustration: Robert Hooke, the famous experimenter of the early Royal Society of London, was interested in processes of memory, though neither he nor anyone else in the seventeenth century described it as a psychological interest. Since Hooke discussed memory as a distinct topic relevant to natural philosophy (and in this he followed Aristotle), it is at least intelligible and it may be pertinent to say that this was a contribution to psychological thought. But

Hooke was not a 'psychologist' and, in fact, his work had no influence. Similarly, it can be argued that seventeenth-century English pamphlet writers are of interest because they started to describe economic activity separately from other aspects of collective life. This may be pinpointed for modern purposes as the beginnings of a form of economic thinking, but it was not 'economics'. The historian's guiding principle is to represent accurately the way people in the past divided up and categorized their own experience and knowledge. This then leaves the practical problem of how to picture their world with our modern means of expression.

The philosophical question concerns continuity and discontinuity in history. The once dominant mode of writing about the history of the sciences stressed continuity. Where it constructed breaks, as it did when it described the scientific revolution, these were pictured not so much as breaks but as leaps – leaps forward within science or philosophical leaps that created the intellectual conditions for science. There are three bodies of work which, by contrast, have argued for discontinuity in history, and they have strongly influenced studies of the history of the psychological and social sciences. The first is the book by the US historian of science Kuhn, *The Structure of Scientific Revolutions* (1962); the second stems from the mid-nineteenth-century work of Marx, who owed a great philosophical debt to G. W. F. Hegel; and the third is associated with the writings in French in the 1960s and 1970s of Michel Foucault.

Kuhn described the history of science (all his examples in fact concerned the physical sciences) in terms of periods of normal science interrupted by shorter and abnormal periods of revolutionary science. The latter, he argued, lead to new problems and frameworks. The relevant point is that he pictured revolutions as breaks with past ways of thought, and he implied that new and old ways of thought are not commensurable, which means that they cannot be fully compared with each other. This idea challenged theorists of knowledge because it apppeared to suggest that standards of objectivity and truth change in scientific revolutions and that particular standards therefore hold only as long as a period of normal science holds. This, at least, was the direction in which a group of sociologists of knowledge took the argument in the 1970s, interested to show how social realities explain what counts as knowledge and how knowledge changes. Kuhn's interpreters put on to the intellectual map the possibility that bodies of knowledge in the human sciences are separated from each other by changes in meaning and conditions of truth rather than by time and progress.

Marx and Engels argued that material conditions, expressed as the social relations of property, capital and labour, foster one rather than another way of thought. 'The sum of productive forces, forms of capital and social forms of intercourse, which every individual and generation finds in existence as something given, is the real basis of what the philosophers have conceived as "substance" and "essence of man" . . .'[15] It followed, they thought, that as material conditions change so new ways of thought become possible: a revolution in material conditions prompts a revolution in ideas. Some historians have adapted this argument and tried, for example, to correlate modern notions of the self, the idea of individual economic man and the development of market relations in seventeenth-century England and the Netherlands. Taking the point further, it might be possible to argue that there was no psychological thought in the modern sense before this time because material conditions had not created the subject, the self, which psychological thought addresses.

Foucault claimed that the conditions of thought precluded the existence of the human sciences before the nineteenth century: 'the very concept of human nature [in the eighteenth century], and the way in which it functioned, excluded any possibility of a Classical science of man'.[16] He believed that radically new ways of knowing human beings appeared around 1800, so that it was possible only thereafter to assert scientific truths about man in the way that the twentieth-century human sciences have done. He also correlated novelty in the conception of man with institutional and administrative changes in schooling, the penal system and medicine – with the government of microsocial life rather than with the macrosocial changes that Marx thought underlie history. It was, Foucault claimed, new patterns of power in everyday social arrangements that created the subject, 'man', of the modern human science disciplines. Few historians are attracted by the idea of such a radical break; but Foucault nevertheless prompted a major reassessment of how to conceptualize the subject matter of the human sciences.

In his first major book (1961), which was a history of madness, Foucault issued a profound challenge to the human sciences, a challenge that few scientists felt it was in their province to take up but which held a fascination for a wider audience. Foucault's proposition was that the label 'mad' hid from what counted as reason the limits of reason's rationality; it followed, in his view, that madness is a state that can teach reason something about itself, something fearful, a knowledge of 'the other', a condition to which reason has no response. This was the position of the court fool who speaks wisdom no one else dare utter. Such explorations of the boundaries of what is thinkable about ourselves fostered scepticism towards the truth claimed by the orthodox human sciences and in the late twentieth century generated excitement about literature, the arts and language as routes to truths not touched by the existing human sciences.

Different views about historical continuity and discontinuity, or about the limits of scientific reason, will not be reconciled independently of agreement about the nature of knowledge in the human sciences, including historical knowledge itself, and this is absent. In addition, historians now generally agree that the history of science should not be subservient to the outlook of modern science disciplines; they call such subservience 'presentism' and believe it offends against basic principles of objective historical scholarship. It is also appreciated, however, that as historians write in the present it is a condition of intelligibility that they use present-day language. It is important to translate past thought into modern ideas, with whatever qualifications are necessary. The discussion of words like 'science', 'soul' and 'passion' will make this clear. Further, the way historians select and structure historical questions necessarily reflects their own intellectual and material situation, which most directly means that it reflects their training and academic experience. How could it be otherwise? Anybody who studies the past does so for a purpose. The social scientist who awards favourable marks to eighteenth-century writers who believed, like later economists, that there is a relationship between the utility of a product and its value, expresses her or his purpose in the study of history. All the same, many people who work in the human sciences, and not only historians, have become sceptical of such purposes crudely grounded in the present: at best they make history a decoration for some other discipline and at worst they are self-serving. Instead, there is the purpose of excitement, enlightenment and liberation in seeking to know how other people understood themselves and their circumstances and how other people gave their understanding a voice in the most profound and expressive ways open to them. More deeply, if it is accepted that human beings are in some sense self-creating, then to write history – the history of that self-creation – is essential to our knowledge of what we are.

The development of this thought in the concluding chapter leads me to write in the past tense about what, as this is written, is the present. The purpose of this somewhat unconventional style is to suggest that what we call the present is only one moment in a continuing story or quest, that the present has no more permanence or completeness than the past, and that when we write about ourselves in the present we also write history – just as in writing history we write about our present selves.

One further general question about the writing of history must be discussed. From the late medieval period to the present, writers referred to 'man' (often capitalized as 'Man'), to *'l'homme'* and to *'der Mensch'*. A well-worn eighteenth-century phrase referred to 'the science of man'. All such usages sound contentious in the late twentieth century because they elide the question of gender. Did references to 'man' include woman and, if so, how? The usage is probably most a problem to modern English readers since 'man' became gender specific in a way, say, that the German *'Mensch'* did not. It is in fact a fascinating historical topic in its own right to trace the settings in which 'woman' is and is not differentiated from 'man' and the senses in which 'man' has gendered connotations. 'Gender' itself, however, is a category introduced in the second half of the twentieth century to distinguish differences between men and women as they exist – or are perceived to exist – without the prejudgement that these differences are sexual or biological. In many cases, there can be no doubt, reference to 'man' in the past presupposed norms of reasoning and mental capacity that were, as a matter of historical fact, identified with men. When the topic concerned what was universal in human nature, the subject of woman simply did not appear. We now wish woman to appear. The answer to the silent erasure of women from history through the establishment of men as the norm, however, is not for historians to change the language in which historical actors characterized human beings. The answer is to write historically about gender and sex: to place on the agenda the intellectual and cultural history of belief about human differences. I shall refer, as I have done in this chapter, to 'man' where this is appropriate historical language; it would be forced and anachronistic to do otherwise. It could be said that the whole purpose of the history of the human sciences is to write the history of what the word 'man' has meant, and this history will include belief about human differences – of class and ethnic identity as well as of gender.

To conclude this introduction, it is helpful to draw a comparison between historical writing on science and religion and on the human sciences. It is easy to see why conclusions about the historical relations between science and religion differ since belief in the truth content of science and of religion differs among those who write. Someone who believes that science establishes uniquely objective truth will write different history to someone who believes that science and religion equally express truth. Historical research, in fact, points to the extraordinary variety and complexity of historical relations, and the lack of boundaries, between science and religion. The historical record neither confirms nor denies modern religious or scientific belief. The historical achievement does not resolve questions each person and each age must confront anew in new terms and for new purposes. Without understanding the historical achievement, however, we would be immeasurably poorer in language and imagination.

The parallel with the history of the human sciences should be apparent. There are major divergences of opinion about the nature of the human sciences – especially about whether the study of human beings should follow the natural sciences or create separate disciplines. There are deep divisions between those who take human nature to be given by biology and those who consider it to be the product of culture continuously recreated through reflective language. Each position leads to a different approach to the history of the human sciences, and the writing of history is therefore in the last analysis inseparable from debates about the human sciences themselves. The following chapters give priority to the range of intellectual positions, to the diversity of the human sciences. We need the history of this diversity of belief about human nature to give us the expressive and imaginative life to create our own beliefs. In fact, there is no choice: if we do not do it consciously, we will assuredly do it together unconsciously.

II

THE 16TH AND 17TH CENTURIES

2

The Dignity of Man

And God said, Let us make man in our image, after our likeness: and let them have dominion over the fish of the sea, and over the fowl of the air, and over the cattle, and over all the earth, and over every creeping thing that creepeth upon the earth. So God created man in his *own* image: in the image of God created he him; male and female created he them.

Genesis (in the King James version, 1611)[1]

i *Ancient and Modern Learning: the Renaissance*

Harmony and dignity, as well as unbounded technical confidence, lived in the buildings of the cities of Rome, Florence or Venice of the early 1400s, and the builders judged their success in relation to the ancient world. Scholars and political leaders in the fifteenth and sixteenth centuries envisaged a rebirth; they set themselves against the more recent past and, inspired by Greek and Roman examples in poetry, philosophy, art, military glory and civil society, proudly set out to learn and then to surpass ancient achievements. The generations that followed believed that they had broken free from the medieval world and had overtaken ancient culture. By the early eighteenth century, it was natural philosophy – the sciences of nature – that appeared to have climbed furthest above the ancient world and, by the revelation of natural law in the universe, to have separated the modern mind from the pagan or Christian superstitions of the Dark Ages. Meanwhile, nascent industrial capitalism, with roots in the trading city states of Italy or the cities of Hamburg,

Amsterdam or London, laid the basis for material conditions about which the ancients had not even dreamed. Nineteenth-century historians, who named the Renaissance, believed that this age inspired and founded modern civilization. The rebirth of ancient learning was interpreted as the first-born child of modernity.

This image of a decisive break with the medieval world was inspirational but it was not true. Few late twentieth-century historians confidently equate the Renaissance and the beginnings of modernity. The beginning of the modern age was gradual and firmly rooted in the diverse and flourishing Christian culture of the thirteenth century. In that century, Saint Thomas Aquinas (*c.*1224–74) and other scholars assimilated ancient philosophy, above all that of Aristotle, to the Christian theology of the Church Fathers. The early universities – Paris, Bologna, Salamanca and Oxford, and later those like Leipzig, Cracow and Vienna – established patterns of training and critical commentary that instilled in their scholars a respect for learning and argument. Christendom was not monolithic in faith, in political life, in economic activity or in culture. And, in particular, medieval Europe was not static nor its expressive life limited. The poet Petrarch, followed by Dante and Boccaccio in fourteenth-century northern Italy, turned language, based on Roman example, into what many considered a divine art. Further north, Gothic architecture and its derivatives had flourished since the twelfth century and blended engineering and spirituality in cathedrals like Cologne, Reims and Salisbury. Roman law was taken up by Italian and then French commentators who sought a unifying and rational basis for civil authority. All this, and much more, fed into what nineteenth-century scholars called the Renaissance and, through that period, into the modern world.

There were also profound changes. Johann Gutenberg devised movable type in Mainz about 1450 and launched printed communication; Columbus reached what Europeans called the New World in 1492, with momentous consequences for the European imagination as well as for the economy; and, beginning in 1517, the Protestant Reformation tore apart any semblance of a unified Christendom in the West. We must also undoubtedly add to this catalogue the new natural philosophy – the scientific revolution – though this was a complex series of processes rather than a defining moment.

Printing, discovery, Reformation and science were the outward events, collective achievements, that transformed a shared literate culture. At the same time, they were inward events, part of a transformation of people's belief, sensibility and imagination. The doyen of Renaissance cultural historians in the second half of the twentieth century, Paul O. Kristeller, believed that there was something new in human experience, a 'tendency [for writers] to take seriously their own personal feelings and experiences, opinions and preferences'. He discerned 'an air of subjectivity' in humanism, the literary movement at the heart of cultural change devoted to the recovery, re-translation and imitation of ancient texts and scholarship.[2] As the word 'humanism' implies, this was a movement that placed man at the centre and gloried in his ability – inspired by ancient example – to discover in himself great qualities.

A large question mark hangs over the continuity of ideas, beliefs and values from the ancient to the modern world. Can we be sure that when the Greeks wrote about the soul, about science or about virtue, they meant what Renaissance writers – let alone later generations – meant? There was no ancient or Renaissance equivalent to the term 'the sciences' in modern English. There was competition for status between different kinds of learning, there was an increased concern with method, and there was debate about the classification of the different branches of systematic knowledge. But the divisions and classifications were not ours, and disciplinary categories like psychology, economics and sociology were not branches of knowledge. The traditional curriculum consisted of the seven liberal arts or sciences – both terms were in use: the foundational *trivium* of grammar, logic, rhetoric; and the more advanced *quadrivium* of arithmetic, music, geometry, astronomy.

Enraptured by the elegance of ancient Latin and the moral rhetoric of Virgil or Cicero, the humanists like Lorenzo Valla (1407–57) propagated a body of learning that spoke to the education of a man as a civil leader and created glory both for himself and for his people. Humanism flourished in the city states of fifteenth-century Italy, where Venice and Florence, for a time, had republican governments, where other cities at least desired their rulers to govern well and where all sought to unite state and citizen in virtue and justice. The written genre of advice to princes wove together the learned man and the just and glorious advancement of the state. By the late fifteenth and early sixteenth century, civic humanism had an influence across Europe from the court of Mathias Corvinus in Hungary to Henry VIII's court in England. Bankers and merchants also, in cities like Prague and Augsburg, fostered the new learning, employed scholars, put their sons to study and had their dignified selves portrayed in oils.

Students were drilled above all in Latin composition and exercises, less often in Greek, after the most respected ancient models. Alongside grammar, there were the inseparable arts of poetics and rhetoric, the means to give elegance and persuasiveness to language and hence to enhance a man's social presence. The moral content of classic texts highlighted questions of sensibility and conduct. Humanist learning thus promoted knowledge of individual men and women of a certain class as responsible agents, men and women who know what it is natural and right to do or feel in different circumstances. Such learning was about people in this world, but knowledge of man always also led ultimately to theological questions and to matters of faith. At the same time a practical literature attempted to communicate more widely what was valued in a man's or a woman's life. All this learning and moral concern built up a picture of human nature, and it is across the full range of Renaissance scholarship and cultural life that we must seek for sources of what later became the human sciences. Enthusiasm for rhetoric or moral philosophy gives the hint that what was novel in the Renaissance was not so much ideas as forms of life, a gradual change that tended to emphasize the value of subjective experience and the relation of that experience to responsible agency in civil society. We might perhaps describe this as the basis for psychological and sociological ways of thought. Key elements of such a basis are discussed in the following sections.

Education and ideas were not static, and there were innovations in the teaching of late medieval and early Renaissance logic, arithmetic and music. What did remain fixed was the assumption that education is text-based, that the core texts are ancient and that the teacher's prime role is exegetical. The logical works of Aristotle (384–322 BC), notably the *Prior Analytics* and *Posterior Analytics*, were known and taught in the West from the twelfth century; by 1400 they were accompanied by a series of commentaries and, at higher levels, the study of contrasted non-Aristotelian logics. A definitive printing in Greek of Aristotle's works appeared between 1495 and 1498, and the logical works were codified as the *Organon*; this edition became the basis for later Latin texts, commentaries and teaching. Students studied the texts systematically and with an eye to the way logical argumentation could be applied in natural philosophy, moral philosophy and theology. Debate about Aristotelian logic and its relevance to the method by which knowledge is acquired reached a high point in the sixteenth century, especially at Padua, the university that educated the patrician sons of Venice. The humanist movement tended to shift the focus from logic, though it had its greatest effect in preparatory teaching in 'grammar' schools. At university level there was pressure to support a cycle of studies called the *studia humanitatis* or humanities, which included *grammatica, rhetorica, poetica, historia* and *philosophia moralis*. The humanities were most distinctive of civil life in the courts and cities of Europe rather than university life, but in both these settings, moral philosophy became perhaps the most important of all the branches of humanist learning for new conceptions of knowledge of man.

In the medieval and Renaissance periods alike, study prepared the most serious students for advanced work in the three university faculties of theology, law and medicine – the three original professions. Here again teaching was primarily exegetical, though this technique did not exclude critical commentary and debate. As the foundation of learning, theology was a living force. Academic theologians, for example, constantly engaged in debate about the proper spheres of knowledge according to faith and knowledge according to reason. One point must be stressed again and again: what modern people refer to as 'the church' was never in fact a monolithic institution and did not impose religious belief; rather, human nature was constituted in and through Christian categories of understanding and practice. There neither was nor could have been something called 'science' independent of this culture. Few people, even among sixteenth-century humanists, considered that they might elevate reason over faith. This step was taken only in the seventeenth century, and only then can we discern elements of thought that some later natural philosophers hoped to turn into knowledge independent of theology.

Humanist lawyers attempted to move behind the layers of commentary added by medieval scholars to recover Roman law. This activity was conceived to be part of 'civil science' or 'civil wisdom', an attempt to order the grounds of right government founded in a notion of the *ius gentium* – the common justice of civilized people. There was also ecclesiastical or canon law, and in addition the effects of custom or local traditions that encouraged law to develop through practice, as in English common law. Debates in jurisprudence about such matters as evidence and responsibility – which related questions about a person's knowledge, a person's nature and his or her agency – contributed greatly to systematic representations of human nature. Further, we must note that the notion of natural law, discussed in the next chapter, which is at the basis of modern scientific understanding, was, in origin, legal as well as theological.

Last but certainly not least, medicine existed as the profession most obviously and directly focused on man's nature. Medicine, like law, related scholarly textual learning to the secular, material and practical affairs of the everyday world. As we will see, a combination of learning and practical action is characteristic, over and over again, of the way the human sciences have developed. For the great majority of people, of course, healing was not a learned subject but a domestic activity that relied on local oral knowledge. Medicine, in the form of systematic knowledge, was taught in the universities through Aristotle's texts on man's nature, especially through the text widely known in Latin as *De anima* (*On the Soul*), and through the works of

Galen (who flourished in Alexandria and Rome in the second century of the Christian era). Arab scholars, especially Avicenna (ibn Sīnā, 980–1037), added substantial commentaries and new studies. Debate in the sixteenth century between humanists who turned to revised Greek texts and physicians who defended the medieval and Islamic inheritance created a rich environment for thought about the proper relation between sensory experience and textual authority as the means to obtain knowledge. There was also debate about whether the brain or the heart is the centre of the vital spirits, and such debates used a language full of reference to individual human character. This language was still present in the twentieth century in phrases like 'cool headed' or 'warm hearted'. Physicians were expected to know the body's humours, parts and temperaments, the disorders to which they were subject and also something of the underlying causes. Medicine, as both a philosophical and a practical subject, by its very nature placed man at the centre of its attention. Popular medicine did the same, though without systematic and formal presentation of what it knew of man's and woman's nature.

Scholarship and humanism possessed no one discipline, nor even one set of disciplines, within which to examine human nature in the manner of the modern human sciences. Rather, 'man' was a ubiquitous subject, and it is in the thought entwined with the daily lives of men and women that we must search for the origins of modern knowledge. Before the development of modern human science disciplines, belief about man's nature was scattered across the subjects of the *studia humanitatis* and of the three higher professions. Such belief was also implicit in practical affairs. For good reasons, then, the historian can find no clear-cut boundaries to the relevant subject matter. Learning there was in abundance, and it ranged across the material, the moral and the spiritual worlds. Sometimes education directly addressed man's nature, as in medical debate about the humours, sometimes views about that nature remained implicit, as in commentaries on the foundations of law, and sometimes there was an informal and ad hoc conjunction of implicit and explicit comment, as in texts on rhetoric.

If we compare the Renaissance and the twentieth century, even though the Renaissance was a deeply religious age we can discern something that looks like a decisive step towards the establishment of secular opinion about human nature and towards the acceptance of modern science as a way to understand that nature. This step was a stress on the dignity of man, an enthusiasm for a way of life that values worldly human accomplishments and a confidence that ideals are achievable. There was a turning to knowledge of what people are as a worthy subject. None of this was entirely absent in medieval society, but it had heightened significance in the fifteenth century. It reached its apogee in the celebrated oration by Giovanni Pico della Mirandola (1463–94) on the dignity of man, an introduction to a set of theses he hoped to offer for pubic disputation in Rome at Epiphany 1487. In fact, Pope Innocent VIII intervened and some of the theses were condemned as heretical. Pico, an ardent Florentine philosopher and follower of Plato, made man central in significance, in responsibility, in freedom and in beauty. This was a vision of man's place in God's creation, a place which was irradiated by God's light but also shone with its own brightness. Thus, Pico had God address man: 'Thou, constrained by no limits, in accordance with thine own free will, in whose hand We have placed thee, shalt ordain for thyself the limits of thy nature. We have set thee at the world's center that thou mayest from there more easily observe whatever is in the world.'[3]

For all Pico's rhetoric, the stress on human dignity was fraught with ambivalence. In the medieval and Renaissance cosmos, humanity occupied a lowly position, bound to what is earthly, changeable and subject to corruption. The sin of Adam and Eve tied men and women to the flesh and to death. Alongside Pico's vision of the grandeur and majesty of man there were images and writings that vividly portrayed human folly, melancholy, subjection to harsh fate and inevitable death. For every line in praise of glory, there was an engraving of the reaper with his scythe, the hourglass and the skull that mocks man. Yet man was also believed to be the most significant of created beings, poised between the spiritual and the merely material, between the eternal and the merely temporal. Most profoundly, the Son of God had become man and had promised eternal life even beyond the ending of the world. When Copernicus placed the earth in orbit, philosophers both feared he had displaced man from his central position in the universe and enthused about the elevation of man to the heavens.

Emphasis on the dignity of man, in itself, explains little about the advent of modernity. What does stand out from a modern perspective is the dignity accorded to the soul as the vehicle of knowledge, particularly knowledge gained through the senses or, as we would say, experience. This idea needs to be expressed carefully; it was, after all, the Aristotelians who were the source of the much repeated dictum: 'There is nothing in the intellect that was not first in the senses.'[4] By the second half of the seventeenth century, however, sensory experience had become, to a new extent, a standard for authoritative knowledge. This could not have occurred without confidence in human capacities and without attention to what those capacities are. At first glance, it appears paradoxical that there was also a substantial rise in scepticism, best expressed by the French essayist, *seigneur* and eminent mayor of Bordeaux, Michel de Montaigne (1533–92), who conducted readers on a tour of contradictory claims to knowledge. But attention to the senses as sources of knowledge went hand in hand with heightened awareness of the difficulties that face the ideal of certainty. Sixteenth-century writers, when they dealt with this problem, turned to accounts of the soul and its relation to the material world as the means by which knowledge can be assessed. The activity of the individual soul, concretely engaged with the world, was placed at the centre of scholarship.

ii *On the Soul*

When we turn from the general characterization of human dignity to what was written on the specific parts and qualities attributed to men and women, we must begin with the soul, for all writers the essence of human nature, the principle which gives man his dignity.

The soul was not only a theological category that could or should be ignored in a history of science. Medieval Christendom regarded the drama of man's immortal principle as of transcendent significance; no philosophy could be countenanced that denied this principle. All the same, argument about the soul drew on the ancient pagan philosophers, and questions that concerned the soul's relation to spiritual goals and to immortality did not circumscribe the intellectual debate. The key texts were Aristotle's *De anima*, as well as the collection known as the *Parva naturalia* (*Lesser Pieces on Natural Things*), which included, among other matters, discussions of perception, memory, divination in sleep and senescence. The analysis of *De anima* remained a core element of academic training through the sixteenth century. The text was used to exemplify the Aristotelian mode of explanation, and it also (along with *De sensu* from the *Parva*) provided terms for debate on the manner in which the mind acquires knowledge. It was proper for philosophy to debate such things as long as it was clearly understood that the outcome offers no purchase on questions that concern the immortal soul. Ultimately, theology remained the superior discipline.

By the sixteenth century, *De anima* existed in scholarly Greek editions and in new Latin translations, all surrounded by different traditions of commentary. Even at this time there was debate about the proper translation of key concepts, and it may be almost impossible in modern English to recover the meaning of Aristotelian and early modern usages. The once standard English translation (1931), for example, added a list of contents to *De anima* that described the first section as on 'The dignity, usefulness, and difficulty of psychology'.[5] But the choice of the word 'psychology' as a term is deceptive. After a brief rhetorical flourish which placed the study of the soul in the front rank of scholarship, Aristotle asked what we would think of as philosophical questions about how the soul should be understood, such as whether it can be thought of as acting without body. He did not announce something called 'psychology' (a word he did not use) but wrote that 'the soul is the cause or source of the living body . . . all natural bodies are organs of the soul'. Indeed, a more recent and widely used revision of the translation of *De anima* silently removed references to psychology.[6] Book II of *De anima* discussed the soul as the principle of life, the form, in Aristotle's terms, that combines with substance to make the entity known as a living thing. The topic which then flowed naturally from this was an examination, in late medieval terms, of the faculties that the soul must have to make possible the nutrition, reproduction, sensation, motion and reason characteristic of human life. This discussion was also one of the bases of medical education, since an understanding of the soul was necessary to understand health and ill-health.

Aristotle then systematically introduced the senses, and he argued logically from their capacities evident in common experience to the necessary attributes of the soul. Finally, in Book III, he moved on to consider the relationship between the soul's activity – the nature of action was itself a major topic – in sensation and reason, and in this context he considered what in English translation is rendered as 'mind'. This was an area that provoked considerable dissension as it raised the vexed problem of the relationship between logical reasoning, generalization and knowledge and the contingent, particular material reality of sensation. The problem was often considered to be the question of how the organic (sensory) and the intellective (rational) soul are related. The further Christian question of the soul's relation to immortality was not dealt with primarily as a logical or empirical topic but rather as a question of how to comprehend the proper relation of what is known by faith to other forms of knowledge.

The *De anima* was the last and highest text for the Bachelor of Arts degree in most Renaissance universities. In one way it tied study of the nature of man to the nature of animals: it envisaged man as in possession of an organic soul perhaps superior in form but not different in kind from that found in animals. In another way, it turned the study of the nature of man into difficult and, in practice, highly technical philosophical and theological questions about the intellective soul's unity, capacity for reason and immortality. Thus, the intellective soul itself was commonly analysed in terms of separate faculties of reason and judgement.

The soul was the topic in terms of which scholars sought to mediate between knowledge of the body and faith in the immortal principle and to debate man's place in the universe, a place suspended between the earthly and temporal and the heavenly and eternal. The American historian of ideas Arthur O. Lovejoy made this vision of man's nature familiar to modern readers when he described how the ancient conception of the great chain of being was revived. This created a picture of the world as a hierarchy of entities from the most material up to the most spiritual. The human soul, divided between organic and intellective parts, lies at the mid-point. The human body is 'the noblest entity in the category of bodies' and the human soul 'occupies the lowest rank in the spiritual order', as an Elizabethan edition of writing by the medieval monk Ranulf Higden expressed it.[7] The study of the soul was therefore pivotal in every sense.

Philosophers and physicians left questions about immortality to the theologians and concentrated on the soul as a natural entity. They did not pose the question of the relation between body and mind in the modern sense but tried to understand, in terms of the Aristotelian explanatory categories of substance, form, cause and purpose, how the soul makes possible all of actual life, from logical reasoning to nutrition. Their arguments dealt with many problems, some of which we recognize as also our own, not the least of which was the question of how the sensation of external material objects enters imagination and mental reasoning. It was also difficult to know how the soul moves the body. In response to this, it was common to use a metaphor which likens the soul to the captain of a ship: he is not the substance of the ship but the ship is lost without his presence. Thus, Francesco Piccolomini (1523–1607), a philosopher at the university of Padua, supposed that the soul possesses innate principles of reasoning that judge of sensible images, rather as a captain's knowledge steers a ship.

Renaissance writers advanced a multitude of claims and counter-claims about the soul. They inherited the principal medieval commentaries from Averroes (ibn Rushd), the twelth-century Iberian Islamic scholar, and Aquinas, and the latter's thought flourished anew in the Counter-Reformation in the later sixteenth century, especially at the Jesuit College of Coimbra in Portugal, as a sophisticated response to both Protestantism and scepticism. Humanist learning added other sources and emphases, notably the Neoplatonic enthusiasm for the soul as the means through which man becomes one with the universe, one with God, and – since man in some degree reflects the creative act of God – effects improvement in the human condition. Renaissance Neoplatonists, backed by Duke Cosimo de' Medici's patronage of the philosopher Marsilio Ficino (1433–99) in Florence, translated and studied Plato's own texts, the early Christian Neoplatonic writers themselves and an exotic mélange of Jewish, Arab and other sources. This study supported magical views of nature in the sixteenth century, belief that the universe is crossed by a web of correspondences that tie man's nature and the fate of individual men and women to the natural world. A disciple of Ficino, who portrayed the human soul as an exhibition of both divine permanence and material corruptibility, called it 'truly the bond and knot of the universe'.[8] Astrology flourished and co-ordinated human fate with the motions of the heavens. A rich moral and intellectual rhetoric tied the macrocosm of the world to the microcosm of man. Ficino's fellow Florentine Pico della Mirandola held that 'God the craftsman blended our souls in the same mixing bowl with the celestial souls and of the same elements . . .'[9] A widely reproduced figure, well known in Leonardo's drawing or in the published work of Robert Fludd, points man's outstretched limbs to the four corners of the world, thereby placing man at the centre of the universe in which he lives but also in contact with the circumference. This figure, which goes back to the Latin writer Vitruvius, symbolized the harmony between man and the world, for the proportions of man match those of the world. These same harmonious proportions also underlay Renaissance architecture; there was a desire for human buildings to recreate the aesthetic principles of the creation of the world.

Philipp Melanchthon (1497–1560), the powerful scholar and administrator who carried Luther's reforms into the universities of Central Europe, perpetuated the Aristotelian agenda in his often republished Protestant texts. Nevertheless, in contrast to

Aristotle himself and more firmly than many commentators, he not only asserted the immortality of the soul but discussed the soul more in theological terms than in the terms then current in natural philosophy. Human knowledge, he argued (as some Catholics had also done), is restricted by original sin to sensory experience; it is therefore necessary sharply to distinguish such limited knowledge from the certainties of faith. With this clearly understood, he still provided a comprehensive account of the workings of the senses and the body, the account of the body based on Galen, setting out a practical moral philosophy for the direction of the passions.

A later section discusses connections between the passions, freedom of the will and moral philosophy. In relation to such issues, learned knowledge about the soul intertwined with ordinary life. Indeed, it was a leading theme with the humanists that learning should be practical, and they turned to language and rhetoric to achieve this end. In this they also followed Aristotle, who had argued that all human conduct, like animal motion, requires appetition, the exercise of the soul's potentiality to initiate movement according to reason, imagination or desire. This was the abstract basis for the attempt to understand what we call motivation and to study the moral and immoral conduct of daily life, and it formed part of the academic field of study then called moral philosophy. The discipline sought to integrate Aristotelian description of the soul's faculties with both pagan and Christian judgement regarding what it is right to do.

Aristotle's authority – he was often called simply 'the Philosopher' – to command the agenda, even while he stimulated critical commentary, came under devastating attack in the seventeenth century. A new natural philosophy argued that the physical world cannot be understood in Aristotelian terms, and there were necessarily implications for understanding the soul. The attack on Aristotle and on the scholarship for which he was the figurehead took many forms. Two of the most celebrated for the origins of modern natural science were those of Francis Bacon (1561–1626) and Galileo Galilei (1564–1642). Bacon's critique was methodological and founded in his belief that learning had become sterile, devoted to false 'idols' rather than to the clarity that comes from arguments based on deduction from experience and the construction of general laws from specific instances. Galileo's was substantive as well as methodological and framed his famous arguments for the truth of the Copernican system of the world in which the earth moves around the sun, a belief that struck at the heart of medieval Aristotelian natural philosophy. In the following generation, René Descartes (1596–1650) attempted systematically to replace Aristotelianism with a new metaphysics, or set of fundamental claims about reality, a metaphysics integrated with a new and mechanistic natural philosophy.

Before these attacks, however, Renaissance Aristotelians had struggled in sophisticated ways to describe how knowing – including knowing about knowing itself – relates to what is known. This was not the problem of knowledge as it is understood in modern philosophy but the problem of how to comprehend the soul in a way that helps to explain how the senses, the memory, imagination, cognition and judgement relate, and also how the intellective soul determines motion in the body. The latter problem required answers to questions about how knowledge of the world (as opposed to intuitive certainties or faith) is possible, how minds interact with the world in conduct, and even how minds relate to minds – the state of love. Some time after the seventeenth century, these questions became central problems in Western philosophy under the headings of epistemology (the theory of knowledge) and the mind–body problem. These are modern terms. For Aristotelians, such questions belonged on the agenda of the moral and natural philosophy of the soul, which is not exactly the same as the agenda of modern philosophy. Renaissance scholars divided up topics into questions about the organic soul and the intellective soul, the soul as the form of living processes and the soul as the form of reason, and it is difficult for us to find equivalent concepts. There is, notably, no clear way to map the modern concept of consciousness on to the Aristotelian descriptions. The soul, the central concept of Renaissance views of man, therefore presents the historian with striking problems in a search for the origins of the human sciences.

iii *The Eye*

One way to carry the search forward is to examine in more detail Renaissance theories of the sense of vision, partly because scholars then and later linked vision to knowledge and partly because the topic relates views on the soul to what to modern people is best known about the Renaissance, its art. Descriptions of the eye, of the eye's nervous connections to the brain and of the manner in which the physical parts are connected by an animated *spiritus*, were inherited from Galen, Avicenna and their medieval commentators. Another body of scholarship represented the relationship between the eye and the visual object in geometrical terms. By the late fifteenth century, the geometrical approach was associated with an elaborate theory of perspective and with the work of artists who, in bravura displays, exploited perspective in painting and architecture. This art was so successful that later generations learned to believe that perspectival painting reproduces the natural visual image in the eye, or, in a parallel way, to think that the eye creates a perspectival picture for the mind to look at. Art created expectations, which became embedded in Western culture and were still extremely strong at the end of the twentieth century, about what a 'real' visual representation of the external world should be like.

The great creative innovators in accounts of physical reality in the early seventeenth century, like Johannes Kepler (1571–1630) and Descartes, were fascinated by the eye. The eye was at the centre of the most concrete and enriched representations of physical reality in Western culture. Through the eye, it appeared, the soul acquires a literal picture of reality. To understand how the picture forms was therefore a pivotal issue for the new natural philosophy and for new conceptions of the mind as a conscious state of being.

The Aristotelian approach to vision discussed as one topic the manner in which external objects create a physical effect in the eye and the means by which the physical effect becomes a mental perception. There was also no clear differentiation between what we think of as conceptual (or philosophical) and empirical (or scientific) questions. There was extensive debate about what derives from the external object and enters the eye to make there the object's image; it was also asked what medium between object and eye makes this possible. The discussion was primarily logical, but it certainly paid attention to common visual experiences. We can note two significant things about the answers provided: they were technical and they were diverse. The learned study of vision supported a high level of intellectual activity that conspicuously failed to produce consensus. These conditions were precisely those which fostered explorations in new natural philosophies in the seventeenth century.

Aristotelian theories of vision exploited the explanatory categories of form and substance, actuality and potentiality. The foreignness of these terms and the difficulties they create for modern readers mark the changes introduced by seventeenth-century natural philosophy. We cannot without effort and imagination slip back into earlier ways of thought. In the Aristotelian world, a substance is only a mere potential for something unless possessed by form – the abstract nature of what a thing is (e.g., the circularity of a drawn circle). Any analysis of the world therefore involved the description of both substance and form. In relation to the eye, this suggested the possibility that the eye receives from objects a representation of the form of the object, technically known as the *species* of the object, which creates an actual object or picture in the eye. Aristotle and his commentators made much of the metaphor of a ring leaving its seal in soft wax; like the eye, the wax retains an image, the form of the object, not the object itself. Yet this metaphor, many generations of commentators thought, created more problems than it solved.

Aristotle attributed colour to an interaction between the object, the transparent medium between ourselves and the object and the eye; he did not identify colour with the abstract form of an external object. Thus Aristotelians, like later natural philosophers who referred to the primary and secondary qualities in our conscious perceptions, distinguished between what is essential to a physical object and what is adventitiously caused by the manner in which the object is perceived. This distinction between what is understood to be essential (primary) to an external object and what is understood to be added (secondary) by the process of perception, was at the heart of conceptual argument about whether sensory knowledge provides true knowledge of what is real and empirical argument about how the sense organs respond to physical objects.

Even when the relationship between the object and the eye was thought to be accounted for, there was still the question of how the soul receives an image. Medieval and Renaissance scholars broke this problem down into a series of questions, though each proved no more soluble than the others. How does the sensible image become perceptible? How does memory recreate an image and imagination create new ones? Is imagination properly assigned to the intellective or to the organic soul? Is the intellective soul active or passive when it possesses an actual image? How does the soul reason from a particular image to the general properties of reality? How does an image give rise to an appetite? These were formidable questions, and the great diversity of answers to which they gave rise contributed to scepticism about the power of Aristotelian philosophy. Aristotle's own account was an attempt to comprehend how the senses and the soul worked, but by the late sixteenth century this question about nature was often merged with questions about how perception makes knowledge possible. This conflation of questions about the workings of mind and the possibilities for knowledge, as well as the difficulties in the way of understanding vision, lasted for the next three centuries.

Mathematicians treated vision as a question in geometry about how lines drawn from the object enter the eye and there recreate an image of the object. This treatment was integrated with the theory of linear perspective and then, in the sixteenth century, with the theory of the camera obscura, the reproduction on a screen or in the eye of a living image, or of an image painted on the inside walls of a box, from light passing through a pinhole. At the same time, alongside the mathematical and artistic studies, there was renewed debate about the eye's anatomy and about which part is in fact sensitive to light – most earlier writers had believed it to be the crystalline humour, the structure we know as the lens.

In 1604, Kepler, the mathematician and astrologer to Rudolph II in Prague, whose work on the elliptical orbits of the planets was of decisive importance in the history of science, published his studies that transformed the understanding of vision. He described the crystalline humour as a lens which focuses rays of light to create the visual image on the back of the eye. His work was a powerful and beautiful exercise in rigorous geometrical analysis, and it was the starting-point for later studies. But to complete his analysis he and others had to accept that the image – which he suggestively called a picture – is upside down and back to front on the retina at the back of the eye. Kepler reported that he 'dutifully tortured' himself to solve the contradictions between this conclusion and what we see, but his answer finally was to exclude the problem from his geometrical studies. He put to one side the questions he could not solve about how the eye 'sees'.

> How the image or picture is composed by the visual spirits that reside in the retina and the [optic] nerve, and whether it is made to appear before the soul or the tribunal of the visual faculty by a spirit within the hollows of the brain, or whether the visual faculty, like a magistrate sent by the soul, goes forth from the administrative chamber of the brain into the optic nerve and the retina to meet this image, as though descending to a lower court – [all] this I leave to be disputed by the physicists [i.e., physiologists].[10]

Kepler separated geometrical analysis of image formation from the attempt to understand how the brain and mind assimilate the image. This was a foretaste of what was to come: brilliant mathematical analysis combined with utter disorder when an attempt was made to relate the world so analysed to the mind. Kepler and his contemporaries were well aware that the optic nerves cross as they enter the brain (in the optic chiasma); but, even when they surmised that this structure had a role in re-inverting the visual image, there was no clear language for talking about *what* is re-inverted (an image? optic spirits? a sensible presence?). And, beyond this, their work contributed nothing to a coherent view of how it can be said the mind 'sees', as opposed to how the retina receives an image.

This specific problem of 'seeing' exemplifies the larger difficulty that existed about how the knowing subject relates to the physical world. On this topic, what was said about the nature of the soul was central to what were to remain core questions in the philosophy of the natural sciences. Further, the whole way in which people came to accept the model of the soul 'seeing' an image, like an observer before a picture, structured the manner in which the mind's relation to the body was thought about over the following centuries. The history of the human sciences, therefore, is central and not peripheral to the history of the natural sciences.

iv *The Spirits and Humours*

Aristotelian studies of the soul and of the senses presumed a high level of academic training. The explanatory vocabulary of form and substance or of potentiality and actuality now rightly suggests the word 'scholastic'. Nevertheless, scholastic vocabulary merged with ordinary everyday language to describe human nature in general and individual character in particular. This was the language of spirits, humours and temperaments, as in she is 'in good spirits', he is 'in a good humour', or she has 'a hot temper'. This was the most widely diffused language for the description of human nature in the early modern period. The spirits and humours were held responsible for the human variety that is the stuff of daily life. Here the pursuit of knowledge met the art of living.

The spirits were understood to be substances, but they were – using modern terms – neither strictly material nor strictly mental. It was held that they mediate between the principle of life, i.e, the soul and its faculties, and the material body built from the four elements of earth, air, fire and water. Philosophers described *spiritus* as the 'first instrument' of the soul. From the ancient world through the Renaissance and, indeed, on into the nineteenth century, references to spirits helped explain how causal agency connects things in nature. Spirits were believed to possess attributes that link action in material and mental realms; for example, it was believed that spirits transmit light impinging on the eye to the intellective soul or convert the imagination into movement of the limbs. In the early seventeenth century, Robert Burton (1577–1640), the famed author of *The Anatomy of Melancholy* (1621), described spirit as a 'most subtile vapour, which is . . . the instrument of the soule . . . a common tye or *medium* betwixt the body and the soule'.[11]

Galen had systematized what we call physiology and his work passed through the Islamic world to the late medieval West. These writings described three major systems in the body, each with its appropriate spirit: the liver and veins, carrying vegetative or nutritive spirit; the heart and arteries, carrying vital spirit; and the brain and nerves, carrying animal or sensitive spirit. In the Galenic scheme, the spirits were thus the medium through which the life, purpose and sensitivity made possible by the organic soul became actual in the body.

The brain and nerves, in this way of thought, are hollow structures containing and communicating the animal spirits. The illustrators who pictured this system incorporated into their work an idea of how the spirits function and showed the ventricles (or brain's internal spaces) as the physical site where the spirits mediate between the body and the mental faculties. They portrayed sensation at the front of the brain, connected closely to imagination and fantasy, and placed the faculties of reason (cognition and judgement) in the middle and memory at the rear. It was also argued, however, that reason, which scholars attributed unambiguously to the intellective as opposed to the organic soul, could not properly be spatially located with quite this freedom. Illustrations of the ventricles and hence of the organic soul were repeated by Leonardo in his anatomical drawings and in the celebrated woodblock plates of Andreas Vesalius (1514–64). Vesalius was the professor of medicine in Padua whose *De humani corporis fabrica* (*On the Fabric of the Human Body*) appeared in 1543, the book which for many physicians marks the beginning of modern anatomy. Such representations of where the soul and the spirits are located continued to appear through the period of the new mechanistic science in the seventeenth century. Descartes attacked the explanatory categories that underlay Galenic medicine; all the same, when he and his followers wrote about soul–body interaction they continued to use terms that had a remarkable resemblance to the language of spirits. As we have seen, Kepler was in a similar position, though he put the problem on one side when he addressed the ancient problem of the formation of the visual image.

The language of spirits is familiar as part of the Renaissance and modern everyday way of describing people; we say that people are high spirited, in low spirits, mean spirited, and so forth. There is a similar familiarity about the related Greek or Hippocratic theory of the humours, the theory that was the mainstay of sixteenth-century opinion about individual character and differences between people and groups, including sexual differences and differences of geographical origin. The humours were discussed as the visible sign of the qualities in the body, associated with the four bodily substances, whose proportion and balance signify temperament and the state of well-being or ill health:

humour	*bodily substance*	*dominant temperament*
hot and moist	blood	sanguine
hot and dry	yellow bile	choleric
cold and dry	black bile	melancholic
cold and moist	phlegm	phlegmatic

This rich and much elaborated language for the human condition united classically educated physicians and the hoi polloi of Shakespeare's theatre. Theatregoers knew, for example, how to understand a description of a woman pining in love: 'And with a green and yellow melancholy, she sat like patience on a monument.'[12] When this language was integrated with astrology and belief in the relation between the planets and human life – which led to descriptions like saturnine or jovial – it created a glittering resource for judgements on human character. The language of the humours, though in increasingly transmuted forms, persisted in medicine down to the nineteenth century. It was the medium for ways of thought that integrated the activity of body and mind, even when physicians had in theory adopted other terms.

The language of humours and spirits was prescriptive as well as descriptive and stressed the relation between moral conduct and health. The key to virtuous and healthy living and temperament was thought to be the proper regulation of spirits and balance of the humours. Aristotelian scholars and popular opinion alike located the passions in the heart and thought of them as an agitation of the vital spirits – and in the heart the romantic passions were to remain. Thus, to be disturbed is to have one's reason, the proper exercise of the soul, agitated by the spirits in the heart, whether for good or bad. There certainly was no assumption in the Renaissance that the passions are necessarily harmful. A 'sensitive soul' was considered to be someone who is easily disturbed, as when (in Burton's words) 'the Imagination be very apprehensive, intent, and violent, it sends great store of spirits to, or from the heart, and makes a deeper impression, and greater tumult'.[13] Moralists inevitably argued that the imagination should be disciplined. When the imagination or senses overwhelm, as Shakespeare portrayed Brutus's feelings in *Julius Caesar*: 'the state of man, like to a little kingdom, suffers then the nature of an insurrection'.[14]

As these lines express, human nature and experience gained added significance from the correspondences drawn between human life, the body and the larger natural and political worlds. It was understood that systems of signs linked the heavens, the political sphere, the well-being of communities, the health of bodies and individual fortune, temperament and mood. This worldview did not simply play with metaphor, though metaphor was commonplace, but articulated belief in a web of reciprocal exchanges of meaning. It was not imagined that any one level of reality, cosmic or human, was grounded independently of any other. To modern eyes, wearied by industry and technology, it sometimes looks as if sixteenth-century people enjoyed an enchanted vision in which human affairs are integrated in the great movement and drama of the world as a living whole. In words translated into English in 1605,

> For in man's self is Fire, Aire, Earth, and Sea,
> Man's (in a word) the World's Epitome.[15]

Mystical and magical ways of thought reinforced such a vision. But, we must remember, this was also a dream to counteract the slaughter, famine, plague and sheer wickedness of the world. Renaissance writers and artists portrayed disharmony as much as harmony. Disharmony in man's nature was understood to be an opposition between the rational soul and agitated spirits, a language that mirrored in naturalistic terms the Christian sense of the duality of spirit and flesh. When writers located the passions in the heart and the faculties of the soul in the brain, they set out in spatial terms a hierarchical evaluation of the lower and higher poles of human action, the opposition of the flesh and the spirit. In the words of the English seventeenth-century religious moralist Sir Thomas Browne, human drama stemmed from man being 'only that amphibious piece between a corporall and a spirituall essence'.[16]

The language of passion had an academic as well as a popular meaning. In Aristotelian thought the word 'passion' (*passio*) denoted the capacity for or actual reception, from outside, of a quality by a substance. Aristotelians considered the experience of passion to be the organic soul's reception of an affect mediated by the vital spirits in the heart. Passion cannot simply be equated with the modern category of emotion – the term 'emotion', in its ordinary modern sense, was not in common use till the late eighteenth century. Though not strictly the opposite of *passio, actio* or action denoted a quality of the intellective soul in thought or judgement.

The correspondence between the spirits, humours and temperaments was also believed to be visible in the outward appearance of a person, in expression in the shorter term and in physiognomy in the longer term. Ficino, in a commentary on Plato written in 1484, stated: 'The soul itself we cannot of course see . . . But we can see the body, which is a shadow and image of the soul: and so surmising from its image, we conjecture that in a handsome body there is an attractive soul. That is why we prefer handsome pupils.'[17] Visual appearance was thought to reveal a person's character, as explained, for example by the Neopolitan natural philosopher associated with the magical tradition, G. B. della Porta (1535–1615), in his *De humana physiognomia* (*On Human Physiognomy*, 1586). A complex system of conventions carried these beliefs into painting and on to the stage. Everyday language, high art and claims about the physical body came together to make possible descriptions of human moods and differences. This pattern, in which scholarly discourse and popular language ran together on the subject of human differences, was one that reappeared in later centuries. The nineteenth-century enthusiasms for phrenology and racial classification are cases in point. This is one very significant respect in which everyday concerns have fed into the sciences of man and those sciences, in turn, have reinforced everyday judgements.

A large number of sixteenth-century books – printed books were increasingly common after 1500 – as well as sermons and tracts were published as guides to everyday life. These practical texts reflected and reinforced existing theories of human nature, and they created a close relationship between classical and popular cultures, especially in belief about the humours and physiognomy. A notable example of a book for the élite was by the Spanish writer Juan Huartes (1529–88), *Examen de ingenios para los ciencias* (1575, translated in 1594 as *The Examination of Men's Wits*). Huarte provided a practical guide to character that linked physiognomy, humour and climate, and he hoped to have theologians, lawyers and even kings among his readers. Not the least of his pieces of advice was on how to procreate clever boys rather than ordinary girls. But opinion about what made men and women distinct needs separate discussion. On this topic, the question of human differences faced everyone.

v *Woman*

Scholars believed in the relative activity of male souls and the relative passivity of female souls. In modern terms, language was gendered, that is, it built evaluations fixed in the culture into the description of sexual difference: softness and passion were female principles; hardness and action, male principles. Each principle derived meaning from its place as part of a couple, such as soft–hard. It was conceived that the harmonious union of male and female principles gives order to the social world and creates new life. There is a striking parallel with the contemporary belief that knowledge, harmony of the intellect, derives from a proper connection between *actio* and *passio*.

Sexual differences appear to many modern people to be timeless, a fixed biological reality underneath, and responsible for, the basic framework of what we are and how we live. Since the 1960s, however, the term 'gender' has become common to describe male and female differences without any presupposition about what in those differences is attributable to culture and what to nature. The fact is, reference to differences is never neutral and free from the controversy about what belongs to nature and what to culture, and historians enter the controversy when they interpret beliefs and practices in the past. It cannot be assumed that the concept of sex (as used in the phrases 'the male sex' and 'the female sex') is universal; at the very least, sex is always known through the lens of a particular culture – even if it is that culture's belief that sex is 'outside' culture. Further, historical evidence suggests that people in the sixteenth century did not experience or refer to sex or sexual difference in the way that twentieth-century people did. Sex was differently constituted and we must look with an open mind to discover what male and female were taken to be. What we find will in turn affect our view of the roots of the human sciences.

We also perceive misogyny in Renaissance and early modern accounts of woman's nature. The language in use to represent human differences was implicitly, and often explicitly, judgemental about what are superior and inferior qualities. It was commonly believed that woman's nature explained, supported and justified her inferior economic position, her legal subordination to her husband in marriage and her absence from political institutions. Rulers such as Elizabeth I or Catherine de Médicis (Regent of France, 1560–74) were very interesting and, for their contemporaries, challenging exceptions. The historians' common phrase 'Renaissance woman' reflects the different status of the sexes, since it signifies that woman is the special topic and that the culture discussed her by comparison to the male ideal of human nature rather than in her own right. Aristotelian authority, once again, provided a language with which to express belief in woman's difference and inferiority. Aristotle had argued that differences between male and female arise in the embryo after conception, and that they are due to the degree of heat and hence animation of the embryo's vital spirit. He supposed that a certain degree of heat is necessary to produce the animation of the male; lesser heat produces the relatively passive female. The outward sign of these processes is the growth of the hot male genitalia in the cool air while the supposedly comparable cold female parts, the womb and ovaries, remain internal. By the same logic it was believed that overheating of the womb produces furore, and it was common to portray the womb as a raging beast, which also explains, for example, the excess of a woman's lust or why women have less control. Representation of the womb – for which the Greek word was '*hystera*' – provided an ample home for men's imagination and desire.

There was scope for much dissension in both learned and commonplace belief about sex and reproduction. For instance, the Aristotelian view that the male contributes the essential animating principle in conception, the female only nutritive substance, was widely opposed by a belief that there is female as well as male semen. Differences of opinion, however, rarely shook the notion of a sexual hierarchy. The literature referred to the relative softness of the female brain (we still say 'soft in the head') and thereby underwrote assumed differences in rational capacity by reference to differences in bodily substance. Similarly, it was argued that the sluggish flow of spirits in women equips them for quiet domestic routine rather than arduous public careers.

The denigration of women was routine. As one early sixteenth-century Aristotelian commentator wrote: 'For a sleeping man is only half a man; similarly, the principle creating woman is only semi-virile. It is for this reason that woman is called an imperfect version of the male by philosophers.'[18] Nevertheless, this denigration was not gratuitous but part of a worldview in which female characteristics were understood to be not low in themselves but low in comparison to their opposites. Active–passive, form–matter, perfection–imperfection, and so forth, were linked terms. Sex was one among many considerations in practical moral philosophy and jurisprudence. Some moralists, it is true, quoted the Latin author Seneca to the effect that 'anger is a womanly vice', and pointed out that since a woman's humours fluctuate, she requires a husband's firm direction.[19] Other writers emphasized the special sufferings laid on women by the Fall in the Garden of Eden. But scriptural authority also pointed to the divine creation of woman as a companion for man – her creation from Adam's rib, not his foot. Women were also believed to possess distinctive virtues – humility, patience, compassion, charity and devoutness. Writers, especially those influenced by Neoplatonism, therefore held up the creation of harmony between female virtues and male capacities as one of the highest ends of the creation. It was common to represent such an ideal harmony by androgynous figures, which exhibit perfection by uniting female and male attributes.

Legal writers who described marriage as natural, i.e., ordained by created natural law, and treated women as 'wives', not as individuals capable of owning property, expressed a wider philosophy rather than simple misogyny. While it was accepted that marriage is a natural institution, and the division of labour within marriage its natural complement, there was a stress on the complementarity as well as the hierarchy of the sexes. The harmonious marriage was believed to be a precondition as well as an emblem of the harmonious social order. The involvement of the sexes in reproduction was pictured as the restoration of a human's whole from two halves. There was reciprocity between material and spiritual representation of this union: anatomists drew the reproductive parts as mirror images of each other (the vagina and womb pictured as an inverted penis), while moralists described man's active virtues as perfected by woman's passive ones. In most theological discussions, woman was understood to be the equal of man, as all men were understood to be the equal of each other, in God's grace. For Calvinists, the question of predestination cut across differences of sex. Yet theological argument usually no more equalized power between the sexes than it did between social groups; the partial exceptions were radical sects like the Quakers in the mid-seventeenth century. Religion sometimes encouraged a revaluation of women in the seventeenth century, as when a French writer argued in 1647 that 'all things are thus equal between men and women with respect to the soul'.[20] Right at the end of that century, Locke simply assumed that education was equally suitable for either sex and that educated women would make better companions for men, though he neither tackled the issue systematically nor married himself.

Further evidence to link characterizations of woman to a wider philosophy, not simply to her inferior social position, is provided by the imagery of female personification. Illustrators, painters, poets, dramatists and essayists all personified Nature as a woman. Similarly, abstract ideals such as science, truth, justice and virtue appeared in the form of women, along with the muses, *Fortunà* and the fates. Reason took the form of the goddess Athena or Minerva armed with a sword and taming a lion. It is also a fact, though unexplained, that abstract nouns have grammatically female gender in the Latin, Italian, French and Germanic languages. Such subjects and personifications imbued the female with dignity. The symbolism and personified images had meaning, however, in a culture in which woman was the object of male gaze rather than herself the valued subject. The dignity of woman was that afforded to her in a world defined by man. It was the male poet who made obeisance to his muse, the male physician or natural philosopher who 'unveiled' the secrets of nature. As 'she' guided 'him', so aesthetic harmony, ordered reason and social justice was thought to become possible. Bacon called for the new science in the seventeenth century to be 'masculine' and virile. Right down through the eighteenth century and beyond, visual imagery portrayed science as the exposure of hidden truths, a process pictured emblematically by Reason removing the coverings of Nature, her breasts bared to signify 'the naked truth'. The frontispiece of the greatest publishing venture of the eighteenth century, the French multivolume *Enclyclopédie*, was an engraving of truth unveiled by reason and philosophy.

What are we to make of this gendered imagery? First, it reveals a rich discourse in which difference was a presumption rather than a conclusion. Renaissance and later belief about the divergent nature of male and female was built into the language. The system of thought associated human differences with nature and envisaged what it is to have a relationship with nature in terms of those differences. Second, though accounts of sexual difference were evaluative in ways that substantially reinforced the social realities of marriage, the personification of abstract ideals indicates that human experience and aspirations did not exist only in that single social dimension – however much personification may have reflected that dimension. The language of human difference linked the everyday world of humours with transcendent ideals of reason and faith. From such a complex culture came the terms with which the human sciences later framed their more specialized studies.

vi *Moral Philosophy*

Renaissance writers admired activity that brought glory to individuals as citizens and glory to the state. They demanded of education that it prepare citizens to exercise a wise authority. The élite families who ruled republican Venice, for example, sent their sons to the university at Padua to be prepared for the arts of government as well as the professions. The faculty there also provided learning and administrative guidance for Venice's affairs in such matters as public health and civil law. An education grounded in the great Latin and Greek authors, along with exercises in the right use of reason, was expected to give a man character, social grace, rational ability and moral discipline. The conception of education as the means to achieve moral and social cultivation affected schools and universities, but it also went far beyond their precincts to affect the lives of those who did not attend university, like Leonardo and like all women, and influenced the circles of patronage that surrounded Europe's many rulers.

Learning and moral cultivation existed at the same time as violent events: moral understanding connected with diverse representations of man's nature and was no remote scholarly indulgence. There was continuous controversy about, and sometimes despair at, humanity's inability to reach the truth. The Italian states were endlessly at war, Spain became enmeshed in genocide in the New World, and by 1520 Martin Luther had launched the Reformation on the back of which secular princes struggled for power and devastated Central Europe for the next century and a half. When scholars considered man's soul, there was every reason to turn to questions of fate, fortune and the powers that man might have to improve his condition.

The civil emphasis in humanism strengthened the place in the university curriculum of moral philosophy, rhetoric and history. The professions – law, theology and medicine – also elaborated concepts and rules about people's activity in their social relations. All in all, it was perhaps along this broad interface between learning and practical affairs that the intellectual developments occurred which were of most significance for the future human sciences. The portrayal in literature, poetry and art of human ideals, the attempt to model action on those ideals by the teaching of grammar, rhetoric and history, the systematic study of right action as a practical contribution to public life, and the elaboration of all these concerns in law, medicine and theology created a culture in which the person became a subject of systematic knowledge.

There was continuity between medieval and sixteenth-century teaching of moral philosophy, a subject divided into three parts: ethics, economics and politics. These respectively addressed the nature of the good man, the good head of the household and the good citizen, magistrate or prince. Moral philosophy directed man to attend both to himself and to others, and it enhanced a sense of subjective and of social identity. Each part of moral philosophy dealt with human conduct as an individual and as a social act, and ethics, economics and politics developed views about what it is natural for people to do just as they developed prescriptions about what it is right for people to do. The question of what is natural did not, however, usually become a subject in its own right. This step was to be much more characteristic of the eighteenth-century search for a science of man. All the same, Renaissance moral philosophy was rich in insights into the practical life of the human soul. Moral and Christian preoccupations were not in opposition to science. Moral philosophy was central to the origins of the human sciences, origins in a world where hopes for knowledge about human nature rested on belief in the divine *Telos* or purpose. There is no direct correspondence between moral philosophy and any of the disciplines of the human sciences, yet the two are connected by categories of explanation like reason, passion, agency and will, however much the meaning of these categories has changed over time. Even where the modern human sciences claim to be branches of natural science, they still deploy terms like 'law' with their roots in Christian medieval culture.

In the *Nicomachean Ethics*, Aristotle had described the purpose of a man's life as the attainment of the supreme good, that which is desired for its own sake. Christians, of course, understood this supreme good to be God, and practical Christian ethics was devoted to the manner in which faith becomes the rule of life. A seemingly endless stream of written and, later, printed instructions and advice, from Catholic confessional manuals to Puritan rules for self-examination, directed attention to what is good. The starting-point for Christian ethics was the story of Adam and Eve, of man's Fall from grace and of the legacy of original sin, the burden of wickedness with which men and women are born. Original sin, and the promise of its redemption through the life of Jesus Christ, was accepted to be the key to the interpretation of human conduct. Different theological views therefore resulted in different practical expectations in regard to human nature. Melanchthon, the Protestant reformer, to take an influential example, argued that original sin vitiates man's spiritual understanding of God, not his faith in God, but leaves his reason capable of judgement concerning what is right and wrong. He therefore described moral philosophy as the study of the natural laws known by reason, and he advocated this study as the means to establish rules about right conduct. Melanchthon then extended this concern with right conduct to encompass government of the household and the state as well as the person.

Aristotle's own pagan conception of the good, perpetuated in many Renaissance versions, stressed man's happiness, understood as the proper exercise of reason in accordance with virtue. Debates about what the virtues are and how happiness can be achieved were thus practical debates about how to live, and in this context there was some description of woman's separate nature and conduct. Writers listed the virtues and divided them into those that belong to the intellective-rational soul, for example, intellect and prudence, and those of the organic-irrational soul, like courage, affability and modesty. Literature, history and common experience provided the models that enabled people to see how conduct and character turn human nature to the good or to the wicked life. Aristotle had claimed that moral virtue lies in the maintenance of a golden mean between actions and passions of all kinds, and it thus followed that the happy person is moderate in all things. This was the substance of Polonius's admonition to Laertes in *Hamlet*:

> Give thy thought no tongue,
> Nor any unproportioned thought his act.
> Be thou familiar but by no means vulgar.[21]

Not everyone agreed with this nor with the Aristotelian characterization of the virtues. Thus the Italian humanist and translator of the Greek historians Valla quoted Revelation rather than Aristotle when he alluded to the bodily humours: 'Because thou art lukewarm, and neither hot nor cold, I will spew thee out of my mouth.'[22] Other writers, however, argued that Valla confused theology and moral philosophy which, founded in practical observation of what is natural to man, should be kept distinct. Conversely, yet others looked to scripture for precepts about the detailed conduct of daily life, a practice which achieved great prominence in many branches of the new Protestantism. At its extreme, Calvinist Protestantism disowned humanism and described man's nature almost entirely within the framework of sin, a description that made man entirely dependent on God's will.

Ancient Stoic moral thought focused on worldly conditions with particular intensity and independence from theology. Renaissance commentators, like their medieval predecessors, were familiar with the doctrines promulgated by Seneca and Cicero. These Latin authors portrayed virtue as the supreme and only good, and they concluded that wise conduct does not allow fortune or misfortune, worldly success or failure, to affect the internal virtue that alone determines happiness. Few people were able to follow such a severe doctrine either in theory or in practice. But it did characterize a notion of the subjective self as the distinct home of what is good, and it concentrated attention on the achievement of *self*-control over the passions as the art necessary before the good life is possible in this world. Many passionate Renaissance critics thought that the Stoics' apparent denial of the passions, of love as well as those that are destructive, was intolerable and unrealistic. In fact, Stoic writers had carefully delineated the passions as part of what defines man's nature, and they had then divided them into those that assist and those that oppose action in conformity with the rational laws of nature. The Stoics did not think it was possible to eliminate irrational passions from man's nature, but they did think it possible to rise above them. This belief, in turn, reflected an assumption that a natural hierarchy relates the higher reason and the lower passions. Some writers, like Justus Lipsius (1547–1606), whose work was a moral response to the devastation war produced in the Low Countries, thought that Stoic ideas therefore provided a route to a practical Christian piety.

Renaissance moral philosophers were also acquainted with ancient Epicurean doctrines that identified what is good with what is pleasurable. Few Christian writers could countenance the acceptance of such thoughts; Melanchthon, for example, sternly observed that we do not seek to know God on account of our pleasure but in order to obey his will. However, the discovery in 1417 of Lucretius' poem *De rerum natura* (*On the Nature of Things*), written in the century before Christ, presented the humanists, who much admired Lucretius' language, with a powerful evocation of man as part of material nature. Lucretius beautifully portrayed what is right as what is natural. But he achieved this by the adoption of a materialist view of nature, complete with mind atoms as well as physical atoms, which was intolerable to Christian writers.

> For when . . . [the spirit] is seen to drive forward the limbs, to arouse the body from sleep, to change the countenance, to guide and steer the whole man, and we see that none of these things can be done without touch, and further that there is no touch whithout body, must we not confess that mind and spirit have a bodily nature?

What we experience as free will, Lucretius argued, arises from the movement of an atom: 'all motion is always one long chain, and new motion arises out of the old in order invariable'.[23] This theory predicated events on the law and order of nature in a way that appealed to many natural philosophers. When, in the seventeenth century, they began to think that some version of atomism might explain physical reality, they first felt the need to distance themselves from ancient atomism's association with the denial of the soul, of free will and even of God. As the next chapter shows, it was a central part of the intellectual challenge presented by Thomas Hobbes to the later sciences of man that he boldly embraced a belief in the soul's material nature.

Moral philosophers provided practical descriptions of the passions as well as philosophical discourses on the good life. These descriptions often took the form of lists which classified important aspects of daily expression, for example, of anger, gentleness, fear, confidence, shame, pity and emulation. Description and classification of the passions was one of the most systematic areas of the study of women's and men's nature; it was also one of the most practical.

It was commonplace to hold up reason as the distinctive quality and crowning attribute of man. The passions were not denied, but they were seen as the irrational part that man shares with the animals. Animals themselves were generally believed to possess an organic but not an intellective soul and hence to be devoid of reason. The limits of reason were represented in the life of many communities which had their days of unreason, of carnival or *Fasching*, when the norms of propriety and nature were inverted, with women dressed as men or the poor free to command the rich. The painter Hieronymus Bosch pictured 'the ship of fools', a rudderless vessel launched on the waters of the world with its all-too-human complement of rudderless raving men and women. It was a metaphor for the dangers of unreason that the good man, good economy and good government sought to navigate around. A mass of practical guides of all kinds, from prayer books to almanacs to instructions for princes, proffered guidance and thereby created a written discourse about the individual agent as both a natural and a moral subject.

Moral argument presupposed that a person has some kind of choice. This was part of the traditional conception of man: Pico's rhetoric on the dignity of man included the striking claim that: 'On man when he came into life the Father conferred the seeds of all kinds and the germs of every way of life.'[24] But it was a difficult matter for theologians and moral philosophers alike to specify what it is that makes man an agent. At the naturalistic end of the spectrum of opinion, some Aristotelians described the will as a natural appetite. The great Netherlandish humanist and theologian Desiderius Erasmus (1466/9–1536), however, conceived of the power to act morally as a power to give inner consent to the rational law in the creation. This conception was in opposition to the non-naturalistic end of the spectrum of opinion, exemplified by Luther, which portrayed man as wholly corrupted by original sin and therefore totally dependent on redemption through Christ to do good. Finally, the sceptic Montaigne, when he turned to individual experience, not least his own, found little of either freedom or glory. As he wrote: 'It is no good our mounting on stilts, for even on stilts we have to walk with our own legs; and upon the most exalted throne in the world it is still our own bottom that we sit on.'[25] To his many modern readers it seemed that when he thus turned to self-experience as a basis for judgement, he showed a different, modern sensibility about human nature.

Pico's admiration for human freedom was often matched by its opposite, a world-weary reference to fate and fortune. When he had advised his prince about every measure that he could take to grasp and retain power, Niccolò Machiavelli (1469–1527), the writer of the most celebrated Renaissance work in practical politics, *Il principe* (*The Prince*, written 1513), conceded that at least half of any outcome was due to *fortunà*. Many people further believed that their fate was bound up with the movements of the heavens and the fortunes of birth. The magical philosophies of the sixteenth century, which tied fortune to the *spiritus* in the universe, strengthened this point of view. Though this line of thought was sometimes fatalistic, it also suggested that there is clear practical value in the search to understand the ways of nature. Pietro Pompanazzi (1462–1525), who spent the end of his life teaching philosophy at the university in Bologna, specifically answered Pico in *De fato* (*On Fate*, completed in 1520), in which he referred to man's place in a universe ruled by law and order, to which man's actions necessarily conform. Aware that this raised the spectre of necessity in everything that a person does, Pompanazzi sought to reconcile his argument with Christian belief through the attribution of all law and order to God's providence. But as he also appeared to argue that the soul perishes at death along with the body, it is not surprising that he provoked a serious theological controversy.

The Christian conception of material and moral order in the creation was the most fundamental intellectual precondition for belief that an order underlies human life. The understanding of human conduct went hand in hand with the understanding of nature. This correspondence was frequently expressed as the intimate relationship of the microcosm, man, and the macrocosm, nature. In the Renaissance and on into the seventeenth century, it was assumed that ordered hierarchies at each level in the created world – the cosmos, the state, the person – were under the sway of natural law and thereby mirrored each other.

vii *Rhetoric*

There was a great breadth of interest in man's nature. This and the following section turn to two subjects once ignored by the history of science, rhetoric and history. These topics confirm how important it is to search for the origins of the human sciences in a culture where abstract beliefs and practical affairs merged. Rhetoric, in the Aristotelian tradition, was an art, but it was an art that concretely expressed abstract notions of what it is to be a civilized human being. History resembled literature, but it nevertheless provided concrete knowledge of how time, custom and human action create – or destroy – civil society; much the same could be said of jurisprudence, the subject of the next chapter. These early modern disciplines were a major part of the background to the eighteenth-century science of man, a science which set out to explain and to advance civilization itself. They were also important as a medium through which people became familiar with naturalistic approaches to human affairs, which explain human achievement in terms of what flesh and blood human beings can do.

Rhetoric was a practical art rather than a systematic science, and it did not articulate presuppositions about man's nature as a subject in its own right. There were practical exercises aimed at pupils and scholars, and there were self-help guides or conduct manuals – as popular in the sixteenth century as they were in the twentieth. The formal discipline with which rhetoric was most closely aligned was logic, as 'logic' then denoted the means to accurate communication rather than the method to generate knowledge. Rhetoric and logic were together central to the scholarly activity which transferred knowledge from one learned person to another or to a student. Later, in the mid-seventeenth century, it was an important feature of the new natural philosophy that rhetoric and logic changed into methods for the advancement of knowledge.

'Rhetoric', in popular late twentieth-century usage, had negative connotations – 'mere rhetoric'. By contrast, the enthusiasm of previous centuries for the subject embraced language as the vehicle for civilized human authority. Rhetoric was embedded in education from the medieval period to the eighteenth century; and it was accepted as an engagement with the highest ends of human affairs, not simply a skill. Pupils at school and university, all future leaders of households if not of political states, trained with Latin models in order to express themselves with grace and power. Orators and poets needed to understand audiences and the effect of words if their rhetoric was to achieve success. Textbooks and manuals which, for example, correlated particular literary tropes with particular psychological effects, supplied the market. In the process discussion of rhetoric propagated opinion about how people naturally respond. Speeches in different settings were understood to require different styles; thus judicial or forensic rhetoric was concerned with proof and disproof while demonstrative rhetoric was concerned with praise and blame. It was taught that the right use of language is essential to doing good since a proper reason cannot sway humanity unless it appears in a form that commands. The Spanish humanist Juan Luis Vives (1492–1540), who lived as a tutor in Bruges and then in London, stated simply: 'Your tongue is the one instrument nature gave you for doing good.'[26] Language, it followed, is the means through which virtue becomes a shared achievement, and rhetoric is its instrument. Since, as Vives also said, 'the bonds of human society are justice and speech' nothing is 'more advantageous to human society than well-formed and well-developed language'.[27]

Rhetoric was thus closely linked to moral philosophy, as ancient authorities like Aristotle and the Roman author Quintilian had argued. Rhetoric mediated between reason and the passions and informed the will. As Bacon wrote: 'The duty and office of Rhetoric is *to apply Reason to Imagination* for the better moving of the will.'[28] Moral philosophers thought that though appetite, prompted by sensation or imagination, directly moves the will, reason is indirectly active in a moral person. But since it cannot be expected that reason will motivate the mass of humanity as it will an educated person, the rhetoric of the man who does reason was understood to be the means by which the quality of a leader becomes the shared quality of those who are led. Elocution was not just a matter of style but the manner in which reason is given its right expression, the means by which the reasoned voice in control of the passions is transferred from the one to the many. And what can be achieved in speech can also be achieved in the other arts, like painting or music. Rhetoric or expressiveness, rightly used, is the guide by which the people are led to an ordered life through their appetites. In the words Shakespeare assigned to Mark Antony when he spoke to the crowd after the death of Caesar: 'Friends, Romans, countrymen, lend me your ears.'[29] How to ensure an attentive audience was subject to much analysis. In the light of the later meaning of words, it is interesting that in the seventeenth century *'pathologia'* described a subdivision of rhetoric concerned with the way to arouse the passions, while *'ethologia'* described a subdivision concerned with the customs and mores of audiences. The existence of such subdivisions is evidence of the degree to which rhetoric fostered an ordered knowledge of man, however strange this use of 'pathology' and 'ethology' is to modern ears.

Memory, also a division of traditional training in rhetoric, was an important topic. Memory had special value before there were printed books, though it continued to be emphasized later. Effective oratory, for example, required an accurate memory. Before and after the sixteenth century, memory was also prized as a faculty of the soul in its relationship with the physical world, and illustrators of the localization of faculties in the brain displayed memory prominently in the rear ventricle. Practical guides to humoral physiology claimed that particular diets or ointments warm and dry out the back of the brain and hence stimulate the memory. These same books also often provided the key to interpreting dreams and reading physiognomy or moods, in which activity they took for granted the expressive unity of the soul and the body, and they thus helped the exercise of self-control in the mêlée of daily life.

The fifteenth century witnessed a revival in the art of memory, especially with the spread of the writings of the Platonist Ficino. This art, attributed by Roman writers to the Greeks, provided the means to enhance recall artificially by the systematic organization of the images with which the memory naturally stores ideas. The basic principle, as enunciated by Vives, was that 'whoever desires to remember things carefully and attentively notes the order in which he commits them to memory . . . For the sight of a place brings to mind what we know happened there or is located in that spot.'[30] The most common art of memory was to envisage a building – later writers often referred to a theatre – in which each image has an assigned place. Memory therefore involves returning in imagination to the building. Among the most famous memory theatres, though uncompleted, was that of Giulio Camillo (*c.*1480–1544), who built his fabric around the seven pillars of wisdom of Solomon's house. This work was part of a magical worldview, and Camillo assumed that there is a natural correspondence between the way internal objects are ordered in memory and the way external objects are ordered in the physical world. Right memory, he thought, is a means to wisdom. The art of memory was the technique to capture this wisdom and thereby, as with rhetoric generally, to fulfil the purposes for which God had placed man in the world. In Bacon's hands in the early seventeenth century, the improvement of the art of memory became one means to make possible 'the advancement of learning' generally. At this point, memory as an accomplishment of the soul began to transmute into the modern theatre of memory, the encyclopaedia.

By no means everyone was committed to the art of memory in the imagistic vein favoured by Camillo, and the use of mnemonics – though more mundane – was more usual. Petrus Ramus (1515–72, born Pierre de la Ramée), who became the most famous reformer of educational practice among Protestants, especially in England, preferred a dialectical method as a way to order knowledge in memory. In his version of dialectics (a classical subdivision of rhetoric), he arranged everything in dichotomies in descending order from the most general to the most specific; each dichotomy differentiated a more general character into two variations. Ramus applied this scheme of logical ordering to the school curriculum, and he laid down rules for teaching that, by virtue of their supposed correspondence to the logical structure of reality, bring into harmony the learning mind and what is learnt. His ideas were extraordinarily popular in Protestant grammar schools, perhaps in part, it has been suggested, because of the appeal of a tabulated method to bourgeois merchant fathers with aspirations for their sons.

The humanist subject of rhetoric also opened up forms of argument and ways of speaking that were exploited by the new natural philosophers of the seventeenth century. Rhetoric, which gives a voice to discourses less rigid and more sceptical than those of the theological, scholastic or mathematical disciplines, was particularly suited to the expression of human affairs – for certainty in human life, most people assumed, is not to be had. When people trained in jurisprudence and studied civil society, they learned to reason in a way that achieves moral certainty, that is, conclusions based on right reasoning and probability, rather than either physical certainty, the direct evidence of the senses, or logical certainty based on deductive argument. It was this standard of evidence and argumentation which then gave seventeenth-century natural philosophers the confidence and authority to demand that their new knowledge should receive public recognition. Once again, what was understood in relation to man affected knowledge of nature as well as of man.

viii *History*

Memory is close to history. Just as there was a desire to enhance memory as the individual's link to knowledge, so there was enthusiasm for history as the collective link to wisdom. In part, history was, like rhetoric, practical, since the lives of Greeks, Romans, saints and more recent civil, religious, military and artistic heroes and heroines provided vivid individual models for the present. At a deeper level, history was embedded in culture, since the perception of what was called the revival of letters was historical through and through, and humanist thought made possible a form of life constituted by recovery, memory and rebirth. The very idea of a rebirth periodized time and created history. Orators compared the modern world to the ancient one and, as learning, wealth, art and literature built up confidence, the comparison was increasingly to the disadvantage of earlier times. History set this comparison within a rational framework. Self-understanding, individually and collectively, looked to memory and history. Origin myths, which reflect a deep need to represent shared identity in terms of a common past, were reconstituted as scholarly study. There was an attempt to recreate the past as objective knowledge rather than to imagine it.

The word '*historia*' denoted the part of grammar concerned with the literal meaning of words, as opposed to syntax, as well as to a story that unfolds in time. To write history therefore meant to write with descriptive clarity and order, with or without reference to time. This usage was retained in the term 'natural history'. Bacon, for example, recommended that histories of the different parts of nature be written as the descriptive foundation for science. The English word 'story' has the same root as 'history', and it denotes an order in words used to describe experience, whether perceived or imagined. When they turned history into a self-consciously disciplined attempt to recreate the past as an objective presence, Renaissance scholars therefore attempted something truly fundamental, the representation of humanity to itself as an objective story.

History as a discipline, equally with rhetoric, was bound up with the fate of the *studia humanitatis* in the late medieval and Renaissance curriculum. George of Trebizond (1395–1484), a Cretan émigré to Italy, in 1434 defined history not simply as the memory or record of events but as their accurate description with reference to causes, motives and consequences. At about the same time, Valla praised history as 'more robust than poetry because it is more truthful' and because it 'exhibits more substance, more practical knowledge, more political wisdom . . . more customs and more learning of every sort than the precepts of any of the philosophers'.[31] Valla himself wrote a history of Ferdinand I of Aragon, and he and others made it clear that they were interested in the worldly acts of men or the fortunes of states as well as with the sublime working of Providence. In the divided and continuously violent political and religious circumstances of the time, such histories of kings, states and empires became important to the rhetoric of political and religious authority. The Reformation in particular called for defence or attack as an event in world history. Did it or did it not fit the pattern of God's providence and the renewal of human civilization? Different histories thus competed for authority when they claimed rhetorical excellence and evidential truth.

The strongest development of historical-mindedness – a sense of the particularity, mutability, relativity and contingency of events – came about through the conjunction of humanist learning and jurisprudence or the theory of law. Renaissance legal thought discovered growth, change and decay in the lives of states and civil societies, from which it concluded that it is necessary to understand the history of human achievements if those achievements themselves are to be understood. As generations of commentators on the inherited Roman law struggled with the contrast between the ideal form of that law and the variety of legal practices and customs in Europe, so they became conscious of the different histories of states and peoples. Jurisprudence laid the basis for comparative history and suggested that the development of states is related to circumstances.

The foundations for historical jurisprudence were spread by humanist scholars from northern Italy, especially by the move made by Andrea Alciato (1492–1550) to the university of Bourges in central France in the 1520s. An influential school of French historical lawyers developed, and its collective contribution was to found modern historical scholarship. Perhaps the most famous of these legal humanists was François Baudouin (1520–73), whose *De institutione historiae universal et ejis cum jurisprudentia conjunctione* (*Institution of Universal History and Its Conjunction with Jurisprudence*, based on lectures given in 1561 in Heidelberg) contained a manifesto of historical method. Baudouin's life and work, like that of many other scholars of his time, was intensely and sometimes dangerously bound up with surrounding events. France, riven by religious and political divisions, saw its kings struggle to consolidate power independent, in one direction, of Catholic Rome and the Holy Roman Empire, and, in the other direction, of Calvinist Geneva and Protestant interests. In these circumstances, the reconstruction of the grounds of religious and legal authority, work such as Baudouin undertook, made memory, embodied as history, an instrument of high politics. When Baudouin sought a 'universal history', it expressed his passion for a reunified Christianity that would bring together custom and morality in common cause: 'If we desire the true and integral knowledge of things human and divine, we must study universal history.'[32] Such ecumenical values died with the massacre of French Protestants, the Huguenots, on St Bartholomew's day in 1572; but the ideal that history should serve political and legal ends had been established, and in this context history meant not just a branch of rhetoric but a science – a systematic subject, with its own method, that sought to understand causes.

History and legal activity shared a concern with evidence. The basic techniques of historical method derived from philology, the linguistic studies of the humanists, and from lawyers who wished to judge the historical provenance and language of legal records. As Baudouin wrote, the scholar's task is 'to determine what is whole and what is diminished, what is old and what is new . . .'[33] The capacity to write history as opposed to recreating memory, to make history a science rather than an art, depended on the development of these skills of textual and linguistic analysis, especially the ability to distinguish primary sources from secondary additions. These skills were also practised by lawyers when they drew distinctions between witnesses who had directly observed events and those who had not. There was debate about the value of judicial torture as a means by which to uncover truth. Further, the law courts deployed rules to reconstruct histories of the cases being tried, and academic jurisprudence constructed these rules as a formal study. The courts sought to reconstruct past events in as persuasive a manner as possible in order to give authority to decisions and to judges when they assigned praise or blame as a guide for future conduct. Similarly, historians sought to evaluate testimony, looked into the causes and effects of events, judged and assigned blame, and reconstructed a story that carried authority.

Jean Bodin (1530–96) took Baudouin's idea of a universal history further, and he related history to the physical circumstances of geography, while he also considered at length historical and legal methods of evidential argument. Significantly, lawyers like Bodin were concerned equally with the scholarly business required to lay sound foundations for learning and with the practical business required to legitimate current political developments or to warn of disastrous actions. Bodin went so far as to proclaim that history is 'above all sciences' since the historian combines 'the narration of facts with precepts of wisdom'.[34] History, indeed, looked as if it was the bar at which the moderns judged and learned from the lives of great men and states. In this way, the study of jurisprudence and history integrated with the study of moral philosophy – and contributed to the examination of right action, right economy and right government which formed the study of man. The methods developed to make these studies more rigorous, such as those that focused on the nature of historical evidence, contributed to a mode of life in which man became a self-reflective subject. The richness of these ways of thought about human life implied that though God had created man in his own image, man is busy at the re-creation of himself.

3

The Province of Natural Law

For the Mother of Natural Law is human Nature itself, which, though even the Necessity of our Circumstances should not require it, would of itself create in us a mutual Desire of Society . . . But to the Law of Nature Profit is annexed: For the Author of Nature was pleased, that every Man in particular should be weak of himself, and in want of many Things necessary for living Commodiously, to the End we might more eagerly affect Society.

Hugo Grotius, *De iure belli ac pacis* (*The Rights of War and Peace*, 1625; in the English translation of 1738)[1]

i *Law and Jurisprudence*

A sense of excitement at renewal and excellence grew from civic life and the generation of wealth as well as from learning and the arts. Cultural change was bound up with the fate of states and empires. The shifting fortunes of politics and war made stability the exception, and scholars – driven by events – changed allegiances to survive in new times. The consequence was a learned culture of great diversity and openness to change. By the seventeenth century, though there were many continuities with what had gone before, the Renaissance had really been left behind. With the consolidation of centralized French and Russian states, the ending of the Thirty Years War (1618–48), the slow decline of Spanish power in Europe, the English Civil War, and the regular settlement by Europeans of the American East Coast and other parts of the world, something recognizably like the modern political order began to emerge. The centre of commercial and cultural gravity shifted towards north-west Europe: England and The Netherlands achieved prominence fuelled by economic prosperity, with their wealth founded in labour, investment, trade and manufacturing as well as agriculture. There was much self-consciousness about these changes. European people in this early modern period thought about their own nature and sought to find in it a cause of the changes that were taking place. Their thought proved to be the foundations for the modern human sciences.

The last chapter took a bird's eye view of the breadth of learning relevant to human nature, the continuities and discontinuities with medieval Christendom, and the ways in which interest in character, conduct and history dignified the individual person as a rational agent. This and the following chapter focus on two specific themes that are enormously significant for the seventeenth century and after. The first theme concerns law. Legal systems and the theory of law, or jurisprudence, are central to the history of *order* in human affairs, and a belief that such order is discoverable underlies the human sciences. A conceptualization of order is a condition for a science of human beings. The science of jurisprudence inspired knowledge about human affairs, and in the seventeenth century enthusiasm for the view that human laws gain authority from divinely created nature was a major stepping-stone to a modern science of man. The second theme, which is taken up in this chapter in relation to Hobbes and carried further in the next chapter, concerns the shift towards a representation of the soul and its activities in terms structured by thought about the material world and sometimes even in material terms. There was a growing willingness to take the mind's relation to nature as the starting-point for knowledge about what it is to be human. The consequence was a strong emphasis on individual human agency in material affairs – moral conduct, economic activity, politics – and from this to draw conclusions about human nature.

It was and is the boast of civil society that it lives according to law. At the most general level of analysis, scholars examined law in two contrasted ways: it was held to be embedded in the nature of things, that is, to be intrinsic to the divinely created order; or it was a human creation, the product of custom and of institutionalized authority. Both conceptions were present in the West. This double articulation of the nature of law paralleled, in a significant way, a basic division of philosophical opinion about the subject matter of knowledge in the human sciences. The double view of law exemplifies the dilemma that faced thought about human action generally: to decide whether order exists and systematic knowledge is possible by virtue of the existence of a divine creation, or a natural reality, of which human beings are part; or whether humanity has constructed its own ordered reality which, since it has created it, it can understand. This dilemma, to which I alluded in the introductory chapter's myth about Tristan and Isolde, reappears in many guises and, if anything, became even sharper in the late twentieth century.

The legal system is of course a practical means to a social end – law and order – and not only a learned occupation. Its history therefore also exemplifies the thesis that, for the human sphere, forms of knowledge have developed hand in hand with the moral, economic and political realities of everyday life. Like moral philosophy in the eighteenth century and psychology in the twentieth, jurisprudence, as a human science discipline, grew as an integral part of the practical governance of affairs. In consequence, it was torn by the contradictory pressures to make itself systematic and to make itself practical. The same situation confronts all human science subjects, and different responses reflect different views of what science is or should be.

The lawyer's occupation achieved high status and some degree of political power in medieval Europe, supported by the demands of property, inheritance, church authority and social or political obligation. There was a strong memory of Roman law, widely held to be Rome's greatest contribution to civilization, known through the *Digest* and *Institutes* drawn up under the Byzantine emperor Justinian in the sixth century. Justinian had made the truly imperial decree that forbade 'all persons now living as well as those who are to come, to write any commentaries on these laws'.[2] Medieval lawyers, however, did exactly that and then, in the fifteenth and sixteenth centuries, the techniques and ethos of humanist scholarship created a vast amount of jurisprudence to accompany these inherited laws. At the same time, in many parts of Europe but perhaps most distinctively in English common law, local traditions affected both the substance of the law and views whether or not jurisprudence should be a systematically rational discipline, i.e., a science.

Commentators who believed that there is a universal basis for law, and therefore that Roman law has universal application, presumed that man has a primary nature given to him by the divine creation. Lawyers therefore referred to custom, the variety of social life characteristic of different times and places, as 'a second nature'. In consequence, the goal of an actual legal system was to find a proper and practical balance between these two 'natures'. As Isidore of Seville had observed in the seventh century: 'The human race is ruled in two ways, by nature and by custom.'[3] In the medieval Christian synthesis, which was given its most systematic form by Aquinas, natural law – the law that rules man's primary nature – was equated with God's law. Yet when commentators studied the relation between Roman law and actual medieval laws, they felt forced to conclude that there are laws created by man through custom as well as natural laws. Beginning in northern Italy, especially at the university of Bologna, fourteenth-century legal scholars differentiated the social and political conditions of Rome from those of their own society, and thus laid the basis for a modern perspective on historical change. A historical awareness made possible an assessment of what in human affairs is owed to nature and what to custom. In the long run, it is possible to suggest, legal scholarship established patterns of thought that became the social sciences. The shift from thought grounded in medieval jurisprudence to thought that we see at the basis of social science began to occur in the seventeenth-century discussions of natural law.

The most basic level of law in the *Corpus Iuris Civilis* of Justinian, in the words of the Roman jurisconsult Ulpian, is 'what nature teaches all animals', namely, the law of self-defence and the desire to mate.[4] Next in generality was the law of nations, the *ius gentium*, the law that all humans share in so far as they are rational beings. Lawyers understood this law in terms of the three basic categories, the categories of persons, things (or possessions) and actions, which were still current in the analysis of affairs in the twentieth century. When they discussed the relations between these categories, the lawyers refined basic terms of social analysis: volition (*voluntas*), authority (*auctoritas*), cause (*causa*), obligation (*obligatio*), and so on. This analysis articulated the concept of the person, the individual who possesses body, property and free will and – since he is also social in nature – obligations and responsibility. Christianity reinforced this articulation of the concept of the person, since faith held that a person is the possessor of an individual divine soul. Christian faith also underwrote belief in a common law of nations as it reinforced belief in the presence of a divine *Logos* or rational spirit.

There was also a body of work, developed in medieval law, which described both the individual and the state as persons on the grounds that both possess property and obligations. It was customary to refer to the state as in possession of a body, the body politic, which – like the body of an individual person – was conceived to have a will and to act as a whole. It was then a matter for open argument, and often enough violent struggle, to decide whether a monarch, a prince or the representation of the people (as republicans believed) rightfully forms the body politic. The terms of argument, which went back to the ancient world, gave politics a vocabulary for judgement about the proper spheres of individual and collective action. Reference to the body politic was not, therefore, 'mere' metaphor but displayed the legal structure in common between the state and the person. In fact, the phrase 'the body politic' was ubiquitous. Baldassare Castiglione (1478–1529), the writer of a famous conduct book for courtiers, claimed that monarchy is a 'more natural form of government' since, 'in our body, all the members obey the rule of the heart': members of the body politic must obey the heart, the sovereign, of this body.[5] He who set the body politic to rights was often called a 'physician'. Over a century after Castiglione wrote, much of the rhetoric about the English Civil War used similar language, especially when parliament cut off in one stroke the head of state and the king's head. This vivid vocabulary of body politics, like the language of natural law, provides clear evidence that the sciences of nature and the sciences of man share the same intellectual preconditions.

Renaissance and early modern lawyers held different opinions about the degree to which actual legal systems should be grounded on comprehensive, rational principles or on custom. Those who favoured the former view aspired to establish jurisprudence as a science; the teacher Alciato, who established French historical jurisprudence, stated that 'reason is the soul and life of a particular law'.[6] He and those who followed him tried systematically to state reasons why the laws are the way they are, and they sought reasons in the nature of things, that is, in the order of the creation. But the history of the development of particular bodies of law, the historical practice discussed in the previous chapter, showed how local customs and circumstances had modified the reasons that derived from man's universal nature in actual legal realities. In this way, the jurisprudential attention to reasons – located in universal realities – and to causes – located in contingent historical or geographical circumstances – were linked.

This was an important if abstract step in the background to the human sciences. As this section has suggested, the lawyers who examined how human law is given both by the nature of man and by the customs that man establishes set precedents for two directions taken by what became the science of man. In the first, philosophers sought the causes of man's nature and social existence in the rational design of the created world, and they then presupposed that, since man's soul is of the world, it partakes of the world's reason and can therefore know the world. This approach was more characteristic of Germanic thought. In the second direction, philosophers were more empirical and tried, through history and the study of civilization, to find out what causes or particular historical or geographical circumstances had made the human world they saw around them. This approach was more characteristic of the French historical school of jurisprudence, and it led to an eighteenth-century search for the physical causes of the rise and decline of civilizations. The culmination of this second trend in jurisprudence was Montesquieu's study *L'esprit des lois* (*The Spirit of the Laws*, 1748). The approach also stimulated a historical imagination, contributed to the methodology of seventeenth-century natural philosophy and underlay the modern conception of natural-scientific explanation. Already in 1689, Jean Domat (1625–96) – a jusnaturalist, as the systematizing lawyers were known – published *Les lois civiles dans leur ordre naturel* (*The Civil Laws in Their Natural Order*), the title of which clearly indicated his hopes for a science of jurisprudence comparable to the sciences of the physical world.

Thought about man's nature received a profound impetus from legal scholarship. Early modern Christian philosophers and lawyers alike assumed that the observable causes of things were also the reasons for those things in what is a rationally created world. The consequence was that the search for causes in jurisprudence and in natural philosophy led at one and the same time to attempts *rationally* to understand history and nature and *empirically* to discover historical and physical agencies. Only later was a search for reasons and a search for causes clearly differentiated. Both aspects were evident in seventeenth-century literature on natural law. During the course of the century, however, theorists of natural law who studied civil society, like natural philosophers who studied physical nature, gradually shifted towards explanations which cited identifiable causes in nature rather than abstract reasons. Virtually all scholars, we must also remember, ultimately referred both reasons and causes to God.

ii *In Search of Moral Authority: Hugo Grotius*

Belief in natural law meant an acceptance that reality has a pattern and order, that order exists equally in moral and physical spheres, and that human beings, who are part of the creation, can obtain knowledge of the natural order and thereby guidance about how to live. Such belief was fundamental for the human sciences from the seventeenth century to the present: twentieth-century social scientists, psychologists and doctors, many of whom disbelieved in a divine creation, nevertheless trusted that knowledge of humankind as part of the natural order leads to a better life. I have argued that the roots of these beliefs were ancient, religious and embedded in the theoretical and practical concerns of lawyers. By the end of the seventeenth century, however, natural law theory gave a new prominence to links between the physical world and human affairs.

The most significant theorist of natural law in relation to civil society at the beginning of the century was undoubtedly Hugo Grotius (1583–1645, born Huig De Groot), a Dutch jurist and statesman. Grotius graduated from the university of Leiden and achieved fame as a gifted humanist while still a young man, though his literary and civic learning alienated him from the local Calvinist church. Religious opinions and political struggles were interconnected in complex ways. Grotius himself was imprisoned, then lived as an exile in France and he ended his career as Queen Christina of Sweden's ambassador to the French court. The continuous wars in the Low Countries between the United Provinces and Spain, followed by the Thirty Years War further east, formed a chaotic and savage backdrop to his great work *De iure belli ac pacis* (*The Rights of War and Peace*, 1625), regarded as the foundation of modern international law. Relentless conflict, widespread civil anarchy and the fact that Protestant and Catholic campaigners fought each other to the ground without victory suggested to Grotius that a new basis had to be found on which to accept a plurality of political states as opposed to a united Christendom. There had to be some limitation to conflict. He did not dream, as did utopian writers like J. V. Andraeae (1586–1654), of building the City of God on earth, but sought grounds in a common humanity to limit damage. This political realism was evident, most famously, in his discussion of the *bellum justum*, the just war.

Grotius can be compared with key figures who founded modern natural philosophy in the seventeenth century, men like Galileo, Hobbes and Descartes, who also felt that the age cried out for certainties on which the human mind could build. These men shared an intense commitment to the ideal that both natural philosophy and moral philosophy should become scientific (to adopt a later usage) – logically consistent, certain and systematic. Their model for science was geometry or arithmetic, and Grotius was inspired by Galileo's use of geometry. He broke with many of his predecessors when he wanted moral philosophy to be a science and not only a practical subject; this step, he hoped, would remove the possibility of scepticism. Natural and moral philosophers alike were disturbed by scepticism. In reponse, they turned to the concept of natural law, which they believed originated with God's will, as the intellectual means to establish science. Thought about physical nature and about human nature went in parallel. Grotius also shared with Bacon an interest in the natural world of man. When Bacon advocated the writing of 'histories', i.e., descriptions of the natural world as a basis for authoritative knowledge, he intended these histories to include descriptions of what man is and how he lives in different times and places. Knowledge of natural law and of natural history, Grotius and Bacon believed, makes possible the improvement of human life.

To establish his argument for a natural law theory of individual legal rights and obligations, Grotius gave an influential account of human nature. Europe might be riven by conflict but, he held, all people, as people, nevertheless share one goal, self-preservation. He argued that when we go below the surface of beliefs, however strongly held, to examine the foundation of action, we find that we do not and cannot act rationally against our self-preservation. Of course, we often act against self-preservation, but such action, Grotius believed, is not informed by reason. Most significantly, we all act in this way; that is, as human beings, we share reason and this reason directs us to self-preservation. 'Love, whose primary force and action are directed to self-interest, is the first principle of the whole natural order.'[7] Here then, he thought, is a common basis in human nature on which to lay the foundations for social agreement.

Through this argument, Grotius claimed empirically to identify a universal human reality, the rational direction of action towards self-preservation. His claim, however, was also informed by theology, by a conception of God's love for his creation. He conceived that God had created man for a social existence and had therefore combined in human nature the desire towards self-preservation and the desire towards sociability. Further, since man is physically weak, Grotius believed that self-preservation in fact requires sociability. Thus, only co-operation can satisfy our nature. On the basis of these arguments, Grotius established what became, when detached from its theological roots, the leading principle of modern liberal moral and legal philosophy: the right of the individual to free action in pursuit of self-interest unless it harms others. 'For the Design of Society is, that every one should quietly enjoy [that which is] his own . . .'[8] He began to direct natural law theory away from its medieval focus on duty and towards what we see as a modern focus on rights.

Grotius made each person's natural commitment to self-preservation, a commitment informed by right reason, the basis for human law whether in the family, government, the legal system or relations between states. It followed that knowledge of man's natural nature is the foundation for the right conduct of his affairs. Grotius took the individual's self-preserving spirit to be primary, and he viewed society as a secondary consequence of these attributes. In his argument, knowledge of what man is therefore primarily concerned his individual nature. These two points, about self-preservation and the primacy of individual attributes, set the agenda for individualist thought about human nature and its government in the West down to the late twentieth century.

All this was not indebted just to Grotius. He is representative of many scholars who grappled with these issues against the backdrop of war and a sense of social crisis. From one perspective, Grotius was a humanist scholar who re-expressed a synthesis of Greek and Christian assumptions already achieved by Aquinas. From the vantage point of the late seventeenth and early eighteenth centuries, however, especially in the work of Samuel Pufendorf (1632–94), Grotius appeared to be the man who had made a new start. What struck his contemporaries and followers – and appalled some religious observers – was the extent to which he attributed the principle of self-preservation to nature; indeed, he even argued that natural laws would bind mankind, 'though we should even grant, what without the greatest Wickedness cannot be granted, that there is no God, or that he takes no Care of human Affairs'.[9] Such a suggestion implied the possibility, exploited later, that the natural circumstances of men's and women's condition, i.e., natural laws in the human sphere, could be studied dissociated from the dependence of law on God. It implied that authority for human actions could be found in human nature rather than in scriptural or theological ordinances. This invites comparison with the controversy precipitated by Grotius's contemporary Galileo, who – whatever the immediate political and legal grounds for the Catholic Church's condemnation of his astronomical views in 1633 – raised the issue as to whether authority for philosophy is to be derived from natural philosophers or theologians. Grotius, in his turn, faced Calvinist and Lutheran critics who argued that human actions must be judged by the will of God, not by standards of nature.

Grotius's treatment of natural law was also an attempt to deal with the scepticism that had become rife in the aftermath of the religious schisms of the sixteenth century. His hope was that if there is disagreement about the direction of God's will or even disbelief that his will can be known, there is still the possibility of agreement about the facts of human nature. Many people feared that scepticism fostered social anarchy and were attracted by his views. Grotius argued that social and moral crisis could be overcome when it is recognized that all people, by virtue of their common nature, have a common basis for moral action in the way they seek self-preservation. It is tempting to see a parallel in the twentieth-century hope that the human sciences will reveal a common human nature and thus make it possible to avoid political conflict.

The search for a common human nature leads to another issue relevant to the contemporary reception of Grotius's work: the dramatic impact on Europe of new knowledge of the sheer diversity of peoples around the world. This experience had a major effect on thought and imagination from the fifteenth to the twentieth century. Not least, moral scepticism was enriched by travellers who provided vivid evidence of the diversity of human values.

iii *The World's Peoples*

Medieval Europe had traded overland across the Middle East and India and, along the silk road, through Central Asia to China; there was also trade with North and West Africa. Arab merchants had gone further south and east across Africa and the Indian Ocean. Sailors had crossed the north Atlantic to Greenland and the North American coast. Yet when the Far East and especially the Americas were opened to European maritime trade, military occupation, settlement and the Christian religion, it was an experience of the highest significance. It occurred relatively suddenly and began in the last decades of the fifteenth century. In India, China, Japan, Central and South America, European adventurers, entrepreneurs and missionaries found established societies complete with customs, laws and religious beliefs utterly alien to Western eyes. Some travellers thought the people they found so lacking in civilization that they doubted whether these people should be considered human at all. Others thought they had stumbled on Arcadia, a mythic land of plenty. The theological problem of human diversity was far from new; medieval Christendom had, after all, lived alongside and often in conflict with Islam for centuries. But the intellectual challenge to Christians to comprehend the variety of humankind as descendants of Adam became intense. So many different peoples, it appeared, were not only totally bereft of Christ but even lacked a semblance of reason, morality or civilization. Where in all of this lay God's providence? Whatever the variety of response, the travellers created a new genre of literature and their reports shaped a new dimension to Western self-consciousness, linking learned and unlearned people in wonder.

The European imagination had slowly incorporated aspects of Africa, India and China over the centuries; but the sudden 'discovery' and extensive colonization of the Americas was a watershed in European history. People who stayed at home as well as those who travelled faced an experience for which there was no precedent. All they could do was draw on ancient authors, on tales of the mythical and the fabulous and on experience of different European social groups – the peasantry, for example – in order to grasp what they saw. European observers looked with the eyes of people well educated in Greek and Roman authors who had written about encounters with others. Those writings gave them a vocabulary with which to grasp the new experiences; 'the others', in our modern language, were therefore often described as 'barbarians'. Classical authors took slavery for granted as a natural and rightful consequence of conquest and superiority, and slavery became endemic in Europe's economic exploitation of its conquests. From a later perspective, it is clear that the early reporters failed even to recognize many differences, so overwhelmed were they by the difficulty of saying anything at all with the concepts available. In their encounter with new peoples, however, the travellers laid the basis for modern comparative anthropology or the study of human diversity.

When they faced 'New World' people, Europeans also began to think that they faced the past of human history, the evidence for what all peoples might once have been. The representation of native American peoples, especially those of the Caribbean and the Brazilian coast, as 'savage', a word first used adjectivally about wild animals or landscape and about cruel or uncouth habits, defined these peoples as isolated by time as well as space from European civilization. Montaigne expressed it vividly: 'Our world has lately discovered another . . . so new and so infantile that it is still being taught its A B C . . .'[10] In imagination, the savage stood at the beginning of social time. The discussion of what savagery meant in order to understand civilization became fundamental in eighteenth-century thought and ultimately to the nineteenth century's belief in historical progress. The European incorporation of the Americas into its worldview required new kinds of knowledge; it could not simply be assimilated to existing beliefs.

European voyages and colonization produced a vast mass of new observations, artefacts and, not least, people, all of which needed description, often from examples, people included, shipped home. The bewilderment this sometimes created is illustrated by the botanist Nehemiah Grew's attempt in 1681 to make some sense of the Royal Society of London's collections which included, among much else, an Egyptian mummy, a Greenland canoe, a pot of poison and a Siamese drum. But it was the people, their appearance, their customs and, especially, their *lack* of what Europeans possessed that aroused most comment. Montaigne, who saw native people from the east coast of South America exhibited at Rouen in 1562, wrote that they:

> [Belong to] a nation . . . in which there is no kind of commerce, no knowledge of letters, no science of numbers, no title of magistrate or of political superior, no habit of service, riches or poverty, no contracts, no inheritance, no divisions of property, only leisurely occupations, no respect for any kinship but the common ties, no clothes, no agriculture, no metals, no use of corn or wine.[11]

A century later, Locke encapsulated the nub of the comparison when he said that the native people and people in a state of nature have no money. Europeans claimed that such native Americans lacked civilization, that is, in Montaigne's words, 'these people are wild in the same way as we say that fruits are wild, when nature has produced them by herself and in her ordinary way . . .'[12]

This claim cut two ways. Some observers, Montaigne included, held up native peoples as an ideal, an example of real humanity unadulterated by civil society and the corrosive effects of too much thought and sophistication. This view appealed down the centuries, not least to a much later period of post-industrial malaise. The savage, in European hands, became a mirror to hold up to civilized men and women and thus to expose the failings and the arbitrariness of civilization. By contrast, many European settlers stressed that what native peoples lacked indicated their inferiority in every way: they were uncivilized, animal-like, without the Christian religion and incapable of progress. For many Europeans, simply to look at them was to know this. One consequence was that reports of cannibalism were almost universally accepted, even though such reports were always about 'other' tribes 'over the hill' and not peoples immediately encountered.

In the light of later racial theories, it is interesting to note that differences of body and colour were not the prime issue; where colour was discussed, it was generally attributed to climate. What gripped the European imagination was the social otherness of native peoples. The spectre this raised, forcibly expressed in the ancient sceptical assertion of Carneades in the second century BC that there is no such thing as natural law but that all law first arose from the convenience and profit of particular states, turned into a principal theme of modern European thought. It was the belief that Grotius specifically sought to disprove. But at the same time as he argued for the universality of natural law and a common humanity, the European encounter with the wider world opened up a perspective that pointed to the relativity of social arrangements, morals and religious faith.

The European expansion also created a problem of moral and political legitimation, and this resulted in a complex literature, especially about the Spanish conquest of the Mexica (1519–22) and Inca empires (1531–2). The defeat, indeed liquidation, of these empires prompted debate about whether Spanish action was lawful, debate which hinged on the Spanish Habsburg sovereign's self-appointed role as the political and military leader of the Christian faith and on the Indians' standing or lack of standing as people comparable with Spaniards themselves. ('Indian' was a contemporary term.) The scholars who took part in these discussions reconstructed a long-standing divergence of view between Aristotle, who had described slavery as the natural condition of a certain class of people, and Christian thought, which had mostly stressed the common origin of all men. It was not obvious to all Catholic lawyers and theologians that either the genocide or the economic enslavement of Indians and the appropriation of their property – their fabulous gold and silver – was compatible with Christian law. No one questioned the actual right of the Castilian kings to conquest; such a step was politically and intellectually unthinkable. Nevertheless, as these kings proclaimed themselves the heirs of the Holy Roman Empire and the secular champions of Christendom, it was an important matter of legitimation to show how their actions advanced Christian faith, morality and law. Members of religious orders and academics entered the debate with different interpretations of the details of Roman and medieval law of rights and property. They used legal and, to some extent, theological categories, like the distinctions between adults and children, husbands and wives and masters and slaves, to describe what they thought is and is not shared by European and non-European peoples.

The leading critic of Spanish policy, Francisco de Vitoria (1483/92–1546), a professor of theology at Salamanca, argued on the basis of Thomist medieval natural law that it is open to any group of people to use natural reason to establish its own rules. This, he argued, the native Americans had done, even though they were deprived of the benefit of revelation; he concluded, therefore, that 'there is no justification whatever for despoiling either their princes or subjects of their property on the grounds that they were not true owners of it'.[13] He used natural law arguments to show that the Indians are legal subjects in the same sense as Europeans. Other scholars answered Vitoria with the argument that the Indians are 'slaves by nature', and exhibit a way of life of 'natural rudeness and inferiority' that not only justifies but requires Spanish rule as a Christian duty. As the Scottish theologian John Mair (Johannes Maior or Major, 1467/9–1550), who worked in Paris, argued in 1519: 'These people live like beasts . . . wherefore the first person to conquer them, justly rules over them because they are by nature slaves.'[14] Yet others wrote that people like the Inca, though socially organized, did not possess a civil society and thus could not be said to possess property in a legal sense. Or it was said that the Indians were like children in law, potentially but not actually able to possess property. This was Vitoria's conclusion, that the Indians have reason but are 'so little removed from the foolish that they are not able to constitute nor administer a legitimate republic in civil or human terms', and hence they should remain under the humane tutelage of Spain.[15] This assertion did not legitimate the *conquistadores* who shipped home precious metals or ruled as despots. The fact was, however, that many native people either died or lived as economic slaves, whatever the theologians back in Spain said about their status.

The potentially disturbing critique of Spanish policy by Vitoria and others did not affect social and economic practice in America. The trade in slaves from West Africa to the Americas, which followed European settlement, was well under way by the early seventeenth century, and this too was accompanied by assumptions about human difference and a quasi-legal legitimation. It appeared self-evident to Europeans that Africans lack reason sufficient to create civil society for themselves and that they are therefore rightly treated as property not as property owners. When a European critique of slavery developed later, it was promoted mainly by evangelical Christians who reasserted a belief in the unity of mankind at the creation. Some secular Enlightenment writers, notably Denis Diderot, also opposed slavery on the grounds of mankind's unity. The nineteenth-century defence of slavery drew more on the assertion, held to be a conclusion of science, that Africans do not belong to the same physical and biological species as Europeans. In early modern Europe, however, the issue was not perceived to be about biological variation but about the representation and legitimation of legal identity and difference.

Two great Spanish scholars in the second half of the sixteenth century began to attempt systematically to compare observations of American and European cultures. The work of Bartolomé de Las Casas (1474–1566) and the much more influential study by José de Acosta (1539–1600), *Historia natural y moral de las Indias* (*Natural and Moral History of the Indies*, 1590) laid the basis for what was to become ethnography. Both men had actually lived in the cultures they described and, in Acosta's words, intended to 'state the truth as certain experience has revealed it to us . . .'[16] Experience, not presumption, they argued, should be the starting-point for the study of foreignness, and hence they described the religions, customs, government and language of the native people. They did not – how could they? – escape from a European mindset. As a Jesuit whose business it was to teach theology to missionaries in Peru, Acosta however recognized problems of translation (though not of structural differences between language groups), faced as he apparently was by the lack of any word for God in the Inca language. Las Casas initiated the argument, developed by Acosta, that the differences between societies are not original but the product of education, custom and time. In order to comprehend cultural difference, these authors turned to history, and they sought explanations in time in the way that was to become characteristic of eighteenth- and nineteenth-century anthropology.

With over a century of debate behind them, early seventeenth-century writers tried to pull together the strands of thought about human diversity, and, significantly enough, they wove the strands together in the narrative form of speculative history. In the English-speaking world, books like Samuel Purchas's *Purchas His Pilgrimage* (1613) and Walter Raleigh's *History of the World* (1614), which synthesized the bible, geography and history into one sweeping interpretation, had enormous influence. Most of these stories accepted the account of the creation in Genesis, which assigned an original unity to mankind, and explained diversity as the consequence of the dispersion of Noah's sons and the effect of time and distance. When the Frenchman Isaac de La Peyrère (1594–1676) suggested in 1655 that there had been pre-Adamitic men, which implied an original diversity of human forms, this was a distinctly heterodox conclusion. It was more usual to hold that all men share a common origin and common qualities; and yet no one doubted that men, women and native peoples share those qualities unequally. Theological, legal and historical thought, which led to accounts of the conditions that make possible civil society, endeavoured to explain why this was so. The experience and study of human diversity lay at the heart of early modern representations of man's nature and of his civilization.

iv *Thomas Hobbes and Natural Philosophy*

There was widespread concern about a crisis of authority in intellectual, political, moral and religious matters in the early modern period. The otherness, for Europeans, of New World cultures added further confusion. While a sense of crisis is common enough, there are times – and this was one – when it becomes the leading force. Grotius was emblematic of his age when he sought a way out in the authority of natural law. Significantly, there was felt to be a crisis about knowledge of the world in parallel with a crisis in human affairs. There was a turn towards natural law as the basis for objective truth about both physical nature and human nature. Writers who used the language of natural law described what reason demonstrates about the whole of nature – human, or moral, and physical. They also believed that the power of demonstration to command authority rests with the quality of reason itself. The seventeenth-century search for natural laws therefore united the study of reason and the study of human beings and of physical nature.

The resolution of the crisis in authority in natural philosophy came with the great achievements in mechanics and astronomy in the period from Galileo and Descartes to Isaac Newton (1644–1727). The Newtonian synthesis of the late seventeenth century set new standards for authoritative knowledge, standards that moral philosophy attempted to emulate in the eighteenth century and which inspired a search for the sciences of man. But the leap forward in scientific knowledge was preceded by several dramatic attempts to integrate moral and natural philosophy. This section discusses the work of Thomas Hobbes (1588–1679), while the philosophies of Spinoza and Leibniz are considered in a later chapter. Hobbes, like Grotius, attempted to find a basis for political obligation in the law of man's nature, but his conclusion was very different. The way that it differed initiated a debate that still ran in the twentieth century about whether human nature is material and, if it is, what consequences this has for social life.

Hobbes was an exceptional figure. His work fits no neat

schemes and has prompted different interpretations, though everyone agrees that his brilliant arguments did much to provoke the long and active stream of liberal political theory. He argued for an authoritarian solution to politics, shared the company of royalists during the English Civil War and yet wrote in such a radical spirit that he was reviled as an atheist and attracted denunciations from across the political spectrum. Even in the seventeenth century, however, he had some covert supporters, and his work certainly provoked responses in continental Europe. He was and remains famous for a materialist view of human nature.

Hobbes spent the whole of his adult life, from when he left Oxford in 1608 with a deep but subsequently despised Aristotelian education to his death in 1679, as a companion and secretary to the Cavendish family, the Earls of Devonshire. Neither Grotius nor Hobbes, nor – as we shall see – Descartes or Pufendorf (for most of his life), occupied university positions. They depended on aristocratic patronage for their livelihoods and freedom to write, and the willingness to grant such patronage was an important value passed on by rulers in the Renaissance. Patronage could, of course, be oppressive, but it could also offer unrivalled protection and intellectual liberty, conditions that Hobbes took ample advantage of. Individual patrons were not necessarily subject, as academic institutions often were, to disciplinary or doctrinal interests – such as those of the Dominicans in Vitoria's Salamanca.

There was a turning-point in Hobbes's intellectual career about 1630, when he was captivated by the idea that Euclidean geometry reaches deductive truths that cannot be doubted if they are rightly argued from the given axioms. 'Geometry . . . is the onely Science that it hath pleased God hitherto to bestow on mankind', by which he meant that geometry is the one means we have to arrive at certainty, and it must therefore be the model to follow in the search for an authoritative basis to human affairs. During the 1630s, inspired by these thoughts, he conceived a large-scale intellectual synthesis to cover, first, the physical world, second, the special characteristics of man's nature, and third, the realm of civil society and government.

The premises outlined at the beginning would, he intended, lead to indubitable conclusions, and would show that human affairs are rightly ordered by nature. 'Nature it selfe cannot erre.'[17] Pressure of external circumstances, the conflict between king and Parliament that led to the outbreak of civil war in England in 1642, encouraged Hobbes to write first on civil society, and he published *de Cive* (*On the Citizen*, 1642) and then his celebrated *Leviathan, Or the Matter, Forme, & Power of a Common-Wealth Ecclesiasticall and Civill* (1651). He later completed his original plan when he published *De corpore* (*On Bodies*, 1655) and *Human Nature: Or the Fundemental Elements of Policy* (1650; Latin version *De homine*, 1658).

Hobbes spent considerable time in Paris, latterly to avoid the English Civil War, a time when he judged that passions ran so high that the voice of reason could not be heard. In Paris he associated with the Catholic natural philosopher Marin Mersenne (1588–1648), who was at the centre of an important network of scientific contacts which supported an intense discussion of alternatives to discredited Aristotelian and magical theories of physical nature. Mersenne and his friends accused existing theories of conceptual obscurity, inconsistency and sheer impotence. By contrast, they were excited about the alternative science of material motion that Galileo had developed using geometrical language. Hobbes even visited the aged Galileo in Italy during his period of house arrest and at the time when Galileo published his foundational work on mechanics and materials, *Discorsi . . . a due nuove scienze* (*Discourses . . . Concerning Two New Sciences*, 1638).

Hobbes's prime concern was to find a basis for an ordered civil society. Since he thought geometry the only true science, he tried to reconceptualize human nature and human relations in terms that incorporated the certain premises and certain deductive conclusions of geometric argument. Such premises could be found, he insisted, once people use language rightly and, in particular, accept that the word 'substance' signifies 'material substance'. He excluded belief in incorporeal souls, spirits or agents as meaningless. Further, he explained the prevalence of belief in incorporeal things as due to priestcraft, the interest priests have in maintaining secular power. It was Hobbes's aim to remove the reasons in terms of which such interests are defended and to find the basis on which to unify power in the interest of peace.

Hobbes argued that nature is corporeal and made out of small particles of matter in motion. His contemporaries Descartes and Pierre Gassendi (1592–1655) came to similar conclusions, though in Descartes' case with different views about material qualities and about mind, and in Gassendi's case as part of a Catholic opposition to magical thought and with a self-conscious emphasis on an ethical reason in man that places him above material necessities. Hobbes's corporeal philosophy, which may well have owed much to his meeting with Gassendi, stood out because he calmly and remorselessly carried through the argument that all the created world, including the realm of mind and human action, is material. This materialism earned him instant and lasting notoriety as an atheist, which he probably was not, and he became a man from whom natural philosophers who sought social respectability were at pains to distance themselves.

Corporeal events, in Hobbes's view, explain human nature. His science of man treated in identical manner the motions of physical particles that cause material change, the motions that cause a living thing like a person to seek its own self-preservation, and the motions in society that, as he believed, cause the natural condition of mankind to be a struggle of each individual against every other. Each particle or person, he argued, is led by a desire for self-preservation to seek power over others. His imagery portrayed particles or individuals, interchangeably, moving as forces against the motions of other particles or individuals. He used the words 'endeavour' and '*conatus*' indifferently in relation to persons and physical objects, to denote the power of motion in persons or things. He described self-preservation as 'a certain natural impulsion of nature, no less than that whereby a stone moves downward'.[18] At one stroke he removed transcendent purposes from the physical and the human spheres. Human acts, which he characterized as acts in pursuit of pleasure or in avoidance of pain, are determined by the same motions that material particles have. 'This motion, in which consisteth *pleasure* or *pain*, is also a *solicitation* or provocation either to draw *near* to the thing that pleaseth, or to *retire* from the thing that displeaseth . . .'[19] Few of his contemporaries were prepared to follow this radical argument, and many later critics of natural philosophy as the means to understand human beings were appalled by it. In Hobbes's world there is no action that is intrinsically right, as Christian moralists constantly proclaimed, but only motions, some of which, in particular social contexts, can be described as having good or bad, pleasurable or painful, consequences. Nevertheless, Hobbes did claim, as a consequence of the way matter moves – that is, as a consequence of the laws of nature – that there is or could be order and law in the human sphere.

Hobbes threw his intellect against the passion of events that overtook his country, one among a distinguished line of philosophers to oppose fanaticism with reason. What is needed, he thought, is the 'mathematical', i.e., deductive, science of man. 'For were the nature of human actions as distinctly known as the nature of *quality* in geometrical figures . . . mankind should enjoy such an immortal peace . . .'[20] It was not utopian, he argued, to seek a peace based on knowledge: men make their own society, and since the basic content of human nature is known to follow from his corporeal being, it should be possible to rearrange men much as it is possible to rearrange the physical world. Conclusions deduced from these premises can be confirmed by observing the causes of actions. Man's corporeal nature and his experience provide political theory with its basic postulates: the natural desire to enjoy for oneself any goods available, and the natural reason to seek self-preservation. On this basis, Hobbes constructed a science of obligation – the laws for a 'Common-Wealth'.

Like Grotius and other natural law philosophers, Hobbes turned to what it is to be human as something, by definition, that all people share and which can therefore provide a base on which to unify human affairs. He stated: 'He that is to govern a whole Nation, must read in himself, not this, or that particular man; but Man-kind.' He demanded, as he said, a 'science of man', a science that is to be a secular guide to what to do, not just idle learning. '*Science* is the knowledge of Consequences, and dependance of one fact upon another: by which, out of that we can presently do, we know how to do something else when we will, or the like, another time . . .'[21] Like Bacon in the previous generation, Hobbes pinned his hopes to belief that the advancement of learning will render mankind, as well as nature, subject to direction and control.

Hobbes's starting-point when he discussed government was an intellectual and moral scandal. In a notorious phrase, Hobbes described the natural condition of people as *bellum omnium contra omnes*, the war of all against all. What he wrote was that 'it is manifest that during the time men live without a common Power to keep them all in awe, they are in that condition which is called Warre; and such a warre, as is of every man, against every man'.[22] At one level, he was describing the circumstances in which he lived: he wrote in self-chosen exile in Paris, where he moved in royalist circles that looked on helplessly as King Charles was beheaded in London. In 1649, rebels against the French king also seized Paris. *Leviathan* was addressed to these dire circumstances, and it argued that the political future depends on individuals, faced by the prospect of perpetual war, contracting to surrender their rights to a sovereign power, a monarch or – it was ambiguous – a collective leadership, in order to establish the stable conditions for a commonwealth. This might have been read as a defence of the royalist cause in England, but royalists as much as their opponents turned against Hobbes because of his materialism and because of his 'atheistical' picture of natural man devoid of God-given morality. At another level, Hobbes's image of natural individuals in conflict and selfishly preserving themselves reflected the explanatory categories of his general natural philosophy. Indeed, his critics reasonably feared that he subsumed moral under natural philosophy. His premise was 'the Law of Nature', the law that binds people to their nature as material beings, not God's will. Civil society, in his view, was an artificial condition, not a natural expression of innate morality, created by men who contract together to accept dominion by a sovereign in order to survive.

If, as Hobbes argued, just and right action means obedience to the law of nature, it followed that:

> The Science of . . . [Natural Laws], is the true and onely Moral Philosophy. For Morall Philosophy is nothing else but the Science of what is *Good* and *Evill*, in the conversation, and Society of man-kind. *Good*, and *Evill*, are names that signifie our Appetites, and Aversions; which in different tempers, customes, and doctrines of men, are different. . .[23]

This equation of good and evil with pleasure and pain was a radical (though not unprecedented) argument of great importance, not so much for Hobbes's own generation – it was often simply dismissed for its immoral denial of God's injunctions – but for the later sciences of man. Hobbes dwelt on the variety of moral standards in the world, attributed all actions to 'appetites and aversions' and described good and evil as what people say they are in the light of their own particular feelings. When he said that the natural condition for a person is unthinking selfishness, a statement that denies innate moral virtues, critics unkindly stated that Hobbes took his own wickedness as a model for the character of other people. Those who actually met him knew he was not a wicked man, though even they sometimes thought his arguments wicked.

Hobbes placed the science of man centre stage in moral philosophy; indeed, Book I of his political classic *Leviathan* was entitled simply, 'Of Man'. Here he discussed what he thought are the primary and universal attributes of being an individual. In practice, to gain knowledge about individual attributes, Hobbes recommended that we compare what we observe in others to what we observe in ourselves. This was a recommendation that assigned great significance to a sense of self and presupposed a capacity for self-reflection. Here again Hobbes's belief in the individual as a primary concept is evident. He did not, however, concern himself with individuals in a modern way that focuses on character and difference.

It is likely that Hobbes's emphasis on the fundamental naturalness of human self-preservation was partly indebted to Grotius.

But whereas Grotius thought that self-preservation leads necessarily to sociability, and thus that sociability is natural in the human condition, Hobbes argued that sociability is artificial. Both in practice favoured some form of monarchical government, but whereas Grotius looked to the King of France or Queen of Sweden to guarantee the order of an existing sociability, Hobbes described in the abstract how sociability is to be established. This description depended in a significant way on the language of mechanical things: knowledge of physical causes makes it possible to construct machinery; the human body is a complex machine created by God; and to reason on the causes of human action makes possible a machine for government – this is the machine he called 'Leviathan'. 'For by Art is created that great LEVIATHAN called a COMMON-WEALTH, or STATE, (in latine CIVITAS) which is but an Artificiall Man . . . in which, the Soveraignty is an Artificiall *Soul*, as giving life and motion to the whole body . . .'[24] He made the link between mechanical technology and political technology. The goal, in both cases, is a stable machine with efficiently functioning parts. Human artifice is modelled on nature, and nature itself is a technology created by God. The physical world, the living body, the machine and the civil state are comparable to each other. There was therefore more than analogy in Hobbes's imagery of the body politic, since this imagery restated the mechanist and materialist categories of a new natural philosophy as the explanatory terms for a science of man. Further, since human beings have made mechanical society – as God has made the mechanical world – so, Hobbes argued, it is possible to have certain knowledge of that society, as God (but not humans, as they are not the maker) has certain knowledge of nature. It followed, for Hobbes, that moral philosophy, one of the roots of what we would call the human sciences, is more certain than natural philosophy.

It is relevant to consider Hobbes's approach to language. He believed that there can be a certain *science* of man once the proper signification of the word 'substance' is clarified. This desire to clarify language was widely shared. Hobbes's Parisian friend Mersenne expressed hopes that language could be revised into a system of clear and rational signs which would replace existing usage, where words are taken to resemble what they refer to, and make possible universal understanding. With Hobbes, this hope for language crystallized into an opposition to rhetoric and an attempt to distinguish the meaning of words by using each word to describe one definite and particular element of experience. He thought that only when words are used to correspond to clearly identifiable things, by which he meant things present or potentially present to the senses, will it be possible to reason clearly with words and thereby construct the rational science that will save men and women from their passions.

Arguments like these were central to the appeal of the corpuscular natural philosophy espoused by Gassendi and by Robert Boyle (1627–97) in England, as well as Hobbes, though Gassendi and Boyle, unlike Hobbes, went to great pains to make corpuscular theories religiously acceptable. This philosophy appeared to make it possible to reason about the natural world in terms of clear and distinct ideas. In a world constructed from particles in motion, it ought to be possible, it was thought, to say precisely what one means by a cause in natural or human affairs, and to define that cause by reference to sensations. That was one reason why Hobbes and later Locke tried in their writings to conjure up for their readers a pre-social world where language directly reported things and hence was clear and true. They distinguished this from the current social world in which, they judged, language is an artifice. The hope that language could become a sign system for the simple elements of sensory experience was important to late seventeenth-century natural philosophy and, subsequently, to the eighteenth-century sciences and the modern scientific worldview. Yet it remained highly questionable as an approach to language and meaning.

Hobbes's views on language were closely connected to his opinion of rhetoric, an opinion formed when he worked on Thucydides, the Greek historian whose account of the Peloponnesian War he translated in 1629. Thucydides had opposed democracy because it opened up the possibility of decisions based on an orator who sways the rabble rather than on the conclusions of reason. Thucydides himself, Hobbes argued, therefore wrote history as a narrative of facts rather than to promote moral or rhetorical examples. With Thucydides as a model, Hobbes praised objective clarity rather than rhetoric. As with history, so later with natural philosophy: the business is to find objective expression for rational conclusions. This, Hobbes stated, requires the authority of an independent rational mind who is able to argue from the sure premises of natural law – not rhetoric and certainly not the rabble. Many later social and natural scientists found this position to be self-evidently reasonable, and it provided authority to place experts and expertise at the heart of the political process.

To conclude this discussion, I turn to the way Hobbes appears to be a figurehead for modern individualism and, in some accounts, for possessive individualism, the position – not in need of explanation in the late twentieth century – that the social world consists of the sum of individuals, their desires and their possessions. The premises of Hobbes's argument about politics concerned the natural action of *individuals*, particular entities that he assumed exist prior to social phenomena. Material corpuscles and individuals, elementary units moved alike by their 'endeavour', interact and thereby form massive substances or societies. More than analogy linked atomism in natural philosophy and individualism in political theory. A complex body in the physical world, such as a living organism, exists when the original coporeal particles are held in a stable pattern. In the same way, civil society (*civitas*), as opposed to crude relations, Hobbes argued, becomes possible when individuals act according to reason rather than passion and surrender their desires to a sovereign power in the interest of their self-preservation. The Spanish debates in the sixteenth century, we may recall, were also about whether the people discovered by Europeans in the Americas possess such *civitas*, and hence legal rights, or whether they are still in a pre-civil situation.

Historians have disagreed about how to relate Hobbes's individualism to social events in north-west Europe where, in the seventeenth century, there was the growth of market economies, an expanded system of capitalist finance, the development of nation states geared to economic efficiency, stress on contractual law in government and civil life, and other marks of modern social existence. Aspects of these changes can be traced back for centuries, while modernization in many parts of Europe was still centuries in the future. Yet one may see conformity between Hobbes's approach to human action and a society organized in material terms on the assumption that individual self-interested action is fundamental. Hobbes found his premises in what could be said of individuals when they act as discrete powers in a state of nature. Later theorists of human action in liberal political systems and capitalist economies also took this as their starting-point, in many cases with conscious acknowledgment to Hobbes. Thus he became an intellectual figurehead for capitalist modernity.

There has also been disagreement about the extent to which Hobbes's political theory was deduced from what he believed about human nature or from the individualist economic activity that he saw around him. It is agreed, however, that when he referred to the state of nature, he referred to a hypothetical and not a historical state – an 'Inference, made from the Passions' which describes 'what manner of life there would be, where there were no common Power to feare'. But it is revealing to note that, like many of his contemporaries, he turned to 'the savage people in many places of *America*' for a picture of something like the state of nature, where 'except the government of small Families, the concord whereof dependeth on naturall lust, [people] have no government at all'. It was in such conditions that, in his famous words, he thought 'the life of man, solitary, poore, nasty, brutish, and short'.[25]

One thing that Hobbes did not discuss was woman. When he argued for the science of man or for knowledge of mankind, he sought knowledge of universal human qualities, self-interest but also reason, which, he thought, dictate the form that a viable civil society must take. Whether he thought these were qualities possessed by women as well as men he simply did not say; the question of sexual difference was not on his agenda. All the same, his language was representative in the way it used the word 'man', and thereby connoted what was taken to be more characteristic of men than women when he defined norms of human nature.

v *Samuel Pufendorf on Natural Law*

Abstract argument about human nature was not thought remote from the harsh realities of day-to-day government – or, in England, the overthrow of government – as Hobbes and the flood of contemporary political essays and pamphlets confirm. Laws and institutions mediated between ideas and political events, while the language of natural law wove nature and daily affairs into a single fabric of understanding. The most articulate and influential reworking of Grotius's studies of natural law in the late seventeenth century is in the books by Samuel Pufendorf. His work was also a response to Hobbes and a legitimation of the consolidation of nation states like France, Brandenburg-Prussia and Sweden. Pufendorf integrated a jurist's concern with legal principles and a statesman's concern with government. His books were the principal resource for discussion of natural law in the eighteenth century when enthusiasm for a science of human nature came into its own.

By the mid-seventeenth century at the latest, it was manifest that Europe was divided permanently on religious lines; endless war had deepened divisions rather than enabled one side to emerge victorious. The Peace of Westphalia in 1648, which brought the Thirty Years War to an end, succeeded because it recognized the status quo, the consolidation of power among different states and principalities. This, therefore, was a key moment in the emergence of the modern system of nation states and for the conception of peace as a balance of power between states. Legal and political theorists took on two main tasks, the first of which was to provide legal authority for each state as it developed; one result was the encouragement of historical studies, such as Pufendorf's history of Charles X of Sweden. The second task was to provide reasons that would hold all people together, even when they were divided politically and by religious confession. To this end jurists placed great weight on natural law – and decisively influenced the search for a science of man.

A politically motivated stress on natural law united Grotius, Hobbes, Pufendorf and many other writers, and it mediated continuity between early modern and eighteenth-century political thought. Grotius conceived of natural law in the human sphere in relation to two innate human needs – self-preservation and sociability. Hobbes rejected the notion that sociableness was innate and argued that civil society requires a rational act of will or contract. Pufendorf accepted Hobbes's position sufficiently to deny that man has innate knowledge of moral laws, and this position attracted condemnation from Lutheran theologians. But Pufendorf also differed from Hobbes in a significant way, a way which anticipated and influenced the direction of argument in the eighteenth century and transformed the human sciences. He argued that though man in a state of nature, i.e., in a state outside civil society, is not sociable, he necessarily acts in such a way that he makes himself sociable. His view on this matter contradicted the sixteenth-century opinion held, for example, by Montaigne and Vitoria, that the native peoples in the Americas are both natural and sociable. Pufendorf also disagreed with Hobbes's opinion that men made a contract in order to survive. Instead, he held the view that the growth of man's sociableness expresses his nature. His conclusion was that sociability is a *historical* and not an original faculty of human nature but, even so, is natural and not artificial.

Pufendorf's youth and education coincided with the Peace of Westphalia and, after he had studied at the Lutheran universities of Leipzig and Jena, he had a career as tutor, professor and then privy councillor to Protestant ministers and kings. He finally received a barony in the year of his death, 1694, from Charles XI of Sweden. His work on international law, on constitutional history, on jurisprudence and on the legitimacy of government added up to an extended analysis of the emerging system of European states. His work for Karl Ludwig of the Rhine Palatinate, Frederick William I and Frederick III of Prussia, and Charles XI of Sweden testifies to the contemporary significance of intellectual and legal defences of the new powers. The major work on natural law was his comprehensive *De iure naturae et gentium* (*On the Law of Nature and Nations,* 1672), which he distilled in the following year into a more accessible compendium.

Humanist learning and the stress on human dignity since man possesses reason were Pufendorf's heritage. Man's understanding and power of 'judgement' – both a legal and a philosophical term – made it possible, he thought, for human beings to know and to obey natural law. He therefore described the purpose of his own work as to 'make Enquiry into that most General and Universal Rule of human Actions, to which every Man is oblig'd to conform, as he is a reasonable Creature. To this Rule Custom hath given the Name of *Natural Law*...' He followed Grotius and interpreted the Roman Law tradition to say that order itself requires preservation and hence that the natural law for man is self-preservation. He then argued deductively that social relations result when men exercise their reason about self-preservation and that sociability is rational as well as useful. 'This then will appear a Fundamental Law of Nature, Every Man ought, as far as in him lies, to promote and preserve a peaceful Sociableness with others, agreeable to the main End and Disposition of the Human Race in General.'[26] Pufendorf believed that this principle established grounds on which everyone, by virtue of a common nature and a common reason, could agree. From such arguments came both modern political rights theories and modern confidence that knowledge of nature, science, provides the authority for agreement on public policy.

Pufendorf's conclusions explained the duties of people to each other, to property and to the state. For instance, though he was politically a staunch servant of autocrats, he argued that all men are in natural law equal, whatever social position they in fact occupy. This is illustrated by his analysis of self-esteem, a quality that common language and opinion treated as a part of human nature.

> We commonly make use of this, as ... Argument against a rude Insulter, *I am not a Beast, a Dog, but I am a Man as well as your Self.* Since then Human Nature agrees equally to all Persons, and since no one can live a sociable Life with another, who does not

> own and respect him as a *Man*; it follows as a Command of the Law of Nature; *that every Man esteem and treat another as one who is Naturally his Equal, or who is a Man as well as he.*[27]

With such arguments, Pufendorf linked reason, jurisprudence and the everyday world of relations between people; he grounded what people commonly thought in general principles of nature and natural law.

The description of man's natural state, often imaginatively informed by reports of non-European peoples, became common in the sixteenth and seventeenth centuries. As Pufendorf rightly perceived, it was a confused and confusing language since the phrase 'the state of nature' could mean at least three things: the state in which the Creator placed humans in contrast to animals (i.e., with reason and will); the state we might imagine a man to be in if he existed by himself alone; and the state men might exist in bound by kinship and not civil society. He also recognized the ambiguity that the natural state 'may be considered either as it is represented by fiction or as it is in reality'.[28] In historical fact, the question of the relationship between a hypothetical state of nature and an actully observed natural man remained open over at least the next century.

Pufendorf, followed by many eighteenth-century writers, dealt with these questions through the medium of a historical narrative that explained civil society and human diversity as the product of time. This was an approach familiar to many earlier students of jurisprudence, but the explanation of civilization by reference to history came into its own in the eighteenth century. Pufendorf himself emphasized marriage as the historical moment when people came together; this was the first act by which man passed beyond a solitary state. 'Marriage may be called the first example of social life and at the same time the seed-bed of the human race.' Indeed, in his view of the natural state, a state which included kinship, marriage appeared 'natural', the cornerstone and exemplar for sociability in general. The beginnings of a well-ordered society lay in the regulation of sex and the care of children in marriage, and these activities thus appeared to be as much a consequence of natural law as sociability itself. It was in this context that Pufendorf discussed woman as opposed to man in relation to natural law. By contrast, when he discussed self-esteem, for example, he did not exclude woman but neither did he explicitly include her. His language about woman was a language about marriage, family and children. It was a small step from this to confirm the legal and moral expectations of marriage of his day and age. 'The duty of a husband is to love, support, govern and protect his wife; of a wife to love and honour her husband, and to be a helpmate to him . . .'[29] Claims about human nature, we shall have much cause to reflect, have never been neutral.

To summarize: though he did not accept, as Grotius had done, that sociability is a premise of natural law, Pufendorf argued that sociability, which begins in kinship and marriage and culminates in civil society, is a necessary historical outcome of the human state. Human beings, he argued, were made sociable by natural law, and on the basis of this knowledge it is possible to establish universal principles of jurisprudence with which to regulate national and international affairs. This was the justification for the hope of peace even though divisions between religious confessions and armed nation states were accepted as a permanent feature of Europe and the wider world.

This chapter has stressed the role of legal activity as both a theoretical and a practical context for the human sciences, a context brought into sharp focus by Pufendorf's influential work. He re-expressed in a modern idiom the position that other writers on jurisprudence had sought to define: what it is by nature and custom for a person to reason and act. Law students in early modern Europe were taught that there are universal moral and social principles that each person knows by the exercise of reason. There are principles that men share with the animals – self-defence and procreation – and there are principles unique to man, which can be divided into universal obligations and those special to particular peoples and institutions. The law applies only where reason exists and is able to discern these principles, and where there is a capacity to will action on the basis of reason. Children, for example, as Pufendorf wrote, 'before the use of reason begins to show itself with any degree of clarity' are 'unable to discern clearly what is to be done and to compare it against a rule' and they can therefore be corrected but not, strictly speaking, punished. 'Similarly in the case of the insane . . . their actions are not regarded as human, since their illness arose through no fault of their own.'[30] Jurisprudential argument therefore undertook to describe human capacities in general and to describe particular differences between human types in individual cases. The legal setting, taken in a broad sense, fostered elaborate discussion of the conditions for human action and conduct, and it thereby built expectations about human nature into the social fabric. Whether the issue concerned man's universal nature or masters owning slaves, the position of women within marriage or troublesome lunatics, legal discourse differentiated human qualities as a basis for political society. Law and legal language were important to the fledgling sciences of man through the way they ordered practical life. And the very concept of law was central to the framework of thought that made a human science possible.

4

Body and Soul

> What then did I formerly think I was? A man. But what is a man? Shall I say 'a rational animal'? No; for then I should have to inquire what an animal is, what rationality is, and in this way one question would lead me down the slope to other harder ones, and I do not now have the time to waste on subtleties of this kind. Instead I propose to concentrate on what came into my thoughts spontaneously and quite naturally whenever I used to consider what I was.
>
> René Descartes, *Meditationes de prima philosophia* (*Meditations on First Philosophy*, 1641)[1]

i *Theology, Pneumatology, Medicine*

The early modern academic study of human existence ranged across introductory topics in rhetoric, intermediate work in moral philosophy and advanced scholarship in law, medicine and theology. In 1651, Hobbes tabulated optics, music, ethics, poetry, rhetoric, logic and 'the Science of JUST and UNJUST' (i.e., jurisprudence) as divisions of the science of man.[2] Arguments about natural law cut across these topics, discussed man's nature as part of the natural world and subsumed moral action and physical change – both inescapable elements of the human sphere – under God's law. Further, the practical conclusions of natural law arguments, for example in a jurist's account of a child's dependency, fostered precise claims about human attributes and capacities.

To see how these arguments developed in the eighteenth century, it is first necessary to return to learned knowledge of the soul. To discuss changing views of the soul in the seventeenth century is to introduce many prominent themes of modern thought about human nature: the enhancement of the concept of mind at the expense of the concept of the soul; the separation between mind and body; the growth of knowledge about the body at the expense of the soul; and the preoccupation with the self. These themes were part and parcel of the new natural philosophy which laid the foundations for modern natural science. Key aspects of modern materialism were outlined by Hobbes. But it was most of all in Descartes' terms that late seventeenth-century philosophers and later generations explored the natural philosophy of the human soul.

There was no sharp break with ancient and Christian thought. Faith continued to give people an absolute conviction in the immortal soul. In the light of reason, as opposed to faith, many scholars continued to understand the soul as the animating principle, the life of the body, and hence the language and subject of the soul was as much the province of medicine as of theology. In the Renaissance, Aristotelians dealt with the soul as an embodied form, and they studied the qualities and divisions of the soul through the soul's relationship to the sensible world. This was in many respects a naturalistic approach, one that joined the study of man and the study of nature. The Aristotelians distinguished between the sensitive or organic part and the intellective part of the soul, and they held that the latter – reason – separates men and women from the animals who possess only an organic soul. This division survived in the new philosophies of the seventeenth century. Bacon, for example, wrote in *De dignitate et augmentis scientarum* (*Of the Dignity and Advancement of Learning*, 1623; this was an expanded Latin version of *Of the Proficiency and Advancement of Learning*, 1605): 'The parts . . . [of the doctrine which concerns the human soul] are two; the one treats of the rational soul, which is divine; the other of the irrational, which is common with brutes . . . [There are] two different emanations of souls, which appear in the first creation thereof; the one springing from the breath of God, the other from the wombs of the elements.'[3] Descartes took over this division and re-expressed it in the terms that are familiar in modern thought. He separated the thinking or rational substance and the bodily substance, and he thereby gave rise to a puzzle about the relationship between animals and humans, and between body and mind, which disturbed philosophers of the new sciences down to the twentieth century.

Aristotelian philosophy continued to occupy a central place in the academic curriculum for most of the seventeenth century. Scholars sometimes classified aspects of the study of the soul as part of pneumatology, the science of *pneuma* or, we might say, of spirit or 'the breath of life', in contrast to somatology, the study of material things. In some schemes, the topic of pneumatology also included study of the angels and higher spiritual beings, the realm of spirit that stretches upwards from humankind to the throne of God. The topic discussed such things as the soul's relation to the body as matters of fact to be found out. The Cambridge Platonist Ralph Cudworth (1617–88) in 1678 praised the ancients for the way they joined '*Corporeal* and *Incorporeal* Nature, *Mechanism* and *Life, Atomology* and *Pneumatology,* and from both these united, they made up one entire System of Philosophy . . .'[4] Pneumatology was the science of spiritual substances and cannot be equated with psychology; nevertheless, when the subject was taught in the first half of the eighteenth century, as it was for example in the Scottish universities, it included the description of what is observed of spiritual qualities as they enter into the everyday human world. In this setting, pneumatology involved the description and classification of the attributes, capacities or faculties of the soul as evidenced by human life. Physicians, theologians and moral philosophers all contributed to this study.

Meanwhile, the deepening and finally devastating attack on Aristotelian thought brought much confusion to the language of soul and body. To make things worse, the language of 'spirit' and 'spirits' used in pneumatology was complicated by inputs from medicine, Platonism, Stoicism, magic and natural philosophy. The consequence was that pneumatologists tried to take a more restricted view of their subject and to occupy themselves with the observable capacities of the human soul rather than with spirit in general. Gradually, but more and more frequently in the eighteenth century, authors referred human capacities to mind rather than soul and in the process they detached claims about mental capacities from claims about the soul's substance or essence. 'Mind' was a common enough word earlier, though it was used more in connection with memory (as in 'to remind') or the action of thought than to denote a psychological entity. There was a gradual substitution of 'mind' for 'soul' in common English-language discourse, and this subtly changed the meaning of what was held to be essential to human beings.

Belief about the soul was deeply embedded in medicine as well as in pneumatology and theology. The Galenic scheme correlated the bodily organ systems (liver, heart, brain) with the nutritive, vital and sensitive capacities of the soul. There was much diversity of opinion, but everyone used the language of the soul as the means with which to describe the healthy and diseased person. William Harvey (1578–1657), the London physician who in 1628 published *De motu cordis* (*On the Motion of the Heart*), the work often regarded as the foundation of modern scientific medicine, was a significant case in point. He gave a powerful logical and empirical demonstration that the heart circulates the blood around the body, a claim that undermined Galenic schemes at their core. But his approach to how the heart works and to the purpose of the circulation was firmly Aristotelian and was couched in terms of the soul's capacities.

As a further point about medicine, it is significant to observe that modern mental and physical categories do not coincide exactly with late Renaissance explanatory categories. In Harvey's world, the soul's capacities were neither mental nor physical. They were, so to speak, soulful. It followed that the moralist and the physician alike approached emotion or illness as states that affect the whole person, alterations in the soul's capacities that naturally affect all aspects of life. Burton's *Anatomy of Melancholy* detailed passions and disturbances of the mind, 'philosophically, medicinally, historically opened & cut up', and he included virtually every aspect of life in his remit.[5] This did not mean that there was a holistic understanding of human nature – an understanding that begins with the person, whole and entire, as a fully grasped reality. To believe that is unwarranted nostalgia.

Dualisms, which divided the immortal soul and its mortal embodiment, the intellective soul and the organic soul, or humans and beasts, were standard elements of the intellectual repertoire. These dualisms were not the same as modern mind–body dualism. Nevertheless, mind–body dualism was not an entirely novel way of thought.

The sheer variety of opinion at the opening of the seventeenth century was emblematic of the unsettled times. Traditional learning was in crisis. The Italian natural philosopher Bernardino Telesio (1509–88), for example, in the 1560s and 1570s criticized contemporary Aristotelians; their knowledge, he stated, was preconceived and too abstract, and he argued that nature should be studied through the senses. Yet he then attributed all change in nature to two forces, heat and cold, described all nature as animated by *spiritus*, and formulated a cosmology that made room for magic. On the topic of human perception, he argued that *spiritus* is a subtle element generated by heat in the brain and is itself able to receive sense impressions. To a degree, he made *spiritus*, which he thought has some material attributes, the substance of the soul, and he thus implied that he explained some things in bodily terms which were previously attributed to the soul. Nevertheless, Telesio also described the existence of a higher *mens* or divine principle which has an inner tendency to seek the divine. It was and is difficult to understand how this added up to a coherent philosophy of man's nature. But the mixture of concepts in Telesio's philosophy was characteristic of his time and symptomatic of increasing divergence among contemporary philosophies.

As the spread of Neoplatonic and magical worldviews in the sixteenth century shows, there was increased doubt about the capacity of Aristotelian philosophy to provide satisfactory explanations of natural phenomena. The outcome, in the middle decades of the seventeenth century, was enthusiasm for corpuscular and mechanical philosophies of nature. These philosophies specifically criticized the vagueness and imprecision of forms, qualities or spirits as explanatory categories and thus drove a wedge between mental and physical levels of reality and between the soul and the body. Telesio's ideas illustrate a shift towards the representation of the soul's agency in terms of spirits and the identification of those spirits with sensible, even material, substances. Descartes and Hobbes took this much further; Hobbes simply considered all human nature to be material.

These changes had implications for theological representations of immortality. Telesio, Descartes and Hobbes, broadly speaking, followed Aristotelian precedent and tried to separate questions that they claimed to be the proper province of natural philosophy from those proper to theology, and they argued that natural philosophical discussion of man's nature did not touch on the question of immortality. Theologians, however, denied that natural philosophers had independent authority to decide what the proper province of any field was. The province of the soul was therefore a contested area. Further, the consequences of certain kinds of natural philosophical argument for faith in immortality had had a high profile since 1513 when Pope Leo X's Lateran Council condemned Averroist arguments for the soul's mortality, and this led to sharp criticism of Pomponazzi's arguments for mortalism put forward in his *De immortalitate animae (On the Immortality of the Soul,* 1516). Later Catholic natural philosophers, Telesio and Descartes included, knew that any consideration of the soul in material terms was likely to provoke accusations of the mortalist heresy. Protestant natural philosophers, like the exponents of new corpuscular ideas in mid-seventeenth-century England, Henry More (1614–87) and Walter Charleton (1620–1707), were equally self-conscious about how the new thought might be judged to affect belief in immortality.

Theological argument and counter-argument about the soul does not translate easily into modern ideas. It is difficult, for example, to grasp the Catholic orthodoxy that required the soul to be understood as a unity which incorporates both the rational principle and, *per se et essentialiter* (of itself and essentially, i.e., without mediation), the form of the body. This requirement followed from doctrine about the incarnation and the resurrection, since it was held that the body as well as the soul of Christ was part of both miracles. No Catholic wished in the last analysis to claim that the soul is divisible, though it is divided for analytic purposes into intellective and organic parts, or material, though the soul has an intimate relationship with the body. Protestant Aristotelian scholars in the sixteenth century, who held a different view of the incarnation and resurrection, were somewhat more willing to describe the intellective and organic souls as separate in substance as well as in analysis. This permitted separation between discussion of the organic soul, a topic about *spiritus*, pneumatology and the study of the animation of animals, and of the intellective soul, a topic about the immortal principle. It is also important that the question of the soul entered into the central Catholic dogma of transubstantiation, the miracle of the eucharist, in which wine and bread become the blood and body of Christ. It mattered theologically – to Descartes and other natural philosophers, as well as to theologians – to describe the soul of Christ in relation to this miracle. Much of this sounds esoteric to modern ears. Precisely so: it signals the differences between a study of human nature framed with reference to the transcendent and a modern science of man.

It was Protestant scholastics who used the word 'psychology'. The teacher Rudolphus Goclenius (1547–1628, born Rudolf Goeckel) in 1590, followed by Otto Casmann (d.1607) in 1594, defined psychology as the study of 'the nature of the human spirit or logical soul'.[6] The word was more common by the mid-seventeenth century, though still in predominantly Aristotelian contexts where it denoted the study of the intellective soul. Classifications of knowledge more frequently referred to pneumatology to cover all aspects of the soul and spiritual being. The outstanding encyclopaedia of the time, produced by Johann Heinrich Alsted (1588–1638), divided pneumatology into two parts: *psychologia*, which discussed the disembodied soul, and *empsychologia*, which discussed the soul as a life-giving principle joined to the body, that is, treated the soul as a part of 'physics'. In the same period, 'anthropology' came into use as a word to denote study of the bodily side of man's nature. These usages continued during the first half of the eighteenth century. As discussed later, Christian Wolff then wrote a *Psychologia rationalis*, a logical account of the soul as the rational and immaterial principle. Pneumatology continued to be taught in the Scottish universities, though with an empirical and naturalistic rather than logical and analytic character. Before this, in the second half of the seventeenth century, the Port-Royal logicians in France and then Locke in England established empirical approaches to what they called the understanding as a basis for logic or right reason. It was not until after 1750 that such empirical approaches were described as psychology.

Much earlier, the Aristotelian intellectual context that gave meaning to the language of the soul was subject to a withering critique. Catholic philosophers, it is true, returned in later centuries to Aristotelian concepts (in Thomist form) as a means to represent man's essential spirituality, and this remained a live tradition. I turn, all the same, to the arguments that were decisive in the seventeenth century for the impact they had, for good or bad, on new ways of understanding the world. It is easy to be sympathetic to one major reason for the critique, the indefiniteness of Aristotelian language, exemplified in the statement in *De anima* that 'the soul is an actuality . . . of a natural body having life potentially in it'.[7] Bacon was particularly scathing about such writings:

> They end in monstrous altercations and barking questions. So as it is not possible but this quality of knowledge must fall under popular contempt . . . And when . . . [the people] see such digladiation [quarrel] about subtilities and matter of no use nor moment, they easily fall upon that judgment of Dionysius . . . 'it is the talk of old men that have nothing to do'.[8]

Bacon was a historically influential spokesman for the new natural philosophy of the seventeenth century, especially when his work was made part of the rhetorical legitimation for the Royal Society of London established nearly forty years after his death. He criticized existing philosophy for its failure to distinguish between clear and distorted reasoning: 'For the mind [*anima*] of man, (dimmed and clouded as it is by the covering of the body), far from being a smooth, clear, and equal glass (wherein the beams of things reflect according to their true incidence), is rather like an enchanted glass, full of superstition and imposture.' He therefore explained what he called the four 'idols' of the mind – the 'corrupt and ill-ordered pre-disposition[s]' that distort its understanding. The first three of these idols are a consequence of human nature: 'Now idols are imposed upon the mind, either by the nature of man in general; or by the individual nature of each man; or by words, or nature communicative. The first of these I call Idols of the *Tribe*, the second the Idols of the *Cave*, the third the Idols of the *Market-place*.' The fourth kind, the idols of the theatre, he attributed to false philosophies. When he referred to the idols produced by human nature, Bacon pointed out: first, how the human mind naturally moves from what it knows to what it does not know and thus assumes uniformity and similarity where none may exist; second, how the individual mind has strange and absurd fancies which derive from dwelling within its own narrow world; and third, how language creates false knowledge, 'Idols . . . which have crept into the understanding through the tacit agreement of men concerning the imposition of words and names', and thus how language leads to the false representation of the world.[9] This discussion was a powerful practical contribution to a new philosophy which made the concrete minds of actual people the place from which to start to create knowledge.

Just at the time when Bacon wrote, Aristotelian natural philosophy was overtaken by a new learning, one that seemed able to resolve the anomalies that had plagued the schoolmen. Kepler's studies of planetary elliptical orbits and Galileo's work on the laws of falling and moving bodies rendered Aristotelianism untenable as a physical worldview. It is therefore time to turn to the major seventeenth-century alternative, mechanical philosophy, which prefaced modern discussions about the soul, the mind and the body.

ii *Cartesian Dualism*

As generally understood, the scientific revolution involved a fundamental change of worldviews that began with the astronomical work of Copernicus in 1543 and ended with the mathematical synthesis of Newton in 1687. Historians currently emphasize a more complex picture in which there is no unified revolution. The same complexity has been introduced into assessments of Descartes, the philosopher at the heart of new explanations of nature in terms of matter and motion. The rest of this chapter deals with pertinent aspects of Descartes' work: his conception of soul and body; his extension of bodily explanations to encompass much of what previous philosophers had attributed to the soul; and his contribution to modern notions of the self. Whether his work and the writings of those who thought like him should be regarded as a triumph or a disaster for the subsequent course of the human sciences is a matter for debate. On the one side, there is the opinion that Descartes' mechanical philosophy initiated the modern study of human beings as part of the natural world; on the other side, there is the judgement that his philosophy made it impossible to formulate a coherent view of people as reflective agents. This debate continues, and it is the debate itself, part of the long-running controversy about what sort of knowledge makes for a human science, rather than any one point of view which gives the history of psychology its character.

The life of Descartes overlapped with the lives of Grotius and Hobbes and, like them, he was obsessed with the search for certainties. His published writings, however, were more concerned with philosophy than with social order, though these matters were probably connected in his own mind. He sought for certainty in the nature of reasoning itself, and hence he came to be regarded as a founder of modern rationalism. His search came in spite of, or perhaps because of, a superior education in the scholastic disciplines, the best available in his time, at the Jesuit college of La Flèche near Paris. His judgement, in the form in which he reconstructed it for public presentation, was as personal as it was devastating: 'For I found myself beset by so many doubts and errors that I came to think I had gained nothing from my attempts to become educated but increasing recognition of my ignorance.'[10] This was the rub: where in the endless arguments of the schools or of worldly battles could there be a basis for truth? Descartes accepted revelation, Catholic authority and scripture, but for an answer he turned in practice to the operations of his own mind. Just as Hobbes was ravished by geometry, Descartes was ravished by arithmetic – in his case by a divine sign that came to his overheated brain after he had locked himself up with a hot stove on one memorable day in November 1619. 'Delighted in mathematics, because of the certainty and self-evidence of its reasonings', he described a method, he believed, that established conclusive axioms and deductive arguments.[11]

Descartes acquired mythic stature in the modern age. In the francophone world, his language served as a model for the clear expression of rational truths and became part of the education of all those who went through the *lycées*. Educated opinion in France held that the qualities of his texts exemplified the qualities that gave France rightful leadership of the civilized world. For historians of science, his philosophy was the purest expression of the mechanistic worldview at the base of modern scientific thought. In the humanities and for modern cultural critics, his account of the duality of mind and body – Cartesian dualism – set the terms for a modern experience of division within the self and alienation between the subjective self and the objective world. Even Descartes' personal life was cultivated as a mystery, and mythic stories rather than confirmed facts dominated biographies. After his study at La Flèche, Descartes took a degree in law at the university of Poitiers, then served in the military forces as a gentleman volunteer under Prince Maurice of Nassau, though we do not know what his motives were. Either then or after leaving the army, probably financed by the sale of property inherited from his mother, he travelled in the Low Countries and in the German and Italian states. He slowly developed a total commitment to the sciences and mathematics, and it is perhaps more in retrospect than at the time that he accorded decisive importance to his vision in 1619 and to the dreams that followed the same night. In his retelling, a sequence of three vivid dreams warned him against past errors, inspired him with the ideal of truth and showed him that his own reason enabled him to reach truth. Ten years later he decided to live in the Dutch United Provinces, where he sometimes sought solitude and sometimes kept the company of doctors, lens grinders and other practical men. All the while, he maintained his Paris connections and through Mersenne and other scholars remained in contact with the rapid and exciting contemporary developments in mathematics, optics, astronomy, mechanics and anatomy. Through connections and correspondence, Descartes' own innovations and opinions in mathematics and philosophy were quickly disseminated, sometimes before they were published. Among the most earnest of his correspondents was the Princess Elizabeth, daughter of Frederick, Count Palatine and Elector of the Holy Roman Empire, who became Queen of Sweden. In 1649 she persuaded him to stay in Stockholm, where early in 1650 Descartes died from pneumonia, by all accounts caught as a result of the Queen's insistence that she be given philosophy lessons, before the day's business, at five in the winter mornings.

While still travelling, Descartes devoted considerable attention to 'meditations on first philosophy', and he arrived at his well-known resolve to 'reject as if absolutely false everything in which I could imagine the least doubt . . .' 'But immediately I noticed that while I was trying thus to think everything false, it was necessary that I, who was thinking this, was something . . .' In short: *'I am thinking, therefore I exist.'* He placed the 'I' that thinks at the centre; from this centre he then argued outwards first to the existence of God and second to the 'laws which God has so established in nature' that make nature as it is.[12] He utilized systematic doubt to isolate an indubitable rational self and a concept of natural law based on God's perfection, in order to establish grounds on which all right-thinking people would be forced to agree. This was his escape from inconclusive arguments and scepticism.

Descartes then deduced that the world consists only of matter, defined as extended substance, and motion. This conclusion was persuasive for those who adopted what became known as the mechanical philosophy because it made extension and motion into *clear* and *quantifiable* modes of existence. Descartes' view of the world turned knowledge into a problem in analytic geometry, a problem resolvable by abstract calculation about the location and relations of points and lines. To explain an event is to describe the redistribution of extended substance, and the language of such an explanation is mathematics. This vision had tremendous depth and impact; it replaced the Aristotelian world of qualities, spirits, formal and final causes and such-like, with precise definitions, extended substance and mathematical demonstration. All this is part of the general history of science, part of the foundations for mechanics and physical science that provided the model over many generations for what scientific knowledge could and should be. Descartes did not succeed well with many of his mechanical explanations of particular things, but he did achieve a fully fledged philosophical argument about the terms in which such explanations should be sought. He originally intended to publish his physical conclusions in a volume on *Le monde* (*The World*) but was inhibited from doing so by the condemnation of Galileo by the Catholic Church in 1633. Instead, he first published the *Discours de la méthode pour bien conduire sa raison* (*Discourse on the Method of Rightly Conducting One's Reason,* 1637) accompanied by essays on the theory of lenses, the weather and geometry in order to illustrate how the method worked when applied to nature.

How did this vision of the world as a mechanical system relate to human existence? Descartes' method of systematic doubt left him with what he regarded as indubitable belief in a perfect God, in extended substance and in a thinking mind – the *'moi'* or 'I'. He described the 'I' in traditional language as a substance distinguished by the capacity for thought and also for being what matter is not – unextended, indivisible and hence unquantifiable. 'I knew I was a substance whose whole essence or nature is simply to think, and which does not require any place, or depend on any material thing, in order to exist. Accordingly, this "I" – that is, the soul by which I am what I am – is entirely distinct from the body . . .'[13] He described the immaterial and immortal soul and the material and temporal world in mutually exclusive terms. This fundamental division, which later generations referred to as Cartesian dualism, had many ramifications; two are of special relevance.

First, Descartes undoubtedly intended in this way to defend the reality of the immortal soul and thus to satisfy his own faith and pacify Catholic theologians who feared the consequences of his method and its anti-Aristotelian conclusions about nature. He intended to support faith in the resurrection and for his philosophy to be compatible with belief in immortality and transubstantiation. The opprobrium which descended on Hobbes when he treated the soul as corporeal like the rest of the world illustrates what Descartes wished to avoid – for reasons of belief, not lack of courage. Second, the combined presence of soul and body is unique to mankind. Descartes was clear and emphatic that men and women alone, not the animals, possess an immortal soul; the presence of thinking substance defines what it is to be human. Any science, therefore, that aims to understand human beings must study the soul, the body and the relationship of soul and body. Though not conceived in precisely the same way by Descartes and later generations, the relationship of soul and body became in later centuries the classic philosophical topics of how mind knows the material world – the problem of knowledge – and of how the mind interacts with matter – the mind–body problem.

Descartes dealt with the problem of knowledge, which we can largely put to one side, through a discussion of God's perfection, and with the mind–body problem through an account of man's bodily life. The second topic requires a separate section. Before turning to it and to an assessment of Descartes in relation to his contemporaries, however, I consider some large-scale claims about his significance for the modern world. This may help explain why the modern human sciences remain divided on basic issues, why the human and natural sciences have such an ambivalent relationship, and why Western public opinion sometimes regards natural science as at best a mixed blessing in relation to questions about human values. One caveat: Descartes' name has often appeared as a symbol rather than as part of a serious historical claim in such discussions; that is, people have cited his work for its canonical status in philosophical argument rather than with historical intent. In historical perspective, Descartes was a man of his time and place, only one man among many scholars who sought, in somewhat scholastic terms, to find authoritative alternatives to discredited worldviews.

News spread rapidly in the 1640s about Descartes' new philosophy and especially about his claim to explain all aspects of the physical world in terms of matter and motion. The attempt to show that his claim was justified, to show that matter and motion do cause the observed phenomena, proved a much more complex task than the early mechanical philosophers thought. But Descartes' programme inspired people like Boyle, Christiaan Huygens and the young Newton, who transformed physical science. At the same time as it led to these triumphs in physical science, however, mechanical philosophy excluded thinking substance, human reason, from the physical world. It portrayed the mind as cut off by its essence from comprehension in terms appropriate to the understanding of nature. It appeared misguided at root to seek a fully naturalistic science of man. The struggle to 'Know thyself' was divided into an effort to know extended substance, the body, and to know thinking substance, the mind. Many later observers thought this was an intellectual and moral catastrophe: incoherent at its foundation and false to everybody's experience of life as a person, not as a separate mind and body. History was sometimes written as if all subsequent theories in the human sciences were either confused by, or were attempts to overcome, this dualism. It was implied that, if confusion continues, something must be utterly rotten in the foundations.

Many modern natural scientists believed that Descartes stopped short, even lost his nerve, when he pushed through the scientific revolution and excluded the human mind from nature. If this is intended to be a historical point about Descartes, it crudely misrepresents his philosophical intentions. It is in fact a point about natural scientists and not about Descartes. It implies that they have a mission, inspired by the explanatory and technological success of mechanical philosophy, fully to carry through an understanding that human beings have only a physical nature. Twentieth-century behaviourists and neuroscientists, in this way, claimed to fulfil the promise of the seventeenth-century scientific revolution. From their point of view,

Cartesian dualism was 'a problem' only because of Descartes' failure to extend the explanatory categories of physical science to encompass the human sphere as a whole. Modern materialists argued that scientific progress shows how all those aspects of human activity previously attributed to mind or soul can in fact be attributed to material processes.

This solution, if such it be, to Cartesian dualism had – and continues to have – many critics. It was opposed by Descartes' own belief in the immortal soul. Non-theological objections, often informally developed, accepted the 'I' or self as the foundation of what people experience in being human. This objection was made dramatic when the Romantics asserted the inherent creativity of the human spirit. More prosaic observers were puzzled to know how consciousness in all its qualitative richness – the sheer colour and depth of the immediately experienced – can be said to be a property of matter and motion. They were even more puzzled to explain conscious judgements of value – moral, aesthetic, romantic and spiritual, or intentions, purposes, passions and visions – in material terms. For those who accepted Descartes' dualism, this was the problem: to explain how experiential richness and evaluative judgements, values, derive from the interactions of material reality with the soul. Descartes' answer appeared to make values subjective, qualities superadded by the mind to nature, in contrast to objective nature itself. Nature appeared to be without value, a dance of atoms. Finally, with arguments close to Descartes' own, critics pointed to reasoning itself, and asked how propositions about material reality can coherently be said to be an activity of material reality. Descartes specifically thought they cannot, and many later philosophers agreed: they argued that the activity of reason must be given priority in any account of the world since nothing can be asserted, let alone known, except as we reason about it.

In the seventeenth century and later, Descartes was understood to have developed a theory of reason based on innate ideas. This frequently misled commentators. Descartes defined the soul as the rational principle, but he did not envisage a psychological theory in which each individual mind at birth possesses a number of fixed ideas about the world. 'I have never written or taken the view that the mind requires innate ideas which are something distinct from its own faculty of thinking.'[14] Locke, who is well known for what is supposed to be his criticism of innate ideas, appreciated this point and, like Descartes, also found it necessary to assume the existence of an innate faculty of mind (that presupposed nothing about innate knowledge), which Locke called 'reflection'. Thus, Descartes and Locke – the supposed exemplars of conflict between rationalism and empiricism – both began with a conception of man as a reflective subject.

Cartesian dualism, like Descartes the man, is emblematic in a modern world that has embraced 'objective' natural science but which is wedded to 'subjective' consciousness. Poets, headed by William Blake, late twentieth-century 'deep ecologists', and some feminists, like the historian of science Carolyn Merchant, have been appalled at the way the mechanical philosophy reduced nature to mechanistic motions: warmth, colour, growth, passion, life itself, appeared dressed in cold and abstract mathematics as opposed to the warm and concrete language of a person's, and especially a woman's, experience. The late twentieth-century degradation of the environment created considerable sympathy with the view that the dominant modern attitude to nature has been that of a male mechanic who exploits a machine. All the same, historical understanding requires other sympathies as well, not least with what it all meant for the historical actors themselves.

One more technical point emerges out of this critical attitude towards natural philosophy. Descartes and later natural scientists drew a distinction between the primary qualities, size and mass (a concept introduced by Newton), which characterize what is supposedly really present in nature, and the secondary qualities, colour, warmth or surface feel, for example, which appear to belong in nature but are claimed to exist only in the mind. This distinction, it seems, made human experience insignificant in the world; modern natural-scientific worldviews have, therefore, been labelled life-denying. Yet as we have seen, the division between primary and secondary qualities existed in earlier accounts of perception, when scholars grappled with the relationship between an object's image in the soul and the object's form or species that impresses itself on the sense organs. The early modern mechanical philosophers even took over earlier analogies, such as the analogy between the way an object leaves an image in the sense organ and a seal leaves its imprint in hot wax. Pre-modern philosophies did not guarantee a match between experiential richness and theories of the external world. Descartes' division between primary and secondary qualities was not an abrupt dislocation of worldviews that integrated humankind with nature. All the same, critical examination of Descartes and of those who followed may provide us with intellectual resources with which to seek for an integration of ourselves in nature.

iii *Les passions de l'âme*

We are faced by an irony when we turn in more detail to Descartes' dualism. In theory, his position was clear-cut: animals and the human body are entirely material and mechanical – there is only one science of bodies; the mind, in contrast, is unique to human beings, and it is known by reason and not by science. In practice, in order to describe something that resembles a person, Descartes wrote about the mind and body together. His philosophy emphasized the distinctiveness of mind and body, but he also wrote about the nebulous zone in which they interact. Given the mutually exclusive terms in which he characterized mind and body, this required some ingenuity.

His strategy was to treat as much of human life as possible in bodily terms. Here he developed a trend already present in the sixteenth century and evident in Telesio's philosophy, to relate all the lesser faculties of the soul, those below intellect or reason in status, to the body. First, however, Descartes had to demonstrate that the human body is indeed the mechanical system his philosophy required it to be. This was the point of the section in the *Discourse on the Method* where Descartes turned to Harvey's argument for the circulation of the blood as 'the first and most widespread movement that we observe in animals, . . . [that]

will readily enable us to decide how we ought to think about all the others'. The movement of the blood, Descartes argued, 'follows just as necessarily as the movement of a clock follows from the force, position, and shape of its counter-weights and wheels'.[15] Such language was inspired by delight in mechanical devices, especially the intricate hydraulically operated figures then to be found in royal gardens, and which Descartes saw at St Germain-en-Laye near Paris. The mechanical imagery of the body was in part a literal representation of technological innovation. After Descartes' death, his full text on mechanical man, replete with marvellous but entirely speculative detail about the body's hydraulics, was published as *De l'homme* (*On Man*, 1664).

However sophisticated man-made machines might become, Descartes argued, they would not look like human beings. He gave two interesting reasons. The first was speech, for though of course machines might be made to emit sounds, 'it is not conceivable that such a machine should produce different arrangements of words so as to give an appropriately meaningful answer to whatever is said in its presence . . .'[16] This argument, latterly refined into dicussion of semantics or meaning, has run and run in relation to the possible distinction between humans and animals. Noam Chomsky, the influential philosopher of linguistics in the 1960s, reiterated the distinction. Secondly, Descartes stated that the variety of human actions and their appropriateness for their purposes shows that they follow from reason and knowledge, that is, from the presence of the soul. When he pinpointed language and reason as uniquely human, he placed his finger on topics that have ever since divided opinion on the relation between the natural and the human sciences. Descartes gave this debate shape when he stated it in terms of a specific question about how the soul interacts with the body.

Descartes never doubted that body and soul do interact intimately: 'it is not sufficient for . . . [the soul] to be lodged in the human body like a helmsman in his ship, except perhaps to move its limbs, but that it must be more closely joined and united with the body in order to have, besides this power of movement, feelings and appetites like ours and so constitute a real man'.[17] Like early modern physicians and in tune with common opinion, Descartes portrayed the soul yoked to the body by 'sensations and appetites'. Accordingly, in *Les passions de l'âme* (*The Passions of the Soul*, 1649), he tried systematically to describe how physical motion in the body acts on and 'moves' the soul, or, to use the language appropriate to the soul, how it has 'passions'. He imagined that motions are induced in the fluid- or, in his word, 'spirit'-filled nerves by mechanical effects such as light that impinges on the retina or agitated motion in the stomach. Mechanical motion in the nervous spirits, he argued, passes to the brain, where he supposed there must be one point at which all these motions come together. This resembled the view of earlier scholars that there is a *sensorium commune*, a common sensory centre, the content of which is *sensus communis* – 'common sense'. Here, Descartes believed, motion 'moves' the soul. In addition, he described feelings caused by the soul itself and not by animal spirits.

When he turned to brain anatomy, Descartes located one structure, the pineal gland, that seemed to him to be uniquely bathed in nervous spirits and centrally placed to receive motions and transmit them to the soul and vice versa: 'a certain very small gland situated in the middle of the brain's substance and suspended . . . [that] the slightest movements on the part of this gland may alter very greatly the course of these [nervous] spirits . . .' On first reading, it looks very much as if Descartes provided a precise spatial location, the pineal gland, for the soul, which is in principle without spatial dimension. He did not intend to make such an obvious error. He wrote that the soul, without extension and without division, 'is joined to the whole body'; that is, the soul is joined in its essence to the whole body, though in reality it acts, and is acted on, at one central point.[18] This sounds almost perversely paradoxical to modern ears, because we think of a cause in post-Cartesian terms as a physical change in space and time. When Descartes considered the soul's relation to the *whole* body, he thought in 'pre-Cartesian' categories of the soul as form or essence and described the soul as the substantial form of the body. But because of what he himself argued so persuasively on behalf of the mechanical reality of nature, he bemused others about what he said on the soul and its relation to matter.

The passage in Descartes' account of man best known to modern scientists is where he apparently anticipated the concept of reflex action. This is an anachronistic misconception. He drew a picture of a crouching boy removing his foot from a fire, and showed a thread running from the foot to the brain; this illustrated 'how external objects that strike the sense organs can incite [the machine] to move its members in a thousand different ways . . .'[19] In his example, a pull on the thread opens a pore in the brain's cavity and allows animal spirits to flow to the foot, the head and the eyes, which leads to withdrawal of the limb. No specific reflex was proposed, though of course his point was to show that such action is machine-like and not willed. The concept of the reflex came later; it was stated clearly only in the 1830s.

His discussion of the soul's passions made considerable reference to the animal spirits. Descartes believed that these spirits are material, like any other part of the world, however subtle their movements and refined the particles from which they are made. He extended the realm of mechanical explanations as far as he thought philosophy would allow, and these explanations encompassed perception, imagination, the passions, dreaming and automatic and habitual activity – indeed, everything except reason and will. In practice, when he imagined how matter and motion result in different passions in the soul, like love, indignation or melancholy, he took these human experiences, metaphorically applied them to the corporeal spirits (e.g., 'agitated' motion) and then claimed that these corporeal motions explain human experiences. At first glance, this appears to achieve little; however, the argument that one could and should redescribe conscious qualities as mechanical events necessarily had consequences for what conscious experience was held to mean. It is not the same thing for people to say they are agitated and to say they are suffering the effects of agitated nervous motions.

It is significant also to note that Descartes described and classified the passions in a way common with late Renaissance moral philosophers and popular moralists. He believed there are six 'primitive passions': wonder, love, hatred, desire, joy and sadness. Other passions, he thought, are species or combinations of these. His descriptions used the same language about the humours and temperaments as his predecessors, with similar references to the warmth or coolness of the blood, vapours and the movements of spirits. Presented with the qualitative richness of the language conventionally used to describe experience, Descartes simply utilized that language. Yet his ideal was that this richness should be translated into a science of quantified motion.

The passions were held to make the soul active and not only to be what the soul 'suffers'; they incite the soul which, in the light of its reason, then moves the spirits to cause speech, bodily motions or other activity. 'Love', for example, 'is an emotion of the soul caused by a movement of the spirits, which impels the soul to join itself willingly to objects that appear to be agreeable to it.'[20] Such passages might make it appear that Descartes believed all our actions to be dependent on the spirits, the line of argument developed by later enthusiasts for a determinist natural science of man. But this was not Descartes' position since, as the same passage went on to say, he believed that the soul can act freely according to its rational judgement as to what is good and evil. In his view, the soul both acts freely and is led by the passion of desire to act. This was a restatement, in somewhat new terms, of a Christian belief in the struggle between the will and the bodily passions.

All moral philosophers wrote about the control as well as the nature of the passions; moral philosophy was a practical as well as a theoretical discipline. Descartes himself included the customary complement of mundane advice about how to suspend judgement at the height of a passion or how to direct attention away from undesirable objects in order to increase the soul's powers. He took a rather instrumental view about how to achieve such control, in accordance with his mechanistic account of what the passions are. While he was conventional in his portrait of bodily forces, in which he included the passions that are movements of the spirits, he took a distinctive line when he described the soul as unambiguously an agent of reason and goodness. The soul, he argued, cannot be divided against itself, a claim that aroused controversy since theologians emphasized original sin as a spiritual fall from grace – a division in the soul. Descartes encouraged belief that when someone is divided against himself or herself, as in madness, 'there is no conflict here except in so far as the little gland in the middle of the brain can be pushed to one side by the soul and to the other side by the animal spirits . . .'[21] Madness, he thought, is physical not mental. Here again, the direction of his argument was to separate, and even to place in opposition, distinct mental and physical categories. His conclusion that the soul cannot in principle suffer unreason or madness was still being reached in the nineteenth century; such calamities, post-Cartesian moralists and physicians thought, have to be attributed to a body that leaves the soul without a properly working instrument, and not to the soul itself.

Descartes' writings had considerable contemporary impact, though few readers, on reflection, went along with everything he wrote about body and soul. By the 1630s and 1640s, scholars looked for certainty in knowledge, sometimes almost desperately, in the wake of the proliferation of worldviews and scepticism rampant at the end of the sixteenth century. Descartes was read along with other philosophers, like Galileo, Mersenne and Gassendi, all of whom raised consciousness about mechanical ideas as the possible way forward. Gassendi, a Catholic cleric, also formulated a comprehensive philosophy, though in his case it took the form of a Christianized Epicureanism or corpuscular theory of nature. Gassendi claimed that there are two souls in man, one corporeal and the other incorporeal, and he assigned to them faculties that resemble those attributed to the Aristotelian organic and intellective souls. By this means Gassendi hoped to reconcile the benefits of explanation by material motions with orthodox belief in man's God-given immortality and reason. As we have seen, Hobbes also took up these new natural philosophies to advance a science of man, and he embarrassed those who followed Descartes and Gassendi when he argued that the soul is entirely corporeal. This shocked contemporaries because it appeared to deny the immortality of the soul and remove the grounds for belief in free and moral action, the possibility of which was thought to underpin social stability. I have indicated how Descartes argued in order to avoid such conclusions, and he met with some success since by the end of the seventeenth century there was an impetus among French Catholic scholars to adopt his arguments. To illustrate the history and difficulties of Cartesian notions of body and soul in the years immediately after 1650, however, I turn to England.

A small number of scholars, mostly well known to each other, distressed by the Civil War and its aftermath, read Descartes avidly from the late 1640s. The enthusiasm of Henry More was quickly tempered by concern that Cartesian dualism threatened belief in the agency of the soul and of God in the world. More concluded that there must be a realm of being that mediates between soul and matter, a 'Spirit of Nature', the agency of spiritual and moral purpose in the world of material motions. More likened the brain to suet or curds, and he thought that the study of matter drew attention to the need for the soul in a living universe and thereby confounded atheists.

More's influence lay mainly in Cambridge. His Oxford contemporary Charleton was attracted by Gassendi's philosophy, specifically by its statement that there are both corporeal and incorporeal souls. Charleton criticized Descartes' dualism as incoherent, and claimed that the pineal gland cannot perform the function Descartes assigned to it. It was preferable, Charleton thought, to accept that the corporeal soul, as Gassendi explained, is something that mediates between body and the immortal soul and shares some attributes with both. Charleton's elaborately titled *Physiologia Epicuro-Gassendo-Charltoniana: Or, a Fabrick of Science Natural, upon the Hypothesis of Atoms* (1654) – in large part a translation of Gassendi – spread these ideas in England. He carried his opposition to Descartes into moral philosophy and medicine; he claimed, for example, that the effort to control the passions reflects conflict between the incorporeal and the corporeal souls rather than the soul's conflict with body. When he came to philosophical problems in his own account of souls – like when he addressed the question how a corporeal entity can be sensitive – Charleton simply answered what he thought of as enigmatic questions by reference to God's omnipotence, which transcends our comprehension.

The physician Thomas Willis (1621–75) was an assiduous follower of these debates in Oxford. While he acquired wealth in his medical practice, his enthusiasm for dissection enabled him to provide an extended description of the brain as the seat of the soul. His book *De cerebri anatome* (*Anatomy of the Brain*, 1664), with its elegant engravings by Christopher Wren, looks at first glance like a founding text of modern neuroscience. But Willis also published other works, especially *De anima brutorum . . . exercitationes duae* (*Two Discourses Concerning the Soul of Brutes*, 1672), which based his descriptive studies in the philosophical world of Gassendi and Charleton. Like them, he attributed a corporeal soul to animals, which he thought explains their purposive and sensitive natures, while he attributed a corporeal and an incorporeal soul to man. He distinguished 'Neurologie', the topic of the brain and nerves, from 'Psycheology', the topic of the 'Nature and essence . . . parts, powers, and affections' of the corporeal soul, and he distinguished both of these topics from theological questions related to the incorporeal soul, with which he respectfully did not deal.[22] Medicine, Willis supposed, is interested only in the body and the corporeal soul; as a physician, he argued that it is the corporeal soul or the body, not the incorporeal soul, which is disordered in madness.

While Descartes and Willis have sometimes been described as the founders of modern scientific approaches to the body and the brain, it is clear that their interests built on sixteenth-century learning. They continued to use the language of spirits and the corporeal soul, as Renaissance writers had done, to mediate between the immortal and mortal poles of the human condition. Both Descartes and Gassendi, in this respect like their Aristotelian predecessors, discussed the soul as a substance with qualities that make possible thought without images ('the intellective soul'), which is linked via the spirits or corporeal soul ('the organic soul') to images created by material changes. When all this is acknowledged, however, there was still something novel and portentous about Descartes' representation of man, minus his soul, as a machine. We must surely think so in a modern world where so much public life presupposes that women and men are machines, and where, if the language of the soul appears at all, it is restricted to private worlds of feeling. There is also no denying the stark difficulties that faced Descartes' philosophy when it tried to do justice to the whole person, the human being as an individual subject. These difficulties remained to confront the late twentieth-century proponents of a mechanistic science of man.

iv *Self-reflection and the Self*

Not the least of the issues thrown up by Cartesian philosophy concerns the reality of the individual self or person. The notion and language of the self has a historical character; it is certainly not unchanging. A new sense of self in the seventeenth century became a crucial part of what is distinctive, modern and Western in the history of the human sciences. As many observers note, twentieth-century people became preoccupied by personal feelings, personal wealth, personal fulfilment, personal health, personal privacy and much else 'personal' besides. This gave much of the twentieth-century human sciences their subject matter. But there is not much agreement about how, why and when these preoccupations arose. It was a long-term historical process. All the same, there are reasons to believe that, in the seventeenth century, there was a considerably increased sense of self connected to developments in natural and moral philosophy as well as to changes in society and culture more generally.

There was a wealth of late twentieth-century literature on the self, some historical and some stimulated by the widespread experience of a fragmentation of self-identity. It is not possible to extract from this work a conclusion about whether or not a distinctive Western sense of self developed in the early modern period. It cannot make sense, for example, to refer, without qualification, to the discovery of the self in this period. It is obvious that Christian belief presumed the category of the person, which denoted someone who possesses an immortal soul, while Roman law presumed that civil society consists of persons endowed with agency and hence responsibility. Since medieval society made the maintenance of social order appear to be rational in terms of Christian and legal notions of the individual, the reality of individuals was in some sense obviously accepted. It is still open to question, however, how far this reality was internalized by people as a subjective sense of self-identity and how far different types of people used a language that represented experience of an inner self.

The category of 'the person' has a very complex history, some political and legal dimensions of which were indicated in the previous chapter on natural law. The dimension least researched in relation to the human sciences is the theological one. There was an extensive medieval and early modern literature that discussed the nature of the person in terms of the individual rational and immortal substance or soul. This brought questions about the human person, an embodied person, into connection with questions about Christ's body as the embodiment of God, a topic relevant to the dogmas of the Trinity and of transubstantiation. When philosophers like Descartes and Locke discussed what they meant by a person or self, their views were understood in theological terms. Much of the controversial character of the new natural philosophy stemmed from this. Locke was notably innovative, and he was much criticized when he detached the question of personal identity from the theology of the Trinity and associated it with the presence of consciousness. When the term 'personality' became current in English-language psychology in the late nineteenth century, the word was already in use in Christology, the theology that concerned the personhood of Christ.

For many reasons, there was a heightened sense of self in the seventeenth century. Whatever the subsequent modifications and intensifications of the sense of self by Romantic writers, by urban society or by modernist art or philosophy, it was possible in the late twentieth century to grasp and identify with the individualizing content of seventeenth-century expression. It is reasonable to suggest that this construction of an expressive language of self, which went beyond the spheres of theology and jurisprudence, created possibilities for psychological experiences, along with appropriate terms and concepts, and hence made possible the modern human sciences. There could, after all, be no psychology unless there was a psychological subject. Part of the history of the human sciences is therefore the history, which in large part remains to be integrated with the history of science, of the subjective self.

Descartes' vivid use of the personal pronoun 'I' establishes a starting point. The *Discourse on the Method* is remarkable not least for the directness and persistence with which sentences begin with 'I'. Descartes' method was a form of autobiography: he stressed what 'I' have experienced in 'my' education and what 'I' have concluded as a new foundation for truth. The point is not that Descartes was egotistic; he chose the 'I' as the hero of the story. He invited readers to reflect as he had done, to find in their own 'I' the grounds for truth. 'For my part, I have never presumed my mind to be in any way more perfect than that of the ordinary man . . . I shall be glad, nevertheless, to reveal in this discourse what paths I have followed, and to represent my life in it as if in a picture, so that everyone may judge it for himself . . .'[23] This language differed from that of academic or scholastic philosophy, in which the personal 'I' was used in disputation only to serve deductive argument or textual exegesis, that is, to serve an impersonal subject. Aristotelian philosophers discussed reason and morality as general conditions of being, not as personal acts. Descartes' 'I am thinking, therefore I exist' rang down the centuries as an individual assertion.

It is significant that when Descartes turned inwards to examine his individual mind as a source of knowledge, he represented this as an individual act not an act characteristic of life in a certain community of people. He stressed self-examination as an individual as opposed to a social performance. It was a taste of what was to come in the nineteenth century, when introspection and the examination of mental content provided psychology with subject matter: the representation of mind in terms that ignore the social constitution of what is represented.

These observations, however, may have an anachronistic

character. It has been questioned whether Descartes really did have a modern sense of the individual 'I'. The style in which he presented himself was heavily rhetorical. Further, when he wrote about his 'I' he referred to the soul as thinking substance, that is, he referred to something universal and characterized by a reasoning nature, and he did not necessarily refer to an individual consciousness. Descartes claimed, for example, that the soul necessarily always thinks, but he did not claim that an individual soul is always conscious. Descartes used the Latin word *'cogitare'* or the French word *'penser'* when he discussed the attributes of the soul, rather than words equivalent to the modern English word 'consciousness'. All the same, even if rhetorically, he portrayed himself as the hero of his philosophical story. Furthermore, he discussed self-control of the passions as the means necessary for the individual to reason rightly, and we should perhaps look for the roots of the modern 'I' in the discourse on the passions rather than the discourse on reason. To picture and train the passions requires a reflective stance, and the 'I' grew in this discipline. The reference to 'consciousness' as constitutive of the self came later; in the English-speaking world, after the work of Locke.

Other writers – such as Montaigne – turned, like Descartes, to the world of reflection as a world in which at least something could be certain to be true. Half a century before Descartes, Montaigne, surrounded by political and religious conflict, wrote the *Essais* which continued to be admired in the twentieth century for their detached, amusing, reflective view of customs and events as elements in the constitution of his own self. In his much quoted preface, 'To the reader', he claimed that 'I want to appear in my simple, natural, and everyday dress, without strain or artifice', and he concluded that 'I am myself the substance of my book . . .' We do not have to believe his rhetoric about the exposure of his 'nakedness' to accept that he made his subjective self the source of knowledge about the human subject. If the object of knowledge is himself, as he said, 'no man ever came to a project with better knowledge and understanding than I have of this matter, in regard to which I am the most learned man alive . . .' He set up the self as the basis for a wider learning: 'Every man carries in himself the complete pattern of human nature.'[24]

Montaigne did not expect women or the ordinary people to possess the same reflective means to knowledge. Rather, he observed, 'the true advantage of the ladies lies in their beauty; and beauty is . . . peculiarly their property . . .'[25] Thus, when he claimed the self as his own point of departure, he denied the same quality of self to others. When he defined beauty he did so as beauty appears outwardly to men and not as it is subjectively experienced by women.

Montaigne chose to express himself in essays, a medium that he developed and popularized as a way to fit words to experience. Since he rejected the possibility of a unified worldview and scorned the religious fanatics who fought around him, he exploited a medium appropriate for the diverse bits of *his* experience that, could, he felt, constitute knowledge. The form of the essay mirrored the disjointed nature of the author's experience. Later, empirically-minded natural philosophers like Bacon and Boyle adopted the essay as an appropriate form in which to describe the plain particulars of nature, to escape the artifice of language and to avoid the grandiosity of worldviews.

Montaigne and Descartes, and indeed Hobbes, were independent scholars whose livelihoods depended on private wealth or patronage. This fact points to another dimension of the heightened or even novel sense of self. Though 'the rise of capitalism' is a phrase no longer in vogue, since that 'rise' appeared in different places in Europe anywhere from the thirteenth to the twentieth century, the use of capital in the seventeenth century did generate wealth, and this wealth did support art, scholarship and moral philosophy. These cultural productions stressed individual attributes and qualities. Wealth, of course, mainly or exclusively benefited élites, though they were as diverse as the courtiers at Louis XIV's Versailles and the members of the Watch in Amsterdam. There was no end to war but, by the last years of the century, there was sufficient political and military stability in north-west Europe to support economic expansion. Economic and political power promoted individual display – of learning as well as of dress – in the courts, country estates and merchant houses of the age.

One sign of the growth of reference to individual status was that the portrait came into its own as a genre of painting, exemplified by the work of Hans Holbein at the court of Henry VIII of England and of Diego de Silva, Velázquez, at the court of Philip IV of Spain. The portrait, with its roots in the Renaissance, certainly represented its subject as king, pope or merchant prince, but it also created the picture of an individual presence, a self. Similarly, the desire for glory and fame led rulers and would-be rulers to grandiose displays of individuality. The Italian Renaissance master Benvenuto Cellini both created evidence of his skill and wrote down what he had done in an autobiography, an example of what the nineteenth-century historian Jakob Burckhardt called 'the most zealous and thorough study of [a man] himself in all forms and under all conditions'.[26] The French historian of ideas Georges Gusdorf suggested that the technology of the mirror, perfected and marketed by Venice in the early sixteenth century, first enabled people literally to reflect on a whole picture of themselves. By contrast, before the sixteenth century, blown glass mirrors magnified what was near their surface, making it difficult for people to see their whole appearance. In all of this there was an enrichment of the sense of self.

Quiet reflection on the 'I' as a subject achieved unparalleled beauty in the Low Countries in the seventeenth century. Jan Vermeer's painting of a young woman reading a letter enables us to see a person who is actively self-absorbed. One self has spoken to another in the letter; the viewer 'sees' intimacy, just as, with the spread of private correspondence, intimacy became a form of refined reflection and the letter a route to self-discovery. The painter took the eye around a curtain to look at the woman's private world, and he illuminated her by light from an open window, the leaded panes of which reflect the woman's self-absorbed face. The painting's subject is the subjective self; we cannot know what is in the letter, but we do know her quiet sensibility. In Rembrandt's self-portraits, created at about the same time, the painter boldly presented not only a picture of what he thought he looked like but a challenge to the viewer to understand what it is to be Rembrandt. These powerful pictures remained a profound emblem of the self. Montaigne, much earlier, put the writer's case for the self-portrait: 'Authors communicate with the world in some special and peculiar capacity; I am the first to do so with my whole being, as Michel de Montaigne, not as a grammarian, a poet, or a lawyer.'[27]

Though portraiture heightened the dignity and value of a person as an individual, it did not cease to represent its subject as a person in a social position, a person who stands for a type of man or woman or even for humanity as a whole. There were portraits of Indians, slaves, women, peasants, madmen and animals, as well as kings and bankers. When art cultivated and enriched the language of individuality it also cultivated the language of social differentiation. The 'I' recognized beneath the clothes and skin was also a social entity. Except in rare instances, such as Rembrandt's self-portraits, the new language of the self was weighted with reference to a person's social position and responsibilities.

The language of Catholic states tended to re-express belief that a person's identity is subordinate to collective political entities of church and state. In contrast, the language of some Protestant communities placed a heavy and even oppressive emphasis on the self as a moral agent; this is the stock image of the Puritan. Protestant values heightened awareness of a direct relationship between each individual fallen soul and God's omnipotent will. They could not be distanced from, and were sometimes in conflict with, the political and social web of individual rights and obligations under natural law. Puritan values directed attention to everyday conduct, business affairs, domestic matters and ordinary things. If, as was believed, God's will had created the world, it was an act of worship as well as an instantiation of God's law for the creation to give full attention to daily life. The Protestant sensibility, notably evident in New England, encouraged belief in the inward self as a responsible agent, and it understood this responsibility to lie equally in relation to the God-given soul and to the God-given material world. The Protestant work ethic united inner-directed and outer-directed responsibilities and enhanced the growth of the individual, self-conscious self. Protestants extolled the individual's prudent management of capital, labour, property and time in the same breath as they preached the individual's duty to manage the spirit. These expectations about individuality, however, were unevenly distributed across the social landscape of Europe and North America. Economic and political power in much of Europe continued to rest with the Catholic Church and Catholic monarchs, with a feudal aristocracy and with new states and their autocratic rulers, not with individuals. The 'I' flourished in spaces between institutions, as in Descartes' meditations or in the domestic order of Dutch merchants.

Responsibility to God's creation, work, economic activity, new technology and new science fostered an instrumental view of the self. The word 'instrumental' denotes a conception of a person as a practical agent, an individual whose identity, subjectively and socially, resides in being able to do things, to act on nature, self or society to achieve practical ends. From this perspective, the new philosophies of nature – most obviously, Cartesian mechanical philosophy – were not just theories about nature but attitudes towards nature. Modern critics argued that these attitudes made for a distanced, objective and sometimes mechanistic relationship with nature and, as hopes spread for a science of man, with women and men as well. For many modern scientists, the distanced stance was a condition of objective knowledge, that is, of science. But critics of the accomplishments of science questioned whether the distanced and instrumental attitude was at the expense of non-instrumental possibilities, such as love and empathy, in the human condition. This is an open debate that gives added meaning to the history of the seventeenth century.

Printed works from the early sixteenth century included many 'conduct books' designed to help readers with self-control. The book itself, like the letter, significantly enhanced a person's capacity to become self-absorbed and self-aware, that is, to become individual. The book and the letter were the material medium of private thought, sensibility and improvement. Descartes wrote about human life: 'These things are worth noting in order to encourage each of us to make a point of controlling our passions . . . [so that] even those who have the weakest souls could acquire absolute mastery over all their passions if we employed sufficient ingenuity in training and guiding them.'[28] Conduct books, especially in Puritan culture, sharpened self-reflection, sometimes to a painful degree. Boyle, who played a pivotal role in the legitimation of the new natural philosophy in England, referred to three 'books' which carry authority – nature, scripture and the conscience. The conscience, he believed, lies in each person and can be known by the use of right reasoning about moral things. From this perspective, the self is a book of truth comparable to the book of nature and the book of God's word.

A sense of self reached its height in the diary, the book written by oneself for oneself as a means of self-reflection and self-control. Serious Puritans recommended the diary as a discipline for the soul's steady contemplation of its proper ends. The parish minister at Coggeshall in Essex, Ralph Josselin, kept a diary through the difficult years of the reign of Charles I. There, amidst the seemingly endless round of his wife's confinements and his own and his family's ailments, he worried about what daily events meant for the salvation of his soul. The diary recorded external events but, more significantly, struggled to make events meaningful by their assimilation into a moral and spiritual narrative of the diarist's self. In the more libertarian atmosphere of 1660s Restoration London, a civil servant at His Majesty's Admiralty, Samuel Pepys, recorded in volume after volume all the vicissitudes and delights of daily life. The grand and the banal lay side by side, each in its own way worthy of record since each experience was uniquely and irreducibly Pepys's own.

Pepys was perhaps more troubled by his bladder stones than by his soul, but whether diarists recorded bodily or spiritual grief and joy they enriched the scope of subjectivity. Some historians suggest that the subjective sphere so marked out and valued for its distinctive and immediate truth characterizes modernity. If so, then it is surely not coincidental that modern science and modern consciousness developed side by side. The shift towards the understanding of what is outward in mathematical and mechanical terms seems also to have involved a shift towards understanding what is inward in terms of a private world of qualitative truth and feeling. But this division between outward and inward, language which is itself interestingly metaphorical, became an insuperable problem when people later pursued a science that would tell them about themselves.

However subjective a diarist's record, it was also a social record; the language, even the choice of the diary as a medium, derived from and shared in the wider culture. An individual who explored her or his own subjectivity did so in a society that valued such sensibility and self-responsibility. An outward sign was the refinement of manners and greater delicacy in public with regard to eating, excreting, cleanliness and the body generally. Society as a whole, bodily expression and subjective sensibility all slowly changed in conjunction with each other, obviously with major differences between women and men, between social groups, between town and country, and between court and plebian culture. The literature on conduct and manners sought to create Christian gentlewomen and gentlemen, to individualize control, to make social control self-control and to cultivate refined subjectivity. The English Puritans, like those who travelled to establish 'New England', and the later German Pietists, who pressed readers or listeners to consider the state of their souls, made the connections between social order and subjective order most clearly. Meanwhile, the theatre of Molière or of Restoration England made much of the same play between social custom and individual character, and comedy and plot relied frequently on the disruptive body.

All this, it appeared in the twentieth century, differentiated and dignified the psychological dimension in human life. Nobody expressed it in such terms in the seventeenth century. Words like 'consciousness' and 'self-consciousness' were unusual in English until late in the century. Nevertheless, the expression 'human nature' came into common English usage, and the phrase took for granted that knowledge of man involves knowledge of a relationship between individual subjectivity and a shared or common nature. This was a different language for reflection on what it is to be human than the scholastic language of intellective and sensitive souls. Modern thought adapted the ancient language of the soul – humours, temperaments and spirits – but it also added a new discourse about human nature, mind and subjectivity. This new discourse stressed self-reflection and self-control, it translated social values into forms of individual refinement and it lay the basis for modern subjective sensibility. The implications of all this became apparent only in later centuries and then often in ways that seventeenth-century philosophers would not have understood. The transitional moment for the creation of a language about mind and consciousness was the work of Locke.

III

THE LONG 18TH CENTURY

5

John Locke and the Natural History of the Soul

> So many thinkers having written the novel of the soul, a wise man has appeared who has modestly written its history. Locke has expounded human understanding to mankind as an excellent anatomist explains the mechanism of the human body ... Instead of defining in one fell swoop what we don't know, he examines by degrees what we want to know. He takes a child at the moment of birth and follows step by step the progress of its understanding, he sees what the child has in common with the animals and in what it is superior to them, he consults especially his own experience, the consciousness of his own thought.
>
> Voltaire, *Letters Concerning the English Nation* (1733)[1]

i *An Essay Concerning Humane Understanding*

The 'long eighteenth century' describes the period from the 1660s to the post-Napoleonic settlement of 1815. It is a useful division for British and French historians, as the Restoration of the British monarch Charles II in 1660, followed by the Glorious Revolution of 1688, settled Parliament's powers in the modern mould; at the same time, the young French king, Louis XIV, achieved full power in 1661 and began the long period of centralized French rule, known as the *ancien régime*, which ended only with the Revolution, and the Revolution led to the modern age. Scientific activity in the period between the deaths of Descartes (1650) and Newton (1727) developed the theories and organizational arrangements that remained dominant in some cases for two centuries. The generation that came of age in the 1660s took for granted what a previous generation had had to argue for, and in its turn it laid the basis for knowledge and activity that later generations did not question. The Aristotelian and medieval worldviews were gone; the business to hand was to construct new knowledge. The philosophers of the Enlightenment, François-Marie Arouet de Voltaire (1694–1778) chief among them, perceived a gulf between their age – an age of progress – and the darkness that had come before. They looked to the future; and we are their future.

The long eighteenth century was of decisive importance for the argument that there could and should be a naturalistic approach to human nature – an approach based on knowledge of the natural world. It contributed concepts and a language to describe men and women systematically as conditioned by the physical world and hence knowable in the same manner as physical events. Nevertheless, there was no one science of man, and some critics questioned the very idea of a naturalistic project to establish such a science. Nor was that project entirely new. Already in the sixteenth and seventeenth centuries, the belief was present that man can transform himself through reason and knowledge of his nature. People aspired to live by reason and experience as well as faith. Human reason reconstituted itself as subject areas – logic, language, law, history, moral and natural philosophy – that were studied in disciplined ways. The language of natural law grew in depth and subtlety, and this made it possible to encompass human action and physical change alike in a vision of the ordered creation. Criticism of the Christian-Aristotelian worldview raised many questions about man's essence, the soul, and new ways of understanding human nature were sought. Descartes' solution redefined what the Aristotelians had called the organic soul as bodily substance and the intellective soul as thinking substance. This created the framework for subsequent debate, but almost no one was happy with what he had done. The long-term consequence was the reformulation of the soul as the mind, and the mind was increasingly characterized by reference to consciousness, language and its relations with the temporal and material world – in contrast to the stress on the soul's transcendent immortality.

If these issues had a certain shape for eighteenth-century philosophers, it was in part because they had all read a single author, John Locke (1632–1704). His book, *An Essay Concerning Humane Understanding* (1690), demonstrated for people in the eighteenth century that progress in knowledge is possible, that human reason and action are subject to natural law and that mind, defined as consciousness, can be the subject of an empirical and, as we would say, scientific discipline. For many modern psychologists, Locke laid the foundations on which was built a science of psychology. Locke's intentions, however, were not necessarily those of his readers or of later psychologists. Modern concepts of consciousness, for example, cannot be retrospectively identified in Locke as if they had clear and unchanged meaning. His own description of consciousness as 'the perception of what passes in a Man's own mind' merely restates what is in need of clarification.[2] This chapter explores Locke's arguments and indicates their significance for eighteenth-century approaches to human nature, approaches with consequences for what people do as social beings as much as for what they understand about their minds. The implications of what Locke wrote run through subsequent chapters; he especially influenced British culture and his ideas culminated in the moral and political utilitarianism to be discussed in connection with political economy. But Locke had critics too, notably in the German-speaking world, and this must be taken into account in the assessment of his significance; in Germany there were profoundly influential alternative views of the soul, which are discussed in the following chapter. In France, Locke was read, but often by those who also looked to Descartes as a model of the rational method in human enquiry.

Locke published his long-contemplated *Essay* after a distinguished career in which he set out to be a physician but became a scholar, state administrator, private secretary and political negotiator. His great book grew out of thoughts that date from his much earlier years as a tutor at Christ Church, Oxford. Like everyone in his generation, he was shaken by political upheaval, violence and fanaticism in Britain and on the Continent, disturbances which did not end for Locke with Charles II's restoration in 1660. In order 'to break in upon the sanctuary of Vanity and Ignorance', Locke wanted to make our understanding '*clear* and *distinct*', so that people can judge better what they know and assess better the relative certainty of beliefs.[3] He hoped it would then become possible to organize society on the basis of sober thoughts and not 'enthusiasm', and in this way to bring about God's purposes.

The project to achieve what Locke called understanding was linked to the search for the right basis on which to act. This was an immediate concern for Locke. From 1683 to 1689, under political suspicion, he chose to live in the Low Countries, and it was in these circumstances that he finally completed his book. He associated with some of the architects of the Glorious Revolution of 1688, and he supported the parliamentary settlement that laid the basis for the remarkably stable British political system. Locke's *Essay* appeared at an opportune time to describe experience and action in terms which justify the hope that educated people can successfully order their affairs.

Locke produced the first draft of the *Essay* in 1671 and it reflected the intellectual milieu of the 1660s, when the legacies of Bacon and Descartes animated the early years of the Royal Society of London and inspired the genius of Newton and Locke himself. Bacon's programme for the renewal of wisdom in human affairs had included the establishment of natural histories, ordered collections of information as a basis for the improvement of man's estate. Locke's concern was the natural history of man, 'this Historical, plain Method' to describe human qualities as they are founded in experience.[4]

Though he possessed some medical education, Locke concentrated his attention in the *Essay* on the mind not the body – and he specifically used the word 'mind' not 'soul'. Though a Christian, he focused his work on this world, and he put to one side aspirations to know, as opposed to having faith in, the reality that transcends the human sphere. 'If we can find out those Measures, whereby a rational Creature put in that State, which man is in, in this World, may and ought to govern his

Opinions and Actions depending thereon, we need not be troubled, that some other things escape our Knowledge.' The measures that ought to guide our actions, Locke thought, can be known; an '*appeal* to Mens own unprejudiced *Experience*, and Observation' was to vindicate his own book. The material for observation, he thought, is everywhere. He was, for example, a great reader of travel books, a major imaginative resource for anyone who wanted to consider the diversity of human beliefs and social life. He was equally interested in reason, the capacity we have to engage with what is in our minds: 'Every Man being conscious to himself, That he thinks, and that which his Mind is employ'd about whilst thinking, being the *Ideas* . . .'[5] His attempt to clarify the nature and origin of these 'ideas' was the technical core of his book, and it was his acumen in his discussion of these issues that continued three centuries later to attract philosophers.

Locke classified 'the sciences', i.e., what falls within the compass of human understanding, into natural philosophy, knowledge of material or spiritual things; ethics, the skill to use our powers for what is good; and logic, the use of signs – above all, words – to consider and to communicate things and actions. His *Essay* was a study in logic, to use the language of his time, not, to use modern language and classification, in psychology or epistemology (the theory of knowledge). For Locke, logic had a double role as a method of investigation and as a method of presentation. He hoped that with the right use of logic people would gain clear ideas about what the mind experiences. His work contributed substantially to the outlook that associated logic with scientific enquiry and culminated, a century and a half later, in John Stuart Mill's *A System of Logic Ratiocinative and Inductive* (1843). One consequence of this new outlook was that rhetoric became clearly separated from logic – exposition became separated from enquiry. Another consequence, of great significance, was that Locke encouraged readers to think of logical enquiry and the operations of the mind, the mind's experience, as closely linked. It is in this regard that he appeared later to be both an empirical philosopher and a founder of psychology. He, however, described no topic as psychology. Even the word itself was little used in English until the time of the philosophers William Hamilton (1836) and William Whewell (1837). As a parallel point, 'epistemology' was not used until the work of the Scottish philosopher J. F. Ferrier (1856).

Locke, then, elaborated a proposal about how we should use words to signify what we can legitimately hold to be true. In his analysis, words signify ideas, and Locke wanted to reveal the sources of these ideas to show that true words relate to true experiences. His exposition contributed to a rejection of Descartes' and other theories of innate ideas, which in turn led to rejection of the belief that everyone shares a natural language. By 'innate ideas' Locke referred to what he took to be common belief in 'some primary notions ... Characters, as it were stamped upon the Mind of Man, which the Soul receives in its very first Being; and brings into the World with it', such as the idea of God, of substance, of identity or of the right words for things. He denied that there are such innate ideas; as evidence, for example, he cited travellers' reports that there are exotic peoples who have no notion of a god. Instead, he argued, all ideas come through experience, a claim that others were to take much further. To convey to his readers what he meant, Locke pictured the mind as a blank sheet of paper, a *tabula rasa*, on which sensation inscribes a record of what takes place. This image captured the eighteenth-century imagination. It was deceptive, however, since Locke took it for granted that the mind also possesses innate powers which transform sensation by reason. He assumed that there is a capacity for reflection, 'the internal Operations of our Minds, perceived and reflected on by our selves'.[6] In his view, this capacity is intrinsic to the mind and not given by physical nature.

Sensation, as Locke described it, followed from the physical world's impact on the mind, mediated by our senses. In this context, he gave the modern meaning to the medieval phrase and Aristotelian argument that there is 'nothing in the intellect not first in the senses'. The physical world gives us ideas (to use his examples) of yellow, heat, soft, bitter and so forth. When we reflect on such sensations, we gain ideas about mental activity such as perception, thinking, believing and willing. The *Essay* then used this analysis to examine the origin of all our ideas, such as those of number, of space, of power, of our own identity and of God. Each aspect of the analysis continued to generate discussion in twentieth-century philosophy. I have highlighted only two main points: his project included a practical natural history, structured by a logical method, and he derived the content of the mind, ideas, from experience, even while he assumed that ideas are acted on by reason.

Locke's approach to experience was connected with the foundations of the modern scientific outlook. His *Essay* appeared only three years after Newton's epoch-making *Philosophiae naturalis principia mathematica* (*Mathematical Principles of Natural Philosophy*, 1687), and many natural philosophers in the eighteenth century linked the two men's achievements. Historians have also portrayed Locke as an underlabourer for Newtonian science, though much of Locke's thought long preceded any concern he might have had with Newton. There are two aspects of Locke's relation to natural philosophy.

Firstly, Newton analysed the physical world in terms of the mechanical forces of bodies that possess mass and motion. Locke, in this regard somewhat like Hobbes, adopted a view of material reality conformable with the Newtonian model, and he assumed that it is the properties of particulate matter that give rise to particulate, atomized sensations and hence to our elementary ideas. He also wrote as if there is a parallel between the way material particles combine to make complex substances in the physical world and the way sensations combine to make complex ideas, i.e., knowledge, in the mental world. The imagery of physical nature reappeared as presuppositions about mental nature. This was to be a recurrent pattern in the modern sciences of mind and at the basis of many claims that the natural sciences are the proper route to a scientific approach to mind. Some scientists even assumed that any research on the mind that does not translate mental events into physical analogues cannot possibly achieve scientific status, and they saw in Locke's analogy the start of a step to make the science of mind a physical science. This was not Locke's position, since he nowhere rejected the view that the mind has distinct existence, however much he implied that analysis of the mind's content can be moved forward by physical analogies.

The second point that links Locke with Newton is the way they both argued for experience as the method to acquire knowledge, the method known since the nineteenth century as empiricism. Whatever Newton's working methods actually were, he legitimated his conclusions with a defence of the value of observation, and this was seized on in public rhetoric about his achievement. Locke's *Essay* belonged to the same culture of experience. His argument implied that Newton's mode of reasoning, and hence the truth of Newton's conclusions, was grounded in the proper, unprejudiced activity of the mind. The British Newtonians therefore argued that their Cartesian rivals on the Continent, who followed both Descartes' philosophy and his physics, were simultaneously deceived about the nature of the mind and the nature of physical reality. Further, the rhetoric of experience helped arguments for a natural religion, for religious truths founded in experience of the creation, which, it was hoped, supported an undogmatic – and certainly non-Catholic – Christianity and the new political order in Britain. Eighteenth-century British writers therefore had good reason to look back on a synthesis between Newton and Locke. Locke, in their view, had revealed the mental mechanism through which experience generates truth, while Newton had stated what the truth is.

In spite of all the talk about experience, Locke's own mode of argument was strikingly logical rather than empirical. Though he recommended the reader to 'follow a *Child* from its Birth and observe the alterations that time makes', he himself rejected innate ideas on theoretical grounds and undertook an *a priori* analysis of simple and compound ideas.[7] A rhetoric of experience coupled with an analytic method continued to feature in British intellectual culture for at least the next 150 years. The importance of Locke's work was that he contributed to a culture that looked to *human* experience in the material world as the measure of all things and the guide to what to do, not that he did in fact study learning and conduct empirically. Locke provided the vocabulary for a new enterprise. All the same, though he claimed to reject much of Descartes' philosophy, he still shared with Descartes a basic belief in the division of reality into soul or mind and body. Locke, too, reinforced a dualistic orientation towards man's place in nature.

ii *Conduct, Education and Politics*

Locke's work is emblematic of a changed sensibility about the soul, a change that in the long run made reference to the soul an embarrassment in everyday discourse, whereas reference to the more secular notion of mind became commonplace. Locke, however, certainly accepted the major tenets of Christian faith, though he insisted that faith should be consonant with reason, and he was a staunch advocate of Christian morals. Substantial sections of the *Essay* considered the respective provinces of faith and reason, the degrees of assent we should give to belief, and the origin of the idea of a god. He endeavoured to make such matters subject to reasoned judgement, founded in knowledge of the mind, since he believed that, if the limits of understanding were clearly known, people would accept toleration in religious matters. Toleration was a major political and religious issue in the wake of the settlement of 1689 and the establishment of a Protestant royal succession, and Locke was the author of a much discussed series of open letters on the subject. The Latitudinarians, who took an inclusive view of Anglicanism, acquired political power, but they continued to perceive threats to both church and state from Catholics in one direction and Deists, who denied revelation, in the other. Locke's belief in toleration paralleled his argument that knowledge is relative to experience – there are limits to our understanding.

In retrospect, from a position distanced from the intense emotions of the 1690s, some historians felt that he introduced a degree of belief in the notion that truth is relative. Locke himself believed that social consensus becomes possible when people achieve clear understanding and hence reach the same knowledge grounded in experience of the same created world. Nevertheless, his argument contained the potential for a much more relativist and socially disruptive view, since it was sometimes concluded, with equal justice to Locke's own statements, that different experiences result in different ideas and hence in incompatible claims about what is true. If experiences do not coincide then knowledge may not coincide either. This is a modern conclusion, not explored by a Christian like Locke. Yet he himself utilized evidence that some people have no idea of a god, in order to attack belief in innate ideas.

What needs to be stressed is that when Locke sought common ground for human understanding in a natural history of the soul, he was – if anything – more concerned with moral accountability than with knowledge. What he desired to show was the basis in man's nature for a shared morality. Yet as with knowledge, so with morals: after Locke's work, a sensitivity to the relativity of belief, conduct, laws and customs was of fundamental importance for thought about human nature and society. Again in the long run, sensitivity to the moral consequences of relative values gave to modern culture much of its anguished edge. Locke himself, however, believed that clear thought would show the basis for agreed morals and conduct as well as understanding. Like Bacon before him, he pursued knowledge for the sake of improvement, a goal relevant equally to the individual and to civil society. Here again Locke's *Essay* was seminal. He articulated an approach to motivation and conduct which reverberated in the topics which, in the twentieth century, were divided into ethics, politics, economics and psychology. He placed his hopes for moral order in the laws of human nature.

Locke's account of the passions was neither extensive nor deep, but it contributed to a naturalistic theory of motivation – a theory that attributes our motives to nature rather than to a force or a reason transcending nature. He described the passions (he did not refer to 'emotions') as ideas derived from the sensations of pleasure and pain. 'For we *love, desire, rejoice,* and *hope,* only in respect of Pleasure; we *hate, fear,* and *grieve* only in respect of Pain ultimately: In fine, all these Passions are moved by things, only as they appear to be the Causes of Pleasure and Pain . . .' It was a small step to argue that the sensations of pleasure and pain give us the mental ideas that precede action and thus to explain our conduct by the experience of pleasure and pain. Action, Locke thought, follows from the uneasiness that comes with an imbalance of pain over pleasure: 'it may perhaps be of some use to remark, that the chief if not only spur to humane Industry and Action is uneasiness'.[8] Action was a consequence of the order and relation of sensations and not of the soul.

The argument was even more radical since Locke carried on, with unbroken flow, from description of mental content to description of good and evil in terms of pleasure and pain. His statement is as deceptively simple as it is famous: 'That we call *Good* which *is apt to cause or increase Pleasure, or diminish Pain in us* . . . we name that *Evil,* which *is apt to produce or increase any Pain, or diminish any Pleasure in us* . . .'[9] He thus referred the standard of right and wrong to individual states of mind. Many of his readers were shaken by this assertion, and some were appalled at the way it apparently denies God-given moral imperatives. In this regard they repeated criticisms already made against Grotius and Hobbes, that these writers had substituted a law of nature for God's will as the source of moral action, and that they denied human freedom. Locke did indeed create a language with which to analyse conduct and civil society as subject to the laws of nature. Later moral philosophers like Joseph Priestley and Jeremy Bentham, the former a Christian, the latter not, seized on the naturalistic possibilities in what Locke had written, and they thereby created modern utilitarian ethical argument – that actions are to be judged by their consequences.

Locke's own account of freedom was more subtle than many readers believed. It was this issue that caused him to write and rewrite by far the longest chapter of the *Essay,* 'Of Power'. He characterized 'power' as the idea of a particular kind of relation, an idea obtained particularly through reflection on desire followed by action. The will, he argued, is the common name for action related to desire, though he accepted that most people, improperly he thought, attribute this will to a faculty of the soul. For Locke, it is the person and not the will which is free, and this freedom consists, he thought, in people's capacity to suspend desire and thereby have the 'opportunity to examine, view, and judge, of the good or evil of what we are going to do'. Locke placed confidence in reason as the foundation for happiness and moral virtue. Freedom lies in our pursuit of happiness through prudence: 'As therefore the highest perfection of intellectual nature lies, in a careful and constant pursuit of true and solid happiness; so the care of our selves, that we mistake not imaginary for real happiness, is the necessary foundation of our *liberty*.'[10] To summarize: pleasure and pain, Locke argued, give people desires which result in action. In addition, however, the mind's ability to reflect, to reason, enables a person to assess where pleasure and pain really lie. He rejected the description of the will as a faculty of the soul and instead explained choice as a consequence of the mind's reflection on its sensations. What others made of this remains to be seen, but no question received more attention in the eighteenth century than how to relate what is good to man's nature, in order to achieve the perfectibility of man. Moral language was the dominant language for the expression of opinion about what it is to be human.

Locke wrote in order to highlight how a person's social responsibility is exercised through reasoned judgement. He therefore also turned his thoughts to education, drew on his experience as a tutor and wrote for 'so many, who profess themselves at a Loss how to breed their Children . . . [and since] the early Corruption of Youth is now become so general a Complaint'.[11] Some eighteenth-century readers took *Some Thoughts Concerning Education* (1693) to be his *Essay* adapted for practical child-rearing. Locke himself appears to have kept these writings apart in his mind, but both the *Essay* and the *Thoughts* influenced reform-minded opinion in the eighteenth century.

To take experience seriously as the source of understanding and conduct certainly placed a premium on education. Two linked principles were of overwhelming significance in the eighteenth century. The first was that knowledge – and hence the capacity to act with reason – is relative to education and experience. Locke therefore argued that education should become more relevant to a child's future and, in the words of the historian Peter Gay, 'thought it wholly inappropriate to beat children for the sake of dead tongues'.[12] He and other educators objected to rote learning; they wanted instead to build on the child's individual interest and aptitude, and at the same time to bring subjects like history and geography into the curriculum. Lest this sound modern, it should be remembered that Locke was no subverter of established social hierarchies, that he had been the tutor of a young man – the future third Earl of Shaftesbury – destined for high office, and that he assumed there should be different patterns of schooling for different social groups. He believed that the pupil experiences the tutor, or indeed parents, as a model. Consonance between experience, the way the mind works and education, it was thought, is the means to incorporate the child into society. Locke also did not neglect the most mundane matters when he pursued the goal (to quote the Roman author Juvenal) of 'a Sound Mind in a sound Body': 'Let [the boy's] *Bed* be *hard*, and rather Quilts than Feathers. Hard Lodging strengthens the Parts . . .'[13]

The second principle followed from Locke's analysis of good and evil in terms of pleasure and pain. At the opening of his *Thoughts*, Locke stated: 'I think I may say, that of all the Men we meet with, Nine Parts of Ten are what they are, Good or Evil, useful or not, by their Education.'[14] This was an extraordinary vote of confidence in the human power to make of humans what we will, though Locke had no doubt that children have to be disciplined to control their desires. In the *Essay* he laid the logical groundwork which legitimated this claim to link will and desire to sensations of pleasure and pain and not to sin. The consequence in the eighteenth century was an emphasis on human malleability, an optimistic belief in human improvement and greater sensitivity to the world and needs of childhood. To be an educator gained new meaning and importance, as is seen in the mixture of dreams about the perfectibility of man and practical projects created later by pedagogues like Jean-Jacques Rousseau, J. H. Pestalozzi and Jean Itard. There were also critics of Locke's opinions – theologians and moralists who were committed to belief in original sin. Locke's work on education, as on the understanding, was therefore also a contribution to contemporary religious and political debate. Locke, better than anyone, knew all this and saw the connections which his theory of mind made between religion, knowledge and society.

Locke's relations with politicians in rebellion against James II in the 1680s confirmed his long-held belief in a theory of political obligation based on natural law. His reasoning resembled that of Grotius and Pufendorf. Locke conceived that people – and in his argument slaves are not people – possess certain natural attributes and needs, the exercise and fulfilment of which satisfies natural law. He described natural man as a free moral agent: people have 'perfect freedom to order their actions, and dispose of their possessions, and persons as they think fit', to pursue pleasure and to avoid pain.[15] In his *Two Treatises of Government,* published anonymously in 1690, Locke spelt this out. He also contributed to the discussion of what he and others called the state of nature as the natural moral state of man, and of civil society as the consequence of the way men come together to form a government for their mutual protection. He believed that there are places where a state of nature still exists – like Hobbes, he looked across the Atlantic – and he hoped that their empirical study would contribute to political thought since, as he succinctly remarked: 'Thus in the beginning all the World was *America* . . .'[16] Natural man, he assumed, possesses property and lives communally, and hence property and community in civil or political society can be shown to have their basis in natural law. What changes natural into political society, he suggested, is money, which makes possible exchange and all the contractual relations that follow from exchange. Once civil society is established in the rational pursuit of the natural 'peace and safety of life', people have, he argued, the right of redress against transgressors by punishment or, if necessary, by rebellion. And it was rebellion Locke had in mind when he published.

Later generations, for whom 1689 secured parliamentary government in England, read in Locke the classic statement of liberal political contract theory, a theory which rationalizes government in terms of individual interests. There appeared to be powerful echoes in the American Constitution. This interpretation is questioned to some extent by historians who link Locke to earlier argument about natural law and recognize the moral and religious content of his work. In particular, they conclude that Locke was not an individualist since, unlike Hobbes, he portrayed natural man as a member of a community. He did not believe God constituted human nature to generate disorder by actions solely individual in nature. Further, Locke portrayed political society as a growth based on reason whereas Hobbes portrayed it as a machine, something manufactured. Locke rejected Hobbes's materialism and accepted the existence of mental powers. Yet they were close in certain respects since both believed that persons in a state of nature have the capacity to reason and 'a right to their preservation', out of which right comes civil society. Hobbes accepted determinism while Locke defended free will, yet both wanted to reason in accord with the laws of nature and held that reason shows that nature, in the form of pleasures and pains, directs our actions. Whatever distance Locke wished to maintain from Hobbes, Locke's stress that sensations derived from change in the material world are the source of action made it eminently possible to argue from his work for causal theories of human nature in which causes are understood to be physical events. Locke, in what he wrote both on mind and on political obligation, created the conceptual framework for a science of man responsive to the new natural philosophy of the seventeenth century. He himself, however, did not specifically derive his leading ideas from that philosophy.

Locke aroused further debate for a remark, seized upon by his contemporaries, that he saw no inconsistency in the belief that God might have endowed certain kinds of matter with the capacity to think. He was immediately accused of materialism, a highly sensitive issue given the close involvement of religious factors in politics. What he said on this subject was taken up in an extensive eighteenth-century literature on matter theory, when natural philosophers strove to give coherent accounts of the basic concepts of matter, cause and mind. The debate about matter linked questions on immortality, the nature of reason, the problem of whether and in what sense animals have souls and the powers of matter. Descartes believed he had cut through all this when he drew a sharp distinction between the body and the soul and stated that animals have no soul, but in fact he caused intensified disagreement. Gassendi, followed by Willis, for example, argued that animals have a corporeal but not a rational soul. Other authors were much more radical. Daniel Sennert (1572–1637), an early seventeenth-century chemist, was accused of impiety for arguments that his critics believed asserted that the soul of brutes resembles that of humans. Pierre Bayle (1647–1706), in his provocative and influential *Dictionnaire historique et critique* (*Historical and Critical Dictionary*, 1695–7) gave a new intensity and depth to scepticism about religion and non-material essences. He too argued that Sennert had implied the immortality of the animal soul, but he used the argument to foster scepticism about theological dogma. Locke was well aware of what was at stake in these debates; yet, while he distinguished animals from humans on the grounds that humans have the capacity for abstraction and language and animals do not, many readers thought he came dangerously close to an anti-religious materialism. When Locke wrote about moral, logical or political questions, his willingness to treat these topics independently of theology and metaphysics helped make the mind a subject in its own right.

iii *Self-identity and Language*

There was an audience in Britain for what Locke wrote: the *Essay*, for example, came out in five revised editions, the last, in 1706, after his death. Early eighteenth-century Britain acquired a self-conscious urban literary culture, in Dublin and Edinburgh as well as in London. Ambitious, politically motivated and often vicious books, pamphlets and poetry jostled for the attention of a public that lived through the political settlement of 1689, the Act of Union of Scotland with England and Wales in 1707, and the Protestant subjugation of Ireland. Joseph Addison and Richard Steele started the *Spectator* and the *Tatler*, magazines which set a widely imitated style of cultural and political commentary. Jonathan Swift's 1729 *Modest Proposal* to solve beggary in Dublin by rearing surplus children for the table was only the most extreme adaptation of style to social commentary.

There was interplay between individual moral or aesthetic sensibility and collective religious or political values. It was a culture with a heightened sense of self, an individualistic mode of expression and a fascination for the workings of consciousness. Into this culture Locke introduced an account of self-identity that both puzzled his readers and seemed markedly at odds with traditional views of the soul as a spiritual entity. Anthony Ashley Cooper, third Earl of Shaftesbury (1671–1713), the moral philosopher whose education Locke had guided, asked the question: 'What constitutes the We or I?'[17] Locke's answer, added to the second edition of the *Essay* (1694), was 'that *self* is not determined by Identity or Diversity of Substance . . . but only by Identity of consciousness'. In other words, he argued that the 'I' is not something concrete but the assemblage of sensations and passions that constitutes experience. This was, to most of Locke's contemporaries, a worry; and it was an argument that later ran together with a modern sense of estrangement and loss of self. When he said that '*Self* is that conscious thinking thing (whatever Substance, made up of, whether Spiritual or Material . . .)', Locke appeared to say that the self disappears when there is no thought.[18]

Locke argued against the view that it is 'the *Identity* of Soul alone [that] makes the same Man'. He was acutely conscious that most Christians considered the salvation of the individual soul, which was understood to have a real identity, to be an article of faith. But he thought that if the content of mind derives from reflection and sensation, then it is necessary to accept that we have no notion of personal identity apart from the idea of the succession of ideas, that is, apart from consciousness. 'For since consciousness always accompanies thinking, and 'tis that, that makes every one to be, what he calls *self*; and thereby distinguishes himself from all other thinking things, in this alone consists *personal Identity* . . .'[19] Many readers thought that this defines man's nature out of existence. Critics were quick to point out apparent paradoxes, such as that a sleeping person, having no consciousness, must have no identity. These were not esoteric matters. In 1708 Londoners thronged to stare at a pair of Hungarian Siamese twins joined back to back. Who could say, 'Whether each of the Twins . . . hath a distinct Soul, or whether one informs both?' Or, 'Could a Man marry the Twins, and not be guilty of Polygamy?'[20]

Locke's arguments, however soberly expressed, touched a raw nerve in contemporary sensibility, and they raised questions even for those unlikely to care about the logical niceties. There was optimism about the way new approaches to knowledge, to the mind and to human nature opened up new visions of connections between the physical, the mental and the social worlds. This optimism was liberating and it upheld human dignity in the face of religious dogmas and repressive social conventions. At the same time, however, a clear sense, linked to clear religious and moral realities, of what the essence of a person is, became lost. Locke himself was at pains to argue for individual moral responsibility but, even in his own lifetime, a few readers relished the possibility that his argument could be turned into a denial that there is a moral as well as a spiritual self. Hedonism, the pursuit of pleasure as the sole goal of existence, seemed to some both a reasonable and a desirable response. To moralists who looked at London rakes or the voluptuous excesses of French aristocrats, it seemed that the self had come under the sway of material pleasure. As one French writer later observed: 'Nothing is more *à la mode* today than to declare yourself an apologist of the passions . . .'[21] Whatever Locke's intentions, his analysis of mind helped shape representations of a new experience of subjectivity, new styles of life and new anxieties about individual purpose and social order.

Philosophical debate about self-identity was closely connected to questions about how subjective consciousness can know something objectively true about the physical world. Locke's analysis implied that knowledge is simply the sum of ideas derived from sensation and reflection; similarly, the self is the sum of those ideas. But Locke supposed that sensations do indeed convey to us something true about the external world. It was then obvious to him that he would have to deal with the nature of language, since it is through words that we represent ideas to others and to ourselves and make claims about what is true. Seventeenth-century scholars in general were fascinated by language. Locke's analysis was different, however, and it led him to what the intellectual historian Hans Aarsleff claimed was 'a critique that laid the foundation of the modern study of language'.[22] Just as Locke upset the notion that self-identity is constituted by and known through a distinct entity, the soul, so he upset the then common notion that words have a natural relationship to objects. He brought the two points together with a question that has a modern ring: whether 'the *Idea*, [individual persons] and those they discourse with have in their Minds, be the same'? Not only did he question a concrete and perhaps comfortable sense of identity but he suggested that when different people use the same words they cannot be sure that they use the words to denote the same ideas. 'Words in their primary or immediate Signification, stand for nothing, but the Ideas in the Mind of him that uses them . . .'[23]

This view of language opposed the belief prevalent in the sixteenth and seventeenth centuries that words are integrally related, through the act of God's creation, to the objects they describe, and even that words resemble the objects to which they refer. The study of the human capacity for language had also had profound theological and political implications. The Protestant Reformation placed language at the heart of scholarship, and Protestants were preoccupied by language, since for them the bible was the *word* of God and as such the key to life and the world. Some scholars and mystics argued that each thing, in its pure state before the fall of man, had possessed an identity and a name which truly and essentially belonged to it. Adam, it followed, had known this language but with his fall and the confusion of tongues which was God's judgement on Babel, the original language had been lost. The great German mystic Jakob Böhme, who believed in what he called the Adamic language, aspired early in the seventeenth century to recover this language in order to achieve purity and salvation. Scholars valued and studied Hebrew because they believed it was the language that, more than any other, retained the original form.

By the mid-seventeenth century, after religious and political violence had cut a swathe of destruction across Europe, scholars dreamed that they might integrate the natural world – the Book of God's work – with the bible – the Book of God's word – and thereby find grounds to reconcile people at war. Students of language hoped that the description of the facts of nature, expressed clearly in a common language, would force refractory minds to reach agreement. But it was first necessary to create a clear language with which to describe nature. Whereas earlier scholars of language had hoped to unite people through the recovery of Adam's original language, the new aspiration was to create a universal language as a transparent and authoritative medium with which to communicate the facts of nature.

This preoccupation was shared by Descartes, Mersenne and Hobbes, all of whom became interested in a philosophical language, a language suitable for the communication of truth, rather than a universal language *per se*, though the two ideals often ran together. The concern about language was also evident among the fellows of the Royal Society of London in the 1660s, the decade in which Locke began to conceive his *Essay*. In this, as in other matters, the Royal Society followed the lead of Bacon, whose analysis of the relation between language and the idols of false reasoning was mentioned in the previous chapter. John Wilkins (1614–72), who under the Commonwealth was warden of Wadham College, Oxford, and a prime mover in the creation of the Royal Society, hoped that 'men might know, and should be obliged unto [i.e., should be obliged to accept], by the mere principles of reason, improved by consideration and experience, without the help of revelation'.[24] He looked to natural religion (as opposed to revealed religion) to unite people, and this inspired his lifetime's commitment to the new natural philosophy, education and the reform of language. In his last work, *Essay Towards a Real Character and a Philosophical Language* (1668), he set out a utopian programme to create a new language, one which would make possible correspondence between words and the objective reality of which they are the sign. He wanted to make language a refined sign system for communication and agreement, and he believed his system promoted this because in it each sign conveys a piece of true information about the world. Wilkins hoped to abolish 'language' and replace it by a 'Real Universal Character' that directly signifies things, something which he believed mankind's original language had done. The new language, he believed, would eliminate misunderstanding between people.

Locke confronted the same issues: 'the Extent and Certainty of our Knowledge . . . had so near a connexion with Words, that unless their force and manner of Signification were first well observed, there could be very little said clearly and pertinently concerning Knowledge . . .' Book III of his *Essay* was 'Of Words', and the opening sentences stressed that the purposes given in common to man require language to make communication and agreement possible.

> God, having designed Man for a sociable Creature, made him not only with an inclination, and under a necessity to have fellowship with those of his own kind; but furnished him also with Language, which was to be the great Instrument and common Tye of Society. *Man* therefore had by Nature his Organs so fashioned, as to be *fit to frame articulate Sounds*, which we call Words.

He argued, however, that words are 'signs of internal conceptions' or 'marks for the *Ideas* within his own Mind', *not* signs for sensations let alone for external objects themselves.[25] He thus drove a wedge between words and the objects or reflections they refer to. Words signify rather than resemble.

Locke was as familiar with ancient languages as Wilkins or Newton and their continental counterparts. He was equally ardent in his desire to find in natural philosophy and in natural religion a new basis on which to reconcile people and create political harmony. His examination of the foundations of knowledge, however, led him beyond the view that others had of language and he concluded that words are a human creation. He attributed the capacity for words to God's creative wisdom, but actual language, he thought, is a matter of convention. This was a radical departure from a worldview in which words were believed to be as natural as the entities they resembled. Locke began to describe language as a social achievement, as something that has changed and will continue to change. It became reasonable to think that it is both possible and desirable for language to improve and not simply to recover. It was possible to look at the existing imperfections of language and, once it was grasped that a word is a sign of an idea, an idea a sign of a sensation and a sensation a sign of a thing, seek to make language more perfect for its function of communication. Locke hoped that an enlightened language would communicate knowledge clearly, separate knowledge from mere opinion and become the instrument of a politically stable and religiously tolerant society.

Readers did not readily comprehend in these terms what Locke wrote about language. It remained commonsense belief that words have a natural relationship to what they signify. His work was opposed philosophically both by the earlier logic of Port-Royal and by his contemporary Leibniz, who had his own plan for a universal language; these arguments are discussed in the next chapter. In the mid-eighteenth century, however, Locke found in the abbé de Condillac an interpreter who turned his approach to language into a systematic basis for knowledge. Condillac then influenced the way the systematic study, or discipline, of languages developed in the German-speaking and French-speaking worlds at the end of the eighteenth century.

v *A New Theory of Vision*

Evidence of Locke's significance is found throughout the eighteenth century: his name was irrevocably linked with hopes to base knowledge – especially knowledge of human nature – on experience. His *Essay*, respected for its analytic power, became a staple part of academic study, notably in the curriculum of the Scottish universities and of Trinity College, Dublin, where it was already being taught in 1692. Locke was also studied in North America and in continental Europe, though with many more reservations. Some topics in his book, like the question of self-identity, aroused intense discussion at the time. Other topics can be compared with the subject matter of the later discipline of psychology as a natural science. The best known of these topics is the account of perception which led to George Berkeley's *An Essay towards a New Theory of Vision* (1709). Berkeley (1685–1753) was an Irish scholar, one of the most profound minds to concern itself with the metaphysical dimensions of the new natural philosophy. His *New Theory* was the incisive work of a young man. In his later career he spent four years in Rhode Island, New England, was associated with a futile attempt to establish a college in Bermuda, and he then returned to Ireland to become bishop of Cloyne.

Locke analysed complex thoughts into what he claimed to be their component parts, simple ideas derived from sensation and reflection. Simple ideas combine, he argued, through an active process brought about by 'perception' (something like awareness), 'the first faculty of the Mind, exercised about our *Ideas* . . .'[26] He described perception as the faculty that distinguishes animals from plants: both receive impressions from the motions of external bodies, but only the former perceive it. In this connection, he understood humans to be animals, but animals with their capacities developed in the highest degree, animals able to combine, retain and reflect on sensation in the most complex perceptions, i.e., able to think about complex ideas. Questions about which complex ideas derive from which combinations of sensations naturally followed. These questions translated Locke's arguments into a series of problems that, at least in principle, were amenable to the empirical study of what children or adults do in fact perceive, with what senses and in what circumstances. In practice, it was a good deal easier to formulate thought experiments than to carry out material experiments on perception. Written analysis remained the dominant style of investigation.

Locke's *Essay* prompted his friend and enthusiastic Irish correspondent William Molyneux (1656–98), whom Locke called 'that very Ingenious and Studious promoter of real Knowledge', to pose a question that still offered challenges centuries later. Locke himself took it up when he revised his work:

> Suppose a Man born blind, and now adult, and taught by his touch to distinguish between a Cube and a Sphere of the same

> metal, and nighly of the same bigness, so as to tell, when he felt one and t'other, which is the Cube, which is the Sphere. Suppose then the Cube and Sphere placed on a Table, and the Blind Man to be made to see. *Quaere*, Whether by his sight, before he touch'd them, he could now distinguish, and tell, which is the Globe, which the Cube.

Molyneux and Locke agreed on a negative answer: newly restored sight would not of itself make it possible to name which shape is which. They thought that the capacity to identify different shapes is learned and that what has been learned for touch has to be learned again for sight. Locke denied that there are either innate visual ideas, e.g., of shape, or tactual ideas that become visual ones. Instead, he suggested that, 'by a settled habit', we judge ideas of light and colour in relation to entirely separate ideas of space and motion, and vice versa. Such judgements, carried out without our notice, enable us to have complex perceptions on the basis of a few visual clues in combination with past experience. Locke concluded that complex perceptions commonly attributed directly to vision occur in us 'by a custom of doing, [which] makes them often pass in us without our notice'.[27]

Locke did not in fact analyse perception in any detail, and it was left to Berkeley to turn his suggestions into a systematic theory about what the perception of space owes to touch. Berkeley's work had considerable impact in a lively natural-philosophical culture concerned with the mind's relationship to the material world as well as with physical phenomena. He was, like his contemporaries, intrigued by puzzles thrown up by the way we perceive, such as the fact that the moon appears larger when close to the horizon. But it was the topics that related vision and touch that had the widest theoretical implications. When he dealt with these matters, Berkeley explored how Locke's analytic approach to consciousness worked out in concrete cases, thus turning Locke's logic into an empirical science of vision. Berkeley's conclusion was that touch, not sight, is the most fundamental of the senses, since it appears to be the source of ideas of solidity and extension – of knowledge of the primary qualities of matter – and hence of the ability to distinguish self from not-self. This argument proved to be a very complex one indeed, and its ramifications were not worked out by Berkeley but by mental analysts who followed his example over the subsequent century and a half.

As paintings with elaborate perspective, the camera obscura, architecture and optical instruments all attest, there was a longstanding and sophisticated fascination with the construction of the visual image. Most seventeenth-century commentators assumed that vision involved an active synthesis of judgements whereby the soul transforms visual signs of distance, depth or size into a picture. It was common to treat the soul as a kind of geometer able to calculate angles between the eyes and objects and thus to work out size and distance. Descartes took this approach in his posthumous *Treatise of Man*, the French edition of which included pictures of the reasoning soul engaged in geometrical exercises to work out the visual image from motions inside the optic nerves.

Berkeley's approach differed fundamentally because he denied that there is anything in common between the ideas derived from the different senses and, separately and specifically, he denied there is any visual idea of depth. The idea of depth or distance, he argued, derives from active motion. In his view, sensations from touch, which we acquire from our movements, the motion of the eyes and from the manipulation of objects, give rise to ideas that are present in the mind alongside but distinct from ideas of colour and light derived from visual sensation. When we reflect on the co-existence of these ideas, the mind – so Berkeley supposed – judges that the co-existence signifies a relationship that exists in external reality. Thus we come to have a picture of the world. The child, he thought, learns by imperceptible degrees that visual ideas exist at the same time as ideas derived from touch; as a result, she or he comes to ignore the original ideas of touch and to think of the picture as the product of vision alone, 'seeing' depth, size, perspective and all the wonders ordinary people attribute to the eye. 'Having of a long time experienced certain ideas, perceivable by touch, as distance, tangible figure, and solidity, to have been connected with certain ideas of sight, I do upon perceiving these ideas of sight, forthwith conclude what tangible ideas are, by the wonted ordinary course of Nature like to follow.'[28] Berkeley rejected the view that perception of distance is a geometrical calculation imposed on the visual image and instead attributed it to the habitual experience of movement of the eyeballs in conjunction with the visual image.

The key theoretical point, which had a lasting influence on theories of learning as well as perception, is that Berkeley attributed a complex visual consciousness to the juxtaposition in time and space of sensations derived from different senses. 'That one idea may suggest another to the mind it will suffice that they have been observed to go together . . .'[29] This statement, that one idea 'suggests' another because the ideas have in past experience been juxtaposed, exploited an analogy that made the categories of the physical world – time and space – the categories of the mental world. The analogy directed study of the mind along the route so fruitful in relation to physical nature, and this approach was important in the later attempt to make the human science of psychology a natural science. When scholars described consciousness in terms borrowed from a discourse about the physical world, they hoped that mind could become the subject of physical science and could be studied by scientific methods. But there were also critics who regarded the analogy between physical relations and the juxtaposition of ideas as false and who dismissed such science as a muddle of what pertains to body with what pertains to mind. Much later controversy about perception, especially when it became the subject of an experimental science in the mid-nineteenth century, was deeply involved with such complex questions.

In order to deal with these and other issues Berkeley devised a philosophy that conceived the material world to exist as ideas in the mind of God. Few people followed his viewpoint. The discussion, however, was by no means all abstruse, and theories of perception were open to correction by experience. At the beginning of the eighteenth century a London surgeon, William Cheselden (1688–1752), operated to remove cataracts and restore sight to the blind. Here was the perfect opportunity to put Molyneux's question to the test (though Cheselden made no reference to the question) and find out what experience, rather than analysis, reveals about visual perception. Unfortunately, experience was not such a cut-and-dried matter since the operation for cataracts was only partially successful or recovery from the operation was slow and reports on what could be perceived muddled. Nevertheless, after his operation on a boy, Cheselden reported a clear-cut result in 1728, and commentators like the French *philosophe* Denis Diderot (1713–84) quoted this as authoritative: 'The young man whose cataracts were removed by this gifted surgeon was for a long while unable to distinguish either size, distance, position, or even shape. An object an inch high placed in front of his eyes so as to cut off his view of a house appeared to him as large as the house . . .'[30] It seemed that Locke, Molyneux and Berkeley were right, even if Molyneux's question did not have the single simple answer he had perhaps hoped for: visual knowledge is a product of learning made possible by the juxtaposition in time and space of sensations. What an imaginative, morally provocative and irreligious writer like Diderot made of this goes to the heart of the eighteenth-century science of man. It was a long way from the sober pieties of Locke and the early Royal Society.

6

The Principles of Rational Science

> To judge men in terms of their power of knowledge (understanding in the general sense), we divide them into those who must be granted *common sense* (*sensus communis*), which is really not *common* (*sensus vulgaris*), and men of *science*. Men of common sense are adept at dealing with rules as applied to instances (*in concreto*); men of science, at rules in themselves and before they are applied (*in abstracto*).
>
> Immanuel Kant, *Anthropologie in pragmatische Hinsicht* (*Anthropology From a Pragmatic Point of View*, 1798)[1]

i *Reason and the Road to Knowledge*

In the seventeenth and eighteenth centuries, what we call natural science was a branch of philosophy, natural philosophy, which occupied a place alongside moral philosophy and spiritual philosophy. The word 'science' was in common use to refer to theoretical or articulate as opposed to practical knowledge, to the branches of scholarship and to connected bodies of demonstrated truths. The word 'science' also had a more precise meaning, to denote knowledge with a rigorously systematic character given to it by its deduction from coherent rational principles. In this chapter, I discuss these rational principles as they affected the human sciences from the second half of the seventeenth century until the age of the German philosopher Immanuel Kant (1724–1804) a century later. Intellectual historians have tended to contrast these arguments about reason with the arguments about experience put forward with such influence by Locke. Yet the contrast was by no means clear-cut. Locke himself made much of 'judgement' and 'reflection' in addition to experience; he also conceived of his *Essay* as a study in clear thinking, in the conditions of knowledge, i.e., in logic. It is therefore necessary to go beyond a simple contrast between reason and experience as sources of knowledge about being human.

Descartes, Locke and their successors believed that clear knowledge requires right thinking about thought itself. They thought this was necessarily a different kind of activity from experience, though Condillac in the mid-eighteenth century tried to explain reason as the mind's activity in the process of having experience. Voltaire described Locke as the author of a 'natural history' of the soul; but though Locke himself was certainly interested in such a natural history, his mode of argument was logical and analytic rather than empirical. Even so, a substantial body of opinion, especially in German-speaking countries but later also in France and Britain, thought that Locke's arguments were unsustainable and that knowledge about being human has to begin with reason, not experience. This analysis of reason was a very important part of the background to the modern human sciences. And a preoccupation with reason itself continued to be a distinctive feature of the continental European human sciences in the twentieth century.

Virtually every writer in the seventeenth century believed man possesses a rational soul. Knowledge of the soul was of paramount interest and importance, and when such knowledge was presented systematically it formed a science. It was judged to be a logical necessity to found the science of the soul in the soul's analysis of reason. Philosophers did not deny the importance of experience for knowledge, but they did argue that the nature of experience could not be comprehended independently of the exercise of reason.

A concrete example of what this meant in practice for the relationship between natural philoosphy and the science of man is provided by the work of Thomas Willis, discussed earlier in connection with the reception of the mechanical philosophy in England. He carried out empirical studies of the brain in humans and animals and speculated about the material relations of the organism's animating principle, the corporeal or animal soul. At the same time, however, Willis, who was a high church Anglican, a royalist and close friend of archbishops after the king's restoration in 1660, hoped that his work on the body enhanced the dignity of the rational soul rather than detracting from it by investigations into matter. He described the rational soul in a conventional way as the reasoning and willing principle given by God to humankind and not animals. When he wanted knowledge of the rational soul, Willis turned away from dissections in order to read sermons by churchmen. When he discussed the soul systematically and sought true causes for man's nature, he drew on a Christian Aristotelianism that characterized man's essence in the traditional discourse of logic and theology. Willis was fascinated by empirical knowledge of the brain; he held that knowledge about human beings, however, requires other disciplines. The rational soul, not the brain, constituted the special subject of the science of man.

Willis was one of many physicians who, because of their occupation and interest, returned to the problem of relations between soul and body and between reason and experience. Half a century later, in the new Prussian university of Halle, Georg Ernst Stahl (1660–1734), better known to historians of science for his chemical theory of phlogiston, took a strongly anti-mechanist line as a professor of medicine, but he certainly did not deny experience. He held that the soul is incarnate in, and works with, the whole body. His views were influenced by Pietism, a form of Lutheran practice which encouraged close attention to inner spirituality and hence made medical views of the soul a sensitive matter. Stahl, whose work was a reference point for later philosophers and physicians in the German-speaking world, therefore made considerations about the soul based on reason and theology the basis of medicine.

It was on the European continent rather than in Willis's Oxford that scholars undertook systematic re-examination of reason and of the causes of human actions. Descartes boasted that he had found an indubitable basis on which to do this, but his picture of how the soul relates in human life to the body was widely judged to be unsatisfactory. The search for alternatives produced much difficult metaphysics, such as that proposed by

Nicolas Malebranche (1638–1715), a priest who played a major part in the French educational system's acceptance of Cartesianism. His metaphysics attempted to solve the problem of causation, which included the problem of the soul's relation to the body, by the attribution of the power of causation to God alone. In his view, events in the soul are the occasion not the cause of interactions. Malebranche also contributed to problems of visual perception, but his method was to use reason to analyse experience. As his studies indicate, logic and theology continued in the late seventeenth century to be core disciplines in the study of man, and they remained so for a long time to come.

Cartesianism provoked many scholarly disputes – disputes no less virulent for being scholarly. When Descartes and Mersenne attacked scholastic learning in the 1630s and 1640s, they precipitated extended struggles over the academic curriculum as well as over metaphysics. In the aftermath of the French Catholic Counter-Reformation and the massacre or emigration of Protestant Huguenots, the Jesuits gained considerable power. Other Catholics tried to resist this and the consequent Jesuit control over education. Some of the best minds to respond to Descartes were in fact Jansenists, a Catholic group held in suspicion in the 1650s because they believed, somewhat like Calvinists, in the need for God to dispose men to receive grace not just for grace to redeem sin. They included among their members Blaise Pascal (1623–62), the brilliant mathematician and natural philosopher of the vacuum. Pascal's profound and anguished exploration of the limitations to human reason and goodness, limitations that he thought were redeemed by grace alone, moved readers in subsequent centuries even though the notes published posthumously as the *Pensées* (Thoughts) were known only in fragmentary or edited form until 1952. Pascal's response to the problems of knowledge was an intensely personal one, but it ran parallel to the concern of some of his Jansenist colleagues to improve learning and education.

In 1662, two Jansenist teachers, Antoine Arnauld (1612–94) and Pierre Nicole (1625–95), published *La logique, ou l'art de penser* (*The Art of Thinking*). This was a handbook on a method for right thinking and an attempt to displace the logic taught at the Sorbonne. It introduced what is known as the Port-Royal logic after Arnauld's connections with the Jansenist institutions at Port-Royal near Paris. Arnauld (with the collaboration of Nicole) drew on Descartes' 'rules' in the *Discourse on the Method* and argued that the soul, through its own activity, conceives, judges, reasons and orders its basic elements and thereby transforms clear ideas into knowledge. It follows, they judged, that to learn about human attributes is to reason clearly and correctly with those attributes. *The Art of Thinking* had an impact because, like Descartes' *Discourse*, it argued for the transformation of a key discpline within the curriculum, logic, into the *method* with which to advance knowledge. Thus its authors thought it necessary to separate part of logic from rhetoric, the art of communication, and turn it into the method of discovery. As their enthusiastic English translator wrote: 'so . . . now Logic may be said to appear like Truth itself, naked and delightful, as being freed from the *Pedantic* Dust of the Schools'.[2] The Port-Royal logicians therefore began their work with the analysis of the mind as it conceives and forms ideas, the task which Locke also set himself.

The new definition of logic established a significant approach to language, one which attracted attention again in the 1960s when Chomsky claimed that language indicated the existence in the mind of universal grammatical structures or rules. The Port-Royal logicians started with the view that language is the means to communicate thoughts, and they therefore proposed that to understand language is first to understand thought; this implied that the structure of language reflects the way the mind thinks. Hence, it seemed to them, all languages are at a deep level comparable and – in any language – successful speech mirrors clear thought. The basic unit of communication, the proposition (such as 'she is a child'), reflects the basic act of mind which affirms or judges two things to be similar or dissimilar. Yet in spite of this theoretically clear position, which explained language by reason, the logicians in practice recognized that much of language, such as figures of speech, derives from experience and cultural tradition rather than reason, and this recognition pointed to the way the empirical and historical study of language might develop alongside its rational analysis.

Arnauld engaged in long-running controversies with his contemporaries, notably Malebranche, about such questions as the relation between our ideas (or way we conceptualize things), perception of the external world and knowledge of God. The two men debated whether, for example, the soul has an immediate representation of physical objects or whether what we think of as objects are only the occasion for the idea present as a perception. From a modern vantage-point, it is possible to tease out questions that we recognize as topics in the philosophy of mind about perception and mental action. The terms of debates then, however, were set by logical and theological agendas – which, in France at this time, also meant political struggles. Yet perception emerged as a topic in relation to which it was possible to have discussions relatively isolated from other considerations. Malebranche, for example, discussed how the size of an object, such as the moon, appears relative to other objects in the visual field. Berkeley's *New Theory of Vision* later confirmed this specialization of interest.

Descartes, Arnauld and Malebranche, much as their scholastic predecessors had done, gave priority to the clarity of concepts as the basis for knowledge. But clarity was elusive, as Descartes' difficulty with the causal relations of soul and body shows. One response was scepticism – to deny that knowledge of any kind can achieve certainty. This was the direction taken by Bayle in his *Historical and Critical Dictionary*, a work that was a major source of information and sceptical argument throughout the eighteenth century. Bayle produced a formidable catalogue of the inconsistencies within and between philosophical, religious and ethical claims to truth. He stressed the historical character of beliefs and hence the impossibility of comparing the knowledge of one age with that of another. Scholars who still equated science and certainty sought to answer this challenge. By contrast, Locke's response to the problem of knowledge (which was not a response to Bayle) was to concentrate his analysis on experience as something which all people can share. Yet many of his contemporaries, notably Leibniz, denied that this dealt with the problem of the origin of concepts, and they argued that certainty must be obtained through the study of the foundations of reason. So reason itself became the basis for the study of man independent of conclusions drawn from experience or natural philosophy, an argument that was very important in the nineteenth- and twentieth-century debate about the proper form of the human sciences.

ii *Deductive Knowledge: Benedict Spinoza and G. W. Leibniz*

All scholars agreed that no proposition could claim to be true that did not conform to basic logical requirements, notably of consistency. In arguments about truth, deductive reasoning – to argue from premises to conclusions logically required by the premises – therefore held pride of place. Hobbes's and Descartes' delight in the rigour and certainty of geometrical and mathematical reasoning inspired their respective sciences of man. They held that the *science* of man is possible since deductive thought can reveal conclusions about human life to be necessarily true. Other scholars shared this presumption, and none with greater intensity and persistence than the seventeenth-century philosophers Benedict (Baruch) Spinoza (1632–77) and Gottfried Wilhelm Leibniz (1646–1716). Their studies exemplify the deductive approach to the establishment of the true causes of human nature. Aspects of what Leibniz argued appeared repeatedly in the human sciences over the next two centuries.

Arnauld died in exile in Brussels in 1694. The Low Countries in the seventeenth century were intellectually and economically possibly the most dynamic in the world. They introduced innovations in government, in the achievement of balance between political and religious powers, in law, in local administration and in natural philosophy. The end of Spanish dominance in the region in the mid-century left what is retrospectively known as the Dutch Republic as a loose association of city authorities, the Calvinist church and the House of Orange. In spite of threats from the Catholic Southern Provinces and the later invasion of the French army, the Northern Provinces, particularly Amsterdam, became a centre for new scholarship and for relative freedom of expression and publication. Locke spent several years there in the 1680s. These conditions also allowed a Jewish community to flourish, a community with many members from families who had escaped from the Spanish Inquisition. Out of this background came the most remarkable of all scholars committed to the power of reason to establish science.

Viewed from the late twentieth century, Spinoza appears to be a medieval figure, part of a great tradition of Jewish learning and religious devotion, a rebellious heir to the twelfth-century Aristotelian scholar Maimonides (Moses ben Maimon). Spinoza's life and thought was a testament to the belief that reason can arrive at certitude and that this certainty of thought makes it possible for man to be one with God. Though reviled in the late seventeenth century as an atheist and expelled from his Jewish community, he appears in the late twentieth century as an awesome combination of rational intellect and religious mystic. This was the man that the German poet Novalis called *ein gottbetrunkener Mensch*, God-intoxicated. Yet Spinoza wrote a powerful defence of secular and constitutional government, the *Tractatus theologico-politicus* (published anonymously in 1670), and he risked himself in dangerous political causes, while his *Ethica* (Ethics, published immediately after his death in 1677) included many conclusions about experience and conduct which have surprised later commentators with their psychological insight. With both Spinoza and Leibniz, however, it is a distortion to pick out part of a text and describe it as psychology or political theory. Their work had meaning as a whole or not at all; this unity of meaning, as they argued, is a consequence of reason itself. They equated the logical with the essentially real. Spinoza was uncompromising: 'The order and connection of ideas is the same as the order and connection of things.'[3] It followed for him that the experience of particular things about a person or a physical state gives rise only to superficial truths. Deep truth is what is known about the eternally real, what is known to clear reasoning by a man undistracted by passing emotion, the chances of experience or political disorder. Spinoza led his life so that he could become such a man.

The *Ethics* was a formal deductive study which argued from definitions and axioms to propositions with the status of necessary truths. It went behind experience to self-evident and clear ideas. Spinoza used technical Latin as the language best suited to his purpose, and his readers then and later faced daunting interpretive problems. In Part II, 'Of the nature and origin of the mind' (he used the Latin word '*mens*'), he dealt with Descartes' problem, how to specify the relation between soul and matter and explain how communication is possible between different substances. Spinoza's argument was that soul and matter are aspects of one reality, not separate entities; he described consciousness of the body, for instance, as the body under the attribute of thought. Part III, 'On the nature and origin of the affections', included an extended description of the feelings. In Part IV, 'On human bondage or the strength of the affections', he described how unclear thought, passion and self-deception enslave us by separating us from the truth. In spite of such human weaknesses, he argued, everything that takes place necessarily gives expression to the absolute perfection of the creation. We do not perceive this because we perceive *sub specie durationis*, within time, and hence have only a partial view. The goal, Spinoza dreamed, is to become one with God, in which condition even human beings, through reason, will perceive – as the eternal Being perceives – *sub specie aeternitatis*, from the perspective of eternity; in this perception, he argued, lies freedom.

Modern English-language translators referred to 'emotion' where Spinoza's Latin was *affectus*. This is misleading since an 'emotion' is a modern psychological category whereas *affectus* was, in Spinoza's words, 'the modifications of the body, whereby the active power of the said body is increased or diminished, aided or constrained, and also the ideas of such modifications'.[4] *Affectus* was a scholastic concept which denotes the effect on a substance, when it has a 'passion', of a cause from without. When Spinoza referred to 'passion', he denoted not 'emotion' but the soul's subjection to a cause not intrinsic to its real nature, just as any entity is subject to such causes. Spinoza defined wonder, contempt, anger, joy, humility and other human states in such language. His purpose was, like other moralists, to show how the passions do not represent our true nature.

Spinoza described each individual thing, including the individualized entity that we call a person, as something that possesses identity through a *conatus*, a striving to realize and perpetuate what it truly is. He then went on to define pleasure as an expression of this *conatus*, pain as an opposition to it; thus, he maintained, whatever gives pleasure is said to be good, and vice versa. Scandalously for his contemporaries, he concluded that a person cannot choose to be good, since belief in choice supposes an impossible contingency in the creation. He described the self as a being on a continuum between an active state, marked by clear ideas and expressive of its *conatus*, and a passive state, marked by unclear ideas and passion, a sufferance of something from outside its essential nature. 'Our mind is in certain cases active, and in certain cases passive. In so far as it has adequate ideas it is necessarily active, and in so far as it has inadequate ideas, it is necessarily passive.'[5]

He discussed the passions, within the framework of his whole deductive scheme, in order 'to treat of human vice and folly geometrically'.

> Thus the passions of hatred, anger, envy, and so on, considered in themselves, follow from this same necessity and efficacy of nature; they answer to certain definite causes, through which they are understood, and possess certain properties as worthy of being known as the properties of anything else . . .

He elaborated propositions from first principles about how the affections, proportionate to the clarity or obscurity of our understanding, express the freedom or servitude of our conduct. Other scholars had considered morality as a form of demonstrable truth but no one had explored action and passion with such (to us, paradoxical) deductive formality *and* insight into the conflicts observable in day-to-day life. This supposed insight is illustrated by his conclusions that we imitate the affections of those things that we imagine to be like ourselves, that our love and hatred is reinforced when we imagine that others share in them and that sexual jealousy provokes obscene thoughts: 'For he who thinks, that a woman whom he loves prostitutes herself to another, will feel pain, not only because his own desire is restrained, but also because, being compelled to associate the image of her he loves with the parts of shame and excreta of another, he therefore shrinks from her.'[6] Spinoza believed that the passions which distract our reason are hidden or at least in shadow. The *Ethics* aspired to lay out the path to make feeling transparent: 'An affection, which is a passion, ceases to be a passion, as soon as we form a clear and distinct idea thereof.'[7] Reason is the road to freedom and in freedom human nature finds its true purpose, an intellectual love of God.

Few readers followed all of Spinoza's thoughts. Much remained obscure, logically questionable and incomplete; later philosophers believed, for example, that he did not adequately deal with the notion of the self or of the other-directed nature of the feelings. Most of his contemporaries were simply scandalized by his reputation for determinism and his view of the creation that identifies God with nature. Only distortions of his work appeared until the late eighteenth-century German Enlightenment, when Johann Wolfgang Goethe (1749–1832) and other philosophers of nature found in Spinoza inspiration for belief that reason reveals the oneness of humanity with the universe.

Towards the end of his life, Spinoza corresponded with and met Leibniz. Though the latter referred privately to Spinoza's monstrous opinions, both believed that science began with logical principles which have the status of necessary truths, and both laboured at metaphysics to overcome the problems bequeathed by Descartes. Leibniz was much more worldly than Spinoza, a servant of secular princes and a polymath whose brilliance was matched by the diversity of his activity: diplomat, academic lawyer, librarian, mathematician, natural philosopher, theologian, linguist and historian. He worked for a while in the philosophical hothouse of Paris in the 1670s and then for successive dukes of Hanover (including Georg Ludwig, who became George I of Britain in 1714). His history of the House of Brunswick was an act of legitimation, comparable to the work of his fellow Protestant Pufendorf, for the ruler's aspirations to advance a modern state. In 1700, under the patronage of Sophie Charlotte, wife of the Elector of Brandenburg, he created the Berlin Society of Sciences, subsequently the Prussian Royal Academy. Sophie Charlotte was one in a distinguished line of eighteenth-century women eminent for both their learning and their patronage of philosophy. Leibniz took a lead in the establishment of intellectual and institutional links between the German states and developments abroad. Behind his formidable activity lay an ambition to reunify the Christian world through the intellect. The principles of reason rightly exercised, he argued, require everyone to agree to the nature of the creation, since the creation has come into being through the rational will and goodness of God. His philosophy was an over-arching attempt to provide the requisite principles of reason. Further, he desired to create a universal language, which he derived from etymological studies of what he supposed to be the common roots of different languages, to provide a transparent medium to assist unification of Catholic and Protestant peoples.

This interest in language was widespread during the seventeenth century – as the previous discussion of Hobbes's, Wilkins' and Locke's views made clear. During the seventeenth century, the hope that a common language would heal man from his fallen nature became the more worldly aspiration that a common language would heal a divided Christendom. The earlier hope that language would reveal what is true about the world became the belief that language can express what the mind thinks clearly and hence knows to be true. Leibniz, an enthusiast for etymology, traced the genesis of words in order to trace the genesis of the human world and of reason, which he thought coeval. The Italian scholar Vico also followed this route. Dreams like these about language remained alive until the twentieth century when, first, philosophers separated logical questions about language from psychological questions about the mind; and second, the Swiss linguist Ferdinand de Saussure stressed again that words are only signs, arbitrarily related to things. Nevertheless, the hope for human unity through language revived with the founding of Esperanto in 1887 by Louis

Zamenhof and, for some auditors, with the spread of English as the *lingua franca* of globalized culture at the end of the twentieth century.

Leibniz's metaphysics, in this respect alone comparable to Spinoza's, argued deductively from logical requirements, the most significant of which is the principle of sufficient reason: 'nothing happens without its being possible for one who should know things sufficiently, to give a reason which is sufficient to determine why things are so and not otherwise.'[8] This principle, Leibniz believed, logically expresses the harmony rather than the chance that reason necessarily assumes is constitutive of the creation. He deduced from this principle, and the logical principle of non-contradiction, that what is exists as elementary units, called monads, which possess self-action and an intrinsic unity. These primary elements, in his analysis, underlie the compound states that we know in common experience as body and soul. It was significant for later psychologists and philosophers of mind that Leibniz attributed two fundamental modes of activity to the monads, perception, the representation within the monad of what lies beyond it, and appetition, the monad's tendency to pass from one perception to another, an activity which is the principle of change.

His statements are notoriously obscure; their full sense emerges only when they are derived logically in the way Leibniz himself attempted to do. What is important to grasp is the logical manner in which Leibniz thought it possible to gain knowledge of the soul and his conclusion that the soul is characterized, first, as an activity, and second, as a unity. Mental activity and unity became central principles in German-language psychology when it later became a discipline. Leibniz did not always directly influence later thought, though he sometimes did; but he provided the classic formulation of the argument that active mental reasoning, which presupposes a unified rational agent – the soul – must logically precede experience if experience is to achieve any order and meaning. This argument requires a science of mind to start out with the presupposition of mental activity, activity that is self-generated and that precedes, and hence cannot be understood in terms of, physical nature. The argument therefore places logical restrictions on the capacity of natural science to originate human science. In philosophy, the argument led to what is called idealism; in psychology, to theories that emphasize the mind's action in capacities like perception and volition.

Leibniz was more directly involved than Spinoza with contemporary developments in natural philosophy. He devised the notation of the calculus adopted in the eighteenth century, and his examination of the conceptual foundations of mechanics was fundamental to the assimilation of Newton's work. It was Locke's *Essay* that provoked Leibniz to address 'the understanding' more systematically. His main response to Locke was written following Pierre Coste's French edition of the *Essay* in 1700, though Leibniz's *Nouveaux essais sur l'entendement humain* (*New Essays on Human Understanding*) remained unpublished until a French version appeared in 1765. These essays therefore did not contribute to the influence that Leibniz had on the reform of the German universities in the eighteenth century, which introduced a secular element into the curriculum. Only at the end of the eighteeenth century was he to become the principal source for the argument that the mind constructs knowledge through its activity. This argument was then contrasted with the theory attributed to Locke, that the world imposes knowledge on the mind through experience. Further, Leibniz was cited for the statement that mental activity determines conduct, Locke for the statement that conduct is a response to external events. In summary, Leibniz became the figurehead for belief that stresses the soul's innate and essential activity when it grasps knowledge and originates conduct.

Leibniz defined a substance as 'a being capable of action', and his description of the soul followed this pattern: the soul has attributes predicated of a substance and does not acquire them through external influences.[9] He analysed a cause as a monad's appetition, which brings about a change that is a development of the monad itself and not a consequence of the interaction between two things. Compounds of elemental things with a certain clarity of active perception and appetition form the souls of animals. A higher level of activity achieves reason and makes the soul of man; activity of the soul as a whole constitutes thought, perception, affection and will.

Eighteenth-century German philosophers favourably contrasted Leibniz's conception of the human soul with Locke's treatment of the understanding, even though they did not accept Leibniz's metaphysics. Leibniz described the soul as a unity able to present to itself in a unified way the sensations Locke supposed the soul receives in the circumstantial order in which the external world presents them. Perception and feeling, Leibniz wrote, 'should be expressed or represented in one indivisible being or in substance which possesses a genuine unity . . . this representation is accompanied in the rational soul by consciousness, and then it is called thought'.[10] Locke described the self as an idea of unity that derives from experience; Leibniz presupposed unity in order to analyse experience. Many later German philosophers and psychologists, such as J. F. Herbart, agreed with Leibniz. It followed that Leibniz could not accept Locke's critique of innate ideas. 'I believe . . . [with] all those who understand in this sense the passage in St Paul where he says that God's law is written in our hearts . . . that is [there are] fundamental assumptions or things taken for granted in advance.' Like Descartes, Leibniz argued that we necessarily reason with certain concepts and that these must therefore be intrinsic to the soul or, to use a later psychological term, 'innate'. He did not deny, of course, that experience is essential as the occasion when reason recognizes that it has concepts. Leibniz pointed out that Locke himself, when he described reflection, seemed to acknowledge that there are innate capacities of reason:

> But reflection is nothing but attention to what is within us, and the senses do not give us what we carry with us already. In view of this, can it be denied that there is a great deal that is innate in our minds, since we are innate to ourselves, so to speak, and since we include Being, Unity, Substance, Duration, Change, Action, Perception, Pleasure, and hosts of other objects of our intellectual [i.e., not sensory] ideas?[11]

In opposition to Locke's image of the soul as a blank slate, Leibniz likened it to a veined block of marble in which the veins represent the pattern that determines the final form of the sculpture.

Leibniz also thought perceptions or feelings might be present in the soul but not actually subject to thought or consciousness, i.e., there may be mental activity of which we are unaware. For Locke, an idea is conscious or it is nothing, and this position led to a puzzle about what self-identity is when ideas are absent, notably while we sleep. Leibniz presupposed continuity in the soul's activity but not in consciousness, and he therefore did not face Locke's puzzle. Leibniz's belief in non-conscious activity had important consequences for views about perception; it led, for example, to his claim that the soul creates a visual image through the dynamic action of a large number of perceptions, each of which we are individually unaware. Belief in non-conscious perceptions appeared to explain much:

> At every moment there is in us an infinity of perceptions, unaccompanied by awareness or reflection . . . alterations in the soul itself, of which we are unaware because these impressions are either too minute and too numerous, or else too unvarying . . . This is how we become so accustomed to the motion of a mill or a waterfall, after living beside it for a while, that we pay no heed to it.[12]

Similarly, just as familiarity leads the soul to become unaware of perceptions, so memory requires attention to perceptions of which we have ceased to be aware.

The historical comparison between early and modern uses of key terms associated with mind and consciousness is fraught with difficulty, and this becomes significant when historians trace the roots of modern psychological views of mind. Leibniz coined the Latin word *'apperception'*, a word directly taken over by German and English, to describe the soul's action, the waking state of a sentient being. It is, in his language, the activity of apperception that brings perceptions into a conscious unity and makes one perception or another the subject of attention and thought. 'Apperception', especially in German, was a term with a long life in psychological theories that analysed experience as a dynamic and constructive as opposed to a passive and reflective process. The French word *'connaissance'* had comparable meaning and denoted an active conscious understanding. By contrast, the Latin word *'conscientia'* and the French word *'conscience'* denoted knowledge shared by a number of people, 'conscience' in the modern moral sense, or simply awareness. Also, Leibniz's references to perceptions of which we are not aware, which commentators sometimes called unconscious perceptions, do not relate to Freud's arguments about the unconscious. Leibniz's deductive science of the soul, with its many scholastic resonances, and Freud's psychoanalytic theory are worlds apart.

It is time to turn from the technical and deductive philosophy of Spinoza and Leibniz to examine the slow liberalization of eighteenth-century German academic culture. While the German-language universities changed, a change that culminated in the first half of the nineteenth century with the formation of modern scientific disciplines, they nevertheless retained a commitment to reasoning about reasoning as the groundwork of science. This philosophical commitment perpetuated a stress on the real presence of the soul's activity.

iii *The German Enlightenment*

The German language united culture across Austria, parts of Switzerland and Bohemia, Bavaria, Prussia – including East Prussia between the Baltic and Russia – and a jigsaw pattern of dukedoms, episcopal principalities, free city states and other political entities, many with feudal societies, over the area which is now Germany and Central Europe. A small, educated, town-based élite developed in the eighteenth century, linked closely to state administration and the churches, but it felt itself to be backward economically, politically and culturally in relation to France and Britain. Both Protestant and Catholic universities were dominated by religious interests and continued to educate in Latin. The philosophy faculty, heir to the medieval arts faculty, had low status and offered preliminary teaching for students before they went on to the higher faculties of medicine, law and theology.

New metaphysical systems and new natural philosophy in the seventeenth century disturbed Catholic theology and Protestant piety. Spinoza was simply outlawed as irreligious. Leibniz's arguments for the necessary form of the creation, like Grotius's or Pufendorf's formulations of natural law, faced criticism from Protestant clergy who believed they detracted from faith in revelation and put restrictions on the will of God. Scholars who emphasized the capacity of reason to pursue science met suspicion from Protestant Pietists who stressed the immediacy and free action of God's grace. This situation changed slowly during the first half of the eighteenth century. Leibniz and other philosophers who stressed reason had some impact but they did not contribute, as Locke's writings did in France, to a secular Enlightenment. Changes in education and scholarship in Germany and Austria were slow; the universities retained many traditions with medieval origins. Real change came only between 1806 and 1810, when the university of Berlin was founded, though this step was possible only because of earlier academic activity.

There were attempts at university reform, associated with the foundation of the Protestant university of Halle (1693) and the university of Göttingen (1734). When Leibniz's work received attention, however, it was mainly in advanced court circles or abroad. In the period known as the German Enlightenment (*Aufklärung*), roughly the last forty years of the eighteenth century, it was court and town culture rather than the universities which provided the driving force. This culture was dependent on royal or aristocratic patronage. Frederick the Great of Prussia brought in French *philosophes* to enlighten his court, and he appointed the Frenchman P.-L.-M. de Maupertuis (1698–1759) to reorganize and give status to the Berlin Academy of Sciences in 1744. The dukes of Saxe-Weimar appointed J. G. Herder and then Goethe, who became chief of state, manager of the theatre, inspector of the roads and almost everything except an academic.

The most influential exponent of the new learning in the first half of the eighteenth century was Christian Wolff, sometimes called the *praelector* or teacher of Germany. He was exiled from Halle in 1723 because his rational religious arguments upset the Elector Friedrich Wilhelm of Prussia – Wolff unwisely wrote that 'reason does not allow itself to be ordered about'.[13] Later he returned and in due course became rector of the university, reflecting the shift of opinion which gradually allowed independent philosophical discourse. Wolff, who wrote first in German and then in Latin in order to have a wider audience, was a systemizer, a classifier of methods and knowledge around logical principles derived from Leibniz and his scholastic predecessors. Wolff's work included the description of two branches of psychology, the science of the soul, in two books, the *Psychologia empirica* (1732) and the *Psychologia rationalis* (1734). The 'rational psychology' argued from metaphysical first principles in order to clarify the concepts that make possible knowledge of the soul. This knowledge was then developed as an ordered deductive science comparable to rational mechanics – Wolff had trained and taught as a mathematician and corresponded with Leibniz about natural philosophy. The 'empirical psychology' described the soul's activity 'by attending to those occurrences in our souls of which we are conscious'; Wolff called these mental occurences *Vorstellungen* (perceptions or mental respresentations).[14] He stressed, like Leibniz, the soul's activity when it has clear perceptions, the activity later studied as apperception. Through empirical psychology, Wolff argued, we arrive at general concepts which join up with the conclusions of rational psychology.

This interest in psychology was practical and moralistic as well as philosophical and logical. Wolff followed Christian Thomasius (1656–1728), a founder of the university of Halle, in believing that it is possible to construct a calculus of the passions which makes possible rational but practical judgements about conduct. Thomasius outlined a programme for 'the discovery from daily conversation of secrets in the hearts of other men even against their wills'.[15] He assigned numerical grades to the passions on a scale from 5 to 60; for example, he described Cardinal Mazarin (the effective power in France during Louis XIV's minority) with

60 points of ambition, 50 of sensuality and 20 or 30 of rational love. Empirical psychology opened out through such projects into the everyday world of the judgement of character, decisions about a child's education and the regulation of the feelings. Moral philosophy, the part of the academic curriculum that linked theory and practice in relation to action, continued in the eighteenth century to be important for the human sciences. Thomasius was also a lawyer, and his arguments for a legal science, with a hierarchical system of rules from the general to the particular, developed later into enthusiasm for comprehensive legal codes, and this also connected the human sciences and the practical social world.

Wolff mapped out a scheme for the scientific study of the soul in terms that remained current with German writers for a century or more. Rational psychology meant the metaphysical and deductive study of the soul's nature, its unity, free will, immortality, purpose in the creation, activity and relations to material substance. Empirical psychology, in practice, meant diverse things, but it tended towards description and analysis of the soul's powers as active faculties. These faculties were grouped under the headings of knowing (cognition), feeling (affection) and willing (appetition). The faculties are the powers of the soul. Significantly, this language about the faculties linked academic psychology to a wider educated world which possessed a rich language with which to describe traits and character. In this way, philosophy connected with ordinary people's views about human capacities for affection, will and the expression of what is beautiful. Some historians refer to eighteenth-century faculty psychology. This conveys too concrete an idea of a single tradition and defined terms; rather, reference to faculties was informal and everyday. The language was in danger of being empty, in that it represented by words such as appetition what needed to be explained. Nevertheless, partly because of Wolff's teaching, writers who adopted this language conceptualized the soul as activity in a manner that gradually became independent of theology. A subject area began to emerge, with its own methods and its own focus on what would later be called mental activity. The assumption remained, however, that mental activity, whatever its variety, derives from a unified soul that carries in itself the designs of the Creator. German writers believed that the mind is a spiritual and formative force, not a mirror of the physical world. This presumption gave a distinctive quality to German-language enlightened culture. The formal foundations for other studies, such as aesthetics, the philosophy of religion, ethics and jurisprudence were in turn all related to the purposive activity of the soul.

The stress on mental activity and the unity of the soul was, all the same, compatible with arguments, common in France or Scotland, that subsumed man under general natural laws and compared men and animals. In 1760, for example, a Hamburg *Gymnasium* (or grammar school) teacher, Hermann Samuel Reimarus (1694–1768), published his *Allgemeine Betrachtungen über die Triebe der Thiere* (*General Considerations on the Drives of Animals*) in which he classified 'drives', or the faculties that give rise to action, into three levels: mechanical drives, awareness drives and volitional drives. Influenced by Wolff's rational approach to religion, Reimarus conceived of animal and human faculties as part of a continuous pattern of created nature. He still distinguished these drives, however, from actions caused by the uniquely human attribute of reason. The German word '*Trieb*', not common in English as the psychological word 'drive' until the twentieth century, was subsequently used widely in the Romantic period by writers like F. W. J. Schelling, a sign of the spread in the eighteenth century of psychological rather than theological language about the soul.

Enlightened but idealist learning is exemplified by the work of Johann Nicolaus Tetens (1736 or 1738–1807), a characteristic figure in the range of his occupations, which included inspector of sea installations, financial official and professor of philosophy at the university of Kiel from 1776 to 1789. By this time it was not necessary, as it had been for Wolff, for Tetens to defend reason against Pietist critics. The increased openness of German society was also reflected in Tetens's response to philosophical developments in France and Britain. He drew on Leibniz's recently published critique of Locke and, in his *Philosophische Versuche über die menschliche Natur* (*Philosophical Essays on Human*

Nature, 1777), he argued that all experience is logically predicated on acts of the soul. These acts, which he classed in the conventional way into thinking, feeling and willing, are, he held, responsible for the assembly of sensations into a meaningful whole. Tetens conceived a relation between two things in experience to be an act by which the human spirit orders sensation and thus constructs truth. By contrast, as explained in the next chapter, writers who followed Locke, like Condillac and Hume, conceived of knowledge of relations as a consequence of the similarity or juxtaposition of sensations.

Tetens analysed feeling as a transformative process in which the soul incorporates experience into a moral and aesthetic unity – feeling is not simply the pleasure and pain that accompanies sensation. Descriptive analysis of the soul therefore related to theories of art and aesthetics that explained how the soul constructs harmony in works of art, a string quartet by Haydn perhaps, and makes the individual qualities of the soul into a shared culture of the spirit. Tetens wrote at a time when there was much concern with human perfectibility, and he was drawn – like many of his contemporaries, in non-German as well as in German states – to think about individual development and the development of humanity in the same way. What later became separated as developmental psychology and cultural history were woven together in this period. Impassioned writers elevated artistic activity into *Kultur*, a word that gave dignity to a collective ideal of the good, the true and the beautiful as the meaning of life. Moses Mendelssohn (1729–86), known to his contemporaries for his wisdom and ugliness as the Jewish Socrates, built a theory of art and culture which used faculty psychology and inspired the generation of Goethe, Friedrich Schiller and the Romantic poets. Mendelssohn argued that there is a specific faculty of the soul, separate from reason and feeling, which is able to respond directly with approval to beauty; hence, he argued, it is possible to know beauty in a way that analysis cannot reproduce. It is, he thought, the gift of artistic genius to create ideal forms that arouse this aesthetic faculty in lesser mortals.

Tetens and Mendelssohn attributed to the soul the power to

transcend particular feelings or sensations and thus to achieve higher values. The study of human nature was therefore the study of the soul's powers, their specific qualities and their mutual relations. This study was also a theory of the perfectability of man and culture. The analytic basis of these theories required, firstly, clarification of concepts (Wolff's rational psychology), and secondly, observation of the soul's activity (his empirical psychology). Analysis of perception pursued knowledge of relations in the world external to the soul, analysis of feeling pursued knowledge of transformative relations within the soul. All analysis, however, presupposed a rational demonstration of the prior unity of the soul; this unity, it was assumed, is the precondition of mankind's aspirations to higher culture.

Tetens discussed the soul's activity in perception as the imposition of forms, for example, the form of causal relationship, on what is perceived. These forms, he argued, make thought possible. This philosophical argument was developed rigorously by Kant, and it is in relation to Kant rather than relatively obscure writers like Tetens (whom Kant nevertheless studied) that these questions were debated subsequently. Kant's contribution was wide and profound. He drew distinctions between different kinds of questions and he did this with such acuity that it made earlier approaches, of the kind discussed in this chapter, appear medieval rather than modern in character. The most basic of Kant's distinctions is that between things as they are in themselves and things as they are as objects of our knowledge. Kant extended the critique of theories of knowledge influenced by Locke and argued that knowledge is necessarily structured by the framework of human reasoning, the framework Tetens called forms. It followed, Kant believed, that absolute knowledge of the kind that Spinoza had sought, is logically impossible. In the *Kritik der reinen Vernunft* (*Critique of Pure Reason*, 1781), Kant undertook to demonstrate that what he termed the forms of intuition, time and space, which constitute the framework of human reason, nevertheless provide an objective foundation for knowledge. Kant denied that either reason alone or experience alone can create true knowledge, and he therefore analysed the relationship between reason and experience to show how knowledge, i.e., science, is achievable. Kant's own discussion of the self presupposed that the unity of awareness, or the unity of consciousness, implies the existence of a soul or, in Kant's terms, a transcendental ego. In line with his theory of knowledge, he denied that we know this ego in itself; rather, we know it as a logical precondition for 'apperception', the word Kant took from Leibniz and used to describe the active process that turns sensory appearances into a unified awareness. It is the business of philosophy, not psychology, to determine how the transcendental ego structures the mind's knowledge of the world or the self.

Kant discussed the forms of intuition as axioms of reason, not as psychological structures. He went beyond earlier philosophers in the way he developed a distinction between what is properly said about reason and what is properly said about mind. He described the forms of intuition as transcendental, universal and necessary, not personal and psychological. A century later, this distinction, especially as expressed by the German logician Gottlob Frege (1848–1925), imposed a division of labour between philosophy and psychology. Twentieth-century readers familiar with this distinction may have judged much of the present chapter as belonging to the history of philosophy rather than the history of the human sciences. The historical fact is, however, that before Kant philosophy and psychology were not separate types of understanding or socially distinct occupations. Wolff, for example, discussed both rational and empirical psychology as part of a larger enterprise; when he distinguished rational and empirical projects, he drew something only tenuously like the modern distinction between philosophy and psychology. The distinction was rarely held to even in the century after Kant by either English-language or German-language writers. Study of how the mind works remained closely bound in the eighteenth and nineteenth centuries to study of reasoning itself.

When Kant analysed the terms in which we reason, he did more than construct a technical philosophical argument. He made reasoning constitutive of the world. The idea of space, like the idea of time, he argued, 'underlies all our outer intuitions . . . It must therefore be regarded as the condition of the possibility of appearances, and not as a determination dependent upon them.'[16] Reflective reason, shared by all men but made concrete in the reasoning of one individual, is the creator of the knowable world and not just its measure. In the generation after Kant, the German generation of *Naturphilosophie* (literally but poorly translated as nature philosophy) and of high Romanticism in philosophy and poetry, this orientation fostered an unprecedented emphasis on the self. Writers and poets sought in their individual selves the truth of knowledge and feeling, as they believed that their own knowledge and feeling transcended any one self and constituted a revelation of what is universal for man. The history of Kant's theory of knowledge is therefore a history of fascination with the self in the groundwork of knowledge about what it is to be human.

Kant also provided a rigorous treatment of what it is for knowledge to achieve the status of science. He contrasted the knowledge of common sense with the knowledge of science, a contrast between knowledge of concrete instances (e.g., the knowledge that leads a mother to feed her crying child) with knowledge of abstract generalities grounded by objective reason (to continue the example, knowledge of physiology). His critical investigation of the basis of reason attempted to lay bare the conditions in which science – formalized abstract knowledge – is possible. These investigations convinced him that not all knowledge can aspire to scientific status. His model science was Newtonian mechanics. But knowledge about mind, he argued, cannot have the character of mechanics; there cannot be a 'science' of psychology because what we observe in our minds does not exist as objects knowable in terms of the forms of intuition of space and time. 'For these inner experiences differ from *outer* experience of objects *in space*, where objects appear juxtaposed and *abiding*. Inner sense sees the relations of its modifications only in time, and so in flux, where the stability of observation that is necessary for experience is lacking.'[17] When psychologists argued that there could be a science of mental content in the late nineteenth century, they did so on the basis of the claim, which opposed Kant's conclusion, that this content can become a fixed object of observation. Though Kant denied the possibility of a science of psychology, he was as interested in psychological knowledge as any of his contemporaries. This interest was primarily practical and moral, and it was through this route that Kant helped establish a subject area for psychology.

As professor of philosophy in Königsberg, the administrative and cultural centre of East Prussia, Kant was a servant of the Prussian state and a leader of the local community's moral, cultural and educational life, not at all the utterly abstracted figure that legend portrays. For most of the last thirty years of the eighteenth century, he gave an annual course of public lectures on 'anthropology from a pragmatic point of view'. These lectures outlined his views about human beings as moral agents whose purpose in daily life is to fulfil the design for which they have been created. He lectured on mundane questions about cognition, the feeling of pleasure and pain, 'the appetitive power' (the passions and the pursuit of what is good) and the character – personal and collective – of the sexes, states and races. He described this field as pragmatic anthropology ('the knowledge of man') to indicate that he was concerned with conduct and experience in their practical orientation towards the world not with a systematic study of the soul's capacities and not with physical qualities. As he wrote: 'Physiological knowledge of man investigates what *nature* makes of him: pragmatic, what *man* as a free agent makes, or can and should make, of himself.'[18] He excluded from his lectures questions about language, aesthetics and other ways in which the soul creates culture, but included generalizations from facts observable in common experience, familiar examples and even the odd ponderous joke. The audience had the benefit of Kant's reflections on what makes for orderly personal life and an enlightened view of social issues like education and mental illness. The lectures helped form a culture familiar with psychological topics even while, as he argued, psychology could not be a science.

Kant distinguished pragmatic anthropology from moral philosophy as well as from science. Anthropology, in his view, is a practical aid to the moralist, while moral philosophy is the formal study of how we know what it is right to do. He articulated in moral philosophy what became a reference point in modern ethics, the claim that there is a moral imperative to do what is good. By contrast, anthropology, in Kant's usage, studies what people do as free and moral agents and the way in which cognition, the passions and character affect action. Man 'has a character which he himself creates, insofar as he is capable of perfecting himself according to the ends that he himself adopts'. Pragmatic anthropology is a practical and moral cultivation of our nature within civil society: 'it gives an exhaustive account of the headings under which we can bring the practical human qualities we observe, and each heading provides an occasion and invitation for the reader to add his own remarks on the subject . . .'[19]

Kant's lectures ranged from philosophical comments on the inability of the inner sense to provide knowledge of the soul in itself to notes on etiquette at dinner parties ('anything that promotes sociability . . . is a garment that dresses virtue to advantage'). Commonplace observations abounded: 'Men are, one and all, actors – the more so the more civilized they are.' He was also fond of regional stereotypes: 'Herr Nicolai [a stock representative of a North German] speaks of the disagreeable *sanctimonious* faces in Bavaria, while John Bull of old England carries even on his face freedom to be rude wherever he may go . . .'[20] Here Kant's comments merged with the language of the theatre, fascination with physiognomy and with regional differences and the small change of sociability between the sexes. The lectures linked formal study of the soul and everyday life. Many of his contemporaries also sought to develop this link and thereby to create an empirical psychology that would take its place in human affairs. Two new journals, the *Zeitschrift für empirische Psychologie* (*Journal for Empirical Psychology*) and the *Magazin für Erfahrungsseelenkunde* (*Magazine for Empirical Knowledge of the Soul*) were founded in 1783 to do just this. One of the editors of the latter, Karl Philipp Moritz (1757–93), was a friend of Goethe, and empirical psychology was yet another of Goethe's interests. Empirical psychology had close links with medicine and the study of physiology. It also took up educational questions, and education was an important part of the social context in which the discussion of distinctively psychological topics became current. Kant and other philosophers, notably his successor at Königsberg, Herbart, brought together the philosophy of the soul with the interest of the Prussian state in education. State interests, pragmatic anthropology and psychology met in the education of the young, with a significant effect on psychology's differentiation as a separate subject.

Pragmatic anthropology included physiognomy, which Kant defined as 'the art of judging what lies within a man, whether in terms of his way of sensing or of his way of thinking, from his visible form and so from his exterior'.[21] This ancient art of reading character from the face enjoyed a considerable vogue in the second half of the eighteenth century, especially in the form in which it was presented by a pastor of the Zurich Reformed Church, Johann Caspar Lavater (1741–1801). Lavater also helped spread the popularity of the silhouette (invented by Etienne de Silhouette), first fashionable in aristocratic circles in the 1760s, considering it a technique of representation that extracts the essential meaning from a face. This essential meaning is clearest of all in death, which 'stops and fixes what was before vague and undecided . . . [and] all the features return to their true relation'.[22] Physiognomic beliefs spread through the nineteenth century, by then often muddled in people's minds with the new subject of phrenology – the study of character based on the bumps on the head. The interest never really died out, and it took modern forms in the popular culture of 'body language' and personality quizzes. Modern Western society placed importance on the public presentation of self; correlatively, there were complex sign systems that made it possible to 'read off' what any particular self is like. This 'reading' was done self-reflectively as well as for others. Physiognomy was therefore a major contributor to modern psychological society. Modern habits of self-presentation owe much to Lavater's generation; it is here, in part, that we find roots of modern sensibilities about the self, especially about the self's feelings, as the source of human values and authenticity.

Lavater contributed to the flowering of individual hopes and feelings of the *Aufklärung*. His religious poetry looked to true human feeling for contact with divinity. In association with friends like Mendelssohn and Goethe, he conceived of each person's soul in terms of an individual character or 'genius', and so attributed intrinsic value to subjective feelings. Poetry and literature, like Goethe's novel *Die Wahlverwandtschaften* (*Elective Affinities*, 1809) – which used chemical analogy to portray the irreducible attractions of the feelings – created a language that gave individualized feeling exquisite expression. Lavater extended this outlook in his physiognomic books, which were beautifully illustrated, to show how the outward face reveals the inner genius or individual character of the soul. His *Physiognomische Fragmente zur Beförderung der Menschenkenntniß und Menschenliebe* (*Physiognomic Fragments for Furthering the Knowledge and Love of Man*, 1775–8), a four-volume study on which Herder and Goethe co-operated, was a sensation. The books included engraved plates (some, in an English edition, by William Blake) of famous contemporaries, whose faces provide ideal models of the human spirit. This work, and many imitations, had an extensive following which cultivated psychological language of the feelings and individual character in daily life and, in a circle of reflections, in the portrait and the novel. For all its idealization and stress on the unique creative aspects of outstanding genius, this language also encompassed everyday relations in a self-consciously cultured middle-class world of domestic sensibilities and friendship. Herder claimed that physiognomy is 'the expositor of the living *nature* of a man, the interpreter as it were of his genius rendered visible'.[23] Physiognomy was also a practical and empirical study, an activity that enabled ordinary people to learn to be psychologists. Lavater's work continued to be republished in English versions for a century.

Physiognomy was a language of domestic life and restated the differences between the sexes. Lavater drew up a series of dichotomies, all of which could be captured in portraiture as well as in words:

> Man is more solid; woman is softer.
> Man is straighter; woman is more supple.
> Man walks with a firm step; woman with a soft and light one.
> Man contemplates and observes; woman looks and feels.[24]

Lavater gave his readers a way to present themselves to each other, and to pursue reflection about their own identity, in a manner that supplied both 'a nature' and dignity. Physiognomy, of course, also naturalized social conventions – it built into the body social values about men and women. In Lavater's writings this pursuit of the self was also bound up with fervent religious feeling and a sentimental belief that God satisfies human needs. 'God and Nature are not botchers: as the eye is, so the ear; as the forehead, so each hair of the head.' Thus, Lavater described the great Goethe's profile: 'How gentle, how utterly without awkwardness, constraint, tension, or flabbiness! How effortlessly and harmoniously the contour of the profile curves from the top of the forehead down to the collar.'[25] When Lavater travelled and preached in Germany and Switzerland as a celebrity and surrounded by people who hoped for such insight, he created a sensibility in which spirit was experienced in human feelings in both their inner form and their outward expression. Romantic writers of the next generation came close to the description of this spirit as Nature herself.

Physiognomy was a language that domesticated the soul. It was a long way from the scholastic fervour of Spinoza to the bourgeois virtues of late eighteenth-century Zurich or Weimar. Nevertheless, belief in the soul's activity – whether in reason or in expressive character – remained a central value. It is, German-language writers stressed, the unity and activity of the soul that makes it possible to aspire to what is true, what is good and what is beautiful. The philosophical culture, with its medieval roots, fostered a language about the soul that made it increasingly possible to describe the Christian immaterial principle in terms of faculties of mind. This stress on mind as activity had long-term consequences for the study of perception and other cognitive processes. It sustained belief in the human spirit as a transformative power, and it rejected the view that this spirit derives from experience and the conditions of physical nature. Descriptions of the soul's activity linked formal philosophy with an everyday interest in individual mental capacities, feeling, morality and strength of will. Concern with education and civil society generally made these discussions relevant to practical affairs. In these complex ways, reference to a subject area called empirical psychology became common in the German late Enlightenment. Whether that subject could become a science was not clear. Meanwhile, however, what historians call the Enlightenment had taken a much more dramatic form elsewhere, and this drama included claims on a larger scale to found the science of man.

7

Human Nature: Natural and Moral Philosophy

> It is not enough for a wise man to study nature and truth; he should dare state truth for the benefit of the few who are willing and able to think. As for the rest, who are voluntarily the slaves of prejudice, they can no more attain truth, than frogs can fly.
>
> Julien Offray de La Mettrie, *L'homme machine* (*Man a Machine*, 1747)[1]

i *Human Nature and Physical Nature*

The period in cultural history known as the Enlightenment had its most developed form in Britain and France. In the east and south, in Poland, Hungary, Russia, Italy or Spain, enlightened learning appeared only in small centres where it was introduced by personal patronage. In the west, in Boston and Philadelphia, there was close attention to what went on in western Europe. The German-language Enlightenment came later and had a different character from that of France, Scotland and England; there was no one cultural centre, with the partial exception of Vienna, with the size and power of Paris or London. German learning was narrower in scope, and when this changed in the generation of Mendelssohn and Kant it continued to be preoccupied with the human spirit and with the principles of rational science. There was no wave of enthusiasm for Locke, no equation of science with natural philosophy, no confidence in the explanation of human affairs by material conditions. Instead, there were the seeds for a powerful critique of these elements of the French- and English-language Enlightenments and for an alternative vision of how to establish a true science of man.

This chapter describes enlightened thought, especially in France and Scotland, and the impact of natural philosophy. Its central theme is what eighteenth-century writers themselves repeatedly called human nature with a rich play on the connotations of the word 'nature'. The language of human nature cut across the real enough differences of politics and culture and concentrated attention on the study of men and women and their social worlds. Many contributors were indifferent to whether or not such study formally created science, but they did pursue knowledge. We will consider the philosophy of men and women as individual beings rather than theories of the social world. The following chapter then stresses the sociable dimension of human nature.

The term 'human nature' (*nature humaine, Natur des Menschen*) was commonplace. Samuel Johnson claimed that 'human nature became the fashionable study' in the late seventeenth century.[2] In the 1720s Joseph Butler (1692–1752), later bishop of Durham, gave sermons 'upon human nature',[3] and in 1739 David Hume (1711–76) published the first two books of what was later considered a classic in philosophy, entitled *A Treatise of Human Nature*. To quote references to human nature in the eighteenth century is a bit like quoting references to God in the bible: it is the subject around which everything else revolves. Above all, the language of human nature satisfied moral purposes; it described the basis on which people claimed we can know what to do, and it thereby elevated nature as the ethical norm. This is the core of what is now understood by the Enlightenment: belief that knowledge will replace revelation as the way to achieve goodness. The French abbé de Mably thus stated, 'Let us study man as he is, in order to teach him to become what he should be'.[4] The power and range of this argument will be apparent again and again in later chapters.

There is a significant ambiguity in references to human *nature*. To refer to a thing's nature denotes its essence or qualities that make it what it is. This was the ancient and medieval usage. In medieval Christian culture the nature of women and men is their God-given immortal soul and their sin inherited from

Adam, and the forms of the soul animate women and men and integrate them with the physical world. In the eighteenth century the word 'nature' also denoted physical reality external to civil society. This double usage provided terms for debate about how far men and women are part of nature, that is, whether their nature derives from Nature, to use a capital 'N' for the physical reality in the manner of contemporary authors. This debate was critical for religious, moral, social and aesthetic opinion. For example, to take a case already present in Hobbes's and Locke's work, reference to a state of nature in which men and women are conceived to exist outside or before civil society, was a commonplace device in the literature on political obligation and a powerful medium for social criticism. The eighteenth century shaped these debates about the relationship of nature and culture as key issues for the modern age, but it did not resolve them. The debates, informed by Darwin's evolutionary theory and scientific Marxism, vigorously revived in the late nineteenth century and were sustained through the twentieth century.

Though historians customarily refer to the eighteenth century as the Age of Enlightenment, neither the age nor enlightenment were monolithic entities. Alongside the new views of human nature affected by natural philosophy, which I emphasize here, Catholic and Protestant conceptions of the immortal soul remained influential. Jonathan Edwards (1703–58), for example, a New England pastor and theologian at Yale College, developed an intellectual enquiry into the mind out of a personal search for salvation. For him, the significant question about man is how it is possible to know whether the fallen soul has been altered by the grace of God, and his answer was that such knowledge is always fallible. These anxious questions, directed to his own problems and to those of his parishioners and students, were a far cry from the enlightened optimism expressed by his younger compatriot Benjamin Franklin (1706–90), who spent much of his time in advanced intellectual circles in Paris.

In the eighteenth century there was a decisive shift of opinion towards the acceptance of nature, that is, the material world, as the basic source of human nature, that is, of man's essence. This was more than a pun on the word 'nature' since to merge the two meanings of the word made the new natural philosophy the source of authoritative ideas about human nature. This was not in any simple sense the advance of science separate from religion, as natural philosophy was itself imbued with religious values and theological concepts. The point is that writers turned to knowledge formulated about physical nature, such as Newton's ideas on the aether, to think about mind and human nature. Physical knowledge appeared to have truth and authority, and it also appeared to be relevant to everyday material matters like health, feeling, conduct and sex. Many authors were at pains to show that natural philosophy applied to man is compatible with religion. A few, however, like the French doctor Julien Offray de La Mettrie (1709–51), became notorious materialists, and they provoked outrage with their treatment of man as a physical thing. If there was a shift towards a modern, secular and materialist view of being human, it may have been helped by the ambiguity of the term 'human nature' since the term itself conflated the physical world and man's essence. All judgements about the secular content of Enlightenment thought are affected by disagreement about whether even late twentieth-century Western culture can properly be called secular. All the same, both supporters and critics of enlightened opinion over two centuries frequently judged the eighteenth century to be crucial to the formation of secular values. These secular values and the enterprise known as the human sciences are closely linked.

There was a further sense, related to the concept of natural law, in which the search to understand human nature was at the foundations of what later became the human sciences. If in the seventeenth century natural philosophers borrowed notions of law in human affairs and applied them to the study of physical nature, in the eighteenth century it was the turn of the laws of physical nature to suggest ways forward for knowledge about human life. It was possible to know human nature systematically, writers thought, precisely because it conforms to laws. The search to uncover these laws, such as Hume undertook in his *Treatise*, was the characteristic pursuit of Enlightenment human science. Whether and in what sense belief that there are such laws implies determinism, materialism or atheism, or merely detracts from religious authority, was debated at great length and with much heat. There was no inevitability that the argument would develop one way rather than another, but most historians agree that the search for the laws of human nature at the very least diminished the prominence given to the spiritual dimension in human affairs.

The search for the laws of human nature did not result in a debate about nature versus nurture, as in the twentieth century. Authors did not sharply separate inheritable and environmental causes in the life of animate beings – animals, plants and people; rather, they referred to physical and moral causes. Physical causes included the body, sexual differences and diet, but they merged with moral causes – obviously in the case of diet. Moral causes are the causes men and women create for themselves; in Hume's words: 'By moral causes, I mean all circumstances, which are fitted to work on the mind as motives or reasons, and which render a peculiar set of manners habitual to us.'[5] To trace moral causes to history and to explain both individual conduct and social customs by reference to man's response to the physical circumstances in which he finds himself became a tremendous intellectual project.

Eighteenth-century writers, artists and philosophers were also fascinated by the body. This was in a sense not new – even denigration of bodily life is only the converse of fascination. Christian confession had been preoccupied with the flesh, and humoral medicine had treated illness as the disturbed balance of the body as much as the soul. There was something novel in the eighteenth century, however, when natural philosophy discussed the material conditions of human life with intellectual range, precision and social authority. The new natural philosophy, or Newtonian philosophy, was valued both for its content and for what was called its experimental method, a method often indistinguishable from observation. Some students of man turned to Newtonian philosophy and applied it to man's body as the way to make progress. They hoped that by the study of the body, women and men would become objects of knowledge as well as, or even rather than, subjects of faith.

One sign that the body came into its own in accounts of human nature was a new stress on sensibility. Writers in the Augustan Age, the reign of Queen Anne, like Joseph Addison, admired a quickness of feeling and a refined sensitivity, and they used a word to describe this – 'sensibility'– that combined faculties of mind such as imagination with qualities of the senses such as delicacy. Sensibility was valued in a person's character because it displays the whole person's response to both mental and physical events. Addison described modesty as 'a kind of quick and delicate Feeling in the Soul . . . an exquisite Sensibility, as warns a woman to shun the first Appearance of every thing which is hurtful'.[6] Over-sensitivity began to be labelled 'nerves' or 'a nervous disposition'; here again the choice of language stressed unity of body and mind. Feeling embedded the soul in the body. A literature of feeling flourished – witness the moral palpitations of Samuel Richardson's heroine in his novel *Clarissa Harlowe* (1747–8). At the same time, medicine was home to a new concern with the properties and disorders of the nervous system. Art and medicine fed each others' imaginations. Johnson was as preoccupied by his nervous condition as by his dictionary. Maidens *blushed*. And in *Tristram Shandy* (1760–7), Laurence Sterne created an ever-more involuted pantomime in which physiology and imagination trod the boards together. 'A Man's body and his mind, with the utmost reverence to both I speak it, are exactly like a jerkin, and a jerkin's lining; – rumple the one – you rumple the other.'[7] Sterne was himself dying from tuberculosis.

The physicians turned sensibility into a physiological property open to investigation by experiment. In Leiden, home to Europe's most famous medical school in the 1720s and 1730s, Jerome Gaub (1705–80) described 'neural man', a phrase that connoted the innate animation in the nerves, separate from the mind's rational faculties, which stirs the body to life and quickens the interaction of mind and body: 'it represents a kind of man within a man'.[8] In Bern, Albrecht von Haller (1708–77), botanist, poet, physiologist, drew a conceptual and experimental distinction between 'irritability' and 'sensibility', terms that described properties of animated substance. The former denoted the responsiveness of substances (such as muscle) to irritation, the latter denoted the special responsiveness, accompanied by conscious feeling, exhibited by the nerves. This language and the experiments it accompanied made it possible to agree that the mind acts in the body through the nervous system, though little beyond this was clear or free from speculation. In Edinburgh, which developed a major medical school in the second half of the eighteenth century, a professor of medicine, Robert Whytt (1714–66), took issue with Haller's principles of physiology. Whytt conducted experiments to show that decapitated frogs or turtles still retain the power to respond purposively to irritation, and he attributed this power to a diffuse soul in the spinal cord. Later researchers discerned here pioneering work on reflex action – though this was not Whytt's language – as the basic unit of nervous function. What is clear, in both Haller's and Whytt's studies, is the attention to the nervous system as the means by which to study the embodiment of the soul.

It was Whytt's younger colleague William Cullen (1710–90) who coined the term 'neurosis' and gave it medical prominence. Cullen had much influence as a teacher of both chemistry and medicine at Edinburgh, partly because he was an obsessive systemizer in the mould of his great contemporary, the Swedish botanist Carolus Linnaeus (Carl von Linné, 1707–78). After he commented that, 'in a certain view, almost the whole of the diseases of the human body might be called *nervous*', Cullen tried to differentiate 'all those preternatural affections of sense and motion' that are directly produced by nervous disorder.[9] These are the neuroses. It was only in the late nineteenth century that the word acquired its modern denotation of psychological disturbance. In the eighteenth century, words like 'neurosis' and 'nerves' did not sharply differentiate mental and physical symptoms or causes. Nevertheless, as Cullen's descriptions illustrate, doctors led the way in the correlation of sensations, feelings and the mind with the nerves and brain. The public learned to describe anxiety, depression or anger as 'nerves'; thought about character and temperament moved away from the language of the humours and towards a language in which mind depends on the well-being of a particular bodily structure. Extreme disorder, madness, was reinterpreted in the same vein, and medical opinion held madness to be a failure of the bodily organ of the mind. Physicians believed that madness was now firmly within the sphere of medical competence and once and for all free from any lingering associations with the unnatural.

Eating, drinking, excreting and sex – celebrated in Restoration drama – reminded everyone of human nature's embodied existence. That the soul is dependent, even unconsciously dependent, on the body, and that the soul's action has bodily effects, no one doubted. Whytt noted, 'As the erection of the *penis* often proceeds from lascivious thoughts, it must be ascribed . . . to the mind, notwithstanding our being equally unconscious of her influence exerted here, as in producing the contraction of the heart.'[10] Whytt was a Presbyterian and respectable society physician. Others, however, took the mind's dependence on the body to be a guide to life and became libertines, critics of what they saw as civilized hypocrisy about sex, and some became supporters of a materialist philosophy of human nature. Materialist values brought the study of human beings fully into the realm of natural philosophy, and those who upheld these values claimed to disclose men and women in their true identity as part of the natural world. The intellectually most radical French *philosophes* scandalized religious opinion and offended good taste when they put this revelation on paper. The marquis de Sade took it to a limit beyond which lay nothing.

Yet, when all is said and done, the eighteenth century remained overwhelmingly a Christian age. This is true in the obvious sense that the great majority of the population, educated and uneducated – though fears were expressed about a godless urban mob – maintained Christian observance and some form of belief. It was also true in the less obvious sense that, even when traditional aspects of belief were questioned, deep-lying assumptions about the created unity of the world and of the purposiveness of human history remained strong. Some modern historians, indeed, have interpreted the hopes invested in reason and progress as a secular flowering of religious values, an attempt to recreate the City of God on earth. Also in the eighteenth century, admiration for Classical Greece and Rome, an enthusiasm for harmony, reason, justice and grace, was accompanied by fascination with pagan myth and religion, the symbols of which dominated the decorative arts. The spiritual life of the ancient world was interwoven in art, architecture and belles-lettres with the Christian life. For most people, even for most of the *philosophes*, the study of man and the study of the world was the study of moral truth, and pagan and Christian language and symbols were used to give truth a voice that transcends history.

ii *L'homme machine*

To claim that the science of man is the science of matter was a bold step, and perhaps it required a doctor, faced in his day-to-day work by the infinitely disruptive body, to take it. The former French army surgeon La Mettrie attracted great notoriety when he published *L'homme machine* (*Man a Machine*) in 1747, and the book has retained emblematic status ever since. He called both for the emancipation of those who work on the body from those who work on mind and for the ennoblement of useful knowledge. As he argued, 'experience and observation . . . are to be found throughout the records of the physicians who were philosophers, and not in the works of the philosophers who were not physicians'. He published in Leiden to avoid censorship in monarchical and Catholic France; subsequently, only the court of the ever autocratic Prussian Frederick the Great offered him a home. He died shortly thereafter, and the whispers of his many enemies claimed that he had succumbed to a surfeit of pâté de foie gras, God's sweet revenge on a materialist.

It must be said that La Mettrie had gone out of his way to provoke such Christian venom: 'Besides it does not matter for our peace of mind, whether matter be eternal or have been created, whether there be or be not a God. How foolish to torment ourselves so much about things which we can not know, and which would not make us any happier even were we to gain knowledge about them!' He belittled the idea that it makes any difference to the way we act, though it might make us happy, socially useful and liberated, if we accept that thought is simply a property of matter, 'on a par with electricity'.[11] La Mettrie's *jeu d'esprit* was reviled by Christian moralists and celebrated by modern materialists. In fact, he fought his own battles on behalf of the surgeons against the physicians of his day, and his essay had little to say specifically about the machine metaphor. On a larger canvas, his work was only a particularly expressive moment in a debate that went back at least to Hobbes in the seventeenth century and continued into modern neuroscience and the philosophy of artificial intelligence. It is necessary to step back to look at this larger picture.

Descartes categorically rejected the existence of the animal soul. Whether he was right became a touchstone for his metaphysics in general and his account of the soul–body relation in particular. The argument about whether animals have souls gave everyone a point of entry into the latest natural philosophy. Descartes himself wrote to More that the denial of a soul to animals is not cruel to them but kind to mankind, since it liberates us from guilt about eating meat. Whether philosophical positions were actually declared by a willingness or an unwillingness to kick dogs, based on belief about whether or not they experience pain, is doubtful. To kick or not to kick was a matter of proper feeling and conduct rather than metaphysics. By the early eighteenth century, English society at least had begun to show a sensibility about animals that gained it a lasting reputation for sensitivity or sentimentality – depending on the point of view. The intellectual stakes concerned both physiological principles, whether a sensitive soul is necessary for the purposiveness of plants and animals, and the doctrinal theology of immortality. Descartes' solution, which reserved immortality for humanity and declared animals to be machines, satisfied few people because the higher animals at least gave every sign of sensitivity.

La Mettrie knew exactly what he was doing when he gave his book the title *Man a Machine*; he was an experienced parodist of hypocritical virtues. As there had been a century of debate about the animal machine, everyone knew that he was signalling determinism, materialism and atheism. To enlist Descartes for his cause, and to deprive Catholics of an intellectual authority, La Mettrie invented a story that Descartes' fear of religious persecution led him to dissimulate his true thoughts on the mechanical nature of the soul. 'For after all, although he extols the distinctness of the two substances [mind and body], this is plainly but a trick of skill, a ruse of style, to make theologians swallow a poison, hidden in the shade of an analogy which strikes everybody else and which they alone fail to notice.' This was a fabrication, but it brought into the open what was, from a Christian viewpoint, the double-edged character of the Cartesian legacy. Dualism sharply delineated the immortal human soul, but its many difficulties encouraged speculation about continuity between soul and body and the extension of mechanistic explanation to humans. This was La Mettrie's line of argument, and he claimed that the equation of human nature with physical nature was implicit all along in the new natural philosophy. 'How can human nature be known, if we may not derive any light from an exact comparison of the structure of man and of animals?' This was a profound question. For many later natural scientists, La Mettrie was a hero, a man who saw rightly that only prejudice held up the advance of physical science, through the comparison of man and beast, from the explanation of human nature. These scientists echoed La Mettrie's sentiments: 'The soul is therefore but an empty word, of which no one has any idea, and which an enlightened man should use only to signify the part in us that thinks.'[12] Nevertheless, he was not a mechanist in the manner of later scientists.

La Mettrie observed that 'one is sometimes inclined to say that the soul is situated in the stomach', but his materialism was a little more subtle than this implies.[13] He contributed to a complex and long-running debate about the properties of matter. Theologians with few exceptions, of whom the chemist Joseph Priestley (1733–1804) is the most distinguished, thought it of paramount importance to defend rational knowledge of the existence of immaterial substances. Philosophical claims about the respective powers of spiritual and material substances featured in debates between religious and political factions as well as in attacks on supposed unbelievers. In Britain, Locke's thought experiment, that the limitation of our understanding means that we cannot exclude the possibility that organized matter might think, provoked a hail of pamphlets for and against. At stake was the post-1689 political and religious settlement, the reliability of knowledge of the Trinity and the unsettling claims from Deists that revelation is not a source of knowledge. One contribution, from John Broughton, defended knowledge of immateriality under the title, *Psychologia; Or, An Account of the Nature of the Rational Soul* (1703), an early use of the word 'psychology' in English.

Parallel debates took place elsewhere in Europe and North America. The growth of an experimental natural philosophy of animals, and struggles over the control of medical education and practice, supported physiological speculation. There was no agreement about how much of this speculation properly belonged to the sphere of Newtonian mechanics and how much to the theology of immortality and the Trinity. No wonder, in these circumstances, that Voltaire praised Locke's book, which he read as a natural history rather than a philosophical romance about the soul. Voltaire's remark had a polemical edge in France where priests exercised secular power and dominated educational institutions. Claims about the properties of material and immaterial substances signified where one stood in relation to political and religious as well as natural-philosophical agendas. It was Voltaire's early hope that France could be reformed from within, if it would follow the light of reason that shone from the work of Locke and Newton.

La Mettrie utilized the idea that natural forces animate matter and therefore argued that matter is not passive and inert but has its own organizing powers. He attributed to matter the causal powers that many Christian writers attributed indirectly to God; but he did not exactly think of bodies as machines, in spite of his book's title. He was not alone in his views; other physicians and naturalists reached similar conclusions as they marvelled at the organized powers of plants and animals. The Genevan naturalist Abraham Trembley (1710–84) provided the fashionable world with a striking case. He showed that the freshwater polyp *Hydra* both produces buds that turn into new polyps and spontaneously replaces damaged parts. As La Mettrie observed, 'does it not contain in itself the causes which bring about regeneration?' He therefore suggested research to examine the properties of organized matter in different parts of the animal or human body. This was a programme comparable to Haller's study of irritable and sensible substances, though Haller was not a materialist. La Mettrie's notoriety stemmed from the fact that he brashly stated what many feared was implicit in the new science of nature, that human nature too is material: 'Man is not moulded from a costlier clay; nature has used but one dough, and has merely varied the leaven.'[14]

The construction of actual machines, automata, to imitate a man was a refined way in which art copied nature and enhanced imagination that man too is a machine. The most famous of all these machines was the flute player constructed by Jacques de Vaucanson (1709–82) in 1737. This was a life-sized model of a man seated on a rock, and the whole was placed on a pedestal which contained a bellows to supply the air which activated the figure. Vaucanson's motive was to understand human movement by imitating it – as well as to display his skill – and the construction required him to study the role of the lips and lungs in the production of sound. Thus, in machinery, man learned to imitate himself.

iii *Feeling and Sex*

La Mettrie's critics accused him of immorality as well as impiety. He did indeed argue that men and women are virtuous only when their bodies work properly, calling into question the significance of moral law as a guide to conduct. He also portrayed the physician as the engineer of social harmony at the expense of the priest as the guarantor of moral order. Commentators suspected that there was a connection between loose thinking and the loose living apparent in urban and court life in Paris and London. Contemporary literature makes it clear that a close link was recognized between subjective feeling and physical sex, even if manners dictated a certain decorum. In Catholic Spain or Protestant Prussia, rigid outward formalities were preserved, but in other settings sex sometimes gained a public prominence that later appalled the Victorians. It is difficult to generalize, but eighteenth-century writers certainly saw sex as a site where mind and body are enmeshed.

It was Diderot who explored with the greatest wit and subtlety what is implied for morals and feeling by the identification of human nature with physical nature. All the elements of what was at issue came together in attitudes about sex (the term 'sexuality' was a nineteenth-century invention). Diderot concluded that we should embrace our sex as an expression of our nature; if human nature is a function of material organization, not of an immortal soul, then bodily life is its own end. As he well knew, such arguments might be held in private, but any attempt to make them public called down the opprobrium otherwise reserved for libertines. In two sparkling essays, *Le neveu de Rameau* (*Rameau's Nephew*) and *Le rêve de D'Alembert* (*D'Alembert's Dream*), Diderot mocked and ran riot with the mores of his day, and he played with feelings of subjective identity as he gave the body its head. Later regarded as classics of modern sensibility, the essays were not published until after Diderot's death; they had some earlier circulation and appeared finally in French in 1821 and 1830.

Diderot's literary games entertained an élite. Lower down the social scale – sometimes very low down indeed – pamphlets, prints, patent medicines and prostitutes all testified to the place of sex in the moral and commercial economy. There was an increased self-consciousness about sexual feeling, and perhaps also an openness in its public expression just as there was an openness about feeling at large. Sex bound the mind and body together in ways apt to make a mockery of philosophers – and common language knew how to mock. It gave physicians, clergymen, pedagogues, traders in leisure and hucksters of every description a role in the description of human nature. The awareness of sexual feeling, in common with sensibility in general, enriched a personal sense of self and a consciousness of difference between men and women, classes and peoples. Sexual feeling, represented in a form with which twentieth-century readers can identify, helped define a psychological sphere of interest, and the content of this sphere was at least as much about women as about men. This happened in a way that tied the mind to the body, linked feeling to life rather than reason and encouraged people in the study of medicine rather than metaphysics. The boundary between what was regarded as public and what private varied greatly, but it is common to see a shift in the nineteenth century away from an earlier freedom of expression. The rise of evangelicalism in the late eighteenth century explains, at least in part, the public sexual repression that gives the Victorian period its unfortunate reputation. But even the social act of repression reveals knowledge of the intimacy of mind and body in sex.

Sexual life created awareness of a contrast between the supposed hypocrisy of polite society and the state of nature. Sex was part of a public debate about the relative contribution of physical and moral causes to man's constitution. This was a central theme for Diderot. In his *Supplément au voyage de Bougainville* (*Supplement to Bougainville's 'Voyage'*, written in 1772, published in 1796) he explored social constraints and moral values in the light of what his critics thought of as a materialist philosophy of nature. Louis-Antoine de Bougainville was the French explorer who prompted Europeans to imagine that the people of Tahiti enjoyed a free and natural sexual life. Appropriately enough, he also gave his name to the riotous flowering creeper *Bougainvillaea*. Diderot imagined a hospitable Tahitian leader, Oron, who is free in his society to acknowledge what he is by nature. Oron speaks to the chaplain who accompanies the socially stunted European visitors: 'What are you ashamed of? Are you doing wrong when you yield to the noblest urge of nature? Man, show yourself openly if you are attractive. Woman, if this man pleases you, then welcome his advances with the same honesty.'[15]

Diderot was well-read in physiology and had thought deeply about Locke's view that knowledge originates with sensation rather than innate reason. Like La Mettrie, he speculated that matter possesses an organizing power and even sentience, and this implied that nothing further is needed, no spiritual principle, to create a man: 'One simple supposition . . . explains everything – to wit, that sentience is a general property of matter or a product of physical structure.' But his position, strictly speaking, was not a materialist one, since he attributed intrinsic organizing powers to matter. His notion of matter was remote from an imagery of inert billiard balls, and closer to the so-called vitalist view of physiological function current at the medical school in Montpellier. It was the moral rather than the material aspects of Diderot's argument that posed issues of lasting significance for self-reflection. He grasped a crucial modern quandary, though he hardly dared pursue it – not just because it was socially impolitic but because it opened an existential abyss from which he retreated. The quandary was implicit in the simple statement: 'Nothing that exists can be either against nature or outside nature.'[16] Diderot made this remark as part of a playful satire on sexual prohibitions, but it lifted the lid on the Pandora's box of demons that the religious critics of the *philosophes* always feared: to deny the immortal soul appears to destroy the foundations of moral order in society as well as the foundations of religion. If taken seriously, Diderot's statement appeared to destroy not only actual moral standards, but the potential for any morality. It was all very well, for example, to turn to the child as mirror of our state of nature, but the mirror reflected a state that is amoral as well as innocent. As Diderot commented, in a passage later cited with joy by Freud: 'If your little savage were left entirely to himself, if his childish ignorance were left intact, if he were allowed to acquire all the violent passions of a grown man while still remaining as deficient in reason as he had been in his cradle, then he would end up strangling his father and going to bed with his mother.'[17] Once human nature is fully equated with physical nature, nothing appears to be left but to embrace what nature teaches. This conclusion was reached very soon after Diderot by Donatien-Alphonse-François de Sade, the marquis de Sade (1740–1814), in a body of work equally infamous, then and later, for its atheism, its immorality and its violence.

The writings of the marquis de Sade are open to many

interpretations. He spent twenty-seven years in prison under five governments, before and after the Revolution of 1789, and yet he read and wrote prodigiously. He particularly admired La Mettrie. I describe the Sade who carried through in his imagination a total identification of human nature with physical nature and thus revealed that the human condition is without any values except those that reflect a selfish force of nature. Sade's notoriety stems from his argument that the honest pursuit of our nature knows no limits except death and that this pursuit, if honest, demands endless and unbounded sexual and physical violence to others. This, he believed, is the law of nature to which humans are subservient. 'The suckling babe that bites his nurse's nipple, the infant constantly smashing his rattle, reveal to us that a bent for destruction, cruelty, and oppression is the first which Nature graves in our hearts . . .'[18] In Sade's imagination, the pursuit of our own nature, our pleasure, depends on causing others to suffer pain and death. The Viennese medico-legal specialist Richard Krafft-Ebing (1840–1902) later introduced the term 'sadism' to describe the enjoyment of sexual violence.

Sade's prose was an obsessive struggle to imagine yet more pain and hence a more intense acceptance of nature. He wrote of virgins whose buttocks are whipped to a pulp before they are split open by anal intercourse, but their flesh is restored so that it can begin all over again. Alternatively, he described ejaculation at the moment of murder only then to invent an ingenious machine to make possible pleasure at sixteen simultaneous murders. It is as if the only thing left when we have fully identified man with nature is pain and orgasmic pleasure, which he treated as much the same thing, and both have continually to be found anew. Perhaps he grasped at the immediacy and all-consuming quality of pain and the orgasm as the only value left once moral injunctions are discarded as dishonest. But even pain and pleasure were denied to Sade, as he himself acknowledged, since it was only in his writing that they could be continuously renewed. Nature, it seemed to Sade, seeks death, and the moment of death offers the fantasy of a moment of pleasure; but beyond this moment and beyond death there is and can be nothing.

His work described an extremity in relation to which it is necessary to decide how to live. Sade was a student of medicine and the materialist philosophy of nature which – as he showed – recreated the image of what it is to be human. Long sections of his works were disquisitions on 'the designs of Nature, of whom we are the involuntary instruments', sections where he developed the logic of materialism.

> Ah, you pedants, hangmen, turnkeys, lawmakers, you shavepate rabble, what will you do when we have arrived . . . [at knowledge of the human constitution]? what is to become of your laws, your ethics, your religion, your gallows, your Gods and your Heaven and your Hell when it shall be proven that such a flow of liquids, this variety of fibers, that degree of pungency in the blood or in the animal spirits are sufficient to make a man the object of your givings and your takings away?

The search for knowledge of human nature, which had its roots in the devout natural philosophy and Christian natural law arguments of the seventeenth century, led in the eighteenth century to La Mettrie's denial of immateriality, to Diderot's advocacy of sexual liberation and to Sade's spectre of natural violence. Clearly, the search for a science of man was not a neutral undertaking. To adopt the new natural philosophy as the basis for scientific knowledge of human beings was to form a new image of men and women. Knowledge and image created a circle of mutual reflections. This was not the truth that enlightened philosophers wanted or expected; rather, they desired knowledge of nature as an objective and authoritative guide to conduct. This was the case even for Sade, though what he found is that 'egoism is Nature's fundamental commandment'. How, he argued, 'are you going to persuade me that a virtue in conflict or in contradiction with the passions is to be found in Nature?'[19] It was a profound and troubling question for the sciences of man.

iv *The French Analysis of Sensation*

Diderot spoke for worldly values when he exclaimed, 'I am a man, and I must have causes particular to man.'[20] He believed these 'causes' are to be found in what human nature makes of feeling and experience. Sade reduced feeling and nature to egoism. Diderot, by contrast, though he sensed that his argument implicitly made values relative, retained a warm imagination in his personal relations and on paper. The French *philosophes*, the philosophers of Enlightenment headed by the caustic Voltaire, desired to free humanity – or as much of humanity as they thought educable, not the mob – from the shackles imposed by priests and absolutist governments. When they turned to human nature for knowledge about how to order human affairs, however, they left an ambivalent legacy for their heirs. There was liberation, of the kind embedded in the grand ideals of the United States Constitution. After military victory was secure, George Washington claimed in 1783 in a letter sent to the governors of the American states: 'The foundation of our Empire was not laid in the gloomy age of Ignorance and Superstition, but at an Epocha when the rights of mankind were better understood and more clearly defined, than at any former period.'[21] No earlier age, he stated, had carried forward so successfully the search for social well-being based on knowledge of the human mind. Yet enslavement also followed: enlightened knowledge was thought to reveal difference as well as identity in human nature and, from the late eighteenth century, what were claimed to be natural differences – of race, sex or class – were increasingly used to legitimate oppression. Even the great statesman and scholar Thomas Jefferson, who drafted much of the Constitution, considered it necessary in the circumstances in which he lived to own slaves.

Ethical and political themes in the human sciences continued in the twentieth century to be debated in relation to the Enlightenment legacy. It is in this context that 'the Enlightenment' was discussed as if it had concrete form. Much of the legacy derived from the way analysis of the mind in the eighteenth century was used to contribute to social criticism. French writers, notably

Etienne Bonnot, abbé de Condillac (1714–80), exploited the potential of Locke's work for a thoroughgoing treatment of human nature in terms of the way the senses mediate between mind, body and the world. Though the *Essay Concerning Human Understanding* appeared in French in 1700, and Voltaire popularized Locke as a lever with which to move established authority in the 1730s, it was in the 1740s that the French began creatively to rethink his ideas.

Diderot was again the author of a lively and provocative essay. His *Lettre sur les aveugles, à l'usage de ceux qui voient* (*Letter on the Blind for the Use of Those Who See,* 1749), a title emblematic of the aims of the radical French philosophers, explored the literal and metaphorical consequences of blindness for knowledge. In the last part of his essay he addressed Molyneux's question about the relationship between touch and sight. It was of course not this that led to Diderot's imprisonment after publication but his suggestion that the presence of the idea of God is relative to the sensations people have had. His imprisonment lent poignancy to his imagined reply of a blind man, also threatened by prison, to his interlocutor: '"Ah, monsieur," the blind man answered, "I've been in one for twenty-five years already." What an answer, madame! And what a text . . .' In his text, Diderot argued that ideas and values, like visual perception, are relative to sensory experience: 'Do we ourselves not cease to feel compassion when distance or the smallness of the object produces the same effect on us as lack of sight does on the blind?'[22] His answer to Molyneux's specific question, however, was not quite what one might expect. He argued that a blind person's visual ability to identify a sphere and a cube, once sight is restored, depends on her or his reason. People without education, he suggested, will just guess, while an educated person will compare new and old ideas, obtained through sight and through touch, to reach a logical conclusion. As this example illustrates, philosophers enhanced rather than diminished the role of reason when they placed so much weight on experience.

The detailed analysis of sensation and the five senses was taken furthest by Condillac. He was, after Locke, probably the writer to whom the *philosophes* themselves owed most because of his analysis of language as well as sensation, an analysis that treated words as signs for ideas derived from sensation and hence regarded language as a sign system for the representation of knowledge. Since knowledge derives from sensory experience, Condillac argued, and since words signify the ideas corresponding to the original elements of that experience, the progress of science depends on clarity and order in language. He stated 'that the art of reasoning, reduced to its simplest form can only be a well-formed language'.[23] This conviction was taken up in the late eighteenth century, notably by A.-L. Lavoisier (1743–94) when he carried out a revision of the language and explanatory concepts of chemistry – the chemical revolution. Condillac also influenced German debates about the nature of language, and these debates in turn initiated historical and linguistic work of great importance in the human sciences. Locke's and Condillac's view, that words consist of a conventional – not necessary – system of significations for ideas, reappeared at the end of the nineteenth century as a basic principle of modern linguistics.

Condillac's analysis has been differently understood, in part because he did not adequately publish aspects of his thought. In his early *Essai sur l'origine des connaissances humaines* (*Essay on the Origin of Human Understanding*, 1746), he announced his intention to 'reduce everything that concerns the understanding to a single principle', the connection of ideas (not sensation).[24] On one reading, his notion of the connection of ideas appears comparable to Locke's notion of 'reflection', that is, it appears to refer to a reasoning faculty of the mind, an intrinsic capacity to arrange ideas. If this text is related to Condillac's later work, however, it appears that when he referred to this ability to connect ideas, he denoted an elemental force, a desire, that expresses a need of both body and mind, and not an intrinsic capacity of mind. He located the origin of ideas in sensation, and he argued that the process by which the mind has sensation, the expression of a basic need of life, makes possible attention to ideas. Attention, in his analysis, is the occasion for connections to form between ideas. Condillac's conclusion was therefore that both the content of knowledge and the mental faculties derive from sensation. This argument was not easy to follow. It was easy, by contrast, to follow his imagination in his *Traité des sensations* (*Treatise on the Sensations*, 1754) when he used the device of an imaginary statue that he equipped with one sense at a time in order to analyse what ideas and faculties we owe to which sensations. This analysis made clear the key role that he and others attributed to touch in the acquisition of knowledge of self and non-self and of spatial notions subsequently associated with vision. Part II was entitled 'On touch, or the only sense that judges external objects on its own.'

At first glance, the image of a statue – used by other writers besides Condillac – pictures human nature as cold and passive. It makes a person appear to be the sum of mechanical interactions between sense organs and the physical world. As Condillac's earlier *Essay* made clearer, however, he did not mean to create such a picture of a person. Though he wrote that 'judgment, reflection, desires, passion, and so forth are only sensation itself differently transformed', he supposed that the mind gains the capacity to reflect and then to attend to as well as to receive sensations of different kinds.[25] 'This manner of successively applying our attention of ourselves to different objects, or to the different parts of one object only, is what we call to *reflect*. Thus we sensibly perceive in what manner reflexion arises from imagination and memory.'[26] His statue was also lively; it had a soul. The idea of an animated human being with a soul underlay Condillac's analysis of the operations of the mind. Sensation, he wanted to say, enables the faculties to grow in the soul; this is the activity of childhood. This conclusion distances Condillac's work from its negative reputation with Romantic critics of mental passivity, a reputation passed on to later generations by the French philosopher Victor Cousin. Commentators claimed that Condillac exemplified sensationalism, the attribution of all mental activity to sensations, and that this led to determinism in moral theory. But this falsely constructs a tradition out of a variety of writings, writings which in Condillac's case did derive the mental faculties from sensation, but in a way that makes sense only in terms of his theory of the soul and his views about the expression of the soul's needs.

The aspect of Condillac's analysis that placed him at one end of the spectrum of opinion about the dependency of mind on sensation was his statement that all mental activity is initially provoked by the pleasurable or painful quality of sensations. 'The principle determining the development of [the mind's] faculties is simple; sensations themselves contain it; for since each is necessarily pleasant or unpleasant, the statue is interested in enjoying some and ridding itself of others.' As it responds with movements to pleasures and pains, the statue acquires further sensations, some of which derive from itself, as when one hand is placed on top of the other, and this gives rise to the idea of itself; while other sensations derive from something else, as when a hand is placed on a table, and this gives rise to the idea of the external world. The statue's 'self' – and by implication each human self – derives from sensations: 'Its "I" is only the collection of the sensations it experiences, and of those that its memory recalls.'[27] It was a religious and political provocation to write in this way, and Condillac's theory of the soul tended to be missed. Further, his belief that sensations generate everything we mean by the self provided the intellectual tools for P.-H. Thiry, baron d'Holbach (1723–89) and Claude-Adrien Helvétius (1715–71) to develop thoroughgoing theories of social determinism.

Condillac's work was analytic rather than empirical; similarly, his analysis of language suggested a way to understand thought itself rather than to observe languages. He described how the capacity to use signs, i.e., words, in place of sensations equips the mind with the ability to reason, to manipulate signs not physical reality. 'In order to develop the real cause of the progress of the imagination, contemplation and memory, we must inquire what assistance these operations derive from the use of signs.'[28] Condillac believed that the soul exists before perception; as it gains knowledge with sensations, it has a certain prelinguistic awareness. But, he claimed, it is only with the development of language, which occurs with sensation, that there is thought. This is another reason why the statue was an inappropriate analogy for a person: it could not develop the ability to speak. As Condillac himself observed: 'A language is necessarily required to acquire theoretical knowledge.'[29] When he analysed knowledge in terms of sensations, he implied that humans and animals are closely related; when he discussed language, he distanced them again. This pattern, which connects people with animals through their shared sensation of physical nature and detaches human culture from nature through language, persisted later and is very relevant to the diversity of the human sciences.

Condillac's contemporaries read him with attention, and his approach was apparent in various authors' articles in the monument to the ideal of systematized experience and applied knowledge, the *Encyclopédie* (volumes 1–7, 1751–7; volumes 8–17, 1765; with later supplements) edited by Diderot, with the earlier collaboration of Jean Le Rond D'Alembert (1717–83) on the mathematical sections. The idea of the encyclopaedia, or *Dictionnaire raisonné des sciences, des arts et des métiers* (*Reasoned Dictionary of Sciences, Trades and Crafts*), to give it its subtitle, embodied the value of systematic and practical knowledge, and the eighteenth century was an age for encyclopaedias: the *Encyclopaedia Britannica* was first published in instalments in Edinburgh in 1768–71. The *Encyclopédie,* although dogged by production problems, censors and disputes about content, was probably the greatest publishing enterprise of the century. There were two editions, the first in Paris and the second (1770–76) in Yverdon in the politically freer atmosphere of the canton of Bern. There were differences in the article on '*Psychologie*' in the two editions which indicate the impact of Condillac's analysis of the mind.

In the first, Paris edition, psychology designated a branch of Wolff's *a priori* and deductive system of knowledge and no reference was made to Locke in this context. The anonymous author assumed that his subject matter was what reason reveals about the capacities of immaterial substance. In the Yverdon edition, the article on psychology was by a somewhat unorthodox pastor, Gabriel Mingard. He, too, defined psychology as the 'part of philosophy that teaches all we can know about the human soul' and the 'first known part of pneumatology [or science of spiritual substances]'; but then, alongside references to Wolff, he cited Condillac and his own Genevan compatriot, the philosopher-naturalist Charles Bonnet (1720–93).[30] Bonnet published two volumes, *Essai de psychologie* (*Essay on Psychology*, 1754) and *Essai analytique sur les facultés de l'âme* (*Analytic Essay on the Faculties of the Soul*, 1760), which, like Condillac's work, related mental activity to sensory experience. Bonnet was also the author of an idiosyncratic philosophy of nature which built self-development over time into the forms of animals and plants. His project was a religious one; he opposed irreligion, especially the belief that matter has its own organizing power, with a view of mind and body that reasserted the dependency of all things on God's power. The Calvinism dominant in Geneva became more liberal at this time, and in this atmosphere Bonnet argued for the embodiment of mental activity in the fibres of the brain but supposed that this activity requires a soul. The result was a kind of religious materialism. His philosophy of nature represented the immaterial soul as activity, but conceived of this activity as a pattern given by the passive organization of matter (the structure of nerve fibres, each predisposed to convey a particular sensation, as the strings of a musical instrument are tuned to produce particular notes). In terms of this broad framework, he wrote a detailed account of the embodiment of sensory experience as the basis for psychology and for knowlege of the world. His appearance in the *Encyclopédie*, associated with Condillac, is therefore an index of the extent to which it had become acceptable to consider mind in terms derived from natural philosophy rather than scholastic metaphysics.

Condillac worked as a tutor to the prince of Parma and published a course of study in sixteen volumes, which included *La logique* (*Logic*, 1780). Pedagogy, or a rational system of teaching, occupied a central place in his work, and it had consequences for the reorganization of education in France which occurred at the end of the century. The revolution in 1789 was initially welcomed by many people in France and abroad as it appeared to create an opportunity for enlightenment in human affairs. Early hopes turned to fear when confronted by the Terror, by the conservative backlash across Europe and by Napoleon's search for an empire and absolute power. For a few productive years in the late 1790s, however, a group of Parisian intellectuals, the *idéologues,* brought together medical interest in the body, Condillac's analysis of knowledge as ordered language and the ideal that knowledge of human nature makes possible a rational society. The revolution swept away the educational and cultural institutions of the *ancien régime,* and it created the opportunity in the 1790s to reorganize higher education and the professional institutions of science, engineering, medicine and law. The *idéologues* provided the new French republic with institutions for professional training to create citizens able to contribute in a rational way to national purposes. Though Napoleon soon closed the second section (moral and political sciences) of the Institut National, which was the *idéologues'* institutional base, they established precedents for the application of the sciences of man to social administration. For example, they influenced the alienist or specialist in mental diseases J.-E.-D. Esquirol (1772–1840), who was at the centre of a network of physicians who wanted to unify health care and social hygiene in the period after the restoration of the monarchy in 1815.

The word *'idéologie'* rather than *'psychologie'* was chosen by A.-L.-C., comte Destutt de Tracy (1754–1836) to stress that knowledge and its application depends on ideas rather than on belief in substances like the soul that are the supposed causes of ideas. It was only after Marx that the word 'ideology' acquired its modern denotation – belief and values that support a political position for which objectivity is claimed but not justified. The *idéologues* had an overt political programme, the rationale of which rested on intellectual analysis of human needs and capacities. Destutt de Tracy spelt this out in the *Eléments d'idéologie* (1801–15), where he argued that well-grounded ideas originate with experience and physiology, and he thus excluded from politics the sort of religious and nationalist sentiment Napoleon sought to revive and exploit. Destutt de Tracy integrated politics and language. He exclaimed: 'Custom is the source of all our progress and all our errors; signs, the most precious invention of men.'[31]

The physical side of *idéologie,* as opposed to the analytic treatment of knowledge and language, was developed by P.-J.-G. Cabanis (1757–1808) in twelve memoirs published in 1802 as

Rapports du physique et du moral de l'homme (*Studies on the Physical and Moral Nature of Man*). He argued that progress in the moral or human sciences depends on the treatment of each human being as a physical and moral unity. He therefore discussed thought and conduct as functions of life related to functions of the body, functions among other functions. In practice, he concentrated his attention on the nervous system, the structure which by then was accepted to be the integrative structure of the body, the physical environment and society. In this context, he made his much cited remark that likened the special organ of the brain, as it produces thought from sensations, to the stomach as it produces gastric secretions in response to food. This was usually abbreviated by commentators to the crude statement that the brain secretes thought (though the idea that the brain is a secretory organ had a long and respectable history). In this way, Cabanis's comment became a classic for nineteenth-century critics who wanted a quick way to illustrate the fatuous consequences of the reduction of spirit to matter. In reality, Cabanis wrote a thoughtful work on the myriad ways in which our experience, well-being and suffering are dependent on physical variables like diet and climate.

The work of Destutt de Tracy and Cabanis reflected as well as influenced a change in institutions and social values. Public opinion began to be sympathetic to an instrumental view of social arrangements and to judge them by the efficiency with which they produce desirable human beings and desirable conduct. The consequences were evident in education, in medicine and in the legal and penal systems. *Idéologie* anticipated a science and a technical expertise – that is, a social science – the purpose of which is to render people happy and moral and society prosperous and well ordered. It exemplified a reforming zeal evident across Europe and North America in education, the treatment of mental disorder, punishment, the care of orphans, legal medicine and public hygiene, poverty and more besides. New practices on the ground generated new expectations of theory, and new theory rationalized these practices. The *idéologues* formed a bridge between the abstract analysis of human nature and the concrete pressures of administration.

v *Moral Philosophy and 'the Science of Man'*

Political conditions in eighteenth-century France encouraged oppositional rhetoric. King and church demanded, as of right, power over secular and religious life. The *philosophes* were in revolt, not on behalf of the people, whom they described as rabble, but on behalf of what they claimed to be reason. Some looked with envy across the Channel to the British Parliament and to liberties embedded in the English common law. Exactly whom Parliament represented and how far liberties were entrenched, however, was questioned in Britain by radicals who supported the Americans in the War of Independence (1775–83) and resisted the conservative reaction after 1789. Nevertheless, English writers on human nature used language which was noticeably less confrontational than that of French writers, and they were sometimes positively eloquent about the harmony of human nature and their social world. Joseph Priestley and some utilitarian theorists, headed by Jeremy Bentham (1748–1832), were the partial exception. Like their French friends, they opposed the established social order with the teachings of reason. The real outsiders were the radicals like Tom Paine, Mary Wollstonecraft and William Godwin (1756–1836), who bitterly attacked their social world because it denied to humanity the goodness of its nature.

There was a pervasive conviction in Britain that God's moral purposes are revealed through the study of human nature and social arrangements and that such study advances wealth and virtue. This optimism supported an impressive body of work in social and moral philosophy. Moral philosophy was part of the academic curriculum from the late medieval period through the eighteenth century, when it covered the study of the soul, human nature and conduct in social relations. It was as much a practical as a theoretical discipline. For most but not all teachers it possessed a religious content which harmonized mundane human relations with God's overall design. In short, it was a subject eminently suited to the education of Christian gentlemen. Moral philosophy formed a distinct discipline in the Scottish universities of Edinburgh and Glasgow, in the colleges of St Andrews, and at Marischal and King's in Aberdeen. It was also taught in the new colleges of the American East Coast, where it continued to be central to the curriculum through most of the nineteenth century. The discipline later became known as 'moral science', a label that persisted in the university of Cambridge into the late twentieth century, though by then it covered philosophy.

Different meanings of the word 'moral' cause confusion. In the eighteenth century, 'moral' signified what pertains to human beings, in the suggestive sense that being human is defined by the capacity for agency in the pursuit of what is good. The eighteenth-century academic discipline of moral philosophy was evaluative as well as descriptive. Values were not added on in the interests of particular social groups, although they may have served such interests; rather, values that were believed to be intrinsic to natural man gave the discipline of moral philosophy, and hence the science of man, its subject matter. To put it succinctly: to have knowledge of man was assumed to give knowledge of what it is right to do. The first edition of the *Encyclopaedia Britannica; Or, a Dictionary of Arts and Sciences, Compiled upon a New Plan* defined moral philosophy as 'the science of MANNERS or DUTY; which it traces from man's nature and condition and shews to terminate in his happiness'.[32]

The seventeenth-century arts curriculum, which preceded advanced study, included logic, metaphysics, moral and natural philosophy and some natural and revealed religion. By the eighteenth century, philosophers rather than theologians taught the moral part, which included the description of man's nature. David Fordyce (1711–51), when he lectured at Marischal College in 1743, told his students that '*Moral Philosophy* has this in common with *Natural Philosophy*, that it appeals to *Nature* or *Fact*; depends on Observation, and builds its Reasonings on plain uncontroverted Experiments, or upon the fullest Induction of Particulars of which the Subject will admit.'[33] He then went on to describe man's moral powers and their relation to conduct, and he emphasized the rules of life and duty. The classification of the sciences was not rigid. George Turnbull, who taught in Aberdeen before Fordyce, described the scientific study of mind or spirit as pneumatology, and he catalogued the 'furniture of the mind' as the basis for moral philosophy.[34] Alexander Gerard (1728–95), by contrast, described the science of mind as psychology. Pneumatology continued to be taught later in the century, and it was a subject division accepted by the celebrated teacher Thomas Reid (1710–96), who was at King's College, Aberdeen, before he became Professor of Moral Philosophy in Glasgow in 1764. In his lectures at Aberdeen in 1752, Reid's syllabus included both pneumatology – 'the History of the Human Mind and its Operations & Powers' – and natural history, with material taken from Linnaeus and Buffon.[35] The natural philosophy and moral philosophy of man were closely related in this Scottish context. Reid divided his subsequently published lectures into those that covered reasoning and understanding, *Essays on the Intellectual Powers of Man* (1785), and those that covered moral activity, *Essays on the Active Powers of Man* (1788).

Scottish moral philosophy related closely to the teachings of the Kirk. The Episcopalian Calvinists held that the moral law had been unmistakably revealed: 'The moral law is the declaration of the will of God to mankind, directing and binding every one to personal, perfect, and perpetual conformity and obedience thereunto . . . promising life upon the fulfilling, and threatening death upon the breach of it.'[36] What was new in the eighteenth century was a willingness to derive the moral law from human nature rather than directly from God's word – though human nature was certainly believed to be God's work. This change was also mediated by English and Irish moral sense philosophers, notably the third Earl of Shaftesbury and Francis Hutcheson (1694–1746), who established an academy in Dublin before he took up the Glasgow moral philosophy chair. Shaftesbury and Hutcheson claimed, in opposition to Locke, that a 'moral sense' (a term introduced by Shaftesbury) is an intrinsic part of our nature. The moral sense, they argued, is a faculty of the soul, a natural guide in the conduct of life. It is known through an inner sense, distinct from the outer senses, and through the affections, which enable a person directly to perceive the virtuous qualities of some thing or action. Shaftesbury compared the affections to the strings of a musical instrument; they are, he thought, designed with a natural harmony, and he likened individual differences to 'different *Tunings* of the Passions'.[37] Butler, in the same vein, described 'a capacity of reflecting upon actions and characters, and making them an object to our thought; and on doing this, we naturally and unavoidably approve some actions . . . and disapprove others . . .'[38] The moral conscience, Butler argued, gives us the power to discern moral truths. Hutcheson understood the moral sense, which he distinguished from an innate capacity for benevolence or virtue, to be the source of feelings of approval or disapproval as we observe the actions of others. These feelings take account of the pain or pleasure generated. In this context, he formulated a statement which was later the mainstay of utilitarian ethical theory: '*That Action is best,* which accomplishes the *greatest Happiness* for the *greatest Numbers.*'[39]

In *An Inquiry into the Original of Our Ideas of Beauty and Virtue* (1725), Hutcheson extended this type of argument about human nature and claimed that there is also an aesthetic faculty, an innate capacity in the mind directly to apprehend beauty. This suggestion initiated a theory of beauty in terms of mental or psychological notions. Beauty was of considerable interest in the self-conscious literary and artistic culture of the eighteenth century. Before Hutcheson, Addison – who for a while dictated London literary taste – considered our sense of beauty to be a natural response to what is harmonious in nature. Hutcheson elaborated this opinion into a systematic theory of the aesthetic sense. In the hands of Joshua Reynolds, the painter, and Edmund Burke (1729–97), later famous for his conservative defence of the English political constitution, these ideas contributed to the conviction that beauty is an expression of subjective feeling. Burke's *Philosophical Enquiry into the Origin of Our Ideas of the Sublime and Beautiful* (1757) described the experience of beauty as a creation of the passions not the reason, a creation that social custom turns into taste in a manner that follows ascertainable laws. Burke's discussion of 'the sublime' captured this sensibility about beauty: it conjured up an obscure and passionate beauty rather than classical ideals – and it alluded to the sensuality of women rather than an ideal of masculine perfection. Such debates about taste and beauty brought theories of human nature into every drawing-room.

Shaftesbury, Hutcheson and their successors lived in a disputatious society. Much of the noise was the small change of a social world ruled by patronage and political ambition. But these authors were surely right when they discerned something fundamental at stake in Locke's attribution of moral actions to pleasures and pains that accompany experience rather than to the soul. As Shaftesbury wrote, 'Twas Mr. LOCKE that struck at all Fundamentals, threw all *Order* and *Virtue* out of the World, and made the very *Ideas* of these (which are the same as those of GOD) *unnatural*, and without Foundation in our Minds.'[40] Moral sense theorists combatted Locke's view with the authority of nature: they turned to what the inner sense teaches about human nature. This inner sense, they claimed, builds knowledge on experience, not of the external world, but of knowing innately what it is right to do. In Butler's words: 'This faculty [of conscience] was placed within to be our proper governor; to direct and regulate all under principles, passions and motives of action.'[41] Further, the moral philosophers, like contemporary natural philosophers, claimed that everyone has in themselves the basis for the same experience, and knowledge of man's nature therefore provides the basis for social harmony. At the same time, it was argued, our experience of the inner sense distinguishes moral knowledge from physical knowledge and shows the groundlessness of materialism.

Moral sense theory claimed a foundation in nature for moral knowledge at the same time as it separated human nature from physical nature. This proved a double-edged defence of established belief. The authority of nature enabled writers who were more interested in moral philosophy than in the defence of religion to bring moral and natural philosophy closer together. This was done by the Scotsman David Hume, known in his lifetime as an essayist and historian, though later mainly famous as a philosopher. The youthful but ambitious Hume anonymously published the first volume of his *A Treatise of Human Nature* in 1739, a book in which he set out to lay the foundations for moral philosophy and thereby, he argued, other subjects as well, in observation of human nature. His grand object was to establish 'the science of man':

> There is no question of importance, whose decision is not compriz'd in the science of man; and there is none, which can be decided with any certainty, before we become acquainted with that science. In pretending therefore to explain the principles of human nature, we in effect propose a compleat system of the sciences, built on a foundation almost entirely new, and the only one upon which they can stand with any security.

Hume set his sights high and found inspiration in Newton. He rejected as vain the hope that it is possible to explain 'the ultimate principles of the soul', just as the public Newton of the *Principia* did not try to explain gravitational attraction. Hume turned to experience and to the four sciences that he thought most relevant to knowledge of human nature: logic, morals, criticism and politics. 'The sole end of logic is to explain the principles and operations of our reasoning faculty, and the nature of our ideas: morals and criticism regard our tastes and sentiment: and politics consider men as united in society, and dependent on each other.'[42]

Hume hoped for fame, or at least learned controversy, as a result of his book, but he was sadly disappointed. After he attempted to re-express his ideas, especially in *An Enquiry Concerning the Principles of Morals* (1751), which brought his anti-religious tendencies into the open, what he called his love of literary fame drew him to essays and history. He lived as a private scholar, political secretary, diplomat and librarian, at the hub of the Scottish Enlightenment in Edinburgh, near to his close friend Adam Smith. His philosophical thought had an impact principally through his critics, Reid and – especially – Kant. This thought, however, also exemplified the way moral philosophy encouraged a science of man, in Hume's case with conscious connections to natural philosophy.

Hume composed his *Treatise* in three books: on the understanding, on the passions and on morals. His starting-point was an analysis, in the manner of Locke, of what is in the mind in terms of elementary 'impressions' – sensations and feelings – and 'ideas', 'faint images of . . . [impressions] in thinking and reasoning'.[43] On this basis he constructed his sceptical argument, which aroused mainly ridicule in the 1740s, that we can never be certain that we have reliable knowledge about either physical or spiritual things. Significantly, his scepticism was directed towards abstract knowledge as the foundation for human affairs and not towards practical matters. Hume never doubted that we are equipped by our nature with the capacities necessary for effective action in the everyday world. We do not, he thought, have to be reflective to know what to do; indeed, it may be better if we are not.

The position is well illustrated by his account of causation. He opposed the common belief that an effect is necessarily related to a particular agent or cause, even though in experience the effect always follows the cause. All ideas derive from impressions, he argued, and we can therefore trace belief that something is a cause to our experience of the way one thing is always accompanied or followed by another. Our ideas of causation follow from the contiguity and succession of impressions. It is therefore our mental habits, which originate with the order of impressions, that lead us to claim knowledge of order in the physical world. What 'we call cause or effect, is founded on past *experience*, and on our remembrance of their *constant conjunction* . . .' In Hume's view, it is our nature that determines the possibility of ordered experience. It also determines our response. He described the passions, our pride, love, ambition, curiosity and so forth, as the result of qualities intrinsic to our nature which lead us to respond in the way we do to impressions. In this context, he claimed: 'Reason is, and ought only to be the slave of the passions, and can never pretend to any other office than to serve and obey them.'[44] This was not a call to freedom of expression – far from it; it was an observation that our actions are, fortunately, founded in our nature rather than in the unreliable conclusions of the understanding. He never doubted that this nature prepares people to live with social propriety.

Hume also took an interest in the physiological side of human nature, rather as Descartes had done. Hume was realistic, not sceptical, about the manner in which physiology underlies imagination, the passions, habits and actions. He wrote that, 'in the production and the conduct of the passions, there is a certain regular mechanism, which is susceptible of as accurate a disquisition, as the laws of motion, optics, hydrostatics, or any part of natural philosphy'.[45] Physiology was not Hume's main concern, as it was for physicians who were his contemporaries like Whytt and Cullen, but he shared with them a culture that linked natural and moral philosophy. This linkage is well illustrated by Chambers' *Cyclopaedia* article (published in Edinburgh in 1728) on 'imagination': 'A Power or Faculty of the Soul, by which it conceives, and forms Ideas of Things, by means of certain Traces and Impressions that had been before made in the Fibres of the Brain by Sensation.'[46]

Hume supposed that 'human nature . . . [is] compos'd of two principal parts, which are requisite in all its actions, the affections and the understanding'. He divided the affections or the passions into those which are 'direct' and 'indirect': the former 'arise immediately from good or evil, from pain or pleasure', the latter 'proceed from the same principles, but by the conjunction of other qualities'. In his view, it is passion and not reason that drives action, and his argument to this effect – that motives are causes not intentions – achieved classic status in the twentieth-century philosophy of mind. In the manner of Locke, Hume thought that the motive element of passion is always analysable into pleasure and pain, and that the feelings of pleasure and pain are the source of our notions of what is good and bad. We must 'pronounce the impression arising from virtue, to be agreeable, and that proceeding from vice to be uneasy'. He therefore argued: 'The mind by an *original* instinct tends to unite itself with the good, and to avoid the evil . . .'[47] At first glance, this looks comparable with the position of the moral sense theorists. But whereas Hutcheson, for example, located the 'instinct' for good in an intrinsic natural faculty, Hume located it in intrinsic natural feelings which are the result of bodily life. Hume accepted that moral distinctions derive from a 'moral sense' (a term he used), but he then analysed the sense, as the moral sense theorists did not, into the elementary impressions of pleasure and pain and (as discussed in the next chapter) of sympathy.

Hume's studies explored the adequacy of experience and human nature as categories in a science of man. He followed Locke, though he diverged in detail, with an analysis of the mental world into elementary units, the impressions, analogous to the supposed material particles of physical nature. He also attempted to apply observation, the natural philosopher's experimental method, to discover knowledge of human nature. While the discipline of moral philosophy originated with a subject matter separate from the subject matter of natural philosophy, its eighteenth-century teachers developed the discipline with terms and methods borrowed from natural philosophy. This was a very significant contribution to the translation of the soul, understood in relation to God, into mind, understood in relation to nature.

Both the moral sense theorists and Hume located the source of right conduct in human nature. The teaching of moral philosophy ensured that the ideas of the moral sense theorists gained a large audience. The literature of the moral sense – reinforced by discussion about the aesthetic sense – was a literature of polite society relevant to everyday personal relations and social obligations. It contrasted favourably as an inspiration to virtue with an earlier harsh discipline of religious injunctions or the dry scholarship of deductive moralilty. Eighteeenth-century society relocated morality as sensibility and thereby constructed part of the modern realm of psychology.

There is one other topic to be considered because it illustrates natural philosophy's contribution to a science of man. Historians of psychology have often given it pride of place in the development of such a science.

vi *The Association of Ideas*

The association of ideas has been thought of as the key scientific achievement of early psychology. The phrase refers to the way sensations, pleasures and pains, ideas and actions were thought to come together as regular patterns. Belief that such associations occur underlies modern theories of learning, behaviour, motivation and perception, and it even underlies psychoanalysis, which developed a technique called free association. Association apparently explained the manner in which the complex content of mind or behaviour is built up through law-like regularities in the relations over time and space of sensations, feelings and actions. It was argued that if simple elements (such as Locke's 'ideas') occur together, or successively, or resemble each other, then they associate to form complex mental content and behaviour. Historians of psychology claimed that psychology achieved a scientific research programme once it was understood that mental content and behaviour are acquired in this way. The work of Hume and of David Hartley (1705–57) was said to be of decisive importance, though some enthusiasts wanted to trace the association of ideas back beyond that. In many histories, the theory featured as the main road towards a scientific psychology.

This is a tunnel vision of psychology's past. There were arguments in the eighteenth century critical of Locke's notion of ideas as the way to conceptualize mental content, and there was antagonism to the scientific claims of associationist principles. Also, contrary to widespread belief, Locke was not an associationist; he added a chapter on the association of ideas to the second edition of his *Essay* in order to account for 'Unreasonableness in . . . Tenets and Conduct' not for the normal work of the mind.[48] The history of association psychology needs to be widened. It is not a history that details the discovery of law-like processes in the mind and pictures the path of scientific psychology. Rather, it is a history that shows the power of metaphor – a social metaphor, 'association', and a physical metaphor, 'ideas' – to create contested versions of scientific psychology. Associationist mental analysis certainly became part of the human sciences, but this was in utilitarian social thought, to be discussed in relation to political economy, as well as in many branches of modern psychology. It was an attractive form of analysis because it represented change and progress in human life as the necessary outcome of past experience. Belief that ideas and feelings of pleasure and pain associate together and determine conduct provided authority for belief that human nature itself guarantees progress through experience – progress at the level of the individual, at the level of society and even at the level of nature as a whole. The history of the association of ideas is one part of the history of optimism and belief in progress.

That said, there is a significant story to be told. Locke's approach to mind required some account of how elementary sensations turn into complex ordered thoughts. Hume saw the problem clearly: 'Were ideas entirely loose and unconnected, chance alone wou'd join them; and 'tis impossible the same simple ideas should fall regularly into complex ones (as they commonly do) without some bond of union among them, some associating quality, by which one idea naturally introduces another.'[49] The 'associating quality', he thought, binds mental parts into a whole, and he picked out resemblance and contiguity in time and place (to which he reduced causation) as the qualities of ideas that lead to association. Later psychologists added to or subtracted from these qualities and gave different accounts of their effects, but the essential structure of the argument remained. James Mill (1773–1836), for example, in his *Analysis of the Phenomena of the Human Mind* (1829), the most thoroughgoing – and dry – of associationist studies, thought that all association can be reduced to contiguity: 'Our ideas spring up, or exist, in the order in which the sensations existed, of which they are the copies.'[50]

Natural philosophy inspired Hume and suggested to him both a general procedure for the study of moral philosophy and a specific metaphor, the 'associating quality' of ideas which, he observed, involves 'a kind of ATTRACTION, which in the mental world will be found to have as extraordinary effects as [gravitational attraction] in the natural, and to shew itself in as many and as various forms'.[51] A Newtonian connection was also important in the other main source for association psychology, Hartley's *Observations on Man, His Frame, His Duty, and His Expectations* (1749). The title of the book revealed its purposes to be unambiguously religious. Hartley had his sights firmly set on the pure happiness that comes from total identification with

God's purposes; his Newtonianism was therefore different from Hume's (and he was unaware of Hume's account of association). These Newtonian connections were present, but they have disguised the fact that the word 'association' had an obvious social meaning. It did not require an esoteric theory of nature to construct imagery of elementary parts in combination as integrated wholes: association was the everyday cement of social relations and a generally accessible source of metaphor.

Hartley's book exemplifies the pervasive British culture of natural theology, with its enthusiasm for nature as evidence of the attributes of God. He was a theologian not in holy orders and a physician without much formal training who brought together religion and physiology in a non-academic study of moral philosophy. His long book set out to prove that the design of the human mind necessarily advances humanity towards virtue and happiness; the association of ideas, coupled with pleasures and pains as the determinants of conduct, are God's means to this end. If, as Hartley supposed, it is a fact that the greatest happiness comes through the pursuit of what is good, then people will be led naturally to that pursuit while they acquire habits based on associations between sensory impressions and impressions of pleasure and pain. 'Since God is the source of all Good, and consequently . . . associated with all our Pleasures, it seems to follow . . . that the Idea of God . . . must, at last, take place of, and absorb all other Ideas, and He himself become . . . *All in All.*' This was a heady vision; but Hartley filled it out with a systematic and detailed discussion of the principles of association, which he thought can be analysed into the simultaneous and successive presence of ideas or sensations. He explained learning and he explained the passions – the passions, for example, 'can be no more than Aggregates of simple Ideas united by Association'. He provided more detail than Hume about association as the working principle of the mind. Everything, he explained, 'results from the Frame of our Natures'.[52]

Hartley, like Hume, built on the early eighteenth-century debate about the moral sense. Hartley was particularly struck by a contribution to this debate by a Reverend John Gay in 1731 that enunciated utility as the explanation of human actions. The very important principle of utility states that it is pleasurable or painful sensations or the pleasurable or painful consequences of our actions that guide our conduct. Hartley's other inspiration was Newton's speculations appended to his work on *Opticks* (1704; Queries 12–16) that vibrations in the fine substance of the nerves might explain transmission and the visual after-effect, the spot that continues to be present after looking at a bright light. The medically-oriented Hartley speculated that such a nervous vibration is parallel to every mental impression or idea, and on this assumption he constructed an elaborate picture of the correlations between vibrations and associations. His imagery of the way vibrations interact or interfer with each other appears to have suggested to him thoughts about how to represent the relations between mental elements. He also accepted necessity, or in modern language, determinism, in the mental as well as in the physical sphere.

Claims for necessity in human conduct appeared to many Christians to be tantamount to atheism. Hobbes and Spinoza were reviled for it. Writers like Hume and Hartley revived such fears. Their critics believed that they did not leave room in human nature for moral choice. Hume in a non-religious form and Hartley in a religious manner, however, presupposed that there is a correspondence between individual desire and the public good, and thus, like their critics, they accepted that benevolence and right action is a natural social reality. While Hume was socially conservative, radical moral philosophers in the last part of the century took up Hartley's arguments, somewhat as the *philosophes* took up Locke, and turned his thoughts into a science of human nature which, it was claimed, underwrote social reform and even revolution. Priestley, the English theologian, chemist and supporter of the American Revolution, had radical political ends in mind when he edited a reprint of the *Observations* in 1775. He thought that Hartley's key contribution, the necessitarian implications of the association of ideas, would reach a wider audience if the argument were separated from its religious and physiological trappings. It was this version, false to its author's intentions, which propagated the image of Hartley as a scientific and not a religious writer held by nineteenth- and twentieth-century psychologists.

The 'essential' Hartley for Priestley and other radical intellectuals was the Hartley who explained mind and conduct in terms of law-like relations between sensations. It was to no avail, Priestley thought, for conservative writers to harp on about moral choice: 'A *volition* is a modification of the passion of *desire* . . . and it is generally followed by those actions with which that state of mind has been associated; in consequence of those actions having been found, by experience, to be instrumental in bringing the favourite object into our possession.' Hartley's theory of mind, Priestley believed, proves that social conditions, rather than an intrinsic moral sense, determine conduct. Priestley, all the same, was deeply religious, in the manner of the radical dissenters, and when he argued for necessity and materialism the argument incorporated his theological beliefs in the resurrection of the whole body and in God's providence. He believed that the association of ideas secures 'provision for the *growth of all our passions*, and propensities, just as they are wanted, and in the degree in which they are wanted through life'.[53] Like Hartley, Priestley thought that human nature guarantees the increase of happiness, but finding this providentially arranged increase of happiness to be thwarted by society, he set out to change society. After Priestley, the radical argument was developed to its logical extreme by Godwin. In his *Enquiry Concerning Political Justice* (1793), Godwin accepted complete necessity in the psychology and affairs of men. When he adapted for his own political ends the view that the mind's content derives from atomic sensations, he came as close as anyone to a picture of the mind in terms of the interactions of billiard balls, the sensations. In his eagerness to justify belief in the progress of humanity, Godwin fell into a reduction of human life to the crudest of physical metaphors.

A further extension of associationist argument occurred in the work of the language theorist John Horne Tooke (1736–1812). He adopted the extreme belief that mental content derives from sensation and accepted the materialist position that all language – and hence thought itself – derives from naming sensations. Tooke reread Locke's *Essay* as a text about language and claimed that it laid the basis for the identification of the sensory origin and meaning of all words. He followed this up with a massive exercise in etymology, an attempt to use the roots of words to reveal their sensory origin. The radicals thought that his overwhelming scholarship, as it appeared to them, drove into retreat the conservative and idealist belief in innate, God-given principles of reasoning. As discussed later, Tooke's reputation survived only for a generation before the inherent weaknesses of his project (e.g., to derive the origin of verbs from the naming of sensations) and the rise of German philology caused its collapse.

Hartley and Priestley understood that theories of the association of ideas take a developmental approach to reason, feeling and action. They described how the child's and hence the adult's character necessarily grows towards virtue and happiness through the acquisition of habits or patterns of conduct based on the past experience of pleasure and pain. The notion of habit was therefore an important part of the natural philosophy of mind. Hartley, in this respect like the French writer Bonnet, correlated habits with physiological changes in the nervous system, and this meant that he believed an acquired pattern of conduct is also an acquired organic structure. That is, he believed, habit becomes represented as a physical structure in the body. At the end of the century, Erasmus Darwin (1731–1802), Charles Darwin's grandfather, a physician, poet and friend of Priestley and early industrialists, extended this idea that acquired and embodied habits can be passed on from the individual to successive generations. In his *Zoonomia: Or the Laws of Organic Life* (1794–6), he extrapolated from Hartley's theory of incremental mental growth to the organic realm in general and portrayed animals in a process of directional change over time. He extended the notion that habits are produced by experience to provide all life with a history – a theory of evolution. In Paris a few years later, independently of Erasmus Darwin, Jean-Baptiste de Lamarck (1744–1829), botanist and invertebrate zoologist at the Muséum d'Histoire Naturelle, took a similar step when he integrated his belief in the inheritance of acquired habits with a view that nature as a whole transforms itself. Most contemporary naturalists rejected Darwin's and Lamarck's theories as speculative fantasies. But they have interest in retrospect, for they show how the directional change over time that Locke described as built into an individual's mental life, in later hands became an evolutionary account of nature. Further, political radicals in the first half of the nineteenth century read into Lamarck's work support for a materialist and necessitarian view that the social world is also subject to change.

Hume's, Hartley's, Priestley's and Darwin's writings provoked opposition from conservative critics who suspected that such work led towards an unholy trinity of necessitarianism, materialism and atheism. A sustained critique was mounted by Reid and his followers, who found in man's shared mental nature, what Reid called 'common sense', authoritative grounds to oppose scepticism and to support belief in moral freedom. 'Common sense', as Reid used the term, denotes the universal qualities in human nature that lead right-thinking minds to rational and moral conclusions. These qualities are 'a part of that furniture which Nature hath given to the human understanding . . . They are part of our constitution . . .'[54] In a similar way, the conservative French philosopher Cousin, who had a preponderant influence on academic philosophy and on the training of teachers in France in the first half of the nineteenth century, referred to 'the permanent good sense of humanity'.[55] Reid described common sense at length in his lectures on the 'intellectual' and 'active powers'. As a teacher of moral philosophy, he wished to defend the grounds of practical judgement: 'Common sense is that degree of judgment which is common to men with whom we can converse and transact business.'[56] He thought practical and moral relations were threatened both by Hume's scepticism and by Hartley's and Priestley's necessitarianism. Reid and his followers in the Scottish universities and North American colleges, where the common-sense school was active throughout the nineteenth century, taught a view of human nature that fitted well with the moral and religious purposes of education. Common-sense philosophy merged with ordinary Victorian discourse about individual lives.

Elements of belief in association and common sense combined in a large number of books on the mind in the early nineteenth century. A much used text was written by Thomas Brown (1778–1820), whose *Lectures on the Philosophy of the Human Mind* (1820) was based on his moral philosophy lectures at the university of Edinburgh. He elaborated what at first glance appear to be associationist principles. But he substituted the word 'suggestion' for 'association', and this did more than change words since 'suggestion' implies mental activity; he moved away from the mechanistic aspects of associationist theory. Brown also described mental elements as 'affections' and not 'ideas', and he described them as linked by relations rather than mechanical juxtaposition. This implied, for example, that one experience may suggest another by metaphorical relation as well as by conjunction in past experience. He also fully integrated feelings into the analysis, and he separated mental analysis from any reference to physiology. In all these ways, he detached analysis of the mind from its links with radical values while he enriched the capacity of analysis to describe the active as opposed to passive character of the mind. Writings like Brown's fostered acceptance of mental analysis as a distinct discipline, a subject soon to be widely known in English as mental science or psychology.

Natural philosophy created a profound opportunity for those interested in a science of man. A few brash materialists claimed simply that this science should become a branch of natural philosophy. This was judged to be crude in the eighteenth century – even if later scientists were to judge it presaged the way ahead. But natural philosophy suggested much besides materialism. It reinforced a search for the laws of human nature and made moral philosophy into a discipline concerned with man as both a physical and a moral being. Further, belief about the material world – its particulate structure and mechanistic causation – created analogies used in theories of mind, and analysis that originated in analogy ended by supplying key concepts to the study of mind. The theory of the association of ideas exemplifies this. As Reid critically observed:

> There is a disposition in men to materialize everything . . . Thought is considered as analogous to motion in a body; and as bodies are put in motion by impulses . . . we are apt to conclude that the mind is made to think by impressions made upon it, and that there must be some kind of contiguity between it and the objects of thought.[57]

Lastly, the pleasure-pain principle evoked a determinist science of human conduct; belief in the necessary sequence of events was transferred from the physical to the human sphere, a prospect that enthralled radicals and appalled the orthodox.

Though we cannot assume that Enlightenment ideas directly caused social change, such ideas were promoted and received in the eighteenth century as if they might. After all, it was the century in which the United States was founded on the basis of a rational and secular constitution, in which the French Revolution showed that long-established royal and Catholic power could be overthrown and in which industrialization – which had no precedent – began to change every aspect of the fabric of life in Britain. It is therefore necessary to examine how the science of man was also the study of men and women as *social* beings.

8

Human Diversity and Sociability

> Cast on this globe, without physical powers, and without innate ideas; unable by himself to obey the constitutional laws of his organization, which call him to the first rank in the system of being; MAN can find only in the bosom of society the eminent station that was destined for him in nature, and would be, without the aid of civilization, one of the most feeble and least intelligent of animals.
>
> Jean Itard, *De l'éducation d'un homme sauvage ou des premiers développements physiques et moraux du jeune sauvage de l'Aveyron* (*The Education of a Savage Man, Or the First Developments, Physical and Moral, of the Young Savage of Aveyron*, 1801)[1]

i *The State of Nature*

The language of human nature cut across occupational boundaries in the eighteenth century. Poets as well as physicians, dissenting radicals as well as academic scholars, women as well as politicians, found hope and meaning for their lives in this nature. The language was rich in claims and counter-claims about differences and resemblances between human nature and physical nature. The negotiation of this boundary gave content to systematic studies of man. The previous chapter discussed these studies in relation to the embodied mind, to human nature within. This chapter turns to the outward form of human nature, to people as social beings and to the many comparisons made between civil society and the state of nature. Studies of the social content of human nature abounded, and they established terms for the analysis of the social world and its institutions in their own right. The content of the discussion is summarized by one question: Is society natural?

There was no subject called sociology in the eighteenth century. There was also, at least at the beginning of the century, no abstract category, 'society', though there was a concrete language about polite society. A discourse about the conditions and history of social life in general, as well as about particular states, nevertheless existed, and it became more distinct during the course of the century. Many writers gave prominence to the sociable nature of man and made the science of human nature a foundation for social thought. A few authors, followers of Hobbes, considered man to be naturally unsocial, but for that very reason in need of social order. The range of opinion was very great. While the question asked in the early seventeenth century was how social arrangements fulfil the kingdom of God, by the mid-eighteenth century it was how social arrangements fulfil human nature. This was not necessarily a fundamental change, since the study of nature was for many authors the study of God's creation, but there was a shift towards study and description of the social world in its own right, a shift eventually marked by the use of the abstract term 'society'. All this was intimately connected to hopes for social improvement and political liberty; the principle that knowledge lays the foundation for achievement of right action, to repeat the point once again, was assumed.

These hopes plunged eighteenth-century students of the science of man into a deep contradiction that neither they nor those who followed them – Marxists, for example – satisfactorily resolved. At one and the same time, authors sought a law-like science of society and upheld belief in individual freedom and dignity. To put it bluntly, they denied *and* asserted individual freedom of action. This contradiction correlated with the subject–object distinction – the split between the knowing mind, the sphere of reason, and the object known, the sphere of physical events – embedded in Cartesian dualism. After Kant, and again in the twentieth century, the contradiction was reformulated as the question of the relation between description and moral judgement or, in modern terms, facts and values. As we shall see, it was a forceful incentive to Romantic idealists at the beginning of the nineteenth century to attempt to overcome these difficulties and to find human freedom and values in nature itself. The argument reached its apotheosis in G. F. W. Hegel's theory of the dialectical development of the Absolute, the starting-point for later Marxist attempts to resolve the contradiction with a philosophy of dialectical materialism. The dominant view in the social sciences in the twentieth century was that these sciences establish value-free factual knowledge, but its proponents then failed to explain how, at the same time, they could claim that such knowledge is the basis for social policy. In reality, over three centuries, the study of objective social conditions, the causes of how people live, was also a practical engagement with evaluating and refashioning the conditions of life.

In the twentieth century it became common to think about the relation of men and women to physical nature in terms of the balance between nurture or upbringing and nature or heredity. The eighteenth century was as fascinated and divided as the twentieth over this question, but its arguments took a different form. Enlightenment authors were concerned with the grounds and origin of morality, virtue and civilization. They did not debate nature versus nurture but asked questions about what the state of nature is and how virtue and civilization relate to it. The eighteenth-century debate was about moral and political philosophy in the broadest sense and not about psychology or biology. The language of this debate referred to the physical versus the moral nature of man. It also referred to the state of nature, the principal conceptual tool for theories about the character of the social world. Writers sometimes treated the state as a real condition in which people live independently of civil society; Hobbes and Locke thought this might be the case for native American people. It was also possible to compare the state of nature with the state of Adam before the fall or the state of mankind before the flood. But such usages merged with the reference to the state of nature for the purposes of argument. The state of nature, finally, had mythic stature: it was a dream of what man truly is or might become.

Mythic qualities radiate from much-loved stories of wild children, of people found living in the wild, alone or with animals, and of animals, apes, monsters or imagined creatures that might or might not be human – Frankenstein's 'creature' is the most eminent, created in Mary Shelley's novel in 1818. These tales interrogated human identity. A life apart from society is like a chemical separation: it distils out what belongs to society and what to nature. It misses the point to argue about the factual truth of these stories, for they possess a deeper truth.

The best-known narrative was a story often read as fact, Daniel Defoe's *Robinson Crusoe* (1719). Defoe placed his hero alone on a desert island, restored to a state of nature, without any social covering – he wore goat skins for clothes. But Crusoe had two inestimable advantages: he salvaged the tools and supplies necessary to improve his lot from the ship on which he was wrecked; and he was able to give thanks to God that 'had cast my first Lot in a Part of the World, where I was distinguish'd from such dreadful Creatures' as the cannibals who visited his island. Yet, face to face with the man he rescues, the savage he named Friday, Crusoe observed with wonder that God 'has bestow'd upon them the same Powers, the same Reason, the same Affections . . . and all the Capacities of doing Good, and receiving Good, that he has given to us'.[2] In the imagined state of nature, man confronts himself as both savage and civilized, good and evil, and – in Defoe's version – dependent on God and the supplies left providentially by his shipwreck.

There were extensive narratives about the boundary between savage and civilized, between beast and man, in the factual literature of natural history. The Swedish natural philosopher Linnaeus, perhaps the most eminent collector and naturalist of the eighteenth century, classified the varieties of mankind along with every other natural object. Under *Homo* he listed special varieties or even species for *Homo ferus* (feral or wild man), *Homo sylvestris* (tree man – which included our chimpanzee) and *Homo caudatus* (tailed man – a category partly mythical and partly designed to include birth disorders). Nothing illustrates better how difficult it was to draw a boundary between animal and human nature than these much debated categories, especially

Homo ferus. Many people were familiar with the stories about children found alive in a bestial condition, and the question of man's relation to the animals was sometimes deepened by belief that animals, such as wolves or bears, had cared for children. These stories were ancient – Romulus and Remus, suckled by a wolf, founded Rome – but at least three new cases had tremendous contemporary impact: Peter the wild boy of Hanover, Kaspar Hauser and Victor the wild boy of Aveyron.

Peter was abandoned by his father and thrown out by a proverbially bad stepmother. He was captured and then escaped several times in 1724 in the forests near Hanover; he appeared to live on and even to prefer the sap from green twigs, and he could not stand wearing clothes. Taken back into society, he was gradually persuaded to dress, to show some sensitivity and to imitate gestures, though he never learned to talk. As a specimen of the state of nature, he fascinated educated society, and he was even presented at the English court of the Hanoverian George I.

Kaspar Hauser was a very different case. In May 1828, a bizarre young man, oddly dressed, with horseshoes for heels on his boots, shambled – he could barely walk – into Nuremberg. In his pocket was a note with his name and the message that he wished to serve the local regiment. He was able to write his name and individual letters of the alphabet but no more, he uttered jumbled sounds rather than speech and when he played like a child this was largely restricted to the movement of his hands while remaining seated. A local doctor became his guardian and began his education, or re-education. He acquired speech but had great difficulty in the use of 'I' or speech in the first person. He also confused dreams with reality and could not at first grasp the idea of his reflection in a mirror. He appeared to be a youth alone and without self-definition. Local knowledge suggested that he had been kept from a very early age in a cellar, fed and given a little attention by someone Kaspar called 'the Man', who had always stood unseen behind him. Rumour spread that Kaspar was the discarded son of the Empress Josephine's niece, and his brutal murder as this story spread served only to confirm it.

Werner Herzog's film *The Enigma of Kaspar Hauser* (1974), immortalized the gentle, unsocialized hero. Victor, too, was translated into film, by François Truffaut, in *L'enfant sauvage* (1970). His case is justly famous as one of the most significant and poignant of all experiments on human nature. Victor was the subject of systematic experiments over several years by a patient, resourceful and innovative physician, Jean-Marc-Gaspard Itard (1774–1838). Itard was of the generation and intellectual inclination of the *idéologues*, and he put into practice the educational implications of Condillac's analysis of mind when he became consultant physician to the Institute for Deaf-mutes in Paris in 1801. Earlier educators had devised sign language and mime and argued that deaf-mutes had an organic disorder and were not stupid. Itard tried to cure dumbness through lip-reading, a strategy which also presupposed that dumbness was not a sign of mental deficiency. With his younger colleague Edouard Séguin (1812–80), he later also developed techniques taken up in the United States, where Séguin moved in 1851, for the education of mentally handicapped children. The Italian educationist Maria Montessori (1870–1952) even suggested in 1898 that Itard's studies 'were practically the first attempts at experimental psychology', a pointer to the way problems of social administration, in this case the deaf and dumb, prompted conceptual and not just practical reconsideration of the human subject.[3] These problems were increasingly filtered through institutional life, and institutions, such as the Institute for Deaf-mutes, became laboratories for the study of human nature.

Local people in the *départements* of Tarn and of Aveyron in southern France sometimes saw a naked child, whom they captured finally when he walked into a farmhouse in the freezing January of 1800. He slept by the sun, murmured but could not speak, loved the rain, ate berries and roots and – like Kaspar – could not identify his own reflection. His character aroused much discussion, and he was eventually brought to Paris. There, Philippe Pinel (1745–1826), as director of the huge Bicêtre asylum and in his capacity as chairman of a committee of the Société des Observateurs de l'Homme (Society of the Observers of Man), diagnosed him as a congenital idiot. Itard opposed this diagnosis, believing that human capacities of the kind which the wild boy so dramatically lacked develop only in society. The boy was put under his care for him to prove the point. Itard named him 'Victor' because the boy responded only to the vowel 'o', and the name thus symbolized the potential humanity in the boy's wild nature.

Over the next five years, Itard and the governess he employed, Madame Guérin, dressed Victor, taught him manners, changed his eating habits, made him sensible, for instance, able to appreciate a hot bath, and equipped him with some powers of concentration – though long lessons or discipline reduced him to animal-like rage. But in his second report (1806, five years after the first) Itard admitted that he had failed to teach Victor language in spite of extreme effort. Itard's critics seized on this as proof of the child's essential idiocy or even lack of humanity. A deeply disappointed Itard acknowledged limitations in Condillac's conception of learning; it appeared that a failure to acquire language at a particular period in a child's development could not be rectified and thus that sensations alone are not the source of the mental faculties. The notion that human development proceeds through successive, qualitatively different steps later reshaped the whole question of the relations between nature and the social world. Condillac's analysis was made to appear abstract and mechanical. Meanwhile, in the encounter between Victor and Itard, nature and society acted out roles as if in a drama on a stage. Victor died in 1828, described as half-wild.

Similar stories, such as those from British India, recurred as late as the twentieth century. Sceptics saw only tall tales or rejected idiot children, but others appreciated the tales for the questions they raised about human nature. As recently as 1973, psychologists in the United States studied a girl, Genie, who had spent thirteen years chained to a potty or confined in a wire-covered crib by her father, kept alive with baby food but denied language. These tales told of a state of nature in which wild but innocent children, with no capacity to identify self and, it was often remarked, an apparent lack of sexual interest, live outside society. The question of language was fundamental. Most but not all eighteenth-century writers believed that a capacity for language in individual human nature makes society possible. It therefore appeared to be vital to decide whether wild children, savages and animals which resemble human beings possess or can acquire language. This question was in turn at the centre of a lively debate about where to draw the line between mankind and the rest of the creation and between civil society and nature. There also began to be unprecedented interest in the distinctive qualities of children and in education as the means by which nature and civilization are united. In the first year of the century, the French libertine Baudot de Juilly wrote: 'For we must draw near the cradle in order to know what nature wants, because it is in children that it acts with greatest freedom . . . so that one may say they are the mirrors of nature.'[4]

ii *Orangs and Others*

Tales of animal-like humans merged with reports of human-like animals. The question of whether there is a clear boundary and, if so, where it is was not easy to decide. In addition, there was the problem of the classification of human beings, who exhibit extraordinary physical and cultural diversity. New evidence crowded in from European voyages and colonization of the Americas, the Far East and the Pacific. Reports and specimens of strange animals and equally strange people were brought back to Europe. Extensive descriptions of native peoples began to appear, and some of these were written by Europeans who experienced native ways of life, learned the languages and admired the customs. Travel books were widely read. A series of bewildering questions about the causes of human diversity, the place of God's providence and the conditions that make civilization possible came to the forefront of attention. Such questions encouraged rash speculation, often on the basis of second- or third-hand evidence that had passed through several geographical and linguistic translations. And the debate about other people was even more complicated as it was also an indirect means to reflect on European economic, moral and political dilemmas. Not surprisingly, there was a vivid consciousness of diversity but no agreed answers.

These topics later gave the discipline of anthropology its subject matter. Anthropology, however, only became a discipline in the second half of the nineteenth century and, when it did so, it was preoccupied, firstly, by the comparative anatomy of physical differences, and secondly, by evolutionary theories of social development. Neither preoccupation existed in the same way in the eighteenth century, and there were then no organizing principles to make possible a coherent science. From the early modern period to the late eighteenth century, the main point of interest was the relationship between the barbarian or savage, the state of nature and civilization. European observers turned to history – to theories of the decline as well as the rise of peoples – as a framework in terms of which to classify information, to make sense of variety and to contrast savage and civilized existence. This was a human history, not an account of biological evolution. Towards the end of the eighteenth century, comparative anatomy started to have a significant impact and it developed a different approach to human diversity, an approach closer to biology.

Linnaeus fussed about whether there were different species and/or varieties of *Homo*, a sign of confusion about the animal–human boundary. The debate among natural historians dated from at least the spring of 1698 when the first anthropoid ape recorded in England was imported. The animal died and it was dissected by the anatomist Edward Tyson (*c.*1650–1708), who published his results in a book entitled *Orang-Outang or, the Anatomy of a Pygmie* (1699). His work, partly a critique of accounts of orang-utans, pongos and other human-like beasts and partly empirical report, influenced naturalists who wanted to separate fact and fantasy. Tyson classified his animal in the light of the chain of being and assigned it a place halfway between the animals and man: '[Though it] be wholly a Brute, tho' in the formation of the Body, and in the Sensitive or Brutal Soul, it may be more resembling a Man than any other animal; so that in this Chain of Creation, for an intermediate Link between Ape and Man, I would place our Pygmie.'[5] Contradictory claims and different names persisted for many years. The animal Tyson dissected was later known as a chimpanzee, while what was later called the orang-utan, though reported in the seventeenth century, was not dissected until the 1770s and not fully observed in the wild by a Western person until the twentieth century. The gorilla, like the heart of Africa itself, was unknown to non-Africans until the mid-nineteenth century.

Early travellers' reports discussed cultural differences at least as much as physical ones. Locke and his contemporaries consumed travel books with an eye to the comparison of savage and civilized people and of the state of nature and civil society. The savage–civilized polarity was fundamental to both natural history and moral discourse. It gave ample scope to Europeans to classify different peoples in a hierarchy given by European ideals. Yet this polarity was not expressed in terms of race in the way that gave nineteenth-century anthropology much of its character. Rather, the language of polarity was the language of philosophy, history and theology. The Latin origins of the word 'savage' connoted a man of the woods who lives outside civil society. It was a word with roots in a contrast made in ancient Europe between barbaric and civilized existence that was later projected on to non-European peoples. It was also a popular theme to contrast savage and civilized worlds, to the disadvantage of the latter. Bougainville's and Captain Cook's celebrated journeys in the Pacific in the 1760s and 1770s, in particular, made possible a rich imagery of the existence outside Europe of an idyllic state of nature. Europeans sensed what they had lost. As Diderot remarked – in an image that endured: 'There was once a natural man, then an artificial man was created inside him. Whereupon a great war broke out inside that man, a struggle that is destined to continue as long as he lives.'[6] At the same time, the reports of travellers grew in detail and complexity, and commentators began, if tentatively, to grapple with the establishment of novel categories appropriate for the description of what is not European, rather than simply subsuming observations under European categories.

The French colonization of North America produced, in late

twentieth-century jargon, some thick descriptions of local culture, as in the self-styled baron de Lahontan's *Dialogues curieux entre l'auteur et un sauvage de bons sens qui a voyagé* (*Curious Dialogues between the Author and a Travelled Savage of Good Sense*, 1703). For all his bombast, Lahontan (L.-A. de Lom d'Arce de Lahontan, 1666–before 1716) had some acquaintance with the Huron and the Algonquin, and he gave his native interlocutor a powerful voice in which to distinguish natural and artificial society. The savages Lahontan described, for example, possess a language that names concrete things in their world but does not enable them to understand abstract civilized words like 'lying' and 'hypocrisy'. J.-F. Lafitau (1681–1746), a Jesuit teacher who knew the Huron and the Iroquois in what became Canada, perceived something of the severe difficulties that face translation, as he appreciated that the native people used a completely different language structure, one, for example, which included kinship patterns, not just different words or grammar. Lafitau's *Moeurs des sauvages américains, comparées aux moeurs des premiers temps* (Customs of American Savages Compared with the Customs of Earliest Times, 1724) was an influential model for the comparison of savage and early societies, one which made sense of cultural difference by means of a theory that describes social development over time. The abbé Raynal (1713–96), in his *Histoire philosophique et politique des établissements et du commerce des Européens dans les deux Indes* (*Philosophical and Political History of the Establishment of European Commerce in the Two Indies*, 1770) provided the most widely read account in the second half of the century of what the colonizers encountered.

Other observers had little sympathy with what they saw as savage society. Simple repulsion was enhanced by European myths of a golden age and, in its Christian version, of the fall from paradise. These myths suggested that savages are the degenerate remnants of earlier and more fully human peoples. Corneille De Pauw (1739–99), for example, took this view in his diatribe against the American continent and the Amerindians in his *Recherches philosophiques sur les américains* (*Philosophical Studies on the Americans*, 1768–9): flora, fauna and humanity alike, he thought, are weak and degenerate. Thus, myth and natural history were sometimes merged in a picture of correspondence between the outward state of savagery in man and nature and the inward degenerate state of a people's nature. An awed fascination with the end of civilization in ancient Greece and Rome brought the theme of loss and degeneration into the context of European history.

A systematic project to incorporate all the new information into a *natural* philosophy of man was undertaken by G.-L. Leclerc, comte de Buffon (1707–88). Buffon was the seigneur de Montbard who rose to eminence as director of the Jardin du Roi in Paris, literally the King's Garden, and who made it into an outstanding institution for plant and animal classification and research. He published many volumes of an *Histoire naturelle* (*Natural History*, publication began in 1749), which assumed that natural history includes man as a natural species. Buffon conceived of natural history in a Newtonian spirit, and this meant that he gave priority to empirical methods – he assembled information sometimes even at the expense of consistency of exposition – and he then attempted to demonstrate general physical causes behind the appearances. Buffon was a powerful voice for a naturalistic science. 'The first truth which issues from this serious examination of nature is a truth which perhaps humbles man. This truth is that he ought to classify himself with the animals, to whom his whole material being connects him.'[7] His project, however, was not identical with a science of anthropology – in spite of his French reputation as 'the founder' of that science. 'Anthropology' appeared in the *Encyclopédie*, published at the same time as Buffon's *Natural History*, merely as a branch of anatomy.

Buffon's human science featured in many articles in the *Natural History* and in *L'histoire naturelle de l'homme* (*The Natural History of Man*, 1749). He discussed man's and woman's physical appearance, and he differentiated species of man-like animals – the orang, the jocko and the pongo – while he quoted reports in full, even if other observers were sceptical, 'because every article is important in the history of a brute which has so great a resemblance to man'.[8] His account tried to explain human variety by reference to geographical and climatic causes and did not delineate a definite number of set types or races. He was as interested in the differences between the sexes as in differences between peoples when he described man's existence in relation to the laws of nature. He wrote at length on 'the ages of man' from birth to death, and he included such information as tables of life expectancy. He always emphasized the effect of the environment. Buffon's tentative account of the causes of skin colour, for example, subsumed this intractable problem under the more general topic of inherited climatic effects and the degeneration or alteration over successive generations of an original stock in animals or man. All this was part of the life of men and women as a natural species; swaddling clothes or puberty – all was grist for Buffon's mill.

> The history of an animal must not be the history of the individual, but that of the whole species of such animals; it must include their begetting, the period of gestation and of birth, the number of young produced, the nurture provided by mothers and fathers, the kind of education received, their instincts, their habitat, their food, the ways in which they obtain it, their customs, their cunning . . . [9]

Many eighteenth-century natural philosophers valued Buffon's natural history of man more than speculative reports of human diversity ('travellers' tales'). Buffon encouraged William Robertson (1721–93), in 1762 Principal of the University of Edinburgh and then appointed historiographer of Scotland, to integrate the history of man and the history of nature, to link man and climate. The result was a major attempt at empirical history, *The History of the Discovery and Settlement of America* (1778–9). Buffon also influenced Bonnet's natural philosophy and psychology. But it was perhaps the *idéologues* Destutt de Tracy and Cabanis, along with contemporary French physicians, who did most to argue for a natural history of the human species, which – integrated with Condillac's theory of knowledge – formed the foundations for a human science. In a contribution to a new dictionary of natural history in 1817, Julien-Joseph Virey (1775–1846) re-expressed Buffon's principles: Man 'remains subject to nature's laws ... How could he not realize that it is in his interest to know himself, together with all that surrounds him, gives him life and causes him to die?'[10] Earlier, Virey was a member both of the Paris Société des observateurs de l'homme, which applied some of Buffon's natural history to man, and an associate of the *idéologues*.

Natural history was closely connected to the study of language. When Tyson dissected his 'orang', he showed that it possessed a voice box, and yet the animal could not talk. He concluded from anatomy that the animal is 'an intermediate link between an Ape and a Man', only then to argue that its lack of speech demonstrates the separateness of human beings, to whom God gave a rational soul and hence a voice.[11] This view, that speech is natural to man and even defines what it is to be human, was widely held. It suggested, for example, that wild children who do not learn to speak are deficient in capacity, not experience. The opinion that language is natural had its critics, however, and they included the idiosyncratic but forceful Scottish judge James Burnet, Lord Monboddo (1714–99). In six volumes *Of the Origin and Progress of Language* (1773–92), he not only argued that there are men with tails but that the 'orang' can be taught to speak. He argued against Buffon that language is a skill acquired in society, not an attribute of human nature. It follows, he thought, that creatures which resemble men might learn to speak; conversely, people who live outside society – as Monboddo thought some savages do – will be speechless. Condillac was another critic of Buffon and of the belief that language is an innate capacity, and it was his argument in the *Traité des animaux* (*Treatise on Animals*, republished 1766) that influenced Itard.

Stories of wild children, the classification of man-like animals, reports of the culture of native people, the natural history of man and the question of the origin of language all contributed to the discourse of human diversity. With such a variety of information and speculation, it was difficult not to be confused. The biblical stories of the descent of human types and cultures from the sons of Noah – Ham, Japhet and Shem, for Africans, Asians and Europeans respectively – and from the dispersion of languages after Babel added yet another dimension. The ancient idea of the great chain of being was pervasive; it reinforced belief that there is a hierarchy among human types and, in a very general sense, it provided a rationale for classification. But, as the subsequent sections of this chapter show, the framework in terms of which eighteenth-century observers began with most hope to make sense of the information was history. The pattern of history appeared to bring order to comparisons of savage and civilized man, of the state of nature and of civil society and of the absence or presence of language. History as a framework was also broad enough to encompass both biblical and secular stories and observation of both progress and degeneration in human existence.

Comparative anatomy achieved prominence in the late eighteenth century when it appeared to offer objective evidence with which to escape the speculation rife in accounts of human difference. In the last quarter of the century, Flemish and German anatomists turned comparative anatomy into what they considered to be a systematic science. Petrus Camper (1722–89), who dissected the orang-utan, argued for the facial angle as an objective measurement which permits the classification of animals and humans on a single scale. Influenced more by traditions of artistic representation than by anatomy, Camper suggested that there is a direct relationship between human quality and the vertical character of the face – animals have snouts while ancient Greeks have high foreheads. In a series of publications, Camper outlined an approach that became standard in the nineteenth century.

> Upon placing beside the heads of the Negro and the Calmuck [Kalmuk, the people of Siberia or Mongolia] those of the European and the ape, I perceived that a line drawn from the forehead to the upper lip indicates a difference in the physiognomy of these peoples and makes apparent a marked analogy between the head of the Negro and that of the ape.[12]

This was one basis for a new empirical science of racial types. A contemporary of Camper's, the German anatomist Johann

Friedrich Blumenbach (1752–1840), assembled the largest collection of skulls in Europe and put the study of human physical variety on a systematic basis. He arranged his skulls side by side in two main scales, but both ran from the most beautiful and symmetrical, the Caucasian, to the most primitive. He regarded the two divergent lines as an exhibition of the way the varieties or races of mankind represent degeneration from the original ideal stock. Shortly thereafter, the influential exponent and manager of French science Georges Cuvier (1769–1832) spelt out the general principles of comparative anatomy and equipped the study of animal and human types with authoritative methods. This helped consolidate the view held by naturalists and physicians that a true science of man must build on foundations laid in the study of physical differences. All this diverged from the eighteenth-century commitment to history as the framework in terms of which to establish this science. Later, linked to evolutionary theory in the mid-nineteenth century, comparative anatomy and history united as the subject matter of a new discipline of anthropology.

iii *Jean-Jacques Rousseau: Savage and Civilized Society*

European thought about native peoples was about European identity at least as much as it was about others. To use non-Europeans as a mirror was a favourite device in literature and political satire. The idea of Europe as an entity, as opposed to Christendom or the civilized world, began to be current in the early eighteenth century. It emerged from a self-conscious contrast between the Asian East and the European West; between nature and culture. The contrast cut in both directions: it confirmed the superiority of civilized Europe, and it criticized European customs as artificial when set against Eastern simplicity and directness. C.-L. de Secondat, baron de Montesquieu (1689–1755), who anonymously published *Les lettres persanes* (*The Persian Letters*, 1721), acquired fame and imitators for his witty views on French society, not least on marriage, projected through the eyes of visiting Persian diplomats. In his version, the East is feminine and erotic – but also prejudiced and despotic. The cutting edge of the *Letters* attacked the artificial content of Western mores. This struck home in the polite society of his day, which was familiar with masks and wigs as well as with the elaborate formal rules of patronage networks, all ultimately centred on the French court. Writers opposed natural to artificial society as the critical vocabulary of social thought, and in the process they created precedents for systematic social analysis. During the course of the century, the comparison of natural and artificial society was, however, increasingly expressed in terms of history as well as geography.

Most natural law theorists in the seventeenth century assumed that man, in a state of nature, possesses basic social relations. Hobbes, who pictured the natural individual in a state of war, was the exception. Locke and Pufendorf restated the belief in man's original sociability, and they focused attention on the relation between natural society and civil or political society. None of these writers particularly sought empirical or historical descriptions of the state of nature – though empirical elements derived from reports of the Americas were present. Eighteenth-century authors went so far as to give the state of nature empirical content, though they never discarded reference to this state as a hypothesis for the sake of argument. This approach is exemplified in the work of Jean-Jacques Rousseau (1712–78), whose description of the state of nature drew self-consciously on scientific and travel literature. His work proved controversial and influential, and it inspired an image of the human being, the noble savage, that echoed in the urban imagination of the twentieth century.

Rousseau was a complex and disturbed person, and his writings have at different times been held responsible for the extreme opposites of the totalitarian state and the self as a transcendent value. Born into the republican civil atmosphere of Geneva, he entered the aristocratic world of the *philosophes* in Paris, only to retreat again, alienated. He wrote with a high moral rectitude which was notoriously absent in some of his personal relations. Later generations distinguished Rousseau's thought from that of the *philosophes* because he gave priority to feeling in human nature and thus grounded life in truths revealed by the heart rather than in knowledge revealed by the head. 'Amidst so many prejudices and factitious passions, one must know how to analyse properly the human heart, in order to disentangle the true feelings of nature.'[13] He therefore appeared to be a forerunner of Romanticism, and later scientists assumed he had turned away from objective methods of research into human nature to stress subjective sentiment. Rousseau's harshest critics said that he projected his own contradictory emotional needs as the parameters of the human condition. Yet, if this is so, it can be argued that he found a voice for existential truth – however opposed to a natural-scientific outlook that truth is. All the same, Rousseau does not fit any of these interpretations very easily, for they play down at least two of his major historical contributions. The first concerns the key concept in eighteenth-century social thought, the state of nature. The second concerns the hopes Rousseau, like many of his contemporaries, invested in education and the perfection of human nature.

Local academies of arts and sciences, which became a feature of civil life in the eighteenth century, often took the lead over universities in the support of enlightenment. One way they did this was to offer prizes for essays on a fashionable topic, and the intellectually ambitious but not necessarily wealthy sought social recognition by this means. In this way in 1755, Rousseau submitted to the people of Dijon an essay on their question, 'What is the origin of inequality among men, and is it authorized by natural law?' He composed a great inspirational vision of human equality. His starting-point was what he claimed to be an authoritative account of life in a state of nature, which – he hoped – showed that natural law theories had presupposed untrue things about man's natural state. Yet he was still ambiguous about whether, when he referred to the state of nature, he meant an actual state or a device with which to clarify man's moral and political position. He observed:

> For it is no light undertaking to separate what is original from what is artificial in the present nature of man, and to know

> correctly a state which no longer exists, which perhaps never existed, which probably never will exist, and about which it is nevertheless necessary to have precise notions in order to judge our present state correctly.

Nevertheless, he then went on 'to judge the natural state of man correctly' with every appearance that this natural man had a historically real existence.[14] Only after this discussion of the natural state did he make his controversial point that the moral life is a consequence of artifice and not natural law.

Rousseau pictured natural man as a solitary animal among other animals, who satisfies uniform and simple needs and is indolent and rudely healthy. Yet he supposed also that man, even in his natural state, is something more: he possesses a spirit, which Rousseau identified not with reason but with the passions that lead to the exercise of will. 'Nature commands every animal, and the beast obeys. Man feels the same impetus, but he realizes that he is free to acquiesce or resist; and it is above all in the consciousness of this freedom that the spirituality of his soul is shown.' In addition, according to Rousseau, natural man possesses both a 'faculty of self-perfection', which 'resides among us as much in the species as in the individual', and compassion. With this seductive but gratuitous claim, he defended the belief that man 'tempers the ardor he has for his own well-being by an innate repugnance to see his fellow-man suffer'. Rousseau's natural man is individual, innocently at one with his feelings, feelings that are firmly feelings of *self* but include a desire for self-improvement and sentiment about others. Thus, the European might envy the savage. Rousseau asked, 'what type of misery there can be for a free being whose heart is at peace and whose body is healthy? I ask which, civil or natural life, is most liable to become unbearable to those who enjoy it?'[15]

Rousseau touched on many contentious issues, as when he linked human language to animal gesture, though he shelved the question whether language or society came first as too difficult to answer. The most resonant of his claims in the longer term was that the human essence is an asocial self in possession of feeling and liberty. This judgement expressed what it is to be human by reference to individual, subjective preferences or feelings. It made it difficult to see how individual selves could ever come together and live in society; Rousseau claimed to answer this problem in *Du contrat social* (*The Social Contract,* 1762), but it haunted subsequent generations.

In the *Discours sur l'origine de l'inégalité parmi les hommes* (*Discourse on the Origin of Inequality among Men,* 1752) Rousseau explained inequality as a consequence of the transition from asocial to social existence. Ironically, this followed from man's faculty of improvement, the faculty, conjoined with sympathy, which separates men from animals and leads men to surmount the difficulties of individual subsistence through interaction with their fellows; the consequence is that man's spirit led to his fall from grace. The moment of this fall is marked by the acquisition of property:

> The first man who, having fenced off a plot of ground, took it into his head to say *this is mine* and found people simple enough to believe him, was the true founder of civil society. What crimes, wars, murders, what miseries and horrors would the human race have been spared by someone who, uprooting the stakes or filling in the ditch, and shouted to his fellow-men: Beware of listening to this impostor; you are lost if you forget that the fruits belong to all and the earth to no one![16]

Rousseau's moral and political purposes required him to explain this transition from animal-like individual but equal existence to human-like social but unequal existence. He drew on contemporary scientific and travel reports, and he swept his readers along with inspirational passages, but consistency escaped him. After that passage on property, he observed more soberly that the idea of property could occur only after it had become possible to reflect on co-operative relations among people. This implied that property need not have such dire consequences. Man's earliest relations, those which stemmed from basic needs, led man, he thought, to conclude 'that . . . [other men's] ways of thinking and feeling conformed entirely to his own'. As he went on to say, this 'made [man] follow, by a premonition as sure as dialectic and more prompt, the best rules of conduct that it was suitable to observe toward them for his safety and advantage'.[17] Whereas natural law theorists attributed moral conduct to reason, Rousseau pointed to feeling as the basis for sociability. Many of his contemporaries took a similar view and believed that men and women possess a social sentiment which, though subjective, is the cement that makes social life possible.

When Rousseau described stages in human progress from the savage to the civilized state, readers were encouraged to think of the state of nature as a historical condition. Human dependency, he thought, created the first elements of civility and *amour-propre,* but it led also to the sentiment of pride, a sense of hurt and a desire for revenge. Some actual savage societies, like the Carib (or Caribbean), Rousseau claimed, preserve a happy balance between 'the indolence of the primitive state and the petulant activity of our vanity'. Other societies developed iron and corn, 'which have civilized men, and ruined the human race'. Manufacturing and agriculture created a division of labour and, through labour, property and inequality. Rather bitterly, Rousseau went on, 'things having reached this point, it is easy to imagine the rest': men became what they once were not – deceivers, exploiters, legislators of inequality, defenders of oppression, tyrants. Where the savage once breathed liberty, the civilized man 'pays court to the great whom he hates, and to the rich whom he scorns . . . proud of his slavery, he speaks with disdain of those who do not have the honor of sharing it'.[18] This was heady language in the absolutist political conditions of eighteenth-century Europe.

Although Rousseau held up the noble savage as a mirror, he did not, in Voltaire's jibe, recommend that people should return to walking on all fours. He did not commend savagery as a way of life since, though he thought savages might be happy, he certainly did not think they are virtuous. He believed that virtue differentiates progressive humanity from savage humanity. When he held up to men and women an image of what they were in a state of nature, he argued that nature and not society, feeling and not artifice, should instruct them on how to live.

He expressed a longing to escape the society in which he lived, a sentiment echoed in Romantic culture, and he turned to the child and his education – he was much less concerned about *her* education – as the means. In *Emile, ou de l'éducation* (*Emile, Or on Education,* 1762), Rousseau wrote didactically about an ideal education with a stress on Emile's innate feelings, his engagement with concrete things and with nature, and his late and guarded introduction to abstract words and social customs. His tutor did not allow Emile books and, when it was necessary to practise reading, the permitted text was *Robinson Crusoe,* a story filled with detail about the material facts of day-to-day existence outside society. Rousseau rejected current educational methods since they 'always seek the man in the child, without thinking of what he is before he becomes a man'; instead, 'We must view the man as a man and the child as a child.'[19] *Emile* was a moral argument rather than an educational manual, and its theme that virtue is created out of the child's nature, if he is kept innocent of artificial society, was seductive.

Belief in childhood innocence contrasted with the widely held belief that the child is emblematic of sin. In an older view, Adam was believed to have been born in possession of innocent moral and natural knowledge. Since the fall, caused by the temptation of knowledge not given to him, Adam's descendants no longer possess this knowledge at birth. Mankind has therefore constantly to be taught moral virtue and there are many opportunities for error. Hence both Protestant and Catholic churches in the eighteenth century, followed at least in this respect even by the *philosophe* Condillac, stressed discipline and a rigid educational programme for children. All these views were criticized in the late eighteenth century.

One of Rousseau's readers, though later also a critic, was a pious Swiss educator, Johann Heinrich Pestalozzi (1746–1827), whose opinions fused the Christian value of love as the essential human quality with a romantic view of nature. He struggled in adverse circumstances to found schools which would educate the child's heart, not impose a soulless discipline of rote learning. The development of the child's potential as a human being depends, he thought, on the child's freedom to sense her or his own loving individuality; this sense alone lays the basis for education. The child should therefore be given the opportunity to assimilate his or her own essential nature before the teacher imposes verbal learning and hence civilization. This argument did not express sentiment only about children. Pestalozzi wrote at the time of the French Revolution and, like many other people at this noontime of hope, he had faith that education is the means to social justice and not only a skill or commodity given to the individual. Education, Pestalozzi believed, has to be taken out of the hands of corrupt and conservative ecclesiastical authorities in order to free the child, the adult and civil society to realize his, her or its own loving nature. The redemption that was once longed for as the grace of God reappeared as faith in redemption through childhood.

iv *Social Sentiment*

Rousseau wanted to determine what makes it possible for people to live virtuously as a political society. The theory of natural law explained the bonds that make a state the outcome of the common interest in self-preservation. Rousseau rejected this argument, denied that savages or children think rationally about their self-interest and instead stressed the role of intuitive feeling, compassion and recognition of the similarity of others to oneself. Like the natural law theorists, Rousseau turned to human nature to discover the basis for social solidarity, but what he found was the feeling self. Feeling was, he claimed, the cause of social relations.

British moral philosophers assumed that man is social when they argued that everyone possesses a moral sense and therefore innate knowledge, often described as a sentiment or feeling, of right and wrong. They presumed that this shared sentiment makes community possible. The majority of moralists, French or British, treated moral feeling as the elementary experience that underpins moral goodness and social stability. As Butler concluded: 'It is manifest great part of common language, and of common behaviour over the world, is formed upon supposition of such a moral faculty . . .'[20] Few authors in the eighteenth century explored the potential in the argument that morals are relative that lay in Locke's attribution of the idea of self to sensory experience. Diderot's brutal question, 'What difference is there to a blind person between a man urinating and a man bleeding to death without speaking?' was – with the notable exception of Sade – not taken up.[21]

Hume is particularly interesting because, though sceptical about the possibility of certain knowledge, he was not sceptical about sentiment. He was socially conservative and committed to the stability of the established order in Scotland. He emphasized that 'morality . . . is more properly felt than judg'd of', and thus – like Rousseau – he attacked the natural law tradition. But, unlike Rousseau, he did not doubt that feeling leads to virtue of the kind his own community in fact valued; he did not experience conflict between personal feeling and social life, as Rousseau did. Though Hume described conduct as the outcome of ideas associated with pleasures and pains, he also argued that there is a natural feeling of sympathy, a capacity to share the feelings of others, 'a proof, that our approbation has, in those cases, an origin different from the prospect of utility and advantage, either to ourselves or others'. Hume was no relativist in morals. He believed that natural feelings constitute the cement of society, rather as fellow feeling held together the cultivated élite of Glasgow and Edinburgh. 'Beside good and evil, or in other words, pain and pleasure, the direct passions frequently arise from a natural impulse or instinct, which is perfectly unaccountable. Of this kind is the desire of punishment to our enemies, and of happiness to our friends; hunger, lust, and a few other bodily appetites.'[22] Hume therefore analysed the passions as the result of complex associations between ideas, pleasure and pain and the 'natural impulses'. Thus, when he explained the love of fame, a passion to which he himself readily admitted, he first explained what he called sympathy. The resemblance that we see other people have to ourselves, and their proximity when they are part of our family or community, causes their expressed pleasures, pains, passions and ideas to arouse feelings in us. As anything that is like ourselves arouses the vivid ideas of ourselves, we necessarily feel what others feel. This is sympathy, the binding force of civil society. When Hume chose the word 'sympathy' he was aware of its physiological connotations, and indeed he linked mental sympathy to the natural bodily appetites. Lastly, sympathy, he supposed, is naturally proportional to the distance of another person from us and hence we need 'artificial virtues' like justice to ensure harmony on a larger scale.

Amongst Hume's friends was the Professor of Moral Philosophy at Glasgow from 1752 to 1763, Adam Smith (1723–90). Smith's lectures covered the study of human action in its social relations, and they included ethics, written up as *The Theory of Moral Sentiments* (1759), and practical judgement – which incorporated material which we differentiate as economics – written up as *An Inquiry into the Nature and Causes of the Wealth of Nations* (1776). It was conventional to teach economic ideas as part of moral philosophy; the writings of natural law theorists, notably Pufendorf, had already linked human nature, jurisprudence and the material or economic conditions of life. Smith grounded ethical and practical judgement in human nature and, like Pufendorf, started from a commonplace of natural law theory: 'Every man is, no doubt, by nature, first and principally recommended to his own care.'[23] This was also a Stoic injunction, a call to lead a certain way of life and not just an abstract idea. It was an injunction to accept what people are by nature. Smith was very self-conscious of his literary style: he believed that a well-formed language enabled the study of human nature, like the study of classics, to guide young men in a commercial society to understand pride, selfishness, ambition, beneficence and, above all, virtue. The Greek and Latin writers, alongside Christianity, therefore provided him with models for conduct in human affairs and the rhetorical resources to give these models force.

Smith elevated sympathy into the human faculty responsible for virtue and sociability. Whereas Hume described sympathy as our identification with the pleasures and pains experienced by others, Smith defined it as 'our fellow-feeling with any passion whatever'. He did not analyse sympathy but assumed it is a natural God-given faculty, just as his teacher Hutcheson had accepted the naturalness of the moral sense. Smith believed that sympathy explains social relations: they are not produced by reason or sentiment, as the natural law or moral sense theorists respectively argued, but simply, as Hume also suggested, by the fact that we enter into the feelings of others. 'Whatever may be the cause of sympathy, or however it may be excited, nothing pleases us more than to observe in other men a fellow-feeling with all the emotions of our own breast . . .'[24] In retrospect, it is tempting to picture a club of Lowland Scottish gentlemen who believed that progress depends on mankind at large adopting its own refined values of mutual regard and obligation. Walter Bagehot, the Victorian writer on the political constitution, quipped that Smith wanted to show 'how from being a savage, man rose to be a Scotchman'.[25] Smith certainly wanted to argue for values which would sustain virtue in the commercial society in which he lived. He was much less concerned with power, the issue that became so important in later social analysis.

Smith's stress on sympathy led him to an interesting argument that individuals acquire a sense of self through social relations. Sympathy with the feelings aroused in others by our own actions, he thought, leads to the notion of ourselves:

> Were it possible that a human creature could grow up to manhood in some solitary place, without any communication with his own species, he could no more think of his own character . . . than of the beauty or deformity of his own face. All these are objects . . . with regard to which he is provided with no mirror which cannot present them to his view. Bring him into society, and he is immediately provided with the mirror which he wanted before.[26]

Later observers commented that the wild boy, Victor, could not recognize his own reflection; as he lived outside society, Smith might have said, he had no objects on which to exercise sympathy and hence no sense of self.

Hume's approach, like Locke's, contained the seeds of a different assessment of sociability based on utility, and these seeds flowered with Priestley and the French followers of Condillac. If all sentiments are species of pleasure and pain, it was argued, then social institutions, indeed sociability itself, result from the individual pursuit of pleasure and avoidance of pain. The social world, like individual action, conforms to a principle of utility. Since men and women are born into at least elementary social relations, every person's pleasures and pains are part of a web of reciprocal pleasures and pains created by everyone's conduct. It was therefore argued that the individual pursuit of pleasure and avoidance of pain necessarily generates social harmony. Priestley, followed in this respect by Godwin, thought that one person's well-being depends on the favourable response of others – it is in a person's interest to give pleasure to others because that pleasure is then returned. In Godwin's elaboration of the argument, only the machinations of established political powers and the corruption of society stand between humanity and its potential for liberty and happiness. He dreamed that men and women, free from constraint, will, in their own interest, do good to others. His belief about human nature underlay his classic text of libertarian and anarchist thought.

Orthodox, Catholic and Calvinist Christianity, in dramatic contrast, stressed the fallen state of human nature, and conservative political thought stressed divine right and inherited obligations within the existing polity. The French *philosophes* experienced censorship, harassment and even prison when they attacked these views. Their political position, however, was a complex one since most of them were aristocrats or enjoyed aristocratic patronage. Their struggle was with the king, his administrators and the church authorities, not a struggle on behalf of the people. All the same, radical conclusions were reached in this context, notably by Helvétius in his volumes *De l'esprit* (*On the Mind*, 1758) and, posthumously, *De l'homme* (*On Man*, 1772).

Helvétius argued that education, by which he meant more than schooling, can do everything. If, as he thought, our individual natures are formed by combinations of sensations, and if our actions are determined by pleasure and pain, then we are what experience makes us. 'Corporal sensibility is therefore the sole mover of man, [and] he is consequently susceptible . . . but of two sorts of pleasures and pains, the one are present bodily pains and pleasures, the other are the pains and pleasures of foresight or memory.' Therefore, 'pleasure and pain are, and always will be, the only principles of action in man'.[27] This view placed him firmly in the camp of those moralists who stressed the power of the passions rather than reason in human nature. As Helvétius observed, 'There is no idiot girl whom love does not make clever . . .'[28] He indeed believed that all men are created equal and that inequality in individual capacity or morality develops as a result of divergent experience. The political potential of this point of view was obvious and became manifest in the American Declaration of Independence.

Helvétius's theory of human nature was rigidly determinist and egoist: men and women, he thought, cannot but pursue self-interest. He also argued, however, that the content of man's self-interest varies enormously with 'the form of government under which he lives, his friends, his mistresses, the men by whom he is surrounded, his reading, and, finally, chance . . .' Thus, 'men, responsive to themselves, indifferent to others, are born neither good nor bad, but ready to be the one or the other' as education determines. He saw clearly that in practice the power of social institutions to inflict pain outweighs the individual's power to pursue selfish pleasure; it is the community that makes a man's or woman's character. Helvétius therefore looked to legislation to impose justice, by which he meant a means to regulate selfish motives by making it in a person's interest to accommodate the interests of others. He envisaged the legislator as the educator of the people. 'It is solely through good laws that one can form virtuous men. Thus the whole art of the legislator consists of forcing men, by the sentiment of self-love, to be always just to one another.'[29] Ironically, it was the absolute rulers like the Russian Catherine the Great, Frederick the Great of Prussia and the Empress Maria Theresa of Austria who, in the eighteenth century, did most to try to mould their subjects through centralized administration and 'education' in Helvétius's broad sense.

Priestley, Godwin and Helvétius shared a theory of human

nature and the belief that this nature is the basis of social harmony. Priestley and Godwin thought that harmony develops naturally from self-interest and that the business of politics is to dismantle government. Helvétius, by contrast, thought that sociability is imposed on self-interest and that the business of politics is to rewrite legislation. This contrast outlined the terms of a lasting debate in social and political thought based on the principle of utility: does human nature itself generate social harmony (the natural identity of interests) or does harmony require appropriate legislation (the artificial identity of interests)? I will return to this debate and utilitarian thought in the context of political economy. This section has highlighted the views of moralists, some of whom were radical and rejected belief in an innate social sentiment, while others were conservative and believed in natural sympathy, but all of whom thought man is necessarily sociable. Writers on the science of man created that science as a practical expression of what they wanted society to be. The science of man and the science of progress were one and the same. As James Madison, a founder of the American Republic, wrote: 'But what is government itself but the greatest of all reflections on human nature?'[30]

v *De l'esprit des lois*

The language of human nature supported an analysis of social life in terms of qualities ascribable to persons. This often but not always obscured the direct study of social realities as phenomena in their own right. There was a literature on the body politic, relations of church and state, trade, poverty and wealth, epidemics, customs, legal systems and much else besides. All the same, there was no conclusion that this literature referred to a unified subject or that it is possible systematically to study social institutions. The categories of 'society' and 'social institution', in terms of which twentieth-century sociologists described and compared the units of collective life, were not present. It was once common, influenced by the modern French sociologists Emile Durkheim and Raymond Aron, to say that Montesquieu founded the discipline of sociology. Though historians no longer find this a meaningful claim, Montesquieu certainly was widely read and admired in the eighteenth century. His book, *De l'esprit des lois* (*The Spirit of the Laws*, 1748), which was the result of twenty years' labour ('I swear that this book nearly killed me'), was considered a masterpiece.[31] That it was placed on the Catholic *Index*, even in France, despite Montesquieu's status as an aristocrat and a Catholic and despite his best efforts to avoid religious or political provocation, shows how far the search for a science of man had the power to threaten the established social order.

Montesquieu accepted that the social world, like the physical world, exhibits regularities; to believe otherwise, he thought, is to accept the absurdity that everything is due to blind chance. His book opened with the statement: 'Laws, taken in the broadest meaning, are the necessary relations deriving from the nature of things; and in this sense, all beings have their laws . . .' Laws of physical nature and laws of human nature exist in the constitution of things. But since man has a rational and sentient nature, he also furthers his ends through positive laws, laws that people establish for themselves, and these laws are no more arbitrary than natural laws. The study of positive law, he argued, reveals regularities in the conditions of social life.

> [Laws] should be related to the *physical aspect* of the country; to the climate . . . to the properties of the terrain, its location and extent; to the way of life of the peoples, be they plowmen, hunters, or herdsmen; they should relate to the degree of liberty that the constitution can sustain, to the religion of the inhabitants, their inclinations, their wealth, their number, their commerce, their mores and their manners; finally, the laws are related to one another, to their origin, to the purpose of the legislator, and to the order of things on which they are established.[32]

All these relations, taken together, form what he meant by 'the spirit' of the laws.

The book that followed described these relations and examined an extraordinary range of topics, from the circumstances that make for frugality to administration by women to etiquette in combat. He self-consciously claimed to observe in what was called the Newtonian manner. Like so many writers on both natural and moral philosophy, he stated proudly that 'I did not draw my principles from my prejudices but from the nature of things' and that 'I began by examining men'.[33] This 'examination' drew on a rich classical education in Paris, wide reading in the legal and travel literature, his stays in various countries – especially England – and his experience as nobleman, judge, landowner and wine merchant in the Bordeaux region. He assembled a compendium of the causes of the variety of social life past and present, and much of what he said was readily accessible by an audience interested in local differences of character. Montesquieu observed, for example, that warm climates expand the nerve fibres, which – given also the bountifulness of nature under this climate – makes people indolent and, in his view, likely to be passive, despotic and Catholic.

The first two parts of his book subsequently attracted most attention since it was here that Montesquieu distinguished different systems of government – monarchies, despotisms and republics – and analysed their different principles and the implications this has for liberty, education and social life generally. In contrast to earlier writers, he supposed that the nature of government follows from the way rule is exercised, rather than the reverse, and he explained how the manner of government follows from the physical and social factors which his book described. His argument was not abstract but continuously exemplified, and in the last two parts of the book he turned to a historical analysis of government in France. This in itself, regardless of his conclusions, appeared to conservative opinion to question the monarch's claim to divine authority.

The scope and richness of Montesquieu's commentary on the law-like patterns beneath the observable diversity of government, law, commerce and custom awed his contemporaries. The Scottish writer Adam Ferguson (1723–1816) wondered whether he had anything to add – though, as we shall see, he certainly did. Montesquieu did not distinguish one topic in its own right, for interesting reasons, and that is the question of the female sex. He did not ignore woman but discussed her as an element in the social fabric only in specific contexts: luxury and sumptuary laws (which regulated conspicuous consumption), marriage, children and population and domestic servitude. There was a similar pattern in Ferguson's writing: only certain topics prompted a mention of woman. Writers on sensation, reason and the understanding, like Hume and Condillac, who constructed a science of mind, did not discuss woman at all. Woman was a significant topic in relation to a restricted range of issues, but not for a universal science of man. Nevertheless, this observation does not tell the whole story. There was also a large amount of writing on woman as a topic in her own right, and some authors even hoped to establish a science of woman. This suggests that writers were not oblivious of woman when they propounded a science of man, but that 'woman' was to them a separable, special subject whereas 'man' had universal connotations.

Many of the contributors to thought on woman were themselves women, such as Louise-Marie-Madeleine, madame Dupin (the great-grandmother of George Sand), who employed Rousseau as a secretary and left extensive notes for a study of the rights of women. The travel literature featured descriptions of the variety of conditions under which women live; a notable example was Lady Mary Wortley Montagu (1689–1762), whose *Letters . . . Written During Her Travels in Europe, Asia, and Africa . . .* (1763) were widely read. Other authors surveyed woman's contribution to civilization and to literature or discussed the special problems of female education or medicine. At the end of the century, and even more in the nineteenth, male physicians directed their physiological approach to human nature towards a specific account of woman and her character built around the reproductive organs and gynaecology. Pierre Roussel (1742–1802), a doctor, wrote a popular *Système physique et moral de la femme* (*System of the Physical and Moral Nature of Woman*, 1775), which was an early example of this genre of books.

The Spirit of the Laws was the most famous contribution to a vast literature that compared social life at different times and in different places and used this comparison to comment on the causes of social and political experience. The literature was practical rather than abstract, and it addressed education, political liberty, moral conduct, the generation of wealth and domestic relations. These comparative and practical aims made space for woman as a subject, just as she was taken into account in the specific detail of legal practice and social administration. She was not absent or invisible in the science of man. Nevertheless, 'woman' as a category was, for the most part, conceived in contrast to 'man'. 'Woman' appeared when discussion of the particulars of passion, government and custom differentiated humanity into its component parts. This interest in social differences encouraged some sharp comments on the conditions which cause woman's social inferiority and dependence, thereby contributing to an argument for emancipation and liberty. Mary Wollstonecraft (1759–97), in her book later celebrated as a feminist classic, *A Vindication of the Rights of Woman* (1792), drew a pointed parallel between women and other dependent social groups, like soldiers and clergymen, and she used the comparison to explain mental habits of dependency and hence the political subjugation of women.

The literature by Montesquieu on woman illustrates the way comparative social description engaged critically with established social institutions, whether they were the law, the family or trade. The comparative imagination crossed space, linking civilized and savage society; it crossed time, linking modern progress and past ages. This historical dimension was also built into the classical education that gave authors on human nature so many of their examples and so much of their rhetorical skill. The search for causes in the social domain became a search for history. This occurred parallel to the search for causes in the physical domain that created the historical sciences of cosmology, geology and evolutionary biology. In the eighteenth century, the novel also created personal history to explain the causes of a person's character and conduct. The historical dimension became a structural feature in the science, literature and social lives of men and women.

vi *The History of Civilization*

Historical writing was well established in the sixteenth and seventeenth centuries. Herodotus and Thucydides were classical models, the bible told the history of creation and jurisprudence invoked the history of custom. It was thus no novelty to turn to the past for knowledge. Eighteenth-century authors, however, recreated this sense of the past in at least two important ways. Firstly, they laid the basis in technique and style for what became an academic subject with its own methods of study and validation. This discipline of history burgeoned in the nineteenth century. Secondly, they expanded historical imagination to include the history of civilization in general and thus to give content to the idea of progress. History became the means by which people understood the relationship between natural man, represented by the past, civil society, represented by the present, and political liberty, represented by the future.

Several of the writers who are thought of in the twentieth century as exemplifying Enlightenment values were best known in their own day for their histories. Voltaire's *Le siècle de Louis XIV* (*The Century of Louis XIV*, 1751) was a masterpiece of literary style and humanitarian moralizing, and it was also a wide-ranging review of progress across a whole age of European history. Hume's *History of England* (1754–62) upset political and religious sensibilities but, in the process, became admired for the way it showed how it is possible for historical writing to distance itself from the narrow interests of particular groups. These histories significantly questioned Christianity as the central theme and causal force of historical change; such questioning was most obvious when historians wrote critically about the Middle Ages and implied it was a hiatus between ancient and modern civilization. Edward Gibbon's monumental *The History of the Decline and Fall of the Roman Empire* (1776–88) ended on a tone of irreparable loss rather than of excitement over the foundation of Christian Europe. His history also disturbed his contemporaries because it implied that it is the historian's business to investigate the causes of Christianity's rise to power rather than to propagate the Christian message.

History, like travel literature, made it possible to compare Christian and non-Christian societies. History and geography together suggested that Christian beliefs and political order are relative to particular times and places. The very act of comparison created a language that could be adapted to systematic and objective study, that is, it created a language for a science of society. D'Alembert, in the 'Discours preliminaire des editeurs' (Preliminary Discourse, 1751) to the *Encyclopédie*, described history as concerned with:

> how men, having been separated into various great families, so to speak, have formed diverse societies, how these different societies have given birth to different types of governments, and how they have tried to distinguish themselves both by the laws that they have given themselves and by the particular signs that each has created in order that its members might communicate more easily with one another.[34]

Montesquieu's *Spirit of the Laws* looked for causes and thereby also reinvigorated historical thought. One chapter was entitled, 'That laws must not be separated from the circumstances in which they were made'; this invoked a maxim of historical scholarship and then used it to explain the apparently appalling Athenian law that 'all the useless people [were] to be put to death when the town was beseiged'.[35] In such ways, reference to historical particulars rather than abstract reasons began to be preferred as the means to understand change, progress and diversity in the existence of man.

A controversial part of Montesquieu's argument was where he discussed physical factors, especially climate, in order to explain different customs. He started with *l'homme physique*, man as corporeal being, and described how man in time became *l'homme moral*, man as cultural being. After Montesquieu, other writers provided accounts of the historical transition from physical or savage to moral or civilized society. The Scottish moral philosopher Dugald Stewart (1753–1828), referring to Smith's history of moral sentiments, called them 'conjectural histories'. Conjectural history equated present savage and historically primitive society and then described the broad history of what was perceived as the civilized West, the static East and the primitive rest of the world. Conjectural history drew into a unified intellectual enterprise the idea of the state of nature, Montesquieu's and Buffon's search for causes, knowledge of the ancient world and travel to exotic places. It also gave systematic form to the Western assumption of its superior values.

The modern historian of economics Ronald L. Meek spread the notion of what he called 'the four stages theory'. Meek argued that a generalized concept of human historical development which became current in the 1750s unified social analysis, with significant consequences in the nineteenth century. The four stages theory, in Meek's description, attributed social change to changes in modes of subsistence, from hunting, to the use of pasture, to agriculture and finally to commerce, and the theory therefore explained the transition from savage to civilized life by material conditions. French and Scottish Enlightenment writers, it appeared, originated the materialist theory of history and initiated the scientific understanding of society through their adoption of a developmental perspective. But other historians qualified Meek's conclusions and pointed out that non-material factors, such as the sentiments of human nature or the political ideals of the legislator were taken to be important causes by writers like Smith who accepted some version of the four stages theory. Montesquieu greatly stimulated a search for causal order beneath the surface of social diversity. A theory of historical stages, many elements of which appeared in Montesquieu's work, made sense of diversity by showing it to be a consequence of the causal story of human history. The theory was taken further by several writers in the 1750s; the earliest in France was the young A.-R.-J. Turgot (1727–81), later a theorist of economic questions and servant of the crown. In an oration at the Sorbonne and in a manuscript on universal history (written *c.*1751–2, published 1808), he explained the advance of the human mind by an account of the origins of ideas in sensations gained in the course of the satisfaction of material needs. To satisfy needs, he conjectured, people had established systems of social organization, which increased productivity by agriculture, the division of labour and commerce. This history, he concluded, is progress. He started with the biblical flood, when he supposed men were hunters, and he went on to describe the achievement of pastoral and agricultural ways of life, and, in his later writings on economic topics, in effect he discussed the causal conditions of commercial society.

This conception of progress became commonplace. A particularly rich discussion was given by the Scotsman Ferguson, in *An Essay on the History of Civil Society* (1767). He was remembered later when Marx cited him with approval for his insights into the historical formation of economic relations, although Ferguson was conservative in sentiment and had served as a regimental chaplain in the British North American colonies before he became professor of natural and then of moral philosophy in Edinburgh. Ferguson was a Christian moralist: he rejected the equation of progress and wealth and denied that progress was inevitable. He wanted to understand civilization in order to fight against moral corruption and decline. To sustain civilization, he argued, appropriately for an army man, requires struggle, hardiness and discipline: 'The virtues of men have shone most during their struggles, not after the attainment of their ends.'[36] It was, he thought, therefore imperative to study the conditions that made civilization possible in the first place.

Ferguson began with human nature, asserting that 'affection, and force of mind, which are the band and strength of communities, were the inspiration of God, and the original attributes in the nature of man'. Sociability, accompanied by reason, language and an upright body, he believed, distinguish man from the animals. Consequently:

> In the human kind, the species has a progress as well as the individual; they build in every subsequent age on foundations formerly laid; and, in a succession of years, tend to a perfection in the application of their faculties, to which the aid of long experience is required, and to which many generations must have combined their endeavours.

Like Smith, he argued that sociability is intrinsic to human nature: 'Man is, by nature, the member of a community . . .'; the history of civilization is therefore the history of communities as much as the history of individuals. Indeed, Ferguson thought that the conventional separation of nature and artifice, especially when the individual is regarded as natural and society as artificial, is misguided. 'We speak of art as distinguished from nature; but art itself is natural to man.'[37] This was an insight that frequently proved illusive in modern individualist societies. The individual, as Ferguson saw, is not definable independently of social existence. The notion of an individual itself represents a social value. To treat the individual and society as independent categories is incoherent, yet the history of the human and social sciences is full of examples of the polarization of the individual and society. Not least, as discussed later, an asocial conception of the individual had deep-seated consequences for experimental psychology as an academic subject.

Ferguson drew on contemporary observations of native North Americans in order to conjecture about the natural art of the savage. He then discussed the history of civilization as the history of the way people have satisfied their basic need for food. He distinguished four main stages – hunting, pastoral, agrarian and commercial – and in this way held together a conception of man as a physical being with a conception of man as a moral agent, since he related moral progress to material changes. Ferguson stressed that advances in the arts and in social arrangements, such as the division of labour, arose spontaneously from the capacities of savages rather than by diffusion from already advanced communities. Local circumstances, like climate or the proximity of a seaboard, he thought, explain the varieties of national character.

Ferguson denied that much could be known about how the transition from savagery to civilization had actually occurred. There was an absence of concrete information, and the problem this caused returned with a vengeance when evolutionary theorists in the nineteenth century in their turn attempted to reconstruct the history of civilization. Both then and in the eighteenth century, many writers were much less constrained than

Ferguson. Two of his compatriots, the judges Lord Kames and Lord Monboddo, published extended, speculative and antagonistic histories. Monboddo's attribution of language to society and not human nature was only one part of his account of the history of civilization in connection with natural conditions. By contrast, in his *Sketches of the History of Man* (1774), Henry Home, Lord Kames (1696–1782) featured the social dispersion and linguistic confusion after God's destruction of the Tower of Babel. In Kames's view, Babel explains the divergent histories of the world's people. More conventional than Monboddo, he regarded language as an original human attribute, and this belief supported his nostalgia for the primitive nobility of Scotland's past and his enthusiasm for the recently 'discovered' and supposedly ancient Gaelic poems of Ossian.

Readers in the second half of the eighteenth century became very familiar with the idea that savagery and civilization are related historically as well as geographically. They learned that man has passed through developmental stages and that historical knowledge, even if critics judged it to be conjectural rather than empirical, is a systematic form of reflection on man. They also acquired a notion of what was later called the materialist theory of history. In the historian Robertson's words: 'In every inquiry concerning the operations of men when united together in society, the first object of attention should be the mode of subsistence. Accordingly as that varies, their laws and policy must be different.'[38] Similarly, the work by John Millar (1735–1801), *The Origins and Distinction of Ranks* (1771; this is the title of the 3rd edition of 1779), discussed social classes and customs and included comments on the condition of women, as part of a search for the causal laws that produce social differentiation ('distinction of ranks') over time. Readers did not find it hard to imagine that Europe had once been savage, especially when this picture invigorated a belief in progress. History established relations between different peoples whilst it still maintained them in a hierarchy. Progress in morality, economy and liberty was understood to mark out the special quality of northwest Europeans and their heirs in North America from savage and static societies around the world. In this way, conjectural history symbolized at the same time as it legitimated the economic and political expansion of Europe.

The link between history and progress reached its greatest rhetorical but also tragic heights in Condorcet's *Esquisse d'un tableau historique des progrès de l'esprit humain* (*Sketch of the Historical Progress of the Human Mind,* 1795). M.-J.-A.-N. Caritat, marquis de Condorcet (1743–94), had faith in man's history as a movement towards truth and freedom. This movement is driven, he thought, by the cumulative observation of the laws of nature. 'Why shouldn't politics, grounded like all the other sciences on observation and reasoning, be perfected accordingly, as more subtlety and exactitude are brought to its observations, more precision, profundity and accuracy to its reasoning.'[39] Knowledge, he believed, inevitably carries humanity as a whole forward as people learn that their happiness depends on the liberty of others. His rhetoric of human progress brought together Condillac's analysis of mind, belief that a natural sentiment of sympathy supports sociability and a political commitment to equality and liberty. Condorcet, however, was not only a dreamer: he wrote what became the first draft of the revolutionary National Assembly's 'Declaration of the Rights of Man and the Citizen', and his reports on education suggested to the French Republic the means with which to educate the new citizens of the 1790s. Carried forward by the *idéologues,* these ideas transformed the French school system. Nearly every administrative *département* established a central *école* to teach the physical sciences and the sciences of history and legislation, while the advanced *grandes écoles* were founded in Paris. The Ecole Polytechnique taught mathematics, natural science and engineering, while the Ecole Normale Supérievre trained future teachers. The tragedy is that Condorcet composed his most visionary writing during the Terror and that he himself was a victim, probably by suicide, of that darkness.

The search for causes, order and progress in the social sphere reconstructed the writing of history. At times, history was the intellectual framework for an understanding of humanity itself. History in the early modern period had provided moral and spiritual insight or legitimation for legal and political power.

In the eighteenth century, history also became the record of sociability and the achievement of civilization, the history of man. This historical imagination went further and deeper than even the French and Scottish writers took it, since Vico and Herder argued in the same century that humanity is what it is precisely because it has a history. With their belief that humanity is its own artificer, they made history the core knowledge of a science of man. But this argument, highly influential in the nineteenth century and beyond, is reserved for a later chapter. I will first say something about the material heart of European civilization and progress, the generation of wealth. At least as early as 1600, commentators desired to control the social transformations they saw about them through an understanding of the economic realities of trade, money, capital and state and individual action. It is possible to suggest that these interests generated the most technical and detailed sciences of human nature in the period before the nineteenth century. Economic events occurred before people's eyes and dug into their pockets; they were a powerful incentive to link theory and experience.

9

Political Economy

> This division of labour, from which so many advantages are derived, is not originally the effect of any human wisdom . . . It is the necessary, though very slow and gradual consequence of a certain propensity in human nature which has in view no such extensive utility; the propensity to truck, barter, and exchange one thing for another.
>
> Adam Smith, *An Inquiry into the Nature and Causes of the Wealth of Nations* (1776)[1]

i *'Economy' in the Seventeenth Century*

The notion of 'economy' implies management, the ability to order things in a way appropriate to their purpose. A medieval steward ran the economy of a household – a vocabulary evident in 'home economics' in the twentieth century. In the seventeenth or eighteenth centuries there were phrases such as 'the economy of heaven', which referred to God's dispensation for the order of the world. From notions of a proper and efficient management came the values of economical action and the careful husbandry of resources. Applied to the physical body and the body politic, it gave rise to the language of the bodily economy and the political economy.

All kinds of economies have abstract features in common, and hence the experience of one kind of economy has frequently reappeared as a way to understand another kind of economy. This analogy has been of great importance for the human sciences. The body, the household, the state and the world were seen to be the ordered wholes of man's material existence, and thought and language about one level carried over to the other levels. In 1628, Harvey compared the circulation of the blood to the water cycle in which rain refreshes the earth; he compared bodily and terrestrial economies in an Aristotelian idiom. In 1860, Herbert Spencer compared the circulation of the blood to the circulation of money; he compared bodily and commercial economies in a capitalist idiom. In both cases, the use of analogies affected the content of what was claimed as knowledge. The language of economy was the language of belief about complex wholes and the way part and whole integrate in the body, the household, the state and the Creation. It was a particularly rich language for the description of relations between individuals and the social economies of which they are part. If, in the twentieth century, 'the economy' denoted only the sum of financial arrangements in a given society, this exemplifies the way modern consciousness acquired a restricted view of social relations.

An economy is not a passive state but something actively achieved. A Renaissance steward managed the household economy or a prince ruled over coinage, labour and the exchange of goods. It was a valued art, indeed a Christian duty, to manage well. Early modern observers, confident in the power of experience, began to collate and analyse relevant information in order to make economy in all its forms a science as well as an art. Bacon declared that 'if books were written of this subject as of the other, I doubt not but learned men with but little experience, would far excel men of long experience without learning, and outshoot them . . . in their own bow'.[2] Day-to-day material relations and the pressures of mundane business affairs shaped the subject matter for sciences of management – the human sciences of practical life that dealt with such matters as food, property, population, labour and exchange. Out of the necessity to manage, whether between master and family, prince and subject, or banker and debtor, came many basic arguments about human nature. European states consolidated centralized power in the seventeenth century, and they were preoccupied by finance, trade, law and order and administration in general. Those involved took it for granted that human affairs are not or at least are not designed by God to be arbitrary, and they tried to give their recommendations and policies authority in terms of belief in the underlying natural order. This belief was religious and aesthetic, signified by the phrase 'the economy of nature', which was used in the sixteenth century and increasingly thereafter. Many writers aspired to praise and worship God with an acknowledgement of the beauty, harmony and efficacy – the economy – of his creation. This view enhanced a conception of the human economy as a seamless part of the natural economy. In the eighteenth century, pleasure and confidence in the design of the created physical world played an important part in the search for the design of the human world.

Many writers went into print to try to influence decisions and policies. In a pamphlet written in 1623 (though not published till 1664), the English merchant Thomas Mun (1571–1641) concluded his advocacy of a free trade policy with the argument that 'this must come to pass by a Necessity beyond all resistance'; he deferred to what he believed to be natural law.[3] A later writer on the English coinage, Charles Davenant (1656–1714), in 1695 restated natural law theory as the foundation for financial practice: 'The supream [political] power can do many things, but it cannot alter the Laws of Nature of which the most originall is, That every man should preserve himself.'[4] By the second half of the eighteenth century, the study of the link between the natural order and material prosperity was known as political economy, the investigation and management of the material and political conditions and laws, physical and social, that underlie wealth. Adam Smith wrote: 'Political oeconomy . . . proposes two distinct objects; first, to provide a plentiful revenue or subsistence for the people . . . and secondly, to supply the state or commonwealth with a revenue sufficient for the publick services.'[5] Political economy developed as a subject hand in hand with commercial society. As a systematic science at the end of the eighteenth century, it both reflected and influenced the social framework. Political economists assumed that individual human nature is a central part of what needs to be studied, and hence political economy was yet another aspect of the eighteenth-century search for a science of man. In the first half of the nineteenth century, in the eyes of some practitioners, it was the most developed of all the sciences, physical as well as human.

This chapter therefore describes the way in which the practical art of the management of material scarcity and prosperity created a human science. In principle, the topic includes politics, war, demography (the study of population), medicine and much else besides. In practice, I describe only the background of political economy as a subject. It is not possible here to say much about the modern disciplines of economics, political science and management science which emerged in the late nineteenth or the twentieth century in circumstances that are described later. Political economy, however, was part of the root and branch transformation of Western life that gave it its capitalist, industrial and urban character, and it was the language in terms of which human nature was understood as this character came into existence.

There was no conception of 'the economy' as a category in the early seventeenth century. In the Aristotelian university curriculum, economy – in the sense of right management – occupied a loosely-defined position in ethics. The separation of a new subject area, which began to occur in the eighteenth century, involved the gradual demarcation of economic from moral questions. Before this time it was assumed that questions concerning trade, taxation or poverty required a decision in terms of Christian and hence ethical conceptions of social order. As a mid-seventeenth-century republican, John Cook, wrote: 'The rule of charity is, that one mans superfluity should give place to another mans conveniency . . . the Magistrate is impowred by God to command every man to live according to the rule of nature and right reason.'[6] Cook, who drew up the indictment against Charles I at his trial in 1649 and was executed for it in 1660, entitled his essay, *Unum necessarium*, necessary unity: each person and thing in civil society possesses a place dictated by God's order as embodied in natural law. Prices, for example, are to be determined by 'the just price' set by guild corporations or royal representatives, not by the market. In this respect like some conservative theorists, Cook believed that it is a Christian duty to control the human sphere in accordance with the natural law. Social realities changed, however, and already in the sixteenth century capitalist economic decision-making and the profit motive were in evidence. Conceptions of what natural law is changed, too, and there was a new stress on the fundamental obligation to self-preservation.

Merchants, farmers, state advisers and householders all tried, of course, to act with practical economic acumen. In the seventeenth century, a desire for systematic knowledge about wealth creation accompanied practical values, just as the interest in the new natural philosophy generally was connected to technology and the practical arts. The centralization of political and economic decision-making in the state's interest was characteristic of the new European order in the seventeenth century. This policy, taken up in France, Austria, Prussia and Sweden, made knowledge of the sources of wealth and security – population, manufacturing, agricultural productivity and a favourable balance of trade – matters of great concern. In turn there evolved a self-conscious attempt systematically to understand economic conditions, that is, to formulate an economic science; and the eighteenth century saw the creation of university chairs in the management of the state. Yet there were no chairs either in the Dutch United Provinces or in Britain, though it was these states that showed the fastest economic growth. As early as 1615, a French writer, Antoine de Montchrestien (1575–1621), who addressed his thoughts to Henry IV's centralized administration, published a *Traicté de l'oeconomie politique (Treatise on Political Economy)*. The economy or efficient management of the state began to be a practical science and economic questions, in the restricted modern sense, started to acquire a degree of autonomy in people's minds.

France, the Low Countries and England were the most dynamic generators of economic activity and wealth during the seventeenth century. Not by chance, they were also repeatedly at war with each other. The French kings consolidated their political power over the whole of France and, in the state's interest, centralized social administration and controls over commerce and manufactures. This policy developed furthest under Louis XIV's minister of finance between 1663 and 1683, Jean-Baptiste Colbert, who pursued the creation of domestic wealth as an instrument of foreign policy. The state took an interest in the health and fertility of the population, in new techniques for the manufacture of goods, in the control of trade, especially the exchange of currency and precious metals, and in the finance of the army. Colbert and his administrators also believed that the state needed empirical knowledge of social and economic conditions and of manufacturing processes, and the Académie des Sciences was instructed to take up such topics when it was founded in Paris in 1666.

The growth of wealth, military power and intellectual and artistic culture in the Dutch United Provinces aroused envy and puzzled admiration. It demonstrated that a small state without resources, a country that even had to create its own land, is able to establish a leading place in the world. Not gold but work, and not traditional social order but finance capital, it appeared, creates wealth and power. Whatever the Dutch rhetoric about Providence or Calvinist deference to predestination, many observers concluded that ingenuity, hard work and financial skill transforms material conditions. The success of the United Provinces therefore focused attention on individual or corporate initiative and on detailed aspects of credit, freedom of trade, the laws of contract, the circulation and source of value of money and so on. These were technical matters and they were debated in relation to specific decisions. The habit of debate familiarized people with economic matters in a narrow sense and developed the terms for more systematic scientific study.

English policy wavered between the state-directed arrangements adopted in France and the greater freedom of individual or corporate decision-making found in the Low Countries. There were debates about whether the national interest requires a favourable balance of trade and whether this in turn requires the retention of currency and gold and silver within the country. The literature on these topics began to isolate particular economic factors, like the quantity of coin in circulation, that could be measured and analysed. It became common to discuss the effect that a certain level of customs tax or rate of interest has on a particular trade – the decline of woollen exports was a recurrent concern. Writers treated these issues empirically, like issues in natural philosophy, as a basis for improvement. They presupposed that the human domain, like the physical world, has a discernible order and discoverable causes. It was accepted that human nature in general, and certain character traits in particular – like those attributed to the Dutch, are relevant causes. In the period from 1660 to the mid-1690s, there was much argument in support of free trade. Its proponents claimed that individual or corporate pursuit of wealth benefits the wealth of the state as a whole; they often described such self-interested activity as natural and inevitable, implying that a free market – like human nature – is a natural condition.

ii *Political Arithmetic*

On the surface, economic and political life appears chaotic. The thronged quays of Amsterdam and Venice or the outbreaks of war and plague exhibited apparent disorder. Merchants and statesmen knew that actions have unanticipated outcomes. The conditions that affect material well-being were so complex that it often looked as if the outcome of particular events could not be known. In consequence, probabilistic forms of reasoning, which make inference possible in conditions of complexity and unpredictability, were crucial to the origins of political economy. In a parallel way, the development of statistical thinking in the nineteenth century went hand in hand with the origins of the modern social sciences. Probabilistic and statistical argument became the technical means with which to reveal order – and hence make science possible – in political economy and social science. It is significant that the writers who in the seventeenth century tackled political and economic questions also developed new techniques to gather quantitative data and to estimate rates of mortality, the quantity of coin in circulation and the likelihood of unforeseen events such as fires. All this had important consequences for natural philosophy and mathematics as well as political economy.

Natural philosophy and early political economy came together in the work both of the Royal Society of London and of the

Académie des Sciences. A fellow of the Royal Society, William Petty (1623–87), like his colleagues, thought that knowledge is reformed through the observation of regular relations among measurable factors. He himself set about the study of the 'political anatomy' of the body of the state; as a physician, he found this analogy between the political and the physical body ready to hand. He began with an analysis of public finances in a *Treatise of Taxes and Contributions* (1662) and proceeded, in *Political Arithmetick* (1690, written 1671–6), to attempt a comprehensive quantification of the country's capital assets and population. Petty knew Hobbes in France in the mid-1640s and, like him, thought about individuals as discrete entities who operate in accordance with their rational self-interest. He therefore made the individual the unit of economic action. This also had the methodological advantage that it became possible to assign a numerical value to each action. Petty dreamed that it is possible to summate and balance individual actions in a political arithmetic. At the same time, he described the market, which he conceived in terms of free exchange between individuals, as a quantitative equation. This form of expression emptied social relations of moral considerations: 'I have taken the course . . . to express myself in terms of Number, Weight or Measure; to use only arguments of sense, and to consider only such causes as have visible foundations in nature . . .' He hoped that his work laid the basis for the application of 'political anatomy' to the country – and he had Ireland specifically in mind – 'without passion or interest, faction or party; but as I think, according to the Eternal Laws and Measures of Truth'.[7] He did not sense any contradiction between this argument and his acquisition of 70,000 acres of forfeited and mortgaged land after he surveyed an Ireland subjugated by Oliver Cromwell.

Petty probably participated in the earlier study by his friend John Graunt (1620–74), published as the *Natural and Political Investigations . . . Made upon the Bills of Mortality* (1662). Graunt used what he himself called mere 'Shop-Arithmetique', to show how the published lists of mortality figures for London could be used to garner information relevant to policy.[8] For example, he teased out of the records the figures for death by starvation or criminal assault in order to provide evidence to counteract public panic that both had become so common that novel methods were needed to suppress population and crime. More ingeniously, he used the figures to estimate London's population and to compare mortality rates between different parts of the city and the countryside. He was sensitive about the unreliability of the information fed into the recorded figures and tried to make reasonable adjustments. Nevertheless, he made questionable assumptions to make up for missing information, especially about the age of people at death. In his work, as so often later in the social sciences, calculative ingenuity outstripped the quality and rigour of measurement.

Graunt would not have been able to proceed at all if the authorities had not required data – or, to use a Victorian term, 'vital statistics' – to be recorded. Formal records of baptisms, marriages and deaths became a characteristic feature of early modern life, first in German towns like Augsburg early in the sixteenth century. English parish records, which Graunt used, began in 1538, and the London bills of mortality, probably designed to record epidemics, in 1519. Much earlier, the Italian cities devised ways to record commercial and financial information, and by the sixteenth century formal accountancy procedures were in use. A quantitative language for social life developed. This language was practical in intent but was also open to more systematic elaboration and hence was utilizable as the language of science. In Britain there was suspicion that government might use information, were it available, in ways inimical to individual liberty. Public opinion associated quantitative reasoning with centralized power and qualitative experience with personal liberty. Nobody then put it in these terms but, in embryo, this was an issue which later dogged the implementation of conclusions drawn from the social sciences. The British parliament in fact opposed the introduction of any national census until 1801 (and then excluded Ireland), and there was no comprehensive census until 1851. Political arithmetic was left to interested individuals.

Graunt's work created interest in France where Louis XIV's administration, in association with the Académie des Sciences, established a bureaucracy to assemble comparable data. His work also stimulated studies of mathematical probability, begun because of an interest in gambling odds, as the means to calculate expectancy. Expectancy was relevant to lawyers as well as gamblers since some forms of contract (aleatory contracts) are forward-looking; for example, a merchant may contract to purchase a crop before it is harvested and its yield known. It is interesting, however, that Graunt's and similar studies of life expectancy did not influence the life assurance industry, though it grew rapidly at this time, for many years. Those responsible for the calculation of premiums in the light of expected risk preferred to do so on the basis of what they knew of local circumstances, even of the particular person to be insured, rather than on the basis of general rules, however rational or scientific. Similarly, nobody used Petty's 'political arithmetic' for financial or commercial decisions.

At the end of the seventeenth century, Gregory King (1648–1712), who had a mixed career as a herald, genealogist, engraver and calculator, extended techniques of measurement to more narrowly economic factors such as prices and demand. He also estimated England's population as a whole and mapped the distribution of people in different socio-economic categories. His work illustrates the way in which study of economic factors and exchange in quantitative terms gave them status as material causes rather than moral or spiritual purposes in a natural order. This contributed to the appearance that social phenomena, from population growth to grain prices, are comparable in terms of order and predictability to phenomena in nature. Petty, for example, objected to government intervention in the export of bullion as contrary to the laws of nature. The expanding commercial market was increasingly structuring human relations in terms of the financial value of what each party had to offer, whether social position, land, labour or goods. It fostered a new type of knowledge about human beings in which they are economic actors. Throughout the eighteenth century, however, other models of human relations – such as the one associated with Renaissance civic humanism – remained influential, even within political economy.

Modern anglophone social scientists celebrated Graunt and Petty as the founders of empirical social research and statistical methods. This judgement ignored both the wider European setting and Graunt's and Petty's marginality to economic life in Britain. Quantitative techniques were common much earlier in business; double-entry bookkeeping, notably, had developed in the sixteenth-century trading city states of Venice and Florence. Quantitative reasoning, which focused on delimited economic topics in King's work, was only the most clear-cut case of seventeenth-century concern with economy in its broad sense. A practical investment in the means to order social life was widespread in early modern Europe, in the entrepreneurial United Provinces as much as in centralized France. All this activity generated a language in terms of which the activity itself was examined. In fact, qualitative moral arguments rather than quantitative language remained uppermost, perpetuating medieval Christian debates about community and property even while economic activity expanded and changed. The display of great personal wealth or conspicuous consumption, called by the generic name 'luxury', juxtaposed to recurrent famine and dire poverty, gave intensity to argument about the proper balance between the rights of property and the rights of community. In concrete terms, governments had to decide whether grain was to be made available at a just price, i.e., at a price that the poor could afford in conditions of dearth, or whether the owner could legitimately pursue the highest price. Moral and political argument, rather than quantitative analysis, took the lead in such debates.

One area where quantification came into its own concerned population, and the debate was able to use the information collected since the sixteenth century. It became common to gauge a state's achievement, even its happiness, by its population size and growth; it also appeared obvious that wealth and military power depended on a large population. It was assumed that a wisely ordered nation has an expanding population, and – as if to confirm this – Britain's population grew rapidly in the eighteenth century along with its power. As Hume wrote, 'if every thing else be equal, it seems natural to expect, that, wherever there are most happiness and virtue, and the wisest institutions, there will also be most people'.[9] There was every incentive to understand the conditions that affect population size. The study of population, widely undertaken in the eighteenth century, spilled over – in ways important for theories of human nature – into the study of the family, sex, woman's place and childhood. It also integrated with medical and public health interests. Right at the end of the century, T. R. Malthus reinforced the joint study of population and political economy and, through his work, it flourished in the nineteenth century.

The attempt to turn all these topics into a discipline that would serve the state was undertaken in German and Austrian central Europe and in Sweden, where small educated élites had an important role in state administration. This was the world that Pufendorf and Leibniz inhabited. In these settings there was little or no separation of economic, social, medical and legal policy; there was little inclination to change the study of administration as a service to the state into a form of analysis distinctively focused on individual economic action. As the council or legislative chamber of the ruler was sometimes known as the camera, the study of ruling and government became known as cameralistics. The discipline became established in higher learning with two chairs, established in 1727, in the cameral sciences at the Prussian universities of Halle and Frankfurt on the Oder. Other universities, notably Vienna in 1750, followed suit and offered teaching in public administration to train state officials. The first chair at Halle was in *Oeconomie, Polizei und Kammer-Sachen* (economy, police and cameralistics) – the economic aspect referred to the conditions of good husbandry, the police aspect to the public order (which concerned food, health, clothing and population rather than crime) and cameralistics to the business of state administration. All areas of concern to state administration were brought together but there was no settled opinion about the classification of topics or the use of terms. The university teaching was intended to provide a rational basis for state policy, and new texts were therefore written to train administrators and not just to guide the hand of rulers. Cameralist scholarship consisted of a mixture of deduction from rigorously demonstrated principles and the compilation of practical data about social and economic life. There were some large-scale projects, especially that projected by Gottfried Achenwall (1719–72), in his *Staatsverfassungen der europäischen Reiche* (*Constitution of European States*, 1752), which assembled tables of descriptive statistics. Achenwall, who was appointed to the university of Göttingen to teach the science of government by the ministry in Hanover, intended that figures should be collected on a regular basis. In 1765 the Austrian academic Joseph von Sonnenfels (1733 or 1732–1817) published the first edition of his textbook, *Grundsätze de Polizei, Handlung, und Finanz* (*Principles of Police, Commerce and Finance*), the most used introduction to *Staatswissenschaft* or the science of government.

An interest in the science of cameralistics and in population came together in Sweden in the work of Anders Berch (1711–74). Sweden, like France, was a centralized state, though it possessed a form of government that somewhat represented four 'estates' – nobility, church, burghers and peasantry. Berch wrote a thesis in which he proposed the use of centralized power to improve the conditions for population growth, trade and manufacturing. This work articulated the views of a group of like-minded scholars at the university of Uppsala who accepted the leadership, in both natural philosophy and political economy, of Anders Celsius (1701–44), the inventor of the centigrade scale. Berch's family connections with commerce and manufacturing in Stockholm, the political centre, helped him in 1741 to obtain the new chair of jurisprudence, economy and commerce at Uppsala. But he had to overcome the opposition of Linnaeus who – though known more widely as a naturalist – understood 'practical economy' to come within his own sphere of interest; for Linnaeus, the natural economy and the political economy ran together in one created world. In *Sätt, at igenom Politisk Arithmetica utröna länders och rikens Hushållning* (*On Studying the Wealth of Nations by Political Arithmetic*, 1746), Berch calculated the size of Sweden's population (the state then included Finland) on the basis of a compilation of the number and size of agricultural units from tax returns. Another compiler of information, Pehr Elvius (1710–49), used figures of births and deaths and assumptions about life expectancy to make a comparable calculation, but his work gave a substantially lower figure. Finally, the state itself thought the issue of sufficient moment to institute a national agency, which began in 1749 to count the population head by head. This project initiated what became the longest demographic time series in the world. Berch also wrote a textbook and attracted sufficient funds to found a *Theatrum Oeconomico-Mechanicum,* a museum for economic science, which exhibited raw materials, manufacturing processes and finished products. Practical natural philosophy and the science of human economy were closely connected in Uppsala and elsewhere in Europe.

Swedish, German and Austrian teachers and officials did not see distinctions between economic and social policy. The modern tendency, in analysis as in policy, to separate ethical, social, economic and political spheres reflects a practical and intellectual division of labour created by later historical circumstances. Most writers in the eighteenth century, in France and Britain as well as in the less dynamic economies of Central Europe or Scandinavia, thought about economic questions as part of a wider concern with the conditions of civil society. This was certainly also the case for Adam Smith who, in spite of his reputation as a founder of the subject of economics, never described himself as an economist and lectured about economic questions as a branch of practical moral philosophy.

iii *The Wealth of Nations*

The English pamphleteers in the late seventeenth century who referred to the laws of nature wanted to free trade, prices and wages from state regulation. They argued against established policy, which in Britain and even more elsewhere made the state the overseer of commercial activity. Men like Davenant argued that human nature rather than the state should direct the social arrangements that produce wealth. The issues came to a head in the 1690s, when there was a pressing need for a recoinage to re-establish the value of money. As it turned out, William of Orange's accession to the throne marginalized the free traders, who were mostly Tory in politics, and instituted a state trade policy that protected home manufacturing. The dominant policy evaluated social and economic transactions in the light of what was claimed to be the collective interest. Concurrently, moral philosophers like Shaftesbury stressed what is in human nature, the moral sense, that builds a representation of the collective interest into a person's feelings and conduct.

A physician and moralist of Dutch origins, Bernard Mandeville (*c.*1670–1733), injected into this debate an outrageous but seductive argument. First in the form of a poem, *The Grumbling Hive* (1705), which he later expanded with essays and a commentary, Mandeville proposed that private greed and desire for luxury is the only basis on which to build national wealth. The final form of his book's title put the point succinctly: *The Fable of the Bees; Or Private Vices, Public Benefits* (1729). He praised burglars, for example, because they keep locksmiths in business and because, as they are profligate, they recirculate money. But he needled his contemporaries with the serious thought that all the talk about a moral sense is so much hypocrisy, that an honest appreciation of human nature acknowledges its selfishness, and that profitable economic activity requires such vices as envy, vanity and love of luxury. In practical terms, Mandeville opposed charitable schools for the poor and supported free trade, and he did not think the state should concern itself with 'private vices'. His views repelled the many moralists who tried to reconcile Christian values with the new consumer societies of London and Paris.

In theoretical terms, Mandeville argued that there need be no match between individual intentions and social outcomes. This is what has been called the law of unintended consequences, the belief, as formulated by Ferguson, that 'the establishments of men . . . arose from successive improvements that were made, without any sense of their general effect; and they bring human affairs to a state of complication, which the greatest reach of capacity with which human nature was ever adorned, could not have projected . . .'[10] Belief in such a law of unintended consequences suggested that analysis should attend to social phenomena in their own right, rather than to the idealized values of a moral sense. Mandeville's contemporary Montesquieu, discussed in the previous chapter, undertook something like such a project.

Mandeville's praise of private vices struck home because he wrote at a time when there was considerable self-consciousness, especially in England, both about what was identified as commercial society and about great disparity of individual wealth. Observers said that something was making English society different from anything that had come before and that this difference lay in its being *commercial*. Smith achieved lasting fame because he provided an analysis of these new conditions, which he summarized in the following way: 'Every man thus lives by exchanging, or becomes in some measure a merchant, and the society itself grows to be what is properly a commercial society.'[11] He meant, as others had suggested before, that every person's place in society is defined by what they buy and sell, whether it be land, rent, capital, labour, goods or services. By the 1770s, when Smith published, the opinion was widespread but not universal in Britain that commercial society is the inevitable and admirable form of modern civilization. It was the last stage in descriptions of the progress of man.

Commercial society appeared to make everything pivot around individual acts of exchange – hence historians have tried to explore connections between the origins of modern society and the modern consciousness of the autonomous self. In the seventeenth and eighteenth centuries there was concern that commercial relations disrupt religious obligations, social hierarchies and the authority of the state, and in consequence there was a sophisticated literature on such matters as the legal right to property, a topic discussed at length by Grotius and Locke. The debate's intellectual context was initially moral, religious and legal; but gradually the dominant discourse became naturalistic and economic. Smith stood astride this change. Earlier, Mandeville mocked the moral discourse of his day and uncoupled moral action from social or economic action. This was initially unacceptable: it was, for example, Hutcheson's duty as professor of moral philosophy in Glasgow to lecture on economic topics like the division of labour, trade, interest and luxury in relation to moral judgement. Smith – unlike Mandeville – did not mock, and – like Hutcheson – he taught moral philosophy, but he did develop a language in terms of which a later generation fully detached economics from morality.

The formation of commercial society, the beginnings of which some historians trace to before 1400 and which was certainly a different process in different parts of Europe and North America, is associated with a key modern view of human nature. The term 'economic man' is a code-word for the opinion that what is called society is only an association of individuals who act in the light of rational self-interest to maximize their material profit and well-being. The term connotes a theory of motivation. In the 1980s and 1990s, a crude rhetoric of economic man had a new political life while state socialism disintegrated in the USSR and Eastern Europe and commercial management values, legitimated as market forces, ran through Western institutions. In the eighteenth century, the term referred to a view of human nature that had to be argued for since critics believed it marginalized ties of moral and social obligation. Smith and other civic moralists in the Scottish Enlightenment, followed by utilitarian political economists like Bentham and John Stuart Mill, systematically discussed motivation and ethics as well as economic realities. Their science of man was a moral philosophy. But they fostered a new self-image of man, the image of a modern person who conceives of his or her self-identity and purposes in terms of his or her position as an economic actor.

The shift towards commercial society occurred first and fastest in Britain (especially England) and its American colonies and in the Low Countries. It was noted with interest that these countries did not experience levels of poverty that turned into starvation. In much of the rest of Europe, by contrast, the efforts to regulate economic life and to subsume it under moral, legal and religious values were also a struggle with endemic poverty and recurrent famine. In France, the state, in pursuit of its own interest, imposed controls over trade, labour and prices. There was no French science of cameralistics, though economic issues were much debated and very visible because of periodic riots over grain prices, a grossly unequal and inefficient taxation system and a fear that Britain's wealth underlay its military successes around the world. In these circumstances a group of well-connected men banded together to press for reform and a more rational system. They shared the practical values embodied in the *Encyclopédie*, and they were influenced in particular by the Irish-born banker Richard Cantillon (1680–1734). Cantillon, whose *Essai sur la nature du commerce en général* (*Essay on the Nature of Trade in General*, written 1730–4) circulated in manuscript for twenty years before publication in 1755, is seen by historians as a link between the seventeenth-century English proponents of quantitative analysis and later authors. The French group coined the term '*physiocracie*' to describe what it hoped would be a new policy founded on rules of nature as opposed to the inherited practices of the state. Physiocracy encouraged a shift, also evident in Britain, towards the acceptance of commercial social relations as natural and attention to them as a central topic for the science of man. In terms of practical policy, the physiocrats tended to favour the monarchy since they thought it offered the best opportunity to implement a rational and scientific form of decision-making.

François Quesnay (1694–1774), the leader of the physiocrats, was a physician, and he approached the financial problems of the body politic as so many symptoms of bodily disorder. His analysis of medical and economic thought attempted to uncover comparable natural laws. Later on, the imagery of disease and healing was important in pro-revolutionary rhetoric and, in the language of the *idéologues* and their heirs in the nineteenth century, informed an influential conception of social administration. *Physiocracie* was thus articulated at a transitional moment in French life when there was a shift towards commercial society, a shift that encouraged a search for natural economic and medical laws as a basis for political decisions.

France was overwhelmingly rural and agricultural, and these conditions were reflected in the assumptions of the two founders of the physiocrat group, Quesnay and the marquis de Mirabeau (1719–89). Quesnay persuaded Mirabeau, during an evening's conversation at Versailles, Mirabeau wrote, that all wealth is traceable to agricultural productivity – not, for example, to population size, as many authors had suggested. Quesnay therefore claimed that the transition from nomadic or pastoral existence to settled cultivation was the crucial moment in the passage from savage to civilized life. History, he thought, follows the increased productivity of the land, driven by the surplus of agricultural goods over the consumption of food required to produce it. This, Quesnay argued, is the material basis for exchange and hence for the expansion of consumption and population. If a state can increase the disposable surplus of agriculture, economic growth follows. This assumption also led him to classify society into the 'productive class' engaged in agriculture; the 'class of proprietors', the landowners including the king and the church, who receive the yield of agriculture as rent, taxes and tithes; and the 'sterile class', which includes manufacturers, who are dependent on what agriculture produces and incapable, he thought, of producing a surplus themselves and hence of promoting other activity.

Quesnay and Mirabeau published a series of books and popularizations of their arguments which brought together the small but active and close-knit group who took the name '*physiocracie*' in the 1760s. They optimistically believed that they had uncovered the causal mechanism that underlies the generation of wealth and established a sound foundation for policy. What they identified as the causal mechanism, agricultural surplus, was described as the agent of historical change and hence the key to civilization itself. In the *Tableau économique* (*Economic Table*, 1758, but repeatedly altered and republished), Quesnay tried in a visual manner to communicate how productivity, rent and exchange generate all the complexities of material life. It seemed – if only one could understand it, and Quesnay was forced to make repeated explanations to puzzled viewers – to lay out the anatomy of society, and this anatomy was material, causal and historical. But by 1770 the school was in decline. Its policies, grounded in the basic requirement to increase agricultural production, upset almost every interest group in one way or another. In 1774 Louis XVI appointed Turgot, an associate of the physiocrats, to the position of Controleur Général des Finances, and the result of his radical experiment was bread riots, violent repression and Turgot's resignation. Removing state controls and allowing producers or middlemen to determine the price of grain did nothing to solve France's problems, but merely illustrated the disjunction between the physiocrats' theories of material reality and the complexities of actual political events.

Most people in Britain as well as France – the issue was not even raised in Central Europe – believed that freeing grain prices would both lead to famine and riot and be a dereliction of the state's responsibility. The great exception was Smith. He was familiar with but critical of the physiocrats' arguments, especially their fundamental premise that agriculture alone is the source of wealth. Smith's starting-point was commercial not agricultural society, reflecting his assumption that civilization had advanced to a new historical stage. With unprecedented thoroughness, subtlety and discipline, he argued for the quantity of labour expended as the basis of economic values and hence for a generalized theory of the origins of wealth that did not depend on the primacy of any one occupational sector. He also distinguished, as the physiocrats had not, the role of the owners of capital who make labour productive. Smith thereby initiated an analysis of the capitalist system of production and at the same time examined the historical development of capitalism in terms of law-like processes. It was his arguments rather than those of the physiocrats that grounded systematic economic thought over the next century, down to and including the work of Marx. Smith's achievement, followed by that of Ricardo, became known as classical economics.

For many modern economists, Smith was simply the founder of scientific economics, the man who showed how it is possible to construct abstract models of economic relations and thereby study the significance of any particular variable. The physiocrats, from this viewpoint, attempted something similar, but they failed because they equated an abstract model with a particular social reality, agriculture. In fact, Smith had no conception of a discipline of economics isolated from the study of moral relations, from the history of civilization or from political questions about how Britain should be governed. He defined political economy 'as a branch of the science of a statesman'.[12] But he did not say that clearly at the beginning of his long-prepared work, *An Inquiry into the Nature and Causes of the Wealth of Nations* (1776). In this book he discussed capital and labour at length, encouraging modern economists to sideline the historical and political aspects of his work, whatever his own intentions may have been.

Smith's work expressed the values of the Scottish Enlightenment, and his earlier activity as professor of moral philosophy was connected with his publication of the *The Wealth of Nations*. He was heir to a civic humanist culture modelled on Greek and Roman example. In this culture, a man's life was assessed by his moral and rational qualities and by the way they contribute to the advancement and welfare of his nation. Smith's lectures to his students, discussed earlier in relation to his published lectures on *The Theory of Moral Sentiments*, cultivated a sense of moral obligation and the correspondence between personal virtue, especially in a public figure, and the common good. Moral philosophy was the basis for 'the science of a statesman'. And, in spite of his later reputation as a proponent of the free market, Smith was convinced that the legislator had a crucial part to play in civilized communal life. He in fact lectured on jurisprudence, though these lectures were published only in the late nineteenth century on the basis of student notes taken in the early 1760s. They make clear how deeply Smith engaged with the natural law debates about property and liberty and how, like his contemporaries, he thought about such matters in terms of a historical relationship between savage and civilized society. From this perspective, his 'science of a statesman' concerned the legal grounds for the right direction of the civil polity.

While Smith was therefore a man with a wide political vision, his work was an intellectual challenge: he argued at length that the vision could not be pursued without an understanding of wealth, that is, the material conditions of history and politics. His discussion of this material dimension pioneered technical arguments that later acquired independence as the subject matter of economics. It was possible for the twentieth-century Austrian-American economist J. A. Schumpeter to make the remarkable claim that *The Wealth of Nations* is 'the most successful [that is, influential] not only of all books on economics but, with the possible exception of Darwin's *Origin of Species*, of all scientific books that have appeared to this day'.[13] Smith, one imagines, would have been flattered, but he would also have wished to know whether society had become more civilized. For himself, his work was moral philosophy not scientific economics.

Most of the issues he addressed had been debated for at least a century in relation to the respective rights and duties of the individual and the state in commercial and in other areas of life. But Smith rightly thought that the new economic activity he saw around him, especially in manufacturing, required new answers, and he took up a position critical of public opinion. He set out to demolish early eighteenth-century protectionist principles, and in the process he repeated thoughts about human nature expressed by his seventeenth-century predecessors. Unlike them, however, he synthesized a discussion of every area of economic activity. He also defined the end of economic activity as consumption, praised the entrepreneur as an accumulator of capital rather than as a moral type and identified mechanisms of equilibrium within the market. Earlier writers thought of the market as an aggregate of human decisions. Smith saw it as the sum of commercial transactions that are as real as physical facts, and he assumed that market relations are governed by general laws. In the long run, though not in Smith's own work, such assumptions made it possible to substitute mathematical models for explanation in terms of historical and social developments. Smith's thought rendered human affairs subject to causal laws, and in this sense his book can indeed be compared to the great works of natural science. And yet, as his critics have not ceased to point out, what Smith described as the market incorporated the values of a particular commercial society, particular human choices and a particular distribution of political power.

Smith is often remembered as an advocate of *laissez-faire*, the belief that the general good is best served by the free economic decisions of individuals. The term *Laissez-Faire*, however, was used by French free-traders in the eighteenth century and became common in English only in the mid-Victorian period. It does not describe Smith's position. His concern was equally with the wealth and the *justice* of civil society, in the tradition of Locke and the natural law theorists. Though there was little novelty in his claim that land, labour and stock or capital are 'the three great original and constituent orders of every civilized society from whose revenue that of every other order is ultimately derived', Smith believed that commercial realities had changed the way these ends had to be pursued.[14] Consequently he attributed a decisively important role to the division of labour, as a result of which, he thought, sufficient wealth is generated to remove physical conditions and poverty as constraints on a state's well-being. In turn, this has consequences for justice and government.

It is not, Smith argued, necessary for the interests of property and the interests of the poor to be at odds, with the balance maintained by the state. Rather, once the principles of wealth generation are understood and applied, wages rise and a high wage economy enhances consumption, production and so on in an upward cycle. The key is the productivity of labour. 'The annual labour of every nation is the fund which originally supplies it with all the necessaries and conveniences of life which it annually consumes . . .' Thus the amount of goods available is 'regulated by two different circumstances; first, by the skill, dexterity, and judgment with which its labour is generally applied; and, secondly, by the proportion between the number of those who are employed in useful labour, and that of those who are not so employed'. Books I and II of *The Wealth of Nations* elaborated these basic principles and showed how 'the constituent orders' of society interact through the self-interested choices of individuals. People, he argued, naturally pursue their own interests; but social interaction, which includes the expression of 'moral sentiment', brings about a natural balance in the state as a whole. He described how this balance is achieved and the conditions that foster or disrupt it. But he thought that equilibrium between 'the constituent orders' is natural.

> By directing . . . [his] industry in such a manner as its produce may be of the greatest value, he intends only his own gain, and

> he is in this, as in many other cases, led by an invisible hand to promote an end which was no part of his intention.[15]

Smith's reference to 'the invisible hand' was a rhetorical flourish, not to be taken literally. Nevertheless, his notion of the balance of interests is not understandable independently of his theory of moral sentiment. He believed that God's providence has so designed human nature that people, on adequate reflection, can and do share the feelings of others. He recreated the virtues of civic humanism as the virtues of liberty and sympathy in a commercial society. Thus he believed that free market relations are the conditions of social harmony, not only the conditions productive of wealth. Though commercial society exacerbates differences in wealth, Smith accepted, it still enables labourers to satisfy their need for subsistence. The market mechanism providentially achieves justice because it reconciles inequality of property with adequate provision for all.

More commentators became sympathetic to the views of physiocracy or of Smith towards the end of the eighteenth century. Condorcet was a notable advocate of the argument that economic and administrative decisions should be based on the laws of nature. Yet there appeared to be irreconcilable pressures to argue with analytic rigour about a narrow issue, such as the balance of trade, while operating in the muddy political world in which actual decisions were taken. This tension faced all the later human sciences when they created an intellectual order which they claimed to be relevant to the disorder of daily life. Smith addressed the problem in moral terms and did not separate economic theory from notions of justice and virtue appropriate for a commercial society. Other students of the science of man concluded that only reform at every level – political, economic, social, legal – would bring a rational and scientific world into existence. And they concluded that if established society resisted knowledge of science then a more amenable society would have to be created. Such thoughts in favour of the science of man did not cause the revolution in France, though that revolution did give them an opportunity. In Britain, such thoughts supported utilitarian argument and made political economy into what some of its advocates claimed to be the queen of the sciences.

iv *The Principle of Utility*

Business in the council chamber of University College London is presided over by an embalmed body that sits upright in a glass case. In earlier years, the body appeared at student lectures. Jeremy Bentham (1748–1832) did not believe in the spirit and he therefore left instructions that his body was to be preserved to inspire the non-religious academic institution founded in the 1820s in his name. The institution stands, and Bentham sits, emblematic of belief that analysis, instruction and reform – applied science – in place of superstition, tradition and sectional interest, unlock human progress.

The intellectual foundations of Bentham's ambition rested on the account of human nature constructed by Locke, Hartley and Hume. They described human action as the consequence of pleasures and pains and they equated what is good with what produces pleasure. Even proponents of moral sense theory, like Hutcheson, helped fashion 'utility' as a criterion of value. During the course of the eighteenth century, the equation of what is best with what causes happiness was turned into the deduction that the sum of human good is the sum of human happiness. If this view was accepted, it appeared possible to calculate the value of a thing or an act by reference to what it contributes to human happiness: its value is its utility. Priestley drew the clear-cut conclusion that the goal of political society should be the greatest happiness of the greatest number. Helvétius reached the same conclusion though, unlike Hartley and Priestley, without the religious inspiration that utility equates with God's purposes for man. The writers who followed this line of thought established a political principle of the greatest generality. 'Utility' became a criterion for the assessment of any act, whether individual or collective, and it appeared to permit the unification of a science of man. Re-expressed in the language of judgements about the function of individual or social acts, it became a central explanatory concept and value in the social sciences in the twentieth century.

Bentham's self-appointed role was to be the Newton of the science of man and pope of the church scientific. He believed that a rational approach to human affairs, which the principle of utility made possible in theory, could not be applied in practice in the irrational conditions of established society. He thought his world was irrational because its activity was the outcome of history and custom, not reason – a point that he and his followers thought the physiocrats had inadequately appreciated. And, unlike Smith, Bentham did not believe that the market generates just social relations. He therefore devoted his life to the elaboration of legal principles that, if instituted, would establish a rational social order. He trained as a lawyer but never practised and instead worked at a theoretical system of legislation and a practical code of criminal and civil law to replace the English common law. He sought to establish legislation and law on first principles, i.e., on utility. The business of law, he argued, is to base social order on human nature, with the expectation that once the basis is established, happiness will necessarily follow. His argument was for the social engineering of human happiness. In the law he identified an institution with the power to translate ideals, so often thwarted during the Enlightenment, into social policy.

Benthamite theory, which contemporaries knew as philosophic radicalism, probably did not substantially affect political events, like the legal reforms of the 1820s and 1830s once attributed to it. But the values Bentham and his followers articulated were fundamental to the modern world. Empirical social data, which the early Victorians called statistics, utility as a criterion of judgement and law as the vehicle for the implementation of policy became characteristic of the modern administration of social life. Bentham's followers were at the forefront of English reform in relation to poverty, public health and crime, and Bentham's language entered the decisions of the republican Convention in Paris and the social policies of the American Republic. France, though neither Britain nor the American Republic, introduced a comprehensive system of codified law during the

Napoleonic period. It was fitting that the Convention made Bentham a citizen of the infant republic in 1792. Indeed, some of his – usually unfinished – writings appeared only in French translation.

The work of the philosophic radicals treated theories of human nature, economic thought, moral philosophy, political analysis, social policy and jurisprudence as linked topics within the overall scope of political economy. Yet of course there was division of labour between authors. The self-made businessman David Ricardo (1772–1823), who came from a Dutch Jewish family settled in England, in fact rethought Smith's economic ideas independently of any interest in the wider concerns characteristic of Bentham. In *The Principles of Political Economy and Taxation* (1817), Ricardo isolated economic factors and treated them abstractly; he took Smith's categories for granted and deployed them independently of their social implications and historical roots. The text that resulted was notoriously dry. He based his largely deductive arguments on an analysis of value, that is, on an analysis of the source and distribution of the units of economic exchange. As he concluded: 'Possessing utility, commodities derive their exchangeable value from two sources: from their scarcity, and from the quantity of labour required to obtain them . . .'[16] Given a measure of value, which this definition provided, it was in principle possible to track all the operations of the market and to make economic predictions. In practice, when he became a member of parliament, Ricardo spoke out for free trade, as he believed this to be the condition in which commodities and labour find their real value and hence the condition in which commercial society functions without distortion. Through his work on the labour theory of value, he became aware of a number of contradictions in his analysis, which he was unable to resolve. This was the starting-point for Marx's later social class-based critique of Smith's and Ricardo's economics; Marx, unlike Ricardo, argued that capitalist activity produces a destabilizing exploitation of labour.

Where Ricardo's work related to Bentham's was in its unspoken assumptions about human nature. Smith, Bentham and Ricardo shared an idealized image of economic man. They believed that fixed attributes of human nature – the pursuit of pleasure and the avoidance of pain and, in Smith's case, also sympathy – explain the fixed reality of economic laws. Since human nature is what it is, they argued, economic reality also has a natural pattern. Even as they wrote, however, the radical Tom Paine, whose work was widely read in France, the American Republic and Britain and did much to create a working-class consciousness, was inspiring belief that supposedly natural economic laws are an illusion conjured up by the rich in their own self-interest.

Two further aspects of Bentham's work are of particular interest for the idea of a human science – though his capacity to complete what he wrote for publication did not match his ability to conceive of comprehensive projects. The first was the hedonic calculus, an attempt to assign numerical values to pleasures and pains and hence to supply a means for the calculation of the desirability of any action. Since, as he declared at the very outset of one major work he did publish (in 1789), 'Nature has placed mankind under the governance of two sovereign masters, *pain* and *pleasure*', it ought to be possible to know in advance what to do. 'It is for them alone to point out what we ought to do, as well as to determine what we shall do.'[17] He devoted much humourless time and trouble in order to assign each pleasure and pain a numerical value, and his reward was ridicule.

They say he cherished men,
Their happiness, and then
Calmly assumed one could
Devise cures for their good,
Believing all men the same,
And happiness their aim.

He reckoned right and wrong
By felicity – lifelong –
And by such artless measure
As the quantity of pleasure.
For pain he had a plan,
Absurd old gentleman.[18]

Given the assumptions with which he started, however, the undertaking was not at fault in principle; nor is it obvious that twentieth-century cost-benefit analyses were very different. Bentham examined the circumstances that affected the quantity and quality of pleasures or pains, such as intensity, duration, certainty and 'fecundity' – the chance of more of the same; and he took into account people's varied sensibility. The more refined his analysis, however, the further away he seemed to get from the practical calculations about the future that ordinary people do all the time. Why should this have been the case? If human actions are part of a law-like order, in which our choices maximize utility, why should Bentham not consult tables to determine the best social policy? Most people could guess the answer: Bentham ignored the variety of feelings and the complex ways they sustain identity and self-worth.

The second topic to seize Bentham's imagination was the application of the hedonic calculus to the specific question of punishment. Punishment, he argued, if rational, should be determined by its utility.

> [Pleasure and pain] will need to be considered in estimating the mischief of the offence. Is satisfaction to be made to . . . [the injured person]? they will need to be attended to in adjusting the *quantum* of that satisfaction. Is the injurer to be punished? they will need to be attended to in estimating the force of the impression that will be made on him by any given punishment.[19]

It is salutary to juxtapose these comments to the penal practice of which Bentham was a lifelong and vehement critic. In 1757, the regicide Robert François Damiens had his flesh torn with pincers, molten lead poured on the wounds and his body pulled apart before the Paris crowd by horses attached to each of his four limbs. Bentham and other reformers thought punishment disproportionate to the crime was cruel; it had no utility; it was even counterproductive when, as sometimes in England, juries acquitted a defendant if they regarded a punishment as too severe. It was commonplace in the English courts to value goods at less than one shilling so that the defendant could be convicted of 'petty' larceny and thus escape the capital sentence which followed conviction for 'grand' larceny. To have value, i.e., utility, Bentham thought, punishment should be certain and swift. He therefore threw all his weight behind a European-wide campaign to reform the established systems of retributive justice.

Penal reform arguments seized the public imagination as a result of Cesare Beccaria's *Dei delitti e delle pene* (*On Crimes and Punishments,* 1764). Beccaria (1738–94) protested against the use of torture to obtain confessions, arbitrary judicial powers and capital punishment for minor offences. As in his later lectures in Milan on political economy, which introduced modern thought about wealth into Italy, Beccaria argued that policy should be founded on human nature. He stated that the business of criminal law is to enforce contract not exact retribution. Each citizen, he held, can be said to have made a contract with the state as a condition of sociability. The best law, it follows, prevents crime and, when this fails, enforces the social contract rationally in accordance with the principle of utility. Bentham's elaboration of Beccaria's proposals was a lifelong obsession. His wanted to classify each and every crime and to assign to it its appropriate type and grade of punishment. Once this was done, judges would then simply administer the rules. Rational administration was to replace arbitrary, and hence ineffective as well as unjust, power. Article VIII of the French National Assembly's 1789 'Declaration of the Rights of Man and of the Citizen' embodied these Benthamite ideals: 'In order for punishment not to be . . . an act of violence of one or of many against a private citizen, it must be essentially public, prompt, necessary, the least possible in the given circumstances, proportionate to the crimes, dictated by the laws.'[20]

Bentham thought he had found not only the principle of rational punishment but also the practical way to implement it. His device was the Panopticon, the design for a building which places each convict, isolated in a cell, on a circumference and observable at all times by a central overseer. Bentham believed that the convict, if always observed, will observe himself and thereby appreciate the justice of his position and the remorseless logic that matches his crime, its punishment and social utility.

Bentham lavished time and money to draw up plans and to lobby the powerful, but his Panopticon was never built. Nevertheless, his reasoning did to a degree materialize in bricks and mortar, as he influenced institutional architecture. Examples are the Pennsylvania Eastern State penitentiary in Philadelphia (opened 1829) and Pentonville prison in London (opened 1842).

The Panopticon had mythic stature in the late twentieth century. It symbolized the dream of the total administration of the human spirit, a state in which the eye of the governor is at all times sensed, though not actually seen, by the governed. When the person who is observed is made to reflect, the governor becomes the internal regulator of everyday life. In the modern world, a critical language suggested, men and women live as disciplined subjects, governed from within. Bentham meant his device to be entirely impersonal: governmental power, he believed, derives legitimacy from utilitarian principles, not from any person or party. His goal was to transcend the temporal and the personal to obtain an economy of social organization grounded in individual discipline. Later critics of the nineteenth-century expansion of bureaucracy and social administration feared that the end point will be reached when society itself becomes a Panopticon. Bentham's dream was a twentieth-century nightmare. Whatever one's views about this, it was political economy, or, latterly, the psychological and social sciences, that made scientific theory out of the technical practices and institutions through which modern administration acts. And the theory, in turn, reinforced the practices and institutions.

The complex knot that tied together social administration and the human sciences is at the centre of the history of those sciences when they became separate disciplines and occupations. I therefore discuss this question later, in relation to 1900 rather than 1800. By 1800, however, political economy was already a complex science that integrated truths of human nature, disciplinary practices, institutional innovation and new forms of government. What this meant in practice is illustrated by the activity of the great pottery manufacturer Josiah Wedgwood (1730–95). Wedgwood became wealthy because he made technical innovations in clays, colours and glazes, on which he worked in his own laboratory. But he also prospered because in the 1770s and 1780s he thought systematically about the organization of his factories and paid attention to the work economy of his labour force. New industrial production methods, he believed, require new kinds of human beings, men who are disciplined to work in time with the schedule of the production process. He redesigned his workforce as well as material nature, and changed his workers' concept of time. Manufacturing made both material and human nature part of a new economy.

Technological ideas were present in new schemes to train children as well as to train workers. The dream of rational education as a production process reached its climax in the Lancaster system. Joseph Lancaster (1778–1838) was a religious dissenter who, in 1798, set up a school in south London in which he trained senior pupils as monitors. The monitors then trained other pupils in the basics of reading, writing, arithmetic and rote learning. By this means, the system appeared able to teach large numbers of poor children sufficiently well for them to read the bible, at low cost and in a short period of time. His school soon had a thousand pupils and attracted the interest of philanthropists and even George III. The system took children who could not pay, equated manufacturing and education and turned out children with basic skills. It was also an investment in the production of ordered lives as it introduced children taken off the streets to discipline.

Bentham had a reputation as a radical since he criticized existing social institutions and argued for new foundations. Radical reform and commitment to the principle of utility continued to be linked in the nineteenth century. Nevertheless, the principle of utility apealed across the political spectrum. One of the best-known writers to apply utility as a criterion of social judgement was the Reverend Thomas Robert Malthus (1766–1834; the first name was not used in his lifetime), who wrote on the general principles of political economy as well as on specific issues of the day. The revolutionary events in France directed everyone's attention to the conditions of political stability and instability. Bentham and the utilitarians argued that stability results from the legislation of an artificial identity of interests. Malthus, who was known to his contemporaries for his political independence, appalled in equal measure by the Revolution and by what he regarded as Godwin's naïve and radical belief in the natural identity of interests, maintained that stability is produced by moral acts reinforced by social sanctions. He argued that enquiry into the natural laws of social progress shows that there are limits to progress, which radicals ignore. Malthus believed that individual moral effort, coupled to economic and social reform, nevertheless makes some progress possible.

'Population Malthus', as he was known, developed his case in terms of a theory of population which was enormously influential. The first edition of *An Essay on the Principle of Population, As It Affects the Future Improvement of Society* (1798) was a polemic against Godwin's and Condorcet's utopias, but in the second edition (1803), which was really a new book, and in subsequent editions he added evidence and moderated his initially exceedingly pessimistic conclusion. Malthus proposed as a 'law of nature' that the rate of population growth tends to increase geometrically while food resources may be increased arithmetically, with the consequence that conditions of scarcity are permanent.

> This natural inequality of the two powers of population and of production in the earth, and that great law of our nature which must constantly keep their efforts equal, form the great difficulty that to me appears insurmountable in the way to the perfectibility of society . . . I see no way by which man can escape from the weight of this law which pervades all animated nature.

This argument, with its appearance of quantitative and empirical reasoning, hung as a spectre over nineteenth-century social thought, an unwanted guest at the industrial feast. It haunted imagination about every kind of economy, notably Charles Darwin's evolutionary approach to the biological economy, and it echoed in late twentieth-century ecological anxiety. Malthus, however, who was a Christian moralist, interpreted his natural law to be a providential arrangement that prompts an indolent humanity to create wealth. He also looked to prudential restraint, the moral law, to mitigate the worst consequences of population growth. 'To avoid evil and to pursue good seem to be the great duty and business of man, and this world appears to be peculiarly calculated to afford opportunity of the most unremitted exertion of this kind . . .'[21]

The *Essay on Population* made Malthus's reputation – Francis Galton later described it as 'like the rise of a morning star before a day of free social investigation'.[22] He moved from his position as a curate to an appointment at the new East India College, an institution where future servants of the East India Company, the chief agency of British power in India, were trained. For the next thirty years, Malthus taught political economy and argued that policy on wealth, labour and trade should be based on natural law. In spite of their political differences, he became friends with Ricardo; they agreed about population and that political economy is a science of natural laws, but they disagreed about the specific definition of economic value and money. Ricardo was more abstract, Malthus more oriented towards concrete policy. Malthus was also more alive, like his eighteenth-century predecessors, to the way social customs and morality condition economic life. But both emphasized the law-like nature of economic determinants and thereby contributed to political economy's reputation as the dismal science, as the Scottish moralist and historian Thomas Carlyle called it.

The arguments of political economy had political implications and were much debated. Industrialization, the scale of the financial and social burden of poverty and sensitivity to political agitation brought about a long-considered revision of the English poor law in 1834. The reformers, who were sympathetic to the principle of utility and familiar with Malthusian political economy, established the New Poor Law as the means to discipline the poor into self-support. As 'a last resort', specially constructed workhouses took in paupers, segregated men from women, provided basic sustenance and inculcated habits of work and order. The new law built a technology of human nature into social administration. The physical fabric of workhouses, like the prisons, was still visible in the late twentieth-century English landscape, the industrial archaeology of the science of man. Malthus claimed that such institutions were needed as a sanction to reinforce individual choices in favour of the public good. His further claim that laws of nature require such policy earned him Marx's bitter accusation that he was 'a shameless sycophant of the ruling classes'.[23] For Marxists, indeed, Malthusian political economy exemplified ideology as opposed to science.

Malthus was a utilitarian in the weak sense that he believed individual action and public policy should be judged by what they contribute to the public good, but he never undertook a calculus of pleasures and pains. At the East India College he was a colleague of James Mill, the most systematic of the utilitarians. Mill's son, John Stuart Mill (1806–73), applied the principle of utility in sophisticated ways; his work was a reference point in Victorian intellectual culture and at the base of the lasting strength of utility as a value in ethical and political debate in the English-language world. James Mill, in his *Analysis of the Phenomena of the Human Mind* (1829), spelt out the foundation of utility in human nature – in the association of ideas, in the pleasure–pain principle and in language conceived as a sign system for sensations. Mill intended to make 'the human mind as plain as the road from Charing Cross to St Paul's' in London.[24] He called his book an 'analysis' to indicate that its methods are like those of chemistry; as one euphoric reviewer claimed, he shows that 'sensation, association, and naming, are the three elements which are to the constitution of the mind what the four elements, carbon, hydrogen, oxygen, and azote [nitrogen], are to the composition of the body'.[25] His son reprinted the *Analysis* in 1869 and confirmed that this psychology is the foundation for utilitarianism as a social and philosophical movement. Yet by then association psychology was vulnerable to criticism, for reasons that the following chapter makes clear. In a celebrated Victorian autobiography, John Stuart Mill described how his father's view of human nature had blighted his own childhood and produced in him a need to balance his father's dry, analytic conceptions with a voice for the feelings. In 1869, John Stuart Mill also felt a need, for political reasons, to counteract the continued appeal of Christian idealism with a philosophy of utility more responsive than Bentham's or James Mill's to the character of history and less abstract in manner. Political economy became a science which applied reason to commercial and industrial society, but many people doubted whether it touched on either what was best in human nature or what was richest in human history.

10

Culture of the Spirit

> There are two ideals of our existence: one is a condition of the greatest simplicity, where our needs accord with each other, with our powers and with everything we are related to, *just through the organization of nature*, without any action on our part. The other is a condition of the highest cultivation, where this accord would come about between infinitely diversified and strengthened needs and powers, *through the organization which we are able to give to ourselves.*
>
> Friedrich Hölderlin, *Fragment von Hyperion* (*Hyperion Fragment*, 1794)[1]

i *La Scienza Nuova*: Giambattista Vico

The light from the mathematical and experimental mind that illuminated nature's secrets in the long eighteenth century shone also on the human sphere. In this picture light spreads from the physical sciences to the human sciences, and thereby man becomes the subject of a science. The previous three chapters gave credence to this view as they began with natural philosophy and ended with Condorcet's or Malthus's claims to knowledge of human progress. There is, however, another view, the basis of which was discussed earlier in relation to Leibniz's and then Kant's analyses of reason. Kant concluded that the mind necessarily shapes knowledge in terms of fundamental forms of intuition. From this point of view, the imagery of reason as a light that illuminates nature's secrets is inappropriate. The image of birth is better, an imagery in which human reason creates knowledge. In this picture theories of man independent of natural philosophy have first place in the history of the human sciences, and the natural science of man appears as one branch and not the whole tree of the human sciences.

This alternative picture is the substance of this chapter. It has been discussed most often in terms of the Romantic reaction against Enlightenment science – a reaction of feeling and spirit against the science exemplified by Bentham's reduction of feeling to a calculus of pleasures and pains. There was such a reaction: William Blake engraved Newton busy with a pair of compasses while his back is turned to the richness of nature. Thomas Carlyle had Bentham or one of his followers in mind when he referred to the 'foolish Word-monger and Motive-grinder, who in thy Logic-mill hast an earthly mechanism for the Godlike itself, and wouldst fain grind me out Virtue from the husks of Pleasure'.[2] All the same, this reaction was only a vivid expression of long-discerned weaknesses in natural philosophy as the basis for knowledge of human nature. The argument continued in the twentieth century, when objections to natural science as a foundation for the human sciences were not necessarily 'anti-science' but committed instead to a different identity for the human sciences. To take seriously writers who were critical of the natural science model in the human sciences, whether in the Romantic period or in the twentieth century, is to make intelligible modern resistance to natural-scientific thought and public enthusiasm, for example, for literature as opposed to psychology as a source of reflective knowledge.

In this discussion of Romantic thought, I stress its interest in what the philosopher Charles Taylor called 'the anthropology of expression'.[3] He used this phrase to denote theories of man that assume an irreducibly active and expressive self. This belief contrasts with the conviction that human nature is constructed by external events in physical nature, by material society or by the structural properties of language. From the Romantic perspective, the self is something in itself and not something that can be known in the way external things are known. Writers at the beginning of the nineteenth century argued that people bring meaning into existence through their spiritual life, which is exemplified by the arts. The imagination turned to the subjective world and claimed conscious feeling as the source of what is most essential to man. At the same time, writers and artists believed passionately that language and the arts – painting, drama, music and poetry – turn subjective meanings into a shared culture. The English poet William Wordsworth defined poetry as 'the spontaneous overflow of powerful feelings'.[4] Christianity, with its stress on the way the soul strives towards the divine, informed every aspect of the argument and its language and symbols. By 1800, however, the arts themselves were held up as the means by which man gives deepest expression to the creative self; they acquired the status that religious dogma once had as the arbiter of what is most of value in man's being. Such cultural shifts may have done more than any amount of new knowledge of physical nature to displace a transcendent by a human-centred frame of reference.

There is another dimension of these beliefs of great importance for the human sciences. The authors of the anthropology of expression also argued that human nature has a historical structure; this was a claim for the historicity, the essentially historical nature, of the human sciences. University professors and the educated public, philosophers like Hegel and writers like Sir Walter Scott, believed that the human spirit has a historical identity: the spirit unfolds over time and hence gives each age, individual and place a unique character. This belief became very important in politics as it contributed to a historical consciousness of national identity, which was especially strong in regions like Germany or Greece where people's experience was of occupation and disunity. A historical consciousness was also built into every aspect of intellectual, religious and artistic culture. German philosophers, led by Schelling, tried to found in metaphysics, i.e., in an account of Being itself, the view that man has a historical nature. The result was the conception that reality is itself expressive, with an expression unfolded over time and apparent in both the physical world and in what man has achieved during the course of history. As the philosophers argued, to understand a person, a nation or human knowledge is to write history – biography for a person, political history for a nation and the history of science for knowledge. They did not think of history as the past but as the living foundation of the present; and they thought optimistically of the future as something in the process of creation. Marx later adapted this philosophy of history into an attempt to change the world. In the natural as well as human sciences in the nineteenth century, historical thought encouraged an understanding of nature as a process that creates new states of being.

Aspects of historical thought went back to the Renaissance and beyond, but history was deepened in the eighteenth century by the theorists of historical stages in mankind's progress, like Montesquieu and Ferguson, by political historians like Voltaire and Gibbon, and also by philosopher-historians who were profoundly opposed to the Newtonian spirit in the human sciences. These philosopher-historians, of whom Giambattista Vico (1668–1744) and Johann Gottfried Herder (1744–1803) were the most original, are the intellectual roots of the belief in history as a formative dimension of the human spirit. This chapter discusses their work before it comments on the philosophy of consciousness in Hegel and describes Romantic beliefs about the human mind held by Coleridge and Maine de Biran.

There were from the beginning critics of Galileo's and Descartes' claims to have provided modes of reasoning that arrive at truth about physical nature. By 1700, however, it appeared to most philosophers that mathematical astronomy and mechanics had put the new reasoning into practice and given authority to natural philosophy. It was therefore a bold step for an obscure rhetorician and student of jurisprudence to argue that it is more possible to create a science of the human sphere than of the physical world, and that the starting-point for science must be man and not external nature. Vico's challenge was to assert that 'history cannot be more certain than when he who creates the things also narrates them'.[5] The reasons why Vico argued in this way are interesting, and they have consequences in twentieth-century debates about the foundations of the human sciences. It is first necessary, however, to untie a historical knot.

Vico is the classic instance of a genius without immediate influence. He spent his whole life in or near Naples where he held an inferior professorship in rhetoric at the university while he vainly hoped for a better position in jurisprudence. He supplemented his income by writing to commission in Latin, and he was humiliated by changes in political power and also perhaps by fear of criticism from conservative Catholics. Naples was no backwater – in terms of population it was, with Paris and London, one of the largest cities in Europe – but Vico's reputation barely rose above that of an erudite local scholar. In 1724 he completed a vast treatise that attacked the leading philosophical schools – natural law theorists, Aristotelians, Cartesians – and propounded alternative doctrines. Unable to raise the money for publication, he cut out the negative portions (now lost) and published his positive ideas as *La scienza nuova (The New Science*, 1725; altered 1730 and again in 1744). The book did not remain unknown but it was unregarded by his contemporaries; it had some Italian commentators in the eighteenth century, was read by Herder – though probably after he had arrived independently at related ideas – and it then belatedly came to prominence through the advocacy of the French historian Jules Michelet (1788–1874). Michelet found in Vico reasons to place the history of national peoples centre stage in the drama of human understanding. Vico was subsequently periodically rediscovered: he was translated into English in 1948; and in the 1960s and thereafter he became the subject of intense scholarly interest and debate.

Vico was a pious Catholic (though there are signs that he privately questioned some beliefs) who nevertheless developed a comparative historical perspective, later central to secular understanding, that viewed each age and state in relation to its time and place. His scholarship was deeply historical: he approached the history of the earth, the history of nations and the history of culture with a common pattern of interpretation. Speculative cosmogony, enthusiasm for the history of the world, interacted in his studies with the legacy of Renaissance history and generated sympathy for a large time-scale. Beyond this, however, Vico claimed that knowledge about human culture is truer than knowledge about physical nature, since humans can know with certainty, and hence establish a science about, what they themselves have created. This was a fundamental challenge to dominant views of science. Vico set down the principles and conclusions of 'a new science' grounded on human nature, and he described the history of culture which, he believed, necessarily follows from that nature.

Like Descartes before him, Vico was rigorously trained in the logical and language disciplines of a Catholic university. Descartes turned in reaction to mathematics for its clear and certain reasoning. He and his followers contrasted deductive reasoning from certain premises with the feeble arguments of literary scholarship, the speculations of antiquarians and the fictional romances of popular belief and poetry. Descartes regarded humane learning as something more suitable for servant girls than scholars. Vico rejected these views, and not only rejected them but turned them on their head when he argued that the Cartesians and those such as Spinoza who thought it possible to achieve timeless truths about physical nature mistook something that had been achieved historically for an absolute that lies outside history. At this point, Vico's modern interpreters parted ways. Some followed the Italian philosopher Benedetto Croce and found in Vico an anticipation of the Hegelian conception of a rational spirit that reveals itself through history; this interpretation is certainly wrong. Other late twentieth-century scholars found in Vico the basis for historical relativism, i.e., belief that each age has distinctive truths. Vico was therefore understood to have claimed for the historian the special role to exercise rhetoric and imagination to reconstruct human truths. Thirdly, there was the interpretation that stressed Vico's pursuit of a *science* that is true because it rests on knowledge of the principles by which culture and language, and hence knowlege or science itself, is formed historically. This third interpetation is historically the most accurate. What was so distinctive about Vico's position was his claim that what human beings have made themselves, their cultures, are known in a way that cannot be the case in relation to what man has not made, the physical world or his own body. He argued that language, poetry and myth, which man has made, are representations of truths with a better claim to be regarded as true than the much-vaunted achievements of mathematical philosophy. He summarized his position in the formula: 'the true (*verum*) and the made (*factum*) are convertible'.[6]

Vico's challenge to the theory of scientific knowledge was made in a striking passage:

> There shines the eternal and never failing light of a truth beyond all question; that the world of civil society has certainly been made by men, and that its principles are therefore to be found within the modifications of our own human mind. Whoever reflects upon this cannot but marvel that the philosophers should have bent all their energies to the study of the world of nature, which, since God made it, He alone knows: and that they should have neglected the study of the world of nations, or civil world, which, since men had made it, men could come to know.[7]

His argument was that the historian-scientist's mind, as it reasons, arrives at general principles of reasoning, and these general principles show why people necessarily construct the culture – the language, the political forms, the law, the art, the knowledge – of their time. Vico also believed that people share a nature and therefore they must construct culture in comparable ways. The principles of reasoning and human nature together explain why there are similarities between different cultures: they have a common historical nature, not a common historical origin. The historian-scientist can reconstruct the thought processes of other ages with certainty and there is therefore no justification for the scepticism about historical knowledge voiced, notably, by Bayle in the 1690s. Since events in the physical world follow from physical laws laid down by God and not created by man, they cannot be known with the same certainty by the human mind. Vico, in the most fundamental sense, gave priority to what I call the human sciences over natural philosophy: he reversed the status of the natural and the human sciences. Knowledge is possible in the human sciences, Vico believed, because here man reflects on his own reasoning and nature as they have been expressed in history.

This theory of knowledge was interwoven with an elaborate

demonstration, which had both deductive and empiricial elements, of how humans – necessarily and in fact – have made the human domain. Man creates history not by art but by the activity of his nature. Vico's new science included an account of man's sociable nature. As he said, man necessarily seeks human contact even though his desires are entirely self-centred, and through this contact he is led to organize social relations in definite ways. Led by their nature, men come together as civil society. It is, Vico thought, a precondition of any civil society that men hold certain beliefs – common sense – which Vico identified as belief in Providence in history, in the immortal soul and in the necessity to regulate the passions. This common sense, he thought, is the foundation for the natural law of civil society described by earlier theorists. He drew together the available evidence about the ancient history of mankind and claimed that each people, each in its own way, passes through necessary and definite historical stages. There is, he indicated, a form of knowledge that expresses truth at each stage and for each people. He wanted his own new science to explain why each stage has the character that, as a matter of fact, it does have.

Some modern interpreters found here a theory of knowledge that makes truth relative to time; if this is so, it is inconsistent with two other positions Vico held. The first was his belief that his own position as a philosopher-historian equipped him with an understanding of the historical process as a whole and hence with the basis for a certain science. Secondly, he never rejected Christianity or criticized the certainty of the Catholic faith, nor did he publicly question the unique spiritual authority of the biblical history of the Jews and of Christ. Christian truth, as he wrote about it, somehow stood outside of time.

Much of Vico's book reinterpreted ancient history as the empirical basis for the science of humanity's self-creation – or, in modern terms, for the science of culture. His data were human symbols. Man, in his nature, is expressive; indeed, Vico argued, there is no thought, no life of the mind, except through symbols, and the record of myth and poetry is the record of human consciousness. The earliest people, Vico supposed, were 'stupid, insensate, horrible beasts'; culture came with the ages first of gods, then of heroes and finally of men. Each age has its symbols, its consciousness and its corresponding social institutions: social order was conceived initially as a matter of divine ordination, then a matter of power exercised by heroes and finally a matter of rational law followed by men. The ancient myths, he thought, tell a social story. Vico believed that primitive language was poetic and consisted only of a few concrete symbols. 'Words are carried over from bodies and from the properties of bodies to signify the institutions of the mind and spirit'; thus the Latin poets Virgil and Lucretius portrayed primitive man as someone who imagines lightning as a god in the sky, demonstrating the literal nature of primitive thought: 'It is impossible that bodies should be minds, yet it was believed that the thundering sky was Jove.'[8]

Vico was well read in the sixteenth- and seventeenth-century debates among natural law theorists who constructed hypothetical history about the natural state of man. Such scholarship encouraged him to think about knowledge as a historical accomplishment. But his argument that since human activity constructs culture it makes possible a science about the civil world was an original direction in which to take historical thought. The fact is, however, that *The New Science* was an impossibly ill-organized book and it was obscurely expressed. Much of the account above has been current only since the 1960s. Earlier interpreters of Vico, such as Michelet or Croce, read into him their own views about the centrality of history in knowledge of the human condition. What was found in Vico at the end of the twentieth century was a surprising precedent for belief that knowledge of human beings is different from a knowledge of nature, for historically relative judgements and for a conviction that we know with truthfulness what it is to be human – regardless of what natural science says – because that is what we are. Nevertheless, Vico was far from the acceptance of relative values in a modern sense, since he tried to establish a science that makes possible true knowledge of the past even while it provides knowledge of each age's historical character.

ii *Philosophy and History*

Vico's thought located meaning in the linguistic and symbolic expression – the culture – of people as they make themselves through history. But virtually no one made much of this idea in the eighteenth century. When such thoughts did achieve prominence, it was in the idiom of the German language and in the distinctive circumstances of late eighteenth-century German society.

In the decade of the 1770s, the decade of *Sturm und Drang*, 'storm and stress', a young generation rebelled against rule-bound education and social conventions to pursue the truths of the heart. Goethe's story of *Die Leiden des jungen Werthers* (*The Sorrows of Young Werther*, 1774) set the values of self at odds with the values of society; Werther was a youthful, romantic individual isolated amid pious people. Some years later, in the first decade of the nineteenth century, France's military successes over Austria, Prussia and many lesser German states signalled the economic, social, legal and political weakness of the German-speaking world. There was a longing for regeneration, to begin, many people believed, in intellectual and aesthetic culture, in the language of Kant and the music of Mozart, a culture to unite German-speaking people across material frontiers. The most visionary saw German culture as an inspiration for the whole of humanity. Such ideals gained institutional support when the universities were reconstructed in the early decades of the nineteenth century. This was a major event in the history of science in general (as discussed in the next chapter). The reformed German-language universities invested in scientific knowledge as a cultural value, a situation possible where a small, élite and highly educated civil service, responsible for the universities as for other areas of administration, was in a position to build its hopes for the human spirit into institutions. The dominant pattern of higher education in Europe and the English-speaking world until the 1980s, which organized both research and teaching in specialist disciplines, was established.

In the 1770s, however, cultural life centred on the local court more often than the university, though, since academics were employees of the state, some professors at least also moved in court circles. The court of the enlightened Duke Karl August of Saxe-Weimar is famous for its patronage and employment of Goethe, and Herder became general superintendent of the Lutheran clergy in the same state in 1776. Herder had been a student in Königsberg, where Kant introduced him to advanced authors such as Hume, Montesquieu and Rousseau. He wrote extensively on theology, art, metaphysics, language and history, but it was his integration of language and history in a universal view of mankind's spiritual journey that was most dramatic. Between 1784 and 1791, he published the four volumes of his *Ideen zur Philosophie der Geschichte der Menschheit* (*Ideas for the Philosophy of the History of Mankind*) in which, like Vico, he portrayed human consciousness and its symbols as part of a historical process. Unlike Vico, however, Herder directly influenced the human sciences. His project was not original – he himself sardonically entitled an earlier essay, 'Another Philosophy of History' – but he vividly expressed the value of the individuality of both the person and people as groups; he gave individual character significance as he connected it with universal values. Herder's readers experienced a heady sense of the manner in which subjective hopes and the aspirations of the German people exemplified the purposes of a common humanity. History, and the future too, in Herder's rhetoric, is a self-creation of the human spirit: 'We live in a world we ourselves create.'[9]

Like other eighteenth-century writers, Herder focused on human nature and – somewhat like Rousseau – fashioned what he said about this nature as a foil to the wooden conventions of his time. 'The more in general I trace the whole sensibility of man, in his various regions and ways of life, the more do I find Nature every where a kind parent.' Man, for Herder, is the most elevated of nature's children. His language was eloquently idealistic. He envisaged nature as the source of man's powers of reflection and symbolic expression, the powers that enable people to seek the ideal through the creation of civil society and culture. The meaning of life, he argued, emerges in the relationships that men create. Each person is 'conceived in the bosom of Love, and nourished at the breast of Affection, he is educated by men, and receives from them a thousand unearned benefits. Thus he is actually formed in and for society, without which he could neither have received his being, nor have become a man.'[10]

Progress 'in and for society' is the shared destiny of mankind, but Herder argued that each stage and path has its own individual character and expresses special truths. He concluded that what is natural for any one group of people is a unique state of cultural development. The expressive power of human nature gives rise to different cultures in response to the diversity of geography, climate, mode of subsistence and history. The study of human nature must therefore be linked to the study of the comparative history of different cultures. History, Herder believed, must recognize the unique value of each culture – its symbols, beliefs, arts and sciences.

> In short, the mythology of every people is an expression of the particular mode, in which they viewed nature; particularly whether from their climate and genius they found good or evil to prevail, and how perhaps they endeavoured to account for the one by means of the other. Thus even in the wildest lines, and worst-conceived features, it is a philosophical attempt of the human mind, which dreams ere it awakes . . .[11]

The student, Herder argued, must enter with sympathy and imagination into the lives of others: the diversity of myth, language and religion requires an empathetic understanding, and this is the way to true knowledge.

For Herder, each people or *Volk* – he had the Germans but not only the Germans in mind – has its own distinctive consciousness and symbols. It is inward consciousness and outward culture, not politics or conquest, that truly unites a people. 'Has a nation anything more precious than the language of its fathers?'[12] Herder's work was in harmony with a contemporary delight in myths, dreams, fairy-tales and folk customs; he himself published a collection of folk-songs and believed that song and language are closely connected in spontaneous emotional life. In this vein, German scholars, like the 'Brothers Grimm' –

the philologists Jacob and Wilhelm Grimm – developed a fascination with language and fairy-tales as the expression of a people's historical identity.

Herder discussed language as the means with which the mind achieves reflective thought; language is not, he argued, the chance consequence of thought. He rejected the view that man constructed language to symbolize emotion or in response to the need to co-operate. Instead, he traced language to human nature itself, and he supposed that a purposive or spiritual capacity spontaneously to express and to reflect on emotion had brought about the development of language. Poetic and religious speech welds together a *Volk*, and the truths of a people are therefore spiritual and symbolic rather than utilitarian. Herder might have agreed with the English poet Shelley who claimed that 'Poets are the unacknowledged legislators of the world': it is the poets who create the symbols in terms of which a people conceives itself to share a common purpose.[13]

When philology, the study of the nature and origins of language, became an academic discipline in the early nineteenth century, Herder appeared to many German scholars – and it was largely a German subject – to be the founding father. His prize-winning essay presented to the Berlin Academy, *Über den Ursprung der Sprache* (*On the Origin of Language*, published 1772), was a response to the Academy's question: 'Supposing that human beings were left to their natural faculties, are they in a position to invent language? And by which means will they achieve this invention on their own?'[14] This was not a factual historical question as later generations would have understood it but a question about how language is possible for man. Herder's answers took up points much discussed at the Berlin Academy since Condillac's seminal *Essay on the Origin of Human Understanding*. They included the argument that language stems from the reflective capacity of the human mind itself, not from animal grunts or material need. In his later writings, Herder developed this claim about language into a form of cultural-historical analysis, which uses the nature of language to reveal the nature, the spiritual world, of a people. He therefore understood language to be a non-arbitrary system of signs that reveals the inner life. This approach proved to be an attractive basis for human science to scholars in the nineteenth century. Like Vico before him, Herder made the study of the humanities, notably history and literature, central to systematic knowledge about society, since he thought that poetic and imaginative language expresses the life of a people. In his own science he portrayed a people as like an organism and suggested that a people's language and culture passes through a life cycle: the human scientist is the morphologist of the human world. He also drew extensively on the writings of travellers and of those who had lived overseas. His work, however unscientific judged by standards which condemn idealization and rhetoric, articulated a powerful alternative to the natural sciences as a way to knowledge of the human domain.

It is necessary to comment on the word '*Volk*'. The terrible path of German history in the twentieth century made it fashionable for a while to seek root causes in peculiarly German modes of thought. National Socialism's propaganda about the mystical identity and destiny of the German people led some critics to attribute to Herder and those who followed some responsibility for the content and strength of German nationalism. Herder certainly encouraged the view that history is about a people united by language and culture and thus about its identity over time. Michelet wrote such history for the French in the mid-nineteenth century and, in a different way, Winston Churchill for the English in the mid-twentieth century. But it is one thing to recognize that Herder believed in the unique characteristics and history of different peoples – 'Each [culture] is a harmonious lyre – one must merely have the ear to hear its melodies' – another to link this belief with events a century and a half later.[15] It is possible that thought about society and history going back to Herder's generation was an element which made later events possible. If this is accepted, however, it is then necessary to explore the marked differences between Herder's conception of the *Volk* in cultural and linguistic terms, as well as his conception of universal history, and the biological and racial notions of human difference which gained in power only in the second half of the nineteenth century.

Herder was a Lutheran pastor who optimistically read God's benign purpose in the overall pattern of history. However tumultuous history has been, 'if there be a god in nature, there is in history too: for man is also part of the creation, and in his wildest extravagances and passions must obey laws, not less beautiful and excellent than those, by which all the celestial bodies move'. God, Herder believed, has put into human nature the quest for its own fulfilment: 'The end of whatever is not merely a dead instrument must be implicated in itself.'[16] The meaning of man's life, Herder implied, is contained in his humanity; his purpose is the achievement of full humanity. He filled such abstract notions with vivid details to show how different peoples and their different values contribute to the rise of humanity. Though the main theme of his story was the rise of European society in ancient Greece and the beginnings of the modern age in the sixteenth century, his imagination swept the world. Himself an unhappy, awkward and unpleasant man, Herder created an inspiring and enlightened vision. In the twentieth century, when history had become an academic discipline preoccupied with evidence and detail, only outsiders like Oswald Spengler or Arnold Toynbee attempted anything on such a speculative scale.

Herder wrote with an expressive historical sensibility. At about the same time, a group of scholars at the university of Göttingen developed history into an academic subject, a science that uses techniques of textual interpretation, the evaluation of evidence and explanations for events related to specific contexts of time and place. After the foundation of the university of Berlin in 1810, as an affirmation of Prussian educational and cultural excellence, a succession of influential scholars taught history there and confirmed history's status as a scientific discipline. Fascination with the ancient world rose to new heights with the studies of the first four centuries of Rome by Barthold Georg Niebuhr (1776–1831), in which he critically examined the reliability of primary sources. When he lectured in Berlin to an audience that had seen its own state collapse, he brought together emotional intensity about politics with a scholarly interrogation of historical evidence.

Historical thought was also a significant force in the Berlin law faculty, where Friedrich Karl Savigny (1779–1861) traced the sources of law to different people's historical experiences rather than to the rational mind's grasp of necessary truths. He thus rejected natural law arguments, as they had been understood by Grotius and Pufendorf, as the basis for social and legal order. Kant had also criticized natural law theory in depth since, in his view, it failed to distinguish legal statements with descriptive content and moral statements with prescriptive form. Between Kant and Savigny, Central European jurisprudence, which had been dominated by natural law arguments, began to be a historical science. Savigny took this further and argued that history alone reveals the order that is in fact present in systems of law. His historical jurisprudence dominated German-language legal scholarship for several generations, and it fostered a different legal culture in Central Europe to either France, with its new and rationally ordered legal codes, or England, followed by the United States, where the common law represented tradition rather than the results of historical scholarship or abstract notions of a people's will.

Savigny's scholarship exemplifies a historical science of the human sphere. When he founded the *Zeitschrift für geschichtliche Rechtswissenschaft* (*Journal for Historical Legal Science*) in 1815, he described law as the expression of the 'second nature' of a people, the means by which a people moves from intuitive relations to national self-consciousness. Later he referred to the *Volksgeist*, the spirit of a people, which gives rise to 'real law' or 'the proper will of the people'.[17] To comprehend and legitimate law, he thought, is necessarily to think historically about what has made a *Volk*. In the earliest period of human history, Savigny claimed, law directly expresses the customs and symbols that give a community its individual character. Complex societies transform this expression into a more abstract form, and this creates the need for lawyers and the search for rational or scientific law. It remains the character of true law, nevertheless, however abstractly, to express the national spirit. Savigny's jurisprudence was therefore an elaborate scientific alternative to the Benthamite science of jurisprudence founded on the principle of utility. It was a conservative alternative, and it was attractive to the established regimes with which most political power rested during the nineteenth century. The concept of the *Volksgeist* also became a key part of the rhetoric in support of struggles for nation status or national independence across Europe. While the utilitarians pinned their hopes on rational deductions from universal attributes of human nature and hence on the equality of man, nationalists placed faith in the unique quality and history of their respective peoples.

Historical writing from Herder to Savigny understood the human domain to express a purposive force or spirit in human actions. Students of human nature looked for knowledge of language, custom, art or law as shared and visible representations of this spirit. Historians thought that man's spirit in itself is universal, but they believed that concrete representations of spirit acquire form in relation to particular circumstances of climate, geography, history and destiny; each pattern of representations – a community's culture – therefore has its own intrinsic qualities and values. Through such beliefs the historian acquired status as a leader who enters sympathetically and imaginatively into the lives of other communities as into the past of his own, guided by scholarly standards of evidential argument and interpretation. History, for its most zealous proponents, is the key to human understanding. In this way, along a path suggested but hardly influenced by Vico, German human science developed a conceptual framework and a method that denied intellectual leadership to the natural sciences. A substantial and self-consciously scholarly body of work found in humanity's historical nature reasons why, in principle, the human sciences and the natural sciences should differ. In the wider world, and unencumbered by weighty academic arguments, contemporary Romantic values also stressed the expressive individual self and the expressive cultural life of communities. The language of poetry, music, drama, painting or the novel rendered feeling as truth, gave the spirit outward form and thereby created an audience for truths about human nature in a way that more academic discourse never could. Romantic values took it for granted that knowledge about human nature begins with reflection on the self, with human relationships and with the expressive culture of the arts.

iii *Conscious spirit*

One strand of German thought suggested a complete revision of what was understood by science, natural as well as human. It was boldly argued that feeling and purpose are primary and provide the terms for a living as opposed to dead understanding of nature. This set up man as a model for nature, rather than the reverse, and it 're-enchanted' the world in precisely the way that Descartes and mechanical philosophy had striven to eliminate over a century earlier. The result was German *Naturphilosophie* (nature philosophy). In an influential version of this thought, *Das System des transzendentalen Idealismus* (*System of Transcendental Idealism*, 1800), F. W. J. Schelling (1775–1854) portrayed nature as the outward counterpart of subjective unconscious activity; human subjective nature, as he understood it, strives to rejoin this universal counterpart, the goal being a higher unity. Lorenz Oken (1779–1855), a biologist as well as a philosopher, declared simply: 'Man is God fully manifested.'[18]

These ways of thought tended to rouse the scorn of matter-of-fact Anglo-Americans, not to mention hard-nosed modern natural scientists. One English response at the beginning of the nineteenth century was that the German writers were philosophically mad. This 'madness', however, expressed a desire that many readers recognized easily enough: a longing, commonplace in the urban nineteenth and twentieth centuries, to identify with something universal in nature. Many people also shared with the German philosophers a passion that individual moral choice and action should make a difference in the world. The experience or fear that the individual does not make a difference, as I shall mention later in connection with Marx, created alienation. *Naturphilosophie* was a full-bloodied attempt to legitimate belief that the universal and the purposive, as they are perceived to be present in the desire of man, are embedded in the essential structure of the world.

The most substantial contribution to a theory that integrates the individual and the universal was the philosophy of G. W. F. Hegel (1770–1831). He achieved fame – though at times he was entirely ignored by English-language philosophy – for one of the most ambitious and comprehensive syntheses of thought ever attempted. Hegel perceived that man's search for unity, for knowledge of what is universal, while simultaneously seeking to assert the individual, involves a contradiction. His philosophy endeavoured to reconcile these apparently irreducible human commitments and held out the promise that thought and action, universal reason and individuality, could become one.

Hegel's career was that of an academic, and he acquired great authority during the 1820s within the reformed Prussian university system. In his own person he represented the cultural value of university life within the state. He taught first at the university of Jena and then in Berlin, and he elaborated a synthesis of daunting technical difficulty that covered natural science, aesthetics, history, jurisprudence, political thought and religion. Each part possessed significance by virtue of its logical integration in a comprehensive philosophy. From a Hegelian perspective – a perspective which in this sense continued in the twentieth century to have authority in Continental Europe – it is philosophy that is ultimately the unifying discipline of science. For Hegel and his successors, it is the articulation of knowledge within a coherent synthesis that makes a subject truly a science. In this regard, his work shared an understanding of what science is with Leibniz and the medieval scholastics rather than with modern Anglo-American natural scientists. Yet, in addition to his influence on Marx, there were twentieth-century approaches in the human sciences affected by Hegel's project. In these schemes, the detail established by a particular discipline such as psychology was made part of science in recognition of its contribution to a comprehensive philosophy of what it is to be rational and human.

Hegel reinvigorated Kant's argument that reflective consciousness makes knowledge possible – and hence that reflection must be the base of knowledge. He believed that it does not make sense to argue that individual consciousness, through reflection, constructs the world; rather, it must be consciousness in general, collective consciousness, at work in a culture and a society, that constructs the world historically. Here Hegel's thought approached that of Vico and Herder. What is true about being human is the truth of human history, the history of human understanding and social organization, which culminates – so Hegel believed – in his own philosophy and in the Prussian political state of his own time. He was well aware of the variety of beliefs but strove through philosophy to establish a unifying idea, the spirit or 'the Absolute'.

> The significance of that 'absolute' commandment, *Know thyself* . . . is not to promote mere self-knowledge in respect of the *particular* capacities, character, propensities, and foibles of the single self. The knowledge it commands means that of man's genuine reality – of what is essentially and ultimately true and real – of mind as the true and essential being.[19]

Each age before his own, Hegel thought, achieved knowledge of a particular character that represents a historical stage of the Absolute; each age therefore has its distinctive conceptual framework. Marx later linked these historical stages and conceptual frameworks to different modes of material production, while non-Marxist philosophers, like Wilhelm Dilthey at the end of the nineteenth century, gave up the search for a universal spirit that transcends history but still conceived of history as different periods of consciousness. A reading of Hegel, of philosophy and of history that makes values historically fully relative informed the culture of the human sciences only at the end of the twentieth century. This reading is undoubtedly foreign to the spirit of Hegel's own time, a time that was seduced by a vaulting ambition to know, once and for all, the truth of man's being.

One technical Hegelian term, 'phenomenology', is important. Loosely expressed, the word connotes an orientation towards what it is to have knowledge about our being that places a consciousness of what we are, as humans, at the centre of how knowledge is obtained. Understanding becomes a philosophical endeavour founded in the immediate qualities of the conscious world, and natural science methods and explanations are secondary. The qualities to which phenomenologists refer are the directly grasped qualities of being conscious – sensible, active and purposive qualities – not the indirect experience of external things. Hegel's first major work was a book on *Das Phänomenologie des Geistes* (*The Phenomenology of Spirit*, 1807). Here he conceptualized spirit as the ground of being human, and he conceived of a universal spirit that apprehends or becomes rationally conscious of itself through the activity of particular minds. By 'phenomenology', he denoted the science or systematic study of the appearance of spirit, that is, 'the phenomena' of spirit as they become known in individual consciousness through the activity of spirit. Though Hegel's relation to Christianity was debated, he structured his arguments in a way that reflected religious values and attempted to relate spirit as a phenomenon to an absolute ground in which it is reality itself. The pertinent point is that Hegel assigned to consciousness the representation of real being – active or purposive spirit – and he therefore vindicated belief that description of what is immediately represented in consciousness is the starting-point for the human sciences. In this context, description of what is known – empirical description – concerns conscious qualities, 'phenomena' in themselves, not phenomena understood as sensations, an external world, the body or the mind.

Phenomenological argument attempted to do justice to the intuition that the conscious self cannot be the object of observation like other objects as it is itself a subject and not an object. Hegel's philosophy was in part an argument about the sense in which the subjective subject can be objectively known. Two aspects of this argument are relevant. Firstly, there is not a conscious self which thinks or observes; rather, the purposive activity of thought or observation is the conscious self as it comes into being. His argument was not structured by the Cartesian categories of soul and body or by Locke's discussion of knowledge as a relationship between the internal observer and the external world. Hegel tried to find an objective way out of the circle in which the human observer tries to observe himself or herself. Secondly, he argued that consciousness inescapably uses signs (most obviously language) in the process of which it becomes a self-conscious subject for itself, and this means that consciousness is a selective process. An attempt to take in everything results in blankness: each and every state of consciousness is selective and has a structure that involves concepts and values. This was a general, philosophical argument, but it was also taken up by psychologists who worked on the topic of perception and who grappled with this same selective feature of consciousness.

In the twentieth century, after the work of Edmund Husserl and, somewhat differently, in existentialist thought, the interrogation of consciousness produced a distinctive phenomenological movement. Phenomenology at times redirected psychological research; it was linked, for example, to the *Gestalt* psychologists' studies of perception before and after World War I. Hegel's influence, however, was more profound in social than psychological thought. His conception of the spirit was a collective and historical one. Within the framework of his dialectical philosophy, he wrote at length about human history – but about its logical essence rather than its empirical surface. His arguments provoked the young Marx in the 1840s and, a hundred years after that, again entered decisively into European social thought.

One of Hegel's arguments in particular, the dialectic of the master and the slave, was re-expressed in many forms. The *Phenomenology* included an abstract account of the master–slave relation as a formative stage in man's history. Hegel claimed that early political society was formed by conquest, the strong and victorious enslaving the weak and defeated. The master's consciousness, he argued, acquired a reflection of its own strength in the subjection of the slave's consciousness. This reflection was a necessity, an affirmation to the masters of their being: they needed the slaves spiritually as well as for their material welfare. Through this necessity, the masters became dependent on the slaves. The slaves, in their reflective consciousness, comprehended the masters' dependency and thus, though physically weak, acquired a position of spiritual strength, strength in being, which was expressed in a desire for freedom. The reversal of strength was enhanced since the slaves, through labour in the mastery of nature, grasped a consciousness of an ability to transform things; the masters, who were merely consumers, lost a sense of their transformative consciousness. Masters and slaves were locked in a historical dialectic in which, through the reflective consciousness of each, positions of strength and dependency reversed, and the outcome was a new collective way of being and a new political order. History, as Hegel portrayed it, shows the old aristocratic, military order, based on heroic virtues, overpowered by the once servile classes with their bourgeois virtues.

This argument was logical rather than empirical in form and open to a variety of interpretations and uses. Marx took the dialectic of reflective consciousness and turned it into a dialectic of classes determined by modes of production. For Marx, history is the overcoming of the slave order by the feudal order, the feudal order by the capitalist order and the capitalist order by the socialist order. Nietzsche in the 1870s and 1880s turned the argument into a violent attack on Christianity, which he portrayed as a life of resentment by slaves against spiritual nobility. In the second half of the twentieth century, intellectuals agonized over man's being in terms of the reflection that one consciousness finds in others and the alienation, resentment and consciousness of otherness thus provoked. In an extreme form, such descriptions reduced the self, even being itself, to an endless dialectic of reflections. This, however, moved away fundamentally from the historically and collectively grounded being that Hegel described as the self-constituting spirit of man.

In the early nineteenth century, a language of conscious phenomena helped delimit a subject area that was more and more frequently called psychology. This language developed largely independently of Hegel's grand philosophical scheme, but, like his language, it made it possible to articulate belief that psychological knowledge originates with conscious qualities rather than knowledge of material nature. The new area of psychology is exemplified in the work of Kant's successor in Königsberg, the philosopher Johann Friedrich Herbart (1776–1841). He was best remembered for his attempt to make psychology a quantitative science; his ambition also lay in two other directions, though in neither was he successful. Firstly, he wanted to construct a comprehensive philosophy to rival Hegel's. This philosophy involved his derivation of psychological claims from metaphysical first principles. Few readers were persuaded. Secondly, as an academic and therefore a Prussian civil servant, he wanted a theory of mind with the capacity to reform pedagogy and ultimately the science of the state. In practice, the gap between abstract scientific analysis and schoolteaching remained too wide and he had little immediate impact on social administration, though educators later in the century found support in Herbart's work for individualized teaching methods and his name, if not his philosophy, became well known in pedagogic circles. Nevertheless, in the space between metaphysics and pedagogy, Herbart created an influential form of psychological analysis, which linked Leibniz's earlier notions of the soul's activity and later experimental research on mental content.

Herbart analysed the mind as a dynamic assemblage of interacting presentations (*Vorstellungen*). Significantly enough, when he described mental phenomena, he borrowed words used about physical nature and referred to the statics and dynamics of mental elements. This reflected a central problem for descriptive psychology: to find a language for mental content that did not simply transfer notions appropriate to the physical world to the mental sphere. It was easiest, and perhaps inescapable, to draw on the most accessible metaphors, the metaphors derived from concrete sensation. This is what Herbart did when he described how presentations interact to create mental content. Much of the obscurity of Hegel's and phenomenological language may be attributed to an attempt to escape metaphorical language.

Herbart treated consciousness as a dynamic unity but nevertheless conceived the 'I' to be composed of mental units each of which possesses a measurable amount of activity and intensity. The possibility of measurement and calculation with these qualities, which Herbart proposed in the abstract, justified the view that psychology can become a science – precisely what Kant denied is possible. In his *Psychologie als Wissenschaft* (*Psychology as Science*, 1824–5), Herbart set out a mental dynamics, replete with algebraic formulae, as a descriptive analysis of mental content, that is, of perception, thought, feeling and volition. The mental world, as he described it, is the sum of interaction, additive and inhibitory, between elements each with its own intrinsic force. These interactions explain the organized form of conscious awareness.

One implication of Herbart's description was that mental elements can be unconscious. The elements achieve consciousness ('apperceptive' awareness) only in combination or through competition with other elements. Herbart's references to unconscious presentations, however, was not an anticipation of later psychoanalytic notions of unconscious processes. As Leibniz had appreciated, it is not possible for every sensation from the body or sense organs to exist in conscious awareness at any one moment, and the achievement of organized awareness necessarily leaves other elements unperceived or unconscious. In Herbart's work, the conscious–unconscious distinction hinged on the competitive intensity of mental elements.

Herbart's contemporaries found his metaphysics incoherent and his psychological dynamics too abstract to make actual measurements possible. Nevertheless, descriptive analysis of the conscious or mental world proceeded apace under the heading of psychology. F. E. Beneke (1798–1854), also an anti-Hegelian philosopher, wrote a series of books, some of which were accessible to a wider audience, which described mental activity in terms of subjectively observable phenomena or introspection. His writings about psychology were independent of a logical analysis of reason and not subsidiary to either religious demands or physical science. Such literature helped establish an audience for psychology as an area with its own distinctive subject matter. M. W. Drobisch (1802–96), a mathematician as well as a philosophical psychologist, looked again at Herbart's goal of quantification which, by the mid-century, did not seem quite so implausible. The medical psychologist or psychiatrist Wilhelm Griesinger (1817–68) adapted Herbart's language of the ego or conscious self, taken to be a field of competing elements, to characterize different pathological conditions. Physiologists trained in specialized university laboratories, which were an important new development in science in the German-speaking world in the 1820s and 1830s, also turned to the detailed experimental study of conscious perception. They began to describe mental content with unparalleled precision, but their techniques appeared to apply only to mental content that derives from sensory perception. It was possible to correlate such mental content with measurable changes in the physical world (e.g., perceived brightness with changes in light intensity). Significant experimental work on sensation included descriptions by J. E. Purkyně (1787–1869), published in 1824–5, of the relative brightness of colours in night vision, and studies by E. H. Weber (1795–1878), published in 1834, of the touch threshold.

Psychology as the description of mental content faced a problem in its goal of being objective. Philosophers like Hegel or, in his different way, Herbart dealt with it by philosophical argument. They sought to make psychological description objective by deriving the concepts used from first principles. By contrast, scholars trained in natural science increasingly turned to experimentation as the way to achieve objective description, even if this restricted their work to limited aspects of the mental world. This path led to the new subject of experimental psychology in Germany between 1850 and 1880 though, as we shall see, it remained linked to philosophy in important respects. Both philosophical analysis and experimental research, especially the latter, moved away from a commitment to expressive feeling as the basis for human truths. But the Romantic notion that subjective feeling is objectively significant remained a deeply-held part of what non-academic audiences expected the subject of psychology to be about. In addition, subjects like history, theology and literature, which presented culture as a shared symbolic achievement of the human spirit, continued to attract interest for what they reveal about human nature. The long-term result was the divergence of scientific psychology and the humanities and, for connected reasons, the divergence of scientific psychologists and the public audience for psychological knowledge.

iv *Imagination and Will: Samuel Taylor Coleridge and Maine de Biran*

This chapter has remained largely within German-language culture where, by the early nineteenth century, two themes were prominent: history as the development of the human spirit, and a descriptive science of consciousness. Romantic values, however, diffused across Europe from Spain, where Francisco Goya's black paintings expressed the mystery of human terror, to Russia, where Alexander Pushkin (classified by historians of literature as a post-Romantic) wrote the verse novel *Eugene Onegin* which provided generations of Russians with models of the feelings. To illustrate these values as a reaction against the Enlightenment science of man, this section discusses the English poet Samuel Taylor Coleridge (1772–1834) and the French statesman and philosopher Maine de Biran (1766–1824). Both men changed their opinion about human nature in parallel with political changes, and both men reflected on the self with new intensity.

When Coleridge's wife Sara gave birth to their first son in 1796, the father – flushed by enthusiasm for events in France – named him David Hartley. He could hardly have found a more emphatic way to identify with radical intellectual opinion: Hartley had deduced how pleasures and pains lead necessarily to moral elevation and social harmony. When he wrote to his friend and future brother-in-law Robert Southey, whom he addressed in 'Health and Republicanism to be!', Coleridge described himself as 'a complete necessitarian'. But political reaction in Britain and the Terror in France, and more particularly his own spiritual longings, expressed in conversations with Wordsworth, caused Coleridge to break his 'squeaking baby-trumpet of sedition' and to seek 'in poetry, to elevate the imagination and set the affections in right tune by the beauty of the inanimate impregnated as with a living soul by the presence of life . . .'[20] His change was deep, and ultimately he became bitter and conservative.

In search of a philosophical synthesis, Coleridge attended lectures in Göttingen and dived into the depths of German thought.

He absorbed Schelling's identification of nature with the ideal and belief that this ideal becomes individual and self-conscious in the human mind. He rejected Hartley and recognized in the mind's activity the source of a distinction between mechanical and creative thought that vindicates the poetic calling. This influential and much-cited distinction was between the 'fancy' and the 'imagination'. By fancy, Coleridge meant a capacity to imitate and to link ideas or feelings together in a mechanical way. In his view, Hartley and Priestley referred only to fancy when they described feelings as pleasures and pains; they treated feelings as elementary sensations that aggregate or associate with other sensations, and then identified the artist as someone who rearranges sensations to create pleasure. Coleridge, in part, had in mind poetry by Erasmus Darwin, who set out to inform his readers through the association of the pleasure of verse with technical information and industrial ideals. In a celebration of industrial and social progress, for example, Darwin wrote:

> How loves, and tastes, and sympathies commence
> From evanescent notices of sense?
> How from the yielding touch and rolling eyes
> The piles immense of human science rise?[21]

This, Coleridge thought, was fancy. By contrast, he argued, we possess feelings that are transformative rather than mechanical, feelings that advance our moral and spiritual condition and, when transformative feeling informs art, such feeling contributes to the progress of humanity. He linked transformative feeling to the faculty of imagination, and he attributed the failings of eighteenth-century ideas and a corresponding inhumanity in social relations to incomprehension of this faculty.

Coleridge distinguished a primary and a secondary imagination. By the former he meant a universal capacity of mind, a spiritual power that graces the finite mind, 'the living power and prime agent of all human perception, and . . . a representation in the finite mind of the eternal act of creation in the infinite I AM'. By the secondary imagination he meant the special poetic capacity, the power to shape the feelings at work in the primary imagination into an organic, living whole – 'it struggles to idealize and to unify. It is essentially *vital...*'[22] Coleridge expressed himself philosophically and portrayed the spirit as a creative unfolding in nature which becomes self-conscious in the human mind. He also expressed himself psychologically and described mental capacities that his readers could grasp in their own subjective experience. Indeed, his use of the word 'psychological' was customarily cited as a precedent in modern English. Coleridge's discussion of the imagination highlighted a profound weakness in theories of the mind that stemmed from Locke: they did not encompass what feeling signifies to a person who seeks meaning in subjective knowledge. The audience for Romantic art turned to the subjective world of love, feeling and imagination as a source of transformative values not informative sensations.

When John Stuart Mill analysed English political culture in a famous contrast between Bentham and Coleridge and described them in 1838 as 'the two great seminal minds of England in their age', Coleridge was made the spokesman for a conservative, organic conception of social order.[23] Though Mill was politically opposed to this position, he nevertheless recognized, as Bentham had not, the place that growth and feeling have in human fulfilment. In later life, Coleridge reached the peculiarly élitist view that a cultivated society requires a state-endowed learned clergy, a view that had little appeal even in class-ridden Britain. By contrast, the value of the imagination, which Coleridge had articulated, was taken up by Victorians like Matthew Arnold who supported the extension of access to education and culture. The education of the imagination became an ideal, the civilized means to address the aspirations of working people.

Just as the young Coleridge linked his thought to Hartley, Maine de Biran, between about 1797 and 1805, associated with the *idéologues*, who wanted to refound social policy on the analysis of human nature into bodily and mental sensations. Biran's prize-winning essay, 'L'influence de l'habitude sur la faculté de penser' (The Influence of Habit on the Faculty of Thinking, 1802), was a study of the effect that repetition or habit has on the vividness of sensation and movement. By 1812, in his *Essai sur les fondements de la psychologie* (*Essay on the Foundations of Psychology*), Biran had changed his views, and he observed the mind's activity – rather than reported on sensations – in order to understand the sources of knowledge. Like Coleridge, he found the word 'psychology' useful as a way to characterize his area of interest. Biran focused on what he called the *'sens intime'* (inward sense). This sense, he argued, gives unmediated experience, first, of a religious character, and second, of the *'effort voulu'* or willed bodily movement. His discussion placed feeling and will attributable to a self at the centre of human science. The novelist Stendhal (Henri-Marie Beyle), his contemporary, placed these psychological values at the heart of literature.

Biran's analysis of the *'effort voulu'* was an original and forceful critique of theories that attribute human nature to sensation and to pleasure and pain. It was also thought to open up a way to reconcile the Catholic faith in the miracle of the soul with psychology as the science of the mind. Biran argued that the earliest primary experience is a sense of effort, a spontaneous will to action. This sense, he claimed, then makes us secondarily aware of something other than our own effort, and this is sensation of the external world. His argument located the active self at the core of psychological analysis on grounds of experience and not on grounds of metaphysics. The analysis extended to language: the experience of effort includes the effort to produce sounds, and the sensations that follow are sounds that result from our own efforts; in this way, Biran argued, language originates in the self-activity of speech. All his later thought therefore emphasized volition as constitutive of a personal self, a position compatible with the Catholic stress on the real agency of the soul. He did not, of course, deny that passion sometimes overwhelms the self but, like Catholic moralists generally, he did not think that this invalidated belief in the primacy of volition. These values achieved a pre-eminent position in academic philosophy and in the teacher-training colleges through the power and influence of Victor Cousin (1792–1867). Cousin controlled philosophy at the *Ecole Normale Supérievre*, the senior educational institution, and hence appointments elsewhere, and he had an influence in the provinces well into the second half of the nineteenth century. He used the term *'personnalité'* to describe an inner awareness of the *moi* or self, and he made the study of its powers through introspection the established position in the science of the mind in France. This was the position that the generation of T. Ribot reacted against when it set out to establish scientific psychology after 1870.

Biran's political career began with his participation as a moderate royalist member of the Council of Five Hundred in 1797, the council that signalled the final end of the Terror. After the restoration of the monarchy in 1815, he became a member of the National Assembly, where he liaised between the assembly and the king on financial matters. But alongside this public life, he led an emotional, depressed and self-reflective existence, which he recorded in his *Journal intime* (Intimate Diary), published long after his death. Just as Coleridge expressed his philosophical views in notebooks and in the partly autobiographical *Biographia Literaria* (1817), Biran's most expressive medium was the essay and the diary. Neither writer achieved the grand synthesis he sought. Their search for self-knowledge indicates the enormous significance of diaries and letters in the creation of the modern self and in the enrichment of language, especially psychological language, in terms of which to express belief in the self's significance. The diary and letters – and letters were often in effect a diary shared among relatives or friends or lovers – cultivated the *sens intime*.

Private writing and the publication of diaries, letters and fiction (often itself in journal or epistolary form) made introspection into both an art form and an education in the construction of a psychology of the self. There were even manuals that taught the proper literary form to adopt in order to express the inward life and to harmonize art and sensibility. The diary was also a means for the observation of others. German pedagogues at the end of the eighteenth century advocated 'the diary method' whereby an observer keeps a record of a child's early development. It was hoped that meticulous observations, when brought together, would provide a standard picture of a child's growth, guide educational programmes and permit assessment of a child's progress. Readers were encouraged to 'collect experimental information through all possible combinations of children ... and objects'.[24] In 1804, the philosopher Dietrich Tiedemann (1748–1803) published a *Handbuch der Psychologie* in which he used systematic observations on his son (the future physiologist Friedrich Tiedemann) to discuss the development of the soul. The diary that recorded the child from without was a mirror image of the diary that recorded the self from within. By such means the educated public acquired habits of thought and self-understanding in terms of an internal psychological life. Philosophers, educators and the reading and writing public rejected or ignored eighteenth-century abstractions about human nature in order to seek truth in concrete experience of the self. They thereby fitted out psychology with a large part of its subject matter.

The Romantics dressed up the artist in finery appropriate to the most real because most creative self. The artist became a model for the life determined by feeling. But material events also dramatically altered Europe. The French Revolution and its conservative aftermath were vividly present in memory throughout the nineteenth century. Meanwhile, the long process of industrialization transformed Great Britain and then other countries. By the end of the nineteenth century, the city not the country had become the place where most people lived in Western Europe and North America. Social and economic change threw outward existence into the melting pot along with inward desire. Disciplines which took man as their subject came into their own and, though coloured by material events and nineteenth-century thought, perpetuated much that was characteristic of eighteenth-century values. It is time to move fully into the nineteenth century.

IV

THE 19TH CENTURY

11

Academic Disciplines and Public Values

> Nothing exists entirely for the sake of something else; nothing is contained entirely in the reality of another. Still there prevails a deep, pervasive connection as well, of which no one is entirely independent and which penetrates everywhere. Freedom and necessity exist side by side.
>
> Leopold von Ranke, from his literary remains, written in the 1860s.[1]

i *The University and the Public*

Educational reform aroused passionate interest throughout the nineteenth century. In the second half of the century, reformers everywhere looked enviously at German higher education. In England, when Mark Pattison argued for the reform of the curriculum of the university of Oxford and Matthew Arnold wanted education to redeem his country from philistinism, they found in German scholarship an example of how to make mental culture an esteemed social value. Arnold wrote: 'What I admire in Germany is, that while there, too, Industrialism . . . is making . . . most successful and rapid progress, the idea of Culture, Culture of the only true sort, is in Germany a living power also.'[2] In the 1870s and 1880s, college presidents in the United States turned to Germany for models of research schools and students travelled to Germany to acquire experience of research. Two decades earlier, idealistic young Russians, who enjoyed some liberalization in their country after 1855, flocked to Germany, Austria and Switzerland in search of higher education, and they returned to westernize their own universities. Humiliated by war in 1870, France assimilated lessons from the German victory, not least about Germany's system of higher education. Italy employed German scientists in its universities in its drive to construct a modern nation after unification in 1870. The German pattern, which organized learning along disciplinary lines and made each discipline into a specialized area of knowledge and research methods, with the purpose of creating new knowledge and new scholars, appeared to be essential to the modern nation state.

All this represented a considerable change, since observers had thought the German universities moribund in the late eighteenth century. At that time only a few academics, notably at Göttingen, were committed to the research ideal and to the establishment of a new role for the university. The transformation came in the first thirty years of the following century, especially after the university of Berlin (later the Humboldt University) was founded in 1810. This new university was the centrepiece of a cluster of curricular and administrative changes that also included the foundation of the university of Bonn, carried through between about 1806 and 1818 by the Prussian minister Wilhelm von Humboldt (1767–1835). The changes were a cultural response to military defeat and disunity in the Napoleonic period. The Prussian state, followed by other German-language states, imposed – and funded – a new regime that required the faculty to teach scholarship, not just train a few future state servants or entrants to the professions. It guaranteed that the students would come because it demanded that future *Gymnasium* teachers gain a degree, and so stabilized a career structure for the academic profession. These measures created a rigorously educated class of people, which included those who took funding decisions about universities, committed to scholarship as exemplary of German values. Competition between the different states for cultural status, for professors and for students – both were free to move between institutions in accordance with the value of freedom of learning – helped university budgets to grow. Within the universities, groups of academics, who emulated the idea of a research community established at

Göttingen, brilliantly showed what full-time scholarship could achieve and vindicated the investment in chairs in new disciplines. In Berlin alone, Hegel in philosophy, Niebuhr in history and Savigny in jurisprudence set a formidable precedent, achieved remarkable social status as individuals and established their subjects with authority.

Eighteenth-century professors were, by and large, relatively unspecialized, with an audience of students and, if they published at all, a small non-academic readership. It was a significant innovation when the new and reformed universities restructured teaching and research along disciplinary lines (in, to use later terms, both the arts and the sciences) under the aegis of the philosophy faculty. The status of philosophy now equalled or even exceeded that of the senior faculties of law, medicine and theology, to which it had earlier been subservient. The new academics were specialists who wrote primarily for their peers and taught mainly to create a new generation of scholars in their subject. Disciplines like philosophy, history, philology, chemistry and physiology came into existence as distinct social entities. The audience for specialist research and the audience for general literature were increasingly separated. The specialists continued to address a general audience – their belief in the place of scholarship in national life, as *Kultur*, required it – but this tended to involve a self-conscious attempt to explain the values and conclusions of research rather than the detailed content of a subject. This chapter describes the relationship between knowledge and values in the interaction between newly specialized research and the public realm.

The new disciplines did not include either psychology or sociology. Scholars who addressed psychological questions were equally likely to be philosophers (e.g., Beneke) or physiologists (e.g., E. H. Weber) and such topics had no distinct constituency or audience. Work on sociological areas was substantially historical and was dominated by debates in jurisprudence over whether legal authority has derived from history or reason. These debates showed little interest in the forms of social analysis influenced by the *idéologues* in France or political economists in Britain.

The German universities established a model of professional scholarship which was vindicated by its results, but it did not automatically lead to the modern human science disciplines. Indeed, there is a strong case to be made that the many roots of psychology and sociology lie elsewhere, in the interstices of social life itself. Though the universities were state institutions and academics civil servants, there was a tension between the academic interest in theoretical refinement, specialization and research as a value in its own right, and the public interest in instrumental knowledge or scholarship visibly relevant to public values. When higher education elsewhere followed the German example, it recreated this tension. Alongside higher culture, given new life by scholarship, poverty, disease, crime, lack of education and harsh conditions of work embraced both the old rural and the new industrial labouring classes. Just as in Britain, France and the United States, German and Austrian social reformers wanted to make progress through a direct engagement with the conditions of the people by educational innovation, new institutions of asylum or punishment, new legislation about health and work, and so forth. When this activity is taken into account, it appears that subjects like psychology and sociology did not originate in the academic setting as much as in the administrative and institutional means developed to manage human beings. From this point of view, it was the schools, prisons, asylums, hospitals, workhouses, families, government reports, charities, church groups, youth movements, friendly societies and factories – the local day-to-day management of human activity – which turned man into a systematic object of study. These sites created much of the subject matter of specialized knowledge in the human sciences – schoolchildren, criminals, the handicapped and so forth. Like the laboratory later, institutions like school or prison placed people in a circumscribed environment and researchers generalized from this environment to truths for which were claimed universal, scientific authority. Knowledge generated in this way was accessible to a general audience, and those who contributed by this route to the human sciences came from every walk of life. Though much new nineteenth-century scholarship was inaccessible to an audience without specialist training, the work in many areas relevant to the human sciences was not only accessible but contentious and had lively non-specialist audiences.

There was, therefore, a complex tension between academic and public sites of knowledge production and discipline formation. Neither in Europe nor in North America was there a simple relationship between public values and the development of scientific disciplines in the academic setting. Scientists did not discover facts which were then applied to particular human problems; in large part, day-to-day social life gave human science its content in the first place. Cultural and social life created views of men and women that then found expression in scientific discourse. What authors constructed as scientific theories of human nature were also prescriptions for one rather than another way of life, and this was as true for Smith as it was for Marx. What the German scholars prescribed as the ideal way of life was scholarship itself; they thought of this ideal as central to national culture and the means to lead humanity towards the highest good. No nineteenth-century scholar in the human sciences separated statements of fact and statements of value, as twentieth-century philosophers of science recommended. These ideals of how to live were ultimately the core of what scholars studied in the human sciences.

Humboldt's educational philosophy, which was built into the institutional framework that became so admired, emphasized *Bildung* (literally but unhelpfully translatable as 'formation'). The word denotes the value in a person of wholeness, integration, a state in which every part of education and life contributes to the pursuit of the good, the true and the beautiful. It is an ideal personal quality, but also the quality that makes possible the high culture of a nation. This ideal, transformed into the value that learning is justified for its own sake, was a powerful ally of university expansion. Once established, the universities gained their own momentum and entrenched interests, and the ideal of all-roundedness, of man as an embodiment of the universal, was diluted by the standards of specialized scholarship. In practice, the different disciplines competed with each other over their respective significance to the ideal of a universal science (*Wissenschaft*). Philosophy claimed the prime role as the subject with the special task of integrating the specialist areas into a whole. By 1850, however, many younger scholars gave priority to the natural sciences because these sciences appeared better able to deliver new and objective knowledge. In the natural sciences, specialization became the norm and scientists concentrated on detailed research rather than the ideals of *Bildung*, except on ceremonial occasions or in public rhetoric. The disciplines of pyschology and sociology, when they did appear, were nevertheless not the result of specialization, though specialization was a consequence, but of decisions to demarcate subject areas and to take up new methods in pursuit of objective knowledge.

Europe and North America cross-fertilized each other in relation to ideas about schools, prisons, homes for the deaf and dumb and all the other sites of social intervention. In this regard, there were the same human sciences on either side of the Atlantic. The Pennsylvania Eastern State penitentiary exemplified Bentham's theories of discipline; Dorothea Dix, an indefatigable mid-century New England philanthropist, toured European institutions for the poor along with those in the States; Itard's younger colleague Séguin realized his institutional hopes for idiots in New York as well as in Paris; the managers of Auburn prison in New York in the 1820s made biographical sketches of their inmates and tried to understand the origins of crime rather in the way that German teachers kept diaries to understand their pupils' development. From the most 'applied' setting, such as the prison in Auburn, to the most 'pure' setting, such as the philological seminar in Berlin, values and the construction of knowledge went hand in hand. The origins of the human sciences lay both in the search for the causes of crime and in the search for the origins of grammatical roots.

The example of the United States illustrates the specific character of the growth of specialized academic activity. In certain respects, the independence and constitution of the American Republic was a great Enlightenment achievement. The new country aspired to be a political community organized on rational principles, free of religious intolerance and responsive to the values of humanity. Political independence from Britain did not break cultural and intellectual ties. Scottish moral philosophy remained the backbone of college education and was designed to fit gentlemen for their place in the professions and in public life. In the first half of the nineteenth century the colleges reflected little of the changes taking place in the universities and in scholarship in Germany, though significant numbers of students travelled to Europe to complete their education. Reid's common-sense philosophy and his description of the intellectual and active powers of man continued to occupy a central place in the curriculum. Teachers equipped students with moral and social notions of good character. One of the most eminent representatives of this tradition of moral philosophy, the Reverend James McCosh (1811–94), a Scotsman who became president of the College of New Jersey (Princeton College) in 1868, taught a curriculum that included descriptive psychology as part of moral training.

Nevertheless, by the 1860s the country's western expansion, civil war, early industrialization and the flood of immigrants from Europe had caused old moral and cultural certainties, embedded in the lives of an East Coast élite, to lose authority. Many educated people were conscious of changes in Europe. Many students, who looked for something more solid from higher education than a moral training, travelled to Germany to experience the new scholarship for themselves. In 1862 the Morrill Act made land grants available to state legislatures and this led to the foundation of state colleges with close connections to local business and social reform interests. New private universities like Cornell, Stanford, Johns Hopkins and Chicago were founded and began to incorporate the German model of the research disciplines. The older private colleges set about wholesale reform of their purposes and curricula in order to respond to new conditions. The result was a major change and expansion of the higher education system, at least in part on German lines, and specialist scientific disciplines developed around postgraduate teaching schools. This was the setting in which there was to be large-scale investment in psychological and sociological research, with the result that psychology and sociology, as social entities, appeared at the end of the century. The new developments, all the same, reinvigorated rather than diluted the interaction of public values and the university.

ii *Philology and Critical Scholarship*

If there is one subject which exemplifies the new rigour of scholarship in the German universities it is philology, the comparative science of languages. In certain respects it was a refinement of the core medieval subject of grammar, the subject at the heart of every student's classical education. A classical education continued to give the educated élite a common identity and sense of civilized values. Doctoral theses, even in the natural sciences, were written in Latin well into the nineteenth century, so that a chemist or a physiologist was to a degree also a humanist. The *Gymnasium* education used rigid methods and strict discipline in the study of Greek and Latin authors in order to inculcate logic, rhetoric and moral philosophy. At the university level, first in Göttingen in the eighteenth century, the study of classics ceased to be directed by theological interests and developed a 'critical spirit' in which textual analysis, interpretation and grammar became a scholarly goal in itself. This method or critical spirit was applied to texts – the bible, historical documents and the law – as the means to reconstruct theology, history and jurisprudence as sciences. Philology was the critical spirit applied to the 'text' of language itself. It became a science that traces the historical development of languages and, it was argued, the historical development of man. Philology was central to the university and it was central to the conception of a human science. The philological *seminarius*, or seminar, in which students met to train in methods under a master, the professor, became a model for other disciplines. Thus 'the philological seminary' on classical languages, held by Friedrich Ritschl (1806–77), 'has become a model for the highest form of university instruction'.[3] This critical teaching of classical languages in the universities influenced instruction and set standards across the school system. Ritschl's seminar at the university of Bonn trained a large number of students to exacting standards, and it is a nice twist that the most famous of these students, Nietzsche, in his turn provided a withering critique of academic scholarship.

Philology, though the most academic of subjects, was also in part accessible to anyone with a classical education, while the conclusions which it reached about human history touched on the interests of the public at large. The connection between academic and non-academic audiences was cemented at the beginning of the century by the educational philosophy of Humboldt which inspired the educational reforms. Through Humboldt and other scholars the technical study of language was linked to emotive questions about world history and national culture.

Humboldt acquired a reputation as the founder of the German school of the science of languages, which, by the development of Herder's arguments, integrated the history of languages and the history of peoples. Nevertheless, just as Herder's early work on language was a response to the Berlin Academy's interest in Condillac, so Humboldt, an aristocratic young man who stayed in Paris between 1797 and 1801 and visited Madame de Staël's intellectual salon, associated with Condillac's heirs, the *idéologues*. Condillac had rejected the traditional belief in an original, universal God-given tongue and considered language to be relative to a people's experience. In Humboldt's hands this approach became the theme of a new discipline, comparative philology, which was claimed to be central to the science of man. As Humboldt stated: 'The *bringing-forth of language* is an *inner need* of man, not merely an external necessity for maintaining communal intercourse, but a thing lying in his own nature, indispensable for the development of his mental powers . . .'[4] In conjunction with other German philologists, such as Jacob Grimm (1785–1863), Humboldt made the analytic study of language advocated by Condillac into a science of man centrally concerned with the *Volksgeist*, the spirit of the people. The history of language, philologists believed, reveals the unique spiritual path of a nation's culture.

This comparative and historical approach was diametrically opposed to that of the contemporary English etymologist, or student of the roots of words, John Horne Tooke. At the end of the eighteenth century, Tooke, influenced by Locke, Hartley and Priestley, hoped to reconstruct the origins of words in elementary sensory experience. His method was to trace the etymological derivation of words to signs for sensory impressions. 'What are called . . . [the mind's] operations, are merely operations of language. A consideration of ideas, or of the mind, or of things . . . will lead us no further than to nouns; i.e. the signs of those impressions, or names of ideas.'[5] Over time, he argued, the signs of impressions are abbreviated as words, and words become used as signs for other words, though analysis can trace all words back to a few simple signs of sense impressions. It follows, Tooke believed, that all languages have the same structure.

German scholars turned instead to history to reconstruct the origins of language. They had a far richer sense of the social nature of language. As Humboldt wrote, 'the predisposition for language is inseparable from the predisposition for sociability'.[6] For Humboldt, language is mental activity – 'it is the ever-repeated *mental labour* of making the *articulated* sound capable of expressing thought . . . language proper lies in the act of its real production' – and thought advances by the expression of this activity and its reflection back to us in the language of others. After he retired from public service in 1819, he devoted himself to research on the patterns in language that he believed to be the record of its evolution. All languages share some properties, Humboldt thought, but particular languages are like nations, which 'can and must be regarded as a human *individuality*, which pursues an inner spiritual path of its own'.[7] European nations, however, especially the German, in his judgement and in the judgement of other German scholars, possess languages more suited for 'higher' purposes. It followed that it is possible to classify languages in relation to the degree to which they are appropriate for the articulation of high culture. Other philologists further developed the comparative study of languages into a science of linguistic evolution.

The study of language, with good reason, claimed a central position in the sciences. It was also at the centre of a complex interplay between academic study and public values. The German philologist August Schleicher (1821–68) stated that 'a knowledge of linguistic relationship is absolutely requisite for anybody who wishes to obtain sound notions about the nature and being of man'.[8] In the nineteenth century this search for 'the nature and being of man' through language entrenched the historical imagination in the public mind. It was a British judge who served in Bengal, William Jones (1746–94), who in 1786 gave new life to an earlier hypothesis, that Sanskrit and the European languages share common roots and that it is necessary to analyse the Indo-European languages as a group.

> The Sanskrit language . . . [bears to Latin and Greek] a stronger affinity, both in the roots of verbs and in the forms of grammar, than could possibly have been produced by accident; so strong, indeed, that no philosopher could examine them all three without believing them to have sprung from some common source which, perhaps, no longer exists . . .[9]

Thirty years later the German academic Franz Bopp (1791–1867) provided a rigorous demonstration of the structural affinity of these languages. He also initiated an argument, presumably physiological at root, which requires the articulation of sounds to evolve in a particular way and hence to impose a set pattern on the evolution of languages. This argument was spread through the second edition of Jacob Grimm's *Deutsche Grammatik* (*German Grammar*, 1822). The Danish scholar Rasmus Rask (1787–1832) defined comparative philology as the science of the development of language, the study of the way language naturally evolves. In his view, this evolution can be traced through laws of etymological and grammatical change, a technique he worked out on a study of Old Norse and Icelandic. The work of Bopp, Rask and their followers constructed a developmental morphology of language comparable to contemporary anatomy. Humboldt and his followers, by contrast, studied language as the activity of the historically developing mind, a more psychological approach.

For the wider audience that followed these debates, the linguistic issues were inseparable from historical questions about the origins of Indo-European peoples. The French historian Michelet called India 'the womb of the world', and belief in a common origin contributed to a fashionable orientalism in the arts.[10] During the first half of the nineteenth century, linguistic affinity became the principal means with which to reconstruct ancient history, beyond Greece and Rome, and hence the origins of modern nations. Significantly, linguistic and historical speculation did not result in a clear decision about the position of the Semitic language and people in relation to the Indo-European group. It became common belief that as many as four waves of people had moved westward, as they followed the sun, from northern India through the Near East to Europe, and that each wave had left a trace in European languages. Such claims were highly speculative but had an almost mythic appeal and, bolstered by the scholarly reputation of philology, satisfied the desire for a science of Western origins. In this connection, the inspirer of the values of German *Kultur* and nationhood, Friedrich Schlegel, in 1819 gave currency to the word 'Aryan' to describe the common Indo-European peoples; its restricted usage came later in the nineteenth century.

The history of languages and the history of peoples were powerful components of the nineteenth-century educated imagination. Scholarly and public attention dwelt on time and change on the grand scale in both the human and the natural sciences. A noteworthy case of work on human history that influenced the natural sciences is in geology. In the 1820s, Charles Lyell (1797–1875), one of Darwin's principal scientific mentors, imagined a physical past with the eyes of a scholar trained in the classics and familiar with recent German scholarship. The search for the fossil record of animal life was not unlike the search for the linguistic record of human life. By the 1830s, when Lyell published his major geological work with its rich sense of time, the immensity of time and change was already a familiar theme in human history. Philologists in their turn exploited an analogy from natural science when they traced the evolution of modern linguistic species from one or more common origins in the distant past. They used comparative morphology, a study of the grammatical structure of languages, to reveal ancestry and establish taxonomy or classification. The biological parallel was clear to all and, in the context of the German search for unifying principles, was at times interpreted as evidence for a deeper structural identity in real developmental processes. Many German scholars in the first half of the nineteenth century thought in comparative and developmental terms about both the organic world and human culture. Thus, when Darwin's work on the evolution of animals and plants became known in German translation in 1860, his audience easily assimilated evolution as a general perspective but was often insensitive to what distinguished Darwin's theory from other theories of development.

Schleicher is a relevant example. He published *Die Darwinische Theorie und die Sprachwissenschaft* (1863; translated as *Darwinism Tested by the Science of Language*), in which he welcomed Darwin's work, drew out the parallels between the evolution of species and the evolution of languages and pictured both types of development with the image of a branching tree. His own studies of language used a comparative method and tried to reconstruct ancestral roots and establish a taxonomy. He described 'language as a product of growth'.[11] Nevertheless, it was unclear in Schleicher's discussion how anything specifically Darwinian, notably the theory of natural selection, related to linguistic evolution. Rather, Schleicher's work illustrates the way historical thought about language and culture was assumed to be related to historical thought about nature. This assumption existed before Darwin and established conditions for the reception of his theory. Schleicher himself, when he linked his work with Darwin's, probably hoped to detach the science of language from its past in speculative idealism and link it with the values of the empirical natural-scientific method.

The relation between Darwin's theory of evolution and theories of the development of languages was complex. Darwin was specifically opposed by the leading Sanskrit scholar of his generation, Friedrich Max Müller (1823–1900), even though Müller exploited the analogy of natural selection in his history of language. He spread the belief in Britain that the past and present destiny of man is bound up with the pattern of migration westwards and the consequent rise and fall of nations. After completing a training with Bopp in Berlin, Müller came to England in 1846 to finish his edition of the *Rigveda* with funds from the London East India Company. Subsequently appointed to the university of Oxford, he eventually occupied a chair in comparative philology, from which prestigious position he created a large English-language audience for romantic and in many regards speculative stories about human history. As he accurately remarked, 'the great problem of our being, of the true nobility of our blood, of our descent from heaven or earth . . . [has] still retained a charm that will never lose its power in the mind and on the heart of man'. Müller's answer to this 'great problem' squarely expressed his belief in a providential Christian God, and, goaded by Darwin's theory of evolution, he applied his science of language to the defence of Providence. For Müller, the study of language is a science of structure and change, like a physical science, not a science of historical evolution. He believed that nouns 'express originally one out of the many attributes of a thing and that attribute . . . is necessarily a general idea'.[12] Language shows that man, unlike the animals, possesses concepts – general ideas. With what he believed was the authority of Kant, Müller argued that concepts are necessarily *a priori* and from this concluded that reason and language are innate to the God-given nature of man. Thus science shows, he believed, that language – in its essence – is not something that can evolve, however much particular languages change over time in response to the 'selection' of one form or another by local conditions. He even concluded that if Darwin had read Kant, he would never have accepted the idea of human evolution.

Origin myths were and are bound up with a sense of identity, and it was no coincidence that linguistic scholarship blossomed during a period when there was intense self-consciousness about national identity. Poets revealed the living spirit in language and scholars equipped that spirit with a past. There was a heady idealism in European struggles for national self-assertion and independence from Ireland to Greece and from Finland to Catalonia. And there was close accord between academic scholarship and public aspiration in the creation of nationalist sentiment.

This accord is evident in a rich body of historical writing about nation states as well as about languages.

History, like philology, became a major discipline within the reformed universities, first in Germany and then elsewhere. The examination of language, law and the bible created techniques for textual interpretation, such as the comparison of divergent manuscripts, which became the method of historical scholarship. Historians defined their speciality as the correct interpretation of primary sources, the direct or immediate documentation of events. Leopold von Ranke (1795–1886), whose historical seminar in Berlin in the 1820s established the pattern for training historians with primary sources, himself trained as a philologist. Ranke intended his method to eliminate false evidence, anachronism and moralism, and thus make it possible for history to achieve a record of past events in its purest form. He took up Niebuhr's critical methods, which had been applied to classical texts, and adapted them to the examination of modern history. In the preface to his *Geschichte der romanischen und germanischen Völker von 1494 bis 1514 (History of the Latin and German Nations from 1494 to 1514,* 1824), he wrote:

> This book attempts to see these histories and the other, related histories of the Latin and Germanic nations in their unity. To history has been assigned the office of judging the past, of instructing the present for the benefit of future ages. To such high offices this work does not aspire: It wants only to show what actually happened.[13]

This was the critical spirit, the replacement of moral tales by scholarship. His often quoted last sentence, however, with its reference to 'what actually happened' (*'wie es eigentlich gewesen'*), did not simply identify history with the recovery of facts. For Ranke, a German of his age, each fact is significant by virtue f its unique place in God's unfolding design for the human world. In the judgement of Ranke and his contemporaries, the key facts of history record the origins of modern European nations – hence Ranke's choice of the period on which his book concentrated.

Ranke exemplifies the historian's predilection to make experience intelligible by the recreation in a narrative of what is individual and particular. He trained students in a critical scholarship intended accurately to portray each age in its own terms. This imposed a commitment to the freedom and individuality in humanity's striving and not only a methodological standard. It permitted Ranke to stress the unique qualities and path of German history and the Prussian state's legitimate claim to leadership of the German people. In this way, academic history became the scholarly core of nationalist opinion. Ranke believed that particular facts reveal a general pattern, a necessary order in reality, which displays itself in the progress of human reason and political society, but that facts are individual and represent the outcome of free human action. This posed yet again the problem of the reconciliation of freedom and necessity in the science of man. In a later debate over the value to general culture of history and the humanities as opposed to the natural sciences, the university of Freiburg philosopher Wilhelm Windelband observed: 'The controversy was at its hottest where it came to the point of finally deciding to what extent an individual person owed the essential worth of his life to his own self, or to the overriding circumstances of his environment.'[14] Ranke and other German historians hoped that history, which describes the individuality of people and nations while at the same time it expresses the necessity of progress in the emergence of nation states, would square this circle.

Academic historians retained a substantial public audience, at least in part because of the support they gave to belief in national progress. In other areas, critical scholarship sometimes produced results sharply at odds with public values; and nowhere more so than in the 'higher criticism'. This body of German scholarship applied techniques for the assessment of the evidential reliability of historical texts to the New Testament. The Protestant Reformation had made scriptural interpretation, exegesis and translation of fundamental significance for generations of scholars and, in many respects, this work equipped both philology and history with their techniques. But the new scholarship was qualitatively different in kind and impact. It dated the gospels to the centuries after Christ's death, detailed inconsistencies in different versions of Christ's life and generally questioned the reliability of the written record as evidence. The scholars who carried out this work, following the example of F. D. E. Schleiermacher (1768–1834), intended to use criticism to make Christian beliefs conform with the conclusions of reason. But because they discussed Jesus as a historical figure and cast doubt on the texts by which he was known, they appeared to most believers to strike at the heart of Christianity. The impact on Christian scholarship was very great, much greater than that of evolutionary theories. The appointment of David F. Strauss (1808–74) to a professorship in Zurich in 1839 led to a peasant riot, with the result that Strauss was immediately pensioned off. As the author of *Das Leben Jesu, kritisch bearbeitet (The Life of Jesus, Critically Considered,* 1835), the book that spread the new scholarship to a wider audience, it appeared as if he personally had dethroned the living Christ. Strauss himself intended to question history, not Christianity; but critical academic work had the potential to disrupt public expectations and to place academic and established truths in opposition. Some members of the public, however, welcomed the new scholarship and believed that criticism stripped away untruths about the human condition, liberated the benighted human spirit and made it possible to find a guide to life in human nature instead of dogma and prejudice. The young Marian Evans, the future novelist George Eliot, nearly drove herself to despair with the 'soul-stupefying labour' involved in the translation of Strauss into English, but she thought it her duty to humanity.[15]

Exponents of the higher criticism referred to theology as a science, and they intended theological scholarship to contribute to the rational understanding of man's nature. Critical scholarship, including theology, contributed to the techniques and conceptual foundations of the modern human science disciplines alongside the natural sciences. The two areas competed for resources and status within the universities and for public attention in the pursuit of higher cultural values. Thus there was no agreement to equate the pursuit of knowledge of man with the natural sciences, though the latter certainly increased in status during the course of the century.

iii *Prehistory*

Enlightenment enthusiasm for universal histories of mankind, Romantic concern with the individual character of a people's history and nineteenth-century linguistic and historical scholarship all demonstrated a preoccupation with origins. The origin myth of the creation and fall of man lost power in its biblical form. Instead, people turned for authoritative conclusions to the historical disciplines, in which critical scholarship and empirical content were believed to create not myth but science. The historical sciences in the nineteenth century included archaeology, a new emphasis on race as the fundamental category of human difference, the reconstruction of the eighteenth-century theory of stages in social development, and the theory of evolution as an explanation of the origin of plants, animals and man himself. The first two developments are discussed in the following sections; the third and fourth require separate chapters.

The disciplines of philology and history were text-based; their notion of history was literally a notion of recorded history. During the first half of the nineteenth century, much other work, often by people outside academic institutions, pushed back the time of history recorded in texts into a more distant past, a past for which the record was artefacts and not language. The outcome in the 1850s was the new concept of prehistory and the beginnings of the modern discipline of archaeology. The work created a new conception of the scale of human history, a scale that merged with the vast aeons of time summoned up by geologists to explain the record of the rocks. Archaeology and geology, indeed, also shared a common background in the activities of collectors and antiquarians whose curiosity had ranged across a past in which human and earth history were not kept separate. The institution that housed the evidence of their activity was not the university but, first, the private cabinet, and then the museum. When archaeology became a university discipline, it retained relations both with museums and the general public to a degree at times unrivalled by any other science. Scholarly conclusions about prehistory were closely followed by public interest in the origins of peoples.

Collecting had been a fashionable if sometimes eccentric passion since the Renaissance, often connected to admiration for classical authors and the world they described. Travellers visited Rome and marvelled at the fabric of a lost civilization; many educated people dreamed of Greece and, in the early nineteenth century, a few began to visit and to bring home the physical artefacts of their Hellenic ideal, like Lord Elgin who shipped the Parthenon frieze, the Elgin Marbles, to London. The standing stones, barrows, legends and sagas of northern Europe became familiar objects of interest and use in Romantic imagery. The French expedition to Egypt under Napoleon, Russian expansion into Central Asia and European interest generally in the Near East brought back vivid evidence of ancient civilizations besides those of Greece and Rome. The Egyptian hieroglyphs were deciphered in the 1820s, adding to the excitement about the ancient world. At the same time, a newly self-conscious science of geology refined techniques to determine the historical identity of excavated objects. Skill in the identification of rock specimens or fossils and in the placement of objects in a time sequence, which was evaluated as progress, was then transferred in the mid-century to the early human record. The great Victorian namer of new sciences and Master of Trinity College, Cambridge, William Whewell (1794–1866), observed in 1837 that 'theoretical Geology . . . has a strong resemblance . . . to philosophical Archaeology'.[16]

'Collecting' conjures up an image of an amateur Victorian in bustle or top-hat studiously bent over in the landscape. Many gentleman and lady scholars, while they did not earn a living through science, were rigorous and disciplined about what they undertook and often active in scientific institutions. The process by which the sciences became professional – monopolized by full-time career scholars in a discipline based in specialist institutions, especially academic departments – was complex and varied. The separation beween professional and amateur was particularly unclear in archaeology, as is well illustrated by the contribution of the French collector Jacques Boucher de Perthes (1788–1868). In the 1840s, Boucher and workmen whom he encouraged by cash rewards, recovered flints shaped by human hands from the river gravel of the Somme at Abbeville in the north of France. Such flints were familiar enough, but Boucher recorded the level in the gravel where they were found and this indicated both their great age and a sequence in the type of artefacts. His scientific and social superiors were initially sceptical of his finds but, over the next decade, verified the location of the flints and accumulated much other similar evidence.

Local enthusiasts elsewhere dug, or employed workmen to dig, down through the layers of a past that long preceded written records and appeared incompatible with biblical authority. In Britain the decisive moment was the excavation of Brixham cave in Devon in 1858, when carefully supervised work recovered human artefacts *in situ* with extinct animals, engraved antlers or stones shaped by hand buried beside the bones of the sabre-tooth tiger. There was little evidence to indicate exactly how old early man was; but the evidence from river gravels suggested at least tens of thousands of years, and some observers speculated about much larger periods of time. A report to the Society of Antiquaries in London, which brought together results from Abbeville and Brixham, concluded in 1859: 'in a period of antiquity, remote beyond any of which we have hitherto found trace, this portion of the globe was peopled by man'.[17] There was new interest in a system established in the second decade of the century by the Danish National Museum of Antiquities under its curator C. J. Thomsen (1788–1865), to order material in its collections excavated from the middens or rubbish tips of ancient Danish villages. The English translation of J. J. A. Worsaae's *Danmarks oldtid oplyst ved Oldsager og Gravhöie* (1842; trans. as *The Primeval Antiquities of Denmark*) in 1849 spread the conviction that early men in general had developed through a sequence of three stages. Worsaae (1821–85) was a student of Thomsen's who developed his teacher's ordering scheme. Early Europe had left a record of its technology in the middens and it was a record of progress, though the Danish scholars did not see this progress in evolutionary terms. The three prehistoric ages of stone, bronze and iron were described as distinct stages.

During the 1850s, English writers began to refer to this past as prehistory, and the term was a useful handle with which to grasp the new sense of man's past. It was crucial that evidence for early man was found in Europe as it required the savage 'other' to be accepted as the ancestor of European people and not just present in some far-away exotic tribe. It became common belief that European people, though now civilized, were through their ancestry connected to primitive human nature. Indeed, the word 'primitive' came into its own. Writers synthesized an image of the living savage described by European travellers with the image of early European man described by archaeologists to picture a historically real primitive man, a man imagined to be ancestral and hence to a degree still present within civilized man. The fusion of studies on the savage, which went back to the Western encounter with other peoples, with the idea of primitiveness, reconceived in the light of the history of civilization and of evolution, created the modern science of anthropology. The science of prehistory attracted public attention exactly at the same time as Darwin's evolutionary theory, and archaeology and anthropology together created a potent historical story about the progress of life from primitive to civilized existence. It will be possible to return to this once the sources of the contemporary interest in race and in evolution have been described.

Also in the 1850s, a limestone cave in the Neander valley near Dusseldorf and river gravels of the Meuse in Belgium yielded human skeletal remains which, like the flint finds, were of great age. These remnants of Neanderthal and the more modern Engis man did not provide evidence of a 'missing link', the half-human, half-animal form imagined as the direct ancestor of *Homo sapiens*; there was no candidate for this position until the 1890s, when so-called Java man (*Pithecanthropus*) was described, only to be rejected in turn as the missing link. The relationship between the ancient bone fragments and modern European man was highly debatable, but it struck everyone that the cranial capacity of the earliest known fossil man, the Neanderthal, was more or less comparable with modern types. This was a puzzle as most observers thought of cranial capacity as a measure of mental capacity. The puzzle was deepened when, in 1868, remains of early man were found in the Dordogne in France, and the leading French anatomists Paul Broca and Armand de Quatrafages deduced that a tall people with fine features had lived there in prehistoric times. As Thomas Henry Huxley (1825–95) commented in *Evidences as to Man's Place in Nature* (1863), his very widely read summary of the anatomical evidence about monkeys and men, ancient and modern: early man must belong to 'yet more ancient' strata than those yet discovered.[18]

iv *Human Differences and Race*

Belief about early man and the history of culture and civilization was inseparable from belief about human difference and the way in which different people were described and classified. At the heart of the values that made reference to the Age of Enlightenment seem appropriate was a belief in the universality of human nature. This belief made possible the search for a science of man. Eighteenth-century writers ascribed human differences to experience and diverse physical conditions. What Locke described as the experience that forms a person's knowledge, Montesquieu and Ferguson redescribed as the experience that forms the different social and legal development of different societies. There was a fascination with difference – Diderot's delight in the Tahitians was shared – but difference was regarded as the outcome of the effect of experience on human nature, something to be explained and not something fundamental with which to explain everything else in human life. Christianity, in theory if not always in practice, also stressed a common humanity, and on this subject religious and secular writers found common ground. Day-to-day life, of course, in relations between men and women, in the interaction of social groups and when Europeans dealt with the rest of the world, was full of assumptions about difference. All the same, there is a sharp contrast between the widespread eighteenth-century belief in a common human nature and the late nineteenth-century conviction that race is a category that divides human nature. In the last decades of the century, vociferous sections of both the scientific community and the public treated the facts of difference as the basis on which to explain everything else rather than phenomena themselves in need of explanation. Further, the biological language of race came to prominence as the way to describe the facts. How and why this occurred was significantly connected to the search for a physical science of human nature, for a natural science of man. On a broader historical map, the prominence of race as a category of human differentiation must be explained by European and United States perception of the world as an arena for competition – if necessary, violent competition – and for the creation of empires. At home, the stress on human differences was also part of the attempted containment of demands for emancipation from workers, women, national minorities and slaves.

There were major studies of human variation in the eighteenth century, such as Buffon undertook in his natural history, and there was debate about where to draw the boundary between man and animals. Herder called for a 'physico-geographical history of the descent and diversification of our species according to periods and climates . . .'[19] The historical and comparative study of languages created a descriptive science of the relations between peoples and nations. Archaeology linked early Europeans and savages with the concept of primitiveness. But it was physical anthropology, the comparative anatomy of human difference, that most directly gave belief in human differences empirical content, and it was the subject area which, so its practitioners claimed, most definitely made the study of man into a science. The stress on race came into prominence in time with faith in scientific progress not in opposition to it. When the systematic comparison of human types on the basis of anatomy became formalized in the late eighteenth century, especially with the work of Blumenbach, there were two strands. One was concerned with individuals and focused on differences within a particular people – differences of sex, character, ability, virtue and so forth. It took the form of phrenology, the science that correlates mental character with the shape of the cranium or brain box, a science which received serious attention from scientists between about 1815 and 1840 but long thereafter continued to have a public following. The other strand was concerned with large-scale human groups, the groups more and more frequently simply referred to as races. This strand was particularly active in the decades after 1850.

Phrenology, once dismissed as a faddish pseudo-science, has interested historians. Between 1810 and 1819, the anatomist Franz (or François) Joseph Gall (1758–1828), assisted by J. C. Spurzheim (1776–1823), published the four volumes and atlas of 100 engraved plates of the *Anatomie et physiologie du système nerveux . . . avec des observations sur la possibilité de reconnaître plusieurs dispositions intellectuelles et morales de l'homme et des animaux, par la configuration de leurs têtes* (*Anatomy and Physiology of the Nervous System . . . With Observations on the Possibility of Identifying many Intellectual and Moral Dispositions of Men and Animals by the Configuration of Their Heads*). This was a major contribution to the anatomical study of the brain. Gall's motive for so much patient observation was to show how the parts of the brain are local sites for the different mental faculties, how the size of the parts reflects the strength of the faculties and hence how physical size, reflected in the shape of the skull, reveals a person's character. He traced his own theories to a schoolboy observation that fellow-pupils with fine memories had protuberant eyes, pushed out, as he later concluded, by the large development of the part of the brain that houses the faculty of memory. Gall elaborated a physiognomical language familiar to painters and sculptors and in everyday life (for example, in the description 'egghead'). He himself interpreted busts of great historical figures. He was also a comparative anatomist in the mould of Blumenbach and like him assembled hundreds of skulls or casts (there are more than 600 items kept at the Musée de l'Homme in Paris) as the empirical basis for the study of human character. Gall intended 'to found a doctrine on the functions of the brain' as a basis for 'a perfect knowledge of human nature'.[20] But it was his commentary on individual differences, very much part of the everyday world, that made his science so accessible. Phrenology dignified gossip about individual character with objective status.

Gall's erstwhile collaborator, Spurzheim, travelled widely in Britain and the United States where he promoted a version of phrenology in the 1820s which made it a guide to life and education. Knowledge of individual strengths and weaknesses, devotees argued, provides the information needed to mould character, including one's own. George Combe (1788–1858) carried the message to the lecture halls and lending libraries of provincial Britain. Prince Albert and Queen Victoria brought cranial measurement into the palace nursery. As discussed in the next section, a culture was encouraged which accepted that human nature is materially embodied in organs, the brain and skull, and is subject to natural-scientific research and to measurement. Phrenology, in its mild way, was a human technology.

Gall thought of character as largely determined by the innate capacities of twenty-seven basic faculties. Education, he supposed, could work with but barely modify an individual's given strengths and weaknesses. Spurzheim and other popularizers took a more optimistic and egalitarian view, advocating exercise as the way to strengthen or control a faculty. In their version, phrenology is the science of human nature that makes it possible for everyone to help themselves. This divergence of view among phrenologists was emblematic of a wider disagreement between those who emphasized innate endowment and those who emphasized education as the cause of differences of achievement. It was very characteristic of nineteenth-century empirical natural science to describe differences between individuals and types and on this basis to claim that differences are ineradicably built into the physical fabric of the human world. The argument frequently implied determinism – and Gall was certainly accused of this; popular phrenology, by contrast, propagated the values of self-help. Other aspects of popular belief, however, were receptive to determinism: a common language linked fortune to individual and national character, or the appearance of babies to that of their parents, and accepted inequality of every kind as a natural condition.

The language of individual differences merged without a dividing line into the language of group differences; phrenologists, for example, described the balance of faculties as different in men and women. And just as phrenology related observed individual differences to the physical body, so anthropology tied observed group differences to physical types. The popular empirical language of human differences was reconstituted as empirical sciences of the physical world; man was reconstituted as nature. But, however persuasive this was to some scientists, the step was controversial and never universally accepted either in the nineteenth or in the twentieth century.

The difficulties that faced the construction of a unified science of physical and cultural human differences is illustrated by the work of an English author, James Cowles Prichard (1786–1848). In his spare time as a young physician, Prichard published his *Researches into the Physical History of Man* (1813), which in subsequent editions became the leading English-language text on what he and his contemporaries called ethnology. He provided a survey of the physical variety of mankind – he used Blumenbach as his main source – yet he also argued as a Christian for man's unity. He gutted travel writers, classical authors, naturalists, biblical historians and every kind of genealogist of antiquity in order to describe man's history and explain his diversity. Where physical observation revealed difference, he used historical observation to explain it as the product of time, climate, custom and the diffusion of peoples. Thus his work carried on the eighteenth-century debate about man's natural history focused around Buffon's work, and Prichard maintained Buffon's emphasis on the role of the environment. The primitive state, Prichard thought, is 'rude and uncivilized' and shared in common: 'all nations who have never emerged from the savage state, are Negroes, or very similar to Negroes'. Civilized people, he concluded, have resulted from 'the evolution of white varieties in black races of men' under the influence of climate.[21] Born into a Bristol Quaker family, though he became an Anglican, and from his earliest years a supporter of the abolition of slavery, his ethnology reflected the values of this background. Prichard also had a flair for languages: as a child he was said to have accosted astonished sailors in the Bristol docks in their own tongues. In the second edition of his *Researches* (1826) he introduced much more ethnographic material in the form of descriptions of exotic peoples, and he used recent German scholarship about language to state the relationship that different people have to each other. His linguistic interests also surfaced in a study of *The Eastern Origins of the Celtic Nations* (1831), a book much encouraged by Bopp, which showed that the Celtic languages belong to the Indo-European family. He continued to defend his belief that all people share a single origin; he referred to race merely as a cluster of characteristics caused by climate, not a rigid quality; and his use of the word 'primitive' connoted man's closeness to Adam rather than the apes. But there was perhaps an air of defensiveness when he rejected the term 'race' as a way to classify human types. In the third edition of his *Researches* (1836–47) the sheer descriptive material increased further, and while his answers to the question of the origin of human types became less clear-cut, he still defended the unity of mankind against a rising tide of contrary opinion about the significance of race. Prichard argued against those who focused on race as a differentiating category on grounds of the psychic unity of mankind. In this way he expressed belief in a common mental inheritance, an idea that was shortly to be given a rich life by evolutionary theories.

Ethnological societies were founded at about the same time in England (1843), France (1839) and the United States (1842) and they co-ordinated historical, geographical and physical data on human variety. These societies also reflected concern about the relations between Europeans, or those of European descent, and people in the rest of the world as the creation of colonial empires accelerated. This overseas expansion itself, often in the wake of Christian missionaries, generated much new information. About the same time, from the 1840s, the argument became prominent that physical facts demonstrate the existence of fundamental differences in human types or races. Some proponents of this view even asserted that the facts demonstrate the separate origins of different races, a belief known as polygenism, which opposed the monogenist belief exemplified by Prichard. There was a general shift of opinion towards physical measurement rather than history or philology in comparative studies of man. The encouragement of the new science was the motive of Paul Broca (1824–80) when he founded the Société d'Anthropologie de Paris in 1859, and matters came to a head in England in 1862 and 1863 when an anatomist, James Hunt (1833–69), who was a sharp critic of current democratic ideals, led many members out of the Ethnological Society and founded the Anthropological Society of London. Broca, whose work on cranial dimensions, such as the difference between men and women, was widely discussed, stated with clarity the general principle that underlay the new investment in systematic measurement of the skull:

> In general, the brain is larger in mature adults than in the elderly, in men than in women, in eminent men than in men of mediocre talent, in superior races than in inferior races . . . Other things being equal, there is a remarkable relationship between the development of intelligence and the volume of the brain.[22]

Hunt and his followers, the 'anthropologicals', presented themselves as scientific radicals as opposed to the unscientific and old-fashioned 'ethnologicals'. They claimed that tangible physical anatomy, not speculative historical studies of languages or travellers' tales, provides the objective method to map human differences. Within a few years there were 500 members of the society, one of whose members responded to Governor Eyre's ruthless suppression of a black uprising in Jamaica in 1866 with the judgement that to hang savages without trial is sometimes 'a philanthropic principle'.[23] After much argument, the ethnological and anthropological societies came together in 1871 as the Anthropological Institute of Great Britain and Ireland, by which time the word 'anthropology' had acquired its modern meaning.

The founding of anthropological societies was preceded by two English-language texts, later notorious, which made physical race pivotal in human history: the first by the Scotsman Robert Knox, *The Races of Man: A Philosophical Enquiry Into the Influence of Race Over the Destiny of Nations* (1850), and the second by two authors from the United States, Josiah Nott and G. R. Gliddon, *Types of Mankind: Or, Ethnological Researches Based upon the Ancient Monuments, Paintings, Sculptures, and Crania and Races* (1854). Knox (1791–1862), whose chapters were based on lectures given in the English provinces to earn a living, adopted an embittered rhetoric that derived from the collapse of his medical career after he received bodies from the murderers Burke and Hare. He stated that race is the limiting condition of all human action: 'race is everything in human history'.[24] For example, though he believed that white races were vanquishing the black races, he also believed that the physical nature of white people would always prevent their acclimatization to the tropics.

Support for polygenism – belief in the separate origins of the different races – was especially strong in the United States, where white progress was in stark contrast to black slavery and the destruction of indigenous culture and people. The popularization of the study of physical conformation as a guide to race was the achievement of Samuel George Morton (1799–1851), an invertebrate palaeontologist turned expert on crania. He assembled a huge collection of skulls in Philadelphia and elaborated a system to make precise cranial measurements as a basis for description of race. He was a close associate of Combe, though not himself a supporter of phrenology. His work was then developed by Josiah Nott (1804–73), a surgeon from Alabama, and G. R. Gliddon, the US Vice-Consul in Egypt, into an anti-scriptural polemic on polygenism and the physical laws of human nature. Nott wrote: 'There is a genus, Man, comprising two or more species – that physical causes cannot change a White man into a Negro . . .'[25] Nott and Gliddon's grossly exaggerated engravings of human racial types became a monument to prejudice in the garb of observation. They shared Knox's belief that physical race determines everything else and that comparative anatomy therefore provides the basic, irreducible data for the human sciences, and wrote a defence of slavery as the natural expression of relations between superior and inferior species of human beings. Their position, though extreme, was not unusual in this period before the American Civil War fought, in part, over slavery.

Some people who claimed scientific status for their work were admittedly a little more circumspect. Researchers devised complex calipers to measure cranial dimensions, or they filled skulls with lead shot to use weight as a measure of relative capacity. The Swedish anatomist Anders Retzius (1796–1860) devised an classificatory scheme based on the cranial index, the ratio between the side-to-side and front-to-back dimensions of the skull. He located original European 'broad-heads' among the Lapps and Basques; elsewhere, he claimed, they had long since been replaced by 'long-heads'. Behind much of this work was the crude assumption that the sheer size of the brain correlates with mental capacity, which related anatomy to prejudices about the relative abilities and contributions to progress of different people, and of men and of women. The French anthropologist Paul Topinard, who was influenced by Broca, wrote that 'the outlines of the adult female cranium are intermediate between those of the child and the adult man; they are softer, more graceful and delicate . . . the crown is higher and more horizontal, the brain weight and cranial capacity are less'.[26] The conclusions of anatomy therefore merged with studies of prehistory and the distribution of language and culture into an evaluation of human differences and progress. All this work also enriched nationalist rhetoric. As the Berlin physician Rudolph Virchow (1821–1902) warned his colleagues: 'The questions which we are dealing with, when taken over by the people, soon become national questions.'[27] Virchow himself directed an extended study of cranial measurements of schoolchildren in the wake of French accusations that the Prussians, who had defeated France in the war of 1870, were Eastern barbarians, a race of inferior people. His measurements showed that the Germans were not a pure type, as far as their heads went, but a mixture of types. In retrospect, this study suggests that there are no correlations between cranial capacity or shapes and different peoples; but, at the time, it seems to have done little to dampen the enthusiasm of anatomists.

There was therefore a shift of emphasis in the description and classification of human differences away from cultural and linguistic studies and towards a comparative anatomy focused on the category of race. In contemporary English usage, this was a shift from ethnology to anthropology. Later, the two approaches came back together, though often in an uneasy alliance, as the social and physical branches of the modern discipline of anthropology. The shift towards physical anthropology, however, was never total, and there continued to be scholars, especially in Germany, who argued that the history of languages was in fact a more rigorous science of human relationships than anatomy. Indeed, in 1860, Moritz Lazarus (1824–1905) and Heymann Steinthal (1823–99) established a new journal, the *Zeitschrift für Völkerpsychologie und Sprachwissenschaft* (*Journal for Cultural Psychology and Linguistic Science*) as a vehicle for research about culture and the expressive activity of different peoples' psychological, not physical, capacities. The journal's editors assumed that the systematic study of man must concentrate on psychology, language and culture, not race. Their subject was *Völkerpsychologie*, 'the psychology of societal human beings or human society', and their journal discussed the way psychological life takes place in culture, making subjective activity a cultural product.[28] Yet another approach was taken in the contemporary, widely read book by Eduard von Hartmann (1842–1906), *Philosophie der Unbewußtsein* (*Philosophy of the Unconscious*, 1869); von Hartmann claimed that culture itself is the expression of unconscious ideas shared by all mankind. Studies such as these on language and symbol formation went on in competition with physical anthropology, and they were continued in the work of scholars as diverse as, for example, Wilhelm Wundt and C. G. Jung.

Even among those who described themselves as engaged in the science of anthropology, there were powerful voices critical of the category of race. The Berlin academic Theodor Waitz (1821–84) gave to anthropology 'the task of mediation between the physical and historical portion of our knowledge of man . . .' He published the first volume of a six-volume study in 1858 in support of monogenism and a logical belief 'as a necessary presupposition of all sciences, [that we] assume that there is a universal and unchangeable human nature'. In an attack on those who claimed the existence of fixed races, he stated, 'it is certain that the absolute permanence of the physical type is nothing but a prejudice, possessing no scientific value whatever to serve as a basis for the assumption of a plurality of human species'.[29] He believed that man's mental endowment was initially more or less identical; the emergence of man from primitiveness was, he thought, due to man's reason, and the timing of this emergence depended on physical and social conditions. This was not just an abstract theory for Waitz since he was very active in a campaign for educational reform. As he believed that the effects of changed social conditions may become inherited, it was an urgent matter to persuade colleagues that people can indeed learn and culture be changed in a positive way. This work subsequently influenced Franz Boas (1858–1942), one of the most powerful critics in the first half of the twentieth century of biological categories as the foundation of anthropology. Significantly, Waitz supported a notion of organic change by the inheritance of acquired characteristics rather than a Darwinian notion of evolution. In his account of human evolution, this reinforced Waitz's argument that cultural change involves the development of a common human nature, a position opposed to the belief that human nature – in the form of different racial natures – produces sharply different cultures.

Lazarus, Steinthal and Waitz were 'armchair' scholars and their notions of human diversity combined theory with travellers' reports. The same could not be said of Adolf Bastian (1826–1905), who did as much as anyone in the 1860s to shape a coherent view of anthropology as a distinct science. Bastian spent more than twenty years travelling over much of the world, and though he travelled light and did not, as he himself recommended, spend long periods in the detailed study of particular peoples, he contributed an enormous amount of information and inspiration to the disciplines of geography and ethnology in Germany. He funded these travels from a private income, but between 1866 and 1875 he worked in Berlin as a scholar and administrator. He then helped to found the Royal Ethnological

Museum (opened in 1886) and to organize the exploration, especially in Africa, that accompanied Germany's attempt to become a major participant in European expansion. Bastian had a clear theoretical purpose for his subject: 'The main focus of ethnology . . . lies in the mental life of peoples, in the research about the organic laws under which mankind rose to a state of culture in the developmental process of history.' He aimed to turn the specifically German concern with *Geist*, man's spirit, into a science. Ethnology was the field that would provide psychology with its data and hence make possible a science of man: 'The goal of modern ethnology is to find an adequate methodology for scientific psychology.' As the individual is part of an organic, historical whole, the study of the whole, of culture, is the empirical basis of psychology. The value of the study of what were called primitive people to this project is that they reveal the basic universals of man's nature: 'The importance of the natural people for ethnology lies in the fact that we find among them the collective thoughts of mankind in their most simple and primitive form, as lucid and clear as if they were a lower cellular form . . .'[30] The study of the world's peoples, he believed, will reveal the universal psychological characteristics of man, the characteristics grounded in his bodily life and basic circumstances that give man a common nature. He judged human diversity to be the outcome of history. Further, in a manner that was common in German thought, Bastian attempted to explain human development in terms of an innate capacity for growth in man's mind, an *entelechy* or final cause, rather than in terms influenced by Darwin. Thus, for example, he explained cultural similarities by a common pattern of development and not by the diffusion of cultures. To summarize: he aimed to turn the characteristic German concern with *Geist*, man's spirit, into a science; the means was ethnology, not the science of race.

In spite of ethnology and its emphasis on the unity of mankind, the subject of race united scientists and public opinion, became commonplace and was ultimately the nucleus of a catastrophe. The degree of prejudice in the late nineteenth-century language of race later horrified observers. Many twentieth-century anthropologists found it perfectly easy to describe human differences without the use of the word 'race'. Nearly every white European and North American in the late nineteenth century, however, believed that the language of race is the language of empirical science. Notoriously, German nationalists gave race mythic status. Extreme nationalists were frustrated even after the unification of their country in 1871 because large communities of German-speaking people still lived outside the borders of the Reich, and they were also obsessed by Germany's lack of world empire and what they perceived as its encirclement in the middle of Europe. German writers began to use the word 'Aryan' to describe a supposed biological type of northern European rather than an Indo-European linguistic group. They dreamed that the character and identity of this northern European people had been forged in the forests independently of the Hebrew and Roman roots of Christianity. On top of this, virulently anti-Semitic values became common across Europe, from Russia to France, and these values were expressed in the scientific language of difference and race.

The Englishman Houston Stewart Chamberlain, an initiate of Richard and then Cosima Wagner's circle in Bayreuth, consolidated many of these values in his wildly successful *Grundlagen des 19. Jahrhunderts* (*The Foundations of the Nineteenth Century*, 1899). George Bernard Shaw, the Anglo-Irish playwright, described this travesty of scholarship as a historical masterpiece. Partly because of Chamberlain's endorsement, the self-styled comte Arthur de Gobineau (1816–82) acquired status in retrospect as the father of racist ideology. Between 1853 and 1855, Gobineau published the four volumes of his *Essai sur l'inégalité des races humaines* (*Essay on the Inequality of the Human Races*), which both scientists and the public virtually ignored. His work was outside both mainstream physical anthropology and ethnology, and few people wanted to hear his absolutely pessimistic pronouncements – 'You are dying' – about Western civilization.[31] The motor of inexorable decline, in Gobineau's view, is the loss through cross-breeding of the pure blood of the three original races of mankind. The mixture of races, he argued, has given rise to human history, but at the cost of inevitable degeneration. His notion of race was a projection on to world history of the feudal categories of nobles, clerks and serfs, which reappeared in Gobineau's work as the white, yellow and black races descended from the sons of Noah. As this projection suggests, Gobineau's interest in race was secondary to an embittered revulsion against an age ruled by ignoble mediocrity at the expense of aristocratic values.

This contempt for modern mass society was also present in Chamberlain's work and in his and Cosima Wagner's zeal for Richard Wagner's music. Many intellectuals feared the advent of participatory democracy, the levelling of culture which they associated with the spread of education, a consumer culture and newspapers for a mass audience, and élites feared what they perceived as a dilution of individual quality in an age of bureaucracy and cities. One sign of these fears was the appeal of languages of human difference, a means to assert, with all the authority of science, the existence of inbuilt hierarchies of nature and hence also of political society. Modern natural-scientific imagery about human nature, which explained quality in terms of body or race, gave authority to an unmodern nostalgia for intrinsic differences of worth. It was a temptation to argue that if society levels people, nature at least does not.

The language of human difference applied to both individuals and groups, and to any social group – of class, age or sex as well as race. The upswing of feminism in the second half of the century, which focused first on access to education and then on suffrage, produced in response a vituperative literature that stressed the natural, physiological unsuitability of women for education and public activity. Physicians took the lead, describing how the reproductive function absorbs bodily energy that in men feeds the brain and makes women different. In a similar vein, throughout the century, a rhetoric about the physiological division of labour between head and hand was used to defend the necessity of social classes. Conservative writers claimed that the higher and the lower classes, like the head and the hand, have separate control and executive functions: 'fine' higher people provide leadership while 'coarse' lower types provide labour. The language of difference also entered into views of childhood and parentage, and belief in paternal responsibility permeated every social level from the family to the empire. All such language, like the language of race, built categories derived from social life into what was claimed to be objectively true about nature – and hence into what was claimed is a natural social order.

It will not do to dismiss all the authors who stressed natural inequality because they were guilty of the misuse of science. Many of them, after all, were scientists. The fact is, the human sciences in the nineteenth century acquired a substantial part of their subject matter through the interest in and language of human difference. This subject matter included values and values entered the subject matter of the sciences. The practical business of social and political life, from teaching good habits to children to the administration of empires, required differential action towards individuals and groups. The sciences of individual, class, sexual and racial difference supplied this differential action with its language, just as the action supplied the sciences with their subject matter. As we shall see, a similar pattern continued to be evident in the psychological and social sciences in the twentieth century. The human sciences were not conceived independently of decisions about how to live and then applied or misapplied to life. They were an expression of the life people lived.

Anthropology, even when it acquired academic standing at the end of the nineteenth century, was a peculiarly difficult field to define. In the hands of the German-American scholar Boas, who had a dominant influence on the way the field was shaped in North America in the twentieth century, it was the subject area that unifies 'the biological history of mankind in all its varieties; linguistics applied to people without written languages; the ethnology of people without historic records; and prehistoric archaeology'.[32] There was no such unified field in the nineteenth century in spite of huge fascination with each of these topics. In continental Europe, anthropology mostly denoted physical anthropology, the study of bodily differences, while ethnology denoted the comparative study of culture. The key preoccupation was the reconciliation of the theoretical unity of mankind with its observed diversity; from this perspective, the polygenist attack on unity, associated with physical anthropologists, was a sideshow. The debate about the relative merits of physical and cultural studies of man was yet another facet of the debate about whether human beings can be the subject of natural science. In the Anglo-American world there was some integration of anthropology and ethnology in an evolutionary framework, but even in the twentieth century sceptics judged that the subject area contained everything left out from the study of civilized, literate humanity, with the result that anthropology in practice became the study of savage 'otherness'. This troubled anthropologists faced by the political independence and independent consciousness of peoples around the world in the late twentieth century.

v *Physiology and Mental Science*

The sciences of human variety, from physiognomy and phrenology to the physical anthropology of race, all correlated mental and physical character and used the latter as a sign of the former. This direction of signification, from the physical to the mental, appeared objective and practical. The result was a substantial discourse, both public and scientific, about human nature that was at odds with the discourse of moral philosophy, the part of academic learning traditionally concerned with man's nature. Moral philosophy described and classified human differences in line with judgements about mental powers, moral virtue and civilized conduct – judgements applied to both people and peoples. Moral philosophy merged with Christian moralism. Moral and religious values and language persisted in the physical significations of difference; Gobineau, for example, linked race loosely with scripture. Yet there is a striking difference between eighteenth-century moral philosophy and the biological racism of the late nineteenth century. This difference relates to the spread of secular values, to what nineteenth-century writers themselves sometimes called 'the rise of rationalism'. Part of this 'rationalism' was the greatly increased authority of belief in the natural sciences as the basis on which to construct the human sciences. Conservative critics of the new descriptions in physical language of human variety certainly believed that the new work was anti-Christian and indebted to materialist science. Anthropologists like Gliddon and Hunt sharply contrasted the virtues of objective physical knowledge with religious prejudice. They attacked or ignored the work of people like Prichard, a Christian and a founder member of an ethnological society, and Lazarus and Steinthal, who wanted to establish an anthropology of the *Volksseele* or soul of the people. There was reason to associate a commitment to physical anthropology with a rejection of the traditional concerns of moral philosophy.

To turn from anthropology to physiology is to appreciate the degree to which the natural sciences lay behind a new discourse about human nature. Experimental physiology exemplifies the way natural science disciplines became separate, specialist large-scale enterprises in the German universities in the 1830s and 1840s and continued to grow thereafter. Physiology seized the imagination of many students as the route by which knowledge of man would finally become objective and true to his real nature. It also became a science in which experimental techniques and the specialist manner in which results were presented to professional colleagues divided trained scientists from public audiences. Clerics and conservatives, many of whom were academics in the disciplines of history, philology and philosophy, looked on physiology's claims to contribute to human culture with suspicion; sometimes they simply attacked it as outright materialism. The development of an academic and a public culture sympathetic to the material explanation of human nature, in physiology as in anthropology, was therefore a slow and complex process. Attitudes were structured by a vast range of local factors. The confused situation is well represented by phrenology, accused of materialism by its critics but propagated in Britain and North America by Combe and his followers supported by the argument that man's ability to acquire knowledge of the brain is part of God's providence. Complexity is also indicated by the spread in the English-speaking world of the term 'mental science' to describe a subject area that brought together knowledge of mind with knowledge of the brain. Mental scientists denied that they were materialists or that physiological explanation detracts from the power of man's will or from his reponsibility. In Britain, *The Journal of Mental Science*, a title used from 1858, was the house organ of alienists – specialists in psychological medicine or, to use the later term, psychiatrists, many of whom were zealous moralists. Nevertheless, as the use of this title indicates, the psychiatric occupation contributed significantly to the acceptance of a physiological orientation towards the mind. In summary, the explanation of human nature by physical facts had a standing and an authority much greater at the end than at the beginning of the century.

Critical scholarship in text-based disciplines and experimental methods in the natural sciences developed in parallel. From the 1820s, with the example of Justus Liebig (1803–73) in chemistry at the university of Giessen at the forefront, there were natural science research schools; the laboratory occupied a place parallel to the seminar in the training of students. Physiology, under the leadership in the 1830s of Johannes Müller (1801–58) in Berlin, promoted a critical, experimental and specialist study of human nature. Physiology was principally but not uniquely a German science, since a large-scale community and career structure was present in Germany long before anywhere else. In Paris, François Magendie (1783–1859) began to carve out a distinct place for physiology in the medical curriculum in the 1830s. He was followed in the mid-century by Claude Bernard (1813–78), who did brilliant experimental work on blood sugar regulation and on nervous control, while he argued forcibly for physiology as the foundation of scientific medicine. In Britain and the United States it was medicine that offered an early home to a physiological interest, and large-scale independent physiological research did not really get under way until the 1870s.

The interactions between medicine, experimental physiology and the search for material explanations of man's nature were no coincidence; as earlier, doctors as a group were markedly sympathetic to such explanations. Medicine as an occupation was also a very significant site where academic interests and public values met; both because medical thought articulated social values and because governments and public opinion supported academic medical activity even when there was some fear of the materialist implications of medical work. The great innovations in urban public health in the second half of the century consolidated a closeness between medical and public values and encouraged the public to accept medical ways of thought in the human sciences. Many physicians, like La Mettrie before them, believed that scientific medicine was truly the science of man, and as the conviction grew that the disciplines of physiology, anatomy, histology (the science of tissues), physiological chemistry and – at the end of the century – bacteriology and biochemistry gave medicine a scientific base, a material science of man, in the view of many doctors, seemed assured.

This cluster of opinions and values was already established in France at the end of the eighteenth century. Medical reformers in the 1790s did not just reform a profession but attempted to reform both the administrative fabric of the country and the very notion of a human science. The hopes invested in medicine were not only for improved health but for the improved regulation and well-being of the nation. Such hopes were present in establishment medicine: Xavier Bichat (1771–1802), a leader in theoretical medicine, classified tissues and also provided analytic terms for the political body. They were also present in unofficial medicine: Franz Anton Mesmer (1734–1805) worked with a healing *banque* or reservoir which stored vital fluids and to which patients were connected to restore their personal harmony. This was 'mesmerism'. At the same time, Mesmer and like-minded people looked to medicine to restore the natural or vital flow of life in the political body, a way of thought that was part of radical culture in the 1780s. Such schemes tied together the material understanding of individual and collective life. The most dramatic implications for a physical science of man came directly from physiology, however, when it claimed new knowledge of the mind's embodiment in the brain and nervous system. It appeared to many scientists that nervous physiology and anatomy together pointed the direction in which natural science would develop to encompass human nature. This argument, too, was evident in early nineteenth-century French medicine, for example in the work of Cabanis; but it was phrenology, argued for by Gall in Paris in the second decade of the century, that forced the issue to the forefront of scientific and public attention.

Gall sowed thoughts about the mind and brain that flowered in natural-scientific approaches to human nature when separated, as they easily were, from his craniology and theory of human differences. After Gall, no one doubted that the brain is the organ of mind; it was firmly established that the brain is the structure through which the mind is grounded in the body. As an Edinburgh professor of medicine, Thomas Laycock (1812–76), observed in 1860: 'All [man's] desires and motives are experienced in and act upon this important apparatus – and all are expressed by it; so that what the man is, in character and conduct, is the expression of the functions of this nervous system.'[33] Phrenology localized mental activity to parts of the brain and attempted to solve the question of the mind's relation to nature by the description of mental activity as the *functions* of the brain. This language was taken up by the new physiology as the means to construct a scientific psychology, knowledge of mind in objective natural-scientific terms. Even scholars who were not physiologists but interested in psychology, like J. S. Mill and his protégé Alexander Bain (1818–1903), argued that the progress of physiology requires old analytic approaches to the mind to be rethought. In 1855, Bain undertook to revise association psychology because 'the time has now come when many of the striking discoveries of Physiologists relative to the nervous system should find a recognised place in the Science of Mind'.[34] The subject area that resulted was in Britain called mental science. Gall himself was accused of materialism, and most of the physiological psychologists in Britain put considerable effort into arguments to avoid the same condemnation. But it was quite another matter to specify clearly how mind and brain relate.

Phrenology was discredited with physiologists by the 1840s. By this time, the German-language universities provided an institutional base for physiology as a discipline in its own right.

A substantial amount of experimental physiology was devoted to nervous functions and, though day-to-day research focused on narrow and detailed topics, all researchers were aware of the wider implications that non-specialists found in the conclusions of neurophysiology. Many physiologists, indeed, were attracted to the subject precisely because of the challenge it offered on the wider intellectual and moral landscape. This situation is exemplified by the work that turned reflex action into a major topic. After experimental studies by Magendie in the early 1820s on the separation of incoming sensory nerves and outgoing motor nerves of the spinal cord, the nervous system was analysed as an organ that links sensation and movement. In the 1830s, after arguments by an English physician Marshall Hall (1790–1857) and the German physiologist Müller, the reflex was accepted as the elementary unit of nervous function. The reflex was a major theoretical innovation because it enabled scientists to describe an organism's achievement of purposive movement, e.g., scratching, as the outcome of material causes, the structural relations of the nerves. It became a model explanation in physical language for a mental attribute, purposiveness. The concept of reflex action suggested a research programme to determine the material correlates of mental processes. Vague psychological questions reappeared as precise physiological ones in this programme. The actual work involved detailed experimental analysis of the conditions that affect the action of reflexes, and the poor frog became the main experimental subject. Scientists reported on the minutiae, for example, of the effects on reflexes after they had cut the spinal cord at different levels or on the alteration of reflex intensity caused by the administration of drugs. Both skilled physiologists and interested commentators, however, thought that the type of explanation advanced in reflex action theory could be applied to many phenomena in human life. The idea of reflex action was turned into an explanation for habits, sleep-walking, hypnotic performance and other automatisms, spiritualist and table-turning experiences, many aspects of madness and much else besides. These were the topics of mental science. They created an audience familiar with explanations of human life in terms of the underlying causal physiology of the body. By this means, mental science came to replace moral philosophy.

The decade of the 1840s was a watershed in the history of physiology as it was then that a group of brilliant students under Müller committed themselves to the exclusive rights of physico-chemical explanation in the processes of life. Emil Du Bois-Reymond (1818–96), Hermann Helmholtz (1821–1904), Carl Ludwig (1816–95) and Ernst Brücke (1819–92) all went on to hold chairs at major universities and to shape the discipline of physiology over the next half-century. At the same time, with a key input from Helmholtz, the clear formulation of the principle of the conservation of energy (the first law of thermodynamics) appeared to contemporaries to demonstrate that only causal processes of a physical character could exist in nature. This conclusion seemed to vindicate the claim that physiology is the only scientific route by which to establish a science of human nature. It was more and more difficult for critics to challenge the authority of natural science. As the Manchester Unitarian churchman and philosopher James Martineau noted with concern: 'To judge from the habitual language of medical literature, the Physiologist considers himself to be treading close upon the heels of the Mental Philosopher, and to be heir-presumptive, if not already rival claimant, to the whole domain.'[35]

These debates had a substantial political dimension, especially in Germany in the 1840s where liberal hopes for representative government and free trade ran high. There was optimism that the realization of these hopes would make Germany a modern, liberal and unified state. At the same time, physiology attracted a public audience since it was an intellectual engagement with the real material conditions of human life. As in the Enlightenment, there were also radical writers who found in the obdurate body a firm foundation of facts on which to build a critique of idealist, Christian and conservative beliefs. The conclusions drawn by the philosopher Ludwig Feuerbach and by Marx are discussed in the next chapter. There was also a substantial body of materialist argument by physiologists themselves, and this was part of the pressure for political change that came to a head

– disastrously, as far as liberal hopes were concerned – in 1848.

One of the most vocal physiologists was a Dutchman who taught at the university of Heidelberg, Jakob Moleschott (1822–93), who also wrote popular studies of the new science. Moleschott's *Der Kreislauf des Lebens (The Cycle of Life,* 1852) which, in his earlier words, explained that 'Life is an exchange of matter', produced a conservative outburst which led to his resignation; he eventually took a post in Zurich. He was an ardent supporter of revolutionary aims in 1848, at a time when he memorably linked diet to the subservient position of Europe's peoples:

> Sluggish potato blood, is *it* supposed to impart the power for labor to the muscles, and the enlivening verve of hope to the brain? Poor Ireland, whose poverty breeds poverty . . . You cannot win! For your diet awakens powerless despair, not enthusiasm, and only enthusiasm is able to blow over the giant [England] through whose veins courses the energy of rich blood.[36]

With the collapse of hopes for change across Europe in the wake of 1848, when conservative regimes repressed liberal as well as revolutionary demands, disillusioned radicals turned with renewed energy to natural science as the authority which, in the long term, would subvert the alliance of political reaction and religion.

Moleschott continued to do some physiological research in Zurich. He also received visitors, like George Eliot and G. H. Lewes, who shared his humanist values. Then, in 1861, he went to Turin to play a part in the modernization of education in what became, in 1870, the Italian state. Colleagues who stayed in Germany fought on against the Protestant and Catholic interests that dominated the school system and resisted the introduction of natural science into the curriculum. One pro-science group, in 1852, founded the popular journal *Die Natur* and, appropriately enough, placed an engraving of a volcano on its title-page. Their picture seemed to say that, whatever the repressive power of ignorance and conservatism, the real material forces of nature and human nature would, in the end, have their say. As Moleschott wrote: 'The natural scientists are the active cultivators of the social question . . . Its solution lies in the scientist's hand, which is guided with certainty by the experience of the senses.'[37] This was the substance of the natural scientists' attacks on the rights of conservative powers.

Conservative power was an oppressive reality in Tsarist Russia. But 1855 saw a thaw; it was a turning-point for the development of the natural and human sciences. The death of Nicholas I and fear of backwardness, symbolized by defeat in the Crimean War, ended an attempt to impose total control over higher education and to restrict belief about human nature in conformity with the faith of the Russian Orthodox Church. Under Alexander II, even after hope faded for political liberalization, students travelled to Western Europe, especially the German-language universities, for an education. Many idealistically believed that knowledge, especially of the natural and the historical sciences, would somehow precipitate reform of the Russian system. The term 'intelligentsia' came into use to describe this class in Russia, a class of unempowered but educated people who believed that objective rational knowledge, science, is the basis for civilized life. Total lack of representative government turned this class either into dissidents, sometimes nihilists, or into scientists and doctors whose political activity was mainly limited to the pursuit of professional interests and the modernization of the social administration of the state. Then, as later under the Soviet system, intellectual commitment to the values of science, as a source of authority that transcends political power, acquired intense significance to individuals as a means with which to sustain personal integrity under repressive conditions.

An influential representation of Western values was created by the journalist N. G. Chernyshevskii before his exile to Siberia in 1864. In his article on 'Antropologicheskii prinzip v filosofii' (The Anthropological Principle in Philosophy, 1860), published in the journal of which he was an editor, *Sovremennik* (*The Contemporary*), which moulded radical opinion, he described man's being, the objective conditions of being human in this material world, as the starting-point for philosophy. This vague formulation was full of meaning to his readers since it was diametrically opposed to the conservative view of human nature that begins with man's spiritual and immortal essence. In an argument built on his principle, much influenced by Feuerbach, Chernyshevskii arrived at the value of humanity as the goal of individual and social existence. In a didactic novel, *Chto Delat?* (*What is to Be Done?*, 1864), written from a cell within the Peter and Paul fortress in St Petersburg, he gave his values fictional expression. With two medical students among his protagonists, the novel described the study of physiology and the recognition of sentimental love as a material necessity, an honest openness to the forces of life, as the basis of a new type of person, 'the new man'. He contrasted human needs with the false needs produced by current society, and he portrayed science as knowledge of true needs. The novel and the values it represented inspired a whole generation with a vision of a human science as a means of liberation. When conservative fears increased in the mid-1860s, Chernyshevskii was exiled to Siberia.

Among the students to travel to Berlin, Leipzig, Heidelberg, Vienna, Zurich and Paris in the late 1850s, was Ivan Mikhailovich Sechenov (1829–1905). He returned to St Petersburg in 1860, where he taught experimental physiology and contributed to the establishment of scientific medicine. His own experimental work, very much on the German pattern, was on the frog's reflexes, and it was at the centre of an international debate about inhibition in the regulation of an organism's movements. But he had a larger ambition and viewed his specialist work as the basis for a scientific psychology, an approach to the human mind that would be objective as it is founded in the material conditions of mind – the brain. He therefore also wrote provocative essays for general audiences. The most substantial of these essays, 'Refleksy golovnogo mozga' (Reflexes of the Brain, 1863), written in Russian and not translated into English until 1935, attempted to delineate physiological analogues for psychological processes, on the model of the reflex, and thereby to found a science of the mind. His argument offered the greatest challenge to orthodox opinion when he developed his physiological concept of inhibition as an analogue for the will, though Sechenov claimed that his theory did not detract from the value of individual responsibility as his critics, and the censors, feared. The essence of the debate – though Sechenov was not the source – was immortalized in Ivan Turgenev's novel *Otzy i Deti* (*Fathers and Children*), published in 1861. The central character of the story, Bazarov, is a medical student and a radical materialist – a 'nihilist' in Turgenev's word, as he accepts nothing not proved by observed material facts, and he has a clear thesis about human nature: 'I shall cut the frog open to see what goes on inside it, and then, since you and I are much the same except that we walk about on our hind legs, I shall know what's going on inside us too.'[38] But in Turgenev's story, Bazarov does not know what is hidden inside himself: love, self-sacrifice and despair.

Soviet physiologists and historians claimed that Sechenov succeeded in founding a science of objective psychology. Sechenov himself, however, probably recognized just how large the gap was between knowledge of the brain and the experiential richness of mental life; certainly, his project to found a physiological science of mind never translated from speculative outlines into concrete physiological detail. This was the general situation in mental science. When he reviewed the hope to advance psychology with knowledge of the brain as the organ of mind, Mill commented in 1843 that knowledge derived through introspective analysis is 'in a considerably more advanced state than the portion of physiology which corresponds to it . . .'[39] This judgement still held when Mill reviewed, very favourably, the books by Bain, *The Senses and the Intellect* (1855) and *The Emotions and the Will* (1859), which systematically brought associationist analysis up-to-date with the new neurophysiology. As Mill wrote:

> Whether organisation alone could produce life and thought, we probably shall never certainly know, unless we could repeat Frankenstein's experiment; but that our mental operations have material conditions, can be denied by no one who acknowledges, what all now admit, that the mind employs the brain as its material organ.[40]

His judgement carefuly accepted the value of natural science and yet avoided the crude materialist conclusion that physiology explains mental life. Similarly, Bain's books in fact discussed body and mind side by side rather than integrating them. What Bain did do was overcome the emphasis on the mind's passivity in British associationist discussions, which treated action as a consequence of sensation and feeling. Bain described activity as preceding sensation, rather than the reverse; this suggested a much more dynamic, interactional view of experience. This orientation entered the mainstream of North American psychology, through the theory of learning, at the end of the nineteenth century. Its most concrete expression was in E. L. Thorndike's 'law of effect' – activity that results in pleasure is repeated, activity that results in pain ceases. But to return to the topic of mind and body, Mill's reservations about the value of physiological analysis continued to be relevant much later. Sechenov and some other believers in physiological explanations in mental science at least implicitly abandoned the belief in physiology as the basis for a viable research programme in the short term. In 1893, Bain concluded: 'Introspection is still our main resort – the alpha and the omega of psychological inquiry: it is alone supreme, everything else subsidiary.'[41]

Meanwhile, the experimental physiology of the nervous system had become a major area of research activity. This occurred in conjunction with clinical studies of brain damage and disease, the medical speciality known from about the 1870s as neurology. The development of antiseptic operative techniques and anaesthesia made possible new precision in experimental studies of the brain. Amongst the most tangible results was the announcement by two German researchers, G. T. Fritsch (1838–1927) and E. Hitzig (1838–1907) in 1870, followed by the English physiologist David Ferrier (1843–1928) in 1873, that experimental stimulation of the cortex of the higher brain justifies belief in the localization of functions. It appeared to be possible to map the sites in the brain where different functions, mental processes like the control of speech, go on. To superficial observation, it appeared that Gall's theory of the localization of mental activity was right all along; but in fact Gall localized mental faculties while the new theories localized sensory-motor functions. Some scientists stated clearly that the new theories localized physical events; there was much speculation, however, about the manner in which these studies translated psychological questions into physiological terms by the attribution of mental activity to a spatial location.

Ferrier, whose initial work was done in a small room in the West Riding of Yorkshire lunatic asylum in Wakefield, brought the new work together in his book on *The Functions of the Brain* (1876). He filled a substantial chapter with comments about the mental function of the cerebrum or higher brain. His hopes for what physiology could do for mental science are dramatically indicated by his claim, also made by Sechenov, that 'thought . . . is in great measure carried on by internal speech'.[42] This implied that rational thought, traditionally held to be central to man's spiritual essence, can be understood as activity in the nervous pathways of subvocal speech. Speech, in this view an integration of sensory-motor acts that culminate in the co-ordination of muscles, can be studied experimentally in a way that rational thought cannot. Natural science offered no empirical justification for the claim, but the claim indicated clearly where the future of science was thought to lie. Brain research for the next century included a complex experimental and conceptual argument for and against the localization of functions. The theoretical puzzles that faced any attempt to deal with the relations of mind to brain through the concept of function took a concrete form in what the French historian of science Georges Canguilhem called the conflict between *les localisateurs* and *les totalisateurs*, those who began with division and those who began with wholeness, in the science of brain.

Political struggles, economic change and religious controversy swept Europe and North America, and – dependent on point of view – there was enthusiasm or horror for a physiological and, as many thought, materialist approach to human nature. The physiological dimension of human science attracted students because it appeared to be the way knowledge about man achieves the same explanatory rigour as the natural sciences, sciences that themselves were transformed in the first half of the nineteenth century into disciplines with high status. The experimental physiology of the reflex and of the localization of mind as brain functions raised hopes that the new physiological psychology of man could have empirical content. Yet the physiological approach to the human sciences was always in a precarious position in the nineteenth century. It had its conservative critics whose views might be dismissed because they opposed the whole enterprise of a science of man. But it also had its sympathetic critics, like Mill, who saw the gap between experience of mental activity and what physiologists describe in frogs. Further, there were other sciences, like philology, the archaeology of prehistory and the comparative anatomy of race, which also clamoured for attention. The rigorous study of language and of history, in particular, established a model for explanations of man's nature that severely restricted physiology's claims. Humanist scientists, who had a secure place within the German universities, believed they exemplified the cultural purposes for which the universities were reformed or founded and held materialism in contempt.

As the study of the physical body as a sign system for qualitative differences of human character developed, however, materialist arguments became more common in science. In spite of the new institutional separateness of academic disciplines, there remained a substantial public interest in the conclusions of philology, history, anatomy and physiology. There was an emotive atmosphere around the interaction between academic and public values over such topics as self-identity and national identity or debate about the balance of social causes and personal responsibility. Every sector of society was fascinated by human differences and their supposed causes. It was also a substantial challenge when first Comte and then Marx, neither of whom were academics, elaborated grand views of historical progress, in Marx's case with a materialist reversal of the contemporary historians' idealism. The study of differences and of physiology was also part of the background that guaranteed an informed and involved audience for Darwin's theory of evolution, which linked man and nature in the most emphatic way.

12

The Science of Society: Auguste Comte and Karl Marx

> I was led by my studies to the conclusion that legal relations as well as forms of the State could neither be understood by themselves, nor explained by the so-called general progress of the human mind, but that they are rooted in the material conditions of life which men carry on as they enter into definite relations that are indispensable and independent of their will; these relations of production correspond to a definite stage of development of their material powers of production. The totality of these relations of production constitutes the economic structure of society – the real foundation, on which legal and political superstructures arise and to which definite forms of social consciousness correspond. The mode of production of material life determines the general character of the social, political, and spiritual processes of life. It is not the consciousness of men that determines their being, but, on the contrary, their social being determines their consciousness.
>
> Karl Marx, Preface to *Zur Kritik der politischen Ökonomie* (*A Contribution to the Critique of Political Economy*, 1859)[1]

i *Reaction to 1789*

Whatever the impact of academic disciplines or scientific specialization in the nineteenth century, many authors who wrote about large-scale themes of human existence – like Auguste Comte (1798–1857), Karl Marx (1818–83), Herbert Spencer and Charles Darwin – worked independently of universities. With the financial support of patrons, journalism, royalties or independent means, they painted on large canvases and attracted criticism from narrower minds unable to raise their sights above detail. Marx and Darwin stand out among them, as they also contributed original, specialist work in economics and biology respectively, which narrowly academic scholars had to come to terms with even if they were unsympathetic to grand theory. These authors also had a large international public audience interested in the implications of scientific progress for national life and private values. Further, Comte, Marx, Spencer and Darwin each wrote history and they created narratives about the past from the origin of life to the industrial revolution that readers seized on to make sense of and take control over the contemporary world. These authors contributed to a sense of progress as a source of meaning which, perhaps fundamentally, separates their time and the late twentieth century.

The significance of the historical imagination in the nineteenth century can hardly be exaggerated. It is a powerful way to make something intelligible, to tell a story about how it became what it is; a person's life history is the archetypal form. This ordinary mode of thought was reconstituted as the historical sciences, sciences that seek causes for particular events and link these causes into a general history of nature and human nature. Nineteenth-century history, both natural and human, was also a narrative of progress, and particular events, people and movements in art or reason were therefore assessed for their value in relation to historical progress. The place of events, phenomena or people in history was taken to give them meaning – the philosophical viewpoint sometimes called historicism. This was a very general feature of arguments, found, for example, in the work of an academic historian like Ranke and in the origin myths of racial theorists. History and progress also formed the framework for the new sciences of society and, as the following chapter shows, for theories of evolution.

Many participants in the events of 1789 believed that they were present during the most momentous days of human history. Certainly, no nineteenth-century politician, whether radical or reactionary, forgot that deliberate human action then produced deep-seated change. Some of the leading proponents of the French Revolution described change as *l'art social*, the reorganization of society along lines dictated by rational knowledge or science. One of the most influential revolutionary tracts, by the abbé Sieyès (1748–1836), *Qu'est-ce que le tiers-état?* (*What Is the Third Estate?*, 1789; i.e., what is 'the commons' as opposed to the monarchy and the nobility as a constituent of the state?), contained one of the first references to *'la science sociale'* (social science) as the study of social questions independently of any one political group's interest. The permanent secretary of the Académie des Sciences, Condorcet, also used the phrase when he linked the study of morals and the study of social organization by common principles. With others, he founded the Société de 1789 to ensure the success of social reconstruction based on *les sciences morales et politiques* (the moral and political sciences). The club did not last but, after the reform of the institutions of higher learning in 1795, the Classe des sciences morales et politiques of the new Institut National included a section called 'Science sociale, et législation'. Condorcet died in prison, but the Classe des sciences morales et politiques became the institutional base for the *idéologues*, who shared his aspiration that what they called *la science de l'homme* (the science of man) should be the foundation for education, medicine, social administration and legislation. Their vision of a *science de l'homme*, an anthropology of men and women in all their aspects integrated with social policy, lived on in Comte and – in certain respects – in Marx. Social science, in some of its recognizably modern forms, was born in the social upheaval of 1789.

The violent aftermath in France, carried abroad by Napoleon's armies, concentrated attention on both the creation and legitimation of new political arrangements and the defence of old ones. There was perceived to be a pressing need to reintegrate people and the state. A large literature grappled with questions about what conditions make for social harmony. The French themselves were faced by the need to define and defend the idea of a 'citizen' of a republic as opposed to a 'subject' of a sovereign. Writers across the political spectrum – the imagery of right and left originated in seating arrangements in the French

Constituent Assembly after 1789 – often drew on the analogy of the living organism, as representing an ideal combination of parts harmoniously at work on behalf of the whole. The language of organic relations achieved more than analogy since the biological concepts of structure and function reappeared as social concepts, in which capacity they greatly influenced social analysis, then and later. This step was especially apparent in the work of Spencer, discussed in the next chapter. Comte, Spencer and many twentieth-century sociologists ('functionalists') thought that a social explanation involves a conclusion about the place – the function – of an action or institution within society as an integrated whole. Biological explanation was transferred to the human sphere. The twentieth-century version differed from that of the nineteenth century in detaching functional analysis from historical reconstruction; Durkheim, for example, rejected Spencer's evolutionism. But nineteenth- and twentieth-century social scientists agreed that social science is the specialist knowledge needed to achieve the functional integration – in twentieth-century language, the adaptation or adjustment – of the individual to society.

The connections between 1789, social science and the search for social harmony were made in the life and work of Henri Saint-Simon (or Claude-Henri de Saint-Simon, 1760–1825), a young French officer who supported the Americans in their struggle for independence and then became a student of human progress. He battled to find the funds for a life of high seriousness; the story, true in characterization if possibly not in fact, is told that he instructed his manservant to wake him each morning with the injunction: 'Remember, monsieur le comte, that you have great things to do.'[2] Like many of his contemporaries, he was deeply impressed by progress in mathematics and the natural sciences, and saw their advance as a corrective to the political turmoil and loss of direction around him. He arranged the sciences in a hierarchy and introduced the term 'positive' to describe those that had fully eliminated reference to metaphysical or final causes. Further, he thought that the science of man would become positive once physiology had made progress, as it appeared rapidly to be doing. He was inspired to assimilate contemporary anatomy, physiology, the science of ideas and Condorcet's notion of progress in order to find 'une science de l'organisation sociale' (a science of social organization). This was to be a science, along the lines of a natural science, which would base knowledge in the observed regularities of the concrete conditions of social life such as climate, health, diet and labour. Saint-Simon's intellectual and moral focus was on what he called *organisation*, the sustenance of the well-being of society as a whole. He was impressed by contemporary French advances in *les sciences de la vie* (the sciences of life), a conception that linked physiology, medicine, psychology and social science into a unified human science through the idea of *organisation*. From this perspective, moral action is a form of social hygiene, a rational intervention on behalf of collective health. A compendium of his views assembled by a disciple in 1825 thus had the title, *De la physiologie sociale (On Social Physiology)*. So emphatically did Saint-Simon think in terms of the organic analogy that he reinterpreted Condorcet's picture of the progress of mankind in the imagery of individual life from childish ignorance to adult experience. He hoped medical physiology would provide a standard of health for the individual body, and that social physiology would provide a standard of harmony for the social body. His ideal of a mature society set the terms in which he and his followers understood actual historical societies. This built values into the fabric of what became the functionalist form of explanation in social theory.

Saint-Simon's science of man focused on organization and, as in physiology or anatomy, on the empirical study of the relations of structure and function that make organization possible. After the experience of turmoil between 1789 and 1815, the new social prophets began to distrust political solutions, and they looked instead to the example of medicine when they tried to intervene in society's ills. They also trusted the decisions of experts, founded on science, rather than public opinion or the popular will. They recognized that emotion and belief have a function in society, that they are the vital force of the social organism like the living force of organic tissues, but had no confidence in the people's emotion and belief as a basis for progress. Saint-Simon himself reconsidered religion and concluded it was an essential binding force; he thought about the need for a religion of humanity and how existing institutions, notably the Catholic church, could be reshaped to guide ordinary people towards social harmony. There were even hints in his work that human progress might not be a matter of reason showing people how to live but might operate in a law-like way independently of human reason. The implication of such hints was that individuals must learn to accommodate themselves to a collective existence that transcends their own lives, as many Catholic conservatives had argued all along. All these themes continued to have an active life through the century. It was Saint-Simon's erstwhile secretary, Comte, who took them up in the most self-consciously messianic way and gave systematic form to Saint-Simon's historical speculations about the progress of the positive sciences. In Comte's hands, progress in positive science became the basis for a fully-fledged religion of humanity.

ii *Auguste Comte and Positive Sociology*

It is easy to mock Comte: the precocious boy brought up among a family of women who in middle age idolized the feminine; the philosopher who dismissed Catholicism in favour of natural science only to devise a calendar of saints; the self-styled lecturer who imposed an encyclopaedic order on knowledge as a whole. Yet the thoughts of this Parisian intellectual – and he was that rather than a scholar – articulated a synthesis of natural science and the study of society that commanded respect and had substantial influence. J. S. Mill, who did not suffer fools gladly, wrote:

> But we know not any thinker who, before M. Comte, had penetrated to the philosophy of the matter, and placed the necessity of historical studies [as opposed to ahistorical 'theories founded on principles of human nature'] as the foundation of sociological speculation on the true footing. From this time any political thinker who fancies himself able to dispense with a connected view of the great facts of history, as a chain of causes and effects, must be regarded as below the level of the age.[3]

It was, in fact, also Comte who introduced the word *'sociologie'* to describe the science of society. His argument to incorporate the study of society into science pointed to two aspects, the static and the dynamic methods, which he adapted from the work of the contemporary French physiologist Henri de Blainville (1777–1850). Both aspects made the intellect pivotal to the achievement of progress. The static method traced the intellect to its organic conditions in the body. The dynamic method, the core of his positive philosophy, traced the intellect historically in the genesis of science:

> It will simply amount to tracing the course actually followed by the human mind in action, through the examination of the methods really employed to obtain the exact knowledge that it has already acquired: and this constitutes the essential aim of positive philosophy . . .[4]

The history of science thus provides the key to future progress.

Commentators during and after Comte's lifetime were puzzled, even disturbed, that this philosopher of sociology as a natural science founded a new religion. Emil Littré (1801–81), the most important of his French followers and the compiler of the great French dictionary, distanced himself from Comte in the 1850s and separated positive philosophy and sociology and the self-aggrandizing absurdities, as he saw them, of Comte's religion. This was an understandable step for a liberal like Littré who was embattled during the conservative years of the French Second Empire (1852–70) and wanted to use reason as a political instrument within a repressive state. Nevertheless, Comte's reason and his religion were all of a piece, and this can be seen when he is related to Saint-Simon and to the deep anxiety after 1789 that there must be a science of humanity if social and moral harmony is to be achieved. In Comte's philosophy, science is not applied to moral ends but is an ethical advance itself, an advance towards the moral goal of an ideal humanity. This explains Comte's appeal in the nineteenth century: he gave hope, where religious hope was questioned, that progress in natural and human science is progress towards social harmony and moral fulfilment. His religion was his attempt to engage ordinary people, whom he thought unable to be moved by science, in a collective reverence towards truth. He therefore imitated Catholicism and devised a clergy, a liturgy, a catechism and a calendar for the new faith in positivism. It was easy to mock. But he also expressed the longing of his age, faced by huge uncertainties, that scientific truths should provide guidance. Comte wrote:

> The great political and moral crisis of our present society is the result . . . of intellectual anarchy. Our most deadly disease is the profound divergence of minds with regard to all the fundamental maxims whose fixity is the prime condition of a true social order . . . if the union of minds in a communion of principles can once be established, suitable institutions will necessarily spring from it . . .[5]

It is legitimate to equate science and progress, as science is the activity that makes consensus possible. The science of society is therefore the medium of social harmony.

Comte was Saint-Simon's secretary for seven years between 1817 and 1824, a period when people took stock of earlier events. After an acrimonious split with his patron, he supported himself from conducting examinations and from an independent course of lectures that he gave to élite members of the administrative and scientific world in Paris; his listeners included the physiologist Blainville, the alienist J.-E.-D. Esquirol and the mathematician J.-B.-J. Fourier. He put together these lectures as his *Cours de philosophie positive* (*Course of Positive Philosophy*, 1830–42). He also received foreign visitors like Mill.

Comte's *Cours* was 'positive' in the sense, derived from Saint-Simon, that it defined knowledge as the laws of 'relations of succession and similarity' among observations: 'It is the nature of positive philosophy to regard all natural phenomena as subject to invariable natural *laws*, the discovery of which . . . is the aim and end of all our efforts . . . we do not pretend to expound the generative *causes* of phenomena.'[6] Comte's positivism expressed opposition to the abstractions of conventional philosophy. As a mathematician (he had been a student at the Ecole

Polytechnique, then the most advanced educational institution in the physical sciences in the world) he took the physical sciences as a model for the social sciences. His ideal of positive science was given by Fourier's contemporary theory of heat, which was expressed mathematically without claims about the nature of heat. Positive philosophy was therefore equivalent to what many English speakers meant by science: the search for 'invariable natural laws'. Positive philosophy and science, in this sense, regard statements that do not have a 'positive' form, like statements that refer the laws of nature to a First Cause, such as the Christian God, as meaningless. The physical sciences are also held up as a model of true knowledge. Comte took this idea of positive philosophy and set it in a historical perspective in which history acquires direction from progress in science. This perspective on history seemed almost self-evidently true in the nineteenth-century scientific worldview. Mill expressed something of this when he observed that Comte's achievements follow from 'a simple adherence to the traditions of all the great scientific minds whose discoveries have made the human race what it is'.[7] Comte himself argued that each science, and hence humanity in general, passes through three stages. In the first or theological stage people attribute events to a living will or desire in things themselves or in a deity; in the second or metaphysical stage the human mind attributes causes to abstract forces or forms; in the final and positive stage, science abandons the search for ultimate causes and seeks law-like sequences in observable phenomena. Comte's history of civilization recorded humanity's achievement of positive science; he believed it had occurred first with the physical sciences in the seventeenth century, followed, he thought, by the life sciences in early nineteenth-century Paris. This vision of the history of science at the core of a progressive civilization dominated the nineteenth century and continued to be central to the worldview of scientists in the twentieth century. The same vision also inspired George Sarton to found the history of science as a discipline after he moved from Belgium to the United States during World War I. In Sarton's view, the discipline is central to the values of humanity as it records the progress of the human mind.

Comte's history of science also provided a classification of the sciences. The sciences of life have a 'more complicated and highly individualised' subject matter than the physical sciences and hence are historically and logically dependent on the physical sciences. But he also emphasized that the sciences of life have their own proper, special subject matter; they are not simply a branch of the physical sciences. The same pattern of dependence in relation to method but independence in relation to subject matter was present in his discussion of the relation between the physical and the social sciences. He distinguished 'two great sections of *organic physics* [the life sciences]: physiology, and social physics [sociology], founded on physiology'. The establishment of social physics, the science that studies the most complex organic relations, is dependent on the earlier elaboration of physiology, but it has its own subject matter, the regularities of the social world, which cannot be translated into the laws of another science: 'Thus social physics must be founded on a body of direct observation proper to it alone, always having regard to its intimate and necessary relation to physiology.'[8] Comte defined social physics or sociology by reference to its distinct subject matter yet made it dependent historically and methodologically on the natural sciences. This definition was widely accepted in later philosophy of the social sciences.

Comte took upon himself the task of founding a positive science of sociology, and thereby laying down the basis for harmony and morality. 'In political philosophy from now on there can be no order or agreement possible, except by subjecting social phenomena, like all other phenomena, to invariable natural laws that will limit in each epoch . . . the extent and character of political action . . .' The methods of sociology, the 'static' and the 'dynamic', are concerned respectively with organization and with change – 'distinguishing between the conditions of existence of a society and the laws of its movement'.[9] But Comte did not carry out his prescription and elaborate the details of the new science. Rather, he turned to his scheme for a religion of humanity as the means to lead ordinary people to live in love on the basis of positive knowledge. He published a *Système de politique positive (System of Positive Politics,* 1851–4) and a

Catéchisme positiviste (Positive Catechism, 1852) to give practical expression to his ideas. Comtean churches were founded in places as far apart as Newcastle in England and Brazil, but most scientists who were otherwise interested could not stomach Comte's egocentric and bizarre notions for a church, modelled on Catholicism, which worshipped human love, embodied in woman, rather than God.

Liberals who were sympathetic to Comte's sociology, like Littré and Mill, detached it from his religion. They admired the Comte who showed how reason and knowledge offer escape from the disasters and arbitrariness of political and social life. In many respects, they also shared his values and hopes for humanity. But Comte perhaps glimpsed what these liberals did not, that reason – however positive or scientific – does not of itself persuade political society to adopt its conclusions. He argued that society also needs sentiment. What neither Comte nor the liberals came to grips with was the social reality of conflict and power. This Marx was to do.

Comte's classification of the sciences was idiosyncratic in one significant respect: it had no place for psychology between the two branches of organic physics, physiology and sociology. Like Saint-Simon, he conceptualized society in organic terms, and he assumed that to observe regularity in society in principle uses the same method as to observe regularity in an organism. He contemptuously dismissed introspection as a source of positive knowledge. 'A thinking individual cannot divide himself into two, one half reasoning, and the other watching it reason ... Our posterity will doubtless one day see these pretensions transferred to the comic stage.'[10] And, as he equated the method of introspection with psychology – accurately enough, given the prominence in France of Cousin's *spiritualisme* or psychology based on inner observation of the spirit and will – he dismissed psychology too. Cousin's *psychologie* exemplified Comte's opinion of knowledge that had not reached the positive stage. Cousin, Comte believed, falsely laid claim to a human science: 'This has been done by proposing as analogous to genuine observation, which must always be external to the observer, that celebrated *interior observation* which can be only a parody of the other . . .' The sharpness of Comte's remarks and antagonism to psychology is easily understood as Cousin dominated educational appointments through his position at the Ecole Normale Supérieure during the period of the July Monarchy (1830–48). He thus had a huge audience, an audience Comte dreamed of, for his view that psychology, or the introspective study of human nature, is the foundation of philosophy: 'Man is a universe in miniature; psychology is universal science concentrated.'[11] Further, Cousin implied and conservatives believed, a science founded on external experience is tantamount to materialism and hence to immorality. Lastly, as an independent scholar, Comte was socially simply beneath Cousin's notice. The space which psychology might have occupied in Comte's classification of the sciences was in fact occupied by phrenology, that is, by a physiological theory of mental functions made externally observable by their location in parts of the brain. Comte believed that psychology requires unscientific introspection and even a metaphysical belief in spirit, while Gall's method for the study of the brain's functions employs the positive method. But phrenology served Comte's wider project badly. As Littré observed, 'Comte prostrates psychology before phrenology', while Mill wrote to Comte: 'I persist in thinking that sociology as a science can make no important progress without relying on a maturer theory of human nature.'[12]

When in the 1890s Durkheim gave scientific sociology detailed content and an institutional base, he too constructed the discipline by the exclusion of psychology. In their concern to establish an independent study of society, Comte and then Durkheim split human nature into the physiological and the sociological: they subsumed what is individual in a person under physiology and what is collective under sociology. Positive sociology preserved a unity of method, the observation of organization and of change, across the life sciences and the social sciences. But the cost was that it explained the human domain without reference to a theory of motivation or conscious intentions. Whether this is valid for social science remained much debated in the twentieth century. When Comte laid out a programme for a positive sociology modelled on natural science, he failed to deal adequately, as Mill observed, with human nature. As he had no language with which to describe the inner world, he achieved only a sentimental comprehension of the emotions both in his own life and in what he thought would inspire humanity. His own sad and intense obsession with Clothilde de Vaux, the woman he elevated into the mother of humanity after her early death, was emblematic of what was missing in a programme for the human sciences that included the organism and society but excluded the person in between. But the criticism should not be exaggerated. Comte's history of the sciences, and the hope that it offered for the progress of humanity through the extension of the sciences to encompass the social organism, had deep power and appeal.

iii *The Young Karl Marx*

Anyone at the end of the twentieth century must think twice before interpreting Marx. His name is attached to too much: to beliefs, political movements, slogans and emotive polemics, and to revolutions, vast social changes and cataclysms. The unpredicted rise and unpredicted collapse of the Soviet Union and the Soviet empire in Central and Eastern Europe was in his name. Much of the trauma of twentieth-century politics cannot be separated from the bearded image of this German-Jewish intellectual. This section describes Marx's view of human nature as he expressed it in the 1840s. I link this view to the long-term consequences of 1789 and, in the following section, discuss the way in which it led Marx towards systematic or scientific socio-economic analysis. He aspired to an objective historical science of society, the science his followers called historical materialism. The purpose of the two sections is to give a historical picture of what Marx thought it is to be truly human and how this related to revolutionary hopes for scientific socialism. The terms in which Marx and Marxists discussed these questions became a reference point in the human sciences.

Marx was a high school pupil in Trier and then studied philosophy and history in Berlin in the 1830s. Socialist, anarchist and liberal strands in radical politics, which all inherited the egalitarian ideals of 1789, began to diverge in this decade. While Saint-Simon conceived society to be an organic whole and dreamed of harmony, Charles Fourier drew up plans for a far more radical experiment in communitarian social organization. A very radical line was taken by P.-J. Proudhon when he developed a critique of property itself, claiming that it is the material root of social conflict. In Britain, Robert Owen spread the ideal of a community – such as the experimental village he built for his own factory workers at New Lanark – in which each person receives goods or services according to need. Liberals, on the other hand, looked to economic freedom of action and the interest in private property to secure progress. Bentham's followers turned to rational legislation and administration to create the conditions to harmonize individual self-interest and social progress. It was a time of rich experimentation in political thought, though – in most of Europe – political conditions were conservative and reactionary. Hopes that established rule would collapse, or at least become more liberal, reached fever pitch during the 1840s. But radicals and liberals paid little serious attention to how the power to effect change was actually to be achieved. After intense and sometimes violent upheaval, which affected cities from Lyons to Budapest in 1848 and 1849, conservative rule re-established itself and hope for change turned to disillusion. The long-term result was a divide between radicals who decided to work within political institutions to bring about change and those who invested their hopes in revolution. Marx's project was a comprehensive science of man to serve the revolutionary goal.

Marx was in the thick of events and ended up expelled from Cologne for sedition in 1849. The fiery *Manifest der kommunistischen Partei* (*Communist Manifesto*) that he wrote early in 1848 with his friend Friedrich Engels (1820–95) was intended to bring people on to the streets. He had lived in Paris between 1843 and 1845, where he met Engels, and they both became well-known activists. After periods in Paris, Brussels and Cologne he settled with his family in London, where he barely kept body and soul together writing journalism – he was the

London correspondent of the *New York Daily Tribune*. He also received support from Engels, who belonged to the German business community in Manchester where he manufactured cotton goods and wrote *Die Lage der arbeitenden Klasse in England* (*The Condition of the Working Class in England*, 1845), an exposé of the new industrial Britain. Engels described the sheer misery and material squalor in which tens of thousands of people lived. This was an all-too-real material state, as the novelist Mrs Gaskell, the wife of a Unitarian minister, also testified. In *Mary Barton* (1848) she created an equally painful picture of these same conditions in Manchester.

Marx's radicalism grew out of a youthful commitment to freedom, a value that at one stage he identified with man's essence:

> Freedom is so much the essence of man that its very enemies realise it in struggling against its reality . . . No man fights against freedom; at most he fights against the freedom of others. Hence every form of freedom has existed since time immemorial, whether as a special privilege, or as a general right.[13]

This notion that to be free defines what it is to be human remained, though buried, in Marx's later writings. His works can therefore be described as an extended study of the conditions in which this inherent freedom has become hidden from people and how it can be restored to them. To give people their freedom is to enable them to become what they truly are. Such a description contrasts with the better-known assessment of Marx as a materialist who pictured human life as a consequence of the iron laws of history. To overcome the tension between a view of man as a free agent and man as the material outcome of history was of fundamental philosophical and political concern to Marx. The depth of his struggle with this problem gives his thought value, even while the failure to resolve it in any actual form of political life called Marxist in the twentieth century destroyed the credibility of the political systems erected in his name.

The terms of this idealist–materialist argument were set by Hegel. Marx was one of a number of young Turks who reacted against Hegel's philosophical synthesis; in his reaction, however, he demonstrated the power of Hegel's analysis of the tension between belief that humans are free expressive agents and that conditions make people what they are. Hegel's resolution was metaphysical: he identified freedom with reality, and he conceived of reality as *Geist* (spirit) which unfolds into consciousness of itself in human existence. Marx's answer was material: he identified *Geist* with human beings themselves, and he found in the history of the material conditions of human life, in labour and technology, the circumstances that deny or fulfil man's freedom. Marx also took from Hegel the concept of alienation and used it to explain how people can in essence be free and yet in fact live in chains.

It was, of course, Marx's passionate belief and political programme that only a communist mode of life would permit labour and technology to serve human freedom rather than limiting it. He indicted other social systems, most of all the capitalist one in which he lived, for what they deny to man. Also like Hegel, Marx looked to the study of history to support his view of the human condition. He rejected all kinds of thought not grounded in the material circumstances of human life and treated thought as the practical means to achieve our full humanity: 'If man derives all his knowledge . . . from the sensible world, and from his experience of the sensible world, it follows that the task consists in so ordering the empirical world that man encounters in it what is truly human . . . that he experiences himself humanly.'[14] Marx turned to politics to create a material world in which people would experience 'what is truly human' as opposed to a lack of freedom, or inhumanity. What people in fact experience, he argued, is the culmulative weight of material history, the long-term consequences of the means by which earlier generations have satisfied their needs by labour and technology. This experience is, for the great majority, an alienated experience, an experience devoid of freedom. Such concrete circumstances, not an idealized view of mind, was, for Marx, the subject of the science of man:

> *Everyday, material* industry . . . shows us, in the form of *sensible, external, and useful objects,* in an alienated form, the *essential human*

> *faculties* transformed into *objects*. No *psychology* for which this book, i.e. the most tangible and accessible part of history, remains closed, can become a *genuine* science with a real content.[15]

Like Vico, Herder and Hegel before him, Marx argued that what humanity encounters is not nature, in the sense of something separate from man, but what humanity itself has created, his industry. He therefore initiated a sweeping, often detailed and, in the case of economic experience, technically innovative historical analysis of the conditions humanity has created for itself. It is, he argued, these historically created – and hence changeable – conditions that have made people what they are. This massive historical and economic project, which Engels, though not Marx, called historical materialism, consumed Marx's energies after 1849. Then, between the 1860s and World War I, Marx and historical materialism merged in the minds of followers and critics alike to create Marxism, a political worldview which holds that there are necessary historical laws of production and hence of every aspect of human experience. It was easy enough to read parts of Marx's work as support for this assertion.

During the 1850s, Marx began the detailed examination of capitalist economics with which he intended to demonstrate the material conditions beneath every aspect, intellectual and institutional as well as productive, of his own society. Over a decade or so, his ideal of human freedom led him, through the historical critique of the alienation that man himself creates, to study the detail of economic mechanisms. This transition in his own work encapsulated the tension that was to remain between the ethical ideal and the determinist stress in Marxism. The transition also enabled later readers and political activists to find the stress – idealist or determinist – that they wanted to find. In addition, Marx's writings were in many respects incomplete. He was a political agitator: he struggled on a daily basis to bring revolution about and did not just write about it. For Marx, practice and not theory alone is the means to resolve tensions.

What in the second half of the twentieth century was regarded as Marx's most valuable work of the 1840s was unpublished and virtually unknown until the 1930s. Indeed, the political impact of these writings was greatest in the West in the 1960s. Written in German, they are known in English as the 'Economic and Philosophical Manuscripts' (a draft written in Paris in the mid-1840s) and 'The German Ideology' (written with Engels in 1845–6). Engels and Marx reflected on the economic logic of the industrial system and studied the way the organization of large-scale production dictates to workers the nature of their labour. The exuberant advocate of the factory, Andrew Ure, had written: the factory 'involves the idea of a vast automaton, composed of various mechanical and intellectual organs, acting in uninterrupted concert for the production of a common object, all of them being subordinated to a self-regulated moving force'.[16] Where Ure celebrated progress, Marx and Engels lamented alienation, the separation of the labourer from himself by the logic of production, so that he becomes an automaton not a man. This argument, linking a person's identity, his or her nature, to the means adopted to satisfy material needs through economic activity, was profoundly influential in the human sciences. It provided a language in which to study how a person becomes what he or she is through material conditions, conditions which are themselves the outcome of history. As Ure discerned, if people work in factories they tend to become factory workers; and, as Marx might have added, if a man owns a factory, he becomes a capitalist. Marx wrote that industry 'is an open book of the human faculties'.[17] Marx and Engels, therefore, did not discuss human nature but what production means for the mode of life. An objective description of a mode of life describes what a person is. All the same, they retained an ideal of what a person is. Thus they evaluated whether or not people have control over the forces shaping what they are; in as far as people do not control these forces, even though these forces are humanly created, they suffer alienation. Alienation is exemplified by the factory worker who labours for money as part of a process of production whose outcome he or she never sees and over which he or she has no control.

Marx and Engels also attempted to explain why people do not perceive the objective, alienated circumstances of their lives but insist on believing that God, natural laws or the state require things to be the way they are. Why do people persist in beliefs that are both false and not in their own objective interest? To address this problem, Marx introduced the term 'ideology'. Ideology, in Marx's sense, refers to false beliefs about the way conditions, though really humanly made, are natural or inevitable. Such false beliefs, and the institutions that maintain them, he argued, originate with the economic interests of a particular social group or class in the maintenance of inequality. Marx, like many other radicals, despised the churches because they taught people to accept God's will rather than to act to secure their freedom. It followed from the analysis of ideology that the real explanation for events in the human sphere lies in the material, historical processes by which labour is used to satisfy human needs. These processes, Marx argued, can be known objectively and are the subject matter of the science of political economy. By contrast, he thought, the sphere of intellectual, religious, legal and political debate or conflict between society's institutions is not the place where one can find basic scientific knowledge about society. These sites are the sphere of ideology; participants believe they debate what is true, but they are dealing with ideas that are representative of a historical reality.

> The distinction should always be made between the material transformation of the economic conditions of production, which can be determined with the precision of natural science, and the legal, political, religious, aesthetic or philosophical – in short, ideological – forms in which men become conscious of this conflict [over the material forces of production] and fight it out. Just as our opinion of an individual is not based on what he thinks of himself, so can we not judge of such a period of transformation by its own consciousness . . .[18]

In his most quoted discussion of these issues, cited at the opening of this chapter, Marx referred to the conditions of production as 'the foundation' or base and social institutions such as the law or the state or religion as 'the superstructure'. These terms gave concrete priority to economic explanations in the human sciences and played a preponderant part in later versions of Marxist thought that stressed the deterministic quality of human history. Conversely, the rejection of the terms was important to the twentieth-century rejection of historical determinism and to the formulation of versions of Marxism that claimed that conscious beliefs and the life of institutions play a causal part in history in relationship with modes of production. As the Polish historian and philosopher of Marxist thought Leszek Kolakowski observed: 'In the history of human thought there are few texts that have aroused such controversy, disagreement and conflicts of interpretation as [the 1859 passage on base and superstructure].'[19]

Marx initiated the modern discussion of ideology distinct from *idéologie*, the science of ideas, present in Paris in the 1790s. He launched what became a major dimension of modern sociology, the study of relations between belief or knowledge and social processes. To unravel the subsequent history of the word 'ideology' is to go into every aspect of what was said about the relationship between people's beliefs and how they live in political society. In the 1920s, especially in the work of Karl Mannheim (1893–1947), this became – in ways much influenced by Marx – the area known as the sociology of knowledge. For Marx and Engels, however, the understanding of ideology, how non-communist society generates false beliefs about reality, was a matter of practical politics and not an academic end in itself. It was therefore a great – and a grim – irony that in the twentieth century Soviet 'Marxism' exemplified what many Western critics meant by an ideology, a belief constructed to defend political power.

Where Enlightenment writers or nineteenth-century liberals started from claims about human nature, Marx sought a historical and economic science that would explain the capacities of human nature as the consequence of human action in the satisfaction of material needs. Together with Engels, he conceived that primitive people satisfied material wants and thus made history. They altered material circumstances, passed from hunting and gathering to herding and agriculture and created a social life mediated by language – 'practical consciousness'. In the process, they also created institutions such as property and legality, and thus human freedom created the instruments of its own denial. Ultimately, they argued, only the common ownership of the means of production could overcome this historical legacy:

> This crystallization of social activity, this consolidation of what we ourselves produce into an objective power above us, growing out of our control, thwarting our expectations, bringing to naught our calculations, is one of the chief factors in historical development up till now. And out of this very contradiction between the interest of the individual and that of the community the latter takes an independent form as the STATE, divorced from the real interests of individual and community, and at the same time as an illusory communal life . . .

Reason, Marx and Engels believed, reveals how people have created societies which are believed to obey natural laws independent of human creation. Thus, reason serves revolution, for it shows that people can remake the world once they see that they have made it in the first place. In the process, they provided a basis for social science, for sociology in Comte's term, as they made the study of man dependent on the study of the historical route along which people collectively have organized production and, in the process, acquired beliefs and institutions that deny the reality of their circumstances.

> The social structure and the State are continually evolving out of the life-process of definite individuals, but of individuals, not as they may appear in their own or other people's imagination, but as they really are; i.e. as they are effective, produce materially, and are active under definite material limits, presuppositions and conditions independent of their will.[20]

Here was a programme for a science of man in the Enlightenment tradition: reason will make man free.

This approach to a science of liberation originated in the radical intellectual circles of the late 1830s and early 1840s. The group known as the Young Hegelians, along with the young Marx and Ludwig Feuerbach (1804–72), reacted against Hegel's argument that grounds man's being in abstract *Geist*. Feuerbach in particular gained a large audience in the 1840s when he detached Christian ethics from Christian faith, relocated the transcendent within humanity and rendered spirit as material activity. Like the contemporary physiologists discussed earlier, and like Marx, he claimed to describe human nature as it really is – concrete, material and motivated by the value of humanity. In a review of a book by the physiologist Moleschott in 1850, Feuerbach adopted a punning German proverb and wrote: '*Mensch ist was er ißt*' – 'We see . . . of what important ethical as well as political significance the teaching of the means of nutrition is for the people . . . If you wish to improve the people then give them better food instead of declamations against sin. Man is what he eats.'[21] This passage was often quoted by defenders of religion to show just how crude materialism could get – and Feuerbach's joke was lost.

Whereas Feuerbach, after 1848, in this way parallel to Comte, rested his hopes on a religion of humanity, Marx turned to the double task of organizing revolution and putting flesh on the bones of the materialist theory of history through detailed economic studies. It was for these activities, not for the writings of the 1840s, that he was largely known from the last part of the century until the 1930s. In his later works, Marx went beyond Feuerbach's anthropology to study how material society actually has determined what people are – their mode of life, beliefs, values and relationships. 'The essence of man is not an abstraction inherent in each particular individual. The real nature of man is the totality of social relations.' Marx and those who followed him committed themselves to the study of 'the totality of social relations'.[22] The claim that Marx had shown how this could be done scientifically and that the results validated the politics of revolutionary communism constituted the core of what, from the 1860s, was known as Marxism. The scientific character imputed to Marx's work therefore needs discussion.

iv *Scientific Socialism*

Marx took a doctorate in a German philosophy faculty and knew the difference between scholarship and sloppy thought. Radical activity in the 1840s taught him bitter lessons, firstly, that radicals lacked a strategy to obtain and wield power, and secondly, that it was incumbent on someone, and this someone could only be himself, to show objectively not just why capitalism is wrong in theory but why it is unsustainable in fact. He was especially critical of the claims made by other socialists to have provided the radical cause with a scientific justification. It was necessary to demonstrate capitalism's limits, he believed, in order to empower radicals and bring people to the communist cause. This was a fateful step, seen in the retrospective light of twentieth-century Soviet history. Marx and those who followed him were persuaded that objective reason applied to the real material world, i.e., *science*, validated their political movement. The contrast with late twentieth-century Western political life, in which government was validated by people's *preference* expressed through their votes, is striking. Marx's attempt to reason objectively about the social system was a political claim of the most emphatic kind for the dominance of science in human affairs. This is why academics and intellectuals whose commitment, occupation or vanity was to reason, turned and returned, even as critics, to Marx's challenge.

Marx published some short works during the 1840s that opened up a materialist approach to history. After he moved to London, and especially after 1857, with a leather chair in a heated reading room and all the resources of the British Museum's library at his disposal, he began the serious and, as it proved, extended study of economics. It culminated in his most famous work, *Das Kapital* (*Capital*; volume 1, 1867; volumes 2 and 3 posthumously edited by Engels and published in 1885 and 1893–4). Proponents of Marx's theories once held that the economic writings came from a 'mature Marx', that this was the work that mattered, and that Marx here escaped from the possible contradictions, which went back to his reference to freedom as a human essence, found in the 'young Marx'.

That is wrong; though the complexity and incompleteness of his work will always leave scope for divergent interpretations, and though he never produced the grand synthesis he perhaps once hoped for, the evidence points to continuity in Marx's thought. As he frequently wrote polemically for specific political purposes, it is no wonder that his comments also exhibit some contradictions.

The youthful Marx read with interest the eighteenth-century Scottish writers Ferguson and Millar who had linked social phenomena like the division of labour or social ranks to the history of material production. He also read their contemporary, Smith, who studied the division of labour and the separation between production and exchange as the key issues for political economy. Smith systematically examined labour as the source of value for what is exchanged, and Ricardo then examined the question of value in much more formal detail. In the early nineteenth century, work such as this encouraged the widely held view that political economy is a science. When Marx undertook economic analysis, then, he did so within the framework of an established science. Marx attempted to make an entirely new analysis of capitalism, but he did not reject the economic concepts Smith and Ricardo had arrived at. He took a broader view, however, and extended the framework of political economy to include both history and social thought. His ambition was to fulfil the scientific promise of political economy, to render it coherent and complete and, as he believed, thereby to demonstrate that the capitalist industrial system necessarily generates instabilities, that these instabilities progressively increase and that they finally produce a revolutionary crisis. Marx's approach firmly linked economics and social change. He discussed the economic reality of the division of labour also as the social reality of the division of classes; economic instability, he thought, expresses itself as class conflict. Economic instability and class conflict, in his discussion, are intrinsic features of the history of production, and ultimately work themselves out in revolution and the transition to communism.

For Marx, therefore, the science of political economy achieves knowledge of the social reality that in objective fact constitutes our mode of being human. His science related society to underlying economic processes and showed how belief like religion, social institutions like the legal system and political activity like imperialism all reflect the same historical forces. He thus laid out a major programme for social analysis, and he is, with Comte, conventionally regarded as a 'founder' of sociology. But most of his detailed work, especially in *Capital,* concerned economic theory, and this work was certainly intended to be scientific, that is, objective, systematic and rigorous. Yet Marx did not isolate economic science from a larger-scale and characteristically German project to establish a science of man. He laboured for a materialist philosophical anthropology, which would present to mankind a true account of what it had created – and hence of what it could change.

The focus of Marx's work was contemporary capitalism, the new industrial society that was turning parts of Europe upside down as he wrote and transformed the United States in the last three decades of the century. His perspective was also always historical since he understood the causal mechanisms of production and exchange themselves to be historically created. What appear in the present to be the natural conditions of social life are, he thought, historical conditions dependent on economic processes. Like his French and Scottish predecessors, he divided human history into stages, and he described each stage in terms of the dominance of a particular means of production. This framework, often rephrased but not entirely discarded subsequently, has underpinned understanding of the origins of the modern world as the transition from feudalism to capitalism.

Marx, who was a voracious reader, was also interested in earlier and more primitive socio-economic systems in Greece and Rome and as revealed by ethnography and the new prehistory. He was notably impressed by a North American study by Lewis Henry Morgan (1818–81). In *Ancient Society* (1877), Morgan proposed an evolutionary sequence in technology as the basis for the emergence of social institutions and social inequality, and this appealed to Marx. A broad interest in the material origins of society, not to mention in philosophical questions more generally, also led Marx to appreciate Darwin's biological theory of evolution. But he did *not* dedicate or offer to dedicate *Das Kapital* (or any part of it) to Darwin, and no English translation of his book appeared until after his own and Darwin's deaths. In 1873, Marx sent Darwin a copy of the second edition of volume one of *Das Kapital,* which Darwin politely acknowledged but never read. Their personal relations never amounted to much, except in the eyes of twentieth-century polemicists for or against Marxism as a science, who used their names as symbols in political arguments.

Yet, at an abstract level of description, both Marx and Darwin explained human existence in terms of causal historical processes. Engels, when he spoke at Marx's graveside in Highgate cemetery in London in 1883, was succinct about this relationship: 'Just as Darwin discovered the law of development of organic nature, so Marx discovered the law of development of human history . . .'[23] Marx read Darwin's *On the Origin of Species* (1859) in 1860 and welcomed it, since, 'although developed in the crude English fashion, this is the book which, in the field of natural history, provides the basis for our views'. Darwin accounted for organisms with a causal history of material adaptation; Marx accounted for societies with a causal history of material production and technology. He legitimately saw that, placed together, Darwin's and his own work, in a general sense, created historical continuity between matter, life and society. At the same time, Marx was contemptuous of Darwin's emphasis on struggle in evolution, which Marx saw simply as Malthus's political economy writ large. 'It is remarkable how Darwin rediscovers, among the beasts and plants, the society of England with its division of labour, competition, opening up of new markets, "inventions" and Malthusian "struggle for existence".'[24] For Marx and his followers, Malthus's law of population exemplified ideology, the elevation of the struggles society has created in particular historical circumstances into the consequences of a timeless law of nature. Engels was blunt: 'the most open declaration of war of the bourgeoisie upon the proletariat is Malthus' Law of Population and the New Poor Law framed in accordance with it'.[25]

During the 1860s, Marx became a public figure in a European

revolutionary movement that looked to his writings for a scientific, irrefutable validation of revolutionary action. He was the leading intellect in the Working Men's International Association, the First International, which lasted between 1864 and 1872 and was the setting where the term 'Marxism' became current. Politics tied Marx's name to a worldview rather than to systematic theory. The exigencies of revolutionary politics, faced by such bloody calamities as the Paris Commune in 1871, demanded attention; the philosophical analysis of alienated consciousness or the economic analysis of the price of cotton seemed secondary.

There was a return to theory with the publication of volumes two and three of *Capital* and of Engels's own studies in the 1880s. Also in the 1880s, some Russians began to use Marx in the long-running debate about whether Russia's development towards industrialization required it to pass through the same stages as the West or whether it could jump forward. Some economists in Central Europe, notably the Austrian Eugen von Böhm-Bawerk (1851–1914), read Marx as a scientific economist and produced in response an extended critique of the labour theory of value and the technical aspects of Marx's economics. By contrast, Marx's work was virtually ignored in France, Britain and the United States. In the English-speaking world, economic analysis during the 1860s and 1870s shifted away from the framework derived from Smith and Ricardo that Marx himself adopted, however critically. What was later called the marginalist revolution of the 1870s was led by academics for whom economics is a science by virtue of the way it uses rigorous techniques to analyse particular aspects of the economy rather than a nation's wealth as a whole. The new economists advanced economic science as a discipline in its own right and detached economic questions from specific social contexts. At the same time, they made their science mathematical and excluded those without training from participation. The economist at Owen's College, Manchester, W. S. Jevons (1835–82), pioneered the argument that economic 'value depends entirely on utility' – not on labour as Smith, Ricardo and Marx had argued. He thought this equation made it possible to develop 'a general mathematical theory of Political Economy' in terms of use values, a theory divorced from specific social and historical circumstances. Jevons believed that the social sciences use precisely the same methods as the physical sciences, and his own goal was to devise a system of economic market mechanics on the model of physical mechanics. This suggested to the next generation of British and US economists the quantitative language and technical arguments in terms of which to establish economics as a separate discipline – and in the process to cut themselves off from economic history. Alfred Marshall (1842–1924), who was appointed to the chair of political economy at the university of Cambridge in 1885, self-consciously worked for the establishment of economics, rather than political economy, as a professsional discipline that could claim status as a science. It was also Marshall who put the word 'economics' into currency. Nevertheless, there was continuity between Ricardo and Jevons and his successors in one significant respect: belief that utility is objectively founded in human nature or in what Jevons called 'the great springs of human action – the feelings of pleasure and pain'.[26] Belief in the reality of economic man carried over from the eighteenth century into the foundations of modern twentieth-century economics.

This belief, that man is by nature a certain kind of economic actor, was opposed by Marx and his communist followers. Marx differed from the new economists in the major respect that he studied economics in order to understand society as a whole – institutions, belief or social conflict as well as markets, prices or labour values. In the long term, his ideas proved more significant to sociology than to economics. Marx's theory, in addition, was not abstract, since he believed that the explanation of human affairs requires a theory of concrete historical development. Finally, of course, Marx was committed to economics as the means to change the world not as the subject matter of a new academic discipline addressed only to fellow academics. His work most stimulated political activists, or scholars with the breadth to think of economic and social questions together, like Max Weber in Germany, or intellectuals who believed history is the key to the human sciences, like Benedetto Croce in Italy. Weber, discussed later in the context of the discipline of sociology, had a profound influence on both sociological theory and political debate about the relation of social science to politics. In both respects much of the stimulus came from Marx. Through the second half of the nineteenth century there was also a flourishing discipline of historical economics in the German-speaking world and German-language academics – most of whom were deeply opposed to the materialist view of history – were challenged to produce different explanations to Marx's for the rise of modern economic conditions.

Marx really achieved prominence, however, with the socialist movements that changed the European political landscape in the late nineteenth century. Marx became the quasi-mythical figurehead of a worldview. He gave a voice to an ethic of human wholeness, freedom and justice; he explained why these values are denied to the mass of people; he made an economic analysis of the division of labour and power – an analysis of every working man's daily experience; he created a picture of history that demonstrated the human origins of current conditions; he held out a vision of a utopian communist future; and as a man he led a life of unimpeachable political activism, which joined heart and mind, labour and reason, in the revolutionary cause. This was a formidable and seductive reputation. Marx also intended to achieve something beyond all this: a science of society. Since, as he wrote, 'Man is . . . not merely a social animal, but an animal which can develop into an individual only in society', it was to be the science of man himself.[27]

Whatever Marx thought about his achievements, many Marxists believed that he had realized his ambitions, and they concluded that Marx's work, supplemented by later writers, added up to a body of systematic, self-consistent and objective truths about the human condition. Many socialists thought that he had achieved timeless scientific results which, for purposes of general exposition, could be represented as a body of historical laws. These laws, which explain history by reference to changes in production and show how inherent contradictions, visible as class conflict, push society forward to new stages, then appeared to guarantee the arrival in the future of socialism or communism. Marx, in various places, used the language of historical inevitability; and, whatever he might have meant, such talk easily became messianic in the context of actual political oppression and revolt. Karl Kautsky, who drafted the Marxist constitution of the fastest growing socialist party, the SPD (German Socialist Party), produced a synthesis of history that reinforced this determinist interpretation. Liberal and left-wing critics alike attacked what they saw as the fatalism inherent in such a view. But it was an interpretation that inspired many rank and file workers to invest their hopes in a socialist future.

The name of Marx thus became synonymous with 'scientific socialism'. This was in ignorance of the young Marx of the 1840s and tended to treat his science of society as a determinist natural science. Naturally enough, political liberals and conservatives attacked the supposed scientific standing of Marx's work in order to discredit the socialist movement as a whole. For some critics it was enough to note that Marx and socialists were ignorant of modern scientific economics. In the period after the end of war in 1918, when events did not go the way scientific socialism had predicted and communist societies did not break out in the industrialized parts of Europe, critics indicted Marx's work as a whole. At the same time, the unexpected success of the Bolshevik Revolution in the less economically developed Russian empire precipitated a new stage in the interpretation of Marx as the foundation for the science of man.

Marx became a household name while Comte sank into relative obscurity. Yet both, at work outside the academic setting and in pursuit of ethical and political goals, gave to the science of society much of its intellectual character. To their contemporaries, they explained and justified historical change as progress. They made sociology, the subject that Comte named, the road to a science of man. Comte discussed methodology and the nature of positive knowledge; Marx analysed economics in relation to historical and social realities. In these ways, they pointed others along the road they perceived. Both were thin in their treatment of psychology and the individual: Comte thought human nature a subject for physiology and phrenology, while Marx referred to man's essential freedom and to specific human natures as the outcome of the historical organization of production. In their different ways, their work was a powerful statement of the human constitution of the human world. In the nineteenth century it carried a message of liberation and hope. To understand how it contributed to the worldviews of the time, it must be placed alongside the theory of evolution. It is necessary to see the extent to which Comte, Marx and Darwin alike used history to link human nature with the physical world.

13

Human Evolution

> It is apparently a truer and more cheerful view that progress has been much more general than retrogression; that man has risen, though by slow and interrupted steps, from a lowly condition to the highest standard as yet attained by him in knowledge, morals, and religion.
>
> Charles Darwin, *The Descent of Man, and Selection in Relation to Sex* (1871)[1]

i *Man's Place in Nature*

George John Romanes (1848–94) is a rare example of a Victorian whose religious doubt derived directly from those arguments in natural science that opposed the evidence of God's design in the world. Romanes was also a deep reader of the works of Charles Darwin (1808–82). Darwin later passed his notes on instincts to the younger man who spread his mentor's ideas in books on comparative psychology and evolutionary theory. Romanes wrote: 'If we may estimate the importance of an idea by the change of thought which it effects, this idea of [evolution by] natural selection is unquestionably the most important that has ever been conceived by the mind of man.'[2] Even allowing for some pardonable exaggeration, this is a strong claim.

A decline of confidence in progress and in ability to find truth made such claims less fashionable. Further, at a distance in time from Darwin, much that once made his work appear revolutionary was found to be already embedded in his intellectual culture. Theories of historical development of language had already explored the evolution of human nature, while physiology had fused nature and human nature in a naturalistic worldview. The point is not that Darwin had predecessors who held exactly the same views (he did not), but that many areas of thought besides his own contributed to 'Darwinism', the scientific, evolutionary worldview to which his name was attached. It is very difficult to maintain belief in one man effecting a revolution in thought. Lastly, what historians know among themselves as the Darwin industry, a veritable empire of historical scholarship, has traced every twitch of Darwin's mind, and twitches are not the stuff out of which to build judgements of the kind that Romanes made.

Yet Darwin was a household name in the late twentieth century, as it was to the Victorians, and if any one name was attached to belief that there is or could be a natural science of human nature, it was his. The reason is straightforward. Darwin made it plausible to believe that human beings, like plants and animals, originate in physical nature and in a manner that accords with causal laws. The evidence that human beings have evolved from physical nature vindicated the conclusion that human nature and physical nature are understandable in the same terms. Put at its simplest, Darwin showed that natural science encompasses man. Evolutionary theory was both an empirical demonstration of the continuity of nature and human nature and the theoretical legitimation of the human sciences. The popular image of Darwin in conflict with religion – an image in need of considerable historical qualification – is at least emblematic of evolutionary theory's wider significance. If, in the words of Genesis, 'God created man in his *own* image', in Darwin's words, 'Man with all his noble qualities . . . with all these exalted powers – Man still bears in his bodily frame the indelible stamp of his lowly origin.'[3] These contrasting words still seemed to many people to demand a choice even after more than a century of commentary and interpretation. The British Prime Minister, Benjamin Disraeli, was glib but he was in tune with public opinion when he enquired, in relation to Darwin's work, 'Is man an ape or an angel?' and declared himself, 'on the side of the angels.'[4]

Three interrelated points in the logic of evolutionary argument are especially pertinent to its standing in the human sciences. The first concerns the authority that evolutionary theory acquired as the proof that humans have an animal ancestry. The evidence for evolution, the facts over which Darwin had a brilliant command, appeared to require people, whatever their previous views, to accept man's continuity with nature. This was how Darwin's champions, such as T. H. Huxley in Britain or Ernst Haeckel (1834–1919) in Germany, promoted the cause. As Huxley said about the origin of species: 'The question is one to be settled only by the painstaking, truth-loving investigation of skilled naturalists. It is the duty of the general public to await the result in patience . . .'[5] This was what made Darwin an apt figurehead for the natural-scientific worldview in the nineteenth century; he was the man who made the facts of nature reveal the basis of human existence. The failure of Darwin's critics to marshall persuasive facts against him enhanced the prestige of the scientific worldview in general. There was, however, much more to Darwin's theory than one claim, or even one set of claims, proved by the facts. His theory exemplified a way of thinking about life and human nature that scientists accepted ultimately because it was for them the only way they could do science. As G. H. Lewes (1817–78), George Eliot's partner, observed, the appeal of evolutionary thought was its presupposition 'that everywhere throughout Nature – including therein all moral and social phenomena – the processes are subordinated to unchangeable Law . . .'[6]

The second point is that the proponents of evolutionary theory presupposed continuity between the animal and the human worlds to justify the extension of their argument to human beings. At the same time, they used evidence for man's evolutionary past as empirical authority for the principle of continuity. This procedure appears to take for granted what the argument sets out to prove. Another way to look at it, however, is to see the argument as simultaneously conceptual and empirical, abstract and concrete. The philosophical and scientific dimensions together made evolutionary theory of such significance for the human sciences. It was also because philosophy and fact were intimately connected that observers thought Darwin brought about a revolution in ideas.

Thirdly, Darwin's impact centred on his picture of men and women as animals. This was certainly his effect on the public imagination, as many cartoons with Darwin himself portrayed as a monkey bear witness. He wrote at the beginning of his account of morality: 'as far as I know, no one [until now] has approached it exclusively from the side of natural history'.[7] But such an approach to morality was precisely what opponents would not concede in the first place; they did not accept that morality is a subject for 'natural history' or that man's essence is a subject for natural science. This was not, ultimately, a matter for *empirical* dispute but a debate about the terms in which it is possible to have knowledge of man. All the same, empirical facts were very important as, in historical reality, they persuaded people to accept one or the other position. The assessment of Darwin's significance for the human sciences cannot therefore be made independently of debate about what sort of human science is under discussion. Many ardent Victorian evolutionists were evolutionists precisely because evolutionary theory unified human and natural science, and their descendants, such as the sociobiologists active in the 1970s and 1980s, shared this position. As Huxley said at the conclusion of a review of *On the Origin of Species*: 'we do not believe that . . . any work has appeared calculated to exert so large an influence . . . in extending the domination of Science over regions of thought into which she has, as yet, hardly penetrated'.[8] This was the confident voice of the nineteenth-century scientific worldview. Nevertheless many critics, then and later, continued to defend other forms of explanation in the human sciences; and there were other disciplines that claimed to be sciences as well as the disciplines of natural science. Beyond this, in the wider culture, many forms of religious faith and argument perceived limits to the capacity of science of any kind fully to circumscribe human existence.

Evolutionary thought and Darwin's name were also inextricably linked with ethical and political debate. Comtean positivism, historical materialism and evolutionary theory alike incorporated an evaluative as well as a descriptive dimension. The analytic distinction philosophers later introduced between facts and values was not present in the nineteenth century. From the outset, evolutionary ideas were part of the value-laden project known in the eighteenth century as the science of man and in the nineteenth century as the science of society. Nor did evolutionary theory separate itself from this enterprise; even Darwin himself believed that, as a matter of fact, human evolution demonstrates the reality of progress. At times, especially at the end of the century, evolutionary language became dominant in ethics and politics and was used to support many different claims that man's natural history determines his prospects. Darwin's name was attached to many different views; and however much these views misused his name, this does not mean that Darwin himself did not also believe that evolutionary thought affects man's values.

Ethics was the explicit end-point of the work of Herbert Spencer (1820–1903) on evolution, which was the most systematic attempt to think through the implications of evolutionary theory for psychology and sociology. His ambition was to show how a system of ethics could indeed be derived from knowledge of the evolutionary facts of nature. Darwin did his level best to avoid thorny matters of philosophy, but even he tried in print to reconcile his Victorian views of moral progress with his account of human evolution. A central feature of evolutionary thought, like Comtean positivism, was that it provided arguments to ground what it is right to do in what it is natural to do. This was part of a shift in cultural values, which historians often call naturalism, that substituted the authority of this world, nature, for the authority of the transcendent in the determination of right action. All the same, many evolutionists, though not Darwin or Spencer, were deeply religious. Others, like proponents of Darwin on the European continent, such as Haeckel in Germany and D. I. Pisarev (1840–68) in Russia, became leaders of Darwinian naturalism in the context of political struggle against conservative Christian forces. Probably the best-selling non-fiction German-language author before 1933 was Wilhelm Bölsche, who wrote romances about evolution as the history of love and reinforced a sentimental view of natural human life. German working-men found inspiration in reading Huxley's book on *Man's Place in Nature*, rather than Marx, well into the 1920s. The idea that evolutionary nature gives man hope had wide appeal. The Russian biologist, and aristocrat in exile, Piotr Kropotkin (1842–1921) found in Darwin's work the basis for belief that man has a natural co-operative nature and that the destruction of existing forms of government will allow this nature free expression. 'Darwinian' theories were as often pacifist as militarist in intent.

The debate on evolution was a debate about biological and geological evidence. But there was nothing remarkable about the fact that naturalists also reflected on Malthus's theory of human population in connection with the life of animals and plants. No educated early Victorian who took an interest in social affairs could have avoided familiarity with Malthusian approaches to poverty and with the belief that society is subject to ineluctable laws. The science of political economy had an established place amongst the sciences of the time, and there were many precedents for the transfer of ideas from one science to another. The banker George Poulett Scrope (1797–1876), for example, was also a geologist, and in the 1820s his thoughts about economy and the balance of finite resources appeared equally in essays on finance and on the earth's surface. Yet it is striking that the two original theorists of natural selection, Alfred Russel Wallace (1823–1913) and Darwin himself, both reported 'eureka' experiences that featured Malthus, and that Spencer's early evolutionary thought was in part a response to Malthus. The political language of 'there is no alternative' was all too familiar to Victorians. The imagery of struggle for a livelihood in human society reappeared in the claims Darwin and Wallace made for natural selection. Their crucial point about why evolution occurs was that animals and plants compete for subsistence and that this competition leads to the survival of one variety rather than another over time. Differential survival, they argued, is the mechanism of organic evolution, the origin of species. Their argument involved many other elements and became very sophisticated. It nevertheless retained a language that linked Malthus's description of human competition as the motor of social progress (whatever the costs), and their own descriptions of the organic struggle for existence as the motor of evolutionary change (whatever the extinction of animals and plants). Even if Spencer had not coined the phrase 'survival of the fittest', the Victorians would still have written and read about natural selection with spectacles coloured by political economy.[9]

Late nineteenth-century polemics on national and international issues frequently invoked Darwin's name or deployed a 'Darwinian' language to portray a particular political value – race or struggle, for example – as a feature of nature. This language was strongest in justification of empire, at a time when European countries and the United States competed fiercely to establish economic and military spheres of influence around the world. Karl Pearson, a pioneer of mathematical biology and statistical analysis, concluded his much read *The Grammar of Science* (1892) with claims about the inevitability of evolutionary struggle and the duty of superior nations to provide world leadership. 'The struggle of civilized man against uncivilized man and against nature produces a certain partial "solidarity of humanity" which involves a prohibition against any individual community wasting the resources of mankind.'[10] Such literature claimed that values are revealed by science itself and did not just draw on the authority of empirical science to underwrite values. The literature of political Darwinism perpetuated the long-established practice that shared language and meaning between the science of man, political economy and natural science. The notion of struggle, the conflict of interests, had been a mainstay of liberal political thought since the seventeenth century. Just as Darwin or Wallace thought around Malthusian notions when they created the theory of natural selection, so Pearson or his contemporary Benjamin Kidd – the author of *Social Evolution* (1894) – thought around evolutionary biology to articulate political goals. All this adds up to the conclusion that evolutionary thought was part and parcel of the human sciences and not independent work that then had an influence on the human sciences.

The phrase 'social Darwinism' does not appear to have been used much, if at all, in the nineteenth century. It alludes in a rather derogatory way to any effort to use evolutionary biology to vindicate a claim about society or a social policy: support for conflict between individuals, classes, nations or races as necessary to progress; the belief that individual or collective action, especially aggression, is natural rather than cultural; or to eugenics, belief that differential breeding is the way to affect human destiny. Little is gained when so many different arguments are lumped together, even though Darwin's name was applied across such a range of views. It is noteworthy, for example, that Darwinian supporters of eugenics favoured central government controls over reproduction, whereas Darwinians in favour of the individual struggle for existence specifically opposed such compulsion.

It is time to turn from this general assessment of the evolutionary debate to the contributions of Spencer and Darwin themselves and to the specific ways in which psychology, sociology and anthropology became evolutionary sciences in the second half of the nineteenth century.

ii *Evolution and Herbert Spencer*

Spencer had the virtues and weaknesses of provincial English Nonconformity. An independent-minded individualist, he made his own way from railway engineer to sage, and he became a scourge of state action, but he impressed many only as a humourless prig. As a young man he was an enthusiast for phrenology; a reading of his head found bumps for firmness, self-esteem and conscientiousness, and the conclusion was that 'such a head as this ought to be in the Church'.[11] He was upset when he was compared to Comte, since he thought his accomplishments unique, but they had much in common intellectually and in personality. Both men constructed philosophical syntheses with the progress of natural-scientific knowledge as the centrepiece of human history. Both were driven by belief that knowledge of the laws of nature shows people how to live and to organize society. They characterized sociology as a distinct science, and they had faith that sociological knowledge would redeem a secular age. Further, they both took the biological language of the structure and function of the parts of organisms and applied it to human society. In both cases, phrenology influenced the notion of function, though only Spencer went on to elaborate a psychology. Both men aimed at the good of humanity and led unhappy personal lives. Where they differed sharply was in their visions of the ideal social order: Comte's was collectivist, Spencer's individualist.

Spencer was original in the way he thought systematically about what an acceptance of human evolution means for psychological and sociological knowledge. He was the evolutionary philosopher *par excellence*. While, unlike Darwin, he contributed no new observations, he provided a conceptual framework for psychology and sociology as subjects comparable with the natural sciences. The framework involved two major principles, continuity and utility, which were fundamental to the subsequent history of the human sciences, especially in the United States. The principle of continuity is the presumption that natural law applies universally and includes every aspect of human existence; for Spencer, the principle is embodied in the evolutionary law of directional change, at all levels of reality, from a state of unorganized simplicity to one of organized complexity. The principle of utility denotes that every phenomenon, be it the solar system or the free market, has a form determined by adaptive integration to conditions. These principles were very abstract – and Spencer's style enhanced this impression: 'Evolution is an integration of matter and concomitant dissipation of motion; during which the matter passes from an indefinite, incoherent homogeneity to a definite, coherent heterogeneity . . .'[12] Nevertheless, his tenacity – he was 'the first to see in evolution an absolutely universal principle', as William James wrote – fostered an intellectual reorientation of great significance.[13]

After his move to London in the 1840s, Spencer was for a while an editor at the pro-free trade *Economist* magazine. His first book, *Social Statics; Or, the Conditions Essential to Human Happiness Specified* (1851) was in the eighteenth-century mould of the science of man; he argued from human nature and experience to an optimistic belief in the natural progress of society, if there is equality of opportunity. He mixed the values of provincial self-help, a phrenological view of human nature and liberal political economy. Spencer read in the literature of science and philosophy, and Mill pointed him towards association psychology. The result was that Spencer brought together an idea of progressive 'individuation' from Coleridge, a 'law of development' taken from the embryologist Karl Ernst von Baer, the language of 'the physiological division of labour' taken from the French zoologist Henri Milne-Edwards, and a theory of organic *transformation* taken from the French evolutionist Lamarck. All this put flesh on his belief that: 'Progress, therefore, is not an accident, but a necessity.'[14] During Spencer's youth, Lamarck's evolutionary theory was linked in Britain with radical criticism of the political and religious establishment and with belief in a law of social progress. Spencer took up Lamarck's notion of structural and functional change in organisms over successive generations in response to the environment and expanded it into a mechanism to produce progress in mind and society. The notion that adaptation occurs through experience, the idea that adjustment of an internal state under the impact of external events produces a stable condition, became Spencer's model for the description of any system whatsoever. In the mid-twentieth century this notion was to become the conceptual basis for the approach to decision-making called systems analysis.

Spencer first applied these general ideas in *The Principles of Psychology* (1855), a book that in the long run transformed the analysis of mind in Britain. Whereas Locke and Hartley had treated experience as the means by which a single mind acquires knowledge and thereby acts constructively in the external world, Spencer treated experience as a continuous historical process, the means by which the minds of animals and human beings evolve over time and integrate animal and human activity with the world they inhabit. This exemplified his supposition that there is a continuous adjustment of inner to outer relations in all living processes. His major insight, that exactly the same could be said of body, of mind and of conduct, laid the foundations for psychology as a science concerned with adaptive *functions*, and its importance is explained in a separate section. Locke had argued that true knowledge, and hence social agreement, is possible because belief grows with experience. In Spencer's hands, that assertion became the argument that mind and culture are products of the adaptive integration (i.e., experience) of previous conditions. He also argued that the natural process of adaptive integration establishes an objective standard for political and ethical judgements, and that – ultimately – humanity cannot but make progress.

Earlier writers who derived the content of mind from experience, like Condillac or Hartley, were in the awkward position of supposing that each individual mind builds up its mental content from scratch. By contrast, Spencer emphasized that all individuals inherit mental functions from previous generations, even though they derive the content of mind from experience. This was an important step in two ways. First, common opinion had always found implausible Locke's image of the new-born mind as a blank sheet; babies, like animals, appear to be born with emotions and instincts. Further, as J. S. Mill argued against his father, James Mill, it is simply indefensible to reduce the emotional life to a calculus of pleasures and pains: 'It is certain that the attempts of the Association psychologists to resolve the emotions by association, have been on the whole the least successful part of their efforts.'[15] Darwin later made much the same point and argued that a theory of the emotions requires a theory of inherited instincts. Evolution, Spencer believed, explains how inherited capacities originate: in experience, but in the experience of the race or the species not the individual. Experience, he thought, becomes embedded in the inheritable structure of the nervous system.

> By the accumulation of small increments, arising from the constant experiences of successive generations, the tendency of all the component psychical states to make each other nascent, will become gradually stronger. And when ultimately it becomes organic, it will constitute what we call a sentiment, or propensity, or feeling . . .

A mental event, over time, changes from a mental process to a nervous structure. This explains, for example, how 'the most powerful of all passions – the amatory passion – [is] one which, when it first occurs, is absolutely antecedent to all relative experience whatever'.[16]

There were more implications to this step than the expansion of the psychology of experience to include instincts and emotional propensities. The instincts and emotions were important evidence for the idealist opponents of empiricist psychology and natural-scientific approaches to mind. British moralists lovingly described animal instincts and human capacities like the moral sense to illustrate God's design in nature and human nature. Conservative academic philosophers argued that the mind possesses *a priori* categories, which of logical necessity cannot originate with experience and are, it was concluded, God-given. Spencer believed that he was able to pull the rug out from under these Christian writers whom he thought bulwarks of the established political order in Britain. Evolution, he argued, demonstrates that emotional and rational capacities are indeed innate, but that they nevertheless derive from experience – evolutionary experience. What had appeared to be convincing evidence for Christian idealism reappeared as evidence for a natural-scientific psychology. Spencer's ambition was to enhance the status of science as opposed to Christian philosophy in the national culture.

Further, Spencer thought he had put to rest the philosophical dispute between idealists (like Leibniz or Kant) and empiricists (like Locke or J. S. Mill) about the origins of knowledge. He accepted that the mind shapes knowledge with *a priori* categories, but held that these categories are nevertheless *a posteriori* in an evolutionary sense.

> Finally, on rising up to human faculties, regarded as organized results of this intercourse between the organism and the environment, there was reached the conclusion that the so-called forms of thought are the outcome of the process of perpetually adjusting inner relations to outer relations; fixed relations in the environment producing fixed relations in the

> mind. And so came a reconciliation of the *a priori* view with the experiential view.[17]

Such a psychological answer to a philosophical problem had exactly the character that philosophers, led by Frege in the 1890s, reacted against when they laid the basis of what became analytic philosophy. In the mid-Victorian setting, however, Spencer wanted his argument to carry naturalistic or scientific ways of thought into the idealist enemy camp.

It took a while for people to grasp these arguments, and Spencer's early work on psychology went largely unread. At the end of the 1850s, just as Darwin's own evolutionary theory appeared in print, Spencer persuaded enough people to back him financially to write a 'Synthetic Philosophy', a systematic exposition of his evolutionary thought, which began with *First Principles* (appeared serially 1860–62) and concluded thirty years later with studies on sociology and ethics. The second edition of the work on psychology (1870–72), in which he reshaped his argument, did attract attention. Meanwhile, Spencer moved towards the ultimate goal of his endeavours, the foundation of ethics and an individualist politics grounded on the natural law of progress. Many readers, to Spencer's annoyance, merged his argument in their minds with what they learned from Comte: that rational political action depends on the systematic study of the natural laws of social development. In the English-speaking world, however, Spencer did more than anyone to inspire intellectual and political investment in the science of sociology.

Spencer's language of 'the adjustment of inner to outer relations' derived from physiology but received its greatest elaboration in sociology. He took the organic analogy further than anyone and, in the process, exposed – almost to ridicule – weaknesses in social science based on biology. This was most evident in 'The Social Organism' (1860), an article in which he systematically compared the structures and functions of an animal – the skin, the digestive system, the blood circulation – with the institutions and processes of production and distribution in society. For example: 'And in railways we also see, for the first time in the social organism, a system of double channels conveying in opposite directions, as do the arteries and veins of a well-developed animal.'[18] He stopped short, however, at the comparison of the brain with government, for Spencer was against centralized regulation and placed a premium on individual moral choice. He opposed, for example, both the government-run post office and the legislation for a system of state primary education passed in 1870. Thus, he drew on organic analogy where it suited his political purposes and ignored it where it did not. Here he was involved in inconsistency, since his analogies were not tropes or figurative phrases to be discarded at will but deductions from principles that, he claimed, underlie organized relations of any kind whatsoever. In theory he could not choose to draw on or to limit analogy because, in his thought, what appears on the surface to be analogy between animals and society is a consequence of a deeper identity. If society is subject to the same natural law as the rest of the universe, the law of evolutionary progress, it appeared to follow that societies, like animals, evolve towards a centralized system – of the kind Comte indeed favoured. Spencer rejected this conclusion, and he argued instead that individual intelligence and moral consciousness, built up through individual experience which becomes a collective inheritance, is the decisive force in society. Forced to clarify his position on government, he withdrew from the implications of his notion of the social organism. Nature did not provide the unambiguous guide to social policy that he hoped for.

All the same, Spencer persuaded a good many people of the need for sociology. His introductory text, *The Study of Sociology* (1873), appeared in Britain and America in E. L. Youmans's 'International Scientific Series', a major publishing venture that contributed to the public understanding of science in the late nineteenth century. Youmans was also the publisher of *Popular Science Monthly*, the premier English-language popular science journal, which also carried Spencer's ideas. Spencer wrote: 'If there is natural causation . . . it behoves us to use all diligence in ascertaining what the forces are, what are their laws, and what are the ways in which they co-operate . . .' From this he drew his basic conclusion: 'And to hold this is to hold that there can be prevision of social phenomena, and therefore Social Science.' He taunted politicians or historians who act or explain action on the basis of common sense or great men: do they mean to deny the reality of natural causation? If they do not deny it, he believed, they cannot deny the need for sociology. He raised people's consciousness that sociology is the prevision society needs to conduct its affairs on a rational basis. In 1851 he wrote optimistically about progress proceeding hand-in-hand with individual improvement; in the 1870s, with Parliament and public opinion in favour of legislation to solve social problems, he became an increasingly pessimistic and embittered critic of the actual social changes around him. He then became known as a leader of what later historians called social Darwinism, in this context, the politics of free competition between individuals based on the belief that such competition is the natural motor of progress. Spencer claimed:

> The study of Sociology, scientifically carried on by tracing back proximate causes to remote ones . . . will dissipate the current illusion that social evils admit of radical cures . . . You may alter the incidence of the mischief, but the amount of it must inevitably be borne somewhere.[19]

Proponents of economic *laissez-faire*, notably Andrew Carnegie, the great Pittsburg steel magnate, took up Spencer in their battle to resist anti-trust legislation in the US (an attempt to prevent entrepreneurs creating monopolies).

Ironically, many leaders of the generation inspired by Spencer's call for sociology, such as the founders, in 1895, of the London School of Economics, saw government as the institution which possesses sufficient power to apply the results of sociological analysis. Throughout much of the twentieth century, applied social science was associated politically with an interest in centralized planning and decision-making. A science of society appeared to its supporters to require implementation by experts rather than the popular vote, and the power of government was thought to be needed to overcome sectional interests. Spencer's utopia, however, was a society in which the education of individuals in the laws of nature gives them the moral power to act for what is good and, over time, to pass down a morally elevated inheritance.

Spencer also assembled a considerable body of comparative information about societies around the world in order to show how his general laws do in fact apply. He tabulated other people's observations to demonstrate social evolution from primitive to advanced, a change marked by increased complexity, specialization of function and adaptive integration of parts to whole. It turned into a wooden exercise in repeated confirmation of his basic principles. More interestingly, he described a change from militant (i.e., military) to industrial society as the crucial step in the rise of the West. With the achievement of relatively stable social order, he argued, Western societies had relaxed the control which had previously been necessary to ensure the survival of the polity. Thus freed, individual initiative generated the material and intellectual prosperity that accompanies moral progress. He projected Britain's experience of change, of the shift of power during industrialization away from an aristocracy with military values to an industrial class with commercial ones, as the general pattern of social development. This theory constituted a major part of the empirical dimension of the account of social evolution he also deduced from first principles. Ironically again, at the end of his life, he watched appalled as public opinion supported military values and Europe rushed to re-arm.

iii *The Descent of Man*

Spencer and Darwin both went to great lengths to control their domestic circumstances so that they could work, and both were often neurotically unwell. But whereas Spencer was an irascible bachelor, Darwin was a much-loved *paterfamilias*. Their paths barely crossed; they divided the labour of philosophy and sociology and of scientific natural history between them and kept a polite but dry distance. Spencer acknowledged that his early evolutionary thought ignored Darwin's central mechanism, natural selection, though he continued to give priority to inheritance of acquired characteristics. Darwin was courteous in print; he wrote that: 'Psychology will be securely based on the foundation already well laid by Mr. Herbert Spencer, that of the necessary acquirement of each mental power and capacity by gradation.'[20] But he was noncommittal in private and regarded Spencer as far too abstract to be a good scientist.

Darwin certainly was a natural scientist in a way that Spencer was not, in the sense that he spent a lifetime in the detailed study of plants and animals, and he was driven by a desire to explain detail. Darwin was a naturalist, but he constantly organized his information in relation to a unifying *theory* of evolution. He understood 'that all observation must be for or against some view if it is to be of any service'.[21] His work not only collated the observed facts into the general claim that evolution has occurred but explored in depth a causal mechanism, natural selection, that explained it in terms of the material laws of nature. He set up what became – with whatever changes – a comprehensive framework for biological research.

Darwin was happiest when at work on some detailed part of the living world, as in his studies of orchids or earthworms. Nevertheless, from at least the time of the world voyage on HMS *Beagle* (1831–6) that made the name of the young naturalist, he was also fascinated by the question of man's place in nature. This interest culminated in two major studies in comparative psychology, *The Descent of Man, and Selection in Relation to Sex* (1871) and *The Expression of the Emotions in Man and Animals* (1872). These books drew out the implications of *On the Origin of Species by Means of Natural Selection, Or the Preservation of Favoured Races in the Struggle for Life* (1859) for human nature. The Victorian public had already been exposed to the idea of human evolution by the time he published on the topic, and this dampened the fire of criticism; his books appeared as contributions to a debate rather than as a shocking novelty. Spencer created a theoretical, comparative perspective for psychology and sociology out of evolutionary ideas. Darwin described human nature in the concrete terms of an evolutionary natural history, which compared human bodily and mental capacities with those of animals. Everyone was able to understand his claim 'that man is descended from some lowly-organised form'.[22] Darwin was sensitive to the implications his work had for ethics, religion, belief in progress, social thought and the philosophy of mind, and, though he tried to keep his discussion at the level of natural history, his views on these matters seeped through in print. But those who opposed his views did not concede that natural history is the appropriate medium for understanding human beings in the first place. As these critics perceived, though Darwin wrote in the language of natural history, his work was the vehicle for a naturalistic philosophy of man.

Mankind was always at the heart of Darwin's experience of nature. On board the *Beagle* with Darwin in 1831 were three Fuegians, indigenous people of the remote southern tip of South America, exposed to Western civilization by a year in England. Left again in their native land, Jemmy Button, York Minster and Fuegia Basket reverted to their native ways. Years later, Darwin reconstructed his experiences at Tierra del Fuego:

> The astonishment which I felt on first seeing a party of Fuegians on a wild and broken shore will never be forgotten by me, for the reflection at once rushed into my mind – such were our ancestors. These men were absolutely naked and bedaubed with paint, their long hair was tangled, their mouths frothed with excitement, and their expression was wild, startled, and distrustful . . . He who has seen a savage in his native land will not feel much shame if forced to acknowledge that the blood of some more humble [animal] creature flows in his veins.[23]

Thus he acquired a vivid image of the primitive as a young man. When he returned to civilized England, he opened a series of notebooks, including those known to historians as the 'Metaphysical Notebooks'. Here, in private, he tried out what it was like to think as a materialist and to take a fully naturalistic approach to human nature. He had no emotional problem in accepting that humanity is the result of causal physical laws at work over time.

After his return to England, Darwin reread Malthus in 1838, and he creatively integrated his thoughts on political economy, human nature and the material he had begun to assemble on the origin of species. The outcome was the theory of natural selection. Marriage to a sensitive and devout woman, awareness that respectable people correlated evolutionary speculation with political radicalism and his identification with the values of the scientific community, which required theories to be proved by facts, kept Darwin quiet on evolution. He also had a mass of research to do to work through the complexities of his ideas. Political economy continued to feature in his thoughts, and the notion of the division of labour helped him to understand divergence (the differentiation of a range of new species as opposed to the simple succession of one species by another). He kept his eyes open to information relevant to human evolution, like the comparison of human and animal instincts. When he finally went into print in the *Origin*, Darwin was at great pains to argue that natural selection could explain instincts since they were popularly believed to exemplify design in nature. Elsewhere in the *Origin* he remarked that 'psychology will be based on a new foundation, that of the necessary acquirement of each mental power and capacity by gradation. Light will be thrown on the origin of man and his history.'[24]

After a decade of debate about evolution, Darwin began the arduous task of turning his notes and thoughts on man into a book. Widespread acceptance of the evolution of animals and plants encouraged him to think that a public statement of his position on human nature would not prejudice sympathy for his cause. He did not want anyone to impugn his integrity and accuse him of dissimulation about his views, though he thought his new work 'contains hardly any original facts in regard to man'.[25] In the work on human evolution he also followed up in detail two topics of great interest to himself, sexual selection (preferential mate selection as a cause of selective reproduction and hence evolution) and the expression of the emotions. Darwin found it difficult to shape his argument, and the *Descent* never achieved the intensity or command over its materials that made the *Origin* so persuasive a book. His argument was necessarily indirect since there was no record of human evolution. He maintained that if *Homo* differs, for example in intelligence, from animals in degree but not in kind, then it is plausible to believe in human evolution. His strategy was therefore systematically to compare animal and human bodily and mental capacities. The result was anthropomorphism: he read into animal nature what is characteristic of human nature, then used what he found in animal nature to confirm continuity between humans and animals. His opponents, by contrast, defined what it is to be human in terms of such characteristics as the moral soul, which animals by definition do not possess, and thus Darwin's argument was intrinsically unpersuasive to them. All the same, by 1871, many parties – except those committed to dogmatic religious beliefs – accepted the evidence for man's bodily evolution.

The *Descent* began with a physical comparison between man and animals, and it sent readers away to check whether friends and relatives had points on their ears. This was relatively familiar and uncontroversial ground; for example, in the 1860s there was a scientific and a popular literature on what, in the West, was the newly discovered gorilla, and much of the excitement derived from the comparison with man. In two subsequent chapters, Darwin compared the mental faculties of animals and humans. Like Spencer, Darwin argued that there are inherited capacities or faculties, and he set out to show how these faculties – including language, the moral sense and intelligence – are present, in however an elementary form, in animals.

> Nevertheless, the difference in mind between man and the higher animals, great as it is, is certainly one of degree and not of kind. We have seen that the senses and intuitions, the various emotions and faculties, such as love, memory, attention, curiosity, imitation, reason, &c., of which man boasts, may be found in an incipient, or even sometimes in a well-developed condition, in the lower animals.[26]

Reason and language were crucial to Darwin's critics. In his approach to the evolution of these faculties, Darwin relied, like

Spencer, on the analyses of mind that had traced its content to experience. He traced intelligence to learning through sensory experience and language to expressive cries about that experience. This did little to address deeper questions that plagued the analysis of reason, consciousness and linguistic meaning (or semantics). But what it did do, again like Spencer's work, was point to the comparative study of animals and the developmental study of children as the route to a science of psychology. Evolutionary theory brought these topics to the centre of the stage, whereas they had often been marginal interests and sometimes even beneath the dignity of learned men since they do not concern the highest forms of reason. Darwin himself published his observations of his eldest son, William, as a baby. This shift of direction in psychological research was very significant in the twentieth century.

The moral sense received special and separate treatment; Darwin, after all, was a Victorian: 'I fully subscribe to the judgment of those writers who maintain that of all the differences between man and the lower animals, the moral sense or conscience is by far the most important . . . It is the most noble of all the attributes of man . . .' He disclaimed ability to deal with such 'deep' questions, but then boldly discussed the topic because 'no one has approached it exclusively from the side of natural history'.[27] This approach was precisely what his opponents would not allow: in one step it collapsed the distinction between culture and nature. Darwin's discussion was a major contribution to the search in human nature for moral reasons for action. He explained the existence of the moral sense, the Victorian conscience, by the imagined evolution of a social animal that also acquired high intelligence. Social instincts have evolved, he thought, because the chances of an animal reproducing its kind increase if it co-operates with animals like itself in the search for food or in self-protection. In the case of the early human animal, he supposed that herd or family instincts increased the chances of survival of individuals who were physically weak by themselves. These instincts then began, he thought, to be accompanied by reflective intelligence, and this caused the comparison of past actions with present outcomes, such as impetuous selfish conduct with long-term suffering. He suggested that this comparison, accompanied by pain, is the basis of conscience. Intelligence meanwhile, he argued, also gave rise to language, and the shared culture that language makes possible, and conscience therefore became reinforced by custom embodied in public opinion. Finally, the inheritance of acquired patterns of reflective activity from one generation to the next, enlarged moral capacities into a fully civilized moral sense. It is therefore possible, he concluded, for altruistic activity to have evolved by natural selection, a process that on the face of it favours 'survival of the fittest'. As Darwin saw it, however, the fittest person is the one who contributes most to society's welfare – and thus contributes to his own. This argument attempted to reconcile the golden rule – do as you would be done by – with an account of the usefulness of moral actions.

Darwin's tentative thoughts on morality did not provide an authoritative basis for naturalistic ethics – ethics based on what happens in nature. Indeed, people to whom the label 'Darwinist' has been attached reached completely different conclusions about what moral instincts man inherits. Darwin's cousin Francis Galton, for example, believed progress requires conditions to be created in which individuals with innately superior capacities are more highly valued than other people, especially when it comes to the production of children. The utopian anarchist Kropotkin argued that people, like animals, have a social instinct and that they are naturally altruistic; progress therefore lies with the removal of present institutions to allow the spontaneous organization of communitarian life.

From the beginning, the theory of natural selection included an element of comparison between the plant, animal and human kingdoms. Just as Darwin speculated about how natural selection might have produced the particular characteristics of human beings, he speculated about natural selection as a force at work in the contemporary world. He never doubted that *natural* selection continued to be a relevant factor in his age – the most obvious example for Darwin and his contemporaries was white colonial supremacy. Indeed, in Darwin's experience, the clearest directly observable evidence of natural selection at work was in human activity. Between the *Beagle* voyage and the publication of the *Descent*, the indigenous Tasmanian people died out (with the qualification that some interbred with other peoples) or were killed off – they were at one time literally hunted down – and white farmers settled their land. For Darwin and virtually all his contemporaries, this was an example of the struggle for existence. Darwin personally was liberal and humanitarian, an opponent of slavery and the mass destruction of people; but he did not clearly distinguish between the elimination of the Tasmanian people and natural events. He certainly thought struggle was important to man's evolution. 'It may well be doubted whether the most favourable [circumstances for advancement] would have sufficed, had not the rate of increase [of population] been rapid, and the consequent struggle for existence severe to an extreme degree.'[28]

Darwin said little about language and culture when he discussed the origins of human morality. Unlike Wallace, his co-theorist on natural selection, he did not see the evolution of language and culture as a revolution in the evolutionary process. Darwin thought that *natural* selection continued to operate in modern society, though modified by custom, law, morality and religion. Wallace raised but barely followed up the profound idea that the advent of rational consciousness transferred the motor of evolutionary change from nature to conscious human action.

> A being had arisen who was no longer necessarily subject to change with the changing universe – a being who was in some degree superior to nature, inasmuch, as he knew how to control and regulate her action . . . Man has not only escaped 'natural selection' himself, but he actually is able to take away some of that power from nature which, before his appearance, she universally exercised.[29]

He implied that once mind and culture had evolved, the evolution that occurred subsequently was a *human* act. The argument followed, as the social evolutionists discussed in the next section claimed, that culture exhibits new laws of change, laws found in human society and not in nature, and which thus require study by anthropology and not biology.

Darwin never assimilated this perspective and continued to think about human evolution in terms of the selection of individual characteristics, not in terms of the cultural evolution of societies in which individual characteristics may be unimportant. The positions were not so clearly contrasted at the time as this analysis suggests – the first generation of evolution theorists, Darwin included, barely got to grips with these issues – but we can see here one of the sources of the recurrent disagreement in the human sciences about the relationship between culture and nature in human affairs. The clearest case of disagreement later concerned eugenics, the policy of selective reproduction for different groups, which was opposed by reformers who thought social problems stem from social conditions. Supporters of eugenics saw continuity between natural selection and human selection as agents of evolution; supporters of change in social conditions saw discontinuity between natural selection and an ethical reform culture.

It is dangerous to read too much into Darwin's brief comments. Nevertheless, because of his status, both in the late nineteenth century and in the late twentieth century, as a figurehead for those who believe that biological evolution is relevant to political life, it is well to be clear about his position. He did not resolve his own philosophical and moral ambivalence. Darwin's work at times described evolution as the consequence of physical laws with no intrinsic meaning and purpose; in everyday language, men and women exist 'by chance'. At other times, he described human history as progress, and he believed that evolution had in fact given rise to moral progress and would continue to do so. His account of the evolution of the moral sense attempted to reconcile these two positions. The warrant to describe human history as progress derived, however, from culture and not biology, and about culture Darwin had almost nothing to say.

There were two further significant topics in Darwin's natural history of man. In the *Descent*, he hoped to clear up the vexed question of the origin of human races and racial differences.

Evolutionary theory, Darwin believed, put the last nail in the coffin of the ancient belief that primitive people had degenerated from a higher condition. He and Wallace also thought it displaced and made redundant the debate between monogenists and polygenists about whether all races have a single origin: from the evolutionary perspective, it was a matter of how far back one looked. Darwin had high expectations that sexual selection would explain the origin of racial differences as well as secondary sexual characteristics such as features of hair or physiognomy which have no apparent relevance to survival. He guessed that sexual selection explains racial features just as it explains the colouration of butterflies. He also used the theory of sexual selection to suggest how human activity like the appreciation of beauty or enjoyment of dance had evolved. In reality, more than half his book was filled with the evidence for sexual selection among animals, the sort of material with which he was most at home.

Darwin was equally fascinated by the expression of the emotions, the study of which, he thought, is a rhetorically powerful but empirically precise way to compare people and animals. His argument had bite. He traced his interest in the subject to the work of the anatomist Charles Bell (1774–1842), whose beautifully engraved plates explored 'his view, that man had been created with certain muscles specially adapted for the expression of his feelings', i.e., that God designed the face as the outward expression for the soul within.[30] It was therefore a delighted Darwin who compared animal and human expressions – with pictures of a snarling dog and a furious child – and suggested the manner in which expressions are explicable by natural selection rather than God's purposes. *The Expression of the Emotions* was an extended study of physiognomy, the practical art of reading human character that had fascinated earlier generations. In Darwin's hands, this art became part of the physical science he knew as physiology. He called his study physiology because he claimed that expression, like crying or shrugging the shoulders, can be explained by three general principles, each of which is a consequence of the physiological organization of the body. These principles summarized Darwin's speculations about the causes of expressions, which are: habitual movements acquired by the satisfaction or relief of sensations in the past – of the individual or of the species; involuntary actions under the excitement of an opposite frame of mind; and movements which result from an overflow of nervous energy. Thus, to illustrate the last principle, an angry person shakes and waves her or his arms about. Tears, Darwin thought, are a habit acquired by the species from the contraction of muscles around the eyes and the engorgement of the eyeballs by blood during pain; and pain, or the thought of pain, now leads to the threat of tears. His book collected together descriptions of animal expression and a mass of observations of children, savages, lunatics, actors and every-day people to build up a picture of the emotional human animal and its innate repertoire of expressions. He discussed anxiety, despair, joy, love, devotion, sulkiness, anger, disgust, surprise, fear, modesty, blushing and much else besides. The result, a natural history of physiognomy, drew human life firmly into the realm of physiological explanation. He observed human behaviour from the outside and correlated expression with muscles, not with mental meanings.

This systematic study was later much admired as a pioneering work in ethology, the biological study of animal behaviour, and as an anticipation of a psychology that observes human life as the exhibition of purposive behaviour rather than mental purposes. It was a fitting capstone for Darwin's work in which photographs of manic lunatics, a diary of his young son, the coloured rump of the mandrill monkey in sexual display, the killing of indigenous people, male nipples, even the nagging Victorian conscience, were grist to his natural-historical mill.

iv *Social Evolution*

It is a source of ironic comment about the English that Darwin, the unbeliever who portrayed man as a monkey, was buried at the heart of a Christian state in Westminster Abbey. His burial was a sign of the stature that evolutionary theory, and the scien-tific community for which it was the flagship, had acquired by

1882. This stature was not solely due to Darwin; there was also the work of Spencer and Wallace, among others. In fact, much of the evolutionary outlook on the human sphere developed independently of Darwin, and it is difficult to disentangle what was owed to biology, philology, history and anthropology in belief about evolution in the human sciences. The conflation of history and evolution in an explanatory framework for the understanding of human nature was central in large areas of the human sciences until about World War I.

Earlier chapters described the richness of historical thought in the Scottish Enlightenment, in German idealism and philology and in Comtean positivism. In the English-speaking world, a new impetus for history also came in part from a reaction against a wooden approach to human nature in terms of pleasures and pains. It was felt that rules deduced from abstract notions, in the manner in which Bentham argued, go against common sense and experience. The essayist and historian Thomas Babington Macaulay wrote in relation to politics:

> We ought to examine the constitution of all those communities in which, under whatever form, the blessings of good government are enjoyed; and to discover, if possible, in what they resemble each other, and in what they differ from those societies in which the object of government is not attained.[31]

Macaulay aimed his remarks at James Mill: he wanted a practical, comparative and historical experience of human nature as opposed to an abstract science of government. The contemporary description of government and society in the United States by Alexis de Tocqueville, which was also a reflective commentary on the implications of democratic government for cultural values, was the sort of study Macaulay might have had in mind.

In the 1850s and 1860s, Spencer and a generation of anthropologists that included L. H. Morgan, E. B. Tylor (1832–1917), John Lubbock (1834–1913) and Henry Maine (1822–88) equipped the comparative study of ancient and primitive society with a theoretical rationale – evolutionary theory. They supposed that all societies exhibit a common developmental pattern, which can be seen in the growth of institutions like government, marriage and law, and that Western societies are the norm of developmental progress to which other societies can be compared. Evolutionary theory appeared to make rational the classification of any society or institution on a scale from primitive to advanced; it legitimated the comparative method as an approach to the interpretation of exotic customs or beliefs; and it made sense of the history of science itself, as it equated science with rational thought and portrayed the manner in which science evolved from primitive superstition and religion. It deeply challenged religous faith by treating religious customs and beliefs as evidence of the stage that a people has reached. The anthropologists implied that monotheistic Christianity, though advanced as a religion, is only one stage on man's progress towards reason, as Comte had earlier argued. Anthropology made religion a subject of scientific study and in the process altered the authority that religious beliefs themselves could command. A fully sociological treatment of religious belief, that is, a treatment of religion as a social institution rather than a set of claims about truth, came in the work of Durkheim just before 1914.

These theories of social evolution, which were developed independently of Darwin, were especially strong in the English-language world. They focused on the description of developmental stages, at the level of social production and institutions like agriculture and religion, to describe human evolution. In German-speaking countries, too, comparative linguistic and historical scholarship discerned an order underneath human diversity and explained it as progress. German scholars, however, were inclined to look for internal developmental patterns as the cause of change. They saw a parallelism between the development of the individual mind, from childlike sentiments to mature thought and emotion, and the development of culture. Scholars then attributed cultural development to the development of universal mental qualities. This was an approach to anthropology mentioned in an earlier chapter in connection with the work of Waitz and Bastian. Mixed in with this anthropology was a commitment to the inheritance of acquired characteristics as the mechanism of biological evolution, for the effects of mental activity appeared to be inheritable, and this supported belief in the existence of directional forces in nature well into the twentieth century. There were in historical actuality many evolutionary approaches to the human past and a good deal of confused and even fantastic elaboration of what the past was supposed to have been. At the end of the century, the speculative quality of the evolutionary outlook brought it into disrepute with scholars who placed rigour in science above everything else.

The German historical school contributed to the idea of social evolution in Britain through Maine's work on historical jurisprudence. Maine developed an evolutionary approach to the central institution of ordered society in his book on *Ancient Law* (1861), which began life as lectures on jurisprudence, philology and history at the university of Cambridge about 1850. He then went to India, as the legal counsellor of the British administration, taking with him an established framework with which to interpret non-Western forms of social order. He argued that ancient or primitive legal practices are intelligible when they are correlated with an evolutionary pattern of development in society. He wrote: 'The undertaking that I have followed in [*Ancient Law*] . . . has been to trace the real, as opposed to the imaginary, or the arbitrarily assumed, history of the institutions of civilized men.'[32] His 'real' history related the legal detail of property, marriage, caste and so forth to a supposed evolutionary pattern; oddities and apparently arbitrary customs acquired meaning when they were seen to be leftovers from earlier developmental stages. Maine was also a political theorist and was interested in the conditions that make one society rather than another open to change and its institutions susceptible to further evolution. Like virtually all Europeans of the period, he contrasted Europe's openness with the closed societies of India and the East.

What Maine did for ancient law, another lawyer, J. F. McLennan (1827–81), did for marriage. In *Primitive Marriage* (1865), he went further than Maine and drew together historical information and anthropological evidence from what he saw as savage societies. As discussed earlier in connection with the idea of prehistory, such constructions of primitiveness out of observation of past and present peoples were a major input to evolutionary thought. McLennan himself began the systematic study of marriage rituals, and he traced an evolution from primitive promiscuity to polyandry to monogamy. The integration of past and present primitiveness was taken further in Lubbock's *Prehistoric Times* (1865), which paid special attention to European prehistory, and in Tylor's *Researches into the Early History of Mankind* (1865) and *Primitive Culture (1871)*. This was the literature Darwin drew on in his own speculations about early human society in *The Descent of Man*. Tylor, when appointed to a readership at the university of Oxford in 1884, became the first specialist teacher of anthropology in Britain. His approach to primitive culture defined the scope of what became social as opposed to physical anthropology. The study of religion was at the heart of his work, which treated religion as a social institution at a particular stage of evolution. He explored animism, belief that there are living forces in the material world, as a primitive social institution that reflects a lack of advance in objective knowledge. He then discussed the continuation of prescientific modes of thought like astrology or folk medicine through their description as 'survivals' of primitive modes of understanding. Evolutionary history explains why such beliefs exist even though, as he thought, they have no utility or empirical content. Tylor concluded that animism is 'a highly rational theory for men in a low state of knowledge', a sort of failed science, which links events and emotions in a relatively direct way.[33] Progress towards a true science, he supposed, was achieved when Western institutions made possible the growth of rational knowledge.

Darwin and the theorists of social evolution reinforced belief in European superiority just at the time when European countries and the United States scrambled for territory in the rest of the world. Political imperialism, popular culture, Darwin's name and belief in social evolution were closely connected. This is illustrated by the connotations of the words 'savage' and 'primitive', words used to describe certain living people, the past of Western peoples and the continuing, even if repressed, presence of early man within civilized man. Belief in recapitulation, or the biogenetic law that individual development from conception to adult repeats the evolutionary development of the species, enriched the picture. The US psychologist G. Stanley Hall wrote in 1904: 'Most savages in most respects are children, or, because of sexual maturity, more properly, adolescents of adult size.'[34] Savages, primitiveness, childhood and sexuality were linked through the evolutionary laws of nature and not ostensibly as moral categories. The opponents of women's emancipation drew on the same language, linking femaleness to the primitive dimensions of human nature.

Theories of social evolution implied that rational knowledge, i.e., science itself, is a crucial evolutionary development in human culture. In effect civilization was equated with the acquisition of a scientific outlook and scientists were the personification of progress. The comparative, evolutionary method was one means by which Western society constructed a social theory of its own nature. At the same time, this theory represented the value of progress actually held in the West as the natural law of social development. Thus, Victorian values were not added to the human sciences but were intrinsic to the framework of these sciences. Tylor concluded *Primitive Culture* with the comment that 'the science of culture is essentially a reformer's science'.[35] From this perspective, the science of culture is the next step in Western man's evolutionary advance. Many social scientists in the twentieth century thought the same. In this, there was much of the Enlightenment: ignorance and superstition will be vanguished by the light of the scientific day.

v *Functional Explanation*

Philosophers as far back as Plato had likened human communities to organisms and drawn attention to the mutual dependencies, order and unified action of civilized existence and animals alike. Saint-Simon, Comte and Spencer filled out the comparison with new ideas about what gives animals their organization. Spencer then reconceived organization as adaptation produced over time, and he argued that evolution theory explains even the most complex integrated system in terms of natural laws. From this perspective, to understand a complex system, like the social world, is to show how its parts have evolved by interaction with each other in ways that have sustained the life of the whole. Psychological and sociological thought at the end of the nineteenth century substantially adopted this perspective, followed Spencer's lead and, inspired by the theory of evolution, attempted to understand individual and social life through the way parts serve organized wholes. This was the basic orientation called functionalism.

Explanation in terms of function was of overriding importance in the United States, where psychology and sociology developed as disciplines. But a commitment to functional explanation was not restricted to North America – as Durkheim in sociology, Bronislaw Malinowski (1884–1942) in anthropology and Marshall in economics bear witness. The human sciences of functional explanation were rooted in evolutionary theory, but early in the twentieth century there was a self-conscious attempt to impose rigorous standards and to demarcate real science from pseudo-science. The new disciplined commitments ruled out speculative evolutionary reconstructions of the kind attempted in Spencer's generation and argued instead for analysis of the function of parts in wholes directly observable in the present – people or animals in laboratories, or living primitive or advanced societies. Malinowski, for example, taught anthropology at the London School of Economics in the 1920s as a science founded on what can be observed in actually existing small-scale societies. The method of the science was observation by the fieldworker who puts aside prejudice in order to see how a society does in fact work. The anthropologist describes institutions like kinship in order to explain how relations between people sustain the social order as a whole. Malinowski's own studies of the Trobriand islanders off the coast of New Guinea established the paradigm of good practice, though a later generation drew attention to the contrast between what he published and what he wrote in his diary when he observed these people.

In the twentieth century, employment in psychology and

sociology in the US far exceeded that found elsewhere, and – especially after 1945 – there was strong American influence over European science which reinforced a widespread commitment to functional explanation. There was also a relationship between functionalist thought in the human sphere and a political culture, dominated by the United States, which equated the common interest with the interests of corporate capital. Political values supported the view that the activity of the parts of society should be understood in terms of the contribution of the parts to the economic efficiency of the whole, measured by the economic wealth and productivity of the whole. Functionalist analysis also became standard practice in the management sciences: personnel and production were assessed by criteria of efficiency, of parts to whole. Political discussions framed social problems in terms of the integration of people and institutions rather than asking questions about whether there is a common goal and what form it should have. Significantly, the desire to question the functionalist orientation, which was particularly strong in the 1960s, accompanied a revival of interest in continental European social theory more indebted to philosophy than to biology. A renewed interest in Marx's notion of ideology and criticism of the academic social sciences in the 1960s led to the sciences of functional explanation being labelled 'ideology'. Critical sociologists then turned the question of how scientific belief about society relates to society itself into one of the richest points of growth in the human sciences.

The commitment to functional explanations was particularly striking in US psychology at the end of the nineteenth century. The Americans took evolutionary principles of continuity and utility especially seriously. Christian philosophers had earlier conceived of the soul as an agent with cognitive, moral and spiritual qualities, qualities in strong contrast with those of material nature. Locke and those who followed him in tracing the content of mind to experience disturbed this position because they correlated mental processes with material circumstances. Physiology and medicine pressed analysis to move in the same direction. The evolutionist theorists then discussed mind as the outcome of the history of nature. They argued that to understand the mind is to determine its place in the history of nature or, in Spencer's language, in 'the continuous adjustment of internal relations to external relations'.[36] This argument was also notable because it compared phenomena labelled 'mental' with phenomena labelled 'physical' as functions of the same complex natural systems – animal or human. In the light of evolutionary theory, it appeared possible to reformulate the philosophical problem of the mind's relation to the body as an empirical problem about the past and present relation of levels of complexity in natural functions. Similarly, it appeared possible to reformulate the problem of the relationship between nature and human culture: culture can be redescribed as the latest, most complex and human stage of nature's evolution of organized wholes. Just as Marx redirected attention from people's conscious beliefs to the economic structures that cause those beliefs to be held, and as Freud later redirected attention from conscious to unconscious determinants of action, so functional explanations turned attention away from subjective reports on the conscious meaning of events to objectively observable functions of actions.

These arguments were of deep significance for the twentieth century. In day-to-day terms, however, abstract vistas about the function of culture in evolution or mental activity in human life translated into detailed empirical work. Such work informed a new comparative and developmental psychology, which involved studies of animals and children rather than the analysis of the rational adult mind. James Mark Baldwin (1861–1934), who founded the *Psychological Review* (with James McKeen Cattell) and who headed psychology laboratories at the university of Toronto, Princeton and then at Johns Hopkins, wrote in 1894: 'The new functional conception asks how the mind as a whole acts, and how this one form of activity adapts itself . . . The mind is looked upon as having grown to be what it is, both as respects the growth of man from the child, and as respects the place of man in the scale of conscious existence.'[37] Functionalist psychology attended to what people *do*, how they behave in the world in which they live, rather than what they think. It was an approach of obvious value to psychologists who wanted to work with animals, since it is hardly possible to ask animals about their thoughts. In many psychology texts at the end of the nineteenth century, such as those written by the British philosopher-psychologists James Ward (1843–1925) and G. F. Stout (1860–1944), psychology was defined as the study of consciousness, and this appeared to make the study of animals, or indeed babies, irrelevant. Evolutionary theory in general, and functional explanation in particular, fostered a different view of what the subject matter of psychology should be. This latter view, which made it possible to have a psychology of animals, became dominant when US psychologists wanted psychology to be a natural science. A commitment to functional explanations which avoid saying anything about consciousness provided the means. Baldwin, as it happens, also brought the behaviourist John B. Watson to Johns Hopkins, and Watson, initially an animal psychologist, reduced the whole of psychology to a study of the functions of behaviour. The commitment was also eminently practical in a social context in which no sharp line was drawn between description and prescription of an individual's integration into the social whole. When they directed attention to the adaptive processes of mind and behaviour, psychologists were able to persuade patrons that they could provide answers to human and social problems. As discussed in later chapters, this was crucial to the large-scale growth of twentieth-century psychology.

William James (1842–1910), the elder brother of the novelist Henry James and son of a wealthy Boston family, was a pivotal figure. His intellectual and emotional life was a struggle to accommodate the truths of physiological science and spiritual ideals in an evolutionary outlook. After an unusually diverse education, much of it in Europe and which included the qualifications for a medical degree, from 1875 he taught a course in psychology at Harvard University. He used Spencer's *Principles of Psychology* as his text. James drew together the threads of psychological experimentation current in Germany, physiology, medicine and functionalist forms of explanation derived from the theory of evolution. Aided by a colourful style and vivid use of metaphor, he produced what for some psychologists remained the masterpiece of their subject, *The Principles of Psychology* (1890). Thereafter, James was more preoccupied by philosophical matters, and he turned to a search for functional explanations in the light of evolution theory into a theory of knowledge called pragmatism. His sympathy with experience as a form of truth if it acts as a guide to life – even if the experience involves apparently unscientific beliefs – led him to studies of psychical phenomena (he worked with the Boston medium Mrs Piper) and to a major study in the psychology of religion, *The Varieties of Religious Experience: A Study in Human Nature* (1902).

Around 1880, James contributed to a lively British debate about what the evolutionary principle of continuity means for the relation of mind to brain. His position contained the seeds of much of his later philosophy. He reasoned that if conscious mental activity is the outcome of an evolutionary process, then it must – like a bird's plumage or any other aspect of an animal or plant – have some adaptive function. It could therefore not be correct to treat consciousness as an epiphenomenon, an adventitious quality and causal irrelevance in nature. This argument also tied in with James's own deep personal need to believe that his choices, indeed his life, made a difference. He therefore opposed Huxley who, carried away by the sound of his own metaphor, likened consciousness to the sound of a bell when it is struck: 'The soul stands related to the body as the bell of a clock to the works, and consciousness answers to the sound which the bell gives out when it is struck.'[38] Instead, James supposed that consciousness has consequences, and he struggled, at one stage, to describe this quality of consciousness as 'conscious interests', a purposeful dimension to our conscious life that makes a difference: 'Mental interests, hypotheses, postulates, so far as they are bases for human action – action which to a great extent transforms the world – help to *make* the truth they declare. In other words, there belongs to mind, from its birth upward, a spontaneity, a vote.'[39] He also described this quality of consciousness as selective attention. 'But there is one thing which [consciousness] *does* . . . and which seems an original peculiarity of its own; and that is, always to choose out of the manifold experiences present to it . . . and to ignore the rest.'[40] James hoped that it is possible to take seriously human values and purposes as real forces in the natural world and thus, so to speak, to re-insert values and purposes into the fabric of nature from where they had been removed by mechanistic science. He was not entirely cogent, but his arguments contributed to the important new philosophy, pragmatism, a way of thought in which the criterion of what is true is set by judgement about the consequences of a claim for our actions. Whatever James's philosophy, he encouraged himself and others, though fully persuaded by the power of natural science, to think that science does not destroy our belief in our conscious agency. Indeed, as he and other psychologists who used functional explanations argued, an evolutionary perspective shows that scientific achievement is the adaptive means by which a conscious humanity turns the cosmic process towards mankind's well-being. This was the intellectual's version of the American Dream, the dream that everything is possible.

James was not the only scientist to argue that human consciousness makes a difference to the evolutionary process. The point was made earlier by Wallace, and two other psychologists in the 1890s took up the theme: C. Lloyd Morgan (1852–1936) and Baldwin. Morgan, who became principal of the university of Bristol, was well known for his philosophy of evolution. He was later remembered because it was believed that he had provided the comparative psychology of animals with a rigorous methodological principle, 'Lloyd Morgan's canon', the aim of which is to advance mechanistic explanation and to eliminate the anthropomorphic form of argument found in Darwin's comparative work. Morgan certainly was encouraged by evolutionary theory to undertake comparative studies. In a much-quoted passage, he wrote: in no case is 'an animal activity to be interpreted as the outcome of the exercise of a higher psychical faculty, if it can be fairly interpreted as the the outcome of one which stands lower on the psychological scale'. Understood in context, Morgan was arguing for the continuity of mental qualities in evolution rather than laying down a rule that psychologists should accept the simplest, mechanist explanations, though many psychologists interpreted the passage in the latter sense. Indeed, Morgan believed that animals experience 'the effective use of the consciousness to which they are heirs' and that consciousness affects the process of natural selection. Experience, he thought, alters behaviour and behaviour is selected for. On this basis, like Darwin before him and like Baldwin, he assumed that animals, even those low down the evolutionary scale, possess adaptive consciousness. Morgan argued that 'our psychological interpretations are inevitably anthropomorphic', that human experience necessarily provides the terms in which we understand animal behaviour.[41] Thus, Morgan's work, like James's, exemplifies the way in which the theory of evolution and the commitment to functional explanation encouraged descriptions of mental qualities as active forces in nature.

The work of two other philosophers and psychologists later stood out for the quality of theoretical argument, though it barely touched day-to-day psychological research at the time. The first was John Dewey's paper on 'The Reflex Arc Concept in Psychology' (1896), which redescribed the central unit for the mechanist analysis of the nervous system, the reflex, as a process. Dewey (1859–1952) argued that the structural elements of the reflex – sensation, central integration, motion – are not distinct and that a proper understanding of the reflex lies with the description of its contribution to the adaptation of the organism as a whole. Similarly, he thought, mental elements cannot be considered independently of the integrated activity of which they are part – the action that gives rise to them and the action to which they contribute. Parallel criticisms were made in the 1920s of the way the analysis of behaviour then treated behaviour as if it consisted of causally linked stimuli and responses rather than as a continuous process.

The second contribution was a series of papers by George Herbert Mead (1863–1931). In 'Social Psychology as Counterpart to Physiological Psychology' (1909), he argued for a social psychology symmetrical with physiological psychology – the recognition that mental activity is dependent for its content on its social nature as much as on the physiological body. In particular, Mead attacked current opinion that imitation, at base a physiological process, is the means by which animals or children become part of a group. Instead, he argued, the existence of a developed self and knowledge of others is a precondition of imitation, and this precondition results from a reciprocal interaction of stimuli and actions, that is, from a social process prior to the differentiation of the self: 'the conduct of one form is a stimulus to another of a certain act, and . . . this act again becomes a stimulus at first to a certain reaction, and so on in ceaseless interaction'. Thus, 'consciousness of meaning is social in origin' and 'reflective consciousness implies a social situation which has been its precondition'.[42] Mead implied that it does not make sense for psychologists to describe their subject matter, mental states, as if they are entities with a natural existence independently of the social whole within which, as he argued, they are social processes, that is, human evolutionary processes. This argument challenged the basis on which most experimental psychology was then carried out, but it was virtually ignored.

Dewey's and Mead's papers illustrate how attention to the theory of evolution led to new approaches to the mind; indeed, they replaced the language of 'the mind' with a language that refers to activity as a process and to the activity's adaptive or maladaptive qualities. It was no coincidence that Dewey and Mead were both familiar with German philosophy, as well as the British theory of evolution, and with its approach to mind through the manner in which mind is formed in a historical process. Dewey and Mead were also both men with high social ideals who looked to their own academic occupation to provide leadership as society evolved and adapted, as they trusted it would, into an enlightened democracy. Dewey went on to become the most influential philosopher of education of his generation, first as chairman of the department of philosophy, psychology and education at the new university of Chicago and then, over three decades, at Columbia Teachers College in New York. Mead was also socially involved in Chicago. Subsequently he gained a retrospective reputation as one of the most productive theorists of a genuinely *social* psychology (a reputation discussed further in relation to the discipline of social psychology).

Between 1830 and 1900, the status of evolution changed from

being a glint in the eye of radical natural historians to being the *raison d'être* of progressive democratic society. What was at one level a debate about the empirical evidence for the origin of species by descent with modification, was at another level about the categories in terms of which to understand what it means to be human. The evolutionary outlook fostered scientific explanation by reference to functions, and this underwrote in intellectual terms a desire to take action on behalf of the future adaptation of humanity. Students of human science did not just believe in evolution but believed in human science as evolution at its highest stage. This theoretical discussion now needs to be matched by a historical account of how psychology and sociology did in fact become institutionalized disciplines, even if they did not achieve unity of outlook, in the half-century before 1914.

14

The Academic Disciplines of Psychology

> What is natural science, to begin with? It is a mere fragment of truth broken out from the whole mass of it for the sake of practical effectiveness exclusively. *Divide et impera.*
>
> William James, 'A Plea for Psychology as a "Natural Science"' (1892)[1]

i *Diverse Origins*

Origin myths create a sense of identity, and this is as true for a scientific as for any other community. A group which struggles to establish itself, whether an oppressed nationality or a science with little institutional standing, may particularly emphasize a moment of birth and a founding father. This happened with psychology, especially in the United States. The psychologist Edwin G. Boring (1886–1968), when he wanted to defend psychology's status as an academic science, published an influential history, *A History of Experimental Psychology* (1929, revised 1950), with a father-figure as its centrepiece. This figure was the German professor of philosophy Wilhelm Wundt (1832–1920), who in 1879 equipped a small room in the university of Leipzig with instruments and created an experimental laboratory to train research students. His use of an experimental method is a symbolic moment for psychology conceived as a natural science. In this origin myth, it is perhaps appropriate that the birth gave life to a method and not to knowledge, as much twentieth-century psychology became preoccupied by methodological rather than substantive issues. Further, the myth portrays psychology with the pursuit of knowledge as its mother and the university as its midwife. There is no hint that ways of life, modern social administration or new conceptions of the self created psychology and its subject matter. Yet seventeenth-century debates about economic self-interest, eighteenth-century writings on civic virtue, or nineteenth-century theories of the expressive self – or of that self's alienation – might be said to have done just that.

Psychology has no 'origin': its identity in the twentieth century is fragmented and its roots are diverse. Psychology is a cluster of activities with a family resemblance but no common identity, and the field undoubtedly has multiple origins. It was sometimes a division of knowledge before the nineteenth century – in Wolff's rational philosophy or in Bonnet's analysis of organic experience – but there was no body of activity conformable with what modern psychologists think or do. Something changed in the second half of the nineteenth century, enabling psychology to flourish in the twentieth century on the scale and with the many identities it then acquired. That twentieth-century history is described in the last part of the book. This chapter discusses the academic dimension of the background, while the broader occupation is explored later. The division is an unclear one. Psychology, like sociology or economics, has simultaneously been an academic subject and a profession which offers services to society at large. Psychologists themselves have continuously debated and re-thought this relationship.

By 1903 there were at least forty psychology laboratories in the United States; the subject awarded more doctorates than the other sciences, except for chemistry, physics and zoology; a professional society, the American Psychological Association (APA), was founded in 1892; and there were several specialist journals, headed by the *American Journal of Psychology* (f.1887) and the *Psychological Review* (f.1894). None of this existed in 1880; at that time, only James at Harvard taught scientific psychology and used some laboratory facilities. Psychology therefore grew remarkably rapidly as an academic discipline. But two qualifications should be borne in mind: firstly, even in the US, there was divergence of opinion about what psychologists should do; secondly, nowhere else was there anything like this growth of an institutionally distinct subject, even where there was similar activity in psychology. The second point is true even for Germany, from where much of the inspiration and training for the American discipline had come. The situation in other countries, like France, Russia or Belgium, needs separate discussion.

The German situation was complex because, by 1900, there was much specialist activity in psychology but few positions reserved for psychologists. What gave the German-language world its initially commanding position in the field was a university culture committed to rational scholarship as a value in its own right and, by 1850, a large institutional investment in the natural sciences, including experimental physiology. Students from Moscow to Chicago came to Central Europe to train, and they took home academic ideals and practical experience of academic organization. This occurred in the area of experimental psychology in the last two decades of the century, in the way in which it had occurred earlier with physiology, and many foreign students studied with Wundt and looked up to him as an example. In Leipzig, however, as elsewhere in Germany, experimental psychology was part of larger philosophical project to provide foundations for rational knowledge. This project did not always survive when students translated their training in experimental methods to other settings. Psychology elsewhere did not always resemble psychology in Germany. This was notably true for the new subject in North America.

The German-language world did not have a monopoly on psychology, though it did have an academic infrastructure able to support experimental activity on a scale not possible elsewhere – until the US established laboratories in the 1890s. Psychology differentiated as a separate subject area in other places, but it did so piecemeal and with distinctive local characteristics. In Britain, Bain's and Spencer's reformulations of the psychology of experience were individual achievements independent of academic support. Bain himself sponsored the journal *Mind* (f.1876), which did not distinguish between psychology and philosophy in its early years but was increasingly dominated by philosophy, a discipline which – unlike psychology – had an institutional base. In 1892, C. S. Myers (1873–1946) extracted funds for psycho-physical apparatus to use in the existing physiological laboratory at the university of Cambridge. An earlier request met with academic and ecclesiastical resistance, even though all that Myers hoped to do was to bring England into line with current German practice and to study perception as a problem in natural science. From 1912, Myers directed a new psychological laboratory which he funded largely from his own wealth, and during the 1920s and 1930s Cambridge became a major centre under his chosen successor, Frederic C. Bartlett (1886–1969). In London, when psychology became a subject during the first decade of the new century, the major inspiration came from Galton, and the chief interest was in individual differences.

The universities in France were, for much of the nineteenth century, examining boards rather than centres of training and research; it was the *grandes écoles* and institutes that had these functions. Aspects of the German university model entered into the educational reforms begun in the 1860s and continued in the Third Republic after France's defeat by Prussia. The reforms, which extended downwards through the school system, were a vehicle for political liberals with secular aspirations for a state involved in a struggle at every stage with embedded Catholic interests. A determinedly secular graduate of the Ecole Normale Supérieure, Théodule Ribot (1839–1916), turned to both British and German activity to find support for a scientific as opposed to a Catholic approach to psychology. He published extensively and his books, which were widely read, promoted acceptance of a distinct science of psychology. *La psychologie anglaise* (*English Psychology*, 1870) introduced British thought, including that of Bain and Spencer, to a French audience in opposition to *spiritualisme*, the Christian idealism then dominant under the legacy of Cousin's patronage, in French philosophy and education. 'The new psychology differs from the old in its spirit: it is not metaphysical; in its end: it studies only phenomena; in its procedure, it borrows as much as possible from the biological sciences.'[2] In the same year, 1870, Hippolyte Taine (1828–93), later famous for his history of France (a significant source for crowd psychology), also attacked French conservative philosophy with a book on British psychology, *De l'intelligence* (*On Intelligence*). The general character of the book, rather than its detailed content, was a call for a secular science of mind.

Ribot and Taine were not much interested in either Darwin or Galton (though Ribot translated Spencer). When Ribot discussed human differences he used clinical examples and medical concepts of abnormality rather than the theory of evolution and variation. This was characteristic of the French style of psychological work, which was oriented towards clinical case studies and to individuals as examples of psychological abnormality. The subject matter of psychology was constructed in close connection with medicine, and French theories of the abnormal mind were modelled in the light of medical views of disease. Ribot promoted the child, the primitive and the madman as the three ideal subjects of a scientific psychology. What separates these subjects from an ideal of mature reason, he believed, gives the psychologist the basic form of the psychological experiment: 'the morbid derangement of the organism which produces intellectual disorders; anomalies, monsters in the psychological order, are to us experiences prepared by nature, and all the more precious as the experimentation is more rare'.[3] His three books on the illnesses of the memory, the will and the personality (published between 1881 and 1885) were studies in the natural experiments of pathology.

Ribot's success in the establishment of the new psychology in France was marked by the establishment of a chair in Experimental and Comparative Psychology at the Collège de France in 1887, to which he was appointed in 1888. Ribot was eclectic and did not undertake original research, but his publications and position as editor of the *Revue philosophique* made him a pivotal figure in the encouragement of a natural-scientific psychology. He advised Alfred Binet (1857–1911) early in the latter's career, and Binet went on to found the journal *L'année psychologique* in 1894, to direct the psychological laboratory (founded in 1889) at the Sorbonne and to formulate the tests that were the beginning of the measurement of intelligence. Ribot was also connected socially and intellectually with the famous neurologist, J.-M. Charcot (1825–93), whose descriptions of hysterics, hypnotized subjects and religious ecstatics were part of a secular campaign against what these Parisian intellectuals saw as ignorance and superstition.

A fascination with abnormal mental states continued to encourage an interest in scientific psychology in France. The psychologist and physician Pierre Janet (1859–1947), who followed Ribot in the chair of psychology among the academic élite at the Collège de France, remarked that the chair existed only because of cases of multiple personality. French conservative philosophers believed in the transcendental self, a reality that could not be the subject of scientific psychology. When Janet suggested that cases of multiple personality proved that there could be more than one self, then, so his point went, psychology achieved status as a scientific subject. The impact of this was made clear by Ribot, who wrote: 'It is but natural that the representatives of the old school, slightly bewildered at the new situation, should accuse the adherents of the new school of "filching their ego" . . .'[4] Janet's own reputation was built on the extended researches, published as *L'automatisme psychologique* (*Psychological Automatism*, 1889), he made in Le Havre on a subject, Léonie, who exhibited remarkable powers when hypnotized as well as a multiple personality. When the first International Congress of Psychology was organized in Paris in 1889, Charcot was persuaded to be its president, and one-third of the papers presented were on hypnotism. Then, in reaction against this situation, and particularly in reaction against the association between psychical research and psychology, psychologists called for a congress on experimental psychology. The debt of organized psychology to an interest in psychical questions was not, however, unique to France; the first society for experimental psychology founded in Berlin in the 1880s, for example, had the purpose of psychical research.

The situation was in some respects comparable in Italy. Supporters of modernization and of the state newly unified in 1870, many of whom were based in Milan or Turin, turned to positivist natural science as a system of belief with which to combat Catholicism. The new professor of legal medicine and public hygiene in Turin in 1876, Cesare Lombroso (1835–1909), launched a fully natural-scientific and determinist programme to deal with social questions. He collected information on what he believed to be innate biological differences between people in Italy, he distinguished a physically distinct criminal class, and he and others devised complicated apparatus to measure differences in head size and physiognomy. He campaigned for legal science to become a branch of natural science, with the implication that an expert scientific administration would then deal with crime as a question of pathology. During the 1890s, debates about degeneration between Lombroso's school, which stressed hereditarian determination, and French doctors and social scientists, who gave more credence to environmental determination, filled the first international conferences on criminology. Topics such as these, as well as some awareness of experimental psychology, attracted students to study human nature as a natural science. Lombroso himself went on to become, in 1906, professor of criminal anthropology in Turin.

The Netherlands, Denmark and Sweden were much influenced by the German academic system, and where there were well-established universities scholars trained in Germany were sometimes able to use the existing institutional base to found psychology as a new scientific speciality. The professor of philosophy in Groningen, Gerard Heymans (1857–1930), founded the first psychological laboratory in the Netherlands in 1892. His appointment was in 'The History of Philosophy, Logic, Metaphysics and the Science of the Soul'; the last of these topics was also taught elsewhere in the Netherlands, but Heymans was distinctive because he treated it as an empirical or experimental as well as a philosophical subject. Special chairs were founded in the 1920s, but – unlike the United States – the system of educational provision gave no basis for rapid expansion because educational theory was dominated by pedagogues committed to a philosophical, moral and religious conception of their occupation. After 1945, however, there was a spectacular growth in psychology and the social sciences generally, attributable in part to the Dutch public's readiness to accept expert guidance in life as a substitute for religious guidance. Alfred Lehmann (1858–1921), who visited Leipzig in order to familiarize himself with psychophysical apparatus, brought Wundt's experimental methods to the university of Copenhagen in 1886. Though he was initially encouraged to relate his work to philosophical questions, he preferred to focus on more narrowly defined psychological topics such as recognition, a problem in perception that he linked to British views about the association of sensory elements.

In other parts of Europe, many liberals put hope in a combination of national liberation or revival, expansion of free economic activity and modern scientific learning. The educated élites often studied in the German universities, acquired specialist knowledge of subjects like experimental psychology and then returned home to spread a scientific approach to social questions. Much of Europe in fact remained rural, remote and economically feudal even at the end the nineteenth century, with only local industrialization. The young men and women who did study abroad, from towns like Helsinki (then Helsingfors, the administrative centre of a Russian princedom) or Bucharest, were able to establish new subjects like psychology only to the degree that local circumstances permitted. To lecture on experimental psychology or on the relevance of psychology to education was one way to push open windows to modern ways of thought. There was also interest in the collective psychology of people, which tied in with attempts to assert the distinctiveness and worth of separate national cultures. An audience for German *Völkerpsychologie* was created. After 1918, with the political independence of countries like Romania, there were opportunities for some institutional growth connected with the practical needs of education and modern commercial operations. In conservative and Catholic Spain, the interest of a few intellectuals in psychology was bound up with the question of Spain's lack of openness towards European culture and philosophy in general. 'Men of letters' took the lead in the attempt to create closer contact. But those who were interested in scientific psychology had little institutional base even at the end of the century.

The Catholic response to the natural sciences and to the associated new psychology was not always negative, especially north of the Alps. After the church's retrenchment against modernity in all its forms in the 1860s, a new pope, Leo XIII, in the encyclical *Aeterni Patris* in 1880, asked for science to be loved for its own sake as a form of truth and asked for a revival of Thomist philosophy (the medieval legacy of Saint Thomas Aquinas) as a source of intellectual leadership. Within this framework, the Belgian scholar and the figurehead of Belgian dignity under the German occupation in World War I, Désiré F. F. J. Mercier (1851–1926), later Cardinal Mercier, established a Thomist institute at the university of Louvain or Leuven in 1889. Here he fostered teaching in modern science, which included teaching experimental psychology. Armand Thiéry and later Albert Michotte became directors of the Louvain laboratory, the first in Belgium, and they supported both precise experimental work and the connection of psychology to pedagogy. Mercier himself wrote general texts that developed a Catholic philosophical anthropology, which defined psychology as 'the philosophical study of life in man' and attacked the supposed mechanist legacy of the science attributed to Descartes. Without a philosophy of the soul, Mercier argued, psychology 'is not a separate science, but only a page of mechanics or of physiology'.[5]

Colonies of Russian students existed in Central Europe in the second half of the nineteenth century. Russia itself was divided between medieval Tsarist absolutism – a form of government legitimated by divine right – and modernization, present in the struggle to establish an industrial economy and to administer a state able to compete on the world stage. An administrative, professional and commercial class possessed considerable education but no political power. In as far as it committed itself to modernization, this class looked to the example of Western European politics and German science. There were concerted efforts in Moscow and St Petersburg to establish institutional bases from which to apply scientific expertise in psychology and the social sciences to society's problems, notably crime. Rivalry or at least a lack of co-ordination between different centres, as well as an element of constraint in public life, caused developments to be patchy. Nevertheless, an institutional base for scientific psychology and for its application was firmly established before 1914.

The origins of psychology as a discipline, therefore, were diverse and complex. The discipline, where it did exist by 1900, had no single identity, and much activity in psychology continued to go on outside anything that can properly be called a discipline. Those interested, however, all looked to Germany, where there appeared to be a blueprint for psychology as a rigorous science. Yet even in Germany there were different claims about the form this science should take as well as developments in psychology that grew out of social, medical and educational questions rather than academic debate.

ii *Wilhelm Wundt and Experimental Psychology*

Elevated values of scholarship made the German philosophy faculty into the home of the humanities and the natural sciences and sustained academic careers relatively free from external pressures. In the first half of the nineteenth century, various scholars were interested in psychology as an area of investigation, but unlike, say, philology, it had no institutional existence as a discipline. Herbart argued at the university of Königsberg for psychology 'as a science', but, although later German academics cited and drew on his work as psychology, his occupation was in philosophy and pedagogy. E. H. Weber at the university of Leipzig experimented on touch discrimination under the auspices of physiology, and at the time his results appeared to be ungeneralizable to wider psychological activity. Society offered no support for psychology as a separate discipline. Nevertheless, philosophy and experimental activity committed to the same values and carried out in the same institutions made possible what was sometimes called 'the new psychology' at the end of the century. As Wundt's career exemplifies, the new psychology resulted from belief that scientific methods lead to better philosophy and not from an attempt to replace philosophy with science as an approach to mind. The natural sciences gained enormous prestige during the century, often at the expense of philosophy; but German natural scientists all studied philosophy and many retained an ideal of the integration of knowledge in a philosophically cogent synthesis. This was certainly true for Wundt.

Wundt was a medical student in Heidelberg in the early 1850s. Even then, but especially after he studied with the physiologist Du Bois-Reymond in Berlin, he saw a medical career as a poor substitute for academic research. He spent some fifteen years as a university teacher in Heidelberg, published textbooks and wrote on psychological, physiological and philosophical topics, while he hoped to be called to a chair. He was also active in Baden state politics as a liberal representative from Heidelberg. For some of these years he was an assistant to Helmholtz in the latter's physiological laboratory. In fact Wundt and Helmholtz had little constructive contact, partly due to Wundt's prickly need to establish himself as an independent scholar; yet Helmholtz was the most brilliant instance of an experimental physiologist to use his expertise to equip psychology with a precise content and a rigorous method. He and his contemporaries in the 1850s and 1860s laid the basis for what became the new experimental psychology. Their interest was in perception, and they took from physiology the terms for an analysis of mind into its elementary constituents. The next generation of researchers, who called themselves psychologists, endeavoured to report more directly on what was in the mind and to create a subject matter for psychology independent of the subject matter of physiology. The new psychology claimed to be an autonomous area of knowledge, but it was physiology that got it launched in the academic setting.

Helmholtz wrote a deeply influential theoretical paper (1847) on the conservation of energy, the principle that underwrote contemporary belief in the continuity of organic and inorganic systems, and experimental papers (1850–52) on the speed of conduction of the nervous impulse. The latter work suggested one measurable physical correlate, time taken, for mental processes. Helmholtz subsequently published a massive and authoritative three-volume *Handbuch der physiologischen Optik*

(*Handbook of Physiological Optics*, 1856–67) and a book on *Tonempfindungen* (*Sensations of Sound*, 1863). They demonstrated how experimental methods and physiological concepts applied to the psychological topics of visual and auditory perception, and became classic works in their field. Scientists were conscious, however, that the psychological topics on which Helmholtz worked were those most directly approachable with physiological concepts and methods. It was not at all clear that topics like thought or emotion could be studied in the same way.

Experimentation on sensory perception was linked to the philosophical theory of the sources of knowledge by Helmholtz's work and his debate about vision with another German physiologist, Ewald Hering (1834–1918). Some experimentalists believed that their work was a natural-scientific answer to the questions bequeathed by Kant about the terms in which the mind knows the world. Physiological studies of perception raised difficult questions about the relation of physical and mental events and the usage of physical and mental language. The authority of conscious awareness as a source of data, and the independence of psychology as a subject separate from physiology, appeared to depend on the acceptance of a mental reality. Yet natural science concepts and methods seemed to presuppose that the mental sphere is in some sense dependent on the physical sphere. Helmholtz and Hering disagreed about whether inherited organization within the brain structures what is perceived. Helmholtz argued that perceptual experience is reconstituted anew in each person, that is, learned by the way the mind forms elementary sensations into a unity. He argued, to a degree influenced by Kant, that perceptions involve central psychological processes in addition to the sensory data of the particular experience, and he called the mental processes 'unconscious inference'. He believed, for instance, that the visual perception of depth is an acquired perception achieved by mental unconscious inference about sensory information provided by binocular vision (in which the two eyeballs move slightly in order to focus on objects at different distances). His position was therefore described as empiricist. Hering opposed all this and argued that an inbuilt structural organization explains why perception is shaped as experiments show it to be. His position was therefore described as nativist. This was a debate of great complexity and involved detailed programmes of research, some of which lasted the researchers' lifetimes. Experimenters became increasingly wary of explanations in terms of psychological processes (like unconscious inference) that lie beyond what experiment actually reveals. The social outcome in general was that psychology, in this area, became a specialist and very precisely focused science. The Helmholtz-Hering debate, which continued in altered form into the 1920s and beyond, exemplifies how philosophers and physiologists constructed psychology as an academic natural science subject. In spite of philosophical differences of opinion, the laboratory became the site where differences were to be resolved. Younger scientists, who were involved in an intense competition to obtain permanent positions, used the laboratory and their experimental training to make their mark. The hope that studies of perception would also address matters of dispute in philosophy made such research doubly attractive. It brought together rigorous methods with the ultimate *raison d'être* of the university.

The same conjunction of experimental work on perception with philosophical debate was evident in the area called psychophysics after the publication of Gustav Theodor Fechner's *Elemente der Psychophysik* (*Elements of Psychophysics*, 1860). Even among German professors, Fechner (1801–87) was an unusual man. His career took him to a chair of physics in Leipzig, where he wrote satirically about natural philosophy under the pseudonym 'Dr Mises'; he proved, for example, that as angels are the most perfect beings they must be spherical. He then temporarily ruined his sight through optical experiments, went through a spiritual crisis and resigned his chair in 1839. On his recovery he criticized materialism and developed a worldview which held the mental and the physical to be alternative representations of one reality. In *Zend-Avesta* (1851; an untranslatable title that connotes 'revelation'), Fechner described the universe as a being with consciousness. At the same time he sought to study scientifically the unity of the material and the mental. He thought this could be done through the correlation of measurable physical changes with mental changes, such as changes in a light's brightness with the mental discrimination of a change in brightness. The quantification of correlations between physical changes in a stimulus and mental changes in conscious perception became psychophysics, and Fechner's book on the subject prompted considerable research. Scientists were excited by the topic because Fechner put forward a specific law of correlation, that could be tested and corrected, and a general programme to make mental events subject to precise description through their correlation with physical changes subject to objective measurement.

Fechner's method of measuring sensory phenomena involved the study of thresholds, a technique also pioneered in Leipzig by Weber, to calculate the amount of change in a physical stimulus which becomes noticeable to an experimental subject. He hoped by this means to deal with what was later known as the stimulus error, the conflation of the measurement of a physical stimulus with the measurement of a conscious sensation. By correlating changes in sensation with changes in physical stimuli, he devised a law which stated that sensation is proportional to the logarithm of the stimulus intensity. Few people were satisfied with Fechner's result and, after about 1870, it was widely argued that psychophysical measurement referred to mental judgements, not to sensation *per se*. But Fechner launched a topic that shaped scientific psychology for a generation because it permitted the establishment of precise techniques. With institutional support, these techniques became the substance of a self-perpetuating specialist discipline; in the process, questions that could not be formulated in quantitative terms tended to get deleted from the agenda. Fechner himself subsequently turned to pioneering work in experimental aesthetics.

The attempt to find measurable correlates for mental processes was also characteristic of reaction-time experiments, which were influenced by a paper by the Dutch researcher on perception F. C. Donders (1818–89) in 1862. The background of the interest in reaction times was in astronomy. Since the end of the eighteenth century astronomers had been aware that individual observers record the time of the passage of a star seen in a telescope's eyepiece slightly differently; the variation appeared to be regular for a particular observer. The phenomenon was initially discussed as observational error but then re-analysed as a personal equation, an individual variable in the sensation of a visual stimulus. Donders and other researchers concluded that the passage of time might provide a measurable parameter for mental events; in this respect, the hope placed in reaction-time research paralleled that placed in psychophysics. The instrumentation and technicalities of measurement absorbed attention and created a specialist expertise; it was another matter to establish clear and authoritative ideas about *what* was being measured. Yet precisely such activity constructed psychology as a natural science.

Instruments had an important role in these studies. There was a close relationship between what instruments made it possible to observe and measure and what sort of experimental work was undertaken, either in physiology or psychology. Helmholtz, who worked in physical as well as physiological optics, developed the ophthalmoscope to look at the retina at the back of the eye. In the 1840s a watchmaker called Mathias Hipp made a chronoscope, an instrument able to measure time intervals to a thousandth of a second; reaction-time research was inconceivable without this precision instrument. The ability to use such equipment separated the scientist from the armchair psychologist. Indeed, instrumentation was such a feature of the new psychology that later observers referred to 'brass instrument psychology', implying in the eyes of sceptics more of a compliment to the proficiency of German engineering than to the psychological results. In the drive to achieve the objectivity of natural science in psychology, the use of precision instruments acquired almost totemic significance. Their use made psychology a scientific occupation even if it did not create scientific knowledge. The major exception to this stricture is undoubtedly the area of perception, where precision tools were appropriate for the questions asked.

Wundt's own early experimental research was on the physiology of muscles and nerves. His ambition went far further, however, and in his *Vorlesungen über die Menschen- und Thierseele* (*Lectures on the Human and Animal Mind*, 1863) he discussed mind in relation to a developmental view of organic nature. Academic success followed the appearance of his *Grundzüge der physiologischen Psychologie* (*Principles of Physiological Psychology*, 1873–4) with his call in 1875 to one of three prestigious chairs of philosophy in the university of Leipzig. The appointment was offered to Wundt because the Leipzig philosophers were concerned that their subject should engage in depth with the pace-setting world of the natural sciences. Wundt responded over the next forty years with what in intent but not in practice was a unified philosophy enriched by natural-scientific methods deployed in psychology. He did not advocate psychology as a subject distinct from philosophy, though he taught specialized psychological methods. He increasingly differentiated psychology from physiology, though it was physiology that provided him with the experimental orientation that he advocated in psychology. He had a definite programme for psychology, and his claim to leadership in the field summarized in successive editions of the *Principles* and reflected in his institutional status and resources in Leipzig was a focus for debate about what sort of science psychology should be.

Much of Wundt's text on the *Principles of Physiological Psychology* discussed the brain and nervous system simply because new knowledge existed about this material dimension. He did not believe the discovery of material analogues for mental events made psychology into a science, only that physiology secured access to one dimension of psychological knowledge. Psychological science, for Wundt, has two methods; the first involves experiment but the second does not. He supported experimental research on perception, psychophysics and reaction times, in which work he thought it possible to correlate a subject's report of changes in conscious awareness with recorded physical conditions. Wundt denied that this research uses introspection, which in his view involves reflection, and he claimed only that it is possible to correlate directly reported change in the size, intensity and duration of sensations with physical variables. In experimental terms, to give an example, subjects were required to release a key at the moment they perceived a change in the brightness of a light source. He adopted an experimental method in order reliably to reproduce identical or near-identical changes of perception by the repeated presentation of the same stimuli. He initiated a *practicum,* or period of laboratory work, and set aside part of the curriculum to train advanced students in these methods and for these students to work up theses. Students used each other or even Wundt himself as the experimental subject since they supposed that the subject needs to be experienced to carry out the activity properly. Experimenter and subject also changed places since the purpose was to study general mental activities, not personal qualities. Wundt secured the institutional support necessary for the expansion of his programme: funding for instruments, space and assistants was made available; he established a journal to publish his students' results, the *Philosophische Studien* (*Philosophical Studies*; the first volume appeared in 1883); and he attracted students from all over Europe and North America as well as from German-speaking countries. The message got around that a training in scientific psychology could be acquired from a period in Wundt's classes and *practicum*.

Nevertheless, Wundt wanted psychology as a science to contribute new and rigorous methods to the solution of philosophical questions. He assumed the reality of mental action, and his *a priori* belief in psychic causality provided the rational basis for his claim that psychology is a specialist science and not a branch of physiology. He took the view that the mind and the body are apsects of the same reality, though each dimension has its own form of causation. No amount of physiology, in his opinion, could make psychology a science because the mind is a psychic principle characterized by a non-reducible rational logic. The aim of Wundt's psychology was therefore to uncover the laws of psychic causality, and the preliminary step was the accurate description of mental contents. Thus he classified action – the will or volition – into three forms: impulsive acts (primitive *Triebe* or drives); voluntary acts, in which one among several motives predominates; and selective acts, in which one motive dominates after a conscious choice. He linked this hierarchy of actions loosely to belief in an evolutionary hierarchy from animal to man. In practice it was extremely difficult to imagine ways to examine volition experimentally. Later experimental psychologists for the most part ignored the will, dismissing it as a remnant of unscientific thought or subsumed the topic within a philosophical discussion of the general active character of mental processes. Wundt himself excluded the will from the scope of experiment and discussed it as one part of the topic of mental activity in which the mind judges.

Wundt lectured, published and supervised students on logic, ethics and the history of philosophy, as well as on psychology. Ultimately, he asked psychological questions in the light of a belief in the mind's rational activity as the precondition of knowledge. He did not think this rational activity itself could be the subject of experiment, and other aspects of his programme for psychology accordingly involved non-experimental methods. To examine complex mental activity, such as thought and language, the feelings and volition, he turned to a kind of cultural psychology widespread in German-speaking Europe, with roots in Herder's vision of the human spirit in history. (This was discussed earlier in the context of anthropology.) Wundt appropriated the word '*Völkerpsychologie*' to denote a psychology that studies mind through the mind's visible activities in the social world of myth, custom and especially language. Whereas experimental psychology does research on individuals, *Völkerpsychologie* – there is no equivalent in English though 'cultural psychology' has been used – researches the collective social world. Wundt thought it possible to infer the operations of the individual mind from its collective products, the results of interactions between minds. This cultural and linguistic side of Wundt's interests dated from his early work on psychology with reference to developmental principles, physiology and man's social nature. He developed the idea of *Trieb* (or drive) as a basic emotional or affective process that leads to expressive movements or gestures. Such gestures are replicated by others and hence both communication and shared mental states become possible; when refined, this communication exists as language. Shared mental states constitute a *Volksseele* – a trans-individual psychological structure that, over time, gives rise to culture. Wundt did not discuss either history or national differences but, beginning in 1900, he published a series of volumes that took a psychological approach to culture and language. For reasons that will become clear, this work had a much smaller audience among the younger generation in search of a scientific psychology than the experimental research. Nevertheless, the notion of gestural communication, Wundt's view of how the individual mind creates a shared culture, was taken up and developed to become an important part of social psychology, for example, through the work of G. H. Mead.

Many academics and students were persuaded that the natural sciences had established truths in a way that philosophy had failed to do. In social circumstances where religious authority was in decline and in which open-minded people perceived the churches to be hand in glove with political reaction, serious young students turned to the new psychology as a subject at the forefront of modern, objective learning. They were naturally attracted to Leipzig, where there was both opportunity to learn the new methods and a programme that placed psychology at the centre of academic ideals.

iii *Psychology in Germany*

Scientific psychology acquired one social identity in Wundt's institute – Leipzig formally established an institute in 1883 and it expanded in 1897 into a purpose-built centre through which many students passed. This did not, however, found the subject, for at least three reasons: Wundt's activity was not unique; contemporaries and even his students had different views of what psychology should be; he remained a philosopher and most psychologists continued to hold appointments in philosophy. This section considers these points in turn as a way to describe the wider picture of psychology in Germany.

G. E. Müller (1850–1934) at the university of Göttingen was Wundt's main rival in training students in experimental technique. Müller was the successor to the philosophy chair of Rudolf Hermann Lotze (1817–81) who, in this chair from 1844 to 1881, wrote at length on natural science and the philosophy

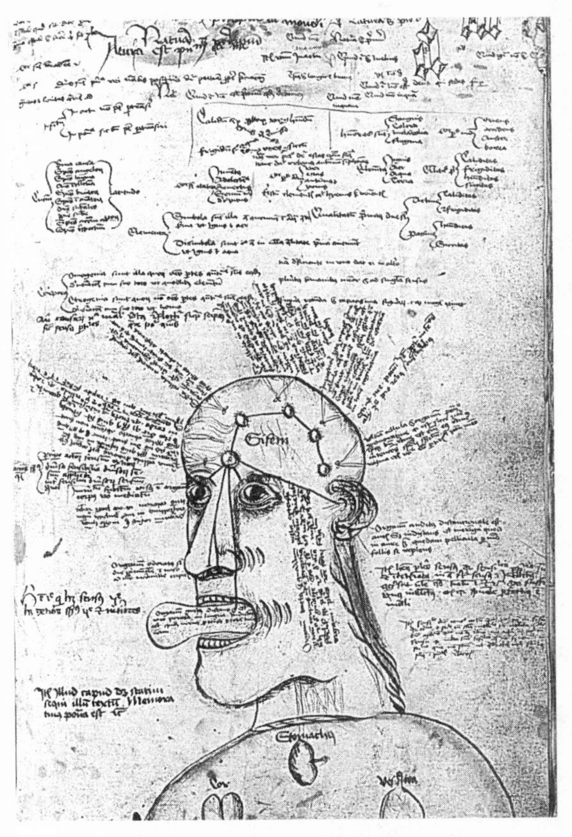

1 Illustration of relations between the senses and the faculties of the soul, from a manuscript of Aristotle's *De anima* [On the Soul].Pen drawing , Leipzig, 1472–1474.

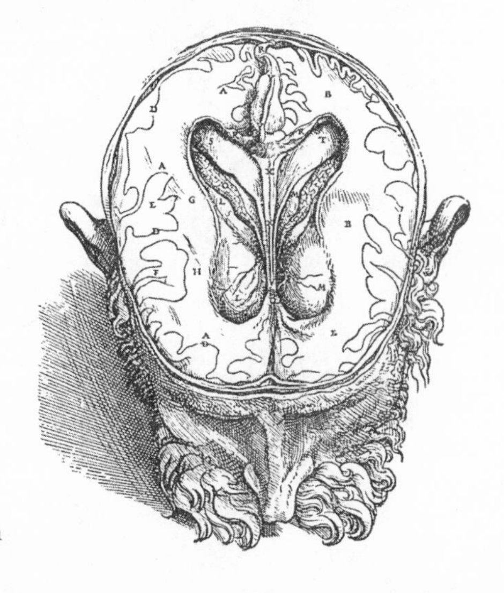

2 Horizontal section of the brain to show the ventricles where the animal spirit is produced; woodblock plate from Andreas Vesalius, *De humani corporis fabrica* [On the Fabric of the Human Body], 1543.

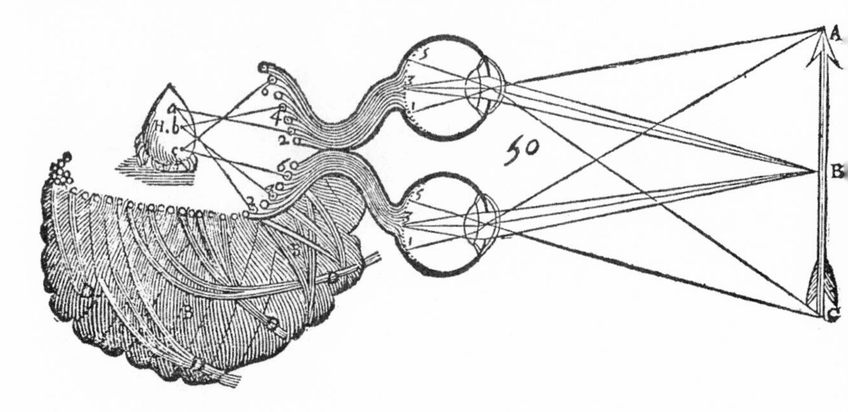

3 Descartes' diagram to illustrate the creation and transmission of the visual image to the pineal gland in the brain; from *L'homme* [On Man], 1664.

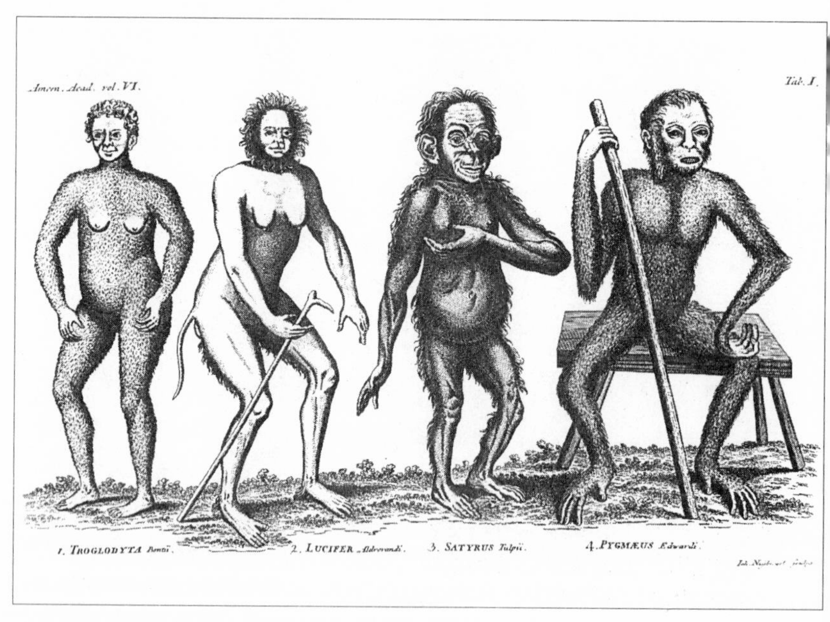

4 Linnaeus's 'Anthropomorphia' – an engraving of beasts in human form, as classified in 1760.

5 The romantic ideal of education: the statue of Johann Heinrich Pestalozzi (1746–1827).

6 The embalmed body and death mask of Jeremy Bentham (1748–1832) in University College, London.

7 Portrait, profile and silhouette of a reflective and quietly mannered soul; from Johann Caspar Lavater, *Physiognomik. Zur Beförderung der Menschenkenntniß und Menschenliebe* [Physiognomy: On the Advancement of the Knowledge and Love of Man] (new edn, 1829).

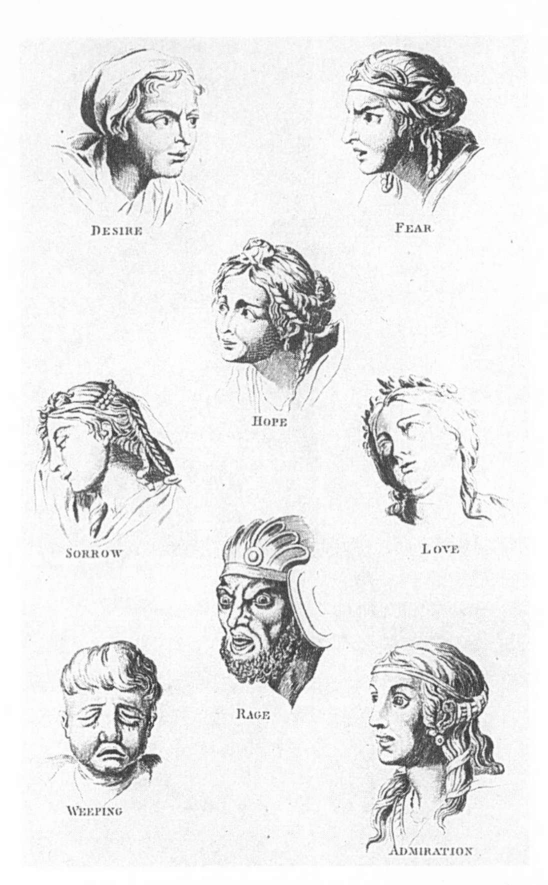

8 Etching to illustrate the physiognomy of the passions.

9 Photograph of I. M. Sechenov (1829–1904) at the Medical Surgical Academy in St Petersburg, 1860s.

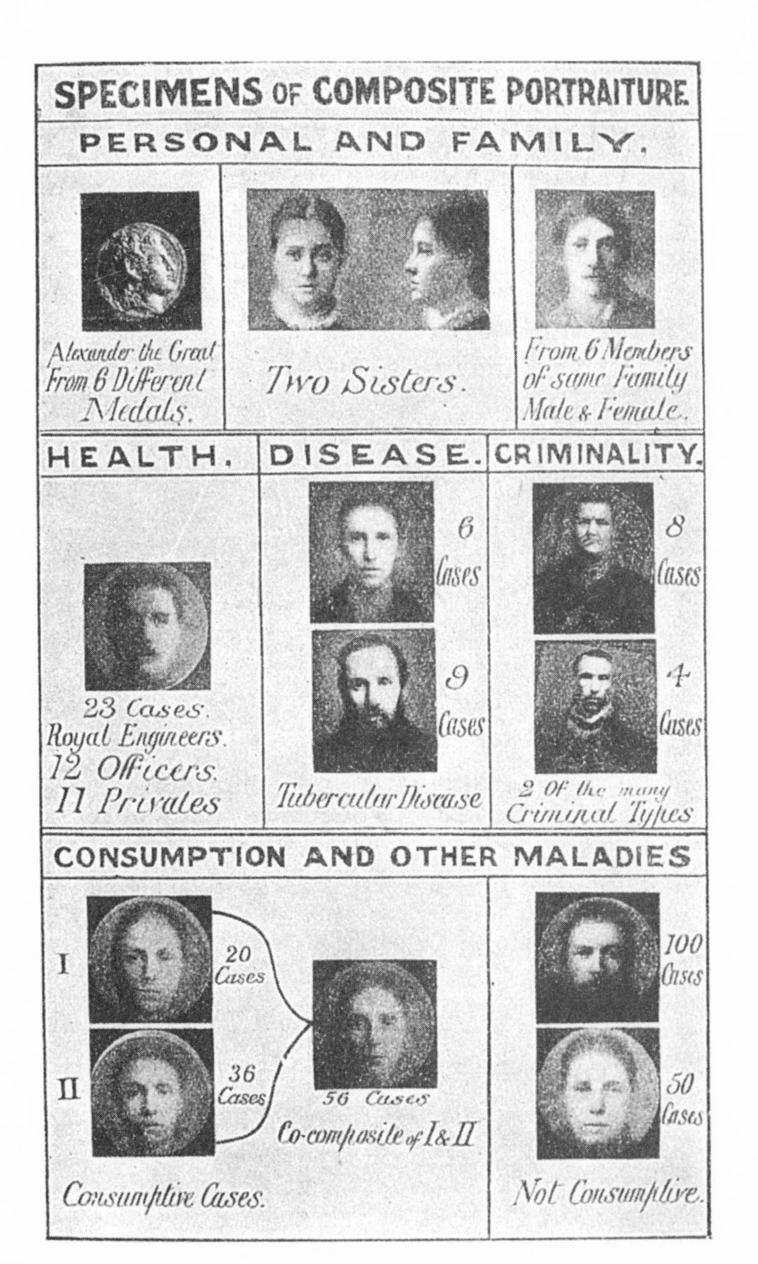

10 Francis Galton's plate to show how composite photographs create pictures of standard human types; from *Inquiries into Human Faculty,* 1883.

11 Lithograph of a Victorian family consulting a phrenologist.

12 Psychoanalysis arrives in the United States: Clark University, photograph 1909. Back row (l to r): A. A. Brill, Ernest Jones, Sándor Ferenczi; front row (l to r): Sigmund Freud, G. Stanley Hall, C. G. Jung.

13 The Pacific Islander's encounter with Europe: statue from Hermit Islands of the Bismarck Archipelago.

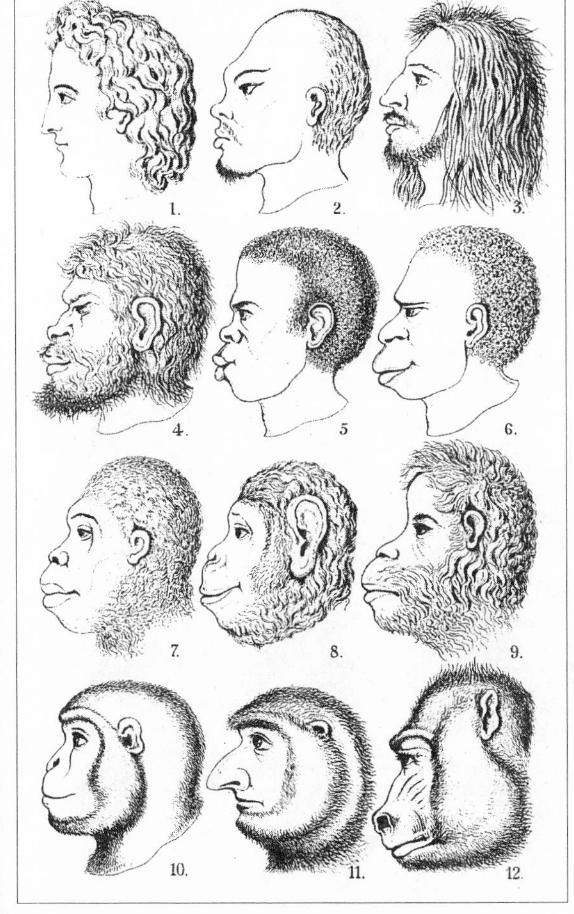

14 A European hierarchy of head forms, from 1 to 12; frontispiece from Ernst Haeckel, *Naturliche Schöpfungsgeschichte* [Natural History of the Creation], 1868.

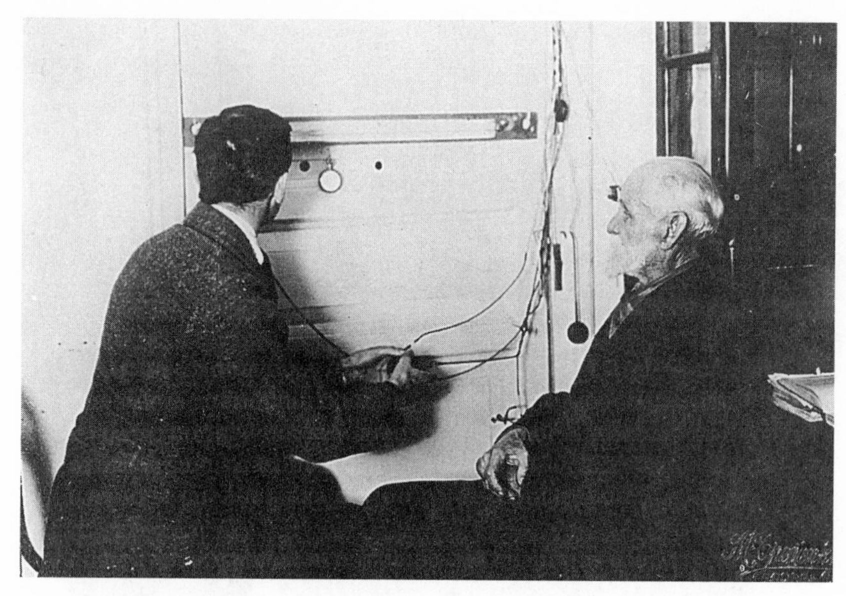

15 I. P. Pavlov and researcher observing a conditioning experiment on a dog; photograph, Petrograd, 1921/3.

16 John Bowlby in child therapy session; photograph.

of mind; his synthesis, *Mikrokosmus* (*Microcosmos*, 1856–64), related human values to cosmic reality. Lotze was a powerful voice for constructive dialogue, as opposed to polarization, between younger natural scientists and older philosophers who feared mechanist views of nature. Müller, by contrast, took for granted the experimental method in a science of mind, and he restricted his work almost entirely to narrow psychological questions, such as the effects that different sensory conditions have on memory. Müller did not have a comprehensive programme to unify knowledge, like Wundt, but gave his undivided attention to experimental activity, and he also taught many students. Müller and not Wundt was the guiding spirit in the foundation of the German Society for Experimental Psychology in 1904.

Müller and other scientists studied sensory perception and psychophysics independently of Wundt. It was only a matter of time before someone made the attempt to adapt experimental methods used in these areas to study the higher mental activities known in the late twentieth century as cognitive processes, such as attention, memory and thought. An independent researcher, Hermann Ebbinghaus (1850–1909), took this step when he devised ways to study memory experimentally. He explored his own ability to memorize lists of nonsense syllables and cantos from Lord Byron's *Don Juan*. He measured the amount of material remembered over time, and he plotted the results graphically as a curve for forgetting. He believed that such research shows how memory works. Subsequently he taught at the university of Berlin and published the study that made his name, *Über das Gedächtnis* (*On Memory*, 1885). The text was preceded by what to philosophers was a provocative Latin epigram, 'From the most ancient subject we shall produce the newest science', which succinctly presented the new psychology's claim to attention.[6]

Though Wundt himself ruled out an experimental approach to complex mental processes, his students and assistants began to go their own ways, and by 1900 there was competition between alternative schools of thought. The most important group was headed by Oswald Külpe (1862–1915), who studied in Berlin and with Müller in Göttingen before he completed a thesis with Wundt. He published on both philosophy and reaction time experiments. Though he concentrated on philosophy and aesthetics after he became a full professor at the university of Würzburg in 1894, he also supported a large quantity of experimental work that considerably changed views on perception and thought. A more expansive view of experiments developed in Würzburg, allowing subjects to make complex and even introspective reports. Ebbinghaus and then the psychologists at Würzburg applied experimental analysis to thought itself. There were sharp exchanges between Wundt and those in Würzburg on the issue and on the nature of introspection. The outcome, a great variety of experimental work, exposed how extraordinarily difficult it was – some were to think it impossible – to devise experiments about the mind's content that would be accepted as valid by other researchers. Different research groups simply could not agree about what they studied in the mind. As Külpe and his associates showed, this situation demanded philosophical clarification as much as new experiments.

German academic activity in psychology began to have a disciplinary nature in the twenty years before World War I, and at the end of this period there were perhaps ten senior experimental psychologists, all of whom occupied positions in philosophy departments. Their main student audience consisted of future *Gymnasium* teachers; but the academic audience interested in their broader conclusions consisted mainly of philosophers, who were increasingly concerned about the number of experimental psychologists who replaced philosophers in posts and teaching. In 1913 they circulated a petition against the trend to faculties and ministries of education across the country, prompting Wundt to respond with a pamphlet called *Die Psychologie im Kampf ums Dasein* (*Psychology in the Struggle for Existence*). This debate about the respective value of philosophy or psychology in the understanding of mind was an emotive one in academic life. Apart from the competition for posts, the two disciplines, in some eyes, symbolized academic commitment respectively to reason or to natural science as the foundation for the unification of knowledge. Some philosophers argued that the new psychology in effect treated thought as a product of mechanistic processes, a step which they found intolerable because, in their view, it is logically incoherent and devalues the status of reason. They did not oppose the new psychology within what they regarded as its proper field, but they did deny that questions about the foundations of knowledge can become empirical, psychological problems. They opposed what was sometimes called psychologism, the extension of psychological explanation beyond its proper sphere. The problem was that there was no agreement on what the proper sphere is.

Many natural scientists at this time were influenced by Ernst Mach's philosophy of science. Mach (1838–1916), who taught at the university of Prague and then at the university of Vienna, systematically criticized physical concepts, such as the concepts of force or absolute space-time in Newtonian mechanics, that cannot be re-expressed as relations between the elements of sensory experience. He argued against what he saw as the metaphysical content of science and for the view that only knowledge based on sense perceptions counts as scientific knowledge. It appeared to follow from this argument that the study of sense perception is the foundation of natural philosophy. This position was later called 'positivism', a term that does not necessarily indicate any debt to Comte. As a philosophy of knowledge, it cut out descriptions of what the self perceives as the physical world in order to describe regularities, in Mach's phrase, in 'the elements of sensation'. It was his hope that this point of view made the mind–body problem a non-issue in science:

> There is no rift between the psychical and the physical, no inside and outside, no 'sensation' to which an external 'thing,' different from sensation, corresponds. There is but one kind of elements, out of which this supposed inside and outside are formed – elements which are themselves inside or outside, according to the aspect in which, for the time being, they are viewed.[7]

His work included an examination of sensation itself, *Beiträge zur Analyse der Empfindungen* (*The Analysis of Sensations*, 1886), based on considerable experimental research. Some psychologists with large ambitions for their subject, such as the young Külpe, were attracted to such a philosophy, in which the empirical study of sensations was made the key to the problem of knowledge. This approach, it appeared, demystified reason: what is needed to understand knowledge is sufficient experimental ingenuity. There was precedent for this attitude in the development of British philosophy after Locke, which analysed knowledge into ideas with a sensory origin and culminated in the mid-nineteenth-century work of J. S. Mill and Bain, but German experimental technique appeared for the first time to make the programme a truly empirical one. As many researchers quickly said, however, one key weakness of Mach's project lay in its treatment of sensations as atomistic elements, a view that itself did not appear to be empirically verified.

At Würzburg, description of the conscious mind in terms of its sensory content had far-reaching consequences, consequences that made Mach's philosophy unviable. Firstly, experiments indicated that thought involves elements, such as attitude, that are not sensory in nature; this precipitated a debate about what in English was called imageless thought. Secondly, Külpe and some of his associates concluded that the content of mind is better described in terms of mental acts than in terms of mental elements, sensations or presentations (*Vorstellungen*). There were philosophical implications for the manner in which psychological knowledge, or psychology as a science, was understood.

The language of mental acts points to yet another dimension of the literature on psychology. This is associated with the German philosopher Franz Brentano (1838–1917), whose significantly entitled book *Psychologie vom empirischen Standpunkt* (*Psychology from an Empirical Standpoint*) appeared in 1874. Although the title stressed the empirical point of view, the book did not adopt the new experimental methods. Brentano aimed to make psychology empirical through accurate description of the reality of consciousness, not by the establishment of a new natural science. He argued that 'the mental', as opposed to 'the physical', is by definition an act not a state. The mental *act*, he thought, is analysable into two parts or objects: the primary object which is immediately observed (the teapot) and the secondary object which is unconsciously perceived but not immediately observed (the seeing). The business of psychology, 'inner perception', is with the second object, and – though it requires special attention – inner perception is the means by which we have knowledge of what distinguishes the mental life of human beings. Brentano understood mental activity to be irreducible – not analysable into something else – and he wanted to establish an empirical psychology that describes this activity rather than the physical or other objects that mental activity is about. In technical terms, which derived from medieval Thomist philosophy, he defined 'the mental' by intentionality: what is mental is an act about or towards an object; seeing is seeing something. This approach was suggestive to psychologists, who were pressed to justify the autonomy of their field, because it was a language that described mind in terms of its own properties and activity, rather than in terms of physical states as studies in physiological psychology appeared to do. The Würzburg researchers found a use for this mental language because their experiments showed it is impossible to analyse thought into sensory elements.

Brentano's philosophical viewpoint had a direct influence on experimental psychology through his student Carl Stumpf (1848–1936). Stumpf, not the pioneer Ebbinghaus or the experimentalist Müller, was the choice of the university of Berlin's philosophy faculty in 1893 when it created a position for the new psychology. Its choice reflected approval of Stumpf's belief that neither Wundt's *Völkerpsychologie* nor the experimental analysis of thought could displace the mind's description of itself as conscious activity. He also held an acceptably philosophical view of science: 'The psychologist must at the same time be a theorist of knowledge . . . as anyone must for whom science is more than artisanry.'[8] Stumpf created in Berlin a major centre for the new experimental psychology, but he did not in any straightforward sense make psychology a natural science. Rather, his psychological seminar, and the experimental work it supported, refined and supplemented a philosophical science to clarify knowledge of the mental realm in its own terms. As we shall see, this viewpoint in psychology was connected to the contemporary development of phenomenology.

Stumpf's vocation was philosophy, his rational conviction psychological experiment and his passion music. He played the cello as well as studying and teaching in a number of universities, and music and psychology came together in his major study on *Tonpsychologie* (*The Psychology of Sound,* 1883–90). Once in Berlin, he successfully organized facilities for psychology, including laboratory space, but he handed experimental work over to others. His students in Berlin included Wolfgang Köhler and Kurt Koffka. Together with Max Wertheimer, they formed what became known as gestalt psychology, which was to be a major influence on theories of perception in the twentieth century, and Köhler succeeded Stumpf on the latter's retirement. Their work perpetuated the close association of experiment and philosophy; and, in this respect like Wundt, they hoped that psychology would make it possible to found philosophy as a science.

The detail discussed in Stumpf's seminar in Berlin led to disagreement with Wundt on many issues, though both men had programmes to integrate the conclusions of experimental science and psychology. This was not the case with yet another scheme for a science of psychology promulgated by the Berlin philosopher Wilhelm Dilthey (1833–1911). His work was a commentary on the nature of knowledge in the human sciences rather than a research programme, and this in itself illustrates the contemporary disunity and insecurity of psychology as a subject area. Dilthey, like other scholars, was impressed with the natural sciences and thought it an urgent matter to examine how they relate to knowledge of man. But he did not accept that the natural sciences establish the route to the human sciences; he drew a distinction, in terms that were still current and the focus of debate a century later, between the *Naturwissenschaften* and the *Geisteswissenschaften,* the sciences of nature and the sciences of mind or the human spirit. He included in the *Geisteswissenschaften* the subject areas that I have described as the human sciences. Dilthey believed that because the two classes of sciences have, in essence, different subject matters they require different forms of explanation. Thus, he argued, to understand nature is to describe law-like causal relations; by contrast, to understand human beings is for the scholar to recreate the mental world, the meanings and reasons, that inform human actions. 'We explain nature; we understand psychic life.'[9]

Dilthey firmly classified psychology as a subject area within the *Geisteswissenschaften*. He was sympathetic to experimental studies of perception, but he denied that such experimental work established psychology, as opposed to physiological psychology, as a science. Instead, he argued that a science of psychology requires understanding, and he used the word '*Verstehen*' to describe the manner in which we can understand and enter into the psychological relation of motive and action in another person or historical figure. When we grasp such an understanding, we make intelligible the reasons or meanings of actions and hence make psychology a science. At the same time as experimental psychology became common in the academic setting, Dilthey reformulated the conviction that to understand man is to understand his history and culture. This argument made psychology a human science alongside history, the study of language, literature or art and the history of philosophy or ideas, a conclusion that involved Dilthey in a series of debates with those who held a different view of psychology. In the 1890s, for example, Ebbinghaus attacked Dilthey's position and set out to refute what Dilthey and others had implied about the new psychology, that it had materialist implications. After 1900, Dilthey moved away from the view that understanding involves a grasp of the psychological world, the world of motives of historical actors, and towards an approach in which he thought understanding possible because of the common structures with which others as well as ourselves necessarily think and establish meaning. The division between understanding and causal explanation in the human sciences, in the terms clarified by Dilthey, re-entered the mainstream of debates in the philosophy of psychology and the social sciences in the second half of the twentieth century.

It is easy to get the impression that every German academic was really a philosopher. This was not so: Müller, a strict experimentalist, was an outstanding exception, and there were psychologists, teachers and politicians who wanted psychology to make a difference to education or social affairs rather than address the theory of knowledge. A prominent example is the physiologist William T. Preyer (1842–97), who broadened his work from studies of reflex control and psychophysics to discuss perception, linguistics and especially child development. In *Die Seele des Kindes* (*The Soul of the Child*, 1882) he combined physiological, evolutionary and psychological information with the clear intention that psychology should affect practical life. This interest developed into recommendations for the reform of the Prussian state school system. Many teachers, in Germany and elsewhere, were interested in the possibility that psychology would provide a scientific basis for education, and this put different pressures on psychology than those which originated with academic philosophy. Wundt's and Preyer's subject matter therefore differed: Wundt was interested in the universal properties of the conscious mind and the mind's expression in linguistic and cultural life; Preyer was concerned with the physical person linked by evolutionary history to nature and by education to society. Wundt's experimental subjects were academics; Preyer's were children.

Many more psychologists became interested in psychology as a practical occupation after 1900. One of Wundt's most successful students, Ernst Meumann (1862–1915), shed his teacher's reservations and directed work in experimental educational psychology. He edited journals on both 'general psychology' and 'experimental pedagogics', terms that indicate the degree to which hopes for practical psychology took psychologists well beyond Wundt's programme. William Stern (1871–1938), who became director of a new psychological institute in Hamburg in 1916, was interested in a great range of experimental and practical topics from the beginning of the century, when he taught at the German university of Breslau (the Polish city of Wrocław after 1945). He hoped that psychology could contribute to the problems of industry, and he contributed the term 'intelligence quotient' to the literature on children's performance. Yet Stern made no attempt in his teaching to separate psychology from philosophy.

All this activity in German psychology demonstrates two things. Firstly, there was great diversity in what was called psychology. Even within the restricted compass of experimental psychology, different research groups did not unite around a common programme; methods and concepts varied from university to university. Secondly, psychology did not take off as an independent discipline in Germany and it gained no secure institutional base. Nearly all the academics who identified themselves as psychologists held appointments in philosophy, and many senior psychologists thought this right and proper. There were those, however, like Preyer or Meumann, who linked psychology to child development and education. These circumstances in psychology changed slowly during and after World War I when both the military and companies, like the railways, who were interested in the rationalization of their activities, began to invest in psychotechnics. A professor in Danzig (Gdansk) in the interwar years went so far as to assert psychology's claims because 'the naked reality of economic life demands exact, natural-scientific objectivity'.[10] Applied psychology was taught quite widely in technical universities during the 1920s, and there were six psychology chairs in German universities by 1930, but the large-scale change in the fortunes of psychology as an occupation came during the period of the Third Reich (1933–45). In the late 1930s, psychologists claimed expertise in personnel selection and so gained access to the resources to establish a distinct occupational identity.

iv *Psychology in the United States*

Psychology was in a completely different situation in the United States, and to a small extent Canada, at the end of the nineteenth century. It had a professional organizational structure, the American Psychological Association, journals, a large academic presence and a clientele that accepted its authority. The concerns of psychologists were with the status of their subject as a distinct natural science, even though it existed, without ambiguity, separate from philosophy. If they had a problem with the boundary of their subject, it was the boundary with physiology, as they worked in circumstances that made it appear natural for psychology to exist as a natural science. Since what happened in the US had great influence elsewhere, overwhelmingly so after 1945, these circumstances of late nineteenth-century North America are crucial to the history of psychology in the twentieth century.

Until the last quarter of the nineteenth century, higher education in the United States was almost entirely centred on the liberal arts colleges. Young men learned moral values and an everyday Christian psychology, sometimes on the Scottish moral science model, that included descriptive analysis of human mental capacities and actions. This was the teaching, for example, of Thomas C. Upham (1789–1872), the college president at Bowdoin in Maine, who published his lectures in 1831 as *The Elements of Mental Philosophy*. One of the best-known teachers later in the century was the president of the College of New Jersey at Princeton, James McCosh. Though he wrote about motivation and emotion as part of Christian moral teaching, McCosh was notably open to both evolutionary theory and German psychology. He perceived no sharp boundary between moral philosophy and the new psychology in the context of the moral purposes of man's improvement. The differences came when new natural-scientific methods were imported, though, as McCosh's interests show, it was easy enough to believe that these new methods served old purposes. Nineteenth-century students who wanted to take their education further were encouraged by their colleges to travel to Europe, especially to the German universities. From the 1870s, these students included some who were dissatisfied with moral science as a science and looked to experimental science for a more modern and objective way to study people. Many of them naturally went to Leipzig, as the most famous centre for the new psychology. One of Wundt's earliest American visitors (though not a student) was G. Stanley Hall (1844–1924), a key founder of the psychology profession and of academic departments in the United States. His work, which focused on children's development and education, exemplifies the continuity of moral purpose between old and new psychology. James McKeen Cattell (1860–1944) was the first student to study at length with Wundt and to spread experimental psychology in North America. His career illustrates what occurred when the new psychology was translated from Germany to the United States; many first-generation psychologists had similar trajectories.

Cattell came from a Presbyterian background reinforced by the courses he took in the liberal arts at Lafayette College in Pennsylvania where his father was the president. He then studied in Göttingen and Leipzig and at Johns Hopkins in Baltimore. He returned to Leipzig in 1883 to work on his doctorate and for three years was Wundt's laboratory assistant: he was skilled with instruments and at recording data on reaction times. Not least, he introduced Wundt to the typewriter. Subsequently he worked in England, set up psychological instruments at the university of Cambridge and acquired from Galton a strong interest in individual differences. In 1889 he became professor of psychology at Pennsylvania, where he established a laboratory and a programme to make mental measurements; in 1891 he moved to Teachers College at Columbia in New York to repeat the pattern. He stayed at Columbia until 1917 when he retired after upsetting many people, both by his pacifism and by his attacks on 'academic servitude', control over the curriculum and research, imposed by college presidents and trustees.

Cattell's programme for mental testing in the 1890s showed what sort of European psychology successfully crossed the Atlantic and what did not. He used his instruments to measure individual differences in sensation and motor performance, hoping to reveal correlations between different capacities. There was a flurry of interest in these techniques because it was thought they would be useful to education; one scientist, for example, claimed to be able to measure 'the physical basis of precocity and dullness'.[11] Joseph Jastrow (1863–1944) headed a team at the Chicago World's Fair in 1893 that carried out a mass of measurements on foreign and native visitors, a venture that again owed much to Galton who had set up an anthropometric laboratory at South Kensington in London. Jastrow, when president of the American Psychological Association in 1900, announced: 'Psychology and life are closely related; and we do not fill our whole function if we leave uninterpreted for practical and public benefit the mental power of man.'[12] Cattell's objectives were also factual and utilitarian:

> To what extent are the several traits of body, of the senses and of mind independent? How far can we predict one thing from our knowledge of another? What can we learn from the tests of elementary traits regarding the higher intellectual and emotional life?[13]

The programme of measurements was a failure; it demonstrated no correlations or insights of lasting significance. Yet the programme exemplifies the way psychologists detached their work from a highbrow philosophical context and relocated it within a practical setting in which they appealed for the value of psychology to everyman. The antagonism to philosophy was expressed forcibly by E. W. Scripture (1864–1945), who took over the psychology laboratory at Yale College in 1892 after he had trained with Wundt. He wrote popular books 'expressly for the people . . . as evidence of the attitude of the science in its desire to serve humanity'.[14]

Cattell taught none of Wundt's philosophy, probably grasped none of his *Völkerpsychologie* and, without qualification, treated psychology as a natural science founded on the collection of facts. He came from a background that thought of education as preparation for Christian social progress, and he assimilated in Germany only what appeared relevant to the development of a psychological expertise applicable to social needs. He was more excited about Galton's techniques for the study of individual differences than he ever was about Wundt's philosophy of mind. Once settled in the US, Cattell became an entrepreneur on behalf of the new psychology; as he argued, it was a valuable resource because research on human capacities is relevant to social performance. He helped establish the *Psychological Review* and increasingly used his organizational and editorial experience on behalf of journals concerned with the public understanding of science and with science education. In the 1920s he was active in the Psychological Corporation, a commercial enterprise to market psychological expertise to business and the public.

Finally, he was for many years a moving spirit in the AAAS, the American Association for the Advancement of Science, the public forum for the national scientific community.

A career in psychology such as Cattell's was a life committed to scientific expertise. Cattell, like many in his generation, believed in expertise as the means to make America modern, civilized and democratic. His interest in science was an interest in the occupations that he believed would contribute to social advance; he was not interested in the development of an academic discipline for the sake of learning. It was precisely because these values were widely shared that psychology acquired an academic base so quickly in the US in the 1890s. Higher education expanded rapidly, there was a new state university system (as at Cattell's Pennsylvania) and new and old schools introduced graduate programmes and research institutes. There was large-scale human and financial investment because academics, administrators and funders shared hopes about what science and education could do for the American people. Cattell's co-founder of the *Psychological Review*, Baldwin, wrote significant studies of child development with moral and social as well as scientific purposes. These studies drew on and contributed to a European interest in psychological growth and education, an interest with a significant institutional base in Claparède's institute in Geneva, which became a famous centre for the study of child development and later home to Piaget. Baldwin was also read in the 1920s by the Russian psychologist L. S. Vygotsky. Young scientists bred on German technique but born with American values put forward a new discipline and they offered their services to make this discipline work. They promised to show how knowledge of individual psychological capacities would smooth the birth of a modern society. This was an offer hard to resist.

Psychologists, however, still found it necessary to fight hard for the scientific credibility of psychology, even though its practical potential began to be accepted. Baldwin told the story that when the Scottish psychologist Bain sent him his books through the mail, Baldwin had to pay tax on them; although tax was not payable on scientific books, Washington bureaucrats stated that 'our experts report that [Bain's] books are in no sense scientific'.[15] While James, when his *Principles of Psychology* was criticized as not sufficiently scientific, happily responded that it was 'a mass of phenomenal description, gossip and myth', most psychologists were unhappy about this image.[16] There was a concerted attempt in the 1890s to advance the cause of psychology as a science. This required a move in one direction against natural-scientific critics who argued that as psychology becomes scientific it becomes a branch of physiology; and a defence in another direction against sceptical lay opinion which felt that psychology was nothing more than practical common sense. In these circumstances, it was important for the development of the discipline that psychologists were able to argue that they did have a unique expertise, the study of human capacities, able to increase the efficiency of a major social activity, education. Once established in academic institutions, psychologists found it easier to argue for the status of their subject as a science.

The link between psychology and education was pressed by the new profession's senior figure, Hall. He published the *American Journal of Psychology* and the *Pedagogical Seminary* (f.1891) and he established close links with those who trained teachers. He became a leader of a systematic observational approach to child growth, the child study movement, and he wrote a large-scale study on *Adolescence* (1904). Hall's career began with a degree in divinity and involved a long struggle to find a permanent academic position in which to bring together his Christian roots and the new standards of scholarship, to which he had been exposed by a visit to Germany. Psychology became the answer. It led Hall in the early 1880s to a post in psychology and pedagogics at the new model university of Johns Hopkins, where he quieted fears about the materialist dimensions of science with the claim that the bible 'is being slowly re-revealed as man's great text-book in psychology'.[17] Johns Hopkins presented a great opportunity as it gave unprecedented prominence to graduate training; other colleges were forced to compete, and psychology as a distinct discipline was the beneficiary. In practice, Hall's rhetoric of religiosity turned into a practical moralism about the life of the child and the activity of the educator.

When he moved to Clark University, Massachusetts, in 1888 as both president and head of a separate department of psychology, he was, in time, able to consolidate an institutional position from which to spread these values.

The largest single constituency for psychological expertise was formed by teachers, public authorities and individuals with the funds to invest in education. There was a desperate need to equip immigrants from Europe and from rural America with the tools for modern urban life. There was also belief in the democratic value, that Dewey did so much to advance, that politics requires individual fulfilment through educational opportunity. The consequence was that psychologists who started off with narrowly focused experimental studies, for example on perception, tended to move towards less rigorous but more obviously relevant work on child development or aptitude testing as they showed funders what their subject had to offer. Once firmly established in the university structure, psychology departments trained an audience sympathetic to their own activities and acquired an institutional momentum of their own. This made possible a considerable amount of experimental research on the German pattern, but a commitment to functional explanation linked to evolutionary theory, added to funding pressures, encouraged research more relevant to the individual's life in society.

This conjunction of academic and public values promoted an institution that had great influence on the social sciences in the US, the university of Chicago. Endowed in the 1890s with massive private funds from John D. Rockefeller, it attracted an outstanding faculty both to give the city a reputation for culture and to address the city's fearsome social needs. Chicago had the nation's social problems writ large: large immigrant groups, many without the English language; exponential urban growth and housing shortages; confrontational labour relations; politics dominated by local bosses; poverty and crime. In these circumstances it was not hard to argue the value of education as the means to achieve social integration and to claim that expertise is needed to assess individuals and social arrangements. Just as the new steel-frame buildings and early skyscrapers created a modern urban landscape, so the psychological and social sciences would create, the thought went, modern urban men and women.

Dewey became a professor in Chicago in 1894 and under his aegis both experimental and educational psychology expanded. He brought Mead with him and then appointed James R. Angell (1869–1949), who had studied at Harvard University under James. These academics shared a functionalist orientation, and they focused on the whole person in her or his adaptive relation to the social environment. Their work was permeated by belief that science is a utilitarian expertise, the highest stage of evolution. In this setting, as elsewhere in North America – both Hall and Baldwin, for example, wrote at length on child development from an evolutionary perspective – evolutionary theory became central for psychology and social science. This never occurred in the same way in continental Europe. Further, it was in Chicago that Watson wrote a thesis on the psychology of rats and began to think that psychology should be a technology for the control of behaviour, a goal that he and his successors saw as the culmination of the evolutionary process. At the same time as Dewey organized the philosophy department to make space for a socially-oriented psychology, Albion Small created a large and influential academic department in the social sciences.

Psychology in the US therefore developed in a different way, and on a different scale, than in Germany, even though American psychologists sometimes had the German example in mind. German psychologists had to justify their existence to philosophers and to the conservative state administrators who had the final word on appointments and budgets. US psychologists had lay, self-styled practical men to satisfy. All the same, the contrast must not be taken too far since there was educational psychology in Germany, while a large academic community of psychologists was dedicated to the pursuit of knowledge in North America. Even in the most academic regions of psychology, however, there were still differences between Germany and the United States, as exemplified by the work of Edward Bradford Titchener (1867–1927). Titchener was an Englishman, more impressed by physiology than philosophy as an undergraduate at Oxford, who wanted to study mind scientifically and therefore went to Leipzig to write a thesis on reaction-time experiments. In 1892 he moved to Cornell University in New York state to head a recently opened laboratory, and there he stayed. Cornell's president, A. D. White, had fought an extended but ultimately successful battle against the religious interests that dominated college life; it was this battle he projected on to the past as a whole in his book later cited to exemplify 'the warfare thesis' on relations between science and religion, *A History of the Warfare of Science with Theology in Christendom* (1895). The result at Cornell was an institution with a secular and scientific approach to mind embodied in a department of psychology. Titchener used German experimental methods and undertook meticulous studies of mental content. He restricted himself and his students to description of what is observable with experimental techniques and instruments. He translated work from the German and published his own *Experimental Psychology: A Manual of Laboratory Practice* (1901–5) and student textbooks. More firmly than anyone else in North America, he promulgated psychology as a natural-scientific *discipline*.

Titchener appeared to his colleagues to be Wundt's representative. In fact, he did not sympathize with either Wundt's philosophy or his *Völkerpsychologie*; he brought from England to the US a theory of knowledge derived from Locke and undiluted by his residence in Germany. He thought that scientific psychology must use experiment to refine objective description of the elementary units whose combination makes up mental content. The psychologist 'takes a particular consciousness and works over it again and again, phase by phase and process by process, until his analysis can go no further. He thus learns to formulate "the laws of connection of the elementary mental processes".'[18] He assumed that the combination of elementary qualities, in an unknown way, reflects causal nervous processes, but he believed that psychology – at least for the time being – must concentrate on empirical description. Wundt, by contrast, built his psychology on a belief in psychic causality. But just as Titchener was distanced from Wundt, he was distanced from his North American colleagues. He defined psychology as the scientific study of the elementary constituents of the conscious mind, and he was not interested in animals or children or individual differences. 'The primary aim of the experimental psychologist has been to analyze the structure of mind . . . to isolate the constituents in the given conscious formation. His task is a vivisection, but a vivisection which shall yield structural, not functional results.'[19] Further, Titchener berated his colleagues when they rushed to apply psychology before they had established the scientific base of the subject. He remained an outsider in this sense, as he gave priority to training rigorous scientists rather than to a psychology concerned with the person's function in society. As a result he was a respected figurehead for an ideal of psychological rigour, but somewhat marginal to the large-scale expansion of psychology as an occupation.

By the time the United States entered World War I in 1917, psychologists had sufficient confidence in their subject's expertise to make a large-scale bid to contribute to the war effort through personnel selection. Their discipline had grown rapidly over the previous thirty years; it was a force in higher education and an occupation with no equal elsewhere. If sheer scale is an index, psychology was an American discipline. Such gross measures are misleading, however, since psychologists were engaged in many different activities and displayed no consensus about theory, the definition of psychology's subject matter or the relation of knowledge to social affairs. W. B. Pillsbury (1872–1960), who established a psychology laboratory at the university of Michigan in 1897, likened early conferences of psychologists to a 'meeting of paranoiacs in a hospital ward'.[20] There was a marked contrast between those who, eager to have psychology accepted as a natural science, restricted what they studied to the laboratory, and those who looked to society itself as the laboratory. This is a major theme in the history of the twentieth-century science.

Before moving fully into the twentieth century, I turn to the parallel development of sociology as a discipline. Psychology's development was part of a broader construction, the human and social sciences. There were manifest differences between a psychophysical experiment and a housing survey, but there were also large areas where the boundaries between subject areas were not clear. The reasons why psychology acquired support, above all in the United States, were also reasons why the social sciences in general were funded. Psychology and sociology alike acquired a social identity as society at large turned to scientific expertise as a way to come to terms with human problems. These sciences are therefore inseparable from the conditions of modernity.

15

The Academic Disciplines of Sociology

> This science, indeed, could be brought into existence only with the realization that social phenomena, although immaterial, are nevertheless real things, the proper objects of scientific study. To be convinced that their investigation was legitimate, it was necessary to assume that they had a definite and permanent existence, that they do not depend on individual caprice, and that they give rise to uniform and orderly relations. Thus the history of sociology is but a long endeavor to give this principle precision, to deepen it, and to develop all the consequences it implies.
>
> Emile Durkheim, *Les règles de la méthode sociologique* (*The Rules of Sociological Method*, 2nd edn, 1901)[1]

i *Social Thought, Social Change and Social Statistics*

There was no sociology before the nineteenth century, but we can identify earlier sociological ways of thought embedded in subjects like moral philosophy, history and jurisprudence. The argument that there is a social realm – a human reality besides individuals and political states, which is to be studied in its own terms – marks out the domain of the twentieth-century science of sociology. As Emile Durkheim (1858–1917) stressed right at the end of the nineteenth century, 'a social fact can be explained only by another social fact . . . Sociology is, then, not an auxiliary of any other science; it is in itself a distinct and autonomous science . . .'[2] In the eighteenth century the interest in 'social facts' was only one part of larger humanistic projects, such as

Montesquieu's study of 'the spirit of the laws'; in Durkheim's time, social facts were authority for an independent science.

There was no sociology as an institutionalized subject before the twentieth century. There were no German chairs in sociology before World War I, and it was only about 1910 that one of the men who has often been described as a founder of the discipline, Max Weber (1864–1920), called himself a sociologist. Only in retrospect has a figure like Marx been described as a sociologist. By 1910 sociology was widely taught in the United States, but even in this setting the discipline was unsure of its status in academic and public life. Certainly, Comte had coined the word many years earlier and argued for a distinct science of the social realm, and he was followed in the English-speaking world by Herbert Spencer. But though both Comte and Spencer had a world-wide reputation, they did not found a discipline with institutional support. Throughout the nineteenth century, political economists and revolutionaries like Marx, political observers like de Tocqueville, race theorists and every kind of reformer and philanthropist felt that they could with confidence pass judgement about social facts. In the German academic world there was a close relationship between the study of history, economic thought and social philosophy. The science of political economy, for example, in the hands of J. S. Mill in the mid-nineteenth century, at times came close to being a comprehensive science of man – of his nature, his economic activity and of his political and social arrangements. Nevertheless, reference to society as an abstract concept was not common. Writers referred, as earlier in the Scottish Enlightenment, to political society, savage society and so on, that is, to the concrete worlds in which people live. There were no clear lines between sociologists, academics in related fields and the concerned public. Even after sociology became established as a discipline in the twentieth century, there was considerable indecision about its boundaries with other subjects (especially social psychology, political science, social anthropology, geography, economics and history) and about the position and authority of social experts in public affairs.

Much more happened between Montesquieu and Durkheim

than the formulation of a distinct argument for a new science of social facts. The social world itself changed: steam power, the factory system, machine tools, steel, gas lighting and the railways changed the conditions of existence, the landscape and human sensibility. The scale and disruption of the shift from rural small-scale communities to industrial urban cities had no precedent. The population of Berlin was 330,000 in 1840; by 1874, only thirty-four years later, it had trebled to 950,000. In the 1860s there began what is sometimes called the second industrial revolution, with industry and life based on new chemical and electrical technology. The world that we know, the world where most Western people live in cities and where consumption and services depend on industrial production, was born. These changes directly and indirectly gave much of modern social thought its subject and defined its purposes. An ambivalent response to change and to the city was at the core of reflection on social life. The novel social world and the city offered action, opportunity and freedom; but they also offered despair, degradation and loss of identity. Change brought wealth and a vast array of goods to many; but it also brought appalling conditions of housing, work and poverty.

Social science, in its modern forms, was an extended dialogue with these changes. This chapter describes that dialogue with a focus on France and Durkheim, on Germany and Weber and on the creation of a professional discipline of sociology in the United States. First we must consider the relationship between social theory and social action in the nineteenth century, when there was no attempt to separate the science of society and social prescription. Only in the twentieth century, when there was a self-conscious attempt to make sociology an academic science, was it claimed that social thought could constitute a value-free pure science of the social world separate from its application. Even then, critics were quick to point out the presence of values built into the questions asked and the theories developed in the science. The question, whether a value-free sociology is possible, remained at issue.

The eighteenth-century writers who created a language for social realities, like Vico, Montesquieu and Ferguson, did not distinguish the study of the social world from the purposes for which men live together. They consciously pursued knowledge in order to contribute to the moral if not spiritual virtue that social progress makes possible. And they consciously desired, in practical and often mundane ways, that knowledge should bring progress in human affairs. This motivated the physiocrats in their studies of agricultural production or the German and Austrian professors in their studies of social police. When in the nineteenth century Comte and Spencer defined the methods and concepts of sociology, they committed themselves to the science because they believed it laid the groundwork for ethical progress. Marx laboured at economic research because he believed that truth would change the world. Social science had many roots, but they all grew in the soil of practical ethics. These commitments changed as modes of life changed. Even before industrialization, in parts of Europe such as the Austrian empire under Joseph II, the attempt was made to create rational social policy. A century before, early modern states were interested in a rational science of government, while the science of policy was elaborated during the eighteenth century. The stimulus for Adam Smith's work was commercial rather than industrial society. As the earlier discussion of political economy demonstrated, attention to population, production and trade, that is, the conditions of wealth and security, created a language about social facts before the industrial revolution. But the change brought about by industry and by city life certainly added a new impetus.

There was change in the immediate conditions in which people lived as well as at the level of the state or the national economy. Change took place in the details of daily life – in the family, in the classroom, in patterns of local community care and control and in the organization of work. Observers with a social conscience attended closely to each site of human interaction, and they argued that the suffering and waste they recorded was the result of the way change had been left to chance, prejudice or custom. There were many attempts to translate the argument into systematic disciplines for the construction of order and morality in human relations. Writers on education in the classroom or on the regulation of labour time in the factory made themselves into experts in a field of practical human science. They also turned people into a subject matter for science. The new human sciences were divided into specialities along the lines of new institutional arrangements, like the school, the asylum and the prison, which separated off distinct classes of people and created the subject matter of the sciences of education, psychiatry and criminology.

The discipline and management of daily life and social relations was the particular interest of the French intellectual Michel Foucault in the 1960s and 1970s. He argued that the very subject matter of the human sciences, the individual as the site of a systematically observable and knowable order – as he expressed it, the individual as a site of truth – came into existence with the new forms of discipline and management that were so prominent in the social landscape early in the nineteenth century. As new practices ordered social relations, he claimed, there was a new subject matter to be observed and not only a new manner of observation. When the prisons, for example, maintained records about their inmates, these records created the question of recidivism or persistent re-offending. The high incidence of recorded recidivism led to belief that there is a *class* of criminals, a distinct criminal type, and it was this class that became the subject matter of criminology, the human science of the criminal, when it was established in the last quarter of the nineteenth century. Thus the subject matter of a science, 'the criminal', and the knowledge produced about this subject matter were a feature of practice not a feature of nature. Institutions like the prisons made possible newly systematic and detailed observation and new forms of knowledge at the same time as they brought into existence new modes of social life. Foucault supposed that these nineteenth-century developments were distinct from the forms of social ordering, and hence the conditions of knowledge, that were present earlier, and this supposition was subject to debate. But the claim that the new human sciences and the new institutions and forms of discipline had to be understood as part of a common world helped transform research in the history of the human sciences.

Great Britain was the first industrial society and, because it achieved relative prosperity and enjoyed relatively liberal politics, it responded to change with some degree of flexibility and self-confidence. Throughout the nineteenth century, part of this response came in the form of work by investigators who sought information about the social world in which they lived as a basis for effective action. Many earnest Victorians studied crime, poverty, housing, insanity, morbidity, conditions of work and other areas of concern. These studies often began as private initiatives but they had long-term consequences for the practice of government, and they created the expectation that the state itself should develop an administration to deal with social questions. Public opinion agreed that a national decennial census (which collected comprehensive information from 1851) was needed to supply an empirical basis for social knowledge and for government policy. Indeed, the fact that the British government was so permeable to an empirical input from the social survey was diagnosed as a reason why sociology as an academic subject developed in a slow and piecemeal fashion: government and administration, rather than academia, soaked up the available resources, interest and manpower in social science. This explanation, however, suffers from an implicit assumption that the 'natural' growth of the subject area should have been towards a unified scientific discipline, whereas unity is not something that can be presupposed of sociological knowledge.

The wide variety of Victorian investigations into the social world is, however, not in doubt. Some – like the report by Edwin Chadwick (1800–90) on provisions for public health (1844) – had a major impact on policy. The pioneering studies of social conditions by Henry Mayhew (1812–87), into the labour and poverty of the metropolis (published in their final form in 1864), and B. Seebohm Rowntree (1871–1954), into the condition of housing in York in 1897–8, both private initiatives, transformed the political and moral consciousness of the country. All this work set out to change the reality it studied. The ideals were well stated by Charles Booth (1840–1916), who followed Mayhew with a more systematic survey of London poverty: 'To relieve this sense of helplessness, the problems of human life must be better stated. The *a priori* reasoning of political economy . . . fails from want of reality . . . We need to begin with a true picture of the modern industrial organism . . .'[3]

The social survey, as it came to be called, grew out of passionate engagement. For example, when the Gloucestershire magistrate Sir George O. Paul became concerned about social order and lunatics in his county, he carried out a national survey of numbers of the insane and of the provision for their accommodation as a calculated means to arouse public opinion. The consequence was an Act of Parliament in 1808 that enabled local rates to be used for the construction of public asylums. The private initiative, the survey, the legislation and the policy to construct special institutions consolidated the idea of lunacy as a social question in people's consciousness, a social reality that requires a social response. Later legislation imposed a national asylum system, a permanent central administration, legal controls and a system of data collection. The systematic collection of data about the insane threw up methodological problems of the kind which were later central to social science: how to define what it is that is to be measured (in this example, insanity); how to create data that can be compared across time and place; how to analyse and present data to reveal norms, correlations and trends; and how to link information to decision-making. The record of information, as Foucault stressed, also created new realities – a certain kind of knowledge of social problems. In the case of the insane, the recorded increase in their numbers throughout the century and the continuously rising financial burden of the asylum programme themselves became anxiously discussed questions and made possible the subject matter of so much late nineteenth-century human science, 'the degenerate'. The rising number of lunatics also became an emotive symbol of fears about what industrial progress and city life had done to the national fabric.

There was a comparable pattern in the history of the response to crime in Britain. Revulsion against the brutality of punishment and against the crowds who watched hangings, fear that severe sentencing rendered prosecutions ineffective, belief that urban settlement cut people off from traditional authority, the spectre of radical rebellion – all encouraged support for a reform of policing and sentencing in the 1820s and 1830s. The most active reformers, wedded to Christian moralism and the principle of utility, wanted the criminal's reform rather than retribution. But the application of utilitarian principles to crime required ordered data collection, regularized administration and a comprehensive police force. As a result, social order and disorder came into existence as topics to be systematically studied. On a broader canvas, changed ways of life, like the housing created for and around the factory in the Lancashire mill towns, transformed the social subject. When Engels, who was himself in the cotton business, conducted his impassioned but empirical survey of Manchester life, he made a subject, social class, visible in a new way. The social world in general was a laboratory where human nature was exposed to new experiences and where the researcher, by means of the social survey, found a new subject.

The observational methods used depended on 'statistics', a term which then denoted the systematic collection of numerical data, as in a census. In England, the Manchester Statistical Society was founded in 1833 and the Statistical Society of London in 1834, the latter in the wake of the initiation of a special section on 'Statistics' in the newly constituted British Association for the Advancement of Science (the BAAS, founded 1831). The members of such societies believed that tables of morbidity and mortality, of the incidence of crime or insanity, of information about working hours or nutrition and so on, are at one and the same time the basis for social policy and social science. The result was a mass of data about the life of the people. Modern statistical analysis, as opposed to data tabulation, developed as a search for patterns and order in this information. It concentrated on two types of regularity: the average of a distribution of measurements of the same phenomenon, and the existence of correlations between two kinds of measurements. This work in social science was of great importance for the development of statistical argument in mathematics and the physical sciences. Statistical argument opened up the philosophical thought that the world itself is in some fundamental sense indeterminate, that is, that individual physical events can never be precisely predicted in the way that earlier mechanist philosophers thought possible. The apparently mundane world of poverty, crime and illness thus became connected with arcane knowledge of the universe. There was a shift during the nineteenth century from the practice of statistics as enumeration to the use of statistics to analyse complexity to reveal order – a major intellectual event. The outcome was a new philosophy of nature, indeterminacy, a new form of inference, statistical mathematics, and – this is the main point here – a new social science that claimed to demonstrate law-like patterns beneath the vast complexities of social life and to provide civilized society with the tools for its own self-regulation.

The work of the Belgian astronomer and servant of the state L.-A.-J. Quetelet (1796–1874) was an important bridge between the eighteenth-century science of the state, descriptive statistics and the new analytic statistics. Sent to study astronomy in Paris in 1823 by the newly unified kingdom of The Netherlands, he became interested in the mathematical theories of probability developed by Pierre-Simon de Laplace (1749–1827) and others. These theories, given systematic mathematical expression by C. F. Gauss (1777–1855), were applied in astronomy as a way to arrive at the correct observations of stellar positions from multiple observations by different observers. It was found that the observations were regularly distributed. Phenomena which exhibited such a regular pattern were later said to show a standard distribution, and in the 1890s the English mathematician Karl Pearson (1857–1936) formalized the description of such a distribution in terms of the normal distribution curve. These ideas made it possible to describe the average of a set of observations. In the early nineteenth-century studies, the variations of a measurement were treated as errors, divergences from an ideal, accurate or true measurement. This way of thinking about the spread of data was perpetuated by Quetelet, who turned the mean of the measurements – the ideal, true measurement – into the notion of *l'homme moyen*, the average man, an ideal of a 'true' man. Through his work, then, a prescriptive norm of a human being acquired status as the mean of variation within the population, and this appeared to give the norm a natural or objective status. In the 1860s, Francis Galton (1822–1911) in England began to study variation as an object of knowledge in its own right, rather than merely as material with which to calculate the average, but the same interdependence of calculations about the distribution of variations and judgement about norms, or social values, was evident in his work.

Quetelet returned to Brussels where he became both director of the observatory and officially involved with the collection of social statistics. After Belgium achieved autonomy as a separate state in 1833, he became perpetual secretary of the Académie Royale des Sciences, des Lettres et Beaux Arts de Bruxelles and the pivotal figure in the country's scientific life. In 1841 he organized the Commission Centrale de Statistique, an agency for the collection of statistics that was regarded as a model by other countries. In a way that was characteristic for statistics at the time, he enumerated data to establish what he believed to be facts and not probabilities. In this intellectual context, he argued that the enumeration of measurements of any one phenomenon, whether of the position of a star or the height of recruits in the army, revealed the average as a fact and showed that differences from the average are regularly and systematically distributed.

Quetelet used the term *'mécanique sociale'* (social mechanics, a term that alluded to the work of Laplace) and then appropriated the term *'physique sociale'* (social physics) to describe the science of social facts established by means of statistics. It was Comte's ire at having a term he himself used taken to refer to something other than his own science, which was unsympathetic to enumeration, that prompted him to coin *'sociologie'*. What really seized Quetelet's and the public's imagination was the way that enumeration revealed patterns beneath the surface complexity and apparent randomness of social life. People were especially struck that a crime like murder recurred at a more or less constant rate over the years even though any particular crime is unpredictable. The winner of a prize for a French book on suicide in 1848 stated what by then was a commonplace observation: 'All the facts . . . tend to demonstrate this remarkable proposition ... that moral facts obey, in their repetition, laws as positive as those that rule the physical world.'[4] It appeared possible to refer with equal factual accuracy to the position of a star, the average girth of soldiers and the murder rate, even though in each case the description of what the fact is came from measurements of many instances of the relevant object. Here was a very persuasive argument that social science could indeed be a social physics.

These considerations enriched Quetelet's conception of *l'homme moyen*, his representation of what is normal and essential as opposed to what is variable about being a person. He accorded the concept immense status in human science:

> This determination of the average man ... may be of the most important service to the science of man and the social system. It ought necessarily to precede every other inquiry into social physics, since it is, as it were, the basis. The average man, indeed, is in a nation what the centre of gravity is in a body; it is by having that central point in view that we arrive at the apprehension of all the phenomena of equilibrium and motion ...

His notion of the average man represented an ideal; but he thought that statistics, not judgement, reveals what that ideal is. Statistics therefore promoted a model of the human figure for the artist in pursuit of beauty to reproduce, a standard of health in medicine and a moral value all rolled into one. The representation of the norm also created a language for the representation of deviations from the norm, and this language was much used in late nineteenth-century theories of degeneration. The language was used about deviations above the norm ('genius') as well as deviations below ('idiocy'). One interesting side path of Quetelet's discussion is the argument that measurement of deviation measures the degree of civilization: higher civilization, he believed, 'more and more contracts the limits within which the different elements relating to man oscillate'.[5] Civilized society, he thought, learns how to regulate and order action around an ideal, and this ideal is a measurable norm.

Quetelet wrote a widely read book that brought together what was known about the statistics of human nature, *Sur l'homme et le développement de ses facultés (A Treatise on Man and the Development of His Faculties,* 1835). His data show how many otherwise unremembered people were involved with data gathering and, by this means, assimilated part of a scientific worldview. His figures on crime, which demonstrated the 'remarkable constancy with which the same crimes appear annually in the same order', deeply impressed his contemporaries: 'Sad condition of humanity! We might even predict annually how many individuals will stain their hands with the blood of their fellow-men . . . pretty nearly in the same way as we may foretell the annual births and deaths.' Many readers thought that this claim raised the spectre of moral determinism and fatalism in human affairs. That, however, was not Quetelet's point. He believed the crime statistics show that man, considered collectively, is subject to natural law like every other aspect of God's creation. There is, he believed, a natural law that links crime and national circumstances, and the outcome is a regular quantity of crime. Human moral will, he held, influences events in particular cases, but it does not alter the natural laws that affect the species.

> All observations tend likewise to confirm the truth . . . that *every thing which pertains to the human species considered as a whole, belongs to the order of physical facts*: the greater the number of individuals, the more does the influence of individual will disappear, leaving predominance to a series of general facts, dependent on causes by which society exists and is preserved.[6]

This was a powerful argument in support of a naturalistic approach to human life, and it interacted in the mid-century with the values of phrenology, physiology, evolutionary thought and mental science to encourage belief that social policies must be founded on the facts of man's condition, not on wishful thinking about moral choice. Nevertheless, this commitment to social science was itself a moral choice, and Quetelet and his fellow collectors of statistics saw their work as part of a new social ethics to guide individual as well as collective moral action.

Statistics therefore appeared to be the means by which the

study of social facts is made as objective and as precise as the study of physical facts, and the means by which social science, like physical science, uncovers general laws. The French physician and writer on social hygiene Casimir Broussais exclaimed in 1837: 'How . . . [statistics] unravel the chaos of particular facts!'[7] The capacity to reveal a hidden order legitimated the large-scale administrative investment in statistics-gathering agencies in both continental Europe and the English-speaking world, and social science was the beneficiary. The official French census forms, for example, carried questions designed by social scientists to answer scientific questions. Quetelet was both the leading light of the Belgian administration's statistical service and the founder of international congresses in statistics whose purpose was to consolidate the subject as a scientific discipline. His work was then taken further in France by Louis-Adolphe Bertillon (1821–83), with Broca a founding member of the Société d'Anthropologie in Paris, who was appointed to a chair in 1877 for his work on demographic questions. With brevity and accuracy, Bertillon wrote in relation to statistics: 'This modest science . . . has become . . . humanity's bookkeeper . . .'[8] One son, Jacques Bertillon (1851–1922), headed the statistical service for the city of Paris and focused concern on the low or even declining birthrate in France. Another son, Alphonse (1853–1914), developed a system of using photographs to record systematically the frontal and profile views of criminals as a means to their identification. His system – '*bertillonnage*', which was applied by the judicial administration in Paris in 1882 – used classification of appearance based on the distribution of variables around a certain number of means. In England, concern with public health reform also strongly promoted statistics. The Registrar General's Annual Reports, the official statistics, were written by one man, William Farr (1807–83), whose techniques were taken up in France. Over a thirty-five-year period from 1843, he created a body of consistent data for the study of morbidity, mortality and marriage patterns.

Another English argument associated with statistics had a European-wide audience, Henry Thomas Buckle's *History of Civilization in England* (1857–61). Buckle (1821–62) wanted to create a science of history, a social science comparable with the natural sciences and based on natural laws of change. He recognized that the study of history must deal with events of great intrinsic complexity and without the benefit of experiment, and he accepted that it is liable to be biased by emotion; but he believed that statistics makes it possible to overcome these difficulties and describe empirical laws. In practice, the sheer weight of material changed what he initially planned as a universal history into a history of England, and that history linked events to the conditions of physical geography, climate, food and soil. Buckle, whatever the weaknesses of his project, had an audience; his book appeared to be the quintessence of determinism embodied in a scientific worldview and applied to the human sphere – and people embraced it or attacked it accordingly. Statistics and history, he wrote, drive us 'to the conclusion that the actions of men, being determined solely by their antecedents, must have a character of uniformity, that is to say, must, under precisely the same circumstances, always issue in precisely the same results'.[9]

The descriptive science of statistics gradually included more analysis; most commonly, the tabulation of the distribution of measurements around what was loosely called the average was the exemplar. The possibilities for analysis were then transformed by Galton who, in the 1860s, began an extended study of the distribution of human physical and mental variation. He was interested in variation itself, not just as a way to calculate an average. Nevertheless, in arithmetical practice, he still treated variation as 'error' and used 'the law of frequency of error' to calculate the average. Further, he tirelessly reinforced his argument that the essence of this human variation is caused by heredity. The older generation in the 1870s initially feared that Galton's work undermined statistics as the empirical basis of social science. In reality, he transformed what was meant by the term in a way that laid the basis for a mathematical theory of the analysis of data, a theory equally applicable to the interpretation of surveys in sociology or variation in animal populations in biology. Galton took up the technique developed by mathematicians to measure and portray the average and the variation in human populations, and this was Pearson's starting-point for the characterization of the normal or bell-shaped distribution curve. Galton also began to study covariance by what became regression analysis, the way to determine whether two or more variables are in fact significantly related. His interest in these techniques derived from a pained consciousness of 'the condition of England', primarily the reality of what seemed to be endemic urban poverty. Galton was committed to the view, which he believed his descriptive and analytic work supported, that improvement must stem from an increase in the reproduction of people with superior inherited qualities rather than from social reform. In the last decade of the century, Pearson and others subjected his statistical ideas to a rigorous treatment and founded mathematical statistics in the modern sense. This mathematical discipline introduced a completely new level of technical expertise into the field of social science. What did not change, however, was the assumption that policy in the modern world requires the empirical and analytic knowledge of the social scientist.

During the early Victorian years of the social survey in Britain, it was assumed that facts, once collected, 'speak for themselves'. Reformers took it for granted that political society would unify around a rational goal and that, if the facts of a problem are laid out for inspection, public opinion will move government to adopt a practical answer. The middle classes began to value the social expert as a figure able to advance a rational form of politics. The late Victorian years saw a partial reversal of the earlier optimism that liberal reason and economics produces social progress, and a rising tide of socialism and 'new liberalism' supported state intervention in social and economic issues. Intervention, it was argued, provides the means by which rational policy, grounded in social science, can deal with the social problems of inequality and poverty that *laissez-faire* fails to solve. Some analysts, like the US campaigning journalist and self-styled political economist Henry George, even argued that *laissez-faire* creates the problems in the first place. The new interest in intervention gave prominence to social scientists as the people equipped to carry through rational policies. Belief in the alignment of government with social science was characteristic of the Fabians (the Fabian Society was founded in London in 1883–4), especially Beatrice and Sidney Webb, who were instrumental in the establishment of the London School of Economics and Political Science, which officially recognized sociology in 1903. This school became the single most important institutional base in Britain for economic and social science. Its founders, and a good proportion of their successors, conceived of it as an institutional link between objective analysis and policy. The beliefs and the school laid the foundations for the post-1945 welfare state, but the involvement with the political process was not especially conducive to the establishment of sociology as a distinct academic discipline. When it came to the appointment of a professor of sociology in 1907, L. T. Hobhouse (1864–1929), who was finally chosen, qualified more because of his commitment to new liberal policies rather than because of past accomplishments in science. Indeed, until 1945, the empirical social survey and scientific sociology remained separate activities in Britain.

ii *Emile Durkheim and the French Third Republic*

An academic discipline of sociology became a reality in France in the period before World War I. This was largely Durkheim's achievement; in the course of his career he established a theoretical, methodological and institutional framework for the subject with remarkable thoroughness. Most of his claims have not survived critical analysis, but the agenda he set – and the intellectual status sociology acquired in the French *sciences humaines* – had a lasting influence. He was inspired by an ethic of patriotism and secular reason, and in his hands sociology was a mode of life not a detached mental exercise. He had vehement critics on the political right and on the political left, but no one doubted that the relationship between scientific reason and moral reason is intrinsic to social science.

Durkheim was the son of a rabbi from Lorraine. From the margins of France geographically and the margins of French society socially, he became professor of the science of education and sociology at the Sorbonne in Paris, adviser to the Minister of Education, a passionate architect of secular republican society and an ardent Frenchman who lost his son in the war. His career was like that of many other brilliant intellectuals with a Jewish background, men and women who fully identified with the secular nation states in which they lived and believed that these states could make rational progress, belief that ended in calamity under Hitler and Stalin. For Durkheim, who did not outlive World War I, this process of political and cultural assimilation was reflected in his fascination with the conditions of social stability and the manner in which the individual acquires a sense of purpose through society. The subject of solidarity was high in French consciousness after the searing experiences of defeat by Prussia in 1870 and of the Paris Commune of 1871, and sensitivity was increased by struggles between Catholic monarchists and secular republicans in the last decades of the nineteenth century. Republican interests slowly consolidated their hold on political power, but Durkheim thought that the Third Republic needed all the support it could get.

Republican policy reorganized education at all levels, including the reform of the universities to create departments committed to scientific research along German lines. With his appointment to the Faculty of Letters at the university of Bordeaux in 1887, Durkheim taught courses in pedagogy and, a new subject, social science. In a climate in which the university in general was restructured, he was able to create a discipline of sociology. Within the space of a few extraordinarily productive years – he meticulously organized his time and his household – he published a study of sociological method, *Les règles de la méthode sociologique* (*The Rules of Sociological Method*, 1895), two major case studies in sociological argument, *De la division du travail social* (*The Division of Labour in Society*, 1893) and *Le suicide: étude de sociologie* (*Suicide: A Study in Sociology*, 1897), and he founded a journal, *L'année sociologique* (1898), to give the new field an identity and to focus the activity of a group of gifted young scholars. He moved to the Sorbonne in 1902, and in 1913 his subject acquired the final seal of approval when his chair in 'the science of education' was renamed 'the science of education and sociology'. From this prestigious platform, Durkheim introduced sociology into the education of teachers and into the school curriculum, developments bitterly opposed by Catholics and conservatives.

If it was Comte who advocated a discipline of sociology, it was Durkheim who brought it into existence. Durkheim developed two arguments with roots in Comte. The first was positivism, the claim that knowledge is restricted to knowledge of the pattern of observations, a position opposed by what was taught in philosophy in Durkheim's youth. He argued that the construction of general laws from observed regularities is the basis of all the sciences, human as well as natural. With the arguments of his teacher, the rigorously scientific historian Fustel de Coulanges in mind, Durkheim set out to show how observation can achieve disciplined results in academic social science. In *Suicide*, for example, he used statistics collected by others to correlate suicide rates with different national, ethical and social conditions. He then used these correlations to explain the likelihood of suicide in terms of the degree of an individual's integration into society and of society's provision of clear-cut rules for conduct. His practice was more like that of the French experimental physiologist Claude Bernard than the social theorist Spencer. Durkheim took up what were originally the biological categories of structure and function as the framework in terms of which to order empirical social observations, just as Bernard ordered physiological observations; for Durkheim, biology did not provide the language – as it did for Spencer – with which to make claims about social evolution. All the same, though Durkheim rejected evolutionary speculation, his conclusions were informed by his prior commitments, by values (shortly to be explained) which had *a priori* rather than empirical standing.

The second point shared with Comte was Durkheim's stress on society and social facts as an independent reality that cannot be reduced to or explained by biological or psychological reality. 'The determining cause of a social fact should be sought among the social facts preceding it and not among the states of the individual consciousness.'[10] This argument again distanced Durkheim from Spencer. Social phenomena, Durkheim believed, require explanation in social terms, and this argument vindicated the rights and value of sociology as an independent discipline. In their eagerness to make this point, Comte (who nevertheless believed in the psychological theory of phrenology) and then Durkheim (who nevertheless held a psychological picture of humans as beings with limitless desires) never adequately related psychological explanation and sociological explanation. Durkheim was understandably scathing about the explanation of social life by reference to great men or to race; his social explanations were developed by means of such concepts as the division of labour and '*anomie*', a condition in which society provides the individual with little or no defined framework in terms of which to make choices. All the same, he presumed that people have a nature that causes them to act in particular social circumstances in one way rather than another. He took this human nature for granted and did not make it a subject for research in its own right, as the later discipline of social psychology was to do. This helped split apart sociology and psychology, which in the short term was perhaps beneficial to the consolidation of the identity of each discipline, but in the long term was detrimental to the achievement of coherent explanations of human action.

Durkheim deliberately chose suicide as a *sociological* topic. People unthinkingly saw it as the most individual and subjective of actions, and so it was the perfect subject with which to raise consciousness about social reality. Quetelet, for example, compared suicide rates as part of a project to measure the chance of any particular individual committing suicide. In Durkheim's view, this was not sociology. The first stage of his argument used statistics to correlate suicide rates with different conditions, such as the existence of Protestant or Catholic beliefs and rural or urban communities. But his ambition went much further: he wanted to explain the likelihood of suicide by general social features, of which egoism and *anomie* were the most important. Egoism is a measure of a society's failure to become the focus of the individual's sentiments and it makes the individual unintegrated and purposeless. *Anomie* signifies a society's lack of norms and it leaves the individual unregulated.

> Both spring from society's insufficient presence in individuals ... In egoistic suicide it is deficient in truly collective activity, thus depriving the latter of object and meaning. In anomic suicide, society's influence is lacking in the basically individual passions, thus leaving them without a check-rein.

The real focus of his analysis was therefore social solidarity not individual suicide. He did not explain, or acknowledge that it might be valuable to explain, why this or that particular individual commits suicide. Like Comte, he implied that this question is one for biology or medicine, as he tended to believe that deviance represents an organic failing, rather than one for psychology. Nevertheless, the concepts of egoism and *anomie*, which were explanatory categories rather than empirical conclusions, presuppose psychological claims whether he intended this or not. He did not avoid assumptions about motivation. Behind his emphasis on *anomie* lay the unprovable assumption that 'the more one has, the more one wants, since satisfactions received only stimulate instead of filling needs'.[11]

Durkheim's interest in social regularities applied to the study of suicide became a much discussed model of sociological explanation. The power and originality of his work can be brought out through a comparison with contemporary studies by Gabriel Tarde on *Les lois de l'imitation* (*The Laws of Imitation*, 1890) or Gustave Le Bon on *Psychologie des foules* (1895; translated as *The Crowd – A Study of the Popular Mind*). This contemporary work put forward what were believed to be psychological capacities, such as imitation and suggestion, to explain social phenomena, as if society could be understood as the sum total of individual attributes. Tarde and Le Bon deployed versions of what historians call a medical model. This model began with an idealized conception of the normal, rational, moral male individual and then explained social disorder as a malfunction attributable to individuals who deviate from this norm – as an ill body deviates from a healthy one. French writers, indeed, devoted considerable space to individual degeneracy in order to explain the social malaise of the *fin de siècle*. In contrast, Durkheim tried to understand social order and its breakdown in terms of what it is that causes the individual to act morally and with a sense of purpose, and he found this cause in society. He argued that people's nature embodies social facts, not simply that people obey social rules. Society is 'something beyond us and something in ourselves'.[12] Without that 'something beyond us', we are vulnerable to egoism and *anomie*.

Behind Durkheim's interest in sociology in general, and behind his focus on suicide in particular, was a political commitment to France's stability and to the contribution education might make to it. His answer to the problem of France's insecurity was to inculcate the values of a certain kind of individualism. Durkheim understood the value in which individuals are held to be a social fact, a reality in a particular kind of social order. He believed that the politics of republican nationalism, but not socialism or conservative monarchism, enables this fact to flourish. Thus his sociology, which studied the conditions in which the value of the individual is increased by the integration of individuals in a certain kind of social order, was also a defence of the Third Republic. This is illustrated by his study of the division of labour. Like Spencer before him, he assumed that the division of labour is a historical inevitability: modern society irrevocably separates men from each other as individual agents. (He simply did not discuss women.) Modern society is therefore faced by the problem of how to achieve social solidarity. His answer was the sociological study of those conditions in which the value of individualism exists, and where the place that individuals have in the state, institutions and occupations is recognized. Hence, during the Dreyfus Affair which divided the country, Durkheim spoke out passionately for Dreyfus, the individual (and like himself of Jewish background), against the representatives of the state who accused him of spying. He defended the state through the defence of individualism, in opposition to conservatives who sought directly to defend the value of the state. Sociology, he argued, is the objective means to enhance the true values of individualism. It should, therefore, have a central place in pedagogy and in the training of teachers who will in turn train young citizens. Belief in individual dignity, in Durkheim's view, derives from the individual's integration in a society that transcends the individual; to defend the individual, as to defend Dreyfus, is to fight against social dissolution. In this way, he hoped to reconcile individualism, which one conservative critic called 'the great sickness of the present time', with moral obligation. 'Doubtless, if the dignity of the individual derived from his individual qualities . . . one might fear that he would become enclosed in a sort of moral egoism that would render all social cohesion impossible. But in reality he receives this dignity from a higher source [in society], one which he shares with all men . . .'[13] The difficulty with this argument is that Durkheim described a social fact – the value of individualism – as both the source of moral action and the authority for moral action. Here he came up against the question of the place of values in social science which caused debate throughout the twentieth century. Weber struggled with this question head-on.

Durkheim further developed his argument about social order with an influential treatment of religion as a social phenomenon and, as such, a source of moral purposes. In his last book, *Les formes élémentaires de la vie réligieuse: le système totémique en Australie (The Elementary Forms of the Religious Life: The Totemic System in Australia,* 1912), he provided a sociological analysis of religious belief as well as religious institutions. His data, which were derivative and later questioned, illustrated the supposedly simplest and most primitive of religious societies, the Australian aborigines. Durkheim attempted to explain aboriginal totemism in terms of the function of totemic belief in the maintenance of social stability. He attributed beliefs about spiritual things to their function in the divisions of the social structure; he explained the parts, such as totemic beliefs, by reference to the whole, social stability. For Durkheim, belief constitutes a *conscience collective,* a social reality, though not a physical one, in which the individual shares. What people hold to be true about the world is a social phenomenon that requires explanation in social terms. In this important argument, Durkheim contributed to what became known as the sociology of knowledge, an approach to belief through sociology rather than either psychology or philosophy.

It is, Durkheim thought, the *conscience collective* and not the physical world that is the source of the meaning of a person's life. The stability of the *conscience collective*, he argued, constitutes a social inheritance that gives human nature its morality and sense of purpose. It is social reality that is real and enduring, not God, the transcendental ego, race or some other metaphysical entity. This was an argument that knowledge of the human being as a subject necessarily comes through the study of social facts, even when the subject is man's nature. Durkheim, in an article published jointly with his nephew Marcel Mauss (1872–1950) in 1903, teased out some of the complex questions prompted by the sociological explanation of knowledge. Their subject was 'primitive classification', the way in which indigenous Australian people order the natural world and assign totemic significance to natural objects. Durkheim's and Mauss's point was that such classifications are not the product of individual minds but a collective ordering of nature in terms of the order of society. As they argued, 'every classification implies a hierarchical order for which neither the tangible world nor our mind gives us the model', and their conclusion was that classification, the basis of our understanding of the world, is not an abstract act but a social one. 'Things were thought to be integral parts of society, and it was their place in society which determined their place in nature.'[14] The argument explored in 1903 in relation to aboriginal society was, much later, in the 1970s, extended to scientific society, when it transformed both debate about scientific knowledge and the practice of the history of science itself.

Durkheim's work, along with further theoretical discussion of the *conscience collective* by Mauss, influenced French social science and, especially through A. R. Radcliffe-Brown (1881–1955) at Oxford in the 1930s, Anglo-American social anthropology. Though a student and colleague of the anthropologist Boas rather than of Durkheim, the US scholar Ruth Benedict (1887–1948) wrote in *Patterns of Culture* (1934) that:

> No man ever looks at the world with pristine eyes. He sees it edited by a definite set of customs, institutions and ways of thinking. Even in his philosophical probings he cannot go behind these stereotypes; his very concepts of the true and the false will still have reference to his particular traditional customs.[15]

Although his study of religion was severely criticized, the way in which Durkheim conceptualized in social terms the categories with which we think, of time and space as well as of the sacred, encouraged a profoundly secular human science. Durkheim glimpsed but did not pursue the insight that his own form of social analysis could be applied to his own science, and that if this is done it describes social science as the *conscience collective* of a particular social order of scholars. This kind of reflexiveness, with its implication that knowledge is relative, was largely foreign to the hopes that he and his generation placed in empirical science and in reason. It was for later sociologists to face the difficulty that the process of investigation in social science is itself a social activity that transforms the subject studied and which can also be studied by social science.

In the years before World War I, *L'année sociologique* was the focus for a group of young men who influenced the development of education, economic history, the sociology of law and anthropology. Sociology itself as an institutionally separate discipline did not make much progress – there were only three chairs in the subject even in the 1930s – but it penetrated other fields as a way to apply scientific methods to the human sphere. The idea of sociology, in Durkheim's thought, perpetuated the Enlightenment ideal of a rational, secular science of human nature. Durkheim, however, derived the rational and moral content of human nature from social reality – though his argument left unaddressed questions about psychology and reason. His ideas were taught in the *écoles normales* and in this and other ways – such as through Durkheim's connections with the moderate left in French politics and his long-standing friendship with the socialist leader Jean Jaurès – sociology entered into the culture of those who looked to science to secure France's place in the modern world. In France and beyond, the functionalist analysis of social realities, which merged in the United States with an evolutionary viewpoint, gave intellectual content to the argument that social and economic policy can meet the requirements of order and stability in the modern world.

iii *Max Weber and Wilhelmine Germany*

Germany was formally unified on 1 January 1871 under Prussian leadership and after Prussia's military triumphs against Austria and Denmark in 1866 and France in 1870. Unification expressed confidence in national strength and purpose. But there was also fear of encirclement, consciousness that many German-speaking people still lived outside the Reich, a desire for a world empire and anxiety about the social disruption of rapid industrialization. The transition from small-town rural society to large-scale urban life was especially sharp. It was all too easy to believe that the old communities represented the 'real' German way of life, rudely broken by entrepreneurial capitalists, if not Jews, and a mass working class. An attempt to understand modernity therefore absorbed intellectual energies and gave sociology its subject matter in the German setting. A few brilliant but not necessarily influential academics hoped by this means to reconcile high culture and modern conditions.

The deepest reflections were those of Weber who, though he died of pneumonia in 1920, exerted a profound influence over the discipline of sociology's self-conception, especially after 1945. Mannheim, a Hungarian sociologist of the next generation, noted that 'Max Weber's whole work is in the last analysis directed toward the question "Which social factors have brought about the rationalization of Western civilization?"'[16] 'Rationalization' was Weber's term to describe what makes modern Western society distinct, to compare the historical development of Western and non-Western societies and to confront his own country with its actual social, political and economic situation. Some liberal observers, from a retrospective viewpoint, felt that Germany's failure in the 1920s to listen to Weber's sober analysis led it instead down the path to National Socialism and ultimately to the division of Europe. But Weber's views were neither consistently liberal nor clear-cut; he was an ardent patriot and his life attempted to hold together deep intellectual contradictions.

Weber was one amongst many scholars to polarize the premodern and the modern age, as a partial list makes clear:

TRANSITION TO MODERNITY

Herbert Spencer: militant [military] – industrial
Karl Marx: feudal – capitalist
Henry Maine: status – contract
Max Weber: traditional authority – rational-legal authority
Ferdinand Tönnies: *Gemeinschaft* (community) – *Gesellschaft* (association)

These terms provided a language for an engaged response to modern conditions. They also related individual psychological conditions to large-scale shifts in material and cultural existence. The German discussions, more than French or English-language social thought, took seriously the contribution of historical knowledge to social science. The German intellectual setting, as the example of Marx as well as Weber demonstrates, encouraged social analysis in the context of academic history, especially economic history, informed by a historical sense of Germany's place in world civilization.

Weber's response to the modern world was intense and anguished. He referred to the modern pursuit of wealth as 'an encasement, hard as steel' (more commonly translated as an 'iron cage'), an inescapable pursuit of what belongs to Caesar in conflict with the values of the Sermon on the Mount.[17] This conflict of material and moral values was also a personal question for Weber – it divided his father and his mother – and he suffered an extended breakdown following his early appointment to prestigious chairs in political economy at Freiburg and Heidelberg in the last decade of the nineteenth century. He took it for granted that the scholar has a high calling and confronts contradictions with absolute honesty, so he must come to terms with the industrial and technological world whatever his values. Weber was therefore frequently at odds with other academics, many of whom yearned nostalgically for an age when the high culture of the universities, they believed, had seamlessly represented the national spirit. He was no less patriotic than his conservative colleagues, but he thought patriotism best served by analysis of the realities of power, economy, social structure and belief. He was fascinated by political sociology, the study of the conditions under which different political arrangements exist. In the context of his search for a democratic order appropriate for Germany, which involved him in party politics as well as academic study, he formulated a much-discussed description of charisma in political leadership. He hoped that charismatic leadership might square the circle of political life and preserve ideal values in a democracy.

In 1872 a group of younger scholars, predominantly economic historians, founded the Verein für Sozialpolitik (Society for Social Policy), which supported research into social conditions and a liberal-reformist response to industrialization. Some members of the Society argued that only a historically and empirically grounded economics, not a deductive science based on an ideal type of economic man – as found in British political economy – will underwrite a social policy able to contain working-class demands and counter class-based Marxist analysis. As scholars they could not accept, as many Marxists accepted, that the economy determines the higher meaning of life. All the same, they believed that the Second Reich in which they lived had to come to terms with industrial conditions and not expect to turn the clock back. They therefore encouraged the state to develop relevant policies, such as a system of national insurance, to deal with working-class poverty. This was the background to Weber's early political involvement. At the end of World War I he went further: he contributed to the draft of the Weimar constitution and helped found the German Democratic Party, a liberal centrist ally of the socialists who came to power when the republic was founded. Weber wanted social science to serve the political state, especially through its analysis of inescapable modern social and economic conditions. This conception of the scholar as rational and autonomous, as well as patriotic, distanced him from both the radical left and the conservative right.

Weber wrote on many subjects. After early work on historical topics, he became involved with modern conditions through a survey of rural labour and the agricultural depression in Prussia in the 1880s. He later studied the history of religion in India and China, to provide a comparative perspective for Western science and Western economic development. He planned a summation, though what was later published as *Wirtschaft und Gesellschaft* (*Economy and Society*, 1922) was cut short by his death. He was closely associated, from 1904, with editing the *Archiv für Sozialwissenschaften und Sozialpolitik* (*Archive for Social Science and Social Policy*) and he helped found the German Society for Sociology in 1909, at about the time when he first described himself as a sociologist. He was a friend and intellectual colleague of Georg Simmel (1858–1918), whose characterization of the city became a classic text in discussion of modernity, a friend of the theologian and social historian Ernst Troeltsch and one of Nietzsche's closest readers.

What became the best known of Weber's studies is *Die protestantische Ethik und der Geist des Kapitalismus* (*The Protestant Ethic and the Spirit of Capitalism*, 1904–5). Weber argued that the crucial economic development in the modern world, capitalism, should be understood historically in relation to mental conditions, especially the meaning of work, saving and investment for seventeenth-century Dutch and English entrepreneurs. He claimed that their interpretation of Calvinist theology associated spiritual salvation with an ethic and psychology of work and reinvestment of material wealth, which transformed commerce and production. The new economic activity, Weber suggested, flourished in social conditions of order and predictability, and this encouraged the rationalization of law and social administration. Power and status shifted away from traditional élites, the nobility and the clergy whose activities and beliefs began to appear irrational, to new classes who exhibited practical, rational activity in economic management and social affairs. Against this background, the understanding of nature shifted from magic to science. Man's knowledge also became practical and rational. Thus Weber attributed a major social transformation to reflective consciousness, though he did not intend to put forward his account of the Protestant ethic as a causal argument and it has not stood up as causal historical explanation. Rather, he outlined what he termed an 'ideal type' of historical action, one that resulted from the coincidence of an ethos found in Puritan sects and capitalist life. This historical study introduced terms for the evaluation of the modern world – 'modernity' – that became basic to the twentieth century's critical vocabulary.

When Weber discussed a historical event such as the rise of capitalism in terms of ideal types, he initiated a model of sociological explanation. The concept of an ideal type is a difficult one. His types did not denote actual social realities, such as the work of a group of merchants, but categories in terms of which the sociologist reasons, firstly, with generality rather than particularity, and secondly, in a way that takes into account the meaning of ways of life (such as the Protestant work ethos). He deployed a theory of knowledge indebted to Kant that makes the rigour of science depend on the rationality of the categories employed while we think about something. 'Although we are far from thinking that it is valid to squash the riches of historical life into formulae, we are still overwhelmingly convinced that only clear, unambiguous concepts can smooth the way for any research that wishes to discover the specific importance of social and cultural phenomena.'[18] Weber's ideal types are concepts intended to make it possible to bring order to the chaos of individual historical facts; they are not the driving forces of history. Weber also hoped that social thought expressed in terms of ideal types would enable social scientists to step outside the value systems to which they themselves are committed. He wanted to find a way to escape assumptions about the purpose of history, notably about Germany's place in history, which he thought distorted so much contemporary social thought, whether conservative or socialist.

As these arguments suggest, Weber was at one and the same time intensely scholarly and passionately political. His attempt to mediate between these two dimensions, which so many twentieth-century scientists, by contrast, claimed should be kept apart, gave sociology a language in which it continued to debate the relationship between facts and values and between social science and the political process. The issues were complex and divisive; it was the debate rather than any agreed answer, therefore, that became distinctive of the sociology discipline.

Weber claimed value neutrality for science in general and

social science in particular; that is, he believed it to be the task of social scientists to state the facts and causes of social phenomena in an objective way free from personal or collective values. He articulated this viewpoint initially as part of a political stance, an intervention in a long-running debate among social theorists about whether a particular group within German society, civil servants – a group that included academics themselves – objectively represents the higher purposes for which the state exists. Weber's view was that science cannot itself give authority for such a claim to objectivity on the part of a social group, and he opposed conservative academics who said it could. Here again he used an argument indebted to Kant, this time to separate moral and evaluative claims from truth claims. Weber, and the liberal theorists who agreed with him, demanded that idealist values – many of which Weber in fact shared – be identified as values and not taken to be scientific truth. In practical terms, this position required that the civil service and universities be treated as part of the political process, not as institutions above it, and therefore questioned the self-image of the university as the guarantor of transcendent values in German society. In his best-known discussion of these issues, Weber described how the choice to study a particular topic, like his own research on modernity, is itself an evaluative act. More deeply, he argued, scientific analysis does not reveal values or an answer to the question, how to live. His lecture to students at the university of Munich in 1919, 'Wissenschaft als Beruf' (Science as a Vocation), was a courageous and subsequently famous attempt to delimit scholarship at a time when defeat, revolution and despair bolstered fantasy about the way forward. In reponse to what, after the Russian novelist Tolstoy, he called 'the only important question: "What should we do? How should we live?"' – Weber said that: 'The fact that science does not give us this answer is completely undeniable.'[19] It was an argument with a bitter message in the German context since many academics, who claimed to be scientists in the broad sense, thought that truth derives from the place of facts or events in relation to the overall purposes of human existence and history. Weber also said that the scholar should not expect to lead national affairs but should equip the nation with rational knowledge of the circumstances in which it pursues values.

Weber also raised questions in the philosophy of scientific knowledge about which there was already extensive argument and which was in turn part of a dispute about the status of the different sciences in the universities and in the national culture. While his work received some attention in the German-speaking world before and after World War I, it was discussed more extensively when it was taken up in the United States and translated into English in the 1940s and 1950s. At this later time, his discussion of value neutrality was viewed as a straightforward argument for the distinction between facts and values, the implementation of which was thought to give sociology authority as a science. This interpretation was possible at a time of high confidence in sociology's standing as a science comparable to the physical sciences, but it did not accurately represent Weber's position. In discussions in the 1970s and 1980s, his arguments were described yet again in more complex terms and it was conceded that they were possibly contradictory – but, if contradictory, contradictory for the very good reason that he addressed deep and unsolved questions about science. He turned his question, 'What should we do?', on science itself and asked why we should study science at all: 'Does [scientific] "progress" as such have a recognizable meaning that goes beyond technical ends, so that devotion to it can become a meaningful vocation? . . . what is the *vocation of science* within the totality of human life and what is its value?' And his answer was that 'it is obvious that all of our problems lie here, for this presupposition [that science is worth knowing] cannot be proved by scientific means'.[20] Scientists who popularized science on the grounds that it is true and that it produces practical results did not face up to this point: the pursuit of truth and the judgement of what is practical are values that cannot originate with science. Weber tried to face this issue, but it left him in the personally harrowing position that he sustained commitment to social science by an act of will, a commitment to reason without reason. In addition, he turned constructively to comparative historical sociology to explain how 'the vocation of science' had become a value in the West, and he turned to politics to make his value choices meaningful in a fully active life.

Weber made substantial contributions to theoretical and empirical sociology, though this subject, as an academic discipline, existed in Germany only after his death. Whereas Durkheim was heir to Comte's positivist conception of social explanation by social facts, Weber was heir to German philosophy's conception of social explanation by reference to the meanings that actions have for rational consciousness. He rejected both Hegelian idealism and Marxist materialism and thought of explanation in the human sciences as the uncovering of the meanings that actions have for conscious persons in interaction with other such actors. He proposed a basic classification of four types of action: purposeful, goal-oriented action, with rational means and ends, exemplified by an engineer who builds a bridge; value-oriented action, when we strive with rational means for a goal that is itself not necessarily rational, exemplified by the pursuit of science; affective or emotional action; and traditional action, action dictated by custom and authority. Armed with this classification, he then argued that modernity exhibits an overwhelming commitment to goal-oriented action, the organization of social affairs by rational means for the rational ends of efficiency, order and material satisfaction. In Weber's view, modern Western societies are marked by rationalization, a history of legal, commercial and bureaucratic structures that have increasingly determined human relationships and purposes. He noted clearly that the means of rationalization has included modern science – social as well as physical.

The modern world, Weber thought, is one in which the gods neither have nor can have a home. Commercial and industrial society, for all its material benefits and the apparent freedom of action it gives to people, brings disenchantment, the elimination of a sense of the magical in ordinary experience and a loss of spiritual purpose for mankind.

> The fate of our age, with its characteristic rationalization and intellectualization and above all the disenchantment of the world, is that the ultimate, most sublime values have withdrawn

> from public life, either into the transcendental realm of mystical life or into the brotherhood of immediate personal relationships between individuals.[21]

We are in a modern age in which a need for value-oriented action cannot be addressed but only massaged by fantasy, consumption and sentiment. We are, as Weber quoted, 'specialists without spirit, sensualists without heart'.[22] This critical judgement challenged people in the twentieth century who were faced by ever more powerful technology, ever more goods for consumption and ever more bureaucratic regulation of life. In contrast to Weber, in both the Soviet Union and the United States, some human scientists – vulgar Marxists and behaviourists at the head – embraced goal-oriented rationality as the final freedom in which humanity discards a superstitious and barbarous past for the brave new world.

The distinction between value-oriented and goal-oriented action relates to another dimension of Weber's philosophy of knowledge in the human sciences. I return to this in the concluding chapter, as it is central to debates about scientific knowledge at the end of the twentieth century; but Weber's basic terms may be mentioned. He used the word *'Verstehen'*, the term Dilthey also stressed, to denote the way it is possible to understand human action because 'the course of human action and human expressions of every sort are open to an interpretation in terms of meaning . . .'[23] This definition drew a contrast with non-human events, which are not actions and have no meaning in themselves (though they have meaning to people who observe or take part in the events). In relation to human actions, Weber then distinguished the manner in which we understand through our description of causal or psychological motives and the manner in which we interpret the meaning of the actions. He examined ways to understand action; he then related action to different forms of life, in one of which values have priority and in the other where practical goals take precedence. This language integrated Weber's thought with a contemporary debate about modernity and especially with the much discussed distinction his fellow social theorist Ferdinand Tönnies (1855–

1936) introduced in 1887 between characteristically pre-modern *Gemeinschaft* (society based on community) and characteristically modern *Gesellschaft* (society based on association). Tönnies defined two types of social structure and two corresponding types of 'will' or psycho-social forms of action: 'natural will', in which action is the result of the organic, historical growth of a community; and 'arbitrary will', in which action is rationally planned. He and Weber sought to describe the sense of difference, then so widely felt, between a person's life expressive of the sacred or unquestioned values of a community, and a life, supported by science, expressive of a rational commitment to the industrial and bureaucratic world. Historical interpretation was intrinsic to their work, and they attempted to be objective about the different forms of life and on this basis to make sociology a science. They both considered that modern rationality is inescapable, but they hoped – Weber also perhaps despaired – that reason might mitigate the ravages of reason. At times, however, especially in the public domain, many scholars who described the difference between the past and the modern age edged close to irrational nostalgia.

Weber, Tönnies and the Berlin sociologist Simmel were preoccupied with modernity as a social, economic, political, psychological and ethical question. Simmel's 'Die Grostädte und das Geistesleben' (The Metropolis and Mental Life, 1903) vividly characterized many of the issues, and the essay was frequently to be culled for its picture of modern conditions. He explained that: 'The psychological foundation, upon which the metropolitan individuality is erected, is the intensification of emotional life due to the swift and continuous shift of external and internal stimuli.' He viewed the city as a new space of human interaction, a space that both excites and alienates, a place that leads to 'the atrophy of individual culture through the hypertrophy of objective culture . . .'[24] In a similar vein, he wrote an original essay on 'the stranger'. As a sociologist, his treatment of the city was part of a more general discussion of social space, an attempt to develop an abstract science of individual interactions. His study set out to fulfil his ambition for sociology, which was 'to describe the forms of human communal existence and to find the rules according to which he or she is the member of a group, and groups relate to one another . . .'[25]

Simmel, who was born in the centre of Berlin and lived there nearly all his life, was in fact, in 1894, the first teacher of a course specifically on sociology. Though he attracted large numbers of students, conservative academics hampered his career because of his Jewishness and his subject – his studies of social forms like money and social differentiation were thought to detract from the higher purposes that make social relations subservient to the state. His most original book was the *Philosophie des Geldes (Philosophy of Money,* 1900) which, from the starting-point that money is the purest expression of exchange between individuals, strove to demonstrate 'the possibility . . . of finding in each of life's details the totality of its meaning'.[26] Simmel's subject matter was not society but social forms, the entities, such as money, produced by individual interactions. His work therefore held together psychology and sociology at the very moment when, like Durkheim, he also claimed for sociology a special content. But, unlike Durkheim, he openly tried to integrate psychology. Simmel was influenced by the *Völkerpsychologie* taught in Berlin, which treated linguistic and cultural expression as the product of psychological activity in social relations, and he also studied ethnology with Bastian. Simmel's sociology was therefore very different from that of Durkheim, who was antagonistic to psychology. Simmel was an influence on G. H. Mead and, through him, on sociology as it developed at the university of Chicago in an academic setting that focused research on the individual in social relations.

Social thought in Weber's world was many-faced. The study of social questions was increasingly undertaken by specialists with their own academic and professional concepts and institutional interests. Weber contributed substantially to this process, which in the German setting occurred in close association with the disciplines of economics and history rather than independently. Social thought also reflected the concern felt by a small but active group of academics, some of whom were liberal or socialist in orientation, to reconcile Germany with its new industrial and urban identity. They expected social science to make a difference to the practical affairs of the state. The discipline was also an intense moral investment, since the understanding of social action required answers to questions about the meaning of modern life, the loss of transcendent purposes and the rationalization of the public sphere; and it also raised basic philosophical questions about the nature of scientific knowledge. Weber's sociology pointed towards a human science that is both a disciplined expertise with social utility and a means to explore the human condition in the twentieth century. Ultimately, as he well saw, science is a moral vocation.

iv *Sociology in the United States*

Weber's lecture on 'Science as a Vocation' prompted critical responses. Critics were primarily interested in the duty of academics and science to restore German culture and to answer the question 'What should we do?', only secondarily in the nature and future of sociology as a discipline. Important social theorists like Robert Michels (1876–1936), who examined the self-serving nature of bureaucratic élites, and Simmel remained on the academic margins for most of their careers because they were socialists or Jewish. Economics and economic history had greater academic support, though sociology departments were established in the 1920s in the new universities of Hamburg, Cologne and Frankfurt. In France, by contrast, Durkheim established a coherent intellectual school, though without many academic positions specifically for sociology, with a powerful institutional base in Paris where it mattered. In even greater contrast, in the period from the 1890s to World War I, sociology in the United States became a distinct academic discipline and an occupation on a large scale. By about 1910 some 400 US colleges taught what they called sociology and there were some fifty full-time professors in the subject.

The situation for US sociology was comparable to the situation for psychology – and for comparable reasons. Democratic culture and progressive values provided fertile ground for the growth of these human sciences as academic subjects and as occupations in the social world. Late nineteenth-century state universities like Wisconsin and Michigan and private ones like Chicago and Columbia put large resources into social science. Social scientists organized themselves, for example with the foundation of the American Academy of Political and Social Science in 1890, around a commitment to the scientific credentials of a university-based response to social issues. The social scientists, in their own word, aspired to become a 'profession', a body of trained specialists who offer a service to society in exchange for society's support for the discipline and deferral to the scientists' own judgement about the expertise on offer. All the same, though the discipline did indeed grow, professional status was not taken for granted since society did not fully defer to sociological expertise or concede that lay people are debarred from knowledgeable comment on social problems. Further, in the period before 1914, sociology did not have fixed boundaries with other social science disciplines like political science and economics, nor was it a unified subject. In the 1920s, sociologists did establish a large academic discipline more and more preoccupied by academic problems of methodology and knowledge accumulation; in these circumstances there was a move to separate pure and applied sociology and to make the former a rigorous science, though the latter remained the discipline's public justification and its primary teaching obligation.

The creation of urban society was rapid and dramatic in the United States. There was massive immigration from rural Europe and a shift of people from rural America into the cities in pursuit of work and a future. Chicago doubled in size in the 1880s. In the decade before World War I – the Progressive Era under the leadership of Theodore Roosevelt – there was a heady atmosphere of unlimited possibility combined with massive disorientation. Forceful figures like Franklin H. Giddings (1855–1931), who was appointed the first American full-time professor of sociology at Columbia College in 1894, and the dynamic president of the university of Chicago, William Harper, seized the opportunity to persuade multi-millionaires like John D. Rockefeller to invest in social science as an answer to the disorder brought about by social change.

A commitment to social science as social engagement went back at least to the middle years of the nineteenth century, to a generation who believed in a disciplined Christian philanthropy and the anti-state arguments of Spencer. A group of concerned north-eastern gentlemen and college teachers founded the American Social Science Association in 1865 and over the next thirty years this society debated policies – perhaps with few practical consequences – to achieve order and control over problems such as poverty and punishment. The Association was a forum for debate about general historical and economic explanations as well as the collection of data from social surveys. By the 1880s, self-consciously scientific scholars, influenced by European example, accused the Association of amateurism and wanted to introduce systematic training into subjects like history, economics and political science. With his appointment at Columbia, Giddings announced that social science was dead and replaced by scientific sociology. He explained that social science had simply concerned itself with unsystematic data and good works; by contrast, sociology is scientific because it seeks causes for social phenomena as a whole.

> 'Sociology', then . . . is the descriptive, historical and explanatory science of society. It is not a study of some one special group of social facts: it examines the relations of all groups to each other and to the whole. It is not philanthropy: it is the scientific groundwork on which a true philanthropy must build.[27]

The result of the new sociology was specialization, more rigorous – especially statistical – methods and firm academic control, symbolized by the foundation of the American Historical Association (1884), the American Economic Association (1885) and the American Sociological Society (1905), in addition to the American Academy of Political and Social Science.

These changes were associated with a new emphasis on academic training in the universities themselves. A key figure was Daniel Coit Gilman (1831–1908), president of Johns Hopkins University, who directed his own university and inspired others to install graduate schools to provide training in scholarship.

The Social Science Association had aimed both to investigate social questions and to lead reform, but by the 1880s men like Gilman thought that it did neither effectively. Like Giddings, he believed that the basis for something better was to make the social sciences academic subjects with the methods and explanatory aims of the natural sciences. Giddings himself, by the 1920s, was openly comparing social problems and engineering problems. Nevertheless, once social scientists acquired an institutional base in the universities, the values prevalent within the university context focused attention on the status of the social sciences as science rather than as the tools of social reform. The new social scientists had a primary interest in the development of knowledge and techniques valued by their peers. This made the application of social science a secondary concern.

The founding editor of the *American Journal of Sociology* in 1895, Albion W. Small (1854–1926), argued that social action should depend on 'the revised second thoughts' of the scientist rather than 'the hasty first thoughts' of ordinary people.

> The conditions of human association are so involved, that it is no longer pardonable to increase present popular sensitiveness and irritation by theorizing about plans for accelerating the rate of human improvement, unless we have reduced all available pertinent facts about past and present human association to generalized knowledge, which shall indicate both direction and means of improvement.[28]

This was a licence for the infinite expansion of academic activity. But Small's allusion to the complexities of modern life and his promise of a more scientific approach to society's problems carried a persuasive message to the governors and funders of the expanding universities. At the university of Chicago, the most dramatic example, president Harper encouraged Small to build up a large department that had lasting consequences for twentieth-century social science as it trained large numbers of people who went on to teach the subject. In 1833, Chicago was a log fort in a marsh; in 1892, the year its university was founded as a graduate and research centre, it was a huge and dynamic city. When Weber passed through Chicago on his way to the St Louis Exposition in 1904, he compared the city to a man with his skin peeled away so that the intestines can be seen at work. The philosophy department at the university headed by Dewey and the sociology department headed by Small shared an ambition to provide objective answers to practical questions. Small's ambition for sociology was that it should 'organize and generalize all available knowledge about the influences that pervade human associations'.[29] Under Mead, who taught social psychology in the philosophy department, and William I. Thomas (1863–1947) and, in the 1920s, Robert E. Park (1864–1944), who taught sociology, successive generations of students trained in social science and accepted scientific analysis as the objective basis for decision-making. The expanding university system provided these students with careers whether or not their science in practice influenced social policy.

Thomas's early interest was in ethnography, and he taught comparative culture rather as his German mentors had done, linking biological and psychological capacities to their collective expression in custom and myth. When local Chicago philanthropists put up the funds, he also undertook research into social problems like delinquency. These interests in culture and in social issues came together in the first decade of the new century when he began a major project on social change based on the experience of Polish immigrants in Chicago. He travelled to Poland to research the peasant communities from which these immigrants largely came, and he there joined forces with Florian Znaniecki (1882–1958), an aristocrat with local knowledge. Thomas and Znaniecki's study of peasants who migrated from rural to urban society and from Europe to America, *The Polish Peasant in Europe and America* (1918–20), which was based on a massive accumulation of data, developed new empirical methods and tackled *the* social question, social change. Their work attempted to do justice to how people experience change and it also delineated broad social forces. Thomas, for example, introduced the notion of 'the definition of the situation', a notion that links the individual's motivation and the social context:

> Not only concrete acts are dependent on the definition of the situation, but gradually a whole life policy and the personality of the individual himself follow from such a series of such definitions. But the child is always born into a group of people among whom all the general types of situation which may arise have already been defined and corresponding rules of conduct developed . . .[30]

As these comments suggest, Thomas and Znaniecki sought to relate the individual to society in ways that would avoid both the Scylla of psychological reduction and the Charybdis of Marxist determinism. But, while their book confirmed the pre-eminence of Chicago sociology and is always cited as a milestone in sociology's history, they did not succeed in the theoretical task they set themselves.

A concern with social reporting continued to be important in American sociology, and the techniques of objective reporting were a significant part of Park's teaching in Chicago in the 1920s. In spite of the academic professionalization of the subject, sociology maintained links with reformist politics and with the ideal of education as the key to the democratic process. Park stated explicitly that 'social control [is] the central fact and the central problem of society'.[31] In the interwar years institutional support of sociology became firmly entrenched, matched by public opinion that accepted the value of expertise, as it accepted the value of technology, in the fabric of social life. What was shared by the discipline and the public was a conception of the nation as an organic whole, a belief that individuals and institutions alike are understandable in terms of their function, and that science and the nation share the goal of efficiency, understood as the adjustment of individuals and institutions to the national life. There was continuity between egalitarian values, nineteenth-century theories of evolutionary progress, a politics of meliorism and the content of sociological research. Optimism in science was such that a teacher of sociology at Chicago in the late 1920s, W. F. Ogburn (1886–1959), told a colleague that 'the problem of social evolution is solved'.[32] This was ridiculous, as the experience of the 1930s painfully showed; but, far from unseating social science, the Depression gave it greater opportunity and suggested that the country needed more rather than less.

With sociology firmly established in academia, some scholars were free to focus intensively on the nature of the subject as a science. Ogburn thought that scientific sociology should separate from social philosophy because a real science should concern itself solely with the verification of empirical truths. He thought that the process of verification requires the use of multivariate statistics. In the 1930s many of his colleagues also became obsessed with statistical methodology, considering that it defined good scientific practice. At the same time, however, other scholars expressed dissatisfaction with the lack of a unified theory in sociology. Theoretical analysis culminated with the work of Talcott Parsons (1902–79), whose book on *The Structure of Social Action* (1937) attempted to provide the field with a comprehensive formal theory. The book suffered from turgid formalisms and unexamined functionalist premises. All the same, it seemed to many sociologists that it was a *summa* of sociology's claims to be a science, since it constructed abstract generalizations – Parsons viewed society as 'a nested sequence of interactive systems' – on the basis of half a century of empirical activity.[33] Whatever was subsequently thought of this, and Parsons became the *bête-noir* of critical sociology in the 1960s, he read Weber extensively and played the major part in introducing Weber to an English-language audience. In the short run, discussion of the concepts deployed in sociological analysis was thereby enriched; in the long run, Weber's examination of the fundamentals of the subject helped open up US social science, Parsons' included, to a critical analysis that called into question both its functionalist premises and its faith in itself as a meliorist expertise.

Comte and Spencer created sociology as an ethical project. They challenged society to achieve moral progress on the basis of positive social science rather than on the basis of inherited customs. Marx and other socialists went further and tied social philosophy to a confrontational rejection of established political power, and they connected science to political events that determined much of the twentieth-century world. Public opinion, which was impressed by natural science and dazed by social change, became receptive to the social analyses profferred by experts. Sociology achieved an identity as an academic discipline, though in ways that varied with local conditions; only in the United States was it a large-scale academic enterprise. In every country, sociology continued in the twentieth century to be a moral as well as a scientific project: it was a modern response to modern social life. When sociology became a discipline, it became a means through which human action became systematically self-reflective. At one and the same time sociology constituted itself as abstract theory and as human technology.

V

PSYCHOLOGIES IN THE 20TH CENTURY

16

Psychological Society

> No revolution changes the essentials so long as it only changes the institutions and ignores the men who live by them. If humanity is a function of institutions, so are institutions a function of humanity. For a transformation of the world to be radical it must grasp things by the root. The root is man. Education changes man. This is the path that is given to us.
>
> Otto Fenichel, after Marx, in a book review (1919)[1]

i *Psychology in the Twentieth Century*

The development of academic disciplines and professional institutions in the human sciences does not immediately seize the imagination as a turning-point in human self-discovery. It is necessary, however, to give it great weight. In the twentieth century the presence of these disciplines and professions was the visible sign of belief in and the special status of specialist knowledge and techniques about human nature. The internalization of belief in psychological knowledge, so that it acquired a taken-for-granted quality, altered everyone's subjective world and recreated experience and expectations about what it is to be a person. The result was an emphasis on 'the personal' in psychological terms, with ramifications in every aspect of life. It became possible to refer to the existence of 'psychological society' in the twentieth century.

The modern human sciences became specialized disciplines with a large-scale investment in their own existence and with distinct agendas, institutional identities and occupations, in this regard like the natural sciences. The sheer scale of psychological activity is significant. A historian cannot construct an overview of the twentieth century by citing a few influential authors, cultural movements and scientific schools. It is necessary to deal with vast numbers of people – there were nearly 120,000 people attached to the American Psychological Association in one way or another in 1992, while in The Netherlands at the same time there were about 20,000 psychologists. In the wider domain of social science, there were people who earned their living as personnel consultants, econometricians, demographers, voting analysts, psychotherapists, development economists, counsellors, forensic psychologists, consultant anthropologists, social workers, philosophers of medical ethics, and so on. The following chapters do not try to deal with all of this, but neither do they draw boundaries round areas that some people would claim are scientific and describe these alone.

Earlier chapters described systematic representations of humanness in relation to the individual body and mind and to collective society and culture. There were no firm discipline boundaries to divide up human nature and culture for study. But from the late nineteenth century to the present, these boundaries, though disputed, were a social reality, and there was a vast amount of specialization. Specialization is a social process, not the natural product of an increase in knowledge. It nevertheless creates a practical problem when we want to gain a general picture, and there is no ideal or easy answer. To deal with this, in a way that conforms with the sort of emphasis given to earlier theories of human nature, the following chapters on the twentieth century concentrate on psychology and the psychological realities of modern life. I say little about sociology, political science, economics, linguistics or anthropology as distinct disciplines, though interconnections between psychology and these disciplines are not ignored. As will be seen, in spite of such restrictions the subject area to be covered is still protean in its nature.

The human sciences generally in the late twentieth century were both disciplinary subjects and activities embedded in everyday life. There were hundreds of psychologists who hardly left the laboratory or academic lecture hall, but thousands who worked in commercial businesses, hospitals, prisons and schools. The academic setting was the place that most often supported systematic and critical enquiry into human nature. Yet it is wrong to assume that knowledge always passes outwards from the university, which pursues pure science, to the wider world where it is applied. While there are different kinds of scientific activity, it is not possible sharply to differentiate those that are pure and those that are applied. Of course, there were sometimes substantial occupational differences between psychologists who worked in academic settings and those who engaged more directly with social policy and administration. Sometimes the differences divided professional organizations and pulled the parts in different directions.

The subject matter of the discipline of psychology, its knowledge, and what that subject matter is about, real people, exist in a circle of interactions. The word 'psychology' itself is ambiguous, as it denotes both a discipline and the state or attributes of a person. There is a reciprocal relationship between psychology and the lives of ordinary people: people are ultimately both the subject matter of the human sciences and makers of psychological knowledge. It is possible to identify a trained or professional psychologist with a description of qualifications and occupation, but there is a significant sense in which everyone in the twentieth century learned to be a psychologist; everyone became her or his own psychologist, able and willing to describe life in psychological terms. The twentieth century was a psychological age, and in this it differed from earlier ages. How people live gave rise to the content of psychological knowledge, and that knowledge suggested changes in how to live. Schooling in mass society, for example, created a need to differentiate individual capacities, and the knowledge that resulted – IQ scores – informed educationl policy and people's self-image. Yet, since academic psychology existed on a huge scale and developed its own rules and practices, it was at times able to achieve considerable autonomy from ordinary life. Much academic psychology diverged greatly from popular psychology. To run rats in mazes to test a theory of learning was, after all, remote from the activity of teaching children and observing the process of learning.

The character of psychological society has attracted a stream of distinguished commentators and critics. The US sociologist Richard Sennett, for example, wrote a history of *The Fall of Public Man* (1977) in which he charted the privatization of the self and the concentration of representations of our existence in the psychological dimension. The consequence, he argued, was the abandonment of the public realm – the crucial realm of political action on behalf of the public good – which is left empty except for people who are significantly known as 'personalities'. In the 1970s and 1980s, there was criticism of 'the "me" generation' and 'the cult of narcissism', a supposed stress on the subjective self as the sole value. For some conservative critics, obsessive attention to the subjective self marked the culmination of the shift from medieval Christendom to modernity, a shift mentioned in earlier chapters in connection with the beginnings of the modern notion of the self and the roots of individualism. It is also possible to think of psychological society as the final triumph of a secular consciousness that banished the soul and the sacred from the human realm. Certainly, the secular language of mind and behaviour dominated discourse where the religious language of the soul and virtue once held sway. But, as with every aspect of relations between science and religion, black-and-white judgements fall down. In the early years of the century, for example, many Protestant clergymen welcomed the new psychology and hoped that it would address the experience and suffering of their congregations. Pastoral concern became as characteristic of psychological society as secular therapy. In the liberal West, it might even be suggested, the formulation of life's problems in psychological terms cut across divisions that separate religious and secular belief.

Twentieth-century Western society was certainly psychological in its values in the sense that individual capacities and differences between people featured in everyone's image of their place in society and in social policy about everyone's place. The language of personality, the everyday language of popular culture and answers to the question 'who am I?', conspicuously signified a social life devoted to the comparison of people. From the cradle to the White House, people acquired an identity constructed in terms of a psychology of differences. The identity acquired was understood subjectively, and confirmed objectively by others, as the self. The identity was material at the same time as it was psychological. The wealth of the West was attributed to the power that individual choice confers, and economic choice, it was thought, made psychological identity; in popular language, people acquired identity through a 'life style'.

Notions of normality and abnormality were also substantially psychological in character. In the jargon of social theory, psychological society is normative – a sense of self in psychological terms is a judgement as well as a description. Standards of normal and acceptable individual, psychological function permeated every aspect of ordinary life and social administration. The language of what is normal originated historically as a practical means to respond to perceived abnormality – in the classroom, the prison or the family, as well as in the clinic – and part of its history was discussed earlier in connection with statistics. The language was a response to what particular social groups perceived to be in need of discipline and regulation. Normality, from this point of view, is a state not perceived to be in need of regulation.

Psychologists claimed to possess objective methods to analyse individual differences. They therefore claimed a special role in a society in which so much hinged on individual capacities. Psychologists acquired authority – in as far as they did acquire it – as the experts who make order possible because they reveal individual capacities and the differences between people. Nevertheless, the relationship between psychologists and the wider society was complex and ambivalent. For instance, since quality and ability were internalized as individual attributes and made central to many people's notion of who they are, public opinion did not always welcome claims by psychologists to superior knowledge. Popularized psychology, quizzes to help determine personality and self-help guides to improve thought or memory flourished alongside professional activity. Psychoanalysis obtained a hold on the public imagination even though many scientific psychologists held it in contempt. All the same, the pressure to enhance the quality of decision-making dependent on knowledge of individual differences was a potent factor in the institutional growth of psychology. The most important area was education. At certain times, and this is particularly clear in Britain, educational applications coloured the whole development of psychology as a discipline.

Nowhere was there a closer connection between the academic discipline of psychology and psychology in public life than in relation to the study of individual differences. Throughout the century, mental tests, especially so-called intelligence tests, aroused enthusiasm and controversy in equal measure. This chapter describes the history of the study of individual differences, gives substance to the notion of psychological society and sketches a picture of the study of individual capacities in industry, education and child care.

ii *Individual Differences and Statistical Analysis*

An interest in differences between people is central to daily life and its history can be traced in many different areas. When this interest turned into systematic research and became a science of differential or individual psychology, that also happened in diverse ways. This section describes the origin of the study of human variation by statistical methods; the focus is on Britain and on belief about the power of heredity – the context in which the concept of intelligence became the centre of attention. The following section then traces the origins of mental tests for intelligence, first in France and then in the United States. Then, after a brief history of the evocative term 'personality', there is an account of psychological practice in the workplace, of approaches to childhood and of the controversy over the respective importance of nature and nurture.

As emphasized earlier in relation to physiognomy and phrenology, everyday arts for the assessment of character, to aid decisions and to guide the passions, were ancient and familiar. Phrenology flourished anew in the United States and Britain in the second half of the nineteenth century when an entrepreneurial family, the Fowlers, carried on an extensive business in advice to clients and in popular psychology. The Fowlers marketed the china busts, the head marked up with the sites of various faculties, that became collectors' items. Phrenology and the wider interest in character and aptitude created an audience receptive to the claims of the new expert psychology around 1900, especially in the US, and later to mental testing. There was also a concentration on individual character or temperament and on a person's inherited disposition (the 'diathesis') in medicine – in diagnosis and in therapy – and medical interest strongly influenced early individual psychology. This too had an ancient background in the theory of humoural medicine and in the physician's response to individual patients.

The inspirational figure for much of the modern scientific interest in individual differences was Francis Galton, child prodigy but disappointed adult until he found meaning for his life in the application of evolutionary biology to the social domain. Galton himself was highly individual. He was, for example, an obsessive calculator, and he recalled a walk where he ingeniously calculated the number of candles, or flowers, in an avenue of horse-chestnut trees. Motivated in the 1860s by Darwin's evolutionary theory and by his fears about lack of social progress, Galton turned his attention to heredity in individuals and its consequences for the evolution of the social world. His key innovation was the argument that the way to study individual inheritance is to look at the distribution of variation in populations, not directly to examine the physiology of heredity. This put the spotlight on statistical argument in biology and human psychology. He believed that heredity is overwhelmingly more important than environment in the determination of individual differences, and he set out to show this and its consequences for human progress through statistical argument. Galton was convinced that biological evolution and human progress are continuous and therefore equally dependent on the quality and distribution of the inheritable variations passed down from generation to generation. Progress, he thought, lies with the spread of desirable inherited qualities in the population. Since he also believed that industrial society and international competition place increasing demands on individuals, he advocated a social policy to increase the numbers of people with the amount of energy and intelligence needed for modern life. He was especially keen to make it easier for young people of good stock to marry. This led to the political programme that he founded, and which he named eugenics, to solve social problems and advance human well-being by the promotion of differential birth rates for groups with different inherited aptitudes.

These arguments came as a revelation to Galton in the 1860s, and it was only at the end of the century that a change in public opinion began to give him a receptive audience. In the intervening years he studied human capacities, with significant long-term implications. In *Hereditary Genius: An Inquiry into Its Laws and Consequences* (1869), he argued that biographical data vindicate belief in heredity, rather than upbringing, as the principal determinant of individual levels of achievement of mind as well as of bodily characteristics. He then abandoned an attempt to understand the physiological mechanism of heredity and turned instead to the statistical analysis of the distribution of individual variation in the population. With the addition of techniques developed by mathematicians, this led to modern mathematical biology and psychology. Galton also devised ways to measure and describe individual variation in the first place; for example, in order to collect data he set up an anthropometric (or human measuring) laboratory at the International Health Exhibition in London in 1884, where members of the public tested themselves for such things as muscular strength and visual acuity. He also studied the variation of fingerprints, leading to their use as a means of identification by the London Metropolitan Police in the 1890s.

Galton's thought was imbued with respect for the laws of nature. He was convinced that a human science is possible once it is recognized that mental variation is inherited in the same way as physical variation, that what gives a person character is heredity not will – let alone the soul. This was his contribution to scientific naturalism, the inclusion of man within a natural-scientific worldview. He took up Quetelet's work on normal distribution and used it to study the distribution and – as he thought – the inheritance of individual differences. Whereas his predecessors had treated individual variation as an 'error' or deviation from an ideal, Galton made variations the focus of investigation. This new approach to individual differences as a quantitative subject with data derived from field research led to a science of psychology not necessarily based on the laboratory: it was possible for the scientist to enter society's institutions and not require society to enter the laboratory.

Galton argued that the study of a *social* phenomenon, the distribution of variation in the population, is the product of the incidence of *individual* mental capacities. This anticipated, as it certainly influenced, a common twentieth-century assumption that social policy issues are really dependent on psychological (or even ultimately biological) matters. The assumption was central to psychological society. Galton himself went further; he assumed that there is a correspondence between an individual's social position and his innate capacities. Indeed, believing that British society was relatively open and hence that natural talent found its natural social place, in *Hereditary Genius* he used the high social position of people in successive generations of the same families as prime evidence for the inheritance of high ability. (He never discussed women's capacities as a significant or distinct topic.) Galton was therefore conservative and élitist in the proper sense of these words: he believed that the existing social hierarchy was desirable, that it resulted from the distribution of individual talent and that attempts to alter it, vociferously expressed by socialists and feminists in the 1890s, could only lower the quality of government and administration. In short, he believed in a meritocracy and thought that was precisely what Britain had.

The technical and statistical side of Galton's work, especially the start he made on correlation coefficients, was developed mathematically and in relation to evolutionary biology by Pearson, the first Galton Professor of Eugenics at University College London (appointed 1911), and in relation to psychology by Charles Spearman (1865–1945), also at University College. Spearman began a military career but switched to science and studied with Wundt in Leipzig and then elsewhere in Germany before he was appointed to the London psychology department in 1906. While in Germany, Spearman published the paper 'General Intelligence Objectively Determined' (1904), which guided all his future work and, after he became head of the department at University College, set much of the agenda for psychology in Britain.

Spearman, who was familiar with Binet's contemporary work in France, postulated that a general factor (g factor) underlies mental performance or intelligence. He believed that this is demonstrated by the way the level of an individual's performance is similar across many activities. Any particular performance, e.g., in musical ability, is affected by other, special factors (s factors), but Spearman and those who followed him argued that there is a correlation of ability across a broad front, e.g., between a child's abilities in English and in mathematics. This correlation reveals the g factor. In *The Nature of Intelligence and the Principles of Cognition* (1923), Spearman set psychology the task of describing human life in terms of two psychological factors, measured as general and special ability. He summarized this 'two-factor theory of intelligence' for a general audience in *The Abilities of Man: Their Nature and Measurement* (1927). Detached from Spearman's formulation, the public tended to refer simply to the level of intelligence as a general guide to a person's ability.

The word 'intelligence' had been used earlier to denote a human faculty, but it was particularly attractive to psychologists in the late nineteenth century as a way to refer to reason as a function of man considered a part of nature. Philosophers from Plato to Hegel took reason as their subject and concentrated on reason as an abstract or logical procedure rather than as an event in the soul or mind. Other thinkers, such as Aristotle and Locke, changed this to some extent since they set out to derive knowledge from experience and discussed reason as the engagement of soul or mind with nature. When some philosophers began also to conceive of what is in the mind in terms derived from descriptions of physical nature, as in the eighteenth-century analysis of elementary sensations, they went further in the treatment of reason as a natural capacity. The representation of mind in fully naturalistic terms was carried through in evolutionary theory, and in this context references to intelligence began to replace references to reason. Intelligence rather than reason denoted a capacity of natural organisms. The new usage is clear in the work of Romanes, who published a series of books to amplify Darwin's evidence that there are no differences in kind, however large the differences of degree, between human and animal mental capacities. In *Animal Intelligence* (1882) he wrote for a general audience that there is no insuperable psychological distance between monkeys and people: 'in their psychology, as in their anatomy, these animals approach most nearly to *Homo sapiens*'.[2] He defined intelligence as the capacity to adapt flexibly to changing circumstances, a definition that made it possible to use the word 'intelligence' equally about animals and humans, which was not possible for the word 'reason'. When the French liberal and reformer Taine published a book to make British psychology available to a French audience and to counteract the *spiritualisme* dominant in academic philosophy, he chose as his title *De l'intelligence*.

Spearman's first study correlated abilities in children in a village school, and he relied on teachers' grades to estimate the level of ability. It was a clumsy method, and tests later transformed the process of measurement. But Spearman claimed to be the first to demonstrate, as earlier researchers like Galton and J. M. Cattell had not, that measures of performance are indeed correlated to a degree that satisfies defensible statistical tests. All the same, his arguments were questioned from the outset. Critics focused on what he had measured, on whether or not his correlations were statistically significant and, if they were, whether they were uniquely explicable by reference to a general factor. Spearman engaged in a running battle with one critic, Godfrey Thomson (1881–1955), who argued that if different tasks drew on partially overlapping sets from a large reservoir of mental abilities one would end up with the same data. The enormous literature on such arguments turned psychology into an esoteric subject that required fluency in statistical methods, as many generations of students painfully learned. Spearman tenaciously defended his two-factor theory, and this defence reflected his identification of g with something real in nature, 'something of the nature of an "energy" or "power" which serves in common the whole cortex [of the brain]'.[3] Though he was less concerned with social questions than Galton, he still characterized g as the innate human capability that a forward-thinking policy of differential reproduction should seek to select. Many later workers in the field, by contrast, whether or not they utilized the notation 'g', in theory treated factors as a measure and eschewed speculative explanations in terms of underlying biological causes. This refinement did not, however, prevent mental factors from being treated as real, natural objects in the practical social settings in which factor analysis was put to work. Schoolchildren, for example, were streamed as if g (or other factors) reveal biological difference. The reduction of intelligence to a single measurable factor, with real political if not natural existence, is exemplified in a British government report of 1938:

> Intellectual development during childhood appears to progress as if it were governed by a single central factor, usually known as 'general intelligence' ... It appears to enter into everything which the child attempts to think, to say, or do ... [and it is therefore] possible at a very early age to predict with some degree of accuracy the ultimate level of a child's intellectual powers.[4]

Factor analysis, then, involved ranking measurements of a characteristic in a population in order to assess whether the variability exhibited by the characteristic can be attributed to one or more distinct factors or variables. In the 1930s, the English psychologist Cyril Burt (1883–1971) adapted the sophisticated statistical procedures of the mathematical biologist Ronald A. Fisher (1890–1962), and brought the results together in *The Factors of the Mind* (1940). Much ingenuity went into the development of psychometric methodology and of the analysis of results. This was sometimes at the expense of the truism that no amount of arithmetical subtlety can replace the need for appropriate measurements across an appropriate range of data in the first place. How mental measurements were in fact made is discussed in the next section, while the irony of all this for Burt's later reputation and for psychology generally is made clear later.

Originally developed in part for the analysis of the distribution of individual abilities, statistical correlational analysis became a key tool of psychology and the social sciences. Faced by the large-scale complexities of the mental, behavioural or social worlds, correlation analysis appeared to be the means to reveal the underlying pattern. Psychologists argued that advanced statistical analysis is necessary to produce reliable evidence in the many kinds of experimental research in which repeated observations are made. Experimental work often involved repeated runs, just as the study of individual differences involved the collection of measurements of one characteristic from a population, and strict procedures appeared necessary if results or correlations were to have authority. The accurate statistical analysis of experimental results became, from the 1930s on, the hallmark of the scientific psychologist. There were at least two significant consequences. Firstly, there was a tendency for the technical side of statistics to become a specialized topic and to develop as a major subject area in its own right. Here, no one could doubt, was a specialist expertise that differentiated the professional psychologist from the lay public. So central was this expertise to psychology that, at times, the discipline appeared to be defined by its methodological preoccupations and the training required to master them rather than by its subject matter. Some scientific psychology journals required research papers to be expressed in a particular statistical manner for publication. It was argued that this condition guaranteed the scientific credentials of the discipline. Secondly, there was a recurrent tension between the opinion that attributed causal significance, often understood to be innate inheritance, to the patterns revealed by statistical analysis, and the opinion that statistical patterns do not tell a causal story. This tension was in fact a feature of philosophical dispute in the natural sciences generally, a dimension of the debate about whether knowledge describes what is real or is instrumental, a means, for our purposes, to order what we perceive. The debate about Spearman's g factor, whether it is a real agent of 'brain power' that varies between people or a mathematical construct useful for certain purposes, exemplifies this tension in concrete terms. Psychology as an occupation was carried on with such tensions rather than through their resolution.

When Burt succeeded Spearman as professor of psychology at University College in 1932, he extended the range of factorial analysis as well as its mathematical sophistication. He applied it – as Spearman had begun to do – to the emotional and active mental powers, that is, to moral character and personality. Burt hoped by this means to develop differential psychology into a general psychology. Before seeing where this led, however, it is necessary to turn back to the early years of the century and to the social questions that Galton had addressed when he studied individual differences. Burt's career itself began with a lectureship at the university of Liverpool followed, in 1913, by his appointment as the first educational psychologist for the London County Council, the body administratively responsible for the education of London's children. In both Liverpool and London, Burt focused on individual differences, and his starting-point was Galton's view that they have a preponderantly innate origin, even though Burt – who had spent his early childhood years in London and had obtained an education through scholarships – was also sensitive to the effects the social environment has on poverty and delinquency. As a child, Burt met Galton and reportedly felt a superstitious thrill that Galton's seminal *Inquiries into Human Faculty* was published in the year of his birth, 1883. While he worked for the London County Council, Burt was at the hub of the administrative machinery set up to solve social problems by education, and his appointment was a sign that society at large had begun to look to experts to carry through its policies. In this respect, Burt's situation can be compared with Binet's a decade earlier. The intelligence test came from the involvement of psychologists with the problems of the classroom, notably the need to distinguish children unable to benefit from the teaching on offer. The psychological test, in origin, was first and foremost a technology to order the child's development in mass society.

iii *Education and Mental Tests*

At about the time evolutionary thought became current, a shift in social and political policy made education and intelligence major matters for the state. In France in the mid-century and then in Britain in 1870, legislation established the framework for universal primary schooling. The principal drive in France, where education for children between six and fourteen became mandatory in 1882, was the political aim of creating a sense of common national identity across a large and largely rural country. In Britain, where elementary education became compulsory in 1880, the stimulus was a desire to enable the people to exercise sound political choice and to become economically self-reliant. Parallel campaigns for education occurred elsewhere, with the result that mass schooling became the norm for modern society. By the end of the century, substantial national and local investment in education in many countries created financial and administrative pressures; it also expanded those occupations with a special interest in the development of human capacities. The significance of this to the expansion and professionalization of psychology in the United States has already been pointed out.

Calls to make education a science were widely heard, and scientific education appeared to require knowledge of the natural development of the child. In the 1880s and 1890s, for example, there was a vogue in the English-speaking world for Herbartian education, named after the Prussian philosopher and pedagogue, which stressed that the child's – not the adult's – interest should be the driving force in teaching and that the child's interest develops in stages. The reality of mass schooling, however, was large classes, a standard curriculum and instruction often indistinguishable from the discipline that was its precondition. There were also expectations that children should reach a certain level of achievement, for which examinations were the measure. The French system was constructed to be a procedure for selecting the best students, though it was conspicuous that it was pupils with lower rather than higher abilities who demanded attention. In this context, the word 'intelligence' denoted in the classroom a child's innate capacity which was thought to be the basis of his or her achievement. In practice, 'intelligence' characterized a child's performance relative to other children and to a set educational standard. Further, in the way in which the word was used by psychologists and especially in public understanding of psychology, 'intelligence' tended to denote a concrete natural object; what was called intelligence became the explanation rather than the measure of performance. This usage reflected and supported a host of everyday assumptions, characteristic of psychological society, about ability as the foundation of social phenomena.

Education and the measurement of intelligence came together in a specific way. During the 1890s, Alfred Binet contributed to the academic infrastructure of the new psychology in Paris by establishing a laboratory, training procedures and a journal. He also became involved with educational research and was the pivotal figure in the establishment of the modern psychology of individual differences and of mental tests in France. Binet was a wealthy young man who found his calling in psychology. After an initial interest in hypnotism, and a painful recognition that he had failed to appreciate the role of suggestion in the investigations, Binet turned medical diagnosis into psychological study as a more reliable way to describe individual capacities. In the 1880s, when there was considerable interest in hypnotism as an experimental and therapeutic technique, it was shown that there were marked differences in individual response. Freud was a significant observer and Binet another, and Binet carried over the interest in individual differences from the area of hypnosis to new work. He made imaginative studies of his own two children, Madeleine and Alice; for example, he noted stylistic and temperamental differences in the way they learned. He also observed differences in the way a child's and an adult's intellect works – differences which he picked out only because he studied higher-level abilities. In 1891 he started working – unpaid – at the new laboratory of physiological psychology at the Sorbonne, and in 1894 he became its director after the retirement of its founder, Henry Beaunis (1830–1921). From then until his early death, Binet was very active in the promotion of an experimental psychology coloured by the interest in individual differences. He began the work that led to intelligence tests after he joined and became the driving force in La société libre pour l'étude psychologique de l'enfant (The Free Society for the Psychological Study of the Child) in 1899.

In the 1890s, Binet and his colleagues and students began to study individual differences experimentally. Though they sometimes drew on the techniques to measure sensory discrimination (by the record of just noticeable differences) or reaction times developed in Germany, their approach was distinctive in two significant respects. Firstly, it tended to consist of in-depth studies, which were analogous to a clinical examination, of people who were special in some way. Binet published a study of *grands calculateurs* and people with a phenomenal capacity at chess (able, for example, to win several games at the same time while blindfolded). Secondly, their work often attempted to compare high-level mental capacities, such as comprehension or aesthetic appreciation; this contrasted with much German laboratory work, which focused on the elements of perception, and it examined the capacities that matter in the social world as opposed to the artificial world of the laboratory. In Binet's words: 'If one wishes to study the differences existing between two individuals it is necessary to begin with the most intellectual and complicated processes . . . however, just the opposite . . . is done by the great majority of authors who have taken up this question.' In 1895, Binet and Victor Henri (1872–1940) published an overview, 'La psychologie individuelle' (Individual Psychology), which gave the field a recognizable identity and proposed a series of topics for research. The authors observed that the driving force of the subject is 'to illuminate the practical importance that . . . [the topic] has for the pedagogue, the doctor, the anthropologist and even the judge . . .'[5] Management of children, the mentally disordered, differences of race and criminals gave psychology both its subject matter and its social significance. All of this reverberated well into the twentieth century.

In 1904 the government established a commission, to which Binet was appointed, to investigate the situation of the mentally subnormal in France. There was a particular problem about the diagnosis of borderline subnormality – individual children were diagnosed differently by different criteria – and borderline children often wasted much of a class's and a teacher's time. There was also a problem when children who were perhaps not really subnormal were stigmatized; sensitivity to the stigmatization of abnormality itself mainly came later. The administrative ideal was clear: if subnormal pupils could be identified early and accurately, they could be taken out of the standard classes; this would free classes to make normal progress and make it possible to give abnormal pupils special education. The classroom problem gained in intensity because of contemporary fears about degeneration in the population, especially in the urban working class, the evidence for which was low reproductive rates nationally and high levels of alcoholism, insanity, crime and prostitution. There was also an interest in eugenics. When Binet set out to devise a diagnostic test with which to identify problem children, the step had significance beyond the classroom. It is also noteworthy that the commission asked a psychologist to achieve diagnostic clarity in an area of medical activity. The division of labour between experts with medical qualifications and those with psychological training, between clinical diagnosis and mental testing, later sometimes became a major issue in the struggle of psychologists for professional status. The mental test, of the kind pioneered by Binet, became the tool which initially legitimated the psychologists' claim to independent expertise and professional recognition.

Binet worked with Théodore Simon (1873–1961), a colleague with experience of an institution for retarded children, and they searched for tasks that would reliably discriminate normal and subnormal children. They succeeded when they realized that the tasks had to take age into account: subnormal and normal children might accomplish the same tasks, but the former did so at an older age. Their standard of what a normal child at a particular age can do, based on a series of thirty tasks of graded difficulty, was published in the journal *L'année psychologique* for 1905. The test was designed to be useful; at the same time, because it examined a variety of separate and relatively complex mental functions, it also appeared to assess a general psychological capacity of theoretical interest, which Binet and Simon called judgement. As Spearman had published in the previous year his argument that correlation coefficients between different mental abilities reveal general intelligence, psychologists tried to integrate the new test scale and the theory of intelligence. Binet himself never offered a unified definition of intelligence.

Binet and Simon revised their test in 1908 and again in 1911, the year in which Binet died, in the light of wider experience and work with larger numbers of children. The uptake of the rather inflexible tests was slow in France, though they were adopted in Belgium and in Geneva (under the direction of Claparède) in 1906. In Paris, Binet's work was continued by his successor Henri Piéron (1881–1964), who turned the laboratory in the university of Paris into an institute of psychology, in effect the first French graduate school in psychology. Piéron, who had broad interests, made it possible for French psychological research to expand and flourish both in the academic world and in the applied setting. With his wife Mathilde, who did the testing while he did the writing, he perpetuated Binet's style of description, which drew up individual profiles rather than constructing a theory of general intelligence. The Piérons' methods were then used, for example, to screen applicants for work on the Paris transport system and on the French railways.

In England, Burt's appointment as a psychologist by the London County Council gave him the same interest in subnormality in schools as his Parisian colleagues, and, as well as routine clinical work, he undertook to develop tests relevant to educational problems. His interest in clinical as well as test work ensured that he never conceived of tests as a mechanical exercise. He undertook an extensive survey of the distribution of the mentally subnormal within London, giving empirical content to the problem then labelled 'backwardness' – a topical matter with the passage in 1913 of the Mental Deficiency Act. This Act called for a national system of special institutions under medical supervision for people known as defectives, though much of its implementation was cut short by the war. By contrast, in the setting of local government responsibility for education and child welfare, the psychologist, Burt, and the many schoolteachers and students with whom he worked, constructed a psychological language about abnormality. The language was applicable to delinquency as well as mental incapacity. Burt also developed a psychological approach to this social problem, which was the subject of his best-known book, *The Young Delinquent* (1925). In the light of his reputation as a hereditarian psychologist, it is worth pausing to note the extent to which he immersed himself in the conditions in which his subjects lived. He saw no contradiction between a policy to improve educational opportunity and belief in the determination of mental capacities by inherent biological causes. What Burt failed to do was to conceptualize the social nature of the interaction between inherited structure and environment during a child's growth; he did not, for example, question whether human heredity can be studied by the same methods as animal heredity. The main point, however, is that Burt established the techniques and pioneered the occupation that tackled social problems in psychological terms. These approaches concentrated on individual variation in mental capacity and conduct.

There was also a German input into intelligence tests from the versatile psychologist William Stern. He objected to the pretensions of a general psychology which fails to make the individual its central concern, and he accused fellow German psychologists of being more interested in abstract notions of mind than in real individuals. His interest in individual differences expressed these values. In this respect he was like Binet and, also like Binet, he appreciated the difficulties that faced the attempt to turn the study of differences into a rigorous and objective science. His interests also included child development and what he himself called applied psychology, for which he organized an institute in Berlin. All these interests were reflected in the work that he summarized in *Die psychologische Methoden der Intelligenzprüfung* (*The Psychological Methods of Intelligence Testing*, 1912). Stern refined the index of a child's intelligence derived from the Binet-Simon test and proposed that the mental-age score be divided by chronological age. He called this index the 'intelligence quotient' or IQ. Devised as a convenient language for the comparison of test results, the concept was easily used to describe intelligence as a real, quantifiable entity in human nature, which reinforced a tendency for descriptions of a test situation to be replaced by causal, hereditarian claims. This, however, was far from Stern's (or Binet's) intent. Stern always assumed that even when the level of two people's performance is comparable, the two people might achieve the performance in different ways. He took seriously his own stress on individuality: 'There never is a real phenomenological equivalence between the intelligence of two persons.'[6] In this context, his use of the adjective 'phenomenological' denoted a person's subjective experience of intelligence in the light of his or her distinctive mental life. Stern elaborated his stance into a position called personalism, on which he lectured to diverse audiences as far apart as the US and Bulgaria. The later usage of IQ, however, tended to go against the purposes that Stern had when he studied and subsumed the individual within a class of an overall scale of general intelligence.

Binet and Simon's work also created interest among American psychologists and doctors concerned with subnormality and what they saw as heredity's contribution to social problems. The key figure was Henry Herbert Goddard (1866–1957), a teacher who retrained as a psychologist under G. Stanley Hall and was then invited in 1906 to take up an unusual position, that of research psychologist at the New Jersey Training School for Feebleminded Boys and Girls in Vineland. Goddard initially imported the experimental apparatus of an academic psychologist, but he quickly realized that what the institution needed was a reliable means to classify the children in its charge. Institutions like Vineland, which were headed by physicians, housed children with what were later distinguished as epilepsy, autism, behavioural problems and so on, not just subnormality. On a study visit to Belgium in 1908, Goddard learned about Binet and Simon's test. He tried it out at Vineland and found that their scale produced results comparable to assessments made by the institution's staff; 'It met our needs', he concluded.[7] That the scale did indeed meet a need is made evident by the willingness of the medically dominated American Association for the

Study of the Feebleminded in 1910 to adopt psychological tests as its main technology for the diagnosis of subnormality.

This insertion of psychology into medicine was Goddard's lasting contribution. He is better known, however, for the grossly simplified hereditarian argument he subsequently developed, in which he attributed human capacities to a unitary intelligence and attributed intelligence to a single genetic factor. In 1912 he published an attention-grabbing book, *The Kallikak Family: A Study in the Heredity of Feeble-mindedness*, in which he described two branches of one family, based in part on real events, one with the gene for feeblemindedness and the other without. He combined the Greek words *kalos* (good) and *kakos* (bad) to construct the word 'Kallikak', and in his story the two sides of the family exhibited a contrast between decent normality and an abundance of idiocy and vice. Goddard invented much and did not even pretend that this was science, but he contributed to a culture in which claims were made that science, and the new instrument of the intelligence test, supported hereditarian arguments and social policies based on them.

At the new private university of Stanford in California, a group of educational psychologists led by Lewis Terman (1877–1956) began to apply the new tests to large numbers of normal rather than subnormal children, enabling them to standardize the scales against which children were graded. The result was the Stanford-Binet test, the mainstay of intelligence testing in subsequent years. Terman saw great significance in these technical developments and anticipated that objective testing would answer two outstanding political questions:

> Is the place of so-called lower classes in the social and industrial scale the result of their inferior native endowment, or is their apparent inferiority merely a result of their inferior home and school training? . . . [And] are the inferior races really inferior, or are they merely unfortunate in their lack of opportunity to learn?[8]

His gratuitous use of 'merely' perhaps revealed his prejudices. Whether or not the results of tests confirmed the natural inferiority of individuals and groups became a hot issue, still debated at the end of the twentieth century; it is discussed below under the heading of nature and nurture.

Earlier than Terman, and to a considerable extent independently of Binet's work, US psychologists like Edward L. Thorndike (1874–1949) and Robert S. Woodworth (1869–1962) had transformed German experimental techniques in order to create laboratory tests to record individual differences. An effort was under way to make such tests relevant and marketable in the world of commerce (advertising), industry (personnel selection) and education (streaming). Thorndike, who had a powerful institutional base at Columbia Teachers College in New York, directed an army of educational efficiency experts, whom he inculcated with his belief that scientific understanding is equivalent to effective measurement: 'Whatever exists at all exists in some amount.' A wave of optimism about psychology's potential contribution to human welfare caused Thorndike to sink to bathetic depths with his exclamation: 'History records no career, war or revolution that can compare in significance with the fact that the correlation between intellect and morality is approximately .3, a fact to which perhaps a fourth of the world's progress is due.'[9] Psychological testing put the scientific outlook at the heart of modern enterprise, supported by the values of the Progessive Era when new specialist expertise in the human and social sciences became part of the social world.

A number of psychologists thought that the entry of the United States into World War I in 1917, with the conscription of over a million citizens, offered special opportunities. Conscription brought into the army men of unknown capacity for soldiering and it was feared that much money and effort might be wasted on those who would turn out to be mentally incapable or unfit for training. The psychologists, led by Robert M. Yerkes (1876–1956), a Harvard professor and president of the American Psychological Association in 1917, worked through the National Research Council to propose to the army a system of mass testing. The proposal was to administer intelligence tests to large groups of conscripts, 'to aid in segregating and eliminating the mentally incompetent' and 'to *assist* in selecting competent men for responsible positions'.[10] Some 1.75 million men were tested; psychology – and especially Terman – thereby gained considerable managerial experience and a large body of data for the comparative analysis of intellectual capacity. Testing on such a scale required the tests to be given to large numbers of men at the same time; Binet, by contrast, took a couple of hours to test just one child. Mass testing changed not just the scale but the nature of the testing enterprise, a change most evident in the use of the recently introduced multiple-choice answer technique. The multiple-choice questionnaire required answers in conformity with predetermined right or wrong choices. It did not give scope to the individual to exhibit his or her capacity in the way he or she understood, which is what Stern, for example, looked for from an individual psychology. In this clear-cut way, psychology introduced into people's lives a technology to shape self-understanding of what it is natural to perceive, feel and think. Through mass testing psychology also gained an audience and a clientele – a public familiar with and willing to accept mental tests – despite the fact that the tests used remained somewhat controversial in the academic literature at the time. There is also little evidence that the army gained much in exchange for the opportunity it gave to psychologists. The army was not convinced that there were gains in efficiency and it remained suspicious that to label its soldiers' intelligence, especially when the label was so often that of low intelligence, was good for morale. Where psychology did make a mark was in government circles. The appointment in 1919 of the psychologist James R. Angell as chairman of the government organization of natural scientists, the National Research Council, has been held up as a sign of psychology's coming of age as a reputable natural-scientific occupation. Angell went on to become president of Yale University and, as discussed later, to preside over a major endeavour in the human sciences, the Yale Institute of Human Relations.

The psychologists interpreted the army test results as support for what they already believed they knew, that white home-grown citizens perform better than immigrants, that the performance of immigrants decreases the further south and east in

Europe their origins and that black citizens are consistently inferior in scores to white. Yerkes, who left the army as a colonel, Terman and others argued that innate mental differences between peoples are real, and this same opinion contributed to the campaign for a restrictive immigration law (passed in 1924). Many other interpretations of the data were possible and expressed; during the 1920s, the support that mental tests supposedly provided for racist values and hereditarian theories was attacked and the value of the army tests questioned. It was, after all, scarcely credible that the average mental age of army recruits should be thirteen, as first uncorrected calculations indicated. There was argument about what it is the tests measured, and the sceptical position was pithily formulated by the Harvard psychologist Boring in 1923: 'Intelligence is what the tests test.'[11] But whatever the differences of view over the tests, the experience derived from the mobilization of a human science on a national scale and from the presentation of the case for such a science within government was important. It was followed in the 1920s by a variety of large-scale enterprises to market psychology as the expertise of efficient human management. Cattell declared in 1922 that the war and army use of intelligence tests 'put psychology on the map of the United States, extending in some cases beyond these limits into fairyland'.[12]

iv *Personality*

The use of tests grew rapidly after 1920 in commerce and industry, often under the heading of psychotechnics, as well as in education. Clients for mental tests outside the classroom were interested in many personal characteristics besides intelligence, and an interest in personality was the background to a major growth area of psychology in the interwar years. The term 'personality' became ubiquitous in the language of both psychologists and the general public, a clear sign of the creation of a psychological society. The psychologists' conception of personality was inseparable from the techniques with which it was hoped to make each person a measurable subject in the complex relations of modern life. The public's conception of personality became central to the way people expressed a sense of psychological individuality and judged individual worth.

The French word *'personnalité'* was much more common than its English equivalent until the turn of the century. It was associated until the 1880s with an idealist view of human character, which attributed character to the individual soul; it was then appropriated by the new psychologists like Ribot and Janet. The psychologists used the word in the context of their work on abnormal states of mind, in debates about hypnotism and in discussion of dramatic cases of psychological individuality. Ribot published *Les maladies de la personnalité* (*The Diseases of the Personality,* 1895), and Janet became famous through his studies on the splitting of the personality, studies that questioned presuppositions about the existence of a unitary self. The concept of personality was therefore intimately bound up with the creation of the French discipline of psychology. In Janet's definition, personality is:

> the reunion of presentations, the rememberance of all past impressions, the imagination of future phenomena. It is the notion of my body, of my capacities, of my name, my social position, of the part I play in the world; it is an ensemble of moral, political, religious thoughts . . .[13]

This definition was vague, but it well described the sort of notion of personality that readers gained from Robert Louis Stevenson's *The Strange Case of Dr Jekyll and Mr Hyde* (1886). Stevenson's fiction captured the mystery of 'personality', a word which he used, and the sense that it can be gained and lost and is a phenomenon or thing in its own right. Shortly thereafter, the interest in spiritualism and in psychical phenomena led British and North American researchers to use the word, personality, to refer to what makes for the wholeness of an individual, the wholeness that believers held persists after bodily death. This work culminated in a book by F. W. H. Myers (1843–1901), the brother of C. S. Myers who established experimental psychology at the university of Cambridge, on *Human Personality and Its Survival of Bodily Death* (published posthumously in 1903). Myers explained psychical experiences by reference to what he called the subliminal self, a region of the mind normally below the threshold of consciousness but which constitutes our essential nature. The Boston neurologist or brain specialist Morton Prince (1854–1929) discussed the pathological splitting of identity, or multiple personality, in a book with the title *The Dissociation of the Personality* (1906). Long before these psychological studies, however, the notion of personality was discussed by the theologians of a new Christology who explored the personhood of Christ. In the context of the argument that Christ's glory lies in the humility with which he became fully human, Christ was characterized as the model for man, the ideal personality. Thus, in its background, the English-language sense of personality connoted a moral and spiritual norm, a wholenesss, with the implication that a true personality is to be found in imitation of Christ.

The early use of the term 'personality', in French and English, had significant religious and spiritual dimensions. Religion and psychology were not independent and opposed preoccupations; at the turn of the century they were often integral to each other. The intellectual background was in part the Protestant argument, powerfully expressed by the German theologian Schleiermacher early in the nineteenth century, that religious *feeling* is an irreducible individual experience and an authentic object of theological discourse. Liberal Protestants at the end of the century hoped to find in religious feeling, that is, in a psychological state, the condition to bring traditional faith back into contact with modern social and economic realities. At the same time, psychologists like William James, in *The Varieties of Religious Experience*, and his Swiss friend Théodore Flournoy (1854–1920) wrote about religious feeling as a psychological phenomenon; they were also intensely interested in spiritualist experiences. For Flournoy, religion is a vital phenomenon, a living human experience, that can be investigated, like life generally, by evolutionary and psychological science. Liberal Protestants also drew on a reconstructed picture of Christ as the model man, a man of true feeling, and an exemplar for ordinary people in daily life. This liberal theology, with its picture of Christ's personality, was profoundly criticized by the Swiss theologian Karl Barth who, in an agonized response to World War I, stressed the absolute transcendence of God. By that time, however, personality was a common psychological term, in most usages divorced from religious connotations.

The idea of personality was in part transformed by the growth of psychology as a technology. Intelligence tests were of much value to administrators of education who dealt with huge school systems and children of great diversity of attainment and background. But when psychologists marketed their services to a wider constituency, especially in business, it was obvious that clients were interested in a wider range of capacities – in character rather than intelligence by itself. One of the pioneers of psychology in advertising, Walter Dill Scott (1869–1955), moved to the business school in Pittsburg in the first decade of the century and explored with businessmen the selection of salesmen. Subsequently, he directed the area of testing that was of most use to the army in the war, personnel selection. This work on selection concentrated attention on character or personality. In the 1920s, many corporations began to perceive personality tests to have instrumental value. At the same time, the mental hygiene movement, which sought to create an organic society through the adaptation of individuals to modern conditions, found tests to be a valuable means to pick out children constitutionally in need, it was thought, of special direction or institutional provision. Large funds became available, notably from the Laura Spelman Rockefeller Memorial, to support research and the application of new techniques. Psychologists who claimed expertise in the newest techniques to assess personality were able to extend their activities.

The word 'personality' described all the individual emotional, motivational and attitudinal qualities that mattered, alongside intelligence, in the market place and in social relations. It was an excellent umbrella term in the shelter of which psychologists became very busy. But there were intellectual costs, since there was no psychological theory behind the measurement of personality. There were no agreed descriptive categories, though people referred to things like emotionality and dependence as elements of personality. The use of the word 'trait' spread in the 1930s as a way to give the descriptive categories a semblance of rigour. It is noteworthy, however, that attempts to improve practice were directed at the measurement of traits rather than at the clarification of the assumptions embedded in the relationships between those who carried out the tests and those who were tested. The traits that were most studied were those that appeared to have a use value in the settings in which the tests were applied. Psychologists and their growing number of clients wanted more precise ways to describe and measure. Burt, when he extended factor analysis, took into his orbit active and emotional aspects of personality, and a large number of US psychologists experimented with test innovations. It appeared helpful to describe systematically a field of personality research, and this is what Stern's American student Gordon W. Allport (1897–1967) undertook in his book *Personality: A Psychological Interpretation* (1937). Stern, who arrived unhappily in the US as a refugee shortly before his death in 1938, hoped that such a field, which he called personalistic psychology and understood differently from Allport, would do something to counterbalance what he saw as the over-reliance on intelligence in descriptions of people. North American personality testing flowered, but it was not the answer Stern sought.

Alongside the expert concern with tests and measurement, the public was fascinated with the qualitative aspects of personality, a fascination that recreated everyday life and language as psychology. Like phrenology earlier, the study of personality made common cause between scientific and public approaches to human nature. Scientific and public values were particularly close in continental Europe, where there were sciences of characterology committed to qualitative methods in the study of individual differences. Scientists who practised character analysis included the Swiss therapist C. G. Jung, who analysed personality in the language of archetypes, the Dutch experimental psychologist Gerard Heymans, who devised a scheme – the Heymans cube – to represent the temperaments in three dimensions, and the German psychologist Ludwig Klages

(1872–1956), who used handwriting to reveal the expressive inner self. In the 1920s and 1930s, Klages' textbook on handwriting and character went through many editions and had much influence; he became the most famous representative of a characterology that linked outer activity to the inward soul. During the 1930s, Nazi rhetoric took over the language of human differences, at the level of both races and individuals, and built on the existing literature about character and the public's ready belief in natural distinctions of personality. A notorious case of continuity between qualitative differential psychology and the Third Reich involved Erich Rudolf Jaensch (1883–1940); he revised his personality theory to conform with Nazi race theory and in 1933 took control of the leading German journal of scientific psychology, the *Zeitschrift für Psychologie*. At the same time, in April 1933, a new civil service law removed those who were classified as Jews or married to those classified as Jews from academic employment.

The European intellectual influence on North American psychology in the 1940s included a radical input from refugees associated with the Frankfurt school and, in New York, with the New School for Social Research. Many of these scholars were interested in the relationship between psychological character and political events, and they contributed a qualitative and overtly political dimension to the study of personality. In *The Fear of Freedom* (1941), Erich Fromm (1900–80) explained to a general audience how the New School's work sought to understand relations between individual psychology and the Nazi or Fascist states. This approach culminated in *The Authoritarian Personality* (1950), a large-scale study co-authored by T. W. Adorno (1903–69), E. Frenkel-Brunswik, D. J. Levinson and R. N. Sanford into what the researchers judged to be the the personality type conducive to authoritarian politics. The study began in 1941 as a response to events in Europe and especially to the power of the German state; in the United States, however, the book attracted attention and criticism for its methodology in the study of personality rather than for its contribution to politics. Adorno, indeed, returned to Frankfurt in 1950, and his subsequent work was theoretical and philosophical rather than empirical.

The investment in quantitative personality research increased during and after World War II, a war in which psychologists were employed in large numbers. All US army officer candidates were tested, for example. Among the most influential studies to follow were those of Hans Eysenck (b.1916). After leaving Germany in detestation of the Nazis, he worked during the war with mentally ill patients in London to show that psychologists who use tests, rather than psychiatrists who use clinical judgement, can produce accurate diagnoses. Eysenck's *Dimensions of Personality* (1947), which was seen to establish quantification in an area notorious for qualitative judgement, developed a two-axis analysis that described a neuroticism dimension and an introvert–extravert dimension. In the 1960s, the psychologist R. B. Cattell (b.1905) located sixteen personality factors; a veritable industry of computer-aided tests and correlational analysis followed. Both Eysenck and Cattell studied under Burt. Ironically, the study of personality, which for lay people was an intensely subjective and personal matter, provided a home for the most mathematically-minded and hard-nosed scientific psychologists. Eysenck's long-term ambition, worked out in a series of books over twenty years, was to show that what he had measured revealed the key innate biological causes of human nature. Yet no agreement was reached among psychologists about the nature of personality, just as no agreement was reached about the factors of intelligence, in the absence of a coherent psychological theory about what the measurements measure.

During the first half of the twentieth century, testing of all kinds gave psychologists an occupation that was simultaneously expert and public. Psychologists engaged in very different activity from the narrowly-focused experimental work that nineteenth-century German academics anticipated would create a scientific discipline. Boring, who was active in the organization of the psychology profession in the interwar period, even referred to 'the schism between experimental psychology and mental testing' and as late as 1957, L. J. Cronbach, in a presidential address to the American Psychological Association, referred to 'the two disciplines of scientific psychology', general and differential.[14] This was a schism created more by differences in methods and training than by the different interests of academic and applied psychology. In the most general way, however, it was because psychologists offered knowledge in the form of techniques for the management of mass society – in the classroom, with delinquents, in the army, in the industrial enterprise, with marketing – that they were able so rapidly to expand their activity. In The Netherlands, for example, in the 1950s and 1960s the government sponsored psychologists to conduct research on the personality of emigrants. Since encouragement of emigration was official policy, the government turned to psychology for information to deal with the 40 per cent return rate within four years. In the course of carrying out such work, psychologists also translated the original managerial tests into academic and laboratory research, achieved relative academic autonomy and enjoyed self-sustaining growth as an occupation. All this underpinned the development of psychology as a science.

v *Occupational Psychology*

The language of energy, efficiency, capacity for work, attention, fatigue and control cut across individual psychology, engineering, the commercial and industrial worlds and national life. Proponents of social science in the late nineteenth and early twentieth centuries looked to expert knowledge to achieve results across this range. Psychologists wanted the new tests to contribute to education and gear teaching to individual capacity, personnel selection to increase productive efficiency and administration to adjust people to be better citizens. Social science turned to a psychological level of analysis and examined the characteristics of individuals in order to make social reality intelligible and manageable. Psychology was a force within the system of production and political order.

About 1900 there was a major shift, which some business historians have called revolutionary, in the scale and nature of business practice. The signs were innovations like the department store – such as the Bon Marché in Paris – large-scale employment of women office workers, the production line and the concentration of capital and market dominance by national and even international corporations. Western industrial society organized itself for mass production and consumption. The jump in scale generated new approaches to management – to finance, production, distribution, personnel and marketing. Management itself became a subject of study; there seemed every reason to create a science of management like any other science of social activity. The new science was obviously goal-oriented and designed to serve business efficiency, but the same goal of efficiency dominated the social sciences generally. This aim also spread the values of pressure groups like the US and European mental hygiene movements in the 1920s. Economic and political imperatives put pressure on science to yield up the secrets of efficient management of the self and of society, as of business. Personnel selection was especially promoted as integral to efficient management. The use of tests to stream children and to allocate educational resources was also, in effect, a form of personnel selection, with the goal efficient society rather than efficient business.

Management was initially conceived in engineering terms. To many contemporary observers it looked as if technology was the drive behind new enterprise. New technology was certainly conspicuous: telephone and electricity supply systems entered people's homes and covered the urban landscape with wires in the 1880s and 1890s. It was an engineer, a man who began as an apprentice machinist, Frederick Winslow Taylor (1856–1915) who systematically marshalled the techniques of industrial efficiency, and a preoccupation with mechanical notions of efficiency was labelled Taylorism. In *The Principles and Methods of Scientific Management* (1911) he brought together his thoughts on how to alter both employers' and employees' conceptions of work, taking it for granted that both parties had a common interest in the maximization of returns, i.e., in efficiency. He popularized techniques to quantify the different elements of the production process, notably time-study – which involved the separation of operations into measurable units and made it possible to measure and hence reward individual performance. The result in practice was the satisfaction of management goals or the adjustment of human capacities to the crude technical demands of machines. One French labour leader referred to Taylorism as the organization of exhaustion. This was the world of human management immortalized in Fritz Lang's *Metropolis* (1926) and Charlie Chaplin's *Modern Times* (1936), films in which the medium showed itself to be an appropriate new technology in which to express art and ethics in mass society.

The crude human engineering of Taylor's methods was widely criticized. In Europe there was more interest in physiological studies of work and fatigue, and this research also engaged with political questions about the efficient expenditure of a nation's energy. These studies were significant for the new psychology, as psychologists thought they could do better than Taylorism. In North America, Hugo Münsterberg (1863–1916) specifically advocated psychotechnics as an alternative. It appeared obvious to psychologists that skill in new research and in the administration of tests, coupled with knowledge about normal development, perception, learning, motor ability and motivation, was needed to mould personnel and workplace or consumers and products into integrated wholes. Much factory production, for example, involved repetitive tasks and output was restricted by fatigue; psychologists therefore asked questions about which movements, with what kind of repetition, produce fatigue, whether breaks in work increase the capacity for work and whether incentives affect tiredness and motivation. The British government sponsored such work during World War I, when the position on the Western Front appeared limited by the production rate of munitions in the factories at home. Another area where psychologists' tests and experimental skills were relevant was in advertising and marketing. During the war, once again, the Johns Hopkins University psychologist J. B. Watson produced a film for the US army to warn recruits of the dangers of venereal disease. He gave thought, not very successfully it seems, how to stimulate desirable responses and how to monitor the results. Watson, famous for his contribution to behaviourism, went on in the 1920s to become an executive of the New York advertising agency of J. Walter Thompson, an acknowledged innovator in the industry. By this time, the advertising industry had almost twenty years' experience of work with psychologists, after the involvement of Scott in 1902, when he was at Northwestern University and told Chicago businessmen: 'If we are able to find and . . . express the psychological laws upon which the art of advertising is based . . . we shall have added the science to the art of advertising.'[15]

The intellectual rationale for the involvement of North American psychologists with commercial and social affairs derived from an evolutionary concept of function, which made human adjustment appear to be the natural continuation of organic adaptation. Psychologists gained confidence from a social role in which they developed new techniques and knowledge as the means to carry human evolution forward. A number of psychologists accepted Watson's polemical view that psychology has as 'its theoretical goal . . . the prediction and control of behavior'.[16] During the 1920s and 1930s, the language of 'adjustment' became common among psychologists and psychiatrists. A. T. Poffenberger began his book on *Applied Psychology: Its Principles and Methods* (1927) with the observation: 'A man's whole life consists in a process of adjustment to his environment . . .'[17]

Imported German experimental techniques were used as a resource with which to devise research relevant to occupations in the commercial world. A notion of society as an organic whole had German roots as well as roots in evolutionary thought and inspired Münsterberg, the psychologist James attracted to Harvard University to take over the experimental topics in which he had lost interest. Like his compatriot Stern, Münsterberg believed that psychological research, especially on perception, was directly relevant both to industry and to legal problems of witness testimony. During the first decade of the century, high hopes were placed in the development of legal psychology, and these developments were supported by reform-minded lawyers as well as psychologists in the German- and English-speaking worlds. But these hopes did not survive the rigours of the adversarial process in United States law. Münsterberg also wrote an introductory text, *Psychology and Industrial Efficiency* (1913), which set out the psychologist's alternative to Taylor's engineering ideal of human efficiency. As he stated, all these expectations of applied psychology, often called psychotechnics until the 1950s, depended on the study of individual differences. Münsterberg's speciality of psychotechnics also flourished in Europe, especially in the area of personnel selection and training. In the interwar period, for example, the Austrian railways ran a special train equipped with psychotechnicians, which was used to test staff. Transport systems everywhere were noteworthy consumers of the new expertise; the People's Commisariat of Labour established a laboratory in Moscow for psychotechnics in 1920, and testing services were subsequently widely used for transport operatives across the USSR.

Experience of government and private business prompted a group of US psychologists, led by J. M. Cattell, to organize the Psychological Corporation in 1921. This was a consultancy, whose practitioners were of a guaranteed standard, founded on a presumed demand for psychological expertise. Its leading personnel, however, kept one foot firmly in academia, and the corporation's commercial viability was marginal in the early years. The large-scale expansion of American psychology in fact took place within the universities, however much expansion was made possible by claims to practical relevance. Substantial employment opportunities for psychologists who did not have academic support came during and after World War II.

Circumstances were different in Britain where C. S. Myers took a lead and founded the National Institute for Industrial Psychology in 1921. Myers was familiar with the work of his former student, the Australian psychologist Bernard Muscio (1887–1926), who had published lectures on industrial psychology given in Sydney and then worked in England for the Industrial Fatigue Research Board. Myers concluded that there was a field of enormous social value to be cultivated, and a small but sufficient number of industrialists, as well as the Carnegie and Rockefeller trusts, agreed with him and made funds available for the new institute. Influenced by his own involvement with the war effort, when he set up tests to select people with the acute hearing necessary to listen for submarines, Myers believed that national efficiency required the informed management of people. He argued that psychologists, not engineers like Taylor, could supply the necessary expertise. Faced by widespread academic ignorance of psychology, and sometimes actual hostility, Myers turned to private initiative and his institution, which he directed full-time from 1922, was integrated from the outset with the world of practical affairs. The Institute provided a range of services, such as personnel selection, production planning and marketing, to industry, government departments and educational authorities, and by this means sustained an income. In much of this work there was no clear demarcation between the aims of psychology and the aims of management. It also conducted research, often in conjunction with the Industrial Research Fatigue Board (renamed the Industrial Health Research Board in 1929) funded by the government through the Medical Research Council. Finally, in conditions where opportunities for academic training let alone academic employment in psychology were rare, the Institute was important as it brought people – like the future professor of psychology at the university of Liverpool and historian of psychology Leslie S. Hearnshaw – into the new area of expertise.

Though developed on the largest scale in the United States, occupational and educational psychology attracted reform-minded élites across Europe and the European empires. As nations struggled to industrialize, to raise educational standards and generally to modernize material and cultural conditions, the general situation in Central and Eastern Europe, a scientific outlook and support for modern ways went hand in hand. Small educated élites in countries like Romania and Bulgaria turned to psychological science for help in the transformation of their countries. Teaching in psychology and education was a focus for groups of intense students and young researchers eager to find in science authority for a better way of life. Since there was a social demand for teachers, psychology obtained some institutional support. The value placed in psychology as the instrument and the embodiment of progress spread by this means.

The career of occupational psychology in Germany had a

special character. Academic psychologists like the aged Wundt (who died in 1920) or the gestalt psychologists studied psychology as an expression of the cultural ideal of *Wissenschaft*, systematic and rationally-grounded science, and hence with philosophical purposes remote from the world of business. They were significantly more interested in a generalized conception of the mind than in issues of differential performance. In addition, students who wished to study psychology had at the same time to satisfy the requirements of the philosophy curriculum. Nevertheless, there were individuals, of whom Stern was the most prominent, who contributed to the technical aspects of differential psychology and sought to integrate academic practice with social and industrial policy. Stern, appropriately, taught at the new university of Hamburg in the 1920s, the sort of institution identified by conservative academics as an expression of the cultural inferiority of the modern age. With the Nazification of higher education in April 1933, Stern, who was labelled Jewish, was removed from his post, even though for a short time he had the illusion that the new regime, which was committed to modernization, would value his ideas for human management.

During the second half of the 1930s, the mobilization of German society for war placed a premium on the efficient use of manpower. Womanpower, at least in theory, was sent back home to breed. The *Wehrmacht* (the German army) listened to university psychologists who claimed the means to select officer material. As a consequence, academic psychology gained a professional identity, and an institutional autonomy from philosophy, that it previously had not had in Germany. In the middle of the war, in 1941, university psychology for the first time acquired a separate programme of study and became a separately examined subject. But its standing with the army did not last, since officers selected themselves at a rapid rate under fire on the Russian front. These circumstances in which professional psychology grew in Gemany left an emotionally and politically divisive legacy. After total defeat in 1945, 'Year Zero', many German academics, psychologists included, distanced themselves from the Third Reich and described the period from 1933 to 1945 as an aberration, a political disaster with which the apolitical ideals of science had nothing to do. There was every personal and occupational incentive to claim complete separation of the objective methods of research from the process in which research is applied. Psychologists rapidly took an interest in North American behaviourism and other experimentally rigorous forms of psychology in the post-war years, further distancing them from belief that they had any special German legacy. A later generation of German psychologists, however, questioned the role that psychological expertise had had in the war effort, pointed out the continuity of psychological personnel before and after 1945 and argued that it was occupational psychology rather than the imperatives of objectivity or the ideals of pure science that established psychology as an academic discipline.

World War II was the scientists' war; it was also the psychologists' war. Technical and scientific experts were employed on an unprecedented scale by the governments of the United States, Britain and Germany. The clearly defined goal of victory enabled all parties within each country to agree to the politics of planning and centralized decision-making, and governments turned to scientific management. Many of the same ideals had already informed the stated policies of the USSR. The modern management disciplines of systems analysis and operations research both originated under the exigencies of war. Psychologists were employed to smooth the adaptation of people to machines and to the vast military organization of war itself, and also to design machines – especially control systems – that people could manage. Much of the impact on experimental psychology in fact came later. In terms of the sheer scale of what was undertaken, testing remained the most important activity; personnel selection, for example, was essential for radar monitors. The US army and navy both developed versions of the General Classificatory Test (GCT) to help invest effectively in people of the right ability and aptitude. These tests built on the methods of factorial analysis developed by the Chicago psychologist L. L. Thurstone (1887–1955), who had earlier analysed human performance in terms of what he called primary mental abilities and which he put forward as a replacement for crude notions of intelligence.

War also employed large numbers of clinical psychologists concerned with motivation, stress and injury. In Britain, the director of the Tavistock Clinic, J. R. Rees (1890–1969), was appointed to head a comprehensive programme of psychological services. This appointment was a coming of age for clinical psychology. The Tavistock Clinic had been founded in 1920 with private funds 'to provide treatment along modern psychological lines' for people with mental problems 'who are unable to afford the specialists' fees'.[18] The clinic, from which a specialist childrens' department separated in 1926, pioneered an individualized psychological response to a wide range of human problems. It established influential training programmes for professional groups such as teachers and doctors, and in this way a psychological technology became woven into the fabric of everyday social and medical arrangements. National validation of the Tavistock orientation in time of war was a validation of the presumptions of psychological society. This positive evaluation of the practical benefits of psychology rose further in the post-war period. The Tavistock opened an Institute of Human Relations in 1948 to conduct research and give training in group psychology methods to people from industry, the civil service, the churches, voluntary care organizations, social work and so on.

In the United States, the war gave many psychologists clinical experience. Psychologists who had trained with rats, test techniques and neobehaviourist assumptions gained a taste for work with humans in need of care. In the wake of the war, the Veterans Administration successfully lobbied for federal government funds which underwrote a large expansion of teaching and employment in clinical psychology. One significant consequence was that an increased number of male as opposed to female psychologists entered the field and there was a shift to accept adults as well as children as clients. The efforts of Carl Rogers and others opened opportunities in health care independent of control by physicians, and psychologists were able to regulate their own clinical field. The outcome was a substantial growth of the speciality of clinical psychology, with the result that the American Psychological Association had to rethink the way the profession as a whole accommodated the interests of both academic and applied work.

In Europe and in the English-speaking world, industry, government, the welfare and health services and education were eager to deploy psychological expertise. Widespread political acceptance of the view that the state has a responsibility for welfare coincided with belief in what social science expertise has to offer. Psychology now expanded rapidly in Western Europe, and achieved a large-scale presence as an autonomous discipline within universities which were themselves expanding in response to demands for educated labour and increased access to higher education. The discipline's new-found prestige also reflected academic and public willingness to view the social sciences generally, and psychology in particular, as forms of expertise outside the political process. The 1930s and 1940s taught what fanaticism and ideology achieve. The public in the West was therefore willing to look to science to advance the common good. The physical scientists appeared to have broken through the ultimate material frontier of the atom and to have created, in the contemporary phrase, the 'limitless horizon' of power and energy; it was a commonplace to point to human management as the remaining frontier. Faced by the continued maladaptation of human beings to modern conditions, psychology offered hope for the future.

Yet, however humane in intention, the spread of psychotherapeutic intervention and techniques for the management of personal relations was neither politically nor morally neutral. The development of new technologies of human management, in which 'the Tavi' played a central role in Britain, transformed power and authority in human relations and built expectations about how people should live into the mundane, day-to-day occupations of the helping professions and thereby into the lives of those helped. Whereas an earlier age held that a person was related by obligations to an external world of social place and contract, the new age located regulation of an individual's relations to others in internal psychological space. The goals of personal adjustment and satisfaction transformed political and ethical issues into topics appropriately managed by psychotherapeutic techniques. These new techniques were operated in every setting – the hospital, school, family, marriage bed, factory or prison – and changed the way people thought about themselves as well as the way they related to others. Psychology therefore contributed to a reconstruction of what was understood as the self and how the power of government in social relations was exercised.

vi *Vienna, the Interwar Years and the Psychology of the Child*

The emphasis on 'the psychological' in human life was not created solely in the English-speaking world. The emphasis in continental Europe, however, differed because of the diversity of cultural and national traditions and because of the shattering impact of war, revolution and political conflict unknown in Britain or North America in the twentieth century. The European quest to understand suffering and its roots in human nature had a special intensity. Some psychologists, faced by dire circumstances, were committed with an almost religious faith to science as a way to save humanity. In the aftermath of two great wars, and again with the collapse of Soviet power in 1989 and the spread of social anarchy and war in Central and Eastern Europe, people appealed to psychology and the social sciences for help. Faced by political and social failure, men and women turned to examine human nature itself, whether to understand how terrible things are possible or, more optimistically, to pull up human nature by the roots and change it. Many scientists held on to an ideal of objective reason and believed that if science could be given a voice in education and the state, it would achieve the humane future that politics had failed to bring about. Many intellectuals, including a good number of scientists, believed that Marxism was itself an objective science, a scientific basis for politics, and hence that the struggle to achieve scientific socialism and scientific knowledge of man went hand in hand. This carried into the twentieth century a version of the Enlightenment project to make women and men free on the basis of knowledge of their nature. Such ideals then became entangled by the confusion of an ideal of scientific socialism with an actual political state, the USSR. In the West there was a parallel entanglement of the dominant values of political individualism and belief about the psychology of human nature. Individualist moral and political values were embedded in what I have called psychological society in Western Europe and North America, and the human sciences focused on the adjustment of the self as a 'natural' activity in the social order. Socialist values in Central Europe and in the Soviet Union, at least for a time, also focused on the adjustment of human nature, but under the Soviet system the science of adjustment was equated with the 'objective' position of the Communist Party.

Some aspects of these complex topics are illustrated in this section, which describes a city at the centre of Europe, Vienna, and the hopes for a new life channelled into psychological studies of children. From its status in 1914 as the capital of an empire that had lasted a millennium, uniting 52 million people of diverse cultures, by 1918 Vienna was an isolated city of a small rural nation. Influenza, lack of food, the destruction brought about by the war, inflation and hope or fear of revolution created terrible living conditions. The new Austrian republic had a socialist government for only five months, but the city of Vienna remained socialist through the 1920s. Until the ultra-conservative dictatorship of Dollfuss took over in 1932–4, the socialist city administration and the conservative provinces, which held a majority in government, were polarized. In these years, indeed to some degree until the *Anschluß*, the Nazi incorporation of Austria into the German Reich in 1938, psychology, especially the psychology of education, flourished. These were also the years in which Freud made Vienna the mecca for psychoanalytic training, and some of his followers looked to his work to redeem human nature. Young, idealistic men and women from Vienna and elsewhere studied human nature, and especially the child, as a way to remake human life and overcome the calamities of the age. There was an outburst of new activity in psychology – investigative techniques, theoretical generalizations, teaching practice and institutions. As it turned out, there was also a tragic mismatch between the power of wider political forces and the energy of creative people.

The university of Vienna had no formal arrangements for psychology until 1922 when it called Karl Bühler (1879–1963) to a chair and at least nominally equipped him with an institute. Vienna had had philosophy professors with a major interest in psychology, notably Brentano and Mach, and Freud had unpaid professorial status in medicine from 1902, but Bühler contributed a new breadth of experience with German theoretical and experimental psychology. Earlier in Würzburg, he was interested in thought processes, contributed an experimental study of the 'aha!' experience of sudden insight and wrote a substantial work on language. A highly cultured scholar, in Vienna he created an atmosphere that supported a broad, humane and scientific vision of psychology's tasks. His appointment was, in effect, a double one, since his wife, Charlotte Bühler (1893–1974), who had been his student, proved to be a formidably efficient manager of the institute and also co-ordinated a major research programme on child development. A vivacious, some say vain, energetic woman, she provided a scientific institutional base for people who saw in the child and in education the way to create a new world. When she attracted funds from the Laura Spelman Rockefeller Memorial in 1926, she turned the child study centre, run under the auspices of the university institute, into a department unique in Europe. Karl Bühler wrote influential studies on the child's development of language and a magisterial survey, *Die Krise der Psychologie* (*The Crisis of Psychology*, 1927), in which he attempted to synthesize competing schools of theoretical psychology.

Charlotte Bühler and Karl Bühler were academics with rigorous notions about what it is to undertake objective scientific research. Karl Bühler was sympathetic with – though not actually a member of – the Wiener Kreis, the Vienna Circle, formally constituted in 1929 but meeting from 1925 until 1936. This circle of positivist philosophers, who accepted the leadership of Moritz Schlick (1882–1936), argued that meaningful propositions are of two kinds: formal, such as logical axioms, or empirical, verifiable by observation. Bühler's advice to his students and assistants to attend meetings of the Kreis showed his concern that scientists must demarcate science from wishful thinking or pseudo-science. Many academics thought that a criterion to demarcate scientific claims, those open to verification, was much needed in the psychological and social sciences. Within the institute, the Bühlers maintained the study of scientific psychology in its own right, without direct reference to immediate practical goals or social values other than those intrinsic to science itself. Charlotte Bühler's child study centre, for instance, had a regimen of data collection, the purpose of which was to advance a general theory of developmental stages and not to formulate research into particular questions posed by education or child-rearing policy. She required a rota of staff continuously to observe babies and young children from a glass-sided corridor. The children in the centre, who were there for observation and not treatment, lived with a minimum of human contact in a sterile environment, a situation that supposedly provided a culture-free baseline for the observation of development.

All the same, many of the Bühlers' students were schoolteachers or worked in training teachers, in experimental schools, in child guidance or with disturbed children. The Bühlers were also hired to teach at the Vienna Pedagogium, the teacher training institute in the city. There was therefore a complex relationship between science and social values. In the first place, the city authorities funded a child reception centre in order to observe children's problems, and this centre supplied the children or subject matter of Charlotte Bühler's researches. The data accumulated at the scientific child study centre derived from a specific group of children who had been brought together as a result of illness, social dislocation and family breakdown. Though the methods of research were self-consciously scientific, the subjects studied were created through society's attempt to respond to a social problem. The pattern was not special to Vienna, and child-study movements existed in many countries. Charlotte Bühler's own ideas became widely known in the US after her emigration and through her later publications on the life cycle.

In the second place, there was a high level of political activism among the teachers and observers who worked with the children. Some, like Marie Jahoda (b.1907), were active in the Austrian Social Democratic Party and were informally Marxist in orientation; others, even if they were not overtly political, were involved with educational programmes the purpose of which was to create new, socialized and fulfilled human beings. From 1920, Freud's former friend and colleague Alfred Adler began an energetic campaign for therapeutic education, and he pioneered a consultation service for parents and teachers, new kindergartens and experimental schools. Vienna provided a supportive context. Though short-lived, the Austrian socialist government pushed through educational reforms designed to create equality of educational opportunity, to modernize the curriculum and, at least for Vienna, to secure a tax system to pay for educational and social welfare for children. The city integrated educational and clinical support, and the collective provision for child welfare was perhaps unequalled until the Scandinavian and Dutch social democracies instituted comparable policies. By 1926 there were some forty different agencies in the city of Vienna concerned with children. In these circumstances, young intellectuals – perhaps especially those whose dreams were unfulfilled because of the absence of revolutionary change – placed their idealism at the service of scientific research and education. Whether the research had consequences for educational practice is questionable; certainly the authorities neither expected nor received direct practical advice from the child study centre. Rather, the scientific research in psychology was part of a general ethos that supported idealism in education. There was a general sense that if the child's nature were known and its needs understood, then its future could be assured. Further, teachers hoped that these new children would remake society as a place fit for a new generation. Even non-socialist psychologists assumed that science should receive community funds and in turn provide the community with information and expertise. In fact, Charlotte Bühler's two paid assistants were relatively conservative in politics, money was desperately short and much of the innovative work was funded by private wealth. But the level of activity and optimism was exceptional.

The career of Lili Roubiczek Peller (1896–1966) provides an illustration. She was born into a wealthy Prague family but trained as a teacher, and she then travelled to Vienna to work with Karl Bühler. There she strove to bring together psychology and the needs of working-class children. Inspired by Maria Montessori, the Italian educationist who extended techniques devised to teach children with learning difficulties to all children, she created the Haus der Kinder (literally, 'House of Children') in the slums of Vienna's Xth district, the first official Montessori school in Austria. With funds from England arranged by Montessori, Peller joined with five women in their late teens to form a collective, and they built the school with help from local craftsmen and architects. Peller went on to become a consultant to the city's child welfare department, to manufacture Montessori furniture and literature, to offer a training course taken by some of the Bühlers' university students and to develop links with child psychoanalysts. She introduced Montessori to Anna Freud, Freud's daughter and pioneer child analyst, in 1930. After she had herself trained as an analyst, Peller worked at the Jackson nursery in Vienna, an observation centre established in 1937 by a wealthy American paediatrician, Edith Jackson. By this date such institutions were under political suspicion; the nursery therefore had no welfare role and its supposed Jewish intellectual debt to Anna Freud was disguised. As she was married to a Jewish physician, and with both Montessori methods and psychoanalysis excoriated by fascists, Peller then emigrated to the United States.

There were other similar careers, many of women. In this context, psychology was a way of life – to practise psychology expressed expectations about how knowledge and technique makes new people and new social relations – rather than an academic speciality. How far any of the new psychological interests directly affected children is hard to judge; but the new interests fostered a middle-class culture that valued psychology. As a way of life, however, this culture was overtaken by a politics that denigrated and sought physically to eliminate what that way of life stood for. The Austrian and the Nazi virulence towards those labelled Jews caught in its hatred large numbers of the Viennese researchers and Charlotte Bühler herself. Those who opposed so-called Jewish psychology equated it with concern for children's self-expression and the welfare of the working-class child. Austrian fascism and then National Socialism repudiated psychological society and, after 1945, the drive to create such a society came from the English-speaking world, informed, however, by the extensive emigration of European psychologists to Britain and North America. In The Netherlands and Scandinavia, indigenous educational and welfare policies built many of the values of psychological society into the institutions of the state. Because these countries achieved high levels of political consensus, especially after World War II, they created circumstances in which psychology and the social sciences burgeoned; these countries in effect put into practice what the Viennese researchers had been able only to hope for, the acceptance of the science of the person as the central plank of a more rational politics.

The Netherlands, for example, in the 1980s probably had the highest ratio of trained psychologists per head of population in the world. Psychology began to expand as a discipline after World War I when church leaders turned to psychology as part of an attempt to address the materialism and spiritual vacuum they saw around them in urban and industrial conditions. Until the 1960s, Dutch society was divided – 'pillarized' – between Protestant and Catholic communities, but though each community established its own institutions like universities, pastoral values and an interest in anti-materialist social science were shared. Both Protestant and Catholic leaders looked to psychology to provide knowledge of the individual's inner world and of his or her orientation towards the wider society and hoped that this would support renewal of a life committed to Christian values. As the Dutch universities reflected denominational lines, an innovation in a Protestant unversity was duplicated in a Catholic one, and vice versa, fostering the institutional development of the discipline. In the 1920s and 1930s there was interest in Stern's personalism as an element of teacher training, and there was a strong interest in characterology, the systematic identification and classification of character types. Characterological schemes, such as Heymans' 'cube', were discussed and there was a particularly strong Dutch interest in graphology or handwriting analysis, adopted from Klages, in the judgement of character. It appeared self-evident that the purpose of psychological knowledge is the achievement of better values, a better orientation in a person's life, and the psychology that developed was therefore at the same time specialist and everyday. Characterology used a language which helped the expert to respond to a person's problems and which enabled that person to reflect on her or his condition. This language was also equally usable with either a pastoral and religious orientation or a therapeutic and secular one. Characterology was a technology of the soul and of the person. It served well to bind together a professional discipline, a subject matter of obvious practical import and the ordinary person's sense of self and values.

Characterology declined in popularity after 1945, partly because some schemes were of German origin and tainted by racist or biological crudities. There was a move in The Netherlands to adopt the more rigorous and quantitative US personality and intelligence tests with their statistical rather than interpretive methods. Nevertheless, there was also a strong defence of qualitative psychology at the university of Utrecht, where F. J. J. Buytendijk elaborated his version of phenomenological psychology, which had an ideal of a moral attitude in human relationships at its core. Very different conceptions of science and of psychology therefore existed side by side in the 1940s and 1950s; one sought to define science as neutral, objective knowledge, the other to define science as a contribution to moral wisdom. But in both cases the outcome was support for psychology as a discipline and, more especially, support for psychology as an occupation in social administration, therapy and education. When Dutch society changed in the 1960s to become more secular and the old denominational division lost significance, shared support for responses in psychological terms to the problems of individual children and adults remained.

Research on children's development, child study and a systematic interest in children's physical, mental and moral welfare was widespread in many countries from the late ninteenth century into the interwar period. This created occupational opportunities for trained psychologists and made psychology part of the expectations of family life. The roots of these interests and of belief in the power of the environment and education to train a child's nature were traced in earlier chapters to Locke, Rousseau and Pestalozzi. A dramatic example of faith in the power of a right upbringing is given by the New York Children's Aid Society which, in 1854, began to collect urchins off the city streets and put them on trains to work on Midwestern farms. Over the next seventy years, some 150,000 children travelled the Heartland Express, with favourable results as measured by the way the children's subsequent careers were interpreted. This policy was the work of idealists and reformers, and it continued through the years that historians often picture as dominated by hereditarian principles of human nature. In the 1920s, these environmentalist values informed a more specialized child study movement led by professional psychologists. One sign of professional activity was the foundation in 1910 of the US *Journal of Educational Psychology*, which was much more experimentally and statistically oriented than the comparable earlier journal, the *Pedagogical Seminary*, founded in 1891 by Hall, who was renowned for his concern that psychology should take up the reins of moral and educational guidance. Even the titles of the two journals stated the difference that science made to the study of the child. The Laura Spelman Rockefeller Memorial funded centres for child research at the universities of Columbia, Yale, Iowa, Minnesota and Toronto before it made funds available to Vienna, and these centres were in communication with each other. As early as 1896, Lightner Witmer (1867–1956) opened a psychological clinic at the university of Pennsylvania in which he examined and treated children with learning problems, and he subsequently began to train psychological experts in children's education. Arnold Gesell (1880–1961) at the Yale Psycho-Clinic, opened in 1911, systematically studied babies and young children within a one-way glass dome, and he used measurement and photography to build up a picture of the stages of normal development. Such accounts were then translated into guides to what every mother could expect from her child, and supplied a growing interwar industry in motherhood manuals or, in a later idiom, parenting guides. The mother was expected to become a psychologist. When she did not, and when the child did not develop normally – physically, mentally and morally – expert psychologists were increasingly available to do the job.

During the 1920s, Burt taught extensively as a professor at the London Day Training College, later the Institute of Education, reinforcing the position of psychology in the mainstream of British educational practice. During the 1930s, the child study movement changed from being the interest of amateurs, those without formal training, to become an occupational movement with psychologically trained personnel, specialized institutions and scientific methods – notably diagnostic tests. One sign of this shift to make the child's life a specialist concern is the establishment in 1933 of a department of child development at the Institute of Education. The department was headed by Susan Isaacs (1885–1948), who had amongst much other activity organized a progressive school in Cambridge. Her own career ran to some extent from practical engagement to academic psychology. The child study movement also contributed to a decline in the emphasis on hereditarian as opposed to environmental forces in the formation of character.

The study of children was the most important factor in the involvement of large numbers of people, especially women, both trained and untrained, in the project of psychology. But neither in the family nor in society at large was the child a natural object who awaited observation and study: it was a projection of aspirations for a better life, and psychology seemed specially equipped to turn these aspirations into reality. The child was the vivid symbol of hope. In Vienna, where pre-modern politics in Europe finally died in 1918, there was an especially close connection between political, educational and psychological ideas and an especially vivacious group of young women and men. But the values of these Viennese researchers were international, and the scientific study of the child brought psychology as a form of politics into every home.

vii *Nature and Nurture*

In a psychological society a person's capabilities simultaneously shape relations with others and become a self-definition. The description of an individual's competence and character – her or his personality – creates an identity and evaluates it in relation to other people and to standards. Psychological tests were developed to provide precise information about competence and character, but because descriptions of identity necessarily contain an evaluation, these tests – however scientific – were not neutral and free of social values. Further, the historical evidence shows that the early proponents of measurement and tests, like Galton, Binet, Spearman and Terman, valued this activity specifically because they believed it would transform the efficiency of social policy and even make new policy possible. The more refined tests of a later period, such as those used by companies to select personnel or by clinical psychologists to determine functional loss after an accident, presumed values appropriate in the settings in which the tests were applied, and the tests assessed people accordingly. Companies valued people with a positive attitude; the community valued people who are independently mobile. Thus, in the century or so during which tests existed, their use and the descriptions of human character that resulted were always bound up with moral and political debate. This is something intrinsic to any enterprise that describes and differentiates people, not an unfortunate accident that stems from psychology's origins.

Debate was most public when it was about the relative contribution of nature and nurture, or heredity and environment, to a person's capabilities. At times it was headline news, as in the United States in 1969, when Arthur Jensen (b.1923) linked intelligence and race and questioned liberal education programmes. Post-1945 British politics involved recurrent conflict over educational policy and resources, with the advancement or retraction of comprehensive education in tandem with the shift of the balance of values attached to a child's natural gifts and social equality. Similarly, debate in many countries about the balance of retributive and reformist approaches to punishment was linked to different views about whether the causes of crime lie in a person's character or in the environment. That public controversy about educational and penal policy focused on the causes of psychological states illustrates again how far psychology had become the language of public life. Few people were so rash as to attribute everything to either nature or nurture. The dispute was about the degree to which natural or social causes in practice limit human potential, and it was about ways in which these causes are subject to change. For these reasons as well, the debate was intensely political, however much proponents on either side cited empirical evidence.

Galton gave shape to the debate through his use of the language of nature and nurture. His beliefs were unambiguous: there are strict and narrow limits set by inheritance on the enhancement of mental or physical performance. He assumed that social eminence genuinely measures innate ability, and used eminence recorded in biographical dictionaries to trace the inherited descent of ability. His critics believed that he did not comprehend the restrictive conditions of life for working people, though Galton thought he took this into account. When he first formulated his views in the 1860s, he saw that they conflicted with the Victorian ethos of self-help which stressed how will-power enables people to do what they wish. He therefore put forward what he believed to be scientific hereditarian psychology in opposition to what he thought of as unscientific moralistic psychology. What was at issue, in his mind, was science versus non-science and not simply nature versus nurture, a conjunction also evident in the twentieth century. Hereditarian thought became very widespread, though by no means universal, at the end of the nineteenth century, and it was associated with debates about race, the struggle for empire, the rise of socialism and feminism and fears of degeneration, alcoholism and mass society generally. Burt followed Galton closely and displayed a faith in the quantifiability of inherited characteristics from early in his career. Then the pendulum began to swing again and the emphasis on heredity declined about the time of World War I. In the 1920s there was extensive concern with the environmental or social determinants of human capacities, exemplified by the young teachers who worked in the Vienna slums. It was in this period, especially in North America, that the modern terms of the nature–nurture controversy became common currency.

The precise ways in which psychological claims and policy interacted varied with time and place. The US debate in the 1920s centred on the restrictive immigration Act passed in 1923 in response to fears about the number and quality of new immigrants. Since the beginning of the century, an increased proportion of people had entered the country from Southern and Eastern Europe, Turkey and the Caucasian region and China and Japan. There was concern that these people could not be integrated into the national life and a belief that they had inferior capabilities to earlier north-west European immigrants. The Act itself did not result from pressure from psychologists, though the army tests were interpreted by some, notably the Princeton psychologist Carl C. Brigham (1890–1943), as evidence for a general difference of intelligence between older and newer immigrants. Goddard also briefly applied intelligence tests to immigrants who arrived on Ellis Island in New York harbour. These tests recorded low ability levels, in as far as they recorded anything, among non-Western European groups for the obvious reason, as Goddard quickly realized, that the tests presupposed a certain level of linguistic and cultural familiarity with the tester's own world. Testing therefore contributed to the atmosphere rather than the substance of restrictive legislation, but it certainly became part of the public debate about nature and nurture and the future of education.

The social anthropologists, particularly Boas and his students in the United States, who worked on cultural differences, encouraged belief in what has been called 'the sovereignty of culture', belief that what divides peoples has little to do with biological heredity.[19] This message resonated with one side of American political life, faith in the adaptiveness of people to new social conditions. There was massive support for public education, and through education some support for psychology, and the faith in adjustment was put to the test during the Franklin D. Roosevelt years and the New Deal of the 1930s.

Boas himself was an immigrant from Germany and he brought with him an interest in the relation between psychological and cultural processes, for example, in language development. In Berlin, Boas was an assistant to Bastian at the Royal Ethnographic Museum, an institution with a commitment to belief in cultural as opposed to biological or hereditarian theories of human progress. In 1899 he became professor of anthropology at Columbia University, and from this position he exerted considerable influence over the development of his subject into a university discipline and away from the museum-based occupation which it had been in North America in the nineteenth century. His book *The Mind of Primitive Man* (1911), which drew on his fieldwork with the indigenous Kwakiutl of British Columbia, as well as on other native peoples, rejected the classification of so-called primitive people in relation to an evolutionary hierarchy. Indeed, he conceptualized cultures in terms of the ways indigenous peoples described the world. He argued for a common human nature. In the same year as his book, Boas published a detailed study of the cranial dimensions of the descendants of immigrants, which showed that head shapes change. Though he did not polemicize in his paper, the message was clear: supposed biological regularities of race vary with local conditions. Such arguments supported the opinion that individual performance cannot be attributed to a primitive mind or a racial type but has to be understood in the context of the culture in terms of which the performance has meaning for the person engaged in it. They also defined a sphere of intellectual activity for the field of anthropology independent of general psychology.

Anthropology gained a large audience in the 1930s through Boas's student Margaret Mead (1901–78) and her book *Coming of Age in Samoa* (1928). She first studied female adolescence in Samoa in the mid-1920s, when she herself was in her mid-twenties, and in her book she contrasted the easy and carefree experience of Samoan girls with the inhibitions and anxieties of her American contemporaries. This study of the contrast between patterns of sexual development, a study designed by Boas, was thought to exemplify the argument that culture rather than fixed biological rules control a child's development. In this context, Mead criticized the earlier work of the psychologist Hall. Many educationalists found support here for their view that a child's development is open-ended and that it is up to society in general and mothers and teachers in particular to create optimum conditions for the new generation. As Mead wrote in the 1930s: 'human nature is almost unbelievably malleable, responding accurately and contrastingly to contrasting cultural conditions'.[20] She continued for many years to relate her ethnographic results to North American preoccupations with child-care, sex roles and personality. In *Male and Female* (1949), however, she showed that she had begun to rethink the extreme environmentalism of her earlier work. Much later, the whole basis of her study of Samoa and the authority of her views on psychology and culture was deeply criticized. But, through the middle years of the century, she was a powerful spokeswoman for belief in the effect of culture on personality.

With the use of mental tests on a routine basis, extended to tests on personality in the 1930s, there were attempts to find indicators for masculinity and femininity and to measure for them. The conclusion that psychology shows systematic differences in intelligence and other capacities between male and female was rejected by the Chicago psychologist Helen Bradford Thompson in 1903, but she thought psychological methods were then unable to settle the matter. In the 1920s, however, psychologists constructed tests specifically aimed to discriminate female and male qualities, and they found what they looked for. A debate resulted in the language of nature and nurture about sexual differences. Many critics of this work, then and later, argued that male psychologists like Terman and Yerkes built their prejudices into their test constructions. Sociologists like Thomas and anthropologists like Boas tended to be sceptical of the existence of innate psychological differences; Mead concluded that 'the personalities of the two sexes are socially produced . . .'[21]

The 1920s was also the heyday of Watson's brand of behaviourism, a subject discussed in the following chapter. Watson took to a crude extreme the argument that human nature results from received stimuli. He claimed that any child – if not physically damaged – can become anything. With his student and second wife Rosalie Rayner, Watson published widely, often in the popular press. General journals and popular psychology took up the theme and, at least temporarily, this opinion influenced advisory pamphlets published by the government. Even when discussion was a little more circumspect and a little more attention was given to the complex developmental stages of childhood, the environmentalist argument placed a heavy burden on parents and teachers. But if this was a burden, it was also the hope for a democratic order in which each citizen is made equal even if circumstances of birth have not made him or her equal.

The practice of racism as a principle of world domination by Germany and Japan in the 1930s and 1940s coincided with and reinforced the US anthropological critique of biological categories of human differences. After World War II, a substantial body of opinion denied any objective validity to the category of race and attempted instead to classify the world's peoples by reference to cultural or ethnic differences. Some human geneticists, interested in the distribution of blood groups, argued, by contrast, that the category of race, defined in strict ways, retains value. Then, in the 1960s, the fate of the terms 'race' and 'intelligence' again became central to politics against the background of the Civil Rights movement in the US, compensatory educational programmes for black US ctitizens and liberation movements worldwide against Western political and economic imperialism. The nature–nurture issue divided psychologists and it divided the public. All sides combined scientific argument with political and social preferences. Hereditarians, who took race and intelligence to be real biological entities described by genetic or factorial analysis, portrayed themselves as the scientific vanguard of a campaign against policies that cannot produce equal achievement. Environmentalists, who took achievement to be a consequence of social arrangements, attacked their critics for mounting right-wing ideology in the language of the facts of nature. The hereditarians, like Galton in the nineteenth century, stressed the hard-nosed quality of scientific argument in contrast to what they perceived to be the woolly wishful thinking dominant in the public sphere. A decision about whether or not the hereditarians practised good science, which the environmentalists denied, was therefore crucial to the conflict.

The controversy started afresh with the publication of a paper by the university of California psychology professor Jensen, 'How Much Can We Boost IQ and Scholastic Achievement?' (1969). The intensity of the controversy reflected the context as much as the content of his paper, a context of student rebellion against the Vietnam war, of hopes for radical social transformation and of substantial commitment to programmes such as Head Start designed to overcome the social conditions believed to disadvantage blacks. The conservative uses to which psychology could be put were taken to a limit in radical eyes when one of Jensen's supporters, the Harvard psychologist Richard J. Herrnstein (b.1930), ignoring the war, attributed student activism to natural adolescent rebellion. Jensen's paper itself argued that 'compensatory education has been tried and apparently it has failed' and then turned to heredity for an answer; it attacked the current 'ostrich-like denial of biological factors in individual differences, and [a] slighting of the role of genetics in the study of intelligence [which] can only hinder investigation and understanding'.[22] The paper reasserted the reality of innate general intelligence as a central factor in human differences and, by extension, the relevance of innate intelligence to social policy. Jensen concluded with a statement of his view that the best evidence supported belief in innate genetic differences in intelligence between black and white Americans.

Jensen cited Burt with admiration, and there was a parallel though somewhat less emotive eruption of controversy in Britain. The 1968 Labour government's attempt to advance comprehensive education, which abolished a divide at secondary level between different schools supposedly appropriate to children of different ability, met strong and successful resistance. The temperature of the debate rose after a series of 'Black Papers' argued against the levelling of education. The authors included Burt and Eysenck. Significantly, Eysenck studied for his doctoral thesis under Burt, though there was soon distance between them, and Jensen worked in Eysenck's department as a postdoctoral fellow. Eysenck's great productivity – he wrote a succession of student texts and popular psychology books – gave him a substantial lay as well as academic audience. His unshakeable self-confidence in his belief that Galton was right to emphasize innate biological capacities enraged left-wing opinion, which accused him of anti-egalitarian politics disguised as claims about the constraints of nature. Eysenck himself viewed his life's work as the development of the techniques of factorial analysis to search for biological causes of human differences.

Even the controversial Eysenck took second place when a storm broke around the reputation of Burt, who died in 1971 as Britain's most eminent differential psychologist. Once again the scientific issue had its roots in Galton. In 1875, Galton published a comparison of the abilities and character of identical twins who had been separated at birth and brought up in different families. Then and later, it appeared that if any evidence is able to clear up the nature–nurture debate, it will come from twin studies. Examples of separated identical twins, however, are rare. Even where they exist, researchers have had to track them down, compare them with proper controls and study them over an extended period of time. Even when identical twins were separated, they might, for example, have been adopted by similar families, since adoption agencies select families by the same criteria. In spite of these difficulties, Burt assembled data on some thirty cases and his conclusions emphatically supported his view that nature and not nurture is the key determinant. He argued that identical twins exhibit comparable ability even when brought up in different social environments, and this argument became part of the standard psychological literature. Though his data were thought to need qualification, it was assumed that his papers showed how scientific psychology can produce objective results on a politically sensitive topic.

All this changed. A mixture of investigative journalism and academic enquiry, the latter by Leon J. Kamin (b.1927) and published in *The Science and Politics of I.Q.* (1974), looked into the remarkable precision of Burt's results and inconsistencies between his papers. It was suggested that he had not only manipulated data but had invented twins and research assistants. It was also known that Burt altered papers by others in journals he edited to increase the authority of his own point of view. His biographer Hearnshaw, who in his own career had admired Burt, systematically – and sadly – confirmed these conclusions: Burt had committed fraud. Hearnshaw qualified this stark conclusion with a portrait of Burt's strengths and weaknesses as a psychologist and as a human being. The Council of the British Psychological Society, much divided on the issue, admitted the fraud. In the eyes of its critics, it preferred to denigrate an individual rather than publicly face up to the weaknesses of a discipline that had accepted clearly questionable data and calculations. In the late 1980s, Burt's critics were themselves questioned, and in the public eye this revived again the nature–nurture polemic. Ironically for the Burt story, the dominant view in the 1990s among psychologists was that the cumulative evidence of identical twin studies, based on a long-term US investigation, confirmed a very significant input from heredity into character and ability.

The constantly polemical character of nature–nurture arguments reflects at least two important things. Firstly, nature and nurture were pivotal categories in everyday language about family, friends, the self and those who make the news, whether merchant bankers or murderers. They were terms with which a psychological society tried to grasp the individual's place in the nexus of social life. What psychologists said on these matters was of public concern: psychological claims were central to the way late twentieth-century Western men and women constructed meaning in their lives. There was no political consensus about forms of life and therefore arguments about the relative significance of nature or nurture ran and ran. Secondly, it is perfectly possible in theory to formulate an account of human development without recourse to abstract generalities like nature and nurture. The genetic endowment, whatever it may be, expresses itself through the environment, the social world, in which a person lives. The lives of children born with Down's syndrome, a disorder that everyone since 1960 has agreed is caused by a chromosomal abnormality, have varied greatly over the last century and still varied in the late twentieth century in relation to local conditions of support. There is always choice about what to do – in education, in welfare or in social policies – and the attempt to delimit choice as a function of psychological determinants, whether attributed to nature or to nurture, is a decision to live one way rather than another. In contrast to this view, most psychologists who emphasized the natural-scientific character of their discipline claimed to describe a subject matter that exists objectively outside the social domain and therefore provides a guide how to live. This expectation, as we will also see in subsequent chapters, was central to the human sciences in the twentieth century.

17

Natural Science and Objectivity

> Before progress could be made in astronomy, it had to bury astrology; neurology had to bury phrenology; and chemistry had to bury alchemy. But the social sciences, psychology, sociology, political science and economics, will not bury their 'medicine men.' According to the opinion of many scientific men today, psychology even to exist longer, not to speak of becoming a true natural science, must bury subjective subject matter, introspective method and present terminology.
>
> John B. Watson, *Psychology from the Standpoint of a Behaviorist* (1919; 2nd edn, 1924)[1]

i *The Status of Psychology as a Natural Science*

Because it became present in every walk of life in the twentieth century, it is possible to describe much of psychology while saying little about it as a professional scientific subject. It is, however, necessary to restore the balance and discuss the ways in which psychologists gave their subject a distinct scientific identity. Identification with natural science, for many psychologists, gave meaning and value to their activity. The history of scientific psychology, as commonly understood, was begun in earlier chapters in connection with physiology, evolutionary theory and functionalism, Wundt and the early history of experimental psychology and the treatment of mental variables as mathematical factors. Physiology, functional description, experimentation and statistics appealed to researchers as ways to give knowledge about mind and human activity a scientific foundation – something they thought psychology had never had. Such hopes justified the phrase 'the new psychology' in the late nineteenth century. This chapter examines the subsequent history of psychology's search to be a natural science.

This is not a straightforward story, not least because the nature of science was variously understood. The pursuit of a scientific identity did not unify psychology – possibly the reverse. Major factors militated against unification. Firstly, the sheer scale and diversity of what psychologists did in twentieth-century society made it unlikely. Secondly, the subject matter which interested psychologists was so varied conceptually as well as empirically – consciousness, behaviour, organisms, the unconscious, persons in society, information systems – that there was no common language, let alone a common theory. Thirdly, it is not clear that other sciences, for example physics, were unified either, so there may not be much precedent for a unification of the kind found in the dreams of some psychologists. Sigmund Koch, a North American psychologist and joint editor of a volume that assessed *A Century of Psychology as Science* (1985), wrote:

> On an a priori basis, nothing so awesome as the total domain comprised by the functioning of all organisms (not to mention persons) could possibly be the subject matter of a coherent discipline. If *theoretical* integration be the objective, we should consider that such a condition has never been attained by any large subdivision of inquiry – including physics. When the details of psychology's one-hundred-year history are consulted, the patent tendency is toward theoretical and substantive fractionation (and increasing insularity among the 'specialities'), not integration.[2]

Nevertheless, there is a substantial history of claims to found psychology as a unitary science. These claims were not at all restricted to behaviourism in the United States, with which they have most often been associated, but were also expressed in continental European arguments that psychology lies at the foundations of the sciences in general and that methods exist to provide an objective account of consciousness itself. Some European science contrasted strongly with the attempt to establish an objective psychology on the model of the physical sciences and to give psychology a scientific subject matter through physical observations. I will refer to European phenomenology in order to throw into relief the dominant view of academic psychologists that their subject was a *natural* science. Further, it is necessary to keep in mind just how little attention differential psychology, discussed in the previous chapter, paid to behaviourist claims. For many researchers it was measurement and quantification rather than explanation in terms of causal laws that made their subject scientific.

Each section of this chapter examines an influential attempt to give psychology a scientific identity: the behaviourism advocated by the intriguing but extreme psychologist John Broadus Watson (1878–1958); the formal neobehaviourism of the 1930s and 1940s that embodied principles later attributed to philosophy of science, logical positivism and operationism; and the gestalt psychology, developed in Germany before World War I, which flourished in the 1920s. These were distinctively twentieth-century movements, but two older claimants to the science of mind continued to exist alongside them. Experimental psychology was discussed in the chapter on the origins of the discipline in the late nineteenth century; physiological psychology, which sought to be objective by taking as its subject matter the bodily instrument of mind, treating it as a function of nervous events, built on the well-established and high-prestige natural science of physiology. The study of the brain as the organ of mind sometimes seemed to threaten the independent existence of psychology as a subject area since, as some scientists judged the matter, it suggested that as psychology becomes objective it in effect becomes physiology. This was a serious concern to the generation of psychologists who, at the end of the nineteenth century, tried to legitimate investment in their subject as a distinct university discipline. To achieve this, they showed that they had objective methods distinct from the methods of physiology – hence the importance to their institutional position of both experiments on mental content and mental tests which defined subject areas and skills that only psychologists possessed. This chapter also includes a section on parapsychology or psychical research, for this was an area of public fascination with the limits of psychology and psychologists' attempts to delineate the margins of what is known reveal much about what counted as scientific objectivity.

The strength of concern about whether psychology is a science sometimes amounted to an identity crisis as psychologists enviously contrasted their subject with physics, chemistry or physiology. Psychologists often desired the objectivity of natural science but feared they did not possess it. What they studied appeared all too often to be either subjective, like mental states, or unmeasurable, like the self-determining person. Physical scientists appeared to be in a more fortunate position: the objects they studied lie outside the observer and are not self-defining. By the end of the nineteenth century, the natural sciences had achieved a position of intellectual leadership within the academic world and considerable public status, and psychologists had an image of this authority at the forefront of their minds. The new psychologists grasped at the experimental method and, later, mental tests as the means to acquire the objectivity of the natural sciences. In the process, the technology of the methods had a strong influence, firstly, on what topics were studied, and secondly, on the content of what was claimed to be known. The methods were not neutral for psychology. Certain topics were amenable to experimental study, such as perception, memory and movement; others were not – Wundt, for example, excluded thought and language from the scope of experiment, though he certainly intended these topics to be included in the science of psychology. Later, however, especially in the United States, the experimental method became a precondition for psychology to be classed as a science, and topics that could not be subjected to objective testing were excluded. At times psychologists concentrated on the learning process – which in practice meant behaviour in rats – almost to the exclusion of human thought, language or imagery. Psychology was distinctive as a science because of the degree to which it defined itself by its methods rather than by its subject matter. Critics doubted whether methods, in the absence of coherent and rational content, are enough to constitute a subject as science. But we must deal with the historical reality of psychology's preoccupation with its methods and note where this preoccupation ruled out topics even while it built in rigour of a certain kind.

Nineteenth-century scholars from Comte to Galton contrasted the state of knowledge about physical nature with the state of knowledge about human nature and society. They looked back and saw a scientific revolution that had swept through knowledge of physical nature in the seventeenth century and moved on irresistibly through knowledge of biological nature in the nineteenth century. They looked to themselves to carry the tide forward and to complete the revolution in human thought with the establishment of the sciences of the human domain. The scientific method was to hand, and they believed that only the accumulated ignorance of the centuries – an ignorance embedded in language, social traditions and conservative institutions – stood in the way of self-knowledge. The prospects appeared to be breathtaking: human beings, after centuries of intellectual wandering and material suffering, had turned to the true path of science. It was an inspiring but grandiose picture. To create an objective science was in practice a mundane matter of minute experimental observation and quantitative analysis. In day to day life, to carry out experimental work or mental tests, to communicate and agree the results as public knowledge, to organize institutional support for the subject and to teach and train new recruits was more than enough to keep an initially small field busy. Many psychologists felt there was a need to get on and catch up with natural science and not reflect continuously on fundamental issues. This 'can do' attitude fitted in well with an important strand of North American cultural values. The two best-known behaviourists, Watson and B. F. Skinner, were both proud of their ability to make things with their hands.

The effort to make psychology a natural science relied on the assumption that psychology has a subject matter – an observable and quantifiable subject matter – that can be known in the way physical objects are known. Any reservation that human beings are in some sense not physical objects, or in essence express qualities not quantities, had to be rejected or else the whole enterprise was in jeopardy. The European gestalt and phenomenological movements tried to deal with this issue through a philosophical debate about the nature of scientific knowledge. By contrast, behaviourists and many physiological psychologists embraced the view that people are indeed physical objects; they saw no intrinsic difficulty in the integration of psychology with the natural-scientific worldview. The search to know and quantify human beings in the manner of physical objects had consequences for psychology as a subject area as well as for everyday attitudes and people's self-image. In the second half of the twentieth century there was greater openness to the view that psychology includes a variety of subject matter. Most psychologists did not worry themselves about questions concerning the foundations of their research, but got on with the business of their particular speciality. All the same, as the concluding chapter suggests, at the end of the twentieth century there were still many divergent views about what the scientific content of psychology actually is.

The preoccupation of psychology with its methods may be explained by its perceived backwardness in relation to the physical sciences, by the peculiar intractability of human beings as subjects and by the practical needs, notably in education, to which much activity was a response. But by no means all research was carried out by formal methods; indeed, a vast amount of work was eclectic in the extreme. In part this reflects a general feature of science, that what scientists say they do, especially in professional academic papers, and what they actually do are different things. But this still leaves historical and sociological questions to be asked about when and why there was a stress on explicit formal method. It is instructive in this regard to describe psychology in Britain, as a contrast with the behaviourist concern with methods in the United States. British psychology sometimes appeared informal, but it was equally rigorous science.

C. S. Myers taught psychophysical experiments at Cambridge in the decade before World War I, and he put his skills to use in the war when he worked on the acoustic location of submarines. Earlier, along with his teacher W. H. R. Rivers (1864–1922) and fellow student William McDougall, Myers joined an anthropological expedition to the Torres Straits, between New Guinea and Australia, where, in the words of the expedition's organizer, A. C. Haddon, 'for the first time psychological observations were made on a backward people in their own country by trained psychologists with adequate equipment'.[3] Rivers and his students mainly studied sensory perception and, owing to the cultural gulf between experimenter and subject, their results had little value. The point here, however, is that Myers's work and the research in the Torres Straits illustrates the way experimental technique developed in Britain – piecemeal, as a craft and adapted to local circumstances and specific intellectual and practical tasks. The same characteristics were apparent in the development of psychology's institutional framework. After the war, Myers moved into industrial or occupational psychology and Frederic C. Bartlett directed the Cambridge laboratory, where he fostered a conception of psychology markedly different from Spearman's and Burt's in London or from that in North America. Bartlett's major book, *Remembering: A Study in Experimental and Social Psychology* (1932), had no statistics and referred constantly to mental states. He later wrote: 'The scientific experimenter is, in fact, by bent and practice, an opportunist . . . The experimenter must be able to use specific methods rigorously, but he need not be in the least concerned with methodology as a body of general principles.'[4] These were the confident words of a man who was for many years the only psychologist to be a Fellow of the Royal Society of London.

Bartlett's work on remembering linked perception, recognition and recall with a stress on the mental schemes and attitudes that people bring to the task. In the final analysis, he understood psychological processes to be social activity, however much he also accommodated to the natural-scientific ethos in terms of which his discipline gained a respected place in Britain. Bartlett thought it obvious that a subjective dimension is part of psychological processes: 'we shall find again and again, *what* is imaged and *when* images occur are both strongly determined by an active subjective bias of the nature of interest'.[5] In his view, what makes psychological study objective is not a method but the experimental ingenuity with which the investigator tests hypotheses in relation to clear questions. His approach preserved conceptual analysis as part of the critical apparatus of science, but it also ruled out grand schemes to make psychology a unified subject. To compare Bartlett's work with what went on at the same time in the United States is to contrast the style of a small but brilliant British academic élite with the style of a large, more egalitarian profession. In Britain what counted as method was a craft a few researchers learned at Cambridge; in the US, formal psychological methods and explicit procedures put everyone on an equal footing.

British research put a question mark against formal methods and the exclusion of the language of mind and consciousness as a prerequisite of science. The varieties of behaviourism removed this question mark and turned to formal methods and non-mentalist language. The behaviourists' claim to make psychology a natural science, however, also required rejection of the credentials of another claimant, physiological psychology. Physiology at the end of the nineteenth century, after all, had unquestionable prestige as a natural science, was a discipline of direct practical value to medicine and had pioneered many of the experimental techniques used by early psychologists. Psychology's intellectual and social relations with physiology were therefore beset with tension and ambivalence early in the twentieth century.

ii *I. P. Pavlov and 'Higher Nervous Activity'*

Somewhat rashly, physiologists like Karl Vogt and I. M. Sechenov in the mid-nineteenth century argued that psychology would become a science by its development as the science of the brain. The difficulty of experiments on the brain, the overly crude argument that knowledge of brain gives knowledge of mind and the unsavoury political associations of materialism encouraged other scientists, like Wundt, to conceive of psychology as a discipline independent of physiology even if it borrowed physiological methods. Any claim that psychology is a science nevertheless provoked questions about how it relates to the science of the brain. The questions were acute when it was claimed that psychology is a natural science, because this claim implied belief that the mind is a function of the brain. Even when psychology was not straightforwardly described as a natural science, as with Wundt, for whom psychology encompassed the domain of psychic causality, there were still complex intellectual and social problems to negotiate about the boundaries between the sciences.

The new psychologists had to justify their claim to independent standing to the established field of physiology. The situation was serious in competition for material support since physiology was part of the medical curriculum and was linked in the public mind with the benefits of medicine, whereas psychology had to establish its claim to be a valuable expertise. In Germany, these pressures were not necessarily great since professors like Wundt or G. E. Müller occupied academic posts that in themselves were of high status; their institutional competition was with philosophy. In France or the United States, however, there were strong pressures on psychologists to show that they offered something in scientific and practical terms that physiology did not. The development of mental tests was decisive in this regard; clearly they demonstrated that psychology was a specialist expertise. Nevertheless, the problem of psychology's relation to physiology, like the philosophical question of the mind's relation to the brain, remained a significant factor in the development of the discipline. It was debated with renewed intensity after 1945, when there was an enormous growth in research on the brain which some proponents believed would create a unified science of human nature. That period is described later; this section illustrates the earlier issues in the light of the work of Ivan Petrovich Pavlov (1849–1936), though an account of the Soviet setting in which, for a while, a Pavlovian approach was enforced in psychology is also discussed later.

Neurophysiology, or the part of physiology which deals with the nervous system, was an area of great activity, both theoretical and experimental, at the end of the nineteenth century. Experimental neurophysiology, which used refined vivisectional techniques, connected closely with the newly differentiated medical speciality of neurology or the study of the pathology of the nervous system. The English physiologist Charles Scott Sherrington (1857–1952) produced a synthesis, published as *The Integrative Action of the Nervous System* (1906), which viewed the nervous system as a structurally and functionally co-ordinated network of neurones or nerve cells. He described the reflex arc as the basic unit of function, the process which turns a peripheral stimulus into a movement while subject to excitation or inhibition by the higher levels of control in the brain. Research on the brain itself, especially studies of the localization of functions like speech, memory, motor control and vision, suggested correlations between physical structures and psychological activities. Most researchers, and in this Sherrington is also exemplary, got on with the rigorously scientific problems of physiology without distraction from the unsolved problems of the mind's relation to the body. A philosophical position known as psychophysical parallelism, which describes mind and brain as separate but parallel realms, often helped as a way to put the problems to one side. Sherrington reserved his interest in the mind for some speculations on sensory perception, but more especially for poetry and for philosophical reflection in retirement.

Pavlov was the son of a priest and, like many ambitious and idealistic students of his generation and background, he turned to the natural sciences for a modern worldview. He became utterly absorbed by the virtues of science. After studying medicine in St Petersburg, he continued with physiological research at the university of Breslau in Germany. These studies culminated in the work on digestion that brought him the Nobel Prize in 1904 and world renown. Only in middle age, as a professor at the Military Medical Academy in St Petersburg, did Pavlov switch his full-time attention to psychological events and to the brain as the structure that underlies them. Even then, however, he continued to call himself a physiologist, since he assumed that scientific explanation requires research to be grounded in knowledge of physiological causes. Pavlov's own account of the emergence of his interests states how he perceived a contrast between objective physiological science and subjective psychological speculation and how this contrast gave direction to his work. The context of his remark was the famous research on dogs trained to salivate at the sound of a bell which they have learned to associate with the presence of food. Pavlov and his students had to decide whether they were physiologists or psychologists. In a later reconstruction of events, he wrote:

> After persistent deliberation, after a considerable mental conflict, I decided finally, in regard to the so-called psychical stimulation [the sound of the bell], to remain in the role of a pure physiologist, *i.e.*, of an objective external observer and experimenter, having to do exclusively with external phenomena and their relations.[6]

As a result, he did not describe the dog's experience of the bell and food, but referred instead to the unconditioned reflex for salivation – the innate response to food – and the conditioned reflex – the trained response after a sound has become a stimulus for the reflex. Thus Pavlov designed a research programme on the psychological topic of learning in terms of the physiological events of conditioning. He viewed conditioning in physiological terms and hoped to connect the conditioning process with nervous events in the brain. This left no space for an independent science of psychology.

Over the next thirty years, until his death in 1936 as a formidably autocratic old man, Pavlov directed a group of scientists who developed this research on the conditioning of an organism's innate reflexes. They soon differentiated the process of conditioning into many sub-varieties and studied the activity of inhibition as well as of excitation. Pavlov inspired younger researchers, who included a significant number of women, with the belief that conditioning was an objective method to study events previously attributed to mental activity. For example, N. R. Shenger-Krestovnikova induced artificial neurosis in a dog, which she trained to salivate in response to a circle and not an ellipse and then presented with intermediate shapes. The neurosis was interpreted as a demonstration of the conflict between excitatory and inhibitory processes in the brain, not as a mental event. Pavlov's programme of research expressed the hope that human beings can come within the scope of science.

As a world-famous scientist, Pavlov campaigned for funds to extend this work, but he had to tread carefully under the Tsars when there was strong pressure not to question the divine essence in man or weaken belief in individual responsibility. He balanced the methodological pressure from his science to push forward physiological analysis and the ideological pressure from his government not to imply materialism. This situation changed in the 1920s, when the Bolshevik government supported the work financially, initially because it wanted the prestige associated with the work of a scientist of international stature, later also because some theoreticians thought it possible to translate Pavlov's theory into the dialectical materialist language of the political system in the USSR. For these and other reasons, Pavlov's programme by the late 1920s flourished at the expense of other research, and physiology had support that psychology, as a separate area, did not.

The research on conditioned reflexes elaborated a language to describe an animal's sensory environment, its behaviour and the way the environment and the behaviour interact. The experimental technique developed initially to study learning opened up objective ways to investigate other psychological phenomena. It became possible, for example, to analyse what a dog sees by conditioning and deconditioning dogs to different stimuli. Where the behaviour does not differentiate between them, it was concluded that the dog does not see the difference. Such experiments enabled researchers to explore differential responses rather than claims about what animals 'see', replacing a mentalist language by a physicalist one and statements about the dog's subjective world by objective observations. Further, though he worked with dogs, Pavlov clearly intended the research to apply to human beings. Around 1930 he introduced the concept of the second signalling system which, he claimed, extends the analysis of activity in terms of conditioning to include the activity of speech. This was an ill-formulated concept, and critics then and later seized on it as a sign of the limits of physiological argument in psychology. It seemed poverty-stricken to describe people's cultural life in terms of second signalling systems.

Pavlov and his students believed that they created science because they studied observable physical changes not mind; experiments on conditioning, they thought, are comparable with experiments by physicists or chemists. The original research, on so-called classical conditioning, showed how a key mental attribute, purposiveness, can be re-represented as habit formation and this in its turn explained in terms of acquired reflexes. In this regard, Pavlov's thought was close to that of Herbert Spencer and the earlier association psychologists who made mind amenable to analysis with concepts informed by descriptions of physical nature. But the science of conditioning was empirical, where the earlier science of association psychology was analytic. Pavlov's ambitions, however, did not stop with empirical description. He wanted to explain conditioning at the level of behaviour by physiological events at the level of the brain. He never called his work psychology and remained critical of attempts to make psychology a discipline independent of physiology. He therefore thought that behaviourism was gravely in error when it claimed to be a new science of psychology and ignored his own claim to have laid the basis for a general science. Pavlov called his science 'the theory of higher nervous activity', intended it to occupy the place filled elsewhere by psychology and identified his programme with the progress of scientific reason itself. His faith in science gave him the energy to maintain a research community through the years of war, civil war and extreme material hardship in Russia from 1914 to 1923. It also stood him in good stead when there was fierce competition for scarce academic resources during the 1920s. But as a comprehensive alternative to the domain of psychology, Pavlov's programme remained restricted to the USSR, at least until it was imposed elsewhere in the Soviet empire in the late 1940s.

Pavlov's technique of classical conditioning was known to Western psychologists from 1906, but it was regarded as a valuable experimental technique, not as the foundation for a whole new science. In 1927, when Pavlov's *Lekzii o rabote bol'shikh polusharii golovnogo mozga* (1926; translated as *Conditioned Reflexes: An Investigation of the Physiological Activity of the Cerebral Cortex*) became available in English translation, the elaborate way in which he correlated the many varieties of conditioning, deconditioning, inhibition and so forth, with cerebral or higher brain activity showed that his physiology, in Western terms, was crude. While anglophone neurophysiologists followed Sherrington and studied the minutiae of inter-neuronal activity, Pavlov speculated about the large-scale irradiation or spread of excitation and inhibition across the cortex (the whole structure of the higher brain). The division of labour in Western science, with few exceptions, fostered highly specialized research and created barriers between the disciplines of neurophysiology and psychology. By contrast, Pavlov and his school, at least in principle, attempted to achieve a unified science of human nature.

Pavlov's ambitions were large. He drew on his medical background to promote his theories as a basis for both psychiatry (though he did not include what was later known as behavioural conditioning in his approach to treatment) and the study of character and individual differences. In this area he was an inspiration to Eysenck, who linked neurosis and character in his research during World War II, and it was Eysenck who created Western interest in the typology or character research of B. M. Teplov (1896–1965) and other Soviet scientists of the generation after Pavlov in the 1950s. Eysenck also recreated in Britain the elements of Pavlov's programme committed to objective experimental technique and the creation of scientific psychology by explanation in terms of underlying biological or physiological events. In the 1950s and 1960s, when there was intense excitement about the potential of the brain sciences to explain behaviour, the memory of Pavlov's enterprise became an important point of contact between US and Soviet science. Earlier, US psychology had gone other ways and established itself as a discipline independent of physiology. That institutional independence, however, did not mean that there was intellectual clarity about how psychological and physiological knowledge relates.

iii *Behaviourism: John B. Watson*

In the 1960s, North American psychologists revived an interest in mental processes, notably cognition, and reported a sense of liberation when they talked about conscious events without disapproval. It suited polemical purposes at this time to picture psychology from about 1910 to 1960 as a behaviourist monolith. This was a false historical picture, though it contained an interesting element of truth: psychology had concentrated on methods as the way to make psychology a science, with extreme consequences for the substance of knowledge. The most obvious consequence was the large-scale exclusion of certain topics from US psychological research, such as thought or imagery, which did not happen in Europe before World War II. The intellectual reasons for exclusion had their clearest expression in the movement known as behaviourism. This was, in fact, never a unified movement and had two clear stages: the early formulations (*c.*1910–*c.*1930) associated with Watson, and the neobehaviourism associated with a positivist interpretation of science (*c.*1930–*c.*1955). I follow that division in this and the subsequent section.

There were at least four sources of behaviourism in the opening decade of the century. Each source has been mentioned in an earlier chapter: evolutionary and functional explanation and the impetus it gave to the study of animal and human action in relation to the whole organism or society of which action is part; the rapid transformation of the United States into an urban and industrial society; the creation of an occupation of psychology with expertise in the classification and adjustment of individuals; and the struggle to make psychology a scientific discipline independent of other fields but objective in the manner of the natural sciences. The four points came together in the promise that psychology would be a science of what people *do*. Young men and women attracted to the new field were not satisfied with a discipline that pursues detached philosophical knowledge. A practical spirit made doing – what the old psychology called conduct and the new one called behaviour – appear much more relevant and objectively observable than mind or consciousness.

Functionalist theory provided the practical attitude with a ready-made vocabulary and underwrote it with the authority of the evolutionary worldview. This was the background of psychologists, like Baldwin at the Johns Hopkins University, who were interested in animal psychology and in childrens' development as well as in the adult human mind. Comparative and developmental psychology, given significance by evolutionary theory, became major resources with which to understand the adult mind, as Darwin, Romanes, Preyer and Ribot, followed in the 1890s by Lloyd Morgan, Baldwin and many others, appreciated. Work with animals also offered experimental possibilities not easily achieved or not ethically acceptable with human subjects. Watson's doctoral research, for example, involved damage to the sense organs of rats in order to study their perception. Simplification, and hence rigorous experimentation, was also more easily achievable with animal subjects. It was not possible, however, to ask an animal to introspect or directly to study its mind. Some researchers similarly thought that it is possible to observe a child's behaviour but not her or his mind; other psychologists, all the same, continued to write about the mind of the child. The overall effect, when comparative and developmental psychology were integrated with functionalist explanation, was to direct attention to what animals or children do, that is, to behaviour rather than to conscious states.

The shift of attention towards the psychology of doing took place in a social context. The emotional experience of sudden social change, which uprooted the rural values of immigrants and people born in North America alike, was disturbing and even traumatic. The new urban way of life required people to make subjective as well as material adjustments to fear and disorientation. A positive response, some psychologists thought, could be promoted by a new science of adjustment, a science of behaviour. This was advocated in a clear-cut way by Watson, whose own life took him from the country to the city and, in leisure and retirement, back again to the countryside. Watson's career exhibited all the elements in the background to behaviourism. As he also issued a concise and pointed manifesto,

'Psychology as the Behaviorist Views It' (1913), he gave behaviourism the appearance of a well-defined identity.

Watson grew up in the narrow atmosphere of a small town in South Carolina, from which he escaped via Furman College to the heady modern world of the university of Chicago philosophy department headed by Dewey. He found it very difficult to adjust to his new life, and as a graduate student he found emotional support for a while in the practical need to look after rats and build experimental apparatus for them. While in Chicago he was influenced by Jacques Loeb (1859–1924), and there was more of Loeb's mechanist explanation of animals in Watson's later psychology than of Dewey's non-mechanist functionalist approach to human action. Loeb was a German-born and German-trained physiologist, famous for his studies of tropisms or movements in plants and animals, which he explained as physico-chemical orientation. Watson completed a thesis on habit acquisition in animals and quickly obtained a prestigious position at Johns Hopkins University. Even so, he felt slighted by psychologists who treated animal work as of secondary importance, valuable only to the degree that it revealed something of the evolutionary origins of mind. Hard-working and ambitious, with secure academic support, he acquired the confidence to articulate his own position. When he replaced Baldwin as head of department he was able to consolidate a programme of action on behalf of his own views. Watson's 1913 paper, based on his opening lecture to a course at Columbia University in New York, was a call to arms. 'Psychology as the behaviorist views it is a purely objective experimental branch of natural science.' Behaviourism would, he argued, replace a psychology that 'has failed signally, I believe, during the fifty-odd years of its existence as an experimental discipline to make its place in the world as an undisputed natural science'.[7] The negative side of this polemic was well grounded. Experimental psychology on the German pattern, which was carried on in the United States as research on the structural units of mental content, had proliferated descriptions about which there was no agreement, and it had lost direction. Only at Cornell University, under Titchener's leadership, was there a well-defined programme to describe the elementary units of mind. Titchener's response to the sprawling diversity of psychology was to distance himself from everything that was not pure science, and by pure science he meant his own programme to describe mental content by rigorous experiments. He claimed that: 'Science goes its way without regard to human interests and without aiming at any practical goal.'[8] Most psychologists, however, moved in another world and wanted a psychology that enabled them to do something for society. Curiously, Watson and Titchener rather respected each other, perhaps because they both saw themselves as strong men who pushed for something more scientific than what was current with mainstream psychology. Both men were also positivist in the manner in which they thought knowledge derives from observation.

The constructive side of Watson's polemic focused on what behaviourism would do for society: 'But until psychology becomes a science and has amassed data on behavior resulting from situations experimentally set up, prediction of behavior resulting from daily life situations will have to be of the hit or miss kind that it has been since the race of man began.'[9] His criticism of psychology's failure as science was inseparable from his aspiration for it as a human technology. The two aspects fitted closely together, as they were to do later for Skinner. Belief that science cannot study an inner mental world was matched by belief that human goals are circumscribed by the search for outer physical well-being and social adaptation.

These values have a place in the wider history of science. Watson and other behaviourists thought that they carried forward the scientific revolution and extended objective explanation from the natural world to the human sphere. This step was, for them, progress pure and simple. What blocks progress, they judged, is the unscientific muddle about mind, consciousness and introspection. In Watson's view: 'We have become so enmeshed in speculative questions concerning the elements of mind, the nature of conscious content . . . that I, as an experimental student, feel that something is wrong with our premises and the types of problems which develop from them.' The answer, he thought, is obvious: psychologists should observe physical variables just like any other natural scientist. 'If you will grant the behaviorist the right to use consciousness in the same way that other natural scientists employ it – that is, without making consciousness a special object of observation – you have granted all that my thesis requires.' In short, he argued that psychologists should make observations of physical stimuli and responses and on this basis correlate behaviour with its preconditions. No reference to mind is necessary. Not least, for the psychologist, 'the behavior of man and the behavior of animals must be considered on the same plane'.[10] Earlier discussions, notably those that analysed mind in terms of elements analogous to physical entities, had laid a basis for this step. Equally, however, arguments existed that made it possible to reject the way Watson and the behaviourists argued for a science of psychology and to suggest that belief in behaviourism was at the expense of both reason and humanity.

Watson contended in 1913 that animal psychology and predictive human psychology require the exclusion of reference to consciousness in order to be objective. Later, he went further and in his book *Psychology from the Standpoint of a Behaviorist* (1919) he effectively claimed that belief in the existence of the mind is a medieval superstition. He expressed himself as a no-nonsense practical man of science rather than as a philosopher, but his position was tantamount to the claim that reality does not include minds. His challenge to fellow psychologists also used the language of masculinity. He rejected the search for a criterion of animal consciousness and talk about animal feelings: 'Such problems as these can no longer satisfy behavior men.'[11] In Watson's world, to reject the existence of the mind itself was a tough-minded adjustment to the needs of progress.

To carry conviction, Watson had to address such problems as the apparent existence of the mental image, language and meaning, problems that returned with great intensity in debates in the philosophy of mind later in the century. Watson believed that language is tractable since it can be treated as verbal behaviour or speech; once this is established, it is possible to treat thought as subvocal speech. Physiologists of an earlier generation, like Sechenov and Ferrier, who were interested in the physical basis of mind, had made a similar suggestion. In 1916, Watson spent some time in an attempt to record movements of the larynx during thought; but, as he was not successful, he had to be satisfied with the argument in principle that thought is a species of behaviour. He cited as evidence the way a young child whispers while apparently thinking, and he suggested that as the child gets older this muscular movement becomes invisible and no sound is made. Watson dismissed the problem of meaning:

> From the behaviorist's point of view the problem of 'meaning' is a pure abstraction . . . We watch what the animal or human being is doing. He 'means' what he does. It serves no scientific or practical purpose to interrupt and ask him while he is in action what he is meaning. His action shows his meaning.[12]

In the hands of later philosophers, notably Gilbert Ryle, this became a powerful argument against the attribution of mental states to people as the explanation for what they are doing. Bertrand Russell (1872–1970) acknowledged an affinity with Watson when he wrote *The Analysis of Mind* (1921), in which he developed a philosophy of mind in terms of sense data, though his analysis focused on the sources of knowledge rather than behaviour. Russell's book in turn stimulated the group of philosophers who became the Vienna Circle and drew their attention to Watson's work. All these discussions about the philosophy of meaning were reassessed in the 1960s, and this is put in context in the last chapter.

Behaviourist polemics challenged psychologists, but they did not bring about a revolution in 1913. Other well-established figures, like J. R. Angell, who taught Watson at Chicago, Pillsbury and Thorndike, who invited him to lecture at Columbia, thought in somewhat similar terms; even James had included prediction in the goals of psychology. Thorndike was of particular importance. By 1913 he had a position at Columbia Teachers College, a major centre for educational research. Like Watson he had worked with animals, and he was also impatient with a psychology that lacked immediate social utility. In 1898 he reported one of the most cited of all experiments, work with cats who learned how to escape from a puzzle-box. The boxes were rough crates and the experiments inconclusive, but Thorndike created a model of research in which learning is the core topic of scientific psychology. Many psychologists besides Watson saw that such experiments had the potential to isolate dimensions of the determinants and outcomes of behaviour and thus to make them precisely observable. They welcomed Pavlov's work on conditioned reflexes for the same reason. But psychologists were not persuaded that Watson had the programme for a new science or that it is possible entirely to exclude references to mental states. When a number of psychologists accepted the label 'behaviourism' as a self-description in the 1920s, it tended to denote their acceptance of the ideal of physical explanation in psychology rather than acceptance of Watson's extreme environmentalism or his denial of the existence of mind.

Watson tried to advance on many fronts at once and the results were anything but systematic. The lack of detail in support of his arguments showed up in his book on behaviourism – much of it was devoted simply to physiology. Medicine also attracted his attention as at Johns Hopkins he had access to the Phipps Clinic headed by Adolf Meyer (1866–1950), an influential figure in the psychiatric profession. Meyer welcomed animal psychology as a source of medical information about 'the conditions under which . . . [pathological reactions] arise and for the ways in which we can modify them'.[13] Watson began to try out some of his ideas with children in the clinic. The most celebrated is Little Albert, whose behaviour, when he was between nine and thirteen months old, Watson and his research student Rosalie Rayner tried to alter. They showed that the child had no fear of a rat; they then clanged a metal bar when the rat appeared and frightened him. Albert, they claimed, became conditioned to show fear when the rat alone appeared. The results, a modern analysis suggested, were not clear-cut and are 'uninterpretable'.[14] Nevertheless, the reported experiment had long-term symbolic value for belief in the acquired nature of capacities: human habits, such as aversion to furry animals, are the result of conditioning. Watson himself used such work to support an extreme belief in the power of the environment to determine a person's nature.

> Give me a dozen healthy infants, well-formed, and my own specified world to bring them up in and I'll guarantee to take any one at random and train him to become any type of specialist I might select – doctor, lawyer, artist, merchant-chief and, yes, even beggar-man and thief, regardless of his talents, penchants, tendencies, abilities, vocations and race of his ancestors.

He did then say, 'I am going beyond my facts . . .', but his unspoken exclusion of the social dimension is therefore even more noteworthy.[15] He viewed people in an entirely individualist way – in his discussion people are formed by stimuli, not by social structure. In a parallel way, his view of habit was modelled on machines not on custom. His writings in the 1920s on bringing up children exemplify the values of psychological society, a society in which everything is considered in terms of individual psychology.

Watson and Rayner together published a popular guide, *Psychological Care of Infant and Child* (1928). To raise a child became a psychological occupation and an activity open to the application of expertise during the 1920s. When he wrote for the popular market rather than for an academic audience, Watson spread knowledge that expertise is needed as well as expert knowledge. He thought he understood why there was a market for psychology: 'The old argument that a good many millions of children have been successfully reared in the past few millions of years has just about broken down in the light of the now generally recognized lack of success of most people in making satisfactory adjustments to society.'[16] This perception recreated social problems as matters of personal adjustment and put forward a new relationship between the mother and the psychologist as the answer. There was an audience: the *New York Times* wrote that Watson's *Behaviourism* (1924) 'marks an epoch in the intellectual history of man'.[17]

There is an intriguing personal dimension to Watson's

approach. He developed a psychology to explain human life in terms of stimulus-response (S-R) links in the physical world. Earlier utilitarian theories had explained human nature by reference to the association of ideas and of pleasures and pains. Behaviourism redescribed ideas as stimuli and, in Watson's version, redescribed pleasure and pain as three basic physiological processes of fear, rage and love. The redescription eliminated language about mind and reinforced belief that the life of people is determined by the causal events of the physical world. The consequences of these beliefs were woven into Watson's personal life, as his medical colleague Meyer argued at the time. After some years in an unhappy marriage, Watson had a sexual relationship with his student Rayner and, in a much-publicized divorce, was forced to resign his position at Johns Hopkins. In correspondence with Meyer, Watson wrote that an unsatisfactory marriage gave him a 'right' to express his feelings outside marriage; Meyer, by contrast, pointed to Watson's denial that behaviour can be understood in terms of its mental meaning as a way of thought likely to foster such (in Meyer's view) unethical conduct. Watson's divorce and remarriage were the actions of a man who exhibited a stimulus-response pattern in his life, in a way that his own theories made true, while the conservative people who disapproved of what he did believed in a psychology that took account of the values of a reflective consciousness. Put simply: Watson understood why he did what he did as a physiological organism; his critics understood that he refused to face the moral meaning of what he did.

In 1921, having become unemployable in an academic institution, Watson joined the J. Walter Thompson advertising company and learned that 'it can be just as thrilling to watch the growth of a sales curve of a new product as to watch the learning curve of animals or men'.[18] He tried to sustain academic arguments, as when he debated behaviourism with McDougall before a huge audience at the New School for Social Research in New York, but his sympathies were with the practical world of instant results not with academic pedantry. After the death of his second wife in 1935 he became isolated, more at home with the animals on the small farm he built for himself in

Connecticut than in a social world. Behaviourism, meanwhile, developed in far more formal directions.

iv *Scientific Method and Neobehaviourism*

The great diversity of psychology in the 1920s was most evident in the United States simply because the scale of the occupation as a whole was greatest there. There was a division between psychology as an academic natural science discipline and psychology as an occupation that offered a service to the community. There were divisions between testers and experimenters. There were popular psychologies of child-rearing and there were esoteric psychologies of sensory perception. Nobody agreed about basic concepts or about the place of theory construction. It was not a satisfactory situation for psychologists who looked over their shoulders at the apparently unified knowledge and communities in the physical sciences.

There were several responses in the second half of the twenties. Boring, for example, published his major study *A History of Experimental Psychology* in 1929 to construct a lineage for experimental psychology and to advance its cause, in opposition to the applied fields, as a rightful claimant to the empire of science. This book influenced the way psychology's history was conceived for the next fifty years, helped by the prominent position Boring occupied at Harvard University between 1922 and 1968 and as a leader of the profession. Carl Murchison (1887–1961), who headed the psychology department at Clark University, Massachusetts, in 1925 and again in 1930, assembled collections of position statements which juxtaposed a dozen different 'schools' of theoretical psychology. In 1931 Woodworth, at Columbia University, published a volume on *Contemporary Schools of Psychology*, and the teaching of courses on 'systems and theories' gained an enduring place in the curriculum. Edna Heidbreder, who went on to a career at Wellesley College, published her much-used text *Seven Psychologies* (1933). Nevertheless, the study of history and systems seemed a poor substitute for a unifying theory to psychologists who had large ambitions for psychology as science. In the 1930s, a new generation, with strict views about what makes psychology scientific, moved into influential positions. The result was a preoccupation with methodology: 'the most sustained attempt ever made to construct a science of psychology through the use of detailed and explicit rules of procedure', as one psychologist and philosopher of science described it.[19] A training in methodology became the hallmark of the qualified psychologist. For several of the field's leaders, however, including Edward C. Tolman (1886–1959), Clark L. Hull (1884–1952) and B. F. Skinner (1904–90), discussed in this section, methods were part of a deeper commitment to a particular view of scientific knowledge that made psychology itself pivotal. They supposed that observation, not logic or metaphysics, is the ground on which to build science. Where better to start, they implied, than with the human observer?

It requires an act of imagination to re-enter the world of neobehaviourism in the three decades from 1930 to 1960. The study of learning in the white rat became the dominant and sometimes sole topic of research, and there was an enforcement of formalized methods to record observations and to analyse them statistically to eliminate error. Some psychologists spent their lives working with rats in mazes, and for over thirty years students trained with such techniques. Pavlov, Watson and Thorndike showed that it is possible to study changes in an animal's behaviour that can be correlated precisely with changes in physical stimuli. These correlations constituted the subject matter of the psychology of learning. Later work elaborated on their descriptions of relations between physical variables with ever increasing subtlety. One retrospective observer concluded that 'it is awesome to contemplate how great a part of the history of fundamental psychology in the United States during the first half of the twentieth century can be seen to relate . . . to the conclusions that Thorndike based on the informal and crude "experiments" he reported in 1898 . . .'[20] A decade later, Thorndike generalized his experimental results as 'the law of effect', a statement about the way the consequences to an organism of its movements determine whether or not those movements become a learned part of its repertoire, i.e., a habit. Thorndike's conclusion, that 'pleasure stamps in; pain stamps out', redescribed in experimental terms the utilitarian pleasure–pain principle, Locke's, Bentham's and the Mills' belief that conduct is explicable in terms of the effects of experience.[21] Other psychologists, who in this respect followed Pavlov, argued that learning occurs simply as a response to the contingent relations between stimuli in the organism's environment. Competition between these two basic models of learning set research tasks for half a century. In the 1930s, the former 'reinforcement' theory was developed in purified form by Skinner, the latter 'contingency' theory by Edwin R. Guthrie (1886–1959).

Watson's early hopes that learning might be reduced to a small set of elementary S-R relations were quickly dashed. There were substantial philosophical critiques of the ability of S-R theory to explain human action in the 1920s, and critics explored alternative but still objective ways to conceptualize the organism as a functional process. Watson debated in public with McDougall the respective merits of mechanistic versus purposive explanations of human action, and the issues anticipated the debates in the 1960s that marked the final demise of behaviourist learning theory. As early as 1922, Tolman presented a philosophically informed critique of Watson's work in which he argued on two main fronts. Firstly, he claimed that the description of an isolated stimulus and an isolated response as the beginning and end terms of a piece of learning is artificially atomistic. Rather, Tolman argued, an animal is in an organic relationship with its environment; hence he substituted the terms 'stimulating agency' and 'behavior act' for stimulus and response. This led to what was later called a molar theory of behaviour, which was necessarily much more complex than molecular S-R theories. Secondly, Tolman retained the concept of purpose (or intentionality), just the kind of mental concept that Watson wanted to eliminate from science. In Tolman's words: 'When an animal is learning . . . a certain *persistence until* character is to be found . . . which we will define as purpose.'[22] He did not, however, reassert the old psychology: purpose 'is a descriptive feature . . . It is not a mentalistic entity supposed to exist parallel to, and to run along side of, the behavior. It is *out there in* the behavior . . .'[23] Tolman described himself as a behaviourist and for forty years as a professor at Berkeley he worked with rats in mazes.

> I believe that everything important in psychology (except perhaps such matters as the building up of a super-ego, that is everything save such matters as involve society and words) can be investigated in essence through the continued experimental and theoretical analysis of the determiners of rat behavior at a choice-point in a maze.[24]

When psychologists studied what is observable during learning, even in the apparently simplified life of the laboratory rat, they still had to decide what concepts to use and whether reference to mental events is legitimate. And then there were still the extraordinary 'exceptions' that Tolman placed in parentheses in his description – exceptions that other students of human nature thought of as the real subject matter of psychology.

As a student at Harvard, Tolman learned a philosophy strongly influenced by James. His teachers, notably Edwin B. Holt and Ralph Barton Perry, turned James's sensitivity about the concrete acting person into a theory of knowledge in which what we can know reflects our life as organisms in a physical world. Tolman constructed his behaviourist psychology in this mould and conceived science to be the extension of natural human adaptation to the physical environment. The orientation, if not the specific Harvard education, was shared by others who were attracted to behaviourism; they were interested in knowledge understood as what animals do. This interest influenced the way they responded in the 1930s to developments in the philosophy of science, and the response became central to debate about how psychology could take its place as a natural science. Both then and later, observers believed that the logical positivism articulated by the Vienna Circle underpinned behaviourism with the authority of philosophy. Consequently, it was argued, when the philosophy was proved unviable in the 1950s and 1960s, the psychology collapsed. The Vienna group, which included Rudolf Carnap, Otto Neurath and Herbert Feigl among its most active members, became internationally known in the early 1930s. It continued in Vienna until 1936, when its leader, Moritz Schlick, was murdered by a probably deranged student. The Viennese philosophers, however, were concerned with the logical structure of a theory of knowledge while Tolman and his colleagues were concerned with how knowledge is acquired, that is, with learning, as a natural process. What American psychology took from European philosophy was its authority rather than its substance; the substance of neobehaviourism had indigenous roots in the commitment to functionalist explanation. All the same, a close connection between psychological theory and the philosophy of science characterized North American psychology over two decades, and this led some people to refer to 'the age of theory'. The Viennese philosophers and behaviourists did both express a vehement rejection of metaphysical and religious claims to knowledge. They also shared belief that meaningful statements about the world have to be made as, or in a form translatable into, statements of physical observations. It followed for both groups that anybody who did not present knowledge in the form of empirically verifiable statements was simply not a scientist, and this made it possible to draw boundaries around the science of psychology.

After publishing a much-discussed synthesis, *Purposive Behavior in Animals and Men* (1932), Tolman spent a sabbatical in Vienna, where his main contact was Egon Brunswik (1903–55), a Hungarian student of the psychologist Karl Bühler. Brunswik, who joined Tolman permanently at Berkeley in 1937, belonged to the Vienna Circle, but his way of thinking, like Tolman's, was naturalistic rather than logical. Both Brunswik and Tolman thought that psychological research had implications for the philosophy of science and not just that philosophy had implications for psychology. Tolman's contacts led him to a new formality in his philosophy of science, and he introduced the concept of the intervening variable, a way objectively to characterize the determinants of behaviour that lie between the independent variable (the stimulus) and the dependent variable (the response). The intervening variables roughly corresponded to what ordinary people meant by mind. As Tolman wrote in

1936: 'the sole "cash-value" of mental processes lies . . . in this their character as a set of intermediating functional processes which interconnect between the initiating causes of behavior, on the one hand, and the final resulting behavior itself, on the other'.[25] 'Cash-value' was James's term, and Tolman's usage indicates how James's qualitative descriptions of functions could be turned into a more formal theory of variables open to algebraic expression. In 1940, Tolman listed ten such intervening variables that influence a rat to turn left or right, and he regarded even this as a simplification. The study of intervening variables made possible a huge amount of detailed research. Brunswik made matters even more complex when he argued that perception and behaviour are probabilistic in form, that they represent an organism's hypotheses about the environment and that there is no one-to-one relationship between psychological variables and observable physical changes in stimuli and behaviour. A decade after his innovative arguments, methods of statistical inference became standard procedure in the analysis of data from psychological experiments. Mainstream experimental psychology, still focused on learning, became a mathematical science.

Tolman loved to compare rats in mazes with humans like himself who try to understand the world. He acted on the belief that all knowledge is provisional, trial and error, open to revision in new circumstances. 'The world for philosophers, as for rats, is, in the last analysis, nothing but a maze of discrimination-manipulation possibilities . . .'[26] He concluded his presidential address to the American Psychological Association in 1938 with a poem written by a friend:

> To my rationcinations
> I hope you will be kind
> As you follow up the wanderings
> Of my amazed mind.[27]

This was not the stuff of heavy formality. But Tolman's scientific papers expressed an extreme formality, and the same was true of the work of Hull.

Hull's neobehaviourist magnum opus, *The Principles of Behavior: An Introduction to Behavior Theory* (1943), is mind-numbing and abstruse. It is therefore all the more significant to ask how Hull came to conceive of the science of psychology in the way he did and why, for a while, he was perhaps the best-known psychologist of his generation. Unlike Tolman, who was born into a well-off and cultured New England family, Hull grew up in small-town Michigan and saw life as a struggle – against religion, against illness (as a student he contracted polio, which required him to walk with a leg brace of his own design) and against his feeling that he came to his life's work only in middle age when his abilities were becoming weaker. Both Tolman and Hull originally planned to be engineers, and Hull retained in his psychology an engineer's imagination about mechanical relationships and belief that it is possible to construct complete working systems. Indeed, in the mid-1920s he invented and made a machine to calculate correlation coefficients, a practical demonstration that machinery can do what mental processes do. He never ceased to think mechanistically – about human beings and about how to organize productive scientific research.

Whereas Tolman relished play and variety – his classes sometimes became free-for-alls – Hull sought certainty and rigour, qualities he found in geometry and in Newton's great work, developed in geometrical notation, on mechanics. He placed an open copy of the *Principia* in his office in the late 1930s. Self-consciously like Hume before him, Hull wanted to be the Newton of human nature. When the full extent of Pavlov's work became known in English in 1927, Hull saw conditioning experiments, which reconceived thought as habits, as the way to study all aspects of thought. On this basis, he argued, it is possible to construct a systematic theory, once a way is found to integrate the two sources of certainty, observed facts and deduction. Psychology, he argued, will therefore become a science when it uses facts about habits to postulate provisional axioms, deduces consequences from these axioms, tests the consequences in experimental settings and finally uses the results to refine the axioms into a conclusive form. In Hull's view, psychological science is a hypothetico-deductive activity.

Besides this, he was, like Watson, a materialist who equated scientific progress with the spread of mechanist knowledge of human nature. As he observed hopefully in a notebook in the 1930s: 'The tide of civilization is running in my direction.'[28]

All this might have remained a bizarre fantasy. In 1929, however, Hull moved from his job as a teacher at the university of Wisconsin to a research position at the newly-founded Yale Institute of Human Relations. This institute, initially funded for a decade to the tune of $4.5 million, had a large impact on the organization and direction of psychology. The Rockefeller Foundation, which put up the money, wanted a theoretical unification of social science research. The plan, which adopted a model established for medical research, was to bring together people from different disciplines to work in groups on tasks set by the goal of integration. Dissatisfied with the lack of progress and with researchers who continued to work individualistically, the Foundation threatened withdrawal. In this situation, Hull took the opportunity to head a co-ordinated programme, in which he shaped the axioms and general theory while groups of researchers organized in a hierarchy worked out the detail through experimental studies. Hull was a man in a hurry, anxious to axiomatize serial habit formation as the basic law of social science, and he generated a mass of empirical questions and points of detail to be answered later. After the Depression, academic funds were short, and Hull's programme therefore stood out as well-organized, purposeful and well-funded.

The interest Hull's programme aroused was a matter of intellectual promise as well as money since there was widespread dissatisfaction in the early 1930s with a situation in which psychology was divided rather than united by theory. Heidbreder recalled that, when she heard Hull talk in 1934, she 'believed – or at least hoped – that such a [logically rigorous] system was possible and that Hull's proposal might be a step . . . toward a comprehensive conceptual scheme within which psychological research might be ordered and integrated, and productively pursued'.[29] She and a few other North Americans were conscious of what the Vienna philosophers were writing about the definition and unification of the logic of science, though little was known of the detail. It appeared that Hull might be the man to bring psychological theory up to the standard that apparently prevailed in the physical sciences. What the Viennese logical positivists were engaged on, however, was the rational reconstruction of the logic of physical theory. By contrast, Hull and the psychologists who thought like him identified logic with the procedures with which to generate knowledge in the first place. In effect, the psychologists were preoccupied with methodology rather than the analysis of knowledge. When the centre of gravity of the European work shifted from Vienna to the United States after Austria came under fascist rule in 1934, Hull took part in the philosophers' unity of science movement. One of his closest supporters, Kenneth Spence (1907–67), helped make the university of Iowa a centre for this movement and worked with the Viennese mathematician Gustav Bergmann to elaborate parts of Hull's programme. All the same, Hull's sympathies with European thought were probably limited. In particular, he was strongly opposed to gestalt theory, a rival claimant to the position of a unified psychology of knowledge. The émigré gestalt psychologists were notoriously cultured, which Hull was not, and he seems to have suspected that metaphysics lurked in their work. Yet the gestalt psychologists Lewin and Köhler had both been members of the Berlin Society for Empirical Philosophy, the Berlin group associated with the Vienna Circle.

A further incentive for psychologists to refine their procedures was provided by S. S. Stevens, a Harvard psychologist, who drew on *The Logic of Modern Physics* (1927) by his colleague the physicist Percy W. Bridgman. Bridgman argued that meaningful statements about physical concepts and unobservable entities (such as the parts of the atom) denote what is done, in terms of experiment and measurement, to arrive at the statements: 'We mean by any concept nothing more than a set of operations; the concept is synonymous with the corresponding set of operations.'[30] Length, for example, is defined by the process used to measure it. This philosophical point appeared pertinent to psychologists and they set out to define the variables of learning in operational terms, i.e., in terms of the operations of measurement used when the variables are studied. Concepts that were not linked to the operational procedure through which they are studied were dismissed as metaphysics, an argument Skinner maintained until his death in 1990. For more than a decade, several psychological journals did not accept papers unless the authors provided formal operational definitions of the terms used. This policy, it was believed, enforced objective scientific standards. Psychologists went beyond anything Bridgman had ever envisaged and physicists required; Bridgman, indeed, repudiated the 'operationism' psychologists constructed from what they thought were his ideas.

In pursuit of deductive rigour and inclusiveness, Hull's system never achieved either. It collapsed because of a mismatch between aspiration and execution and under the weight of its own complexity. In the last years of his life, Hull retreated to more modest and restricted projects. His work survived his death in 1952 only in the less deductive and less inclusive version propagated at Iowa by Spence. In addition to the problems created by the attempt to achieve a unified formal theory, the view that all learning involves reinforcement did not command universal assent. In 1954, Hull's project was subjected to criticism which the author, Koch, himself later described as 'probably the most mercilessly sustained analysis of a psychological theory on record'.[31] This paper initiated a long-running debate about whether learning is a category that can unify psychology; critics believed that learning is not a category but a label for many things. Views about what learning is depended on where psychologists had trained. The irony that it was perhaps experimenters rather than rats who learned a particular performance was not lost. The aim to unify a field through formal theory construction began to look less credible, though the full impact of these criticisms was not felt in psychology until the 1960s. Finally, when the Viennese philosophers migrated to important academic positions in the US during the 1930s, they began to dilute their verifiability criterion of meaningfulness. In particular, philosophers of science became convinced that observational statements cannot be formulated independently of their theoretical content. The dream to purify psychology of everything except statements about observed physical variables began to look misguided in conception. It appeared as if the behaviourist enterprise had emptied psychology of content in order to pursue an image of science that was itself a mirage.

One prominent position, linked to neobehaviourism by methods and ideals but not properly included under the heading, remained unaffected by the psychological and philosophical criticism; indeed, during the 1950s and later it flourished as a distinct scientific school. This was Skinner's operant psychology, which had a firm institutional base first at Harvard University and then elsewhere. Skinner's psychology was not simply another form of behaviourism since he rejected completely the value of theory and the formalization of knowledge. In certain respects this psychology was a return – but with much more rigorously defined experimental procedures and a different theory of learning – to the anti-mentalist zeal of Watson's programme.

Skinner's early ambition was to be a writer, but he realized his real interest was human nature. Armed with books by Watson, Russell and Pavlov he came to Harvard as a student, attracted to psychology rather than physiology by the quality of the psychology department's laboratory workshop as much as anything else. His first major book, *The Behavior of Organisms* (1938), was felt at the time to diverge unacceptably from mainstream theories. Skinner recorded animal movements – later his favourite set-up was to release grain to a pigeon when it raised its head above a certain line – and rejected theory construction for induction, generalization only from observed instances. He attacked the notion of trial and error learning and instead argued that all behaviour involves one basic principle of reinforcement. His view was that an animal makes movements some of which are repeated or reinforced, others of which are not; a pigeon that raises its head, to be followed by the release of grain, as a matter of fact, then raises its head more often. As a result of empirical study, Skinner argued, it is possible to devise schedules of reinforcement, i.e., to construct circumstances that lead to one pattern of behaviour rather than another. He called a movement that is followed by the repetition of the movement (e.g., the raising of the pigeon's head) the operant, and his whole approach became known as operant conditioning. 'Operant conditioning is the making of a piece of behavior more probable.'[32] Skinner's descriptions of behaviour did not refer to mind (pleasures and pains), to drives (hunger), to habits, or even to stimuli (the presence of grain). He recorded sequences of movements and argued that science consists solely in the description of the pattern that such observations reveal. This was indeed radical.

Skinner's work was underwritten by four consistently maintained principles, which he built into his systematic work on *Science and Human Behavior* (1953), and this very consistency made him a marked figure. Firstly, he denied the value of theory and argued that all science needs is systematic observation, a position that drew him closer to the philosophy of Mach than that of the 1920s. He therefore usually declined to debate his position with other people, but presented his empirical observations as his position. Secondly, he clearly separated psychology from physiology, which he thought the source of misdirected experimentation and useless theorizing in psychology. He accepted that there are internal physiological factors which initiate an animal's movement, but he thought this of no concern to psychology, the science of the observable consequences. 'An adequate science of behavior must consider events taking place within the skin of the organism, not as physiological mediators of behavior, but as part of behavior itself.' Thirdly, he refused to refer to mental states and ethical concepts, like freedom, that we associate with mental states. Animals, including people, he thought, exhibit a pattern of movements with operant value, nothing more. As he observed, for example, 'we may regard a dream, not as a display of things seen by the dreamer, but simply as the behavior of seeing'.[33] Lastly, he took biological adaptation to be the core value, and he therefore assessed every action, animal and human, for its contribution to survival. He wrote: 'What matters to Robinson Crusoe is . . . whether he is getting anywhere with his control over nature.'[34] He turned this evolutionary value into a fully-fledged theory of culture – though, for Skinner, it was not a theory but a summary of scientific observation. 'A culture, like a species, is selected by its adaptation to an environment: to the extent that it helps its members to get what they need and avoid what is dangerous, it helps them to survive and transmit the culture.' It followed, he concluded, that 'survival is the only value according to which a culture is eventually to be judged . . .'[35] For Skinner, then, in a more radical way than Tolman and Hull, science is the most evolved strategy for survival. Only a dinosaur, it appeared to him, would not embrace operant science.

The label 'radical behaviourist' accurately describes Skinner's position, and his social values had much in common with Watson's. Skinner, unlike Watson, did not deny consciousness: he 'accepted the existence of private states but as states of the body, the study of which should be left to physiology'.[36] Like Watson, however, Skinner perceived that human thought and language set the greatest challenges for his views, and his twenty-year attempt to deal with this culminated in his book *Verbal Behavior* (1957). He tried to show that language consists of reinforced movements like any form of behaviour. He claimed, for example, that when we say we see the colour red, we have a physiological state and we make movements of the larynx that, in the past, have been reinforced by other people's use of the same movements – if a child says red, and her or his parents smile and say, yes, red, the child says it again in the same circumstances. His book and this argument precipitated intense criticism and in this way contributed to the expansion of psycholinguistics in the 1960s.

Skinner's ability to stir up controversy because he worked out his scientific position to its logical conclusion, a behaviour he thought necessary for human survival, came to a climax with his best-seller, *Beyond Freedom and Dignity* (1971). He attacked notions of free will and individual responsibility as remnants of a pre-scientific past that were leading to social and political disaster. This was not the embittered talk of an old man but the utopian dreaming of one young at heart. In his novel *Walden Two* (1948), Skinner imagined a community run on radical behaviourist lines and with a harmony and efficiency impossible in the outside world. A few communities modelled on Skinner's ideas even existed. Skinner also constructed a baby box, his idea of a good environment, which earned him some notoriety because it appeared so unaffectionate, though the box was little more than an easily cleanable crib with controlled conditions, in which his daughter Deborah lay naked and in comfort.

By the 1960s, psychologists trained in Skinner's experimental methods were a large and productive group but somewhat isolated. His techniques left a mark on machine-learning, of the kind used in language laboratories, and in training people, for example, when token rewards are used to reinforce habits of cleanliness in mentally handicapped children. Meanwhile, criticisms of neobehaviourism became common in many areas of psychology in the second half of the 1950s – in studies of perception, accounts of child development, learning theory and especially work on language. The idea that all learning conforms to a single pattern, and that learning is the key to psychology's secrets, appeared to be chimeras. The behaviourist approach to language found few supporters and a vehement and influential critic in Chomsky. In his dogged adherence to his principles, however, Skinner gave a voice to something implicit in theories of human nature grounded in natural science that had worried critics of such theories since the seventeenth century. This was the way mechanist knowledge appeared to leave no space for values. Skinner asserted the strictly objective quality of the scientific knowledge his research established, and he denied to human nature any value except in so far as that nature survives. Evolutionary science, he thought, gives a value to live by. Yet, in the terms in which he himself expressed knowledge in science, survival is a fact not a value. Values are qualities of cultures, but Skinner's principles did not allow him to say anything about what matters to people as part of a culture. Skinner himself reimported a cultural value into mechanistic science through the door of evolutionary theory, and redescribed the value of survival as the fact of survival. He was not at all alone when he took this step. If Skinner was wrong in the way he developed a natural science of human nature, then much more was at stake in the human sciences than his own particular views. His and other natural sciences did not satisfactorily deal with values.

v *Parapsychology*

Psychologists who made their subject a natural science rejected metaphysical language and subjective methods. Some behaviourists thought that at long last the remnants of spiritual belief in the study of human nature had been eliminated. The demand for operational definitions, for example, had appeal because it both rationalized the psychologists' preoccupation with methods as the last word of science and translated mental notions, once thought essential to psychology's subject matter – such as the mental image – into a language of objective record. Scientific psychology tried to police its boundaries against a pre-scientific past, against pseudo-science and against popular beliefs held to be incompatible with science. Since modern psychology was identified with progress, the campaign against pseudo-science, while partly concerned with the authority of experts also had the zeal of a mission against superstition. This section considers a topic which focused debate about the boundary of scientific psychology. Most psychologists never thought about parapsychology, the study of extra-sensory perception (ESP); they dismissed it to the realm of the impossible or at least the unthinkable, of interest only to cranks or, more generously, to those who like that kind of thing. Nevertheless, a small minority gave it attention and were thereby required to mark out the boundaries of what can claim to be objective. The public, whatever psychologists said, was inexhaustibly fascinated.

The modern history of parapsychology originates in the Victorian enthusiasm for spiritualism and the search for purpose, freedom and comfort from a spirit world and in knowledge of life after death. Trance, table-turning, spirit materialization and other such experiences were then associated by doctors and physiologists with the condition of hypnotic sleep produced by healers and entertainers and extensively studied by doctors and psychologists in the last thirty years of the nineteenth century. The debate about the authority of spiritualist experiences and hypnotic conditions, and about their moral propriety, went back a century to the work of Mesmer, whose activities were condemned by the establishment in reports from members of the

Académie des Sciences, the Académie de Médecine and the Société Royale de Médecine in Paris in 1784. Many scientists took the line exemplified by the physicist Michael Faraday, who showed with a surreptitiously introduced apparatus in 1853 that table-turning, attributed to spirits, is a response to pressure from the hands of sitters around the table. Though this conclusion encouraged scepticism about spirits, it was of great psychological interest because it suggested that participants act unconsciously. Further, a few scientists, such as the co-founder of the theory of natural selection Alfred Russel Wallace and the French physiologist Charles Richet (1850–1935), who won a Nobel Prize for work related to the study of immunity, believed what they experienced when they attended seances. In 1882 a group of British intellectuals unhappy with what they saw as absent in the scientific worldview, formed the Society for Psychical Research (SPR), which was still active a century later and for many years led the attempt to examine psychical phenomena objectively. The first president, Henry Sidgwick, was a Cambridge philosopher and his wife, Eleanor Sidgwick, who was very active in the Society, was the first principal of Newnham College for women in Cambridge. After four years, many of the spiritualist members of the Society, led by E. Dawson Rogers, resigned, unhappy that the Society did not give authority to the beliefs for which they thought it had been formed. This split expressed the divide between those who gave priority to objective methods, the Cambridge-led academic researchers, and those who gave priority to belief as a healing or spiritual power. The same divide threatened every aspect of ESP research, and it was a conspicuous element in the repeated struggles between experimenter and experimental subject for control over the circumstances in which the research is carried out. Psychological investigators wanted to take control so that they could acquire authority to persuade sceptical colleagues and to claim scientific status. But for the subjects themselves, often women (though some famous psychics were men), it was openness to control from another, spirit realm, not control by scientists, that mattered.

The early psychical researchers, led by Edmund Gurney

(1847–88) and Frederick W. Myers in Cambridge, diligently pursued the Society's aims, 'to examine without prejudice or prepossession and in a scientific spirit those faculties of man, real or supposed, which appear to be inexplicable on any generally recognized hypothesis'.[37] These researchers, who included James in New England, were, in an intellectual way, also motivated to search for knowledge that incorporates rather than excludes personal values. This early work culminated in Myers's book on *Human Personality and Its Survival of Bodily Death,* which consolidated the notion of personality, described much of mental content as subliminal, or below the threshold of consciousness, and claimed empirical evidence for contact between the subliminal self and a wider spirit world.

Gurney and Myers and their colleagues set out to be empirical, though many scientists thought they were naïve in their relations with clairvoyants and mediums. By the 1920s, however, the psychical researchers had learned much about deception and illusion and yet they still came up against the inexplicable. One of the most impressive cases was that of a Polish engineer, Stefan Ossowiecki, who exhibited his skills at medical congresses and was examined by the then research officer and later critic of the SPR, Eric Dingwall (1890–1986). Ossowiecki, even under the most stringent conditions, was able to reproduce images and even words securely sealed in envelopes. The problem for the researchers was that whatever evidence was put forward for a psychical occurrence, critics were always able to point out the possibility of deception, even where – as with Ossowiecki – no deception was in fact shown. Given that known physical laws did not permit psychical phenomena, and given that deception had been shown or at least suspected in many cases, most scientists preferred to assume that no paranormal phenomena are real. It was precisely to escape controverted evidence about mental powers that no-nonsense 'behavior men' like Watson propagated their views; such men were not sympathetic to reports of psychical experiences. Psychical research, indeed, was an embarrassment to a discipline that felt inferior in scientific status. Perhaps for similar reasons, James's work on religious experience had little impact on academic psychology and the psychology of religion was, by the 1930s, virtually ignored.

Psychical research and spiritualism, for the most part, went different ways and inspired different enthusiasts; researchers looked for a science, while spiritualists longed for an experience that transcends science as defined by the scientific establishment. The split over whether methods or substance came first recurred. The president of the SPR in 1924, J. G. Piddington, divided the 'High-and-Dry School' from the 'Not-High-and-Dry School':

> The dividing line here has nothing to do with how much or how little of the phenomena under investigation one accepts as supernormal, and concerns only opinions as to methods of investigation and standards of evidence. The Not-High-and-Dries ... take the line that so much has been established beyond cavil that we can now safely relax to some extent the stringent cautions and the very high standard of evidence on which the Society has hitherto insisted ... The High-and-Dries take the line that we cannot ... afford to lower our 'evidential' standard, or modify our methods – our admittedly irritating and meticulously wary methods – of investigation.[38]

There was stalemate: scientists, it appeared, were objective but without knowledge; enthusiasts possessed knowledge but were not objective. In this situation, beginning in the late 1920s, J. B. Rhine (1895–1980) began his work, in which his wife Louise was a constant colleague, that transformed the 'High-and-Dry' methods and institutional standing of psychical research, though the results remained controversial. As a young man trained as a botanist, but with leanings towards a deeper experience of the human condition, Rhine attended a seance in Boston. Appalled by what he saw as trickery, he became convinced that research should begin with the most elementary psychical effects such as card-reading or simple telepathy. Attracted to Duke University in North Carolina by an independent psychical researcher, in 1928 Rhine was offered a post by McDougall who had himself recently arrived from Harvard

University. McDougall had long defended what he called animism, a biology of non-material life forces which he suggested might explain psychical effects. Rhine thus obtained sufficient institutional support to begin systematic work in 'parapsychology', the term he used to describe a field in which rigorous methods would, he hoped, force scientists to pay attention. This initiated parapsychology as an academic subject: other academics had shown interest, but Rhine ran the first laboratory devoted to it on a long-term basis.

Rhine's ambition, informed by the values of the scientific community, was to find a regularly reproducible psychical phenomenon that could be studied by objective methods. The great difficulty in the way of research was that the phenomena in question were exhibited irregularly by exceptional individuals and were also context dependent – the aversion of spirits to the light was notorious. The research method employed previously resembled the method of clinical medicine: descriptive reports of individual cases. To find a new approach, Rhine devised card-guessing experiments, in which the subject identifies cards in a specially designed pack (the Zener cards) of five sets of five simple shapes. The results were then compared in a standard way with calculations of random probability. Many variations were introduced; for instance, the subject was asked to guess cards before the pack is shuffled or was aroused by drugs. It was crucial for the launch and continuation of the enterprise that there were a succession of high scorers in the early trials. But such a succession was never repeated.

This early experience exemplifies parapsychology's methodological problem. The 'falling off' feature, when spectacular non-random results drop away over time, left the field without reproducible results – except for the dropping-off itself – and vulnerable to sceptics. Among the prominent sceptics of the Rhines' work were experimental psychologists, scientists who, as earlier sections showed, were sensitive about the enforcement of explicit objective procedures. But the Rhines did enough to secure some authority. J. B. Rhine's book *Extra-Sensory Perception* (1934), and the results publicized in *New Frontiers of the Mind* (1937), attracted extensive academic and public attention. He popularized the term 'ESP'. The Rhines believed they had finally proved ESP – systematic variation of results with the test conditions was, they thought, particularly impressive – and that research could shift its effort from proof to the process. During the 1930s they persistently argued the case with their psychological colleagues. The very dullness and mechanical repetitiveness – the meaninglessness – of research, in a laboratory setting with quantitative results, was, they hoped, the key to the scientific authority of the enterprise. The Rhines' commitment of so much time and effort, with so many attendant risks, depended however also on values of the kind that had motivated earlier researchers, as the titles of later books, like *The Reach of the Mind* (1947), indicated.

Rhine obtained funds for an independent laboratory because his relations with academic psychologists were difficult. Private funding, indeed, was essential for the development of the subject, and this reinforced a tendency for mainstream experimental psychology and parapsychology to develop along different paths. The kind of work Rhine initiated continued to grow in sophistication, especially when the availability of computers eased the drudgery of counting and the generation of random tables. Rhine founded the *Journal of Parapsychology* in 1937, and other journals followed and published material that extended the range of effects studied well beyond that of the original card-guessing experiments. In the decades after World War II, the field had all the characteristics of a scientific speciality – an independent research programme, experimental and quantitative methods, a social infrastructure and even a founding father – but it was marginal and ignored by most experimental psychologists. This did not necessarily reflect antagonism, but it did reflect indifference: psychologists lacked interest or energy to do research in an area with such a poor record in the production of regular results. Who would spend a career in gathering results that in the end might prove nothing? There were exceptions, notably Gardner Murphy (1895–1979), who was a senior figure in the American Psychological Association and wrote extensively on a wide range of topics. He brought together a reformed American Society for Psychical Research, i.e., a society rededicated to standards of scientific evidence, and academic psychology through his position at the City College of New York. Some work that he supervised, by Gertrude Schmeidler, well illustrates parapsychology's circumstances. She showed that the experimental subject's initial attitude towards ESP – 'sheep' express sympathy, 'goats' scepticism – affects the chances of a high score at card-guessing. Her work was conventional in terms of experimental design. But those who supported parapsychology's findings understood her results to reveal further regular patterns in the phenomena, while sceptics assumed that previous beliefs biased the results. Such inconclusiveness, whatever parapsychologists did, led in the 1940s to their gradual withdrawal from the effort to integrate with mainstream psychology. Psychologists, parapsychologists concluded, were not yet ready to incorporate the new knowledge. The social isolation of parapsychological work was further enhanced by the fact that research papers appeared only in specialist journals, by the separation of Rhine's laboratory at Duke University from the psychology department and by ESP researchers who opened the discussion to a wider range of topics, like survival after death, which interested supporters but were even further from scientific respectability.

Rhine sustained his research programme by force of will, and some of those who worked with him regarded him as dictatorial. At the end, it was apparent that the exhaustive application of a scientific method still had not produced regularly reproducible results, suggesting that any methodology, however rigorous, is open to criticism if a scientific community is sufficiently committed to explanatory concepts incompatible with the results. There is irony here, since many of ESP's critics who were experimental psychologists and simply could not accept the idea of nonmaterial causation, had themselves tried to claim authority for their subject as science by the development of rigorous methods rather than explanatory intelligibility. Moreover, many people who were sympathetic to the possibility of ESP became disappointed in the Rhines' work since the rigour of the methods and the triviality of the results detracted from the living realities that were felt to give meaning to psychical life.

For many ordinary people around the world, the inability or unwillingness of qualified psychologists to take psychical experiences seriously marked out the limitations of a scientific approach to human nature. For most psychologists, laypeople's credulity marked out the work still to be done to persuade the public of the discipline of scientific methods and the truths of the scientific worldview. In these circumstances, it was the authority of personal experience compared to the authority of the expert, and not only the empirical evidence for the parapsychological, that was at stake. The Rhines, however, and the scientists who followed their lead, wanted to overcome this polarization of psychologists and public opinion by the extension of psychology's range of subject matter; but to get the consent of fellow psychologists – still not forthcoming in the late twentieth century – they had to concentrate on their methods.

There were some institutional successes; for example, there were chairs in parapsychology in Freiburg (1954–75) and Utrecht (1974–88), but the only European chair in the 1990s was in the Department of Psychology at the university of Edinburgh. The incentive and funding for the Edinburgh position, established in 1985, came again from outside, in the form of a bequest from the novelist Arthur Koestler (1905–83), who had worked for Stalin's secret service in the 1930s but had turned into a scourge of scientific materialism. Koestler knew exactly what he was doing: he had set himself up as a critic of mechanist science and of the way its boundaries are policed. In the United States, in foundations like the Mind Science Center at San Antonio, Texas, there was considerable activity, for example, with free response techniques, which tried to capture actual life conditions rather than perpetuate the forced conditions of the laboratory. Most interesting in the long run, in many countries outside the Western world, for example, in Brazil and in China, non-physical extrasensory effects continued to have a more embedded place in belief about human nature, and it is in these cultures that we may perhaps best look for insight.

vi *Gestalt Psychology*

From some vantage points it appears that for twenty years or more neobehaviourists occupied the high ground in arguments about psychology's status as a natural science. In fact this was the situation only in certain major centres in the United States. There was very different scientific work by Spearman in London, Bartlett in Cambridge, Pavlov in Leningrad or the Bühlers in Vienna. This section takes the comparative perspective further and describes the work, known as gestalt psychology, of the prestigious and well-supported Berlin institute of psychology in the 1920s. Gestalt psychology did not just form another school, though it did have an unusual degree of institutional and intellectual coherence. It was a full-bloodied attempt to establish psychology as a science by a re-examination of philosophy and hence of the foundations of the sciences in general, and it was not just an attempt to impose the methodological prescriptions supposedly characteristic of the physical sciences. The three architects of gestalt psychology, Max Wertheimer (1880–1943), Kurt Koffka (1886–1941) and Wolfgang Köhler (1887–1967), all experienced the sense of cultural crisis that gripped Central European intellectuals before and after World War I. Their psychology belonged to a debate about science and values thought crucial to civilization itself. A concern with philosophical fundamentals, which contrasted with the temper of much North American psychology, also engaged the movement known as phenomenology. This movement interacted with gestalt psychology and, in the 1950s and 1960s, provided resources for the criticism of behaviourism. It is discussed in the following section.

A lack of agreement about philosophical fundamentals and debate about the relationship of the sciences to each other was of moral and cultural moment to German scholars. Further, when they looked outside the university, they saw a national life that increasingly did not appear to accord academic culture its proper value. They feared a modernity without deeper purposes. Weaknesses in the intellectual foundations of culture had aroused concern long before 1914. Radical innovations in physics, which culminated in Albert Einstein's theories of special and general relativity (1905 and 1916) and in the quantum mechanics of the 1920s, seemed to question the intelligibility of causal understanding itself. Beyond the university, the rapidity and depth of industrialization in Germany after the 1850s brought about a shift of power to business and to the technical professions that disturbed academics. The organization of the working class into a mass Social Democratic Party, and the call from women as well as men for social equality and political representation, raised doubts about the capacity of high culture to continue to inform national purposes. Even within the university, teachers feared that students at the turn of the century were inspired by the harsh critic of academic impotence, Nietzsche, at the expense of rational values.

A constructive response to these fears and perceptions was the call for a new philosophy, a philosophy specifically able to do justice to the natural sciences but which did not empty knowledge of everything except the facts revealed by physical science. The gestalt psychologists continued the response made by late nineteenth-century German-language psychologists, who looked to the natural-scientific methods of experimental psychology to provide philosophy with an objective content to be organized by reason into systematic form. They aimed to describe the conscious structure of knowledge of physical reality, then to use the description as an objective foundation for judgement about reality and to understand the reasoning in terms of which we make statements about it. Where the gestalt theorists differed from earlier philosophical psychologists was in the claim that the basic conscious structures are not elementary, as Mach or Titchener assumed, but organized wholes or '*Gestalten*' (the German word for 'forms'). They then argued that these *Gestalten* are a part of natural reality. Their claims are best understood through the psychological evidence with which the gestalt theorists themselves tried to substantiate their point of view. It is also helpful to picture what they reacted against, the long-standing approach, taken by Locke, that analysed mind in terms of elementary ideas conceived as analogous to physical atoms. The gestalt psychologists specifically denied the mosaic theory of perception, associated with Helmholtz, which presupposed a one-to-one correspondence between local stimulation and local sensation.

The simplest case of a gestalt phenomenon is the well-known line drawing that reveals a duck one moment, a rabbit the next. We do not see a series of lines and on this basis assemble in our minds an image of a duck and then reassemble the lines as an image of a rabbit. We see the whole as duck or rabbit; subsequently, as an analytic exercise, we may break the perception into parts. It was Wertheimer who established an exemplar for this gestalt approach to perception in a paper, published in 1912, on apparent movement. To produce this phenomenon, known for many years, two stationary light sources are flashed at intervals; with a particular timing of the intervals, the observer will see a moving light. What is at first glance an obscure phenomenon presented a great challenge and provided an experimental route into general questions of perception. Wertheimer argued, about what he called the phi-phenomenon, that the observer perceives motion – not motion of a light source but motion as a dynamic psychological event or whole. It is not possible to understand psychological events as the result of the sense organs' or the brain's construction of unit sensations into perceptions; instead, consciousness itself must be supposed to have an organized structure. Experiments will reveal what this organization is. He went on to postulate that these organized structures of consciousness have a temporal as well as a spatial character; the organization of space-time, he suggested, is a property of the conscious act. Finally, though this was worked out as a general principle later, Wertheimer supposed that the psychological organization is paralleled by organization in the brain. The important point for Wertheimer and his colleagues in 1912 was the claim that psychological experiments disclose organization; they also used this point to argue that psychological knowledge has precedence over physiological knowledge in research on perception.

These steps of argument were felt to be very significant. When Wundt was appointed to the university of Leipzig in 1875, he claimed for psychology the status of a science that would reform philosophy. He argued that psychology has its own subject matter distinct from physiology, even though the more elementary aspects of mental awareness are studied with techniques adapted from physiology. At the univerity of Göttingen, G. E. Müller taught these techniques, without philosophical elaboration, as a training in the new psychology. The situation that faced Wundt and Müller remained to face other psychologists: the discipline had to legitimate itself both in relation to philosophy, which reached conclusions through rational analysis, and to physiology, which used well-established objective methods. In order to advance their careers and their subject, psychologists wanted to demonstrate the centrality of their work to the advancement of philosophical science (*Wissenschaft*) and the rigour of its research methods. All the gestalt psychologists, Wertheimer, Koffka and Köhler, approached this situation in the light of their common training with Carl Stumpf in Berlin. Stumpf resisted the tendency, evident in experimental practice if not in theory, to break down conscious awareness into elementary units and to suppose that such units are somehow organized by apperception into larger wholes. He and his collaborators used experiment as a tool to assist the description of conscious contents in terms of the qualities as they appear directly in awareness. This search for a precise language in which to express knowledge of conscious contents 'in themselves' and not by analogy to physical states linked Stumpf to the contemporary development of phenomenology. It also prepared those of his students who became gestalt theorists to consider that the organized quality of awareness might be intrinsic to awareness. Stumpf propagated a philosophy founded empirically in the givens of consciousness, and he did not fall back on the physical analogy of atomistic sensation to make sense of those givens. This gave psychology an autonomous subject matter at the same time as it tackled the central problem of the organized form of the conscious world in our knowledge.

Wertheimer was born in Prague, into a family which assimilated German-language culture in public while it maintained Jewish traditions and learning at home. He rejected Jewish beliefs for humanism and then acquired a deep knowledge of experimental psychology in Berlin from 1904 to 1906. Like

Stumpf, a cellist, Wertheimer was intimate with music: he played the piano and the violin. Music – in which phrasing, rhythm and harmony exist as wholes and as values which are not assembled piecemeal by the player or listener – is a cultural paradigm for gestalt psychology. While behaviourists like Hull and Skinner were constructing mechanical devices, Wertheimer and his friends were playing Haydn quartets. Koffka was born in Berlin and was attracted to Stumpf's psychology as it appeared to him to be a science relevant to daily life; that included his own partial colour blindness, one topic in his subsequent research. After Berlin he went to the university of Würzburg where he learned about complex experimental research on thought and mental acts, research in which the higher mental processes were treated as dynamic events rather than as static entities. Köhler was born in Estonia, but his father became director of a *Gymnasium* in Saxony. Attracted to the physical sciences in Berlin, Köhler studied sound pitch in Stumpf's institute. He also used a mirror ingeniously placed in his outer ear to correlate physical vibrations with consciously perceived sound, and he introduced terms like 'tone colour' into his descriptions.

These three young men, who shared a cultural background, a training in experimental techniques for the precise description of conscious awareness and an exposure to the difficulties faced by current psychology, came together in Frankfurt in 1910. How this occurred involves a classic myth – maybe the events even happened – about scientific genius. Wertheimer got so excited by an idea for an experiment on apparent movement that he left the train on which he was travelling from Vienna to the Rhineland at Frankfurt. He bought a toy stroboscope, a device for flashing light, which can be used for the study of apparent movement, and played with it in his hotel room, then descended on Köhler who taught in the psychological institute of the Academy of Social and Commercial Sciences (subsequently the university) in the city. What Wertheimer showed with his stroboscope, Koffka later wrote, was that 'movement as experience is different from the experience of successive intervening phases . . .'[39] Wertheimer and Köhler were soon joined by Koffka and his wife Mira. They liked each other, had much in common and were fascinated by the same intellectual problems. In this setting, the arguments for gestalt psychology acquired an unusual degree of intensity and integration.

The Frankfurt researchers agreed that there was no systematic scientific psychology because psychologists tried to explain perception, thought or memory by a series of incompatible and *ad hoc* assumptions. They criticized in particular the persistent assumption that the primary mental units are atomic. Wertheimer, Koffka and Köhler published both experimental papers and position statements, the latter in the form of detailed responses to critics, in which they argued that the basic mental structures are organized wholes and that, once this is recognized, a systematic articulation of theory becomes possible. The key principle of gestalt theory was later expressed by Wertheimer in these terms: 'There are wholes, the behaviour of which is not determined by that of their individual elements, but where the part-processes are themselves determined by the intrinsic nature of the whole.'[40] Even before his work on apparent movement, Wertheimer wrote a paper on the numerical capacities of what were then called primitive people in which he suggested that they represent numbers as part of their way of life and that the form that numbers of things have in their mutual relations and context, rather than arithmetic number, determines thought about them. This explains, he argued, Western observers' reports that these people 'fail' with elementary arithmetic. His conclusion was that the whole perceptual situation, rather than presuppositions about the natural status of calculations with elementary entities, has to be considered if mental processes are to be understood.

In 1913, Köhler went to Tenerife in the Canary Islands where he intended to spend a year as director of the recently established Prussian Academy of Sciences animal research station. He had to stay until 1919 because of the war and while there, he carried out experiments with chimpanzees that made his international reputation. Observing the animals while they learned to obtain food, he claimed that they show 'insight'. In contrast to North American learning theorists, who set animals tasks like pressing a bar, which dictates either a unit positive or a unit negative response, Köhler gave his animals situations in which they could act in complex ways. He observed chimpanzees put sticks together or pile up boxes in order to reach bananas that were otherwise out of reach; sometimes he perceived sudden insight before the chimp purposefully solved the problem. For Köhler, this was convincing evidence that animals learn by the comprehension of a situation rather than by assembling habits acquired through trial and error. Given the focus in the United States on learning as the topic with the potential to unify psychological research, and given that this research claimed unique scientific credentials because it linked objectively observable stimulus-response elements, Köhler's work attracted much attention and controversy. It challenged both observational methods and conceptual assumptions in psychology in the US. This contrast and sometimes conflict between gestalt and United States psychology lasted over the following decades.

In 1920, Köhler began to teach at the university of Berlin and then, in 1922, he was called to the Berlin chair on Stumpf's retirement. From this position, in which Köhler became very much the academic aristocrat in the German manner, and with Wertheimer's collaboration in Berlin and then from Frankfurt and Koffka's support from a chair in Giessen, Köhler directed an extremely creative and productive group of students and colleagues. The Berlin group included the brilliant young psychologist Kurt Lewin, who developed field theory (discussed later in relation to social psychology), and it associated with philosophers such as Ernst Cassirer and, later, the theorist of probability Hans Reichenbach. The core of the Berlin work continued to be technical experimental studies, and gestalt-oriented researchers achieved leadership in research on perception. What inspired and gave the group such a sense of common purpose, however, was something larger, the hope that gestalt psychology is a science able to resolve philosophical questions. Koffka and Köhler both systematically addressed this project; and Köhler, who had the competence to work in mathematics and physics, took the argument to the wider community of physical scientists. They claimed that ordinary experience and scientific experiment demonstrate the inherently organized structure of mental content, that this content is isomorphic with (or parallel in structure to) brain organization and that structural organization is characteristic of physical systems ('fields') as a whole. Further, they hoped that a science which takes seriously the structured reality of what we consider ourselves to be directly aware of, the phenomenal field, will make possible the inclusion of moral and aesthetic values in the subject matter of science. This hope addressed 'the crisis of knowledge' which so many German intellectuals attributed to nineteenth-century mechanist science. The psychologists believed that a superficial materialist science, which the gestalt psychologists and others thought was perpetuated by behaviourism, leaves human values without a home in systematic knowledge and – they feared – creates the circumstances in which people turn to irrational beliefs and values.

Such irrational beliefs seized hold in Germany in January 1933. Köhler was a rare example of a senior academic who publicly objected to the dismissal from the universities of people who were classified as politically or racially unreliable, and he wrote a newspaper article in defence of Jewish colleagues. He expected to be arrested, and he sat up all night with friends to play chamber music. The arrest did not happen though his institute was harassed and his assistants dismissed. Köhler, who conceived himself to be a scholar above politics, continued to argue with the authorities and to help his staff; he began his lectures, as required, with what everyone knew to be a token Hitler salute. In 1935 he finally resigned, regarding the conditions as incompatible with scientific work, and left for the United States. Even then, he continued to edit the *Psychologische Forschung*, the academic journal for gestalt research published in Berlin, in order to provide an outlet for his former students' work. When the unpublished backlog of that work was used up in 1938, the journal ceased.

Koffka was attracted to Smith College, Massachusetts, in 1927, and Wertheimer, in exile, taught at the New School for Social Research in New York, the independent institution founded in 1919 for 'educating the educated' which gave much support to refugee academics. Thus the gestalt movement reassembled in North America where, however, it was perceived as a school with a particular theoretical orientation relevant to studies of perception. The gestalt psychologists were not appointed to positions at major US universitites and they failed to attract substantial numbers of graduate students to perpetuate their work. Gestalt psychology competed unsuccessfully with other schools of psychology in the United States and its ambition to be a scientific philosophy was largely unassimilated. Its experimental work was debated without reference to the conceptual framework of which the experimental studies had been part. Whereas in Weimar Germany gestalt psychology was one among several holistic psychologies and, more broadly, holistic philosophies of nature, in North America such thinking went against the grain of the dominant mechanist thought. The contrast is symbolized by the way Köhler and Wertheimer had personal links with physicists involved with twentieth-century field theory and relativity, whereas Hull maintained reverence for Newtonian mechanics. The challenge that gestalt psychology offered to the relationship between natural science and psychology barely crossed the Atlantic.

vii *Phenomenological Psychology*

This chapter has described a series of very different responses to the challenge to make psychological knowledge objective, systematic and theoretically consistent. The divergence between Pavlov's theory of higher nervous activity, Watson's behaviourism, neobehaviourist formalism, Skinner's radical behaviourism and gestalt psychology was so great that we might even question whether they formed a subject in common. In each case, however, and this point is reinforced by the history of experimental parapsychology, the physical sciences were the authoritative point of reference as psychologists tried to construct a science. The gestalt psychologists were different because they also strove to incorporate psychological knowledge into the philosophy of the physical sciences.

That psychology, to pursue science, should take natural science as its point of reference was obvious to most English-language psychologists in the twentieth century. This 'obviousness' needs to be put in cultural context. The high status of the physical sciences at the end of the nineteenth century, and the public authority of the truth achieved by physical science, did not mean that philosophical problems about the relationship between reason and the physical world had been solved. Many natural scientists, and more especially the public, took a realist view: scientific knowledge gives a true picture of what is real. Nevertheless, such realism was widely questioned at the turn of the century. Intellectuals reconsidered the conditions of reason itself and it was argued that knowledge is limited to the phenomenal awareness that we call consciousness. This led some philosophers to argue further that psychological science cannot in any straightforward way take natural science as its point of reference. They claimed that reason and conscious awareness, not knowledge of the physical world, constitute the foundations on which an objective psychology must be built up. This point is difficult for non-philosophers to follow. But in this section I indicate the direction of the argument; it is important to do this as a corrective to anglophone histories that simply assumed that there was and is agreement about the identity of psychology as a natural science.

The view that psychological analysis, made rigorous by refined observation, provides data for the philosophy of knowledge was shared by such otherwise different philosophers or scientists as Helmholtz, J. S. Mill, Wundt and Mach. Frege subjected such approaches to a logical critique in the 1890s, though his work became central only later; he accused them of psychologism, the misrepresentation of logical issues as psychological events. An influential group of German neo-Kantian philosophers were similarly opposed to the philosophical claims made on behalf of empirical psychological data, and at the same time they expressed concern about the growing occupancy by psychologists of academic positions in philosophy. Lastly, Edmund Husserl (1859–1938) published his *Logische Untersuchungen* (*Logical Investigations*, 1900–1), a book dedicated to Stumpf, which launched a revision of fundamental philosophy with major long-term implications for psychology.

Husserl's philosophy, which he called phenomenology, responded – like the systematic work of the gestalt theorists – to a sense of intellectual and cultural crisis. He analysed procedures to examine the conscious or phenomenal world in terms proper to this world rather than in terms that presuppose these phenomenal qualities represent something else, something such as a mind or a physical reality. He wanted to escape from the subject–object distinction, not to provide a more objective form of introspection, and studied what he held to be the necessary structure of consciousness. Husserl inspired others, for example, Martin Heidegger and through him J.-P. Sartre, to treat phenomenal consciousness as a subject in its own right and to believe that objective statements can be made about it. This inspiration had a distinct humanistic resonance: it made the values – moral, aesthetic, spiritual – present in conscious awareness, values many people wanted to be made into the proper subject matter of the human sciences, fundamental to a theory of knowledge. Concern with values was evident in the language Husserl chose; as a description of sensation, for example, he wrote: 'Experienced sensation is besouled by a certain act character, a certain grasping or mean-ing . . .'[41] This expression of values – 'besouled' awareness – attracted people to phenomenology who feared that the human reality of values is excluded from knowledge by the natural sciences. But the core of phenomenology as a philosophy was an attempt to re-lay the foundations for knowledge in general through a manner of reasoning that goes behind the distinction of the knowing subject and the known object to the phenomenon of being.

Husserl's interest in the description of the phenomenal world, or world of conscious awareness, derived from the lectures of Brentano in Vienna which made claims about psychology as an empirical science. Brentano defined the subject matter of psychology as mental events and believed that this subject matter makes psychology, though an empirical science, inherently different from the natural sciences. Husserl rejected Brentano's characterization of 'the mental' but retained the search for a descriptive language appropriate to the qualitative bearing of consciousness, a bearing intrinsically expressive of an attitude towards or a valuation of the intentional object in conscious awareness. Brentano also influenced Stumpf and, though Stumpf taught through the medium of concrete particulars rather than abstract theory, thus gained a powerful institutional voice in Berlin on behalf of a psychology of phenomenal qualities. All these approaches contrasted strongly with contemporary differential psychology and behaviourism. The same contrast existed even in the North American setting, where James's colourful writing evoked the discordant split between the riches of the directly experienced world and the desiccated terms in which psychologists – his target was principally the Wundtian experimentalists – turned that experience into science.

Husserl's contribution was a technical one: he provided a method by which, in his view, phenomenological description becomes objective. He thought it possible with this method to describe consciousness without reference to the truth of what we are conscious of or any psychological presuppositions. He wanted, in his term, to 'bracket-off' the conventional attribution of conscious awareness to mind – and here he sharply criticized the Cartesian dualism of mind and body – in order to describe conscious awareness itself. To those who were not primarily philosophers, he made it appear possible to defend belief in knowledge that reflects the significance of the world as directly experienced. Phenomenology sought to shut out the contingent reference of a conscious quality (that, for example, a colour is the colour of a physical thing) to describe the quality in its intrinsic terms and as a value (the colour as a state of being). Phenomenologists did not therefore predicate objective knowledge, as natural scientists did in the twentieth century, on the separation of facts and values; instead they set out to describe values objectively as part of the intrinsic subject matter of what may be known about human beings. Husserl offered hope that such an approach could be rationally grounded and systematically developed, i.e., that it could become a science, and that could thereby overcome the crisis in which nineteenth-century natural science had left the human spirit. Husserl developed his own work in a philosophical direction and went beyond the method of reduction to make claims about what necessarily exists for conscious awareness to have the qualities that it does. He also argued that direct intuition requires us to posit the self as the agent of reason. Other intellectuals, sometimes influenced by Husserl, were more directly interested in the possibilities of a science that includes human values in its subject matter, and they were primarily responsible for the development of phenomenological psychology. This development took place along two main lines: firstly, in conjunction with the refinement of qualitative descriptions in experimental psychology; and secondly, in German-language psychiatry as a way to comprehend the disturbed mind's attitude towards the world as the extension of a shared human condition, e.g., anxiety. Later, in the 1950s and 1960s, phenomenology was a resource for humanistic critics of behaviourism and of natural science models of human life more generally.

Husserl's work was philosophical and non-empirical, but, as psychology was taught in Germany as part of philosophy, psychologists were trained to deal with the issues. In the decade before 1914, a large amount of experimental work, such as that done in Würzburg by Karl Bühler on the analysis of complex thoughts or by Narziss Ach (1871–1946) on the will, rejected the correlation of elementary conscious states and simple external variables. The experimenters instead tried to describe consciousness in a way that reflects its qualitative activity. In this connection, they also used the word 'phenomenological' to indicate their interest in the qualitative structures intrinsic to consciousness of something; this usage contrasted with the 'phenomenalism' of Mach, a position that restricted knowledge to the sum of sensations. The word 'phenomenology' was used whether or not psychologists cited or attempted to use Husserl's work.

The philosophy of phenomenology developed parallel to and had common ground with experimental work on perception of the kind in which the gestalt psychologists engaged. The connection was sometimes direct, notably in the work of David Katz (1884–1953), a psychologist trained at the university of Göttingen where Husserl taught. Katz himself taught at Rostock and then in Stockholm, and he worked at length and with originality on the richness in experience of colour and touch. His originality was that he found ways to describe this richness as part of an experimental programme. The titles of his main publications convey a sense of what, during the interwar period, he called a phenomenological method: *Die Erscheinungsweisen der Farben und ihre Beeinflussung durch die individuelle Erfahrung'* (*The Modes of Appearances of the Colours and Their Modification by Individual Experience*, 1911; translated as *The World of Colour*, 1935), with a second edition, *Der Aufbau der Farbwelt (The Structure of the World of Colour*, 1930); and *Der Aufbau der Tastwelt (The Structure of the World of Touch*, 1925). His description of colour, for example, distinguished film colour and surface colour; the latter is attached in conscious experience to a surface, the former not. In Katz's account, these two experiences are not modes of appearance of one and the same colour but are different colours; which colour is seen depends on context, a matter that experiment explores. His language for colours opened up a range of qualitative distinctions – about lustre, glow, luminosity and the like – all familiar enough to painters but previously outside the range of objective research. In his book on touch, Katz pointed to the 'almost inexhaustible richness of the touchable world'.[42] This language that referred to consciousness as a 'world', a realm of meaningfully structured relations, was the characteristic language of phenomenology.

By the 1920s the word 'phenomenology' was widely used but with little of the precision Husserl attached to it. A key figure in this wider usage, and one who gained an audience outside the universities, was the Munich philosopher Max Scheler (1874–1928). He directly linked description of consciousness to values and hence appeared to reclaim science from materialism and fulfil a human need for meaning. Scheler's over-arching interest was in a philosophical anthropology, a philosophy of man's nature that starts from each person's irreducible and essential value, and this implied that knowledge of human beings must start from each person's conscious representation of values. His main phenomenological studies were on the emotions. He detached emotions from their association with the subjective world and described them as structures of objective consciousness, objectively or meaningfully connected to the values to which the emotions refer. It followed, in his argument, that ethics is not a matter of subjective sympathy with others, of feeling alone, but of the objective orientation of consciousness, as our essential nature, in an authentic manner. Many readers before and after the moral and material catastrophe of 1914–18 looked to authors like Scheler to renew hope. He was a philosopher who appeared to legitimate recourse to authentically experienced emotions and values as the precondition for knowledge about being human. Scheler was willing to develop phenomenology as the basis for a redemptive worldview, with all the loss of precision in language and philosophy this implied. But he also contributed to psychology in more specific ways: for instance, his descriptions of self-deception and *ressentiment* (an attitude of envy and resentment and hence alienation from others) influenced psychiatry in Munich. This psychiatry, developed systematically in the *Allgemeine Psychopathologie* (*General Psychopathology*, 1913) by Karl Jaspers (1883–1969), started from a description of the psychology of consciousness in which the patient expressed experience and an orientation towards the world. During the interwar period, phenomenological psychiatry burgeoned, though it was almost entirely restricted to the area of German cultural influence.

Phenomenology originated in German-language philosophy but it was taken up with interesting consequences in French, Belgian and Dutch cultures. The French-language Belgian psychologist Albert Michotte (1881–1965) was a significant influence who, while not a phenomenologist, did experimental psychology in a way that made possible some links with phenomenology. Michotte directed an influential school at the Catholic University of Louvain until 1946 after he had become familiar with German experimental work – especially at Würzburg, where he did research in 1907–8 that was critical of belief in sensory elements. His later experimental work explored the concrete nature of actual experience. This is most evident in his much cited studies published as *La perception de la causalité* (*The*

Perception of Causation, 1937). He concluded that we have a direct and irreducible experience of causal influence, of one thing as the agency of another. Though Michotte did not claim to solve philosophical questions, his experimental results on the perception of causation struck against Hume's classic philosophical analysis in the mid-eighteenth century. Hume, who analysed mind in terms of ideas, described what we call causation as a habit of thought which follows our experience that particular ideas in fact always follow other particular ideas. This approach exemplifies the atomist view of consciousness so widely criticized by German psychologists. Michotte's work gave empirical content to the criticism and suggested that what we call causation is a structural feature of conscious awareness. Like the work of the gestalt psychologists, Michotte's studies had critical implications for the learning theories then dominant in North America. He did not employ a phenomenological method, but he and others recognized that his experimental arguments and phenomenological description were complementary.

Dutch psychology was distinctively influenced by phenomenology through the work of Frederick F. J. Buytendijk (1887–1974), who was a professor at a large institute for psychology in Utrecht after 1946. Buytendijk was a physiologist and a physician who was also interested in Scheler's aspirations for a philosophical anthropology. The links between these interests were not obscure: Buytendijk wanted to observe animals as they express themselves in a way of life, and he wanted to know man in his 'innerness'. Such qualitative realities cannot be understood, he thought, in mechanical terms. An openness to the expressiveness of animal existence, for Buytendijk, comes by the study of animals in their natural habitats rather than by experiment. This was an attitude shared in the 1920s by other researchers such as the English naturalist Julian Huxley, who established ethology as a subject area. Buytendijk believed that knowledge of animals is predicated on our experience of the human situation as an expression of life. Animal studies must have a foundation in general psychology, and general psychology must address the human existential situation – the meaning that our being in the world has for us. Knowledge of people's innerness comes through the 'encounter' – a method which, in Buytendijk's description, involves 'disinterested and desire-free yet personally interested participation in each other'.[43] The flavour of this is illustrated by his remark that 'fashion is not a habit, but a significant behavior, the expression of a value-project on the part of human being as being-in-the-world . . . We can only guess at the influence which the fashionable expression "O.K." is having on contemporary European society.'[44] His psychology was a philosophical anthropology, and he looked to a phenomenological method to describe the qualities of lived experience.

Buytendijk desired 'the rehabilitation of the great tradition of German anthropological meditation which is still anchored in reverence before the human in all its manifestations and in the unconditional love for everything that bears the human face'.[45] He translated his values into a wide range of detailed descriptions of human movement (e.g., the special meaning of self-movement and the difference between touching and being touched), pain and the 'encounter' – a term which indicated that even basic sensory relations exist as part of an attitude or orientation; and, partly in response to Simone de Beauvoir's study *Le deuxième sexe* (*The Second Sex*, 1949), he recharacterized 'the feminine' in terms of a caring orientation towards the world. Dutch psychologists influenced by Buytendijk described everyday psychological events like the handshake and driving cars. Their description of the everyday in terms of what such situations mean, structured by values intrinsic to a person's awareness, created a psychology in the 1950s remote from the psychology dominant in the English-speaking world. But Buytendijk and others nevertheless thought that phenomenological psychology is objective because it starts from the objective reality of values in the human situation. The Utrecht psychologists considered their psychology to be more objective than any psychology based on natural science, which denies man's inner nature, could ever be. But by the late 1950s, the tide in The Netherlands had turned against this view of human science.

Phenomenology was widely taken up by European intellectuals as it seemed to bring together the search for knowledge and the search for values. It was, for example, a significant channel through which modern existential and psychological values were propagated in Spain. Before the period of the Franco regime, which began in 1938, the philosopher of 'vital reason', José Ortega y Gasset (1883–1965), devoted himself to making his contemporaries more open to a wider European and especially Germanic culture. The group of people who formed around Ortega, the so-called school of Madrid, associated itself with Husserl but rejected Husserl's attempt to characterize being by the radical reduction of consciousness to its basic essences and instead characterized being in terms of life acts. This attitude elevated the individual's subjective awareness of valued action to the key position in a vague and idealistic, but psychological, form of analysis of the human condition.

Phenomenology entered French philosophical life through a variety of routes, though it had little direct influence on those who were primarily psychologists by training and occupation. In the late 1930s both Maurice Merleau-Ponty (1908–61) and Jean-Paul Sartre (1905–80) wrote about topics like the emotions and imagination from a non-experimental, abstract point of view, and in their work they characterized universal features of awareness as they are lived through in the human situation. As Sartre's own turn to the medium of fiction indicates, this approach to the psychological world was close to literature. Merleau-Ponty's work was more rigorous and abstract and he responded at length to current work on perception and behaviour. His concern was impersonal, with the universal structure of consciousness not the particularities of personality: 'In so far as . . . something has meaning for me, I am neither here nor there, neither Peter nor Paul; I am in no way distinguishable from an "other" consciousness . . .'[46] This argument contributed to the French stucturalism of the 1950s, a cluster of intellectual commitments to the description of the supposed universal formal features of mind, reason and language. Despite the appearance of lengthy texts by Sartre and Merleau-Ponty, however, no distinct school of phenomenological psychology developed out of their work.

Many things were labelled phenomenology and there was no

single intellectual movement. But though there was no unity, there was a common orientation, most evident in the negative reaction to the kind of natural-scientific psychology taken to an extreme by behaviourism. Phenomenologists started at the other end, so to speak, from behaviourism; they started with the values and meaning of conscious being rather than with the measurable relations of physical stimuli and responses. Their movement developed at a time of perceived intellectual crisis, when the nineteenth-century scientific worldview appeared to deny or to devalue a human search for meaning. Phenomenological psychology was made known in the US through the proselytizing of Robert B. MacLeod (1907–72), but it had little impact on the main streams, with the exception of the humanistic psychology to be discussed later. British and North American psychology, whether through factorial analysis or behaviourist research, went to great lengths to give content to psychology conceived as the extension to the human sphere of the scientific worldview. This trend was followed in Europe after 1945. Research on parapsychology did not break the mould but sought to bring a new range of phenomena within the scientific worldview. By contrast, the gestalt theorists and even more the phenomenological psychologists argued that the form of our conscious awareness demonstrates the false premises on which nineteenth-century mechanical philosophies of nature – and twentieth-century psychologies that followed them – were founded. Instead, gestalt theorists and phenomenologists argued, though in different ways, consciousness intrinsically discriminates qualities and values, and it is therefore possible and necessary to refound scientific psychology in a way that starts from the qualities of our awareness.

This chapter has brought together North American and European psychology in order to put the well-known behaviourist claims to establish psychology as a science into a broader context of debate about the identity of science itself. Much of the European work discussed was unknown or ignored in the English-speaking world. Where it was known, it was in a partial and specific way, as was the case with gestalt experiments on perception. The North American psychologists' preoccupation with science, defined as the application of rigorous objective methods to observe physical variables, simply excluded much continental European work from being considered psychology at all. But this exclusion was a social phenomenon, and it resulted from the development of psychology in historically different contexts. Psychologists may have legislated boundaries around what is and is not science, but their boundaries were humanly created and there were always settings in which they were questioned. The search for objective science created a huge investment in psychology as a natural science, whether this was in terms of Pavlov's physiology, Hull's hypothetico-deductive scheme or Skinner's atheoretical operant research. Yet this investment presupposed many things about the nature of science itself, and these presuppositions were subjected to a critique as part of a renewal of philosophy and human values. Both behaviourists and phenomenologists had the future well-being of humanity in mind, but they chose radically different means, and these means represented completely different choices about how to live.

18

The Unconscious: Reason and Unreason

> In the course of centuries the *naïve* self-love of men has had to submit to two major blows at the hands of science. The first was when they learnt that our earth was not the centre of the universe but only a tiny fragment of a cosmic system of scarcely imaginable vastness . . . The second blow fell when biological research destroyed man's supposedly privileged place in creation and proved his descent from the animal kingdom and his ineradicable animal nature . . . But human megalomania will have suffered its third and most wounding blow from the psychological research of the present time which seeks to prove to the ego that it is not even master in its own house, but must content itself with scanty information of what is going on unconsciously in its mind.
>
> Sigmund Freud, *Vorlesungen zur Einführung in die Psychoanalyse* (*Introductory Lectures on Psychoanalysis*, 1916–17)[1]

i *The Tensions of Enlightenment*

Sigmund Freud (1856–1939), the son of Jewish migrants from Moravia to Vienna and in the last year of his life a refugee from Hitler, acquired mythic stature. The way he wrote about human nature became inseparable from the self-reflective world of individuals in twentieth-century society; his work, however open to contention, informed thought about what it is to be human throughout the century. Yet, as the loud silence about him in the previous pages has implied, his position in relation to academic and professional psychology was at best anomalous and at worst scandalous. Freud claimed to establish a new science, psychoanalysis, as the basis for a new psychology; yet scientific psychologists habitually denigrated this claim and regarded his contribution as at best peripheral, at worst pernicious.

Though the previous chapters have barely alluded to Freud, they make possible an appreciation of his work in a long-term perspective which reveals a dramatic tension. From one viewpoint he was a figure of the high Enlightenment, engaged on a science of man in order to describe the natural laws of our being and guide how we must live. He was an enemy of superstition and a liberator from ignorance about the intimate world. From the opposite viewpoint he was a prophet of the impossibility of enlightenment. He uncovered in human nature the power of the unconscious, an irrationality deeper than reason which determines individual and collective life: 'The unconscious is the true psychical reality.'[2] Reason, in Freud's work, revealed the limits of reason; he concluded the process of enlightenment with knowledge of dark powers. Freud's message was pessimistic, and outward events in Central Europe through most of the century accompanied his work like the chorus of a Greek tragedy. Yet the source of this pessimism was also the source of an optimism that enlightened reason, which, as Freud saw it, must live in the modern world without benefit of spiritual grace, can improve the human condition. It is in relation to this tension, rather than in the debates about the details of his psychology, that we may seek to understand his contribution. In his own unguarded estimation, Freud asserted that 'I am actually not at all a man of science, not an observer, not an experimenter, not a thinker. I am by temperament nothing but a conquistador – an adventurer . . . with all the curiosity, daring, and tenacity characteristic of a man of this sort.'[3]

Freud acquired mythic stature because he placed about his shoulders the mantle of the hero, the man who quests for the light of truth though aware that the journey may end in darkness. At the same time, he was of course an imperfect human being. He was a brilliant intellect of liberal scope who devoted much emotional energy to the enforcement of distance between himself and those who developed work that differed from his own. He was a man of reason and culture who was nevertheless sometimes overwhelmed by petty spite and ambition, and who deceived himself as well as others about the origins and development of his work. As a man who emphasized the power of the unconscious, he attracted biographers who want to turn the tables and find the sources of his work in his own unconscious. By the way his life made possible such a circle of reflections, as Peter Gay's biography put it, it was 'a life for our time'.[4] If, as Freud suggested, all language and action is structured by the unconscious, irrational elements of the mind, on what basis can we accept as truthful anything that we say or do? Freud straddled the nineteenth and twentieth centuries; he was educated in the certainties of nineteenth-century natural science but developed a psychology of the uncertain inner life. He became a figurehead for the late twentieth-century loss of scientific innocence and absence of faith in progress.

Freud himself, however, intended to contribute to the progress of science. He made a series of psychological claims for which he became famous: the presence of a dynamic, continuously active unconscious mind; the sexual theory of the neuroses and, by extension, the sexual theory of motivation in general; the interpretation of dreams as wish-fulfilling fantasies; the Oedipus complex as the basis of both individual character and moral culture. His readers, however, did not agree about what these claims amounted to, in the sense of either what they mean or whether they are credible. This was not a failure to understand the 'real' Freud; it was a reflection of the power of his work to make possible different interpretations and for debate about these interpretations to be a key resource of modern psychological culture. With Freud and with the psychologies and psychotherapies that called on his name, if only to reject him, the human sciences acquired much of their reflective language. Psychoanalytic notions about emotionally laden forces became part and parcel of public opinion in modern society. In Vienna or Berlin in the 1920s, in New York or London in the 1940s, and in Paris or San Francisco in the 1960s, critically engaged people looked to psychoanalysis as a humane technology of the self, a means by which the individual person comes to terms with both the social and the existential conditions of life. Psychoanalysis suggested means to deal with what are often called eternal questions of birth, death, love, hate, rage and loss. It was also a way to manage individual suffering – phobia, anxiety, loneliness, obsession, emptiness and inadequacy of every kind. Elaborated in countless ways since Freud's death, the psychology he initiated entered into every aspect of human relations as part of the chosen means to deal with human problems.

All this suggests that Freud's contribution must be understood in relation to twentieth-century cultural values and not only as a body of would-be truths. Freud claimed, however, to have established new knowledge, and he claimed scientific status for that knowledge. In discussing this claim we must again ask what the word 'scientific' means in this context. The next section outlines the historical development of his thought, while later sections consider psychoanalysis once it had become an established activity and the major alternatives to Freud's own views, especially the analytical psychology of Carl Gustav Jung (1875–1961).

Freud's initial theory came together between 1885 and 1900. During this time he worked mostly in private practice as a specialist in nervous diseases. His patients' symptoms and his attempts to find cures for them supplied the empirical foundations for his work. He then added dream analysis, including analyses of his own dreams. From this clinical and dream material taken from individuals, he constructed a general psychology, a theory he believed to be a true science of human nature. The empirical authority of his views depended on argument from a small number of cases, that is, not from experimental studies but from people subject to the social conventions of his middle-class consulting room. He was self-conscious about the fact that his manner of writing had much in common with the short story; he also appreciated the comparison between Sherlock Holmes and his own form of detection. He wrote to persuade readers that his own stories make human actions intelligible and, further, to persuade them that the fact that they do so requires belief in an elaborate theory of the unconscious. His form of argument was therefore in marked contrast to the ways contemporary psychologists attempted to make their subject scientific through experiments on mental content, studies on the physiological basis of conditioning, factorial analysis or observations of behaviour. His approach was comparable, however, with much late nineteenth-century French psychology which argued from intensive studies of individuals. Over the period from about 1950 to 1980, when psychoanalysis aroused interest among trained psychologists in the United States, Freud's ideas were studied by experimental methods and the results were largely negative. The consequence of the distance between Freud's ideas and anything that experimental studies demonstrate was the reassertion of the commonplace, Freud's protestations notwithstanding, that psychoanalysis is simply not a science.

Yet Freud was scientifically and medically trained in anatomy and physiology to the highest standards. He did research for some years, when he worked on the micro-anatomy of nervous tissue in the eminent Ernst Brücke's physiological institute in Vienna. It was an enthusiasm for science and an ambition to make an impact as a scientist that led him from clinical particulars to general laws when the need to earn a decent living forced him out of the laboratory and into medical practice. Freud's new psychology was strongly coloured by his physiological background: he held to determinism, belief that everything has a specific cause; and he broadly conceived of mental structures and functions in terms of a physiological model of the economy of energy. He also discussed the subjects of individual development and culture in the light of evolutionary theory. Given all this, it has been thought valid to assess psychoanalysis as a natural science and, since it does not meet the standards established by experimental psychology, to dismiss its claims. Freud's fame has then been attributed to public gullibility, passing intellectual fashion and the seductions of persuasive writing.

None of this begins to explain Freud's historical importance. When he wrote about his patients, about the unconscious and about sexuality, he was extremely aware – sometimes painfully so, since he ardently wished to be original – that he was rephrasing insights embedded in ancient myth, in literature and in critical thought. The precise nature of his relation to Nietzsche's work, to take one significant instance, is a vexed question. Freud was extremely well read – he was, for example, intimately familiar with Shakespeare's English – and he was a great stylist in the German language who received the Goethe Prize in Frankfurt in 1930. He collected antiquities, and a sense of archaeology as the recovery of the past by the excavation of the covering layers of history is an important image of his psychology – the archaeology of the psyche. He brought these interests as well as his natural-scientific training to his approach to a patient's symptoms and to his general claims about the mind. When he used the language of energy, for example, the language ambiguously connoted physical and mental power. He drew on the quasi-mystical philosophy of nature of Fechner as well as the physiology of Brücke. He observed his patients as clinical specimens, but he also collaborated with them to tell stories that made meaning out of their lives and their suffering. To many readers and followers he held out the hope that he had constructed an objective science, but he also appeared to make subjective life intelligible through the removal of the layers of the mind that hide and deceive. It is therefore not valid to assess psychoanalysis only as a natural science; it must also be assessed, as subjects in the humanities are assessed, in terms of intelligibility and meaningfulness.

This argument leads again to the historical ambiguity of the word 'science'. Freud has been read and judged as a natural scientist and he has also been read and judged as a scientist in the wider sense – as a man who searched for a systematic, consistent and objective understanding of the psyche. On the latter reading, Freud's work is judged as a language, as a subject like literature or history and – above all – as a practical response to the suffering of individual people. Freud himself endeavoured to be a scientist in both senses, something that helps account for the tensions in his work and its openness to diverse interpretations. He had followers who thought psychoanalysis could become a natural science, but for most late twentieth-century people interested in his work questions about the topic were simply misguided. Those who sympathized with psychoanalysis argued that it is a way of thought making the particulars of human life intelligible and thereby potentially more richly livable. Sympathizers of a theoretical bent found parallels in the practice of phenomenology and in exegetical textual scholarship to support belief that interpretive procedures of the kind they thought Freud undertook can be rigorous and lead to science.

This complex of issues about Freud's status as a scientist reappeared in a debate about his language and its translation from German into English. What is known as the *Standard Edition* of his psychological works (in twenty-four volumes) appeared in English between 1953 and 1973 under the editorship of James Strachey (the brother of the literary writer Lytton Strachey). In many respects it is a monument to careful scholarship; it was a major contribution to making Freud accessible to non-German-language readers; and it was substantially responsible for the enormous amount of English-language commentary on Freud. There were critics, however, who stated that Strachey and his translators deferred to natural science and found words with a natural-scientific coloration in English where Freud used German words with humanistic or even spiritual connotations. The central examples are '*Seele*', rendered in English as 'mind' whereas it might have been rendered as 'soul', and '*das Ich*', rendered as 'the ego' whereas it might have been rendered as 'I'. It was also pointed out that Freud's discussion of psychic energy sounded mechanistic in English, although he may have had in mind a philosophy of will like that of Schopenhauer. It was even suggested that there should be a new translation. More was at stake than words: there were questions about whether knowledge of people must conform to natural-scientific knowledge to be knowledge. Given the emotional and evaluative load carried by Freudian psychology, which points to deep powers within the psyche and to socially unsettling sexual forces, the argument about the form of knowledge of ourselves to which we have access emerged with special intensity over psychoanalysis.

Even the roots of psychoanalysis lay in controversy – in the world of marginal medicine practised by Mesmer and, later, by nineteenth-century hypnotists, in the bourgeois world of *fin-de-siècle* nervous illness and in the cultural crisis engendered by the scientific worldview. There was an extensive pre-Freudian literature about unconscious mental events; conventional Romantic imagery, for example, sought to call up the hidden powers of the soul. There was a notable Victorian preoccupation with sex; the repression of public discourse about the subject was matched by private obsession. When Freud's work began to appear about 1900, sexology was already an established medical domain. Dream interpretation was an ancient art, a means with which men and women endeavoured to confront what has been called the inner other, a strangeness within that is yet intrinsic to self-identity. Dreams were also a source of the tension between the desire to know and the fear of unreason. Freud himself claimed that all his assertions could be found in literature. Yet it was his work that gave form and language to all these issues in the twentieth century.

ii *Sigmund Freud and Early Psychoanalysis*

As a self-confessed conquistador of the inner life – a hero who descends into the dark night of the soul to return bearing the truth – Freud has attracted more than the usual complement of stories about his youth. A suitably Germanic inspiration was found in his own report that he desired to pursue natural science when he heard a lecture on Goethe's lyric essay on nature. Some biographers, by contrast, stress his Jewish identity, the sense that he was marked out for special insight. His ancestors took part in the large-scale movement of Jewish people from Eastern and Central Europe westwards, into provincial Austro-Hungary and on again to cities like Vienna. Along the way, these people became assimilated to the German-language way of life to varying degrees. Like many other bright students, Freud took the path of upward social mobility opened by a medical training. His ambition lay with research and he worked for a while on the description and staining of nerve tracts in simple fish. But as there were no career openings and he needed secure prospects in order to marry, he shifted to medical practice, specialized in various Viennese institutions as a neurologist and became a private consultant in nervous diseases. He also rashly experimented with cocaine in the hope that a quick medical breakthrough would make his name. His marriage to Martha Bernays, the daughter of a Hamburg Jewish merchant, was delayed for five years while he established himself. The cost of bourgeois marriages and professional careers remained a sombre theme throughout his work.

Freud chose to specialize in the borderland where mind and body interact. His research involved the anatomy of the nervous system; his medical practice covered disorder where there was obvious physical injury, for example, in cerebral palsy, and cases where the disorder was clearly 'in the mind' as, for example, in the case of a young woman who obsessively washed her hands. In between the extremes, however, doctors were faced by an utterly confusing hotchpotch of hysterical and nervous symptoms caused by unknown combinations of mental and physical factors. Often enough there was little agreement whether the symptoms were 'real', 'put on' or even 'imagined'. There was much debate, because of the legal and financial implications, about railway injuries. People involved in accidents frequently suffered serious pain, though no physical cause was visible and the pain showed a remarkable capacity to lessen when compensation was forthcoming. It is easy to laugh, but even sceptics had to face the patients who suffered. Male and female patients also exhibited 'neurasthenia', a then recently coined term for a malaise that exhibits lassitude, weakness, loss of will, with such symptoms as sensitivity to light or noise and headaches. Hysterical symptoms were particularly dramatic, and women and men exhibited partial paralysis, hallucinations, compulsive actions and prostrating anxiety.

The medical response to this borderland world was as polyvalent as the symptoms themselves. Many medical authorities, like Theodor Meynert (1833–92), Freud's head at the psychiatry department of the Vienna general hospital, took a physicalist approach. They believed that physiology is the research edge of medical science; as a consequence, they assumed that all disorder, if genuine, is ultimately physical in nature, even when the symptoms are functional, i.e., when they appear in the disordered activity of the body and are not visibly due to organic damage. Freud adopted this manner of argument with some sophistication in his study of aphasia, or loss of speech, which developed an evolutionary orientation towards brain function. He correlated type and degree of loss of speech with progressive loss in the evolutionary sequence of levels in the brain. At the same time, however, in 1886, Freud opened his private practice; and to attract patients he needed to make sense of and to offer cures to the bewildering disorders that came through his door. His response was marked by a willingness to experiment: he used hypnosis and cocaine, for example, as well as more conventional rest cures and electrotherapy. He also had a driving intellectual ambition, the ambition of a scientist, to reveal causal order beneath the surface confusion of symptoms. The therapeutic and intellectual resolution of this experience in his work became psychoanalysis.

When he started out as a doctor, Freud was helped emotionally, intellectually, financially and practically, in the form of referred patients, by Josef Breuer (1842–1925), a well-established Viennese physician. Among Breuer's patients was Bertha Pappenheim, celebrated as the first psychoanalytic patient and known in the literature as Anna O., whose case Breuer and Freud wrote up in their joint *Studien über Hysterie* (*Studies on Hysteria*, 1895). Anna O. had two personalities, a normal but sad person and a morbid, agitated and disturbed person who hallucinated black snakes, suffered paralysis in one arm and spoke an ungrammatical jargon. At one stage Breuer was able to change her from one state to another when he showed her an orange. She exhibited a kind of self-hypnosis during the disturbed condition, and Breuer found that if he then encouraged her to talk about memories this had some therapeutic effect. This creative patient called the talk 'chimney-sweeping' while her doctor described it as catharsis.[5] In the version of the story propagated by Freud, but perhaps a false version, it dawned on Breuer that his patient renewed her symptoms because she was in love with her doctor, and he therefore broke off the relationship. Whether true or not, the story makes sense of the way in which, out of the recovery of Anna O.'s memories, Freud constructed the theory of the unconscious, and out of the patient's feelings for her doctor he elaborated what he called the transference, the utilization of emotional energy in the medical relationship as therapy. These steps, however, took Freud well over a decade, a decade in which he gradually shifted his attention and his language from physical to psychological explanations of his patients' worlds.

Another major input came from his experience of hypnotism. If Meynert's physicalism represented one wing of medical opinion in the 1880s, belief in hypnotism as a form of therapy represented the other. Freud, ambitious to explore, obtained a travel grant to study in Paris in the winter of 1885–6, and his base was the Salpêtrière hospital where Charcot was then at the height of his fame as a neurologist and protagonist of hypnotism as an experimental tool. The experience was decisive for Freud, but it is necessary to recall the history of hypnotism to see why this was so.

Though Mesmer's claims were rejected by the Parisian medical and scientific establishment in the 1780s, what was called animal magnetism continued to have a lively existence as an alternative medicine. In the mid-nineteenth century there was a renewal of interest, in Britain and the United States as well as in France and Germany. Though this interest was closely linked to spiritualism, attention was given to the psychological characteristics of trance and automatic states as well as to healing forces. In the French centre of Lorraine, Nancy, a local doctor, A.-A. Liébeault (1823–1904), and then an academic physician, Hippolyte Bernheim (1840–1919), used hypnotism regularly as a therapy. They claimed it had a healing power with ordinary people for everyday illnesses. The transformation of the standing of hypnotism with the medical establishment occurred in Paris, however, with its use, though in a special way, by Charcot. The Salpêtrière, where Charcot worked, was more like a small town than a hospital and it provided him with a vast array of nervous disorders to study. He imposed descriptions and classifications, just when neurology became a self-conscious medical speciality, with a flair that made him a medical celebrity. He boldly included hysteria within his remit, persuaded his colleagues that it is a real disorder and then used hypnotism as a technique to reproduce and study hysterical symptoms. There was considerable interest in all these events throughout Europe and North America. Hypnotism moved in from the medical margins, and physicians began to engage with it without risk of censure.

Charcot's whole approach, meanwhile, was attacked by the Nancy physicians, and this dispute raised hypnotism's public profile. Charcot, like Meynert, looked for physical explanations. He assumed that hysteria is a pathology with a physiological basis and an inherited component, and he used hypnotism as an experimental and demonstrative technique to reproduce the symptoms. This was all of a piece with the theatrical way in which he staged his teaching and his life generally. He thought a susceptibility to hypnosis is itself pathological and not, as Bernheim believed, a feature of general psychology. Nevertheless, Charcot's hysterical patients exhibited symptoms that followed ideas and not anatomy: hysterical paralysis, for example, affected the parts used in a particular bodily expression rather than the parts linked by the same nerves. For Charcot, this was of interest; for Freud, it became key evidence that symptoms are the symbolic expression of something in the mind of which the patient is unaware. In Breuer's and Freud's famous phrase: 'Hysterics suffer mainly from reminiscences.'[6]

Also during the 1880s, Pierre Janet investigated a phenomenon apparently related to hypnotism and trance, the splitting of personality. The idea of the double, the *Doppelgänger*, a shadow that parallels, opposes or complements the ordinary self and is a source of danger or of inspiration, has a long history. Janet brought this idea firmly into the medical arena after he spent much of the 1880s teaching in Le Havre, where he experimented with a young woman who exhibited a split personality. Léonie, who appeared able also to fall into a hypnotic sleep at a distance, was a sensation in the world of psychical research. Janet himself returned to Paris to study medicine and to pursue his interest in psychopathology. He attempted to understand Léonie by the description of three conditions, or three personalities, which he correlated with her symptoms and with those of other subjects who exhibited phenomena like amnesia (loss of memory) or a capacity for automatic writing. He posited the existence of a subconscious realm. Thus, when Léonie said, for example, 'I am afraid and I don't know why', Janet interpreted this to show that 'the unconscious has its dream; it sees men behind the curtains, and puts the body in an attitude of terror'.[7] Janet therefore described different dimensions of psychic life as split parts of the personality endowed with subconscious ideas. When he practised as a doctor, he took these ideas further and developed, independently of Freud, a fully-fledged dynamic theory of the subconscious. In 1902 he replaced Ribot as professor of experimental psychology at the Collège de France, where he lectured for the next thirty years on the normal and abnormal mind and was sharply critical of Freud's sexual theory of the neuroses. But while Janet's early work attracted attention, his later systematic theories did not. His institutional position gave him few students, but his presence in the Collège contributed to the substantial absence of psychoanalysis from French culture before the 1930s.

Freud opened up his practice and expanded his ideas after he returned from Paris. He later visited Bernheim and showed sympathy with the belief that the susceptibility to hypnosis is a general feature of psychic life. Freud was also willing to accept that the rapport or emotional relation between patient and physician is central to treatment – as Breuer's experiences with Anna O., told to Freud, confirmed. At the same time, Freud reflected on Janet's notion of the split personality as a way to make sense of symptoms as the expression of hidden memories. He did not find hypnotic techniques very satisfactory in practice and instead devised what he called free association and then dream analysis in order to gain access to hidden memory. This introduced the couch into the consulting room and obliged each patient to report 'whatever comes into his head . . . not being misled, for instance, into suppressing an idea because it strikes him as unimportant or irrelevant or because it seems to him meaningless'.[8] On this basis, during the 1890s, Freud elaborated his distinctive theory of psychoanalysis: a method to gain access to the unconscious; a theory about what drives the unconscious: 'the mechanism of defence'; and a therapeutic practice – 'the transference' – that draws out the emotional energies of the unconscious through a psychological relationship between analyst and patient.

This creative decade in Freud's life, which in retrospect he portrayed as a solitary heroic journey, involved many debts: to Jewish culture, to evolutionary biology, to a physiological training, to his close friendship – which sustained him through a self-perceived isolation – with a Berlin nose and ear specialist, Wilhelm Fliess (1858–1928), to cocaine, to the literary sources of a practice based on talk, to the ferment of ideas in the *fin de siècle*, to the political alienation of Vienna's intellectuals, and – not least – to his patients, many of whom were well-educated nervous women. Beginning in 1894 and intensified by the death of his father, Freud undertook a self-analysis, a procedure made possible by the way he made associations in his mind to material in his dreams. This clarified his beliefs about what drives the unconscious and how these energies affect ordinary life as well as pathological symptoms. Freud stressed that the purpose of studying pathology is not to label a separate category of people, as was so common in contemporary theories of degeneracy. Rather, he argued, the study leads to conclusions about what is normal: normal and abnormal people share psychic mechanisms. In the interpretation of dreams, Freud found what he believed to be a reproducible method with which to carry out analysis; he called it 'the royal road to a knowledge of the unconscious activities of the mind'.[9] Dream analysis also helped Freud to elaborate his key concept of resistance, the process whereby the mind goes to great lengths to prevent the entry of unconscious material into the conscious world. When we dream, Freud argued, a censor operates to alter and filter unconscious material before it passes into awareness, and this process results in the surface meaninglessness of dreams. Dream analysis exposes this activity and restores meaning to the dream content. Freud therefore compared dream analysis to the process of translation.

After the publication of the book on hysteria, when he pressured Breuer into co-authorship, Freud brought out his major work on *Die Traumdeutung* (*The Interpretation of Dreams*, 1900), with its key theoretical chapter seven, *Zur Psychopathologie des Alltagslebens* (*The Psychopathology of Everyday Life*, 1901), and *Drei Abhandlungen zur Sexualtheorie* (*Three Essays on the Theory of Sexuality*, 1905). These works gave psychoanalysis a public life. A small group began to meet every Wednesday evening in Freud's apartment at Berggasse 19 (which became a Freud museum in the 1980s) and formed the Vienna Psychoanalytic Society. Freud's publications were not ignored, as he and other early analysts sometimes liked to claim in order to enhance their self-image as lonely explorers, but neither was his work recognized as a new science as he thought it deserved to be. *The Interpretation of Dreams* inspired a small number of intellectuals, mostly not physicians and all Jewish, who formed the early Vienna group. In the wider world it was admired by the Swiss psychiatrist Jung and his younger colleague Ludwig Binswanger (1881–1966); their visit to Freud in 1907 was followed by the foundation of a psychoanalytic group in a medical university context in Zurich. This marked the movement's expansion on to the international scene and away from its Jewish context in Vienna. With Freud's and Jung's visit to the United States in 1909, and the foundation of the International Psychoanalytic Association in 1911, its public and medical visibility began to grow.

Psychoanalysis was launched outside the academic setting, and the contrast between psychoanalysis and contemporary academic psychology is part of the history of the debate about Freud's claim to establish a science. The contrast also explains much of Freud's impact on wider twentieth-century culture. His books were unquestionably influenced by the mechanist worldview and academic physiology. Freud took it to be axiomatic that any action, however trivial, like a slip of the tongue, has a cause, and that knowledge of causes makes possible general laws. His work is noteworthy for the seriousness with which he assumed that determinism applies in daily life, not least to humour. Nevertheless, there were moments when Freud appeared to acknowledge a mystery at the heart of human experience, and his discussion of dreams and jokes delighted in the very unscientific play of Viennese language.

Freud initially wanted to understand the unconscious activity of symptom formation in terms of energized relations between the brain's structural elements. In 1895 he spent an intense six weeks writing a draft paper, a description of a physical model of the mind, subsequently known as the 'Project for a Scientific Psychology'. Significantly, he abruptly abandoned the project and thereafter used psychological and not physiological language, a language appropriate for the clinical and dream evidence that he actually had. As he explained in a letter to his friend Fliess: 'But apart from this conviction [about a physiological basis] I do not know how to go on, neither theoretically or therapeutically, and therefore must behave as if only the psychological were under consideration.'[10] He still retained elements of physiology in his theory: firstly, the determinism; and secondly, metaphors to describe an economy of energy and the dynamic or even hydraulic interaction of mental structures. Freud conceived the unconscious to be a cluster of energies formed by the libido (his word for sexual forces) and modified by the libido's encounter with the conditions of life, that is, by repression. He explained repression as a defence, a mechanism to avoid the pain of mental conflict. From one perspective, then, Freud gave new life to the association of ideas and the pleasure–pain principle of conduct, and in this way he perpetuated eighteenth-century psychology as well as enlightened ideals. But when he described association and motivation as unconscious and repressed and made them the hidden causes of pathological symptoms like giddiness ('losing one's balance') and everyday mental events like dreams, Freud emphasized an irrational content in human life that people committed to a life of reason ignored. He concluded that though the unconscious cannot be known directly it is everywhere and at all times the power behind the throne of reason. Psychoanalysis, he believed, is the key to unlock its secrets.

When Freud adopted psychological language and left brain research to the future, he did more than simply adopt a language of convenience. His use of the language of the psyche aligned his work with the thought of poets, philosophers and healers of the spirit. It built ambivalence into his work understood as a science. Freud felt cut off from the medical establishment by the direction of his work, though he had a respectable if unspectacular reputation as an expert on hysteria and nervous diseases. With *The Interpretation of Dreams* and other books, however, he gained a non-medical audience, one appreciative of the wit, style and insight with which he observed ordinary life, and some readers were prepared to receive his work as a revelation. Freud opened up a psychology that spoke to cultured readers who looked for a science of human nature which was neither impotent before emotional needs, like experimental psychology, nor unbelievable, like theology. He wrote stories about individual cases like a novelist, and these stories made sense of what his patients said and provided a language with which readers made sense of their own lives. Ebbinghaus's experimental study of memory in the 1880s presented the subject, initially himself, with *meaningless* lists of syllables; Emil Kraepelin (1856–1926), the author of the leading psychiatric textbook of Freud's generation, wrote down his patients' statements but treated them as symptoms of disorganization and not as meaningful communications. Freud, by contrast, thought all language has meaning: we never remember anything meaningless or say something that does not communicate. The choice of the word 'interpretation' in the title of Freud's book on 'dream work' also signalled an ambivalence about natural science. The book was replete with proposals about causal mechanisms to explain the form and content of dreams; at the same time, it was a commentary on the meaning that dreaming has for individuals in particular and the human condition in general. At one and the same time, Freud explained dreams and pointed to psychic depths that lie beyond the scope of causal language. 'We become aware during the work of interpretation that . . . there is a tangle of dream-thoughts which cannot be unravelled and which moreover adds nothing to our knowledge of the content of the dream. This is the dream's navel, the spot where it reaches down into the unknown.'[11]

Freud followed an established method in clinical science when he generalized from particular cases. Nevertheless, his work was exceptional for the extent to which he constructed a general psychology on this basis. He was not satisfied to put forward only a medical theory of the neuroses, but used such a theory as the starting-point for an all-encompassing account of mind. Hysterical symptoms led him to the unconscious, the unconscious to concepts of resistance and repression, and the explanation for these phenomena to the theory of sexuality – his argument about the early development and decisive role of sexual feelings. Later he even speculated further about the most general structure and dynamics of the psyche, and he used these speculations to account for morality, culture and religion. It was a bravura performance. The important historical point is not a judgement whether it is all true or false, but recognition that it gave to *psychology* an ultimate seriousness in the description of the human condition. Freud combined stories about a governess who smelt burnt pudding – Miss Lucy R. – with references to sex and with philosophical questions about the meaning of life; and he gained a general audience that no amount of studies on nonsense syllables or of rats in mazes was able to do.

The popular reputation of Freud's psychology from early on in the century was that it emphasized sex at the expense of everything else. This reputation doubtless suggests in part that changed mores in the twentieth century were projected on to a father figure, Freud; at the same time, the term 'sexuality' itself became current. The changes were certainly not all due to Freud, though his name became synonymous with a view of life which makes sexuality the basis of motivation. Many physicians, teachers, anthropologists, writers and moralists focused on sexuality in the late nineteenth century: Freud's psychiatric colleague Albert Moll emphasized childhood sexuality, and another colleague, Richard von Krafft-Ebing, classified sexual abnormalities; Arthur Schnitzler, the Viennese writer, portrayed bourgeois marriage as a tragedy which frustrates sexuality within its confines and makes happiness in sexuality impossible outside; Gustav Klimt, the leader of The Secession, a group of young artists who left the official art academy, brought direct eroticism into painting; in 1908, Richard Strauss staged his opera *Salome*, with its violent portrayal of sexuality. And that was just Vienna. In Britain, Havelock Ellis (1859–1939) began to publish his encyclopaedic series of *Studies in the Psychology of Sex* (7 volumes, 1897–1928); in the United States, G. Stanley Hall focused parents' and teachers' attention on adolescence, and brought Freud and Jung across the Atlantic for a series of lectures.

Breuer observed in his studies on hysteria with Freud that 'the great majority of severe neuroses in women have their origin in the marriage bed'.[12] As Schnitzler's stories and Freud's patients confirmed, this was no idle joke. Freud's sexual theory developed when he tried to understand what sort of material is repressed and forms the unconscious content of the mind. His search led him to trace symptoms back to childhood rather than to current circumstances. When his patients on the couch associated with ideas, they regularly showed resistance, a refusal or inability to follow an idea in a certain direction. Freud found that this direction had a sexual content. For some time he thought this content derived from a traumatic experience, and he concluded that his patients had experienced a childhood seduction, abuse, the pain of which caused its memory to be buried in the unconscious until adolescent or adult experience released the energy of this buried memory as symptoms. Then, in 1894, Freud began what has been called the creative illness that led to his self-analysis. From the examination of his own sexuality, as well as the reconsideration of his patients' reports, he became convinced that many memories of childhood seduction are in reality memories of fantasies. This implied that young children possess sexual feelings and that these are expressed and then repressed in their relationships with their parents. In order to make sense of dream material, Freud constructed a general theory of sexual development, which emphasized what he named the Oedipus complex, to characterize the child's relation to the parents, and the successive location of pleasure in the oral, anal and genital parts of the body. Freud intended to be objective and scientific even while he wrote on such emotive subjects as a boy's desire to sleep with his mother. He rejected the accusation made against him that he was biased towards sex; he believed, when he claimed to reveal sexual forces behind ordinary life, that – like any scientist – he traced back what is observable to its causal roots.

At this stage of Freud's thought about sex, several things are noteworthy. Firstly, he presupposed the existence of instinctual energy but gave little attention to what it is; his focus was on the repressed content in the unconscious. Secondly, while he stressed childhood sexuality, his notion of its energy, the libido or 'the power behind the sexual drive', was rather wide, and it involved a desire for pleasure rather than sex in a narrow sense.[13] Thirdly, speculative notions such as the inheritance of acquired characteristics, belief in which was widely supported in late nineteenth-century biology, and theories of collective psychic ancestry underlay his ideas. Freud retained these notions but they became increasingly dated in the second half of his life. Fourthly, he was overwhelmingly concerned with male sexuality, even though a substantial number, perhaps a majority, of his patients were women and he was intimately familiar with problems of women's sexuality. Freud never seriously questioned a male-oriented norm of sexuality, a weakness painfully exposed in the 1960s and 1970s. Finally, his concentration on sexuality was not a call for its liberation. He remained a bourgeois man of his time, probably wedded to monogamy (though there is a minor biographical industry that claims this was not so), and he demanded sexual restraint as a condition of civilized social life.

Historians have debated the significance of Freud's Jewish family background, and this debate repeated in microcosm discussions about the relations between Jewishness, tradition and modernity in Europe. Though without religious faith, Freud identified with his Jewishness in a way that went beyond reason. 'But there remained enough [of my Jewishness] to make the attraction of Judaism and the Jews irresistible, many dark and emotional powers all the stronger the less they could be expressed in words, as well as the clear consciousness of an inner identity . . .'[14] This encouraged the view that he interpreted dreams or neuroses rather as the rabbis interpreted scripture, or even the view that he was a secret mystic. There is little evidence to support the latter position, however much there is a need to read Freud as the author of an interpretive rather than causal science. When he lent his name to support the

Jewish university in Jerusalem in the 1920s, he continued, in new political circumstances, to hope for the enlightened emancipation of his people.

Freud had an insatiable intellectual curiosity and ambition. Many of those attracted to his work in the early years, like Adler, Ferenczi and Lou Andreas-Salomé, looked for something more like a guide to life than a psychology. For Helene Deutsch (1884–1982), before she modified her views on femininity and motherhood in the 1930s, Freud was a guiding light in dark times. If psychoanalysis began as a medical practice to heal the individual, many of its followers hoped it would become a movement to heal the world. It thereby made an inestimable contribution, especially in the 1920s, to the formation of a society in which people reflected on their lives in psychological terms. Whatever Freud's protestations about the scientific character of his work, psychoanalysis was perceived to have a mission. The mission, which Freud also contemplated, was enlightenment, to hold up a mirror to human nature so that people can live in the light of what they truly are. Some analysts dreamed of redemption for modern men and women for whom religion had lost the power to redeem. All analysts felt that they had truths to foster and nurture in a hostile and resistant world. This had, as a negative consequence, the formation of analytic sects dogmatically loyal to a mythic Freud. But, as a positive consequence, it also motivated the personal resources to build a better world. Some of this was evident in the radical Viennese experiments with education in the 1920s, discussed in an earlier chapter. Lastly, however, if psychoanalysts had a mission, so did many other psychologists – even if the mental testing movement, Watson's or Skinner's behaviourism and social psychology took less colourful forms.

iii *The Id, the Ego and the Psychoanalysts*

There was a rapid growth of interest in psychoanalysis just before 1914 in both the English- and German-speaking worlds. Freud broadened what he wrote to include culture and civilization: he published on the prevalence of nervous illness, on Leonardo and on the anthropology of primitive taboos. The analysts consolidated themselves into a movement and Jung, at Freud's insistence, became the first president of the International Psychoanalytic Association in 1910. Schism immediately followed: Alfred Adler (1870–1937) went his own way in 1911 and Jung left in 1913. These painful events, 'defections' from Freud's point of view, led to the establishment of 'the committee', an inner circle of followers to each of whom Freud presented an intaglio ring. Then came war and Freud, a loyal Austrian with two sons in the army, turned in on his own reflective resources. He also gave his first 'Introductory Lectures', in which he explained his ideas in the form of an adult education course.

Though long anticipated, and in many quarters ardently desired, war shattered emotions as well as bodies. Many observers portrayed war and the proletarian uprisings that followed as the outbreak of the animal, the instinctual and the irrational in human nature, the stripping away of a veneer of evolutionary progress and civilization. Their language also resonated with Christian belief in original sin. Freud responded with a powerful painting in the sombre tones of the unleashed unconscious. He delineated more precisely what he meant by the instincts and how he thought they relate to culture as the source of conscious moral values and civilized achievement, and he counted the cost in individual unhappiness, guilt and anxiety of the effort to maintain civilization. As his leading British supporter Ernest Jones (1879–1958) wrote: 'how well the facts of the war itself accord with Freud's view of the human mind as containing beneath the surface a body of imperfectly controlled and explosive forces which in their nature conflict with the standards of civilization'.[15] The results of Freud's reflections appeared in *Jenseits des Lustprinzips* (*Beyond the Pleasure Principle,* 1920), *Massenpsychologie und Ich-Analyse* (*Group Psychology and the Analysis of the Ego,* 1921) and *Das Ich und das Es* (*The Ego and the Id,* 1923).

These short books clarified a structural theory of the psyche. Freud described the id, the instincts or energizing power of the mind, as entirely unconscious in nature. The ego, which he held to be both conscious and unconscious, arises as a precarious and always tense integration of the instincts and the world in which one must live. The material realities and values of this wider world, imposed on the child through the mediation of her or his parents, form the mind's superego, which utilizes energy taken from the instincts and turns against the ego. The ego's assimilation of the superego is experienced as guilt, which is often a destructive process but also a source, through sublimation, of creative power. The human ego, in Freud's portrait, derives its power from redirected instincts; he described the instincts as at one and the same time the source of repression, neurosis and anxiety and the the energy for social existence and civilized values. Freud's large audience in the 1920s found in his work rich resources with which to understand how civilization may break down and irrational forces rule. Analysts close to Freud developed similar themes. Paul Federn (1871–1950), for example, who became acting head of the Vienna Psychoanalytic Society with the onset of Freud's illness in the 1920s, explained the political upheavals in Central Europe in 1918–19 by reference to 'the fatherless society', but in relation to these events he gave a positive description of a society of brothers that has overcome its old leaders.

Freud's contemplation of instinctual forces led him to a startling conclusion about their nature: they are both life-affirming and the source of sexual libido, *Eros*, but also destructive, life-denying and the source of death, *Thanatos* (a word sometimes used by Freud's followers at the end of the 1920s). Four factors help elucidate a dimension of Freud's work that readers then and later found counter-intuitive and controversial. Firstly, war and social upheaval gave Freud and his contemporaries an experience of the sheer scale of the human capacity for barbarity. Secondly, Freud interpreted the pleasure principle, which he accepted, as the outcome of an elemental desire to restore equilibrium. He thought that the ultimate equilibrium is the state of death: 'the aim of all life is death'.[16] Thirdly, Freud was puzzled in his clinical work by the tenacity of certain disorders, like obsessions, and he hoped to explain the self-destructiveness that this tenacity involved with an enriched conception of the instincts. Fourthly, Freud himself faced death – he feared for his sons in the war and for himself as he aged and suffered from cancer, and he lost a dearly-loved daughter, Sophie, and grandson.

Freud reached into the heart of myth as well as into science to write his books and the result, to many observers, resembled a religion. But he specifically argued the wish-fulfilling and essentially childish nature of religion, especially in *Die Zukunft einer Illusion* (*The Future of an Illusion,* 1927). In *Das Unbehagen in der Kultur* (1930; published in English as *Civilization and Its Discontents*) he described the tenuous condition of civilization, a state that is, he believed, always threatened by the repressed and sublimated powers that make it possible in the first place. We have a stark choice: to be civilized and unhappy or to destroy ourselves. The one hope, to which Freud was committed, is reason, the capacity to establish systematic, objective knowledge – i.e., science – which enables us to ameliorate or at least understand our condition. Through enlightenment, to which Freud believed psychoanalysis makes an exemplary contribution, men and women achieve dignity and a means to lessen suffering caused by ignorance, stupidity and superstition. In a memorable and oft-quoted admonition to an imagined patient, which Freud used about the ends of therapy: 'But you will be able to convince yourself that much will be gained if we succeed in transforming your hysterical misery into common unhappiness.'[17]

There was a personal dimension to this view of life. In 1923, Freud was operated on for cancer of the jaw and lived thereafter with a prosthesis and in daily pain. He refused analgesics and continued to smoke cigars in a conscious decision to preserve his alert mind – to pursue the dignity of enlightenment though faced by the prospect of death. After 1933 he witnessed the destruction of his books in Germany, for his work was reviled as 'Jewish science', but he refused to emigrate. With the *Anschluß* in 1938 he finally left Vienna, after negotiations and payments to the Nazis, and spent the last year of his life as a refugee in London. Freud himself did not comment much on the phenomenon of fascism, but other writers found in his work the terms with which to describe the politics of unreason and hate.

Freud was the revered founder of a movement that generated its own momentum and diversity. Two priorities for analysts were the attempt to deepen the therapeutic process and the need to train future analysts. Therapy, in the tradition of the hypnotic healers Freud visited in Nancy, involved a psychological relationship between therapist and patient. The Freudian analyst endeavoured to become the object of the emotions, whether of love or hate, of the repressed unconscious, and then sought to redirect those emotions on to a more appropriate person, occupation or mental activity. This placed a heavy burden on the analyst, who was required to be objective and unemotional whilst the object of the strongest emotions. The training therefore consisted of an analysis of the future analyst, to make it possible for her or him to have an objective self-understanding and hence the detachment required to be a therapist. Freud, uniquely and heroically, was thought to have begun the process with his self-analysis. Further, before there were formalized training procedures, he began to analyse his own youngest child, Anna, an extraordinary act given the nature of the repressed contents presupposed by his theories.

The issue that proved most divisive was the question whether a medical training should be required for anyone who wished to undertake an analytic training. Freud resisted this, and indeed many of his earliest and closest supporters – like Anna Freud herself (1895–1982) – were not medically qualified. His defence of non-medical therapy created major problems with the medical profession, the occupation to which he looked for recognition, and it raised – as mesmerism had done – the whole vexed issue of the legitimacy of non-medical forms of healing. In the United States, doctor-analysts imposed a medical requirement, with the consequence that when a refugee like the Berlin analyst Otto Fenichel (1898–1946) arrived in middle age, he had first to become a medical student in order to continue to practise. Questions of organization and training proved in practice to be inseparable from theoretical disputes. In spite of Freud's claim that he had established the foundations, fundamental disagreements continually surfaced, often with an emotional intensity that was a source of wonder to outsiders and one reason for the cliché that compared psychoanalysis to a religion.

The interwar period was creative for psychoanalysis, though the end result was not what Freud desired, as the many divergences from his views laid the basis for the post-World War II spread of a huge range of psychotherapies, many of which used watered-down analytic ideas. Disputes focused on therapy and training, areas in which Freud's informal methods, appropriate perhaps for a brilliant individual, were inadequate as a basis for large-scale organization. A member of the committee, Karl Abraham (1877–1925), a physician, had the abilities needed in an organizer, and he turned Berlin into a major centre for training alongside Vienna. Already by 1920, Abraham and his colleagues had established a polyclinic, which brought treatment and research together in a single centre. During the 1920s several Viennese and Hungarian analysts moved to Berlin and Anna Freud taught there in 1929. Karen Horney (1885–1952), Helene Deutsch and Melanie Klein (1882–1960) all practised in the city, though Klein left for London in 1926 in reaction against the rigidity of the methods used by her Berlin colleagues.

War turned Freud's and his followers' attention towards the social and political conditions in which people develop into adults. In Vienna and Berlin, young and sometimes radical analysts worked uncomfortably alongside a more conservative older generation. A significant political argument was constructed to link personal repression and misery to the conditions of labour imposed by capitalism, and there was an ambition to make analysis available to those who could not pay. Most explosively, Wilhelm Reich (1897–1957) who, though young, had an important position in analytic training seminars in Vienna during the 1920s, turned to analysis as a technique of sexual liberation. He argued that capitalist interests literally become embodied in the workers, as he thought could be seen in the repressed posture and personality of workers who feel guilt at the thought of bodily pleasure and hence accept poor material conditions. He concluded: 'The character structure, then, is the crystallization of the sociological process of a given epoch.'[18] Reich established clinics in working-class districts of Vienna, and later Berlin, to restore to workers their sexual desire and hence their desire to control their own lives. Earlier, Freud's close colleague Sándor Ferenczi (1873–1933) held a university position under the short-lived communist government in Budapest in 1919. During the 1920s, Ferenczi's therapeutic practice caused increasing disquiet among other analysts because he questioned the taboo on the analyst showing any emotional reaction to the person on the couch. The analyst's neutrality was thought to be the way to defend objectivity and propriety and hence to be crucial for the movement's public image. Ferenczi, however, called for an analytic responsiveness dangerously close to emotional involvement. Reich and Ferenczi, in their different ways, thus tilted the balance – always precarious – that psychoanalysts maintained between the pursuit of science and the pursuit of salvation and liberation.

Most of the early analysts were atheists. Nevertheless, hope that religion could recover its place as a form of spiritual healing caused some Christians to turn to Freud, as they were later to turn to Jung, in order to understand psychic needs and processes. Behind this was a deep concern among Protestant pastors, before and after World War I, that the churches had little to offer men and women in the modern world. One of Freud's most loyal supporters in Zurich was a pastor, Oskar Pfister (1873–1956), who believed that psychoanalysis shares a goal with religion as it seeks the sublimation of the natural instincts to higher ends, and he thought the techniques of psychoanalysis are valuable as a means to help achieve this goal. In The Netherlands, psychologically oriented pastoral concern created an environment receptive to new techniques, psychoanalysis included. Many psychotherapeutic interventions began to flourish under the rubric of 'healing', a word with both material and spiritual connotations.

In France, psychoanalysis had only a precarious position. The rapid collapse of Charcot's reputation with his death in 1893 left a lasting suspicion about hypnotism and suggestion. His onetime protégé, the neurologist Joseph Babinski (1857–1932), insisted that all psychological symptoms are caused by and can be cured by suggestion – an argument that rendered psychological symptoms of little medical interest and marginalized Freud's response to the neuroses. Janet's prominence as a medical proponent of the psychology of unconscious forces also had the effect that Freud's work was excluded from discussion. Parisian intellectual culture, in any case, was not inclined to look beyond itself and exhibited a degree of haughty isolation from the wider world. Nevertheless, a small but intense analytic community established itself during the interwar years. An interest in hysteria also continued to exist in the 1920s, focused on people like shell-shocked soldiers and the new woman, the woman who tried to escape her traditional roles. Most dramatically, the surrealists celebrated the unconscious in art, theatre, cinema and writing, and they assigned to the unconscious positive attributes precisely where polite society found negative ones.

Though Freud avoided active politics, he was not indifferent; he held scathing views about mass democracy and expected the mass of the people always to desire a leader. Whatever his own opinions, in the 1920s his work contributed to the rethinking of Marxist theory, the questioning of historical determinism, which followed the failed revolutions in Central Europe after 1918. Max Horkheimer (1895–1973) assembled a group of social theorists interested in psychoanalysis at the university of Frankfurt, a group subsequently famous and known as the Frankfurt School, which included Adorno and Herbert Marcuse (1898–1979) and had links with socialist Berlin analysts like Siegfried Bernfeld (1892–1953) and Erich Fromm. In a way that can be compared with Reich's arguments, though expressed in more sedate terms, these theorists found what Freud had written a means to link material conditions, social structure and the individual's psychological identity. They argued that earlier self-styled scientific Marxists had erred because they ignored the different psychological constitution of groups and hence the way in which political power, mediated by family and class, is recreated as the internalized world of the individual. Freud's theory of culture described sublimation of libido as the necessary means to a civilized existence. By contrast, 'the left Freudians' argued that sublimation, whatever else it might be, is a political process which reproduces the existing structures of power in society as part of the individual personality. These radical arguments were overtaken by events in the 1930s, but they had their day in the 1960s, when they were encapsulated in the slogan, 'the personal is political'.

Claims about the effect of social structure on individual psychic structure were particularly pointed in feminist critiques. Freud gave belated and then never fully committed consideration to female as opposed to male sexuality and to the position in the family of the female child. His work exemplifies what was then a convention of male discussions of sexuality: he took male nature as the norm. He described femininity as founded in the young girl's realization of her lack of a penis and consequent feelings of inferiority: 'The discovery that she is castrated is a turning-point in a girl's growth.'[19] With maturity, he argued, the focus of a woman's sexuality moves to the vagina and, when she desires and gives birth to a child (especially a male child), a woman finds satisfaction through a substitute for what she has previously not had. Already in the 1920s, Horney and Deutsch made a more positive evaluation of femininity and motherhood. Later, more radical critics followed the lead of Simone de Beauvoir (1908–86) who in 1949 described how the man 'is the Subject, he is the Absolute – she is the Other', encapsulating the essence of feminist criticism.[20] These critics launched a derisive rejection of Freud's male characterization of sexuality. It is therefore at first glance somewhat paradoxical that the practice of psychoanalysis in the interwar period and after offered special scope to women and that their contributions were substantial. Freud attracted and enjoyed the company of brilliant women, such as Andreas-Salomé, though his wife Martha ensured that their theories stopped at the nursery door. His consulting room also taught him women's secrets to an unusual extent, though it has been remarked that his patients responded to him as a man and thus confirmed the assumptions about sexual difference that he already held. He was less than clear when he commented that 'after all, the sexual life of adult women is a "dark continent" for psychology'; that is, like darkest Africa, it had never been explored.[21] When it came to generalizations about the social position of women, apart from the élite few, he was largely uninterested. Yet some later feminists found in psychoanalytic work a way to provide women with the terms in which to understand the reproduction of social power along lines of gender within the mind and the body.

Freud's clinical studies resulted in a theory of childhood, and teachers and child psychologists took analytic training or at least sought analytic insight. Psychoanalysis appeared to be a potent means to reveal the true nature of the child, to determine the role of parenting and to grapple with the irrationality of disturbed children. Since Vienna was a hotbed of educational experiment, there were cross-links between teaching, psychology and analysis. The analysis of children in itself required new techniques, since the talk of the analytic sessions, on which everything else pivoted, had to be adapted to the language of children. Anna Freud was an authority because of her innovative technique with children and her elaboration of the theory of defence, published as *Das Ich und die Abwehrmechanismen* (*The Ego and the Mechanisms of Defence*, 1936), as well as because of her unique personal position. When she moved with her father to London, she continued her work and cared for child victims of the Blitz.

When the Freuds arrived in London in 1938 they were met by a small but thriving analytic community, with Jones, a member of Freud's committee, the senior figure. As in Vienna, there were close connections with education and child welfare; Susan Isaacs, for example, who was head of the department of child development at the Institute of Education, undertook an analytic training. There was also an independent theory of child development, proposed by Klein, which emphasized the mother's role. Klein delved further into childhood than either Sigmund Freud or Anna Freud thought possible, and she used patterns of play to interpret how children relate to objects in their world. She decided that the child's attitude towards food is the best way to understand the earliest months. Her interpretations depended on a theory about how babies in the first weeks and months establish relations between their need for food and the object, the mother's breast, that satisfies it – or not, as the case may be. This basic contact, she argued, is the prototype of all later relations. Violent love and hate experienced at this stage, she believed, establishes unconscious patterns which the child projects into every kind of attitude as she or he grows up. 'The analysis of very young children has taught me that there is no instinctual urge, no anxiety situation, no mental process which does not involve objects, external or internal; in other words, object-relations are the *centre* of emotional life.'[22] Klein developed the creative tension between reason and unreason at the heart of psychoanalysis: she used a sophisticated study of human identity to reveal the most elementary irrationality. During the war, the followers of Anna Freud and of Klein engaged in an extended and bitter controversy, centring on training, which determined the future shape of analysis in Britain. The outcome was the existence of an orthodox Freudian school, a Kleinian group and an independent group. The Kleinians and independent analysts like Donald W. Winnicott (1896–1971) and W. R. D. Fairbairn (1899–1964) became known as object-relations theorists. Links between child welfare and psychoanalysis were reinforced by the institutional support of the Tavistock Clinic, where the work of Michael Balint and John Bowlby (1907–90) stressed the mother's role. Through the Clinic's activities, analytic thought spread widely among people in the management, social work and caring professions. The object-relations theorists' focus on mothering and on the way the early bond between mother and the child of both sexes, indeed between the breast and the child, led to a theory of personality and integrated psychoanalysis with the post-war political hope that welfare provision would achieve a civilized society.

Nazi power had dreadful consequences for the analytic community. Freud and many of his followers were, in Nazi terms, Jewish, and it was easy for the Nazis to brand the negative side of what analysts portrayed in human nature as the preoccupation of a peculiarly Jewish science. There was a large-scale migration of German, Austrian and Hungarian analysts, especially to the United States, and they had a marked impact on views about the explanation of social life in psychological terms. It is not true, however, that analysis disappeared in the Third

Reich. A cousin of Field Marshall Göring, Matthias Heinrich Göring (1875–1945), headed the renamed German Institute for Psychological Research and Psychotherapy in Berlin, dubbed the Göring Institute, and offered psychological treatment to the Nazi party élite. In this bizarre situation, an attempt was made to rephrase psychological concepts formulated to deal with neurosis as a language fit for heroes who remained in Berlin during the war.

Early psychoanalytic theory placed considerable weight on the instincts. When Freud elaborated his structural theory of the ego and the id, he attributed the driving power to the id. Public interest in psychoanalysis immediately before and after World War I reflected excitement about psychology's capacity to reveal in human nature a buried, primitive, sexually active and aggressive inheritance. The rapid popularization of analytic ideas was part of a wider reaction against Victorian culture and against hypocritical public denial of what was common private knowledge. In the words of a flapper in a play from 1927: 'Don't you even know, mother, that everybody's thoughts are obscenely vile? That's psychology.' A 1917 article, similarly from the United States, was entitled, 'Is the Psycho-analyzed Self a Libel on the Human Race?'[23] Though Freud did his best to distance his movement from such popularizations, his own stoic pessimism did little to dampen down conviction that a beast lurks within the human breast.

This emphasis on what was thought to be vile changed with the migration of analysts to the United States, especially through the work of Horney and Heinz Hartmann, along with the contributions of an American psychiatrist, Harry Stack Sullivan (1892–1949). They created what became known as ego psychology, a psychological theory that described the ego as an original mental structure with its own positive powers. They argued, with an optimism in tune with US meliorist values, that the psychic core of personality is a power with the capacity in a mature individual to integrate innate drives and social pressures in a genuinely self-fulfilling way. Horney's work built on a significant critique of Freud's view of femininity, and she created a positive image of the special, nurturing qualities of the female ego. Sullivan, whose theoretical disagreements with Horney divided analysts in the US and influenced the institutional growth of the field, endeavoured to develop analytic work with psychotics. Freud had concluded that psychoanalysis is impotent before psychosis, as psychotics lack the linguistic capacity necessary to the analytic process, but Sullivan suggested ways in which the therapist can engage with even the highly disturbed psyche and seek its integrative core. Hartmann argued, in *Ego Psychology and the Problem of Adaptation* (1939), that there is a 'conflict-free ego sphere' responsible for conscious psychological functions like perception, learning and memory, which performs the normal activity through which the self adapts to the social environment. This argument was influential in part because it made it possible to relate psychoanalysis to the academic experimental psychology of learning and adjustment. Hartmann also stressed social circumstances, which many critics thought Freud had ignored. 'We do not believe that one can completely explain the total behavior of an individual from his instincts and from his phantasies . . . one cannot under any circumstances ignore or neglect the part played by the economic or social structure as partially independent factors.'[24] This encouraged an attempt in the 1950s and 1960s to synthesize the psychoanalytic emphasis on psychological forces with the social science emphasis on social forces and to create a unified theory as a basis on which to improve people's lives. Hartmann made his own influential contribution in a paper on 'Psychoanalysis as a Scientific Theory' (1959). The analyst David Rapaport (1911–60) constructed natural-scientific models of the psyche, which compared mental and physical energies and encouraged attempts systematically to integrate analytic and non-analytic theories.

Another émigré psychoanalyst from Europe, Erik Erikson, also stressed the existence of an autonomous ego, a conflict-free area of the personality, which he approached through a study of the conditions that nurture or damage an emerging autonomy during children's lives. Erikson, who had Danish parents though he was the stepson of a German-Jewish paediatrician, trained as a child analyst with Anna Freud and then emigrated to the United States in the 1930s. His work, which gained a large public audience, expanded on the idea that there is a natural pattern of ego development to produce a theory of the life-cycle. He popularized the concepts of identity and of the adolescent identity crisis in *Childhood and Society* (1950). This book supported belief that there is a normal pattern of adolescent development that ends with a natural adjustment to society when the child achieves maturity. Erikson focused attention on society and on the conditions that foster the strength and integration of the young ego. By the time he published his book *Identity: Youth and Crisis* (1968), psychoanalytic and psychological approaches to development had become closely connected in the academic setting. In the public domain, psychoanalysis and psychology were connected by the rapidly growing interest in psychotherapy. All this work had psychological rather than medical authority. At the same time, US analysts and analytic ideas achieved a position of prominence within the medical speciality of psychiatry to a degree that happened nowhere else, a situation that lasted until the 1970s.

Psychoanalysis achieved prominence in the US in connection with scientific psychology, in the medical profession and as part of a culture of personal fulfilment. In each area, analytic theory and activity fitted in with what experts and lay people alike perceived as the need for personal development and adjustment. The ego psychologists' belief that there is a natural pattern in the self's development, which makes possible the integration of all aspects of the psyche, valued aggression as necessary for competition but rejected Freud's notion of the death instinct. The psychologists encouraged optimism that maladjustment is tractable through therapy. In popular versions – incompatible with the spirit of Freud's work but in keeping with US ideals – psychoanalysis was equated with a search for personal growth and for the true self, a fantasy of a personality independent of culture. This was one of the expressions of the 'me generation' of the 1970s and 1980s. There was enormous private middle-class prosperity during the interwar years and especially after 1945, and psychoanalysis sometimes became the means to supply that wealth with a soul. Though controlled by the medical profession, it nevertheless aimed at a full prescription for life. In the brash words of the influential physician and analyst Karl Menninger (1893–1990): 'There are two fundamentals in life. One is the business of making love, the other is the business of making a living.'[25] In *Man against Himself* (1938), Menninger acknowledged the presence of an aggressive instinct but described how the adjusted person turns it towards his or her own success and the social good.

The extraordinary spread of psychoanalytically-informed culture in North America was not without its critics. It was the butt of endless jokes and contempt from the populist heartland, helped by the concentration of analysts in the cities of the east and west coasts. There were also individual critics like Norman O. Brown (b. 1913), who in his book *Life against Death: The Psychoanalytic Theory of History* (1957) attacked ego psychology because he believed it disengaged from the tragedy of the human condition and the personal autonomy he considered to be fuelled by the death instinct. The Frankfurt School's views about the connections between capitalist culture and personal repression were also reworked in the US, especially in Marcuse's *Eros and Civilization* (1955), and this bore fruit in the 1960s in radical critiques of what psychologists and psychoanalysts conceived to be normal development. Lastly, it was argued that psychoanalysis is not a natural science at all but nevertheless a fundamental form of knowledge and crucial to human self-understanding. This challenge was most vocal in France, especially in the work of Jacques Lacan, who created considerable interest in psychoanalysis in France. But this challenge was part of debates that concerned the human sciences in general in the late twentieth century, and it is therefore discussed in the last chapter. It is necessary first to return to earlier divisions within the Freudian world.

iv *C. G. Jung and the Collective Unconscious*

Schism is a regular feature of life among marginal social groups, whether they are religious sects, radical political parties or analytic societies. It is no coincidence that splits occurred in the

Freudian community just when it changed its nature from being a local, Jewish group in Vienna to an international society with pretensions about medicine and culture in general. Freud thought he had discovered truths of major importance, and he therefore perceived the divergences of erstwhile colleagues as defections that endangered his enterprise; naturally, his accusations of disloyalty were deeply resented by those who pursued their own path. These ungrateful children, as Freud saw them, justifiably claimed that they had never fully accepted his views in the first place. Freud tolerated difference in the context of a small group; when it came to the presentation of psychoanalysis on a world stage, it was another matter.

Adler and Wilhelm Stekel (1868–1940) were two of the first members of Freud's group in 1902. Adler's departure in 1911 and Stekel's a little later were therefore major blows. Both men had backgrounds in general medicine; Stekel had an inexhaustible interest in many subjects, while Adler was probably attracted to Freud through a special interest in the neuroses. Committed to social medicine as part of welfare, he hoped through a life in medicine to contribute to the good of humanity in the broadest sense. He did not fully accept Freud's sexual theory of the neuroses and recognized the existence of an autonomous aggressive drive, and he linked the qualities of this drive to functional relations between bodily organs and to family and social relationships.

Few of Adler's views became public until after the break with Freud, when he published *Über den nervösen Charakter* (*The Nervous Character,* 1912), and more especially until after the war. He believed that psychic forces work in the interest of the unique individual, as if there is a distinct goal for every human being. When a person deviates from this goal, he thought, a neurosis is produced. He also developed views about the feeling of inferiority, which he claimed to be a major psychological force. This argument began life as an orthodox medical theory about how, in any particular person, one bodily organ may have a congenital weakness and hence be a likely site of pathology. Adler translated this physiological idea about the existence of inferior organs into a psychological theory that feelings of inferiority are the core of the neuroses and responsible for much ordinary motivation, for example, the desire to exert power over others. Neurotic people, he claimed, have an 'inferiority complex'. With his own feelings as the middle one of three brothers in mind, Adler linked strong feelings of inferiority to sibling rivalry; but he also linked such feelings to social position and to the lack of power experienced by employees and women. Adler worked in social and occupational medicine before he met Freud, and this interest continued to inform his psychological thought. Later, his psychology became a theory of pedagogy and his psychotherapy a concentrated means to educate people in community feeling.

During the 1920s, Adler developed his work into what he called individual psychology, a way to understand character, neurosis and aim in life as a consequence of the interaction of psychological forces and social environment. For Freud, this was just woolly thinking, caused by resistance to the unpleasant truths of the libido. The post-war educational reforms in Vienna provided Adler with a setting in which to propagate his ideas, especially after he became a professor at the pedagogical institute in 1924. He set up child guidance clinics and he was intensely committed to adult education. Later he emigrated to the US, where he had lectured widely in the 1920s, as he saw a future for his ideas there. He lectured and wrote in an accessible and practical way about the links between personal mastery and an individual's early experience, and this was something many ordinary people wanted to know about. His book *Menschenkenntnis* (1927; translated as *Understanding Human Nature*) conveyed this practical concern in its title. Indeed, the very practicality of Adler's notions meant that they became absorbed into North American culture and into ego psychology without much awareness of their source.

Freud was even more reluctant to admit that Jung held views irreconcilable with his own than he had been with Adler. Jung's final separation from Freud in 1913, after a relationship that reveals much about the character of both men, was fateful for the history of psychoanalysis. Jung was an energetic and imposing man, a generation younger than Freud, a Swiss citizen and not Jewish – as Freud 'nearly said . . . it was only by [Jung's] appearance on the scene that psycho-analysis escaped the danger of becoming a Jewish national affair'.[26] He was a trained psychiatrist who, when he met Freud, worked at the internationally known Burghölzli institution, directed by Eugen Bleuler (1857–1939), in Zurich. Jung was Freud's chosen heir-apparent, appointed – in spite of Adler's seniority – the first president of the International Psychoanalytic Association. Both men wrote self-serving accounts of their relationship; with the benefit of hindsight and access to their correspondence, it is possible to see that there always were different agendas beneath their apparently close friendship.

Jung was born into a prominent and cultured Basel family. One grandfather was a physician, the other a theologian who was said to have had visions, and his father was a Protestant pastor whom the son, at least, believed had religious doubts. A distinctive mixture of the medical and the spiritual, a mixture that Jung and his followers claimed makes possible true psychological knowledge, entered all his work. As he later said: 'A psychology that satisfies the intellect alone can never be practical, for the totality of the psyche can never be grasped by intellect alone . . . the psyche seeks an expression that will embrace its total nature.'[27] While still a medical student, in the last years of the nineteenth century, he worked with his younger cousin Hélène Preiswerk, who was a medium. In 1900 he took an appointment at the Burghölzli where he soon became clinical director, managed the outpatient service, which involved hypnotic therapy, and began to teach at the university in Zurich. In 1906 he published the results of word association tests and this spread the notion of a psychological 'complex' or cluster of emotively linked ideas. He also began to correspond with Freud after he was inspired, like his Zurich colleagues, by *The Interpretation of Dreams*.

Their extensive letters and intense meetings reveal the stimulus Jung gained from Freud, to an extent a father-figure, and the deep hopes Freud invested in Jung as an intellectual son and hero who would carry psychoanalysis to the international medical profession and the wider world. Jung, though the younger of the two men, came to the relationship with a strong sense of his own personal and professional worth, with a belief that he and Freud were pooling resources in an empirical programme to understand the neuroses, and with his own speculations about the psyche. He made it plain that Freud's stress on the libido as entirely sexual was never fully acceptable, though Freud believed that he would eventually find the sexual theory forced on him by the evidence. Jung used the language of will or life force to describe libido or psychic energy, in contrast to Freud's language which often described the libido as analogous to a physical power.

By 1911 even Freud had to face up to the continued divergence of view, and in 1913 the rift became complete. Jung meanwhile had begun to devote his clinical time to private practice. His work with psychotic patients in the hospital and with wealthy neurotic clients impressed him with the power and constancy of symbols produced by the mind as the expression of emotional pressures and also found in dreams and fantasies. He conceived these symbols to be representations of the positive goals of the psyche – of its intrinsic constructive as well as destructive nature – and opposed Freud's view that they represent the negative consequences of sexual repression. Whereas Freud thought deterministically about the causal formation of unconscious contents, Jung thought about the psyche in terms of purposive forces known to consciousness in symbols and images. Freud went through a creative illness, a period of neurotic exploration of his own unconscious, and Jung undertook the same journey, in his case with danger to his own health and with emotional damage to others. Freud's route involved free association with the material produced by his dreams, Jung's encouragement to his inner self to produce visions and symbols. In his own imagery, Jung tried to construct shafts and tunnels into the depths of his mind. Out of this dangerous experience, with painful consequences for his wife and other women, Jung emerged as a mythic hero, a bearer of light from the dark world of the human unconscious: the comparison with myth is Jung's own.

Jung wrote his major study, *Wandlungen und Symbole der Libido* (publ. first as articles then as a book, 1911–12; translated as *The*

Symbols of Transformation; originally published in English as *The Psychology of the Unconscious*), while still in touch with Freud. It was only as he ended the book that both parties had to confront the reality of Jung's disbelief in and not just indecision about the sexual theory of the neuroses. Jung thought that the theory 'cherishes the belief that the essential cause of all disturbances is sexuality . . . [and shows] a confusion here between cause and effect'.[28] His book had all the excitement and disorder of a creative mind in full flow; in it, as the psychiatrist and historian Henri Ellenberger wrote, 'Jung devoted more than four hundred pages to a mythological interpretation of a few daydreams and fantasies of a person he had never met'.[29] He used reports published by Flournoy about a young American woman, known as Miss Frank Miller, who produced rich dream and fantasy images. In an interpretation of this material in which he drew on myth, philology and religious experience, Jung introduced a new language for the dynamic unconscious powers or libido. He did not deny the value of Freud's insights into repression but, all the same, he painted a picture of the unconscious with its own archaic, pre-individual structure. After the book, he set out on his journey to discover these inherited powers in his own psyche. The result was the theory of the collective unconscious, which Jung made the basis of a systematic theory of personality in *Psychologische Typen* (*Psychological Types*, 1921). In this book he introduced the categories of introvert and extravert to describe whether the sources of motivation derive primarily from inside or outside the psyche. These terms were subsequently taken up by the psychologists Spearman and Eysenck in Britain who were interested in personality tests.

The foundation of human nature, in Jung's view, is an inherited psychic energy, unconscious but continuously active in a purposeful manner: there is 'a common psychic substrate of a suprapersonal nature which is present in every one of us'.[30] This energy is the source of intuitions that have no words; but it also creates a language through symbols. References found in different cultures to the wind as 'the breath of life' and as spirit, for example, indicates – in Jung's view – the collective nature of the unconscious. He discerned recurrent patterns in the ways symbols are used and found these patterns in myths and stories, as when Orpheus travels to the world of the dead in search of the true prize. He discussed these patterns as archetypes or permanent structures of the collective unconscious. During the 1920s, Jung devoted considerable energy to non-Western cultures – for example, he visited the Pueblo Indians of New Mexico – and to Eastern religions, mythology and alchemy, in order to demonstrate the shared archetypical structure of the psyche. He also collaborated with the leading German sinologist and translator of the *I-Ching* or Chinese book of changes, Richard Wilhelm.

Jung resigned his university position at about the time he separated from Freud to give himself the freedom of private practice. Much of his work, which was often with wealthy and cultured patients, pointed to a split between the conscious realm with its ethical motives, respectable emotions and rational knowledge, and the unconscious realm, which is amoral, asocial and non-rational. This split, he thought, emerges symbolically as pathological symptoms, in dreams and in our glimpses of unachieved ideals beyond the surface of life. His approach to therapy followed from this. 'The pathological element . . . [lies] in the dissociation of consciousness that can no longer control the unconscious. In all cases of dissociation it is therefore necessary to integrate the unconscious into consciousness. This is a synthetic process which I have termed the "individuation process".'[31] His belief in the possibility of integration supported an optimistic way of thought that was attractive in the confused years before and after World War II. Jung, however, stressed that the unconscious energies are irrational and in many respects destructive and threatening; their successful integration into the modern mind can by no means be assumed. Jung, like Freud, in his work reflected a sense of imperilled civilization, and in an age of war there was an audience ready to believe in the tenuous nature of reason compared with the undiminished powers of the primitive psyche. Whereas Freud offered his readers a stoic message that scientific reason might lessen the intensity of suffering, Jung discussed the positive power of the collective unconscious to heal and transform.

A striking number of people attracted by Jung's work were educated, professional and middle-aged – outwardly successful but inwardly without direction. For Jung, the existence of such people supported a generalization, that this sense of split between outward and inward life provides a diagnosis of Western culture. The West, he suggested, is preoccupied by rational-scientific knowledge and material mastery, but it lacks collective purpose and fails to give individual spiritual satisfaction. The culture as a whole is unintegrated with the collective unconscious. Jung moved on from his role as a young hero who journeys into the underworld, to become a wise old man in a world eager for prophets. He viewed political events in the 1930s in terms of a whole civilization's denial of unconscious forces. As a German-speaking Swiss citizen, sympathetic to German culture but not directly vulnerable to German politics, he was slow to appreciate the nature of Nazi power. It seemed to Jung that Hitler's appeal resulted from the dictator's genuine responsiveness to unfulfilled interests of the unconscious, interests unfulfilled because of the failure of post-Reformation Europe to achieve a collective symbolic life that gives expression to spiritual needs. The result, he feared, is that the European mind is overpowered by dark but irresistible forces. Jung himself viewed such discussions as objective contributions to the understanding of the modern psyche, that is, to psychology. Given the political events of the 1930s, however, they have sometimes been represented as sympathetic to National Socialism, and debate on the issue is not closed. This emotive matter is further confused in two ways. Firstly, Jung described inherited unconscious powers and believed that their particular character reflects the history of a people or a race; it therefore seemed natural and objective to him, as it did to many of his contemporaries – including some Zionists – to characterize the psyche in racial terms. He thought there is a Jewish psyche structured by what, like many writers before him, he described as the rootless culture of the Jewish people. He assumed also that differences between inherited Jewishness, Swissness, Germanness or whatever pointed to the need for different psychotherapy for different people. In the 1930s, whatever Jung's stated intention to be objective, all this was to play with political fire. Secondly, the Nazification of German institutions in April 1933 resulted in the forced resignation of Ernst Kretschmer (1888–1964) as president of the German Society for Psychotherapy and that of Jewish members. Jung, the vice-president, in response took over as president of a newly constituted International Society for Psychotherapy. His supporters argued that this put him in the best position to help those expelled and to preserve therapeutic ideals in the German-speaking world; his detractors argued that it betokened his willingness to work alongside National Socialism and its racial policy. Jung faced a moral dilemma; but we may conclude, at the very least, that a man who dismissed political activity as superficial in comparison to psychological understanding, was ill-equipped to comprehend the political nature of his role. A stronger view is that he was slow and reluctant to distance himself from belief in the politics of racial stereotypes which appeared to him to express an inner reality.

Jung, like Freud, described his work as science and claimed to have discovered objective truths about human nature. Biological and mechanist notions ran through Freud's work, often in tension with the psychological language; Jung's thought was from the beginning freer, with a non-mechanist language about mental purposes. Jung accepted religious belief as a natural human expression of archetypal psychological forces, which, some readers thought, opened his theory to a religious interpretation; Freud was an atheist. Both Freud and Jung made extremely speculative use of belief in the inheritance of acquired characteristics. This belief, commonplace in 1900, became unacceptable during the 1920s, though more slowly in the German-speaking world than elsewhere. The idea of psychic inheritance simply lay outside the new science of biological genetics. Jung's orientation was shared by other so-called philosophers of life, though Jung did not develop a systematic philosophy as a context for his theory of psychic inheritance. His work acquired authority because of the insight and order he brought to clinical material, to comparative mythology and to seekers of an integrated and fulfilled subjective life, not for a contribution to natural science. Indeed, for most of Jung's psychotherapeutic followers and his large public audience, the question whether his work was scientific was beside the point.

This situation indicates once again the divergence in the twentieth century between academic and public expectations about what psychology is for. It also illustrates again the conflict in Western culture between scientifically authenticated knowledge and the authority of personal or experiential meaning. Jung confirmed public belief that personal meanings are significant, especially after the publication of his posthumous, widely-read autobiographical essays (to which his long-time personal secretary Aniela Jaffe substantially contributed), *Erinnerungen, Träume, Gedanken* (*Memories, Dreams, Reflections*, 1962).

This chapter started with the apparent paradox that Freud was an Enlightenment scientist who, through reason, brought the Enlightenment to a close. While he and – much more – his followers hoped that truth moderates suffering, outward events confirmed that progress cannot be taken for granted. Jung, because he attributed a creative power to the unconscious, appeared a more optimistic writer; yet he too stressed the dark side of human nature, and he did so in a way that presented Enlightenment reason as the symptom rather than the cure of modern woes. Psychoanalytic work faced reason by its limitations. Jung took religious experience seriously as an irreducible human need, whereas Freud regarded faith as a regression to childhood and primitive fantasy about the omnipotent father. Many readers thought Jung integrated modern psychology and religious faith and that he thus overcame the crisis in human existence attributed to the rise of the scientific worldview and provided psychological evidence that modern people can believe in the existence of eternal values in the human spirit. Jung perhaps enabled ordinary readers and religious believers to project what they wanted to find into his work, but he was hardly a Christian psychologist. What he did do, as Freud also did if his work is read as an interpretive human science, was give psychological language the richness, and independence from mechanist conceptions, which made it possible for healers and those in need of healing to respond to human needs. This language of healing was neither material nor spiritual: it was psychological.

Freud, initially when he projected the Oedipus complex on to human prehistory and then – confronted by war – in his examination of the ego, began to consider the individual as a social being. His followers, like Federn, the leader of the Vienna Psychoanalytic Society in the 1920s, went further, linked individual and social repression and sought a psychoanalytic key to society's failures. Adler and others associated with Freud took the same route. Jung, by contrast, was avowedly uninterested in politics and regarded all action that does not originate with the transformation of the individual as superficial. Yet he also, with sad consequences for his reputation, projected psychological categories on to the social and political stage and treated social events as the outcome of psychological forces. Yet this elevation of 'the psychological' at the expense of 'the social' was not restricted to any one form of psychology. Psychology in the US achieved institutional and financial support because it offered individualized adjustment and control rather than a sustained analysis of personal experience and character in relation to social structures. Twentieth-century psychology, both analytic and non-analytic, had an awkward relationship with the social dimension of human existence.

19

The Individual and the Social

> Social psychology began to flourish soon after the First World War. This event, followed by the spread of Communism, by the great depression of the 1930's, by the rise of Hitler, the genocide of the Jews, race riots, the Second World War and the atomic threat, stimulated all branches of social science. A special challenge fell to social psychology. The question was asked: How is it possible to preserve the values of freedom and individual rights under conditions of mounting social strain and regimentation? Can science help provide an answer?
>
> Gordon W. Allport, 'The Historical Background of Modern Social Psychology' (1954)[1]

i *The Individual and the Crowd*

Specialization is visible as the human science disciplines developed in the late nineteenth century. Disputes about the proper methods and content of knowledge reappeared as arguments about where the boundary between different disciplines lies and the autonomy of each discipline. All this is conspicuous in discussion about the relation between psychology and sociology. Earlier chapters described the separate establishment of the two disciplines; yet, given that individuals live in societies and societies consist of individuals, there was no self-evident boundary between their respective subject matters. This was a sensitive matter because each discipline was vulnerable to the criticism that its basic concepts – like 'individual' or 'society' – were poorly conceived, and that neither really had the intellectual independence it claimed for itself. This chapter discusses social psychology, the area where psychology and sociology most obviously met and where the question of the individual's relation to society had most obviously to be confronted. The history of social psychology is a microcosm of larger intellectual difficulties which faced the human sciences.

The way individual life is conceived in relation to collective existence is fundamentally a political and moral as well as scientific subject. The history of social psychology is inseparable from much of the political history of the twentieth century and from argument about power, justice, freedom and obligation. Given the common identity of the subject matter of social psychology, the social person and the political person, social psychology has not been politically neutral. The history of the area forces us again to confront the question whether it is either possible or desirable to pursue a non-evaluative science of human beings. This connection between social psychology and politics was most obvious in Soviet history, which is discussed in the chapter's concluding section, though it applied equally elsewhere. It was no coincidence that the social sciences expanded rapidly in the English-speaking world and in Western Europe in a reform-minded political context from the 1940s to the 1960s. The US psychologist Gordon W. Allport (1897–1967), the author of a standard essay on the history of social psychology, believed, for instance, that: 'Whether social science, under proper ethical guidance, can eventually reduce or eliminate the cultural lag [between material and socio-political progress] may well be a question upon which the human destiny depends.'[2] It is hard to imagine a more political judgement about science. Similar sentiments were also forcibly expressed at this time by Skinner.

The history of social psychology has most often been conceived in terms of the emergence of a specialist area within the overall discipline of psychology. This definition prejudges the way to think about the individual's relation to society. If, as some theorists argued, it is not possible even to formulate a concept of the individual independently of social categories and, conversely, if it is not possible to describe social existence without presuppositions about the physical and moral autonomy of human beings, then social psychology might be thought, in principle, to be the general discipline and the foundation of specialist disciplines like psychology and sociology. This question of how to think about the individual–society phenomenon was shared by psychology and sociology in common with anthropology, economic and political science, history, linguistics and sociology, and psychology was by no means always the dominant science.

Social psychology existed as two separate specialities, one located in psychology and the other in sociology. In North America, these two specialities, with the same name but with different personnel, institutional support, methods and content, existed side by side for most of the century – latterly distinguished by the appellations 'psychological social psychology' and 'sociological social psychology'. The psychological version of the discipline generally took natural science as a model, and so became a laboratory science focused on the individual which explained outward social relations by reference to inner states like motivation or attitude. The sociological version, by contrast, was more open to the non-natural science disciplines, focused more on collective entities like culture or class, and sought explanations in the linguistic and symbolic structures of historically constituted society. In some settings outside North America, as in Sweden after 1945, social psychology developed as a speciality first within sociology. In France, perhaps the most interesting work on the individual and society was done by mid-century historians, the *Annales* group, who envisaged a historical social psychology of which the subject matter is shared *mentalité* or belief about the basic dimensions of human existence. The history of social psychology is therefore more a history of the contingent boundaries between disciplines and the contingent division of intellectual labour than of the construction of a speciality in its own right.

The history of social psychology, if it is to do justice to thought about the social individual, becomes a very complex story indeed. I focus on psychological social psychology and on the way 'the social' in human life was understood by psychologists. But that is only part of the picture. The following sections discuss, firstly, the central position of the crowd in late nineteenth-century imagination about relations between people, and secondly, the interest in instincts as the foundation of social life. These attitudes provoked a reaction, discussed in the third section, which informed experimental social psychology in the United States, and this psychology was influential in Western Europe after 1945. In the last section I consider the special history of psychology in the Soviet Union, the state that claimed a unique, objective understanding of the social nature of people.

Values and facts of nature were intertwined in the nineteenth-century background. A liberal political economist like John Stuart Mill as well as the German theorists of the *Volksgeist*, studied human social nature, as it had been studied in the eighteenth century, as a basis for the right management of public affairs. This intimate connection of facts and values is particularly colourful in a late nineteenth-century cluster of writings about crowd psychology, predominantly French and Italian. Crowd psychology flourished in the 1890s and directly preceded a self-conscious attempt to formulate social psychology. At the same time, Durkheim set up sociology on an independent intellectual basis, and to do this he formulated social categories that were supposedly free of psychological content. The contrast and conflict between crowd psychology and Durkheim's sociology over the explanatory role of psychological as opposed to social categories particularized the general question of the relation between knowledge of the individual and knowledge of the social. Durkheim's arguments remained a reference point for later sociologists who critically evaluated the limits of a psychological approach to human relations.

French observers saw the crowd as an unpredictable but powerful force in politics. Taine wrote a celebrated history of his country, rich with feeling about the crowd in relation to the events of 1789 and its Napoleonic aftermath. More immediate was the experience of the Paris commune in 1871 and the violent attacks on order and property, followed by a bloodbath as the communards unsuccessfully resisted the restoration of government. An image of 1871 symbolized a loss of control, personal and social. Horrified bourgeois vividly pictured women from the commune, *les pétroleuses*, who were supposed to have burned their city rather than surrender. Taine, followed in this respect by Emile Zola's novel *Germinal* (1886), portrayed the crowd as an unleashed beast made up of people who have lost control and moral restraint and instead act automatically like others around them. The crowd continued to be a visible presence in turbulent French politics – in strikes, May Day demonstrations, the brief but meteoric rise to political prominence of General Boulanger and in the extended Dreyfus affair (1894–1906), which divided the country over anti-Semitism and the meaning of nationalism. Behind these events lay an indeterminate fear of the vast cities of Paris and London, the trauma of lives affected by urbanization in Europe and the United States and aspirations for democratic government – seen by many as government by the crowd. The sharpened focus on people as members of a mass accompanied the transformations that created mass society, and the outcome was enthusiasm for a human science that claimed to reduce the mass to order. Taine did not just wring his hands in horror but argued in favour of support for moral and political science so that public life could be efficiently administered. His friend and fellow historian Emile Boutmy (1835–1906) commented that 'it was the University of Berlin that triumphed at Sadowa' (where the Prussians defeated the French in 1870, events followed by the Paris uprising and the commune), and he implied that France's future depended on the education of its people, 'to give the people back their heads'.[3] Taine introduced British empirical psychology in a book on intelligence in 1870 and, with Boutmy, founded a school for political science as an institution that would follow the example of Mill and Spencer in England and found politics on the science of human nature. The belief spread that the systematic study of rational action, when the individual acts as an individual, and of irrational action, when an individual is subsumed in a crowd, is of utmost importance to modern mass politics.

There was also a literature in Italy on the crowd, though authors like Scipio Sighele (1868–1913) were liberal-radical rather than liberal-conservative in politics. Lombroso, Enrico Ferri (1856–1929) and other doctors and lawyers wanted to recreate Italy as a modern state after its unification in 1870, and they believed that to achieve this it was necessary to study the biological and material condition of Italy's people and hence of its problems. In the wake of Lombroso's book on *L'uomo delinquente (The Criminal Man,* 1876), this group, centred on the Turin and Bologna law schools, advocated legal reform on scientific lines. The lawyers argued that retributive justice should be replaced by a science of the criminal type, and that such types should be treated as a biological not a moral problem. Lombroso and the theorists of the crowd divided, however, in their science: Lombroso accepted physical determinism while the crowd theorists attributed crime to a person's situation in relation to other people en masse. In *La folla delinquente (The Criminal Crowd,* 1891), Ferri's student Sighele argued that as part of a crowd a person turns into a criminal type and hence the crowd itself can become a criminal entity, though he did not imply that crowds are necessarily criminal. In the background there was an experience of agricultural unrest and strikes by labourers, and the Italian lawyers (unlike the French theorists) even cited the crowd as an extenuating circumstance for rioters when they appeared in court: the individual as a member of a crowd, they argued, is less responsible. Social rather than jurisprudential writers then undertook a more systematic comparative study of different social groups, and, besides dangerous groups, they studied those that they viewed positively as the expression of the people's collective life.

The French and Italian work, like that of the British political economists, began from the premise that the individual is the key to knowledge of social disruptions like crime, strikes and political upheaval. It was argued that when individuals assemble they are more likely to act in the way other people act than to make an autonomous choice; this creates the appearance of a group or crowd mind. The contrast drawn between the two types of action was highly evaluative – superior individual moral choice was contrasted with inferior collective conduct. Two French authors, Gustave Le Bon (1841–1931) and Gabriel Tarde (1843–1904), enriched the psychological content of this analysis during the 1890s, Le Bon in his hugely popular book *The Crowd* and Tarde, somewhat more academically, in *The Laws of Imitation*. Le Bon and Tarde were influenced by the public debate in the 1880s between Charcot's Paris school and the Nancy school over the subject of hypnosis. This debate made everyone familiar with suggestion, the supposed power of external forces or people to influence a person's action without that person's awareness. At times the idea of suggestibility became a key metaphor for the modern urban man or woman, the person buffeted back and forth by the bedlam of the city. In striking and open contrast to Le Bon and Tarde, during the 1890s Durkheim developed an approach to sociological analysis that aimed to exclude individual psychology and to treat collective life, such as shared values, as a social reality in its own right, not to be explained as the sum of individual realities.

Le Bon was eccentric and egocentric, a misogynist and rabidly élitist in politics, but all the same one of the most important popularizers of science in France. While there was little originality in his work, his formulation of crowd psychology became the best known. For Le Bon, the crowd represents everything inferior in the modern age:

> It will be remarked that among the special characteristics of crowds there are several – such as impulsiveness, irritability, incapacity to reason, the absence of judgment and of the critical spirit, the exaggeration of the sentiments, and others besides – which are almost always observed in beings belonging to inferior forms of evolution – women, savages and children, for instance.

He absorbed work by Taine, Sighele and the French psychologist of evolutionary progress and dissolution Ribot, and he reflected contemporary fascination with hypnosis. The outcome was a description of the individual in a crowd as a person who is abnormally suggestible. The individual en masse, he argued, loses her or his capacity to chose. 'All depends on the nature of the suggestion to which the crowd is exposed.'[4] Crowds are like women, constitutionally inferior and easily led. This explains the power of leaders and the fickleness of crowds, just as it explains the power of men over women. The outstanding leader can sway the crowd. Contemporary politicians noted with interest that the crowds who excitedly called for General Boulanger to take power in the mid-1880s were unaware that he had been launched into prominence by a small group of wealthy tactitians. The Paris newspaper *Le Figaro* wrote: 'Instinctively, the crowd acclaims the man which it feels in possession of a strong will . . .'[5] Observers also noted that the radical May Day crowds which terrified conservatives at the beginning of the 1890s were, within a few years, baying against the liberal Dreyfusards. These observers assumed that the same proletarians made up the crowds and followed whatever suggestion was the order of the day, though historians doubt whether this was true.

Out of such experiences and prejudices, Le Bon constructed a crowd psychology that, he believed, enables the astute leader to mobilize mass support. He turned an élitist fear of mass politics – 'The populace is sovereign, and the tide of barbarism mounts'[6] – into an opportunity: since the mental quality of people is inferior en masse, the superior man who stands outside can achieve control. His work was translated into Romanian, Swedish, Turkish, Japanese and other languages, and he became a correspondent of the élite and powerful. After Mussolini led the fascists to power in Italy in 1922, Le Bon sent him autographed copies of his works, which the leader claimed to have read already. Even before 1914, a down and out would-be painter in Vienna, Adolf Hitler, perhaps absorbed Le Bon's message. Le Bon helped shift the perception of the crowd from uncontrollable rabble to disciplined mass. When, in his grandiose manner, he ran a salon in Paris at the beginning of the century – he was held in almost total disregard by the academic establishment – he gained an audience in the higher reaches of the French army, where officers were interested in the means by which a mass army could be made to retain its order and effectiveness.

Le Bon generalized his ideas to cover any kind of social gathering, such as parliament, not just the Paris street crowd. The quality of rational intellect declines, he thought, wherever men attempt to reach collective decisions. He regarded socialism as fundamentally incompatible with the natural conditions for rational decision-making and he wrote virulently against it. By contrast, Tarde, though also conservative in politics, developed the general psychological arguments with more subtlety and greater liberality of purpose. Impressed like Le Bon by the evidence of hypnotic suggestion, Tarde elevated imitation into the basic principle of social relations. He attempted to understand social life by the attribution to people of a psychological capacity to imitate others that leads to a shared social existence. In his later work, he focused on the idea of the public, understood 'as a purely spiritual collectivity, a dispersion of individuals who are physically separated and whose cohesion is entirely mental', and he then discussed the formation of the public as a process of interaction rather than imitation.[7]

Tarde's interest in social issues began while he was a provincial judge in the Périgord (Dordogne). His work on criminology led to an appointment in Paris in 1894, when he became head of the statistical department of the Ministry of Justice. His best-known book, *The Laws of Imitation*, developed out of his interest in the explanation of abnormal or criminal conduct, by individuals or crowds, but, like Le Bon, he saw that his explanation could be generalized as a psychology of how people behave in groups. He summarized his position: 'Society is imitation, and imitation is a kind of somnambulism.' In *L'opinion et la foule (Opinion and the Crowd*, 1901), however, he defined social psychology as the study of 'the mutual relations between minds, their unilateral and reciprocal influences', and he detached it from preoccupation with crowds and distanced it from vague notions of a collective mind.[8] Instead, he suggested the value of investigations of 'the public'. Though Tarde discussed psychological factors, he always had in mind the achievement of a social science. He defended his position against his contemporaries on two fronts. In one direction he argued against the biological reductionism common in his day, the belief that social structure and social relations result from biological laws. He thus opposed both racial theories and the evolutionary sociology of Spencer. By the end of the 1890s Tarde was also compelled to defend his views in the other direction, against Durkheim's sociology. Durkheim accused Tarde of psychological reductionism, the explanation of what is truly social by psychological factors, and especially disliked the explanation of social solidarity by the theory of imitation. The historical outcome was that Durkheim's approach rather than Tarde's acquired an institutional position in France. All the same, Tarde initiated suggestive work on social interaction and on the formation of public opinion, and Durkheim, in spite of his intentions, in fact made implicit assumptions about individual motivation.

Durkheim succeeded in his bid to secure the academic position of an autonomous sociology but, as the popularity of Le Bon's *The Crowd* attests, the notion that individual qualities explain social relations had a wider audience. Public belief was also influenced by the evolutionary outlook and the assumption that biological inheritance, embedded in instinct, race and sex, as well as in individual differences, profoundly affects the public sphere and a person's place in it. At the same time, less biological and more rarified notions of inherited psychic structures were also common. In this context, there were ubiquitous – if vague – references to a collective mind. It was a severe intellectual challenge to say something precise about men and women as social beings, which required an academic division of labour, without an artificial or incoherent division of the subject matter. Writers were tempted to allow notions like the crowd, the group mind, imitation or instinct, which themselves required explanation, to serve a shallow explanatory function.

ii *Instincts and Group Psychology*

Evolutionary theories of human nature and the powerful values associated with race, sex, empire and élite rule reinforced each other in the late nineteenth century. Many theorists equated the historical order of evolution with an existing social order and a desired hierarchy of social relations. Sometimes the arguments were crude: Le Bon, for example, equated children, savages and women as primitive in an evolutionary sense. There were, however, more sophisticated theories that claimed to understand human nature by reference to the evolution of instincts or innate psychic structures. The range of such theories is illustrated by the work of J. M. Baldwin, William McDougall (1871–1938), Freud and Wundt. When an experimental social psychology developed in the 1920s, the step often regarded as the beginnings of the modern speciality, this happened as a reaction against the theories, discussed in this section, that emphasized instinct and innate psychic structures. Anthropologists in the same period reacted in a parallel way against what they believed to be the inherently speculative nature of evolutionary approaches to society and engaged instead in empirical field studies of non-Western societies, like Malinowski's on the Trobriand Islanders.

Baldwin, one of the founders of the psychology discipline in North America, brought elements of German, French and British work together in his studies of mental development, especially in *Mental Development in the Child and the Race* (1894). His approach reconciled the demand placed on North American psychology that it be both scientific and practical. Research on mental evolution and development dealt with the question of how children become educated and functional members of society. Baldwin praised Tarde's laws of imitation, which he claimed to have understood independently, as basic to knowledge of the process of socialization. He distinguished three main stages in early development: a passive stage in which children receive images; an active stage in which they assume the character of what they perceive – they imitate like 'veritable copying machine[s]'; and a more mature stage in which they comprehend the world separate from themselves.[9] This description clearly related individual psychological growth to the process by which an individual becomes a social person. The self is formed in the image of others, Baldwin argued, and the image of others is formed in the light of the development of consciousness. Though he did not found a research tradition, his work was one source for the later very influential studies of development by Piaget. Baldwin's career ended abruptly, however, in 1909 when he resigned from Johns Hopkins University, apparently because he was named in a police raid on a brothel, and he went to live in Mexico.

Baldwin's work was assimilated in an academic world concerned with education and social reform, processes understood as the socialization of the individual. A professor of social science and social reformer who spent most of his career at the university of Wisconsin, Edward A. Ross (1866–1951), published widely-read books on *Social Control* (1901) and on *Social Psychology* (1908), which reflected current views on the crowd and on imitation. Both books advanced the opinion that there should be a social science speciality concerned with individual relations. These books also exhibited the characteristic aspiration of the Progressive Era that the social sciences should help to restructure industrial, urban and mass society. There was no sharp division between sociology and psychology in this connection since psychological laws were perceived to be relevant to social relations. Ross himself referred to 'psycho-sociology'. The absence of a clear intellectual boundary between the disciplines was also evident in the work of earlier sociologists like Giddings and in many later texts on social psychology written by sociologists as well as psychologists.

The psychological approach to social relations was strong in Britain as well as in the US, especially in the work of McDougall. He became interested in psychology as a medical student and then at the university of Cambridge, where he took part in Haddon's expedition to study the primitive mind in the Torres Straits Islands, before he obtained an appointment at Oxford. Frustrated by conditions and attitudes there, he left for the United States in 1920. 'The scientists suspected me of being a metaphysician; and the philosophers regarded me as representing an impossible and non-existent branch of science.'[10] Later, after he also failed to settle in the US, he moved from Harvard to Duke University, and he enabled J. B. Rhine to launch experimental parapsychology. While at Oxford, McDougall published *An Introduction to Social Psychology* (1908), one of the best-selling English-language psychological texts of the interwar years. During the 1920s he maintained that he had established a systematic psychology, which he called hormic psychology, but though he had a large public readership he was isolated in academic circles and his ideas were more characteristic of the pre-1914 period. McDougall's work amounted to a philosophy of nature rather than a narrowly-based psychology, and he developed an evolutionary philosophy of nature that fully incorporated man within the natural order as a basis for ethics and social science. He believed mechanist arguments cannot explain the evolutionary process and he argued instead for animism or belief that change in nature comes from mental forces. This philosophical position underpinned McDougall's psychological claim that experience and conduct are fundamentally determined by instincts or innate predispositions. He elaborated a theory of human nature, in a way that popular audiences found accessible and attractive, and he used this theory as a framework for social psychology. It is, he thought, inherited forces that shape human life. His *Introduction to Social Psychology* described the principal instincts and emotions of human nature and then explained individual character and capacity on this basis. That was what the public wanted and expected from psychologists, and it made McDougall – who was himself an unqualified political élitist – into a spokesman for the commonplaces in terms of which ordinary people voiced their views on human differences.

It was not easy to define clearly what an instinct is or to arrive at a definitive list of the instincts, and criticism of the vagueness of the term became common after a paper by the US psychologist Knight Dunlap (1875–1945) in 1919. In McDougall's woolly definition, an instinct is 'an inherited or innate psycho-physical disposition which determines its possessor to perceive, and to pay attention to, objects of a certain class, to experience an emotional excitement of a particular quality upon perceiving such an object, and to act in regard to it in a particular manner . . .'[11] His object was to determine the dispositions that form character and thereby to understand interaction between people and hence social relations in general. Psychology, he wrote, leads to the study of the 'group mind' or 'a mental life which is not the mere sum of the mental lives of its units'.[12] The highest ambition of the social psychologist is to study the group mind, especially in its highest form as 'the national mind'. It appears in retrospect that when McDougall studied the group mind he succeeded only in re-expressing the national and racial prejudices of his own class and background.

Beliefs of the kind McDougall articulated encouraged sympathy in the 1930s for irrationalist forms of politics legitimated by supposed innate differences of national or racial character. He came to regret he had ever used the phrase 'the group mind' since it was so liable to misunderstanding. He never intended to say that there is a consciousness other than the consciousness of individual minds, merely that there are 'similarities of structure of the individual minds which render them capable of responding in similar fashion . . .'[13] All the same, he and others wrote that this common structure makes possible a mentality that transcends the life of any one individual. By contrast, anthropologists and sociologists, some influenced by Durkheim, described this mentality as culture and argued that it exists in historical, linguistic and symbolic forms rather than as a psychological phenomenon.

Interest in the relations between instinct, political life and national character was widespread. In Britain, for example, Graham Wallas (1858–1932), an active member of the left-wing Fabian Society and, much later, professor of political science at the London School of Economics, published *Human Nature in Politics* in the same year as McDougall's *Social Psychology*. More subtle in his understanding of politics than McDougall, Wallas nevertheless also included knowledge of the primitive and instinctual forces in human nature on the agenda of social science. He coupled such studies with high expectations about what social engineering will contribute to ordered political life, whereas McDougall placed his hopes in a biologically or naturally superior élite. Just as World War I began, a London surgeon, Wilfred Trotter (1872–1939), published *The Instincts of the Herd in Peace and War* (1914, based on articles which first appeared in 1908–9), which discussed his belief that a gregarious instinct predisposes people to join and act in groups. For Trotter, this instinct is a positive force which makes it possible for people to live together; without it, 'the egotistic reason would . . . have rapidly carried the race to destruction in its mad pursuit of pleasure for its own sake'.[14] He argued that human science must include a comparative perspective on social animals (bees, sheep and wolves were his favorite exemplars). In contrast to McDougall and Freud, however, he invested his hopes in human life in a 'herd', a group in which the leader is one among equals, and which he distinguished from a 'horde', in which a group is subservient to a leader. The outbreak of war and the shared emotions this aroused were felt at the time to be powerful evidence for the reality of the instincts he described. Only with further reflection were questions raised about what an instinct is supposed to be and about the role cultural values have in the expression of emotions.

Trotter married Ernest Jones's sister, and the two men practised medicine together in the 1900s at the time when Jones became fascinated with psychoanalysis and first acquainted with Freud. It was also at this time that Freud began to venture out of the consulting room and into anthropology and social psychology. Freud attributed social relations to instincts which are re-expressed in a form re-created by the socialization of every child in the family. 'It seems to me a most surprising discovery that the problems of social psychology . . . should prove soluble on the basis of one single concrete point – man's relation to his father.'[15] He attempted to understand relations in general and group solidarity in particular in terms of speculative evolutionary reconstructions, the evidence for which he derived from clinical material, dreams and the theory of childhood sexuality. In *Totem und Tabu (Totem and Taboo,* 1912–13), Freud located the origin of civilization, of ordered social relations, in the rebellion of the sons against the original father, the father's murder and the subsequent suppression of fratricide – the suppression of a hypothetical state of social anarchy in which the sons kill each other in pursuit of the father's wives – by the internalization of guilt and by the institution of the incest taboo. In this way, as Freud conceived it, the primal horde of mankind acquired a social order and the basis for law. Elements of this ancestry, he thought, persist as an inherited psychic structure in the modern mind. When he referred to the regression to childhood that we experience when we dream, he wrote: 'Behind this childhood of the individual we are promised a picture of a phylogenetic childhood [that is, of the childhood of our collective evolutionary descent] – a picture of the development of the human race, of which the individual's development is in fact an abbreviated recapitulation influenced by the chance circumstances of life.'[16] Freud accepted the then widely held – but subsequently judged false – belief that an individual's development repeats the evolutionary development of the species ('ontogeny recapitulates phylogeny'). Later readers often found Freud's theory bizarre, though it had a context in late nineteenth-century anthropological speculation on the origin of religion. It was, indeed, precisely this kind of speculative reconstruction that came under attack from anthropologists on the grounds that it had no empirical content. Freud's anthropology remained the least influential part of his work.

Freud also tried to understand the collective psychic forces that, he thought, found expression in the war. In *Group Psychology and the Analysis of the Ego,* he reviewed the literature on group conduct, including Le Bon and Trotter. He rejected belief in the existence of an autonomous social instinct and instead attributed social solidarity to sublimated sexuality, to 'Eros, which holds together everything in the world'.[17] 'Masses are lazy and unintelligent', and crowds remain stable, he thought, as long as people in them are bound by shared love of a leader and by the illusion that the leader loves them.[18] But when groups collapse, as German, Austrian, Hungarian and Russian society did at the end of World War I, repressed feelings of envy and aggression break out. If a shared identification with the father is lost, the sons will fight each other, unless the civilized order proves strong enough to withstand the psychological onslaught. Freud's fear was that civilization has so severely repressed the instincts that when the time of trial comes, as it appeared to do during the turbulent years from 1917 to 1923, everything will break loose.

Both Freud and Jung assumed that there exist inherited psychological structures acquired in the common evolutionary past. Such beliefs were common, though there was a range of opinion about what is supposedly inherited. There was in 1900 no clear-cut view that inheritance takes the form of material instructions for a particular character or particular patterns of behaviour. It was customary in German-language publications to refer to life forces or mental inheritance in a way that left vague or open the question whether the inheritance is material or psychic in nature. Freud's and Jung's notions of inheritance demonstrate this imprecision. Theories of inherited psychological structure also entered into Wundt's science of the higher mental functions of thought and language. In different ways, the work of Freud, Jung and Wundt integrated a theory of the instincts as inherited mental structures with the analysis of social phenomena like law, language and religion. Like other German-language writers, Wundt considered that these structures have been affected by the experience and history of particular peoples, and in this context he referred to the *Volksgeist*. In 1860, the cultural psychologists Lazarus and Steinthal had defined the *Volksgeist* as 'that which is common to the inner activity of all the individuals [of a people]'.[19] The relation of this orientation to anthropology, in the work of Bastian, and sociology, through the ideas of Simmel (who studied psychology with Lazarus in Berlin), was discussed in earlier chapters. In his work after 1900, Simmel did not demarcate between sociology and psychology since human beings are in their nature social. Weber and Tönnies were other German scholars who discussed the psychological dimensions of social action. In the ten volumes of his *Völkerpsychologie* (1900–20), Wundt studied myth, custom and language in order to reconstruct the collective structure of mind. He included an account of the stages through which he believed mankind has passed in its cultural evolution and used this to account for such social phenomena as totemism. This part of Wundt's work, unlike the training in experimental psychological methods which he offered in Leipzig, remained largely unknown in the English-language world. Yet a similar belief that mental inheritance explains culture was widespread also in the anglophone world. It was encouraged by certain anthropologists, like James G. Frazer (1854–1941), a Scotsman who moved to Cambridge and was one of the supporters of the Torres Straits expedition. His lifelong work on *The Golden Bough: A Study in Magic and Religion* (twelve volumes, 3rd edn, 1907–15, supplement in

1935), a study of classical mythology and religion in terms of the ancient and primitive psyche, was designed to illustrate 'the gradual evolution of human thought from savagery to civilization'.[20] This and similar work gained a substantial public following as it appeared to offer insight into the primitive foundations of human nature in the rites and beliefs of early peoples.

Scholars, however, steadily differentiated the respective provinces and methods of classical studies and anthropology, isolating a figure like Frazer who crossed the boundary. In addition, such studies as Frazer's produced a strong counter-reaction because they lacked empirical rigour and fabricated general psychological explanation without reference to specific social contexts. The British anthropologist and physician W. H. R. Rivers (1864–1922) expressed this critical attitude posthumously in 1923: 'A great deal of so-called "social psychology" consists in the direct application of the conclusions of this psychology of the individual to collective behaviour on the assumption . . . that since society consists of individuals, what is true for the individual must necessarily be true of the group of individuals.'[21] Rivers wanted to separate psychological facts from social facts and rejected the idea that the former explained the latter. Anthropologists turned to fieldwork rather than reconstructions of the primitive mind. Biologists added to the critical atmosphere as their work made notions of psychic inheritance appear intolerably vague if not just plain wrong. As biologists began to examine inheritance statistically as an effect of the material properties of genes and denied the possibility of the inheritance of acquired characteristics, they threw the weight of scientific opinion against belief in psychic inheritance. The interests of psychologists turned in new directions and they gave people tests or put them in the laboratory to see how their conduct altered under observable conditions; they stopped writing in general terms about the group mind. The empirical approach in psychology was especially strong in the United States, where in the attention given to experiments on individuals it is possible to see a reflection of that country's political individualism and the belief that individuals hold the key to social phenomena. There was also an expansion of the occupation of psychology in the interwar years, and work on the individual as the subject matter of social psychology was one means by which the occupation laid claim to an autonomous sphere of interest and justified its growth. At the same time, however, sociology also expanded its activity, and this was also often justified to public audiences on the grounds that it served the integration of the individual and society. The outcome was an unclear, and sometimes uncomfortable, relationship between the disciplines of psychology and sociology, since both were concerned with a whole that disciplinary interests kept recreating as the abstractions, 'the individual' and 'society'.

iii *Social Psychology as Psychology in the United States*

Both sociology and psychology acquired a modern disciplinary and professional form in the United States in the closing decade of the nineteenth century. There was no obvious intellectual reason why the study of relations between people should belong to psychology rather than sociology or vice versa. There was no unambiguous division of subject matter, yet approaches to the individual–society relation differed in significant ways as social psychology developed separately in the two disciplines. For the most part, psychologists acted *as if* individuals have a non-social dimension of existence, e.g., in the activity of perception, memory or emotional expression, and presumed that this gives psychology its core subject matter. Social psychology was therefore understood to be the study of this 'non-social' activity in settings that psychologists labelled social, such as in friendships or in school. This is well illustrated by the content ascribed to social psychology by Floyd H. Allport (1890–1978) in 1924 in what was widely cited as the first psychology textbook on the subject. Allport described social psychology as 'the science which studies the behavior of the individual in so far as his behavior stimulates other individuals, or is itself a reaction to their behavior; and which describes the consciousness of the individual in so far as it is a consciousness of social objects and social relations'.[22] It might be thought that, since there is and can be no human behaviour outside a social context, social psychology thus described simply amounts to the same thing as psychology in general. The definition gave no idea where the boundary between the social and the individual is supposed to be. Nevertheless, psychologists in the twenties and later accepted Allport's definition, or something like it, as a useful one.

North American sociology, though its interests initially overlapped with the psychologists' social psychology, increasingly left the study of activity at the individual level to psychology. On certain topics, however, like the study of personality, there was little clear separation between sociology and psychology. Personality research became a major area of activity for both sociologists and psychologists during the 1930s, driven by the common interest in knowledge to make possible the individual's social adjustment. Yet, with time, the research effort became concentrated in the new sub-speciality of psychology, social psychology, and this sub-speciality grew into a huge field in its own right. There was then specialization even within the new field, which again reflected divergence of view about what is properly 'psychological' and what is properly 'social'; for example, there was a division between psychologists who studied and measured supposedly universal traits of human nature, such as assertiveness, and those who studied interaction between people, such as aggression and deference.

The absence of a predetermined boundary between psychology and sociology was evident in the early texts of social psychology. This was a constructive intellectual situation since, as the sociologist C. H. Cooley (1864–1929), who taught at the university of Michigan, observed, 'a separate individual is an abstraction unknown to experience . . .'[23] It was a psychologist, Baldwin, who wrote a preface for the English translation of Tarde's work on *Les lois sociales* (*Social Laws,* 1898) in 1899. Baldwin himself referred to social psychology in *Social and Ethical Interpretations in Mental Development* (1897), a book in which he developed his views about the value of psychological knowledge for social reform. Ross, by contrast, was not a psychologist but an economist, though he popularized the particular phrase 'social psychology'; his book of that title was an eclectic jumble of history and social commentary. The division of intellectual labour was especially blurred at the university of Chicago where, under the philosophical leadership of John Dewey and the sociological leadership of Albion Small, there were high expectations about the practical benefits to follow from the study of people in society. In 1917, Dewey, when he had moved to Columbia Teachers College and developed a philosophy of education, called for a new, co-ordinated discipline of social psychology across the boundaries of sociology and psychology, to produce 'that ordered knowledge which alone enables mankind to secure a larger and to direct a more equal flow of the values of life'.[24] It was also the Chicago sociologist W. I. Thomas who, with Florian Znaniecki, in their study of peasants who moved from rural Poland to Chicago, introduced the concept of attitude to describe personal orientation towards social conditions and values. Thomas and Znaniecki both initially tried hard to find in universal human characteristics a psychological basis on which to erect a theory of social systems. They hoped for a social psychology built on the study of attitudes, but they treated the notion of attitude in a way that differed markedly from the usage of later experimental social psychologists. For Thomas and Znaniecki, attitude is expressed in the language and reflects the experience of immigrants as they move from Polish to US culture and engage in the social process of assimilation. For later psychologists, attitude is a measurable variable in a laboratory context. Znaniecki, like many other sociologists, subsequently decided that a social-psychological programme is not viable and used only social categories in his theoretical work.

Sociologists continued to show interest in the fundamentals of motivation through the 1920s. There was discussion, for example, of the work of the Italian émigré scholar who was known as a sociologist in the US, Vilfredo Pareto (1848–1923). He drew a distinction between real motives, those ingrained in human nature, and fictitious motives, intellectual systems of justification, and used this in his studies of the manipulation and control of the public by élites. Such elaborations of psychological categories declined with the rise of the attempt to make sociology rigorously scientific, since psychological theories of motivation themselves lacked rigour and authority. Sociologists instead turned to terms like 'role', which were defined by reference to social function not psychological content, to describe the aspects of a person relevant to the examination of social systems. The term 'attitude' was taken over by psychological social psychology, and for a time studies of attitude almost circumscribed the field. In 1954 it was stated that 'this concept is probably the most distinctive and indispensable concept in contemporary American social psychology'.[25]

A profound assessment of what was implied for psychology if it took seriously the social framework of human life was made by G. H. Mead in Chicago. Elements of his analysis, in broad terms, were also developed by a range of social theorists, for instance by Marx and by the Soviet Marxist Vygotsky, discussed below. In brief, condensed and difficult papers, Mead argued that all psychological activity, even activity such as perception or emotion that clearly has physiological content, also has social content. His point was that it is not possible to talk about human experience or action independently of that experience's or action's social existence. In his view, psychological categories are in their very nature social. When a psychological social psychology developed in the 1920s, it conceptualized the biological individual as an autonomous entity and then studied that individual's interactions with social conditions or other individuals. By contrast, Mead argued that what is understood subjectively as well as objectively to be the individual, or a representation of the self, is the outcome of a social process. For Mead, there is no self to become socialized; the self is formed by socialization. When Mead stated that 'for social psychology, the whole (society) is prior to the part (the individual), not the part to the whole', this seemed to many psychologists to define the subject of social psychology in a way that put it outside the psychologist's domain.[26] Mead's view, in fact, failed to gain a substantial audience among psychologists. It did not help that he made his argument in an unsystematic and inaccessible form. His ideas diffused slowly through his teaching and through posthumous publications based on his lectures.

Mead was strongly influenced by Dewey's 1896 paper on 'The Reflex Arc Concept in Psychology', the theoretical masterpiece of functional explanation in psychology. This attempt to escape from dualism, whether of mind and body or of centre and periphery in the organism, and to achieve a sense of persons in environments as a whole process, underlay all Mead's work. His writing was theoretical, and he wanted a science of action as 'a dynamic whole . . . no part of which can be considered or understood by itself . . .'; all thought and action has to be understood in a social context.[27] His work was also practical, and he understood science to be a refinement of developmental processes in the most general sense. 'Our reflective consciousness as applied to conduct . . . researches its highest expression in the scientific statement of the problem and the recognition and use of scientific method and control.'[28] Mead put this philosophy into practice and was committed to the social use of rational science. He attempted, for example, though without success, to provide objective assessment and conciliation in the large garment workers' strike in Chicago in 1910.

Although Mead had an intellectual influence, it was mainly after his death and then principally among sociologists who derived from his work a theory, called symbolic interactionism, about the social construction of the self and of the individual's sense of self. His work diffused back into psychology through sociology only in the 1950s and 1960s. He became known for the phrase, 'the conversation of gestures', which denoted the social process of communication – not necessarily by language – in which the entities of self and other get constructed. The gesture 'reacts upon the individual who makes it in the same fashion that it reacts upon another', and this develops consciousness and the individual as part of a social process.[29] The continuous accommodation of individuals to each other, which begins in a young child as adjustment of movement, is, Mead argued, in maturity primarily a function of language and what he called significant symbols. It was an abstract theory, a fact that in part explains its lack of impact on psychologists. Mead tried to find a language to describe how the identity and subjective world of the self is constructed through a person's dialogue, in movement and language, with the surroundings at every moment of life. Those who elaborated Mead's ideas dismissed the laboratory as a setting in which to understand such questions; instead, they studied the language and meaning of social actions.

Social psychology became a major speciality within academic psychology during the interwar years, and, like its parent, the offspring adopted a behaviourist orientation and a statistical methodology. Allport's book in 1924 gave the field a shape, showed where the subject fitted into the psychology curriculum and suggested an experimental research programme on the behaviour of individuals in interaction with other individuals. He showed how social psychology could be constructed as a branch of individual psychology. There was considerable overlap between this new speciality and work in occupational psychology, educational psychology and advertising. There were studies of individual attitudes and opinions, often related to current political concerns about race and labour disputes, and about how attitude and opinion varies when one or more other people are present. Thurstone introduced a technique to scale attitude in 1928. During the 1930s, the threshold for aggression was studied in research which provoked children to show aggression under controlled conditions. The methodological pressure to be more precise and objective about physical variables, and hence to do experimental research on narrow topics, set the agenda.

Research in experimental social psychology was beset by antagonistic forces. There was pressure to be methodologically rigorous and to give the science credibility through experimental research. This brought social psychology into the laboratory and translated human relations into behavioural variables that can be studied under controlled conditions. The irony is that this step did not only simplify social relations in order to examine them empirically, as the researchers intended, but created a new kind of social relationship, the person in a laboratory. The irony had a bite because the other pressure on social psychology was to provide information of use to businessmen, politicians, educators and other people interested in guiding or changing people's behaviour in the wider social world. The search for precise knowledge created a new subject matter isolated from the wider society; but the justification for the whole research was supposedly its value to this wider world. Rigour and relevance worked in opposition. This situation is comparable to that later identified by the Dutch psychologist Johan T. Barendregt as the neurotic paradox: a methodologically correct project is irrelevant to life; a project relevant to life is methodologically incorrect. But this did not stop the growth of the subject.

The high level of interest in the interwar period in the harmonious integration of American society through personal adjustment made it important to understand the formation of opinion and the conditions under which people co-operate at work or in the community. The journalist Walter Lippmann used the expression 'public opinion' in 1922, and in 1936 George Gallup initiated a technique to measure public opinion by random sampling. This kind of elementary social analysis was separate from what went on in laboratory studies. Social psychologists studied individual capacities, not public opinion, as is shown by the content of articles published in the leading journal, the *Journal for Abnormal and Social Psychology* (the 'and social' was added in 1923). This direction of work was encouraged by the mental hygiene movement of the 1920s, which supported studies on child development and personality in order to provide tools for the individual's healthy adaptation to life. By the 1930s, the journal carried many papers on personality and this conveyed the clear message that personality is the key to social relations. Floyd Allport's younger brother, Gordon W. Allport, became an authority on both the trait theory of personality and on attitude and he set the agenda for much of the later work in psychological social psychology.

Many psychologists and social scientists were aware that disciplinary divisions were artificial and that laboratory studies were of limited value to practical industrial or social problems. External patrons who wanted results were frustrated. This led the Laura Spelman Rockefeller Memorial (subsumed under the Rockefeller Foundation in 1929) to make a multi-million dollar investment in the Yale Institute of Human Relations, founded in 1929 after a pilot project, which then ran for twenty years, though for the last ten on a much reduced budget. The motivation within Yale University came from the deans of the law and medical schools who wanted to resist what they saw as over-specialization, a goal shared by Rockefeller. The Institute was itself a large-scale social and psychological experiment, as its sponsors searched for a way to integrate psychological, sociological and anthropological expertise into a unified social science. There was a widespread feeling that a co-ordinated research strategy, modelled after the strategies of large business corporations, was needed. The Institute was to be a 'symbol of that synthesis of knowledge, for which need is now so widely recognized'.[30] As a result, a distinguished group of researchers, including anthropologists like Malinowski and Edward Sapir and the psychoanalyst Erikson, worked there for a time. It is striking, however, that integration was very difficult to achieve, and where it was achieved it was under the leadership of Hull, who was ambitious for a formal science of psychology. Psychological concepts and methods dominated most of the research agenda. Sociologists kept their distance from the Institute, partly because of internal politics at Yale, partly because they sensed an incompatibility between social and psychological levels of analysis.

The Institute's activities were overtaken when the United States entered the war in 1941. Psychologists and social scientists of all persuasions eagerly redesigned their research in order to contribute to national goals. The war was a major stimulus to social psychology; for example, it spurred large-scale studies of both soldier and civilian attitudes and morale, and it provided support from the intelligence services for studies of attitude change. It also gave academics a taste for the organization of their work at the corporate level rather than as an individual pursuit.

The spread of psychology to business had begun before World War I, but the link between academic studies and corporate interests achieved great prominence from the Hawthorne experiments conducted between 1924 and 1933. These studies took place in the Hawthorne plant of Western Electric in Chicago, then the manufacturing division of the American Telephone and Telegraph Company (AT&T). In the better-known version of what happened, academic researchers set up experimental production facilities within the works and surprisingly discovered that physical changes in working conditions, like the lighting level, do not correlate with productivity. This led to the discovery that productivity is a function of the workers' attitudes, while attitude itself was thought to be formed by the way workers follow the person accepted to be the group leader. In this version of the story, the experiments demonstrated the practical value of research in social psychology and led to a revolution in business practices that introduced personnel management and a worker-oriented redesign of work. Management at the Hawthorne works instituted an extensive programme of interviews with workers and a counselling service that lasted for more than two decades. As a result of the publicity given to the experiments by an academic at the Harvard Graduate School of Business Administration, Elton Mayo (1880–1949), who presented himself as the organizer of the studies, they became widely known and discussed. Mayo interpreted the results to show that productivity can be increased when workers get to know and like each other. Management, he argued, must look beyond engineering to personal relations. Through such books as *Human Problems of an Industrial Civilization* (1933), Mayo influenced the future course of academic business studies.

This version of events, apart from Mayo's influence on later business, does not stand up. The original experiments were part of a programme conducted under the auspices of the National Research Council and funded by electrical companies to set standards of high lighting levels in the interests of industrial efficiency. The experiments were started by company engineers who were not at all blind to the effect of psychological factors but set up a test situation to try to eliminate psychological variables. Managers had for some time attempted to manipulate productivity by psychologically-oriented schemes, such as payment by group rather than individual results. The fact that women workers engaged on repetitive tasks responded enthusiastically when given special attention as an experimental group and led by observers to expect higher productivity was hardly a discovery of social psychology. Before Mayo even visited the works in 1928, the company's technical manager had expanded the research into lighting levels from the shop floor to a special test room where he measured the effects of different rest periods and different attitudes among the workers. Mayo subsequently imposed his own interpretation of experiences in the test room, a version that emphasized the women's irrational as opposed to rational motivation, particularly in the way that their attitude was affected by leadership. In his reports he transformed the test room into a type of laboratory, treated the workers' views as behavioural variables, took for granted the expert's superior understanding and assumed the rationality of management as opposed to workplace goals. These reports found an audience just when there was a new militancy in the labour unions during the Depression in the 1930s, a situation that gave personnel management its opportunity. Mayo's work was attractive to management as a theory that higher productivity results from good supervision by a supervisor who is accepted as the team leader. The social psychology of business turned to the skills that encouraged workers to accept the rationality of management goals personified in a well-chosen leader. The Hawthorne experiments were another aspect of what, in earlier chapters, was referred to as the psychology of adjustment.

North American social psychology treated people's traits and attitudes as real natural objects that can be observed. Researchers favoured studies designed to eliminate interaction between experimenter and experimental subject and in which the psychological factors under study can be treated statistically as independent variables. But the psychological factors thus studied did not escape a socially constructed and historical nature. Further, the experimental settings devised by social psychologists were not the same as the settings in which people ordinarily acted. Yet most psychologists in the 1930s and many thereafter assumed that they studied natural variables relevant to ordinary life. A major exception was Kurt Lewin (1890–1947), a Berlin gestalt psychologist who migrated to the United States in 1933.

Lewin was a member of the university of Berlin Institute of Psychology during the 1920s, where Köhler was the professor, and he attracted an unusual group of students – women, East European and Jewish – to study human relations. The experimental work of Lewin's group was reflective and innovative, and the researchers accepted that the construction of an experiment is itself a social act that creates psychological circumstances for the purposes of study. Their work therefore replaced the model of the scientist as the observer of nature with the model of the scientist as a social actor who studies other actors with whom she or he is in dynamic relation. Lewin's student from the Ukraine, Tamara Dembo, who later also became well known in the US, exemplified this approach in her studies of anger, which required her subjects to tackle a difficult problem while she observed their conduct when she offered them no help. She wrote up her results as a qualitative description of the social dynamics of anger in a paper that illustrates a strong contrast to statistical studies of traits or attitudes. Lewin generalized such methods as a means to understand human action through the experience of group dynamics.

Dembo's and Lewin's descriptions of actions somewhat resembled informal reports by phenomenologists of the meaning that psychological action has for a person's place in the world, and their descriptions therefore went against accepted notions of objective reporting. Lewin, however, was not a phenomenologist; like the other gestalt psychologists, he believed that scientific knowledge goes beyond description to find general causes. He therefore tried to analyse psychological situations, like persons at work in groups, to locate their most general features. Further, he tried to combine these general features in a formal theory of the psychological force field, using the metaphor of the physical field of forces. To this end he used the mathematics of spaces known as topology, a mathematics suitable for the description of whole-part relations, in such books as *Principles of Topological Psychology* (1936).

There have been different views as to whether Lewin had an immediate impact in the United States or was at first marginalized. He found it difficult to get a suitable permanent position in North America and his methodological studies remained untranslated into English. Yet, while he did not have a prestigious university chair, he worked in situations, well-funded by Rockefeller, that gave him good research opportunities, high-quality doctoral students and access to a network of friends and private funding agencies. From 1935 to 1944 he held an untenured position at the university of Iowa, where he developed experiments on group behaviour that began to attract the interest of social psychologists. He also became involved with other socially concerned psychologists and together they founded the Society for the Psychological Study of Social Issues in 1936. The Society's activities, like Lewin's own work, were soon redirected as part of the war effort. Already in 1939, Lewin put his scientific study of group dynamics to practical use when his students worked on problems of productivity with the Harwood textile company in Virginia. This work led to techniques of participatory decision-making; it brought those who have to carry out decisions into the decision process and thus motivated them in relation to the outcome. This was part of what was hailed in the post-war years as a business revolution. His interest in methods encouraged action research, an attempt to advance knowledge through the involvement of the researcher with the people who are being studied. The new ideas about management paralleled the logic of Lewin's approach to experiments with human subjects in the 1920s. Both the manager and the scientist, he thought, need to recognize the continuously changing dynamics of the social relations that form the context of individual action.

The significance of Lewin's ideas for the management of social relations was acknowledged when he moved from Iowa to the Massachusetts Institute of Technology (MIT) in 1944 to establish the Research Center for Group Dynamics. This group and its students exerted a major influence over social and industrial psychology. During the war, Lewin became an adviser to the Office for Strategic Studies and the Office of Naval Research, and he was able to attract large funds. He gave thought to practical problems about how the individual relates to social change, and this became part of a discussion about how to re-educate German Nazis after their defeat in 1945. All this built on work Lewin had initiated in the late 1930s, especially in studies of the contrast between authoritarian and democratic group structures. A paper by Lewin, R. Lippitt and R. White, 'Patterns of

Aggressive Behavior in Experimentally Created Social Climates' (1939), provided a model of research in which complex social phenomena are brought into the laboratory, and experimental work on groups remained a major theme in social psychology until the 1960s. After Lewin's death in 1947, the MIT group moved to the university of Michigan where it became part of an Institute for Social Research, which was the largest centre for academic studies of organizational practice in the United States. Lewin also established a technique with which to teach group dynamics to future group leaders, and researchers explored techniques of leadership, motivation and conflict resolution within small groups. In 1947, these techniques were formalized in the programme, co-sponsored by Lewin's Research Center, of the National Training Laboratory at Bethel, Maine, where leaders were trained in the means to effect organizational and attitudinal change. The training groups became known as T-groups, the forerunners of many of the group techniques familiar in the second half of the twentieth century, and used, for example, to train nurses. The US innovations were followed closely in Europe. In London, the Tavistock Institute of Human Relations was founded immediately after the war; the Institute, where Wilfred Bion explored leaderless group dynamics from a psychoanalytic perspective, joined forces with professional bodies, especially social workers, to study and train people in group relations. Through these training routes, significant numbers of people became familiar with psychological principles in social life.

The US psychology profession grew rapidly after 1945. This expansion added to the pressures to divide psychologists into independent academic and applied areas, and even social psychology and occupational or organizational psychology tended to separate from each other in this way. Psychological social psychology became an independent academic field, later with separate experimental and applied journals and values that pulled even the speciality in opposite directions. A pattern of research developed in which a few major American studies created conceptual frameworks in terms of which a mass of detailed work was conducted, at least for a few years. These studies included those of Leon Festinger, *A Theory of Cognitive Dissonance* (1957), Fritz Heider, *The Psychology of Interpersonal Relations* (1958), and H. H. Kelley, whose innovative paper was called 'Attribution Theory in Social Psychology' (1967). Each of these general theories proposed an explanation of what, in the context of social relations, drives people to act in one way rather than another.

The work by Festinger was part of a wider argument in social psychology that relations between people depend primarily on cognitive understanding, on belief and evaluation about situations and other people. Festinger, who had studied with Lewin, argued that motivation derives from dissonance between belief or aspiration and experienced achievement. The underlying notion that people seek cognitive consistency appeared applicable in many situations. Another cognitive theory of motivation, the theory of personal constructs, was developed at more or less the same time by George Kelly (1905–67). Festinger and Kelly attributed behaviour to thought and expectation rather than motive and need, and they thus added to the opinion that was turning against mechanistic notions of human action of the kind exemplified in behaviourist psychology. Heider, who trained at the university of Graz in Austria before he worked with the gestalt psychologists in Berlin and then in the United States, translated some of Lewin's work into English and contributed to the idea of cognitive balance behind Festinger's work. He elaborated his ideas on what he called interpersonal relations – relations between a few, usually two, people – in his 1958 book, and contemporary psychologists were struck by the qualitative language with which he discussed the relation between the inner and outer dimensions of human contacts. Heider, whose book was unusually conceptual, seemed able to reconcile the demand for science with respect for the meaning that an orientation towards things or people has in an ordinary person's experience or action. He was an articulate voice in the orientation of academic psychology towards a non-mechanist, non-quantitative representation of action. H. H. Kelley, who was influenced by Heider, was among a number of psychologists who founded attribution theory, an approach to action that takes seriously a person's own explanation of behaviour in terms of causes classified into those located in the self and those located in external agencies. Researchers tried to understand social relations by the study of how people attribute causes and assign and take responsibility for what occurs. Among other things, this encouraged psychologists to take seriously how ordinary people explain events in psychological terms rather than simply to assume that scientific psychology makes ordinary psychology uninteresting.

One particular experiment aroused enormous interest and continued to be debated much later. In a paper on the 'Behavioral Study of Obedience' (1963), Stanley Milgram reported how college students were instructed to give electric shocks of increasing and finally of great severity to other participants who were out of sight but audible and whose task was supposedly to memorize words. The second group of people were in fact accomplices of the experimenter and received no shocks. What the experiment showed was that the naïve student participants were willing to obey an authority figure, even though they showed distress while administering the shocks. At first glance, what the scientist had done was show how easy it is to induce acts of cruelty despite countervailing values and emotional distress. Milgram, who had by then moved from Harvard to the City University, New York, discussed the implications of his experiment at greater length in *Obedience and Authority* (1974) and included other data. Two main criticisms were levelled at the experiment. Firstly, it was argued that it exploited the respect in which science is held and revealed the ordinary student's trust in institutions and experiments rather than anything about obedience and cruelty. Secondly, it raised questions about the ethics of experimentation with human subjects. Milgram in fact talked after the experiments with his naïve participants, informed them about the procedures and dealt with any emotional consequences; he called this process debriefing. Subsequently, ethical concerns entered more and more into the design and implementation of work in social psychology. This involved some recognition of the social dynamics of the experiment in a way that was foreign to many earlier behavioural studies.

All this work gave an academic speciality an identity but it did not provide a comprehensive basis for the understanding, let alone the prediction, of how individuals behave as social beings. Much else was written by social scientists, alongside the statistical and experimental work of psychological social psychology, on the individual as a social being, but different areas of the social sciences were often barely in communicaton with each other. Some anthropologists became interested in psychology as a way to understand human drives as part of the explanation of social activity. In the 1930s and 1940s, a 'culture and personality' group, influenced by theories of personality and by psychoanalysis, formed within US anthropology, and members of this group argued that it is possible to infer cultural characteristics from individual personality traits. The best-known member, Margaret Mead, was a prominent spokeswoman for a viewpoint that linked child development, sex roles and personality to the national culture. During the 1940s, she and other social scientists did a considerable amount of work, funded by the military and intelligence services, on national character and on attitude change as a contribution to US political and military dominance. Beyond the ranks of anthropologists, David Riesman, a sociologist at Yale University, wrote a bestseller (in an abridged version) about character and culture, *The Lonely Crowd: A Study of the Changing American Character* (1950). He argued that there was a shift in the American character from inner-directed to outer-directed motivation and he correlated this with long-term social and demographic change. Such work was in the spirit of studies on character and modernity by Weber and Tönnies. The study referred to earlier, which resulted in *The Authoritarian Personality*, focused on family structure and discipline in early childhood in order to explain what was thought to be a special German openness to authoritarian politics. The study illustrates yet another way in which social commentators assumed that individual personality and culture enter into the formation of each other.

In the 1950s there was a renewed attempt to unify the human and social science fields and to create a unified social psychology, this time under the banner of the behavioural sciences. It was not a success and by the early 1970s there was widespread reference to the crisis in psychological social psychology. Only within the narrow confines of particular branches of the psychology profession was there any agreement about the key terms for the psychological analysis of social realities. On the wider stage, there was variety of opinion, the divergent interests of different occupational groups and all the messiness of political life. Yet an excessively general set of psychological assumptions about economic man, an idealized respresentation of human nature in terms of free economic individuals, was part of the intellectual rationale for the political and economic policies that dominated Western countries in the 1980s and then spread eastwards in Europe after the changes of 1989. Human nature, the political argument implied, creates social relations through acts that consciously and rationally maximize material goods for the individual. This viewpoint had its roots in liberal political theory that went back to Hobbes and Locke. The irreducible basis of society, on this view, is the free and independent individual, and society, it was believed, can be formed and reformed in accordance with the individual's wishes and preferences, as expressed through the market and representative government. At the same time it was assumed that individuals have a 'nature', natural sentiments about family, community or nation, that bind people in communities through shared natural emotions. Sharp divisions of wealth, gender, ethnicity, life-style and moral values, however, in practice exposed 'common sense' belief about human nature and common natural emotion as extremely fragile. North American and European economic individualism, its critics argued, correlated with loss of commitment to social and community structures and hence to the sources of values. The individualism seemed to be of a piece with a psychological social psychology that posited autonomous individuals as its given subject matter. The politicians of the free market tended to denigrate the social sciences, that is, they denigrated institutional support for the argument that 'the individual' is a socially constituted category. In these circumstances, where political and psychological assumptions were intertwined, there were many divisions in the broad area of the human sciences. In one dimension, there was a contrast between formally elaborated science and political rhetoric about human nature; in another dimension, there was a contrast between psychologists who started from the individual and sociologists who started from society as irreducible categories.

With the rise of a new political consciousness in Western Europe and North America in the 1960s, there were moves towards a different kind of social psychology. Sometimes this drew, like the work of the Soviet theorist S. L. Rubinshtein discussed in the next section, on the Hegelian dimension of social thought in the young Marx. Small but intense groups of theoretical psychologists, led by Klaus Holzkamp in Berlin and K. F. Riegel (1925–77) in North America, attempted to refound psychology within a Marxian dialectical framework. A sense of crisis about the lack of direction in social psychology in the early 1970s also resulted in an influential non-Marxist philosophical critique, influenced by Wittgenstein, in Rom Harré and P. F. Secord's *The Explanation of Social Behaviour* (1972). This book argued that explanations of human action require reference to socially situated purposes expressed in the ordinary language of actors themselves. The argument fostered an emphasis on people's active construction of their world, which also implied recognition of the way experimental subjects actively shape experimental tasks. Harré and Secord also founded the *Journal for the Theory of Social Behaviour* in 1971 in order to provide a forum for theoretical debate. Kenneth J. Gergen (b.1934) reconceived social-psychological knowledge as historical knowledge. His paper on 'Social Psychology as History' (1973) argued that social psychologists, whether they know it or not, study a historically specific subject matter and not the isolated actions of people in experimental settings: 'social psychological research is primarily the systematic study of contemporary history'.[31] Psychologists and historians of psychology, when they developed this and related points, began in the 1970s to argue that the history of subjectivity and of the emotions, which draws on resources like biography, diaries and imaginative art, are of central importance to psychology. As a result, researchers began in the 1980s to take the traditional humanities subjects, like literature and history, and transform them into human science subjects, the sources of knowledge about men's and women's historical nature. Gergen later elaborated a position he called social constructivism, which was linked to contemporaneous developments in the sociology of knowledge. He argued that scientific knowledge must be understood as a social action rather than in terms of a correspondence betweeen its statements and a supposed asocial 'reality'. The German-born Canadian social psychologist and historian of psychology Kurt Danziger described in detail how experimental social psychologists historically constructed their subject through the social relations embodied in the experimental methods of their own occupation. Such arguments supposed that people express the form or 'habitus' of a culture in which they work, to use a word favoured by the French intellectual observer of intellectuals Pierre Bourdieu, and that this habitus becomes part of the knowledge they produce.

This orientation in social psychology connected with work in historical anthropology, for example by the Romanian-born writer Norbert Elias, who published a study, originally in the 1930s, on 'the civilising process'. Elias suggested ways in which social customs, personal manners, bodily expressions and belief about the world all interacted to form the early modern European character. Once again, this added to the weight of opinion in the 1970s and 1980s that considered both subjective feeling and objective expression of feeling in bodily acts or language as historical constructions and therefore pointed to cultural history rather than natural science as the proper home for a theory of *human* nature. A pioneering study of the psychology of past people, Zevedei Barbu's *Problems of Historical Psychology* (1960), brought the principles of the French *Annales* group of historians, encapsulated in Lucien Febvre's call for a history of 'sensibility', into contact with English-language interest in character and personality. The French historians themselves, who had considerable influence in the mid-century and in the following decades, envisaged studies of the large-scale *mentalité* of peoples about time and geography, childhood and death. Such work, which took for granted Barbu's starting-point that 'of all living creatures man alone is truly historical', tended to equate the subject matter of psychology and the subject matter of history.[32]

This conclusion was a long way from what most psychological social psychologists understood their subject to be. That subject, by and large, continued in the 1980s to presuppose that 'the individual' and 'the social' can be defined independently of each other and hence that psychology and sociology may properly specialize, each with its respective subject. Yet a considerable body of literature, some of which was written by social psychologists themselves, called this presupposition into question and hence questioned the basic orientation of the subject area in the direction of natural science. In the 1990s, social psychology was, with good reason, a field open to debate about fundamental issues of explanation in relation to human nature. The next chapter discusses debate in the second half of the twentieth century about psychology's scientific identity, a debate much concerned with how to conceive the social individual. First, however, I turn to an area of the greatest interest, though usually ignored in the context of social psychology, the history of psychology in the Soviet Union between 1917 and 1991.

iv *Soviet Psychology*

Soviet psychology belongs in the history of social psychology for the reason that the Marxist-Leninist theory of the Soviet state claimed to be an objective social science of human action. In theory, the Soviet state was uniquely constituted and possessed unique legitimacy because it was the political vehicle of objective progress, in the manner first understood by Marx. The Soviet state contrasted itself with Western states in which the legitimacy of government rested on the right of citizens to express preferences, which Soviet theorists held perpetuated non-progressive power structures. The distinctive way Soviet power was legitimated by theoretical claims meant that the theory of the state was simultaneously a human science, an argument about the social and historical constitution of being human. It is in this context that I discuss Soviet psychology in relation to social psychology. Further, Marxist, Marxist-Leninist and dialectical theorists of various persuasions, not only in the Soviet Union, engaged with many of the difficult conceptual questions about 'the individual' and 'the social' when they attempted to construct a historical and materialist human science.

There is a harsh irony in this section. For all that may be said about theory in the Soviet Union, for much of the time psychology was dominated by the model of the physical sciences and substantially ignored people as social actors. The fact was that the close identification of the state, which meant the exercise of centralized power, with Marxist science had the consequence that the science could not itself be examined seriously, least of all by empirical social studies of human action under the Soviet system. At its Stalinist extreme, the state tolerated no theory which the state itself had not formulated. 'All efforts to think of any theory, of any scholarly discipline, as autonomous, as an independent discipline, objectively signify opposition to the Party's general line, opposition to the dictatorship of the proletariat.'[33] This was the voice of the party boss. The very argument which, in theory, made the Soviet system uniquely able to create a social psychology, in practice prevented serious study of such a subject. Exactly this kind of paradox or mismatch between theory and practice in the long run removed any intellectual legitimacy from the Soviet exercise of power.

The history of science in Russia is a record of both identification with the West and pride in the unique path of Russian achievement. The 1890s saw limited but rapid industrialization and Moscow and St Petersburg became huge conurbations which bred radical political agitation, poverty and crime. Educated liberals responded in a manner that paralleled the Western response and turned to the social and medical sciences for the tools with which to tackle the new conditions. The conservative administration of Tsar Nicholas II (reigned 1894–1917), however, gave little encouragement to the growth of the academic social sciences, and it held psychology in suspicion as a potentially materialist approach to human nature that detracts from the soul as the agent of individual freedom and responsibility.

A significant part of the public turned for inspiration to literature and the great writers rather than to politics or science. All the same, the liberal position achieved some standing in, for example, the psychiatric services and especially in the area of education. In Russia, as in North America, liberal opinion hoped that a scientific approach to pedagogy might improve the capacity of children and hence improve the conditions of life.

The attempt to establish psychology as a separate discipline sometimes met with resistance. As early as 1885, however, V. M. Bekhterev (1857–1927) established a psychological laboratory in the psychiatric clinic attached to the university in Kazan. N. N. Lange (1858–1921), when he returned from study with Wundt in the 1890s, argued for an experimental approach to psychology. There were other individual initiatives, but the first formal university institute was founded in 1912 when G. I. Chelpanov (1862–1936), who had also studied with Wundt, opened an institute for psychology, independent of philosophy, in Moscow University, and this institute attracted large numbers of students. Chelpanov was an adept organizer who was able to hold together an eclectic cluster of interests and tendencies. He questioned whether psychological science was ready to help practical pedagogics, and though his institute did take an early interest in mental testing, it was as a theoretical activity rather than as a practical intervention in the poorly developed state school system. Russian liberals, in contrast to liberal professional opinion in the West, took it for granted that low performance in intelligence tests reflected the poor social conditions in which the majority of the people lived.

Several scientists who became important after the Revolution trained or worked with Chelpanov, notably K. N. Kornilov (1879–1957), who was his senior research assistant. Kornilov was a son of what was called the rural intelligentsia, which in his case meant that his father was a bookkeeper, and he projected himself as a model of the hard-working student who raises himself and then returns, via educational psychology, to help his own class. 'Cast up by a fortunate wave out of the mass of the people's teachers, I understood that I am obliged to devote all my energies to the service of those who have, in the past, been chastised by many but caressed by no one.'[34] His work, which involved studies of reaction times, was of little service to the people, but his conventional rhetoric reveals something about the social context in which young and idealistic people turned to psychology. Another researcher to work with Chelpanov was P. P. Blonskii (1884–1941), who was imprisoned as a student member of the Socialist Revolutionary Party during the uprisings of 1904–6. The first psychologist to support the Bolsheviks in 1917, he saw in revolution the opportunity to realize pedagogic ideals and to create the conditions in which people can be saved from their ignorance.

There was no institute comparable to Chelpanov's at St Petersburg University; instead, a unique organization grew out of the medical interests which did much to encourage acceptance of Western natural science in Russia. The neurophysiologist, neurologist and psychiatrist Bekhterev, who had moved to Petersburg, was an entrepreneur who managed to put together private and government funds for an independent psychoneurological institute in 1908. This institute became a huge, eclectic enterprise, and different people worked on everything from experimental psychology to sociology, from social psychology to criminal anthropology – everything with 'a clear connection with the psychology of man'.[35] Bekhterev himself tried to systematize this work under the general heading of 'objective psychology' and later 'reflexology'. He used the word 'reflex' as a metaphor to explain action as the consequence of organic connections between a person and surrounding conditions. In his definition, 'Reflexology . . . is the science of human personality studied from the strictly objective, bio-social standpoint.'[36] However loose conceptually, this work was widely cited and also translated as a significant contribution to a biological attitude towards the study of human beings.

Bekhterev's claim to leadership in objective science competed in St Petersburg with that of Pavlov, institutionally well entrenched at the Military Medical Academy with the status of a Nobel Prize winner. While his work concentrated on the experimental study of conditioned and unconditioned reflexes, Bekhterev had wider interests that included more neurophysiology and clinical neurology. Bekhterev rejected Pavlov's methods as a way to uncover knowledge of the brain, but neither accorded to psychology the status of a subject independent of physiology. In 1916, in a rare comment on his general stance, Pavlov dismissed those who were concerned with mind and consciousness: 'We are studying all the organism's reactions to events of the external world. What more do you need? If you find it nice to study, so to speak, the poetry of the problem, then that is already your business.'[37]

The collapse of Tsarist power during Russia's war against Germany and Austro-Hungary, followed by revolution and civil war, was accompanied by huge material deprivation and virtual anarchy in public life. Many people in the professions, including natural scientists and doctors, left with the Revolution, but a certain number, such as Chelpanov, Bekhterev and Pavlov, stayed on and sustained scientific work under appalling conditions. Some radicals during the early years of Bolshevik rule thought that everything had become possible: the Revolution had created entirely new social conditions and these conditions made possible a new human nature. A. K. Gastev (1882–1941) headed the Central Institute of Labour in the 1920s, and he tried to use the organization to apply scientific knowledge that man really is a machine. Radicals believed that the truth of a materialist theory of human nature means that men and women can be re-engineered. They thought that there need be no hypocrisy or frustration over the satisfaction of bodily needs, and each person will learn to act for the public good when she or he understands what these needs are. Blonskii threw himself into utopian plans to reconstruct not just academic culture but human nature itself. This transformed the engineering ideal into social psychology. With different goals in mind but in the same spirit of infinite possibility, the radical feminist Alexandra Kollontai advocated free love to liberate women from the economic dependency of marriage. As early as 1920, Blonskii wrote that we must 'take our stand on the Marxist viewpoint as the only scientific one . . . not only in economics but also in social science generally . . . and in psychology, and also in philosophy, and in all of science'.[38] As a philosopher of education, he associated with Nadezhda Krupskaya, Lenin's wife and a force in the creation of a new educational system.

From 1923, radical plans for pedagogy and human engineering succumbed to the contingencies of political and economic life in the new state. Theorists paid more attention to systematic relations between the sciences and Marxist theory, and Kornilov announced at the 1923 congress of pyschoneurology a mission to establish a Marxist psychology. Since it was held that Marx had uncovered the objective reality of the human condition, and since the USSR was the only state to reconstruct itself in the light of that objective knowledge, the human sciences were expected to acquire a new and distinctively objective character under communism. The Socialist Academy of Social Sciences was founded in 1918 in order to train party cadres in a Marxist understanding of human conditions. Renamed the Communist Academy in 1924, it was repeatedly in conflict with the prerevolutionary Academy of Sciences, especially when it tried to extend a Marxist evaluation of knowledge into the natural sciences – for example, when it opposed the new quantum mechanics. There were intense discussions among psychologists as to what the new character of science is and how communist psychology relates to bourgeois psychology, that is, to a science constructed in societies without an objective relationship to the historical process. This was the philosophical and ideological side of debate. The other side was a bitter struggle for scarce material resources; researchers fought for access to the Party as the source of power and dispenser of funds. In the background there were debates in Marxist philosophy, which resulted in the late 1920s in the replacement of Nikolai Bukharin's historical materialism by A. M. Deborin's dialectical materialism when the former was judged to have too determinist connotations. All theoretical debate was then cut short in 1930–31 by the Communist Party's subsumption of theory and practice at every level in society to the transformation of the country known as the Great Break. Many of those active in this step were trained in Marxist practice at the Communist Academy. The agronomist T. D. Lysenko, who led the attack on scientific genetics, became notorious in the West.

Several factions competed to dominate psychology and neuropsychology during the 1920s. Bekhterev re-expressed reflexology as a materialist theory and attributed human actions to 'associative reflexes'; he intended his theory 'to break the yoke of subjectivity in the scientific appraisal of those complex activities of the human organism which establish man's correlation with the environment'.[39] He envisaged a programme for psychology, analogous to behaviourism, that translates mental processes into observable correlations of sensory input and movement. Non-specialists found it difficult to distinguish Bekhterev's and Pavlov's arguments, though they were competitors for resources in Petrograd, renamed Leningrad in 1924. Pavlov's programme was founded on experimental studies of conditioned reflexes and the assumption that it is possible to move from such experiments to a theory of brain action, while Bekhterev's programme, which described reflexes as associated rather than conditioned, was more loosely articulated, more an imagined unification of the sciences than a definite plan of work to achieve this goal.

The call to establish a Marxist psychology came from Moscow, not Petrograd, from a physician involved in the mental health movement, A. B. Zalkind, and from Kornilov. Both were party members and committed to the communist cause in a way that Bekhterev or Pavlov were not. Chelpanov's Institute of Psychology, where Kornilov worked, continued to function after 1917. Then, in 1923, there was a breach and Chelpanov was ordered to pass the directorship to Kornilov. As the Russian proverb puts it, 'when you chop wood, chips fly'. Kornilov and his supporters argued that Chelpanov resisted psychology's reconstruction on Marxist lines and that his position in effect treated Marxism as one possible philosophy rather than the objective foundation for the sciences in general. Chelpanov fought back, though he recognized – as all scientists did by this time – that he had to accommodate to the communist regime or emigrate; he argued that Marxist philosophy supports an experimental psychology that links the brain and introspective reports, and he accused Kornilov of vulgar materialism. But he could not shake off the accusation that he was an unreconstructed idealist from pre-revolutionary times. Kornilov headed the institute for the rest of the decade. In the early 1930s he was branded 'eclectic' in his turn; it was stated that he put together a mishmash of theories and methods under Marxist labels rather than an objective scientific practice that carries history forward, the history then in the making under Stalin.

At first glance it is surprising that Pavlov's work benefited most in the long run from these conflicts. He initially had nothing but contempt for what he called the 'barbarians' who seized power, but he was so committed to natural science research that he was prepared to judge the new regime simply by whether or not it supported his work. The Communist Party had its own reasons to back a scientist of Pavlov's stature. After a direct response from Lenin to his threat to go abroad in 1920, Pavlov received special resources, and then, because of his standing as a scientist with an international reputation who continued to work in the Soviet Union, he began to receive large-scale support. In 1924, Bukharin, the party's chief ideologist, broadly endorsed Pavlov's work as 'a weapon from the iron arsenal of materialism' and, as long as the meaning of such slogans was not examined, this served well to legitimate Pavlov's programme.[40] Consequently, by the early 1930s, when Pavlov was in his eighties, he was head of two purpose-built institutes, one in Leningrad and one, the Koltushi Biological Station, in the nearby countryside, which together employed some forty scientific workers. His comments about communists mellowed, and in 1934 he offered public support to the state, though the content of his science never included any reference to Marxist principles. By the time of his death in 1936, there was a bizarre situation where Pavlov was described publicly as a hero of socialist labour and yet had only the scientists of his own school as an audience. Pavlov's history is strange because it was intimately connected with the history of the Soviet state while detached from the larger Marxist project.

The ideological battles of the 1920s presupposed that it is possible to construct a unified science that integrates man's biological and historical nature, and that only Marxists can achieve this because only they have freed themselves from idealist prejudices about man. Marx, it was believed, had shown the true relationship between the individual and society, the historical constitution of human nature through the material organization of labour. Marxist-Leninists of the 1920s, within this general framework, tackled the specific problem of the recreation of psychology by reference to Marx's and Engels's writings on the historical nature of human consciousness and action. This emphatically was not Pavlov's programme – whatever material support he received and whatever the position later attributed to him.

It was, however, the programme of a figure who was to achieve posthumous fame, Lev S. Vygotsky (1896–1934), from the 1960s a significant influence in Western psychology as well as in the liberalization of East European science. Vygotsky was then variously called 'the Mozart' and 'the muffled deity' of Soviet psychology, phrases that alluded to the brilliant versatility of a man who died young and whose reputation survived the darkest years of Stalinist brutality.[41] Vygotsky's career is dramatic and moving. As the separate Russian-language and English-language publication of his collected works in the 1980s and 1990s attests, he is also a remarkable example of delayed influence. Born into the family of a Jewish bank official, Vygotsky had to overcome quotas on Jewish students to get an education. He was involved as a student with the artistic avant garde in Moscow before and after the Revolution, a time of wonderful inventiveness in the arts. His earliest work was as a literary critic, and in all his later work he attempted to accord a place to the expressive and aesthetic human consciousness. He probably welcomed the Revolution; certainly, within a few years he had thoroughly digested the Marxist classics. For reasons that remain obscure, in January 1924 Vygotsky appeared not in the guise of a psychologist of art but as a spokesman for a psychology that takes consciousness as its subject. At the second psychoneurological congress in Moscow, the young man delivered an electrifying lecture that took apart (though he did not name names) the pretensions of both Bekhterev's and Pavlov's programmes to be the objective science of man, and he implicitly criticized the sympathy of party ideologues with such work. 'A human being is not at all a skin sack [a sausage] filled with reflexes, and the brain is not a hotel for a series of conditioned reflexes accidentally stopping in.'[42] Instead, Vygotsky turned to Marxism as a philosophy potentially able to reconcile a psychology of consciousness with the materialist science of the body. This defence of a psychology of consciousness was welcomed by Kornilov's Moscow institute as it challenged the claims of the rival Leningrad schools, and Vygotsky was invited to join the institute. Both his lecture and the invitation may have come at the instigation of two young researchers, A. N. Leont'ev (1903–79) and A. R. Luria (1902–77), major figures in the revival of non-Pavlovian work in the 1950s. Leont'ev became known for his studies of child development, carried on through the 1930s, and Luria for his research on brain damage (an area known as defectology in Russia) on which he worked extensively during and after the Great Patriotic War of 1941–5.

Luria's reminiscences of the mid-1920s give a vivid idea of the turbulent social and intellectual setting.

> Instead of cautiously groping for a foothold in life, we were suddenly faced with many opportunities for action – action that went far beyond the confines of our own tiny circle of family and friends . . . This atmosphere immediately following the Revolution provided the energy for many ambitious ventures. An entire society was liberated to turn its creative powers to constructing a new kind of life for everyone. The general excitement, which stimulated incredible levels of activity, was not at all conducive, however, to systematic, highly organized scientific inquiry.[43]

In the midst of this Vygotsky took seriously dialectical philosophy and hoped to integrate a theory of biological development and of historical consciousness. In a short number of hectic years, years increasingly disrupted by tuberculosis, he worked in pedagogy and clinical psychology, designed a psychological field study of Uzbek peasants and debated philosophical questions about methods, and did research on developmental psychology. He wrote a large-scale theoretical text, *Istoricheskii smysl psikhologicheskogo krizisa* (*The Historical Meaning of the Psychological Crisis*) in 1926–7, which included a cosmopolitan review of competing philosophies but which remained unpublished until 1982. His studies of developmental psychology culminated in a series of essays, edited into what became his best-known work, *Myschlenie i rech'* (*Thought and Language*, 1934; translated into English in 1962, again in 1986 and in 1987).

Vygotsky divided the child's early development into a prelinguistic stage, when he thought the child possesses only sensory and emotional consciousness, which can be understood biologically, and a linguistic stage in which he thought the child interacts with historical culture and thereby acquires both language and the capacity for thought. As a way to communicate his ideas, he argued against the developmental theories of the Swiss psychologist Piaget, and it was in this specific context that Vygotsky's ideas became known in the West, marginally in the 1930s and then more widely in the 1960s. His broader theory concerned the interaction, mediated by language formation, between the individual as a biological organism and historically specific culture. He was unable to do much systematic socio-psychological research and, while he dreamed of a Marxist social psychology that would unite psychology, a theory that would integrate biological and linguistic processes in actions and conscious meanings, almost none of this appeared in *Thought and Language*. Even though Vygotsky's work was cut short by his death, it was the most serious attempt during the interwar years to achieve a Marxist psychology. This aspect of his thought was either not perceived or not accepted either in the Soviet Union or by his US translators in the 1960s; the English-language abridged version of the text made it appear to be a contribution to Western debates on developmental psychology.

Meanwhile, conditions that went far beyond misrepresentation dominated life in the USSR. For several years before his death, Vygotsky, like other psychologists, faced criticism that his work did not adequately embody party-defined goals. Any serious intellectual or artistic undertaking was subject to such criticism in the 1930s: the very act of standing outside the party, even for purposes of debate or aesthetic expression, was taken to be opposition to the party and its claim to have achieved an objective hold on the human condition. More prosaically, party activists, faced as they were by the demands of the Five Year Plans for accelerated industrialization, were impatient with theory of any kind and disdainful of highbrow theorists educated before the revolution. These officials tolerated theory less and less, and they questioned the practical expertise of psychologists and even the value of psychology as a distinct discipline.

Pavlov's school, with its well-supported infrastructure and its claim to be objective physiology, a subject which was politically more neutral than psychology, stood out as a relatively thriving area of research, and it was able to maintain continuity in its work. Most other areas suffered. For example, while there was some interest in psychoanalysis during the 1920s, since it was claimed to reveal true instinctual, material needs beneath repressive bourgeois values, it was politically incorrect by 1930. One area that expanded for a while was 'pedology', a term which denoted the subject area of children's development and included the use of tests in schools. Pedologists argued that the tests were valuable because they make it possible to pick out intelligent future leaders amongst the uneducated children of the people. In practice, the tests picked out children with incapacities and led to an alarming rise in the number of children described as defectives. As in every other country, the practical consequence of testing was to favour children from the more privileged social strata. This and other arguments shut pedology down in 1936.

Party activists did not think that the Soviet 'new man' needed a psychology. The rhetoric of the 'new man' went back at least to Chernyshevskii in the 1860s and carried utopian connotations of how human nature itself will be transformed in new social conditions. By the 1930s, the rhetoric referred to the ideal of a man who transcends what were previously thought to be the constraints of nature, who is also a moral agent with objective consciousness who, when he carries out party decisions, is the objective agent of historical progress. Political pressure equated this ideal man with the loyal supporters of Stalin and the party. Stalinist proponents of dialectical materialism argued that it is possible for labour to leap beyond existing material circumstances through a consciousness of historically constituted human nature. There was no place for a psychological science or a social science in their scheme for progress, and they branded any defence of these sciences an attempt to replace the dialectical laws of human history by the mechanical laws of nature.

Historical evidence about the factional struggles beneath the vague but coercive public language is hard to uncover. Some things are known about individual lives. Vygotsky died in 1934, aged thirty-seven, distressed by the marginalization of his life's work. His colleague Luria moved sideways from psychology, went to medical school and worked in the politically less contentious area of neurology. Though these scientists suffered from the political changes, a new philosophical voice was heard in psychology during the 1930s, that of S. L Rubinshtein (1889–1960). How he achieved prominence is not clear, but he became the spokesman for a dialectical psychology that links man's biologically evolved body and his historically situated consciousness in a manner that proved politically acceptable, at least in certain quarters. Rubinshtein was a professor in Odessa before promotion in 1932 to head the psychology department in the Herzen Pedagogical Institute in Leningrad. From this base he published 'Problemy psikhologii v trudakh Karla Marxa' ('Problems of Psychology in the Works of Karl Marx', 1934) as well as a textbook which presented psychology in an appropriate Marxist language and kept alive teaching in the subject. Interestingly, Rubinshtein, though perhaps in somewhat vague terms, turned to the recently published manuscripts of the young Marx for his basic concepts. As Rubinshtein wrote, 'the point of departure for the reconstruction [of psychology] is the Marxist concept of human activity', and he then went on to quote Marx's view that Hegel had revealed 'objective man, true, actual man, as the result of his own labor'.[44] Western Marxists later grappled with the same issues from the same starting-point.

Rubinshtein won a Stalin Prize, was promoted further and founded a department of psychology in Moscow in 1942. Between 1946 and 1949, however, nearly all forms of psychology, including Rubinshtein's, were criticized. The only exceptions were psychologies that built in name on Pavlov's work. The party's support for the status of Pavlov's science as the foundation of the natural science of man was dogmatically confirmed in three All-Union congresses, in physiology (1950), psychiatry (1951) and psychology (1952). These congresses were characteristic of the final paranoid years of Stalin's rule (he died in 1953). In the aftermath of the war to end the German invasion, a new wave of terror and anti-Semitism gripped the USSR, and show trials paralysed the Soviet-dominated half of Europe. Isolation, suffering and the Cold War fostered an extreme chauvinism about Russian science. Public speakers stressed Russia's unique historical contributions to scientific progress. A genealogy was constructed to link the nineteenth-century physiologist Sechenov to Pavlov, and Pavlov to a dialectical science of being and consciousness, and this was elevated into dogma that had a lasting influence on Soviet writing about the history of psychology. In practical terms, psychologists feared that the power given to the heirs of Pavlov's physiological research threatened the survival of psychology as an independent subject. This extreme consequence was resisted, however, and psychologists attacked 'the nihilist attitude toward the psychological legacy' at the 1952 psychology congress.[45]

Within a few years of Stalin's death, psychologists were able to broaden the research agenda, though appropriate obeisance to Pavlov in print or in public was still required. Rubinshtein returned to prominence in the second half of the 1950s with a dialectical study of *Bytie i soznanie* (*Being and Consciousness*, 1957). B. M. Teplov (1896–1965) elaborated part of Pavlov's thought into a systematic typology of human character, but he also published a textbook that clearly defended the autonomy of psychological topics from physiological ones. Leont'ev published the studies of child development on which he had been engaged since the 1930s and, under the heading of action theory, gave new institutional life to psychology in Moscow. By 1960, Soviet scientists had re-established regular contacts with the West, and in the area of brain research there was a real sense of discovery and excitement on both sides. Many researchers later commented on the stultifying effects of the official support of

Pavlov's science. Most damage was probably done in Central Europe – especially in East Germany – where ideological controls were tight and where scientists, distant from the élite centres of Moscow and Leningrad, were unable to decide what if any leeway there was in the research agenda. Nevertheless, psychologists found ways and means of doing much non-Pavlovian work; the Central European capacity for constructive survival under repression became legendary.

The history of Eastern Europe in the twentieth century is particularly open to revision. Marxism was elevated into principles that purportedly made objectivity in the human sciences possible – and yet the exercise of power destroyed objectivity and even, at times, almost eliminated the sciences themselves. Soviet science, which was rhetorically erected on a platform of Marxism, singularly failed to research relations between 'the individual' and 'the social' or seriously to examine how these categories might be conceptualized as abstractions in a continuous historical process. The great exception was Vygotsky, and his work belatedly reinforced the search for a truly social psychology, that is, a psychology that takes seriously the way that society, as a cultural and political presence, enters into the development of individuals and hence of what we denote by an individual. There were other Russian scholars, notably the literary theorist Mikhail Bakhtin, whose work had a delayed but profound impact on topics that connect language and reason and are central to the human sciences.

That Soviet Marxism consumed its own children in the human sciences cannot be explained solely as the aberration of a totalitarian regime. Intellectual, disciplinary and political difficulties, in every national culture, faced the creation of a theory able to unify and make intelligible individual and social agency. Consider what social psychology in the United States was for much of the century. It seemed natural to US researchers to presuppose the biologically independent person as the starting-point for psychology: they rarely attempted to study the socially and historically constituted nature of their subjects. The political and economic individual lauded in the public sphere in North America reappeared as the experimental subject. Experimental science examined individuals, usually young students, in the laboratory setting, and recorded results as statistically significant variables. This work was not itself discussed as historically situated action. Scholars like Mead, who questioned the intelligibility of description of the individual independently of the social world, or those who questioned the value of quantitative experimentation governed by statistical models, remained marginal to social psychology as it developed within academic psychology. The thought that social psychology, or some discipline concerned with people as the embodiment of historical processes, should underlie both psychology and sociology, was alien in both the USSR and the USA.

20

The Past and the Present

'So you are saying that human agreement decides what is true and what is false?' – It is what human beings *say* that is true and false; and they agree in the *language* they use. That is not agreement in opinions but in form of life.

Ludwig Wittgenstein, *Philosophical Investigations* (1953)[1]

i *The Behavioural Sciences*

The variety of psychology and the human sciences becomes almost limitless the more closely the past approaches the present. Two metaphors suggest themselves. In the first, a many-branched tree grows from a single trunk with a complex but common network of roots. Knowledge of human nature, in this picture, grows in the soil of human experience over the ages and gives rise to a common structure of concepts and objective methods, which in the light and air of the modern age bursts into lush greenery. The second metaphor derives from myth: Penelope awaits the return of her husband Ulysses, lost after the Trojan war. She is surrounded by suitors and promises to remarry when she has woven the cloth on which she works by day but which, unseen, she unpicks by night. Knowledge of human nature, in this picture, is woven in daylight only to be unravelled by the critical night. The weaving of the cloth begins afresh each day; claimants to the throne of psychology, like suitors, come and go.

Earlier chapters tell a story more in the spirit of the rewoven cloth than of the tree. The modern scientific disciplines of psychology and sociology have been traced to nineteenth-century developments, and knowledge and methods in the psychological sciences have been described in the twentieth century. The specialization of knowledge and the diversification of occupations has been explained by reference to local cultural and national contexts and to practical engagement with the conditions of life. Each chapter on the twentieth century traced a strand of psychology from the nineteenth century into the post-1945 period. Within defined areas, for example, in research on colour perception or in the elaboration of personality tests, psychologists experienced the accumulation of knowledge and expertise. Yet there was no overall agreement about what progress in psychology had achieved. Moreover, even in specialist areas, scientists encountered many difficulties, and some of these appeared to raise questions about assumptions built into the foundations of the subject. The sheer scale of psychological activity, the variety of its roots and patterns of growth and the range of contexts in which it was undertaken, made it an exceptionally complex enterprise. As two commentators observed in 1985: 'After a hundred years of ebullient growth, psychology has achieved a condition at once so fractionated and so ramified as to preclude any two persons agreeing as to its "architecture".'[2] Between 1920 and 1974, 32,855 doctorates were awarded for psychology in the US, 5,000 more than in physics. Finally, as the reference to something called psychological society has stressed, the complexities of psychology's history reflect the complex fabric of social relations and subjective life as well as the activity of scientists.

This concluding chapter sketches developments and debates in the immediate past in order to complete a picture of the background to the human sciences in the late twentieth century. It is exciting to see the weaving of a fabric with many of the strands used earlier but in new ways – and perhaps with new strands. We, in our turn, refashion belief about human nature.

Psychologists who located their subject in natural science and believed in the inherently progressive character of science sometimes claimed that, once and for all, they had founded psychology as an objective science. These claims occasionally became messianic; J. M. Cattell, for example, in 1930 said that 'it is . . . for psychology to determine what does in fact benefit the human race'.[3] The claim to have established an objective science was especially associated in the 1940s with the neobehaviourist Hull, and later with Skinner, whose research programme, though increasingly isolated, continued after his death in 1990. In the period after 1945, the claim was also associated with three major areas of activity, biological psychology (and sociobiology), neuroscience (and neuropsychology) and cognitive science. I will deal with each in turn. Researchers in these related areas enthusiastically argued that, finally, their subject had become a science, like physics, chemistry or biology, and these enthusiasms dominated academic psychology. The areas shared a background in the 1940s – in the neo-Darwinian evolutionary synthesis, in the brain sciences and in the sciences of communication, organization and computing. Much of the impetus for their development came from the United States and much of the motivation came from war or the threat of war. The description begins, however, with the idea of 'behavioural science', the term used in the 1950s and after to denote an ideal unification of psychology and the social sciences as a single field.

The Rockefeller Foundation funded the Yale Institute of Human Relations in the 1930s in order to unify science and to provide the basis for an expert, scientific response to public affairs. The result was Hull's formalized psychology, largely because his research programme resembled the sort of corporate effort envisaged by the funding body. US intervention in World War II, as well as the internal contradictions of Hull's work, shifted everybody's commitments. The war also made many more people sympathetic to the corporate organization of large-scale research – 'big science' – in circumstances where there was an easily agreed political objective, the defence of freedom against Germany and Japan and, later, the USSR. Belief that the United States pursued objective values like human rights and freedom, and a search to reduce social and racial tensions at home, continued after the war. At this time, social scientists reached a high point of optimism about how they could provide the tools to achieve social and political objectives. It was therefore thought to be a priority to integrate knowledge and research on human nature and public affairs and create a unified science. The result was the behavioural sciences. The catalyst was the Ford Foundation, which decided in 1952 to put several million dollars into academic social science. The Ford Foundation, like the Rockefeller Foundation earlier, wanted a co-ordinated mission, not piecemeal, individual research. The Behavioral Sciences Program was articulated as a business plan. Many researchers identified with its broad aims – to attract Ford funds, but also because they shared the aspiration to unify expert endeavour for the social good.

The word 'behaviour' retained the loose meaning that G. H. Mead, for example, gave it in the 1920s: 'Behaviorism in this wider sense is simply an approach to the study of the experience of the individual from the point of view of his conduct, particularly, but not exclusively, the conduct as it is observable by others.'[4] The word was current in the social, economic and political sciences as well as in psychology. It indicated that what was of interest in research on human beings was neither physical phenomena, the subject matter of the physical sciences, or imaginative and subjective experience, the subject matter of the humanities, the arts and religion. The word, it was understood, denoted the distinctive province of the human scientist – individual action in society. A leading researcher on government, Arthur R. Bentley, wrote:

> [Behaviour is] that specifically separate field of scientific inquiry, set over against the physical and vital, within which both 'social' and 'psychological' research must be carried on. It is that great type of activity which cannot be held within a physical description . . . nor within a vital, but which requires a directly psychological and social form of research, with whatever better descriptions and techniques we may secure to replace the two very imperfect words 'psychological' and 'social'.[5]

The word 'behavioural' marked out an area to be studied by observation in the manner of the natural sciences, but which could not be claimed by the biological and physical sciences. It also focused on the individual as social actor, in the hope of side-stepping the dilemma whether to assign the individual or society explanatory priority and thus overcome competition between psychology and sociology. The latter point was a particularly attractive one given that the goal was to establish a *practical* science of human affairs in a situation where training and specialization divided psychologists from sociologists.

A key figure in the Ford initiative was Bernard Berelson (1912–79), director of the Behavioral Sciences Program from 1951 to 1957 when it ended. As Berelson noted, 'the effort to build knowledge cumulatively require[s] that general categories descriptive of the behavior be set up and used more or less systematically'. The behavioural sciences represented the hope that 'behaviour' could systematically be made the subject matter of a unified social science. Berelson went on to say: 'The ultimate end is to understand, explain, and predict human behavior in the same sense in which scientists understand, explain, and predict the behavior of physical forces or biological factors or, closer home, the behavior of goods and prices in the economic market.'[6] One concrete result was the establishment in 1952 of the influential Center for Advanced Studies in the Behavioral Sciences at Stanford University in California, an institution which brought together scholars from different disciplines to pursue a common theme.

Berelson himself was a professor of library science who worked during the war on the analysis of German morale and opinion. He took this interest in opinion further and developed techniques to study voting behaviour; he worked with the Bureau of Applied Social Research to produce *Voting: A Study of Opinion Formation in a Presidential Campaign* (1954). Such studies were models for behavioural science. One of the co-authors of this study, Paul F. Lazarsfeld, was a lifelong promoter of empirical studies of social life, from the time of his innovative work, done with his colleagues Marie Jahoda and Hans Zeisel, on a community with high unemployment near Vienna, *Die Arbeitslosen von Marienthal* (1933; translated as *Marienthal: Study of an Unemployed Community*). There remained a gap, however, between such work in what was known as applied social science and what psychologists undertook when they studied individuals, a gap social psychology filled in only limited ways. The weakness of Berelson's own research, represented by a co-authored volume on *Human Behavior: An Inventory of Scientific Findings* (1964), was that neither science nor policy emerged from his lists of empirical propositions about what people do, however clearly the lists were formulated or however relevant they were to pressing social problems.

In spite of intervention by Ford and many gestures about behavioural science, the academic cores of scientific psychology and sociology remained distinct. Psychologists and sociologists sometimes joined together in applied work, but that work was not matched by the development of integrated theory. Within the broad field of psychology, however, there was a reinvigorated approach to behaviour in biological terms, which sought to understand observable activity as an organic process with a causal basis in the nervous system. I discuss in turn two substantial areas of research, the first of which built on basic principles of evolutionary biology, the second deriving from experimental studies of the nervous system.

ii *Between Biology and Culture*

Evolutionary theory appeared to establish continuity between nature and human nature and between natural science and human science. This principle was always of decisive importance to the theory's proponents, but it was not clear how to apply it. Throughout the twentieth century there was much disagreement about whether this belief requires that human nature be treated as a part of nature and human science accepted as simply a branch of natural science, even though all participants in the debate believed in evolution.

Early in the twentieth century, a reaction against speculative nineteenth-century reconstructions of anatomical, psychological and social evolution distanced much of the human sciences from evolutionary ideas. Durkheim's work in sociology, Malinowski's in social anthropology and Saussure's in linguistics directed their respective fields to the empirical study of present structures and processes to the exclusion of the evolutionary dimension. Nevertheless, at least two other major strands of the nineteenth-century worldview continued to run through the human sciences. Firstly, belief in the power of empirical methods retained its hold; Malinowski, for instance, drew on the views of the positivist philosopher of science Ernst Mach when he advocated field-based anthropological research. Secondly, functionalist explanation, explanation by the analysis of parts in relation to their place in a whole, behaviour in relation to mode of life or people in relation to society, continued to link the biological and the human sciences. The general result was the pursuit of detailed empirical research, with considerable attention to the achievement of objective methods, within a framework of explanation that presupposed particular norms of wholeness. By the mid-century, however, the dominant view separated the natural and the cultural spheres, and biology and the human sciences, on both intellectual and moral grounds. As Theodosius Dobzhansky, an architect of the neo-Darwinian synthesis of the 1930s, commented:

> Over and over again, some biologists made themselves ridiculous by urging solution of human social and political problems based on the assumption that man is nothing but an animal. How dangerous may be such false keys to human riddles is shown by the fruits of one of these errors – the race theory.[7]

Such comments did not hold back repeated attempts, backed by evolutionary theory, to support exactly the assumption Dobzhansky criticized, but the argument that the human sciences are a branch of biology was unorthodox until the 1970s.

Natural science and human science in the middle decades of the century broadly divided their labour between nature and culture; but these categories begged every question that was ultimately of interest. What indeed was 'culture' other than an umbrella term for everything that human scientists studied independently of natural science? The US anthropologist A. L. Kroeber (1876–1960), in an influential textbook, defined culture as 'a set of phenomena that invariably occur in the world of nature wherever men appear in it . . . to be studied comparatively, with complete equality of regard for all such phenomena, and without preappraisals among them'.[8] This statement described an area of study rather than clarifying the concepts of nature and culture. Kroeber, with Clyde Kluckhohn, made a brave attempt in 1952 systematically to review the meanings of the word 'culture', separating culture and individual behaviour as the subject matter of different levels of explanation. They were very willing to admit, however, that, though the concept is indispensable in the classification and explanation of human activity, there is no general theory of culture. Whether or not the concept of culture was a clear one, its use did signal agreement with the dominant view that the sciences of nature and the sciences of man (or culture) are separate. Another anthropologist in the US, Ashley Montague, wrote that man 'has no instincts, because everything he is and has become he has learned, acquired, from his culture, from the man-made part of the environment, from other human beings'.[9]

The emphasis on the cultural origins of human nature, linked to belief in the meliorist possibilities of the redesign of culture through social policy, was strong in the mid-century. This was an age when political social democracy flourished in countries like The Netherlands and Sweden and made possible the implementation of policies of planned welfare. The German Third Reich brought such horror to the biological language of human differentiation that it almost silenced non-cultural theories of what makes people different. In Britain, for example, eugenic argument was common in professional circles in 1930; outstanding scientific supporters included the statistician and biologist Ronald A. Fisher, the mathematician Pearson and the psychologist Spearman. The arguments then disappeared from public view during the 1930s, and when the leader of the Eugenic Society, C. P. Blacker, tried later to revive the society's fortunes, he bent over backwards to distance the study of heredity from a policy of state intervention. All the same, support for the hereditary basis of human capacities and the biological basis of human nature was present, for example, in Burt's and Eysenck's long-running research on the biological underpinnings of intelligence and personality.

Biological arguments about human nature began again to interest human scientists and to grip the public imagination in the late 1960s. Inspiration came from natural history and the deeply-embedded habits and traditions, which preceded but were reinforced by Darwin, that compare humans and animals. While the study of animals and plants became the subject matter of academic natural science in the nineteenth century, at the same time a great amateur and sometimes an academic interest in natural history persisted. There was a rich appreciation of animals and plants in their natural settings and – with self-conscious pleasure in the human comparison – in the personalities of animals, especially as pets. The academic and the public interest came together in the zoo and in the garden. In the 1940s a new science of animal behaviour, ethology, brought the natural history dimension, with its patient observation of natural animal behaviour, into contact with university, laboratory-based science. Then, in the 1970s, a group of evolutionary scientists argued for sociobiology, the integration of natural selection theory, ethology and human science, intending to subsume the human sciences into biology. Sociobiologists believed that the unity of knowledge, so conspicuous by its absence in the human sciences, could be achieved through the acceptance of the unity of man and evolutionary nature, that is, by the reconstruction of culture as biology.

The roots of ethology go back at least to the period before World War I. Laboratory-based research set the pattern for scientific biology, but individual scientists as well as natural historians continued to seek a less analytic, more direct experience of living animal nature. In England, Julian Huxley (1887–1975), the grandson of Darwin's 'bulldog', Thomas Henry Huxley, made subsequently famous studies in the wild of the dazzling display of the great crested grebe. The director of the Berlin zoo in the 1920s, Oscar Heinroth (1871–1945), pioneered a critique of a zoo's functions and stressed the differences between wild or natural and captive or artificial animal habits. Huxley and Heinroth wanted to observe what they significantly and positively valued as 'natural' animal behaviour, a motive far removed from those of US animal behaviourists and which distanced their work from what was called comparative psychology. This desire to know the natural animal was close to a public moral and aesthetic sensibility that favoured the natural over the artificial – an ever resonant contrast in an urban and industrial age. Heinroth's colleague Jakob J. von Uexküll (1864–1944), director of the Hamburg zoo between 1925 and 1944, introduced the concept of the *Umwelt*, the world to which an animal is bound by its sensory and motor capacities. He then conceived of research as the scientist's creative reconstruction of the animal's world. Such ideas were developed by the Dutchman Nikolaas Tinbergen (1907–88) and the Austrian Konrad Lorenz (1903–89). They devised rigorous ways to observe animal behaviour unaffected by man; they also refined the concept of instinct and initiated research on inherited behaviour patterns. Tinbergen's connections with England – he arrived after a period in a concentration camp during the war and was given a personal chair in zoology at the university of Oxford – led to the establishment of a distinct discipline of ethology, a discipline that subsequently interacted with comparative psychology in North America. Lorenz gained a huge audience with his stories of animals, *Er redete mit dem Vieh, den Vögeln und den Fische* (1949; translated as *King Solomon's Ring: New Light on Animal Ways*) and for his studies over many years of the greylag goose.

Lorenz's career exposed the moral and aesthetic ambivalence of the critique of what he and many others felt to be the unnatural qualities of modern civilization. When he compared domestic animals unfavourably with wild creatures and observed a loss of instinctual vitality, he also commented on the human values supposedly lost in modern life. He felt revulsion against industrial civilization in the 1930s and, perhaps idealistically persuaded by the rhetoric that the German peoples might make a radical return to their natural roots, he joined the Nazi Party. In 1940 he took up an academic appointment in Königsberg and in academic papers linked his biological theories to Nazi interests, particularly the concern to purify the *Volk* of degenerate tendencies. He evidently drew connections between the purity of nature and the purifying ideals that some intellectuals found in National Socialism, though there was little or no support for his work within the party. In 1939 he compared the domestication of animals to the deleterious effects on people made to live in large cities; in 1963, he discussed political activity in terms of twisted aggressive instincts.

> Aggression ... is an instinct like any other and in natural conditions it helps just as much as any other to ensure the survival of the individual and the species. In man, whose own efforts have caused an over-rapid change in the conditions of his life, the aggressive impulse often has destructive results ... For behavioural science really knows so much about the natural history of aggression that it does become possible to make statements about the causes of much of its malfunctioning in man.[10]

Lorenz's book on aggression was joined by a cluster of other studies by authors like Robert Ardrey, Desmond Morris and the opportunely named, and much more circumspect, anthropologists Robin Fox and Lionel Tiger, who attributed human psychology – aggressiveness, territorial imperatives, emotional expression – to an inherited animal nature. Their work found a receptive public audience in spite of – or perhaps because of – critical comment from social scientists who stressed the social and political determinants of human action and experience. The desire to find a basis for human action beyond politics, a ground in what popular language called real human nature, had strong appeal, which was reinforced by the rhetoric of objective biological observation used by the scientists. The new biological anthropologists accused their critics of knee-jerk left-wing politics and an unthinking rejection of the biological dimension. The critics accused them in turn of using biological determinism to legitimate political inequality and social injustice. There were clear parallels, as well as shared participants, with the contemporaneous debate about inheritance and IQ.

It was in this context that Edward O. Wilson (b.1929) published *Sociobiology: The New Synthesis* (1975), followed by a polemic aimed at the wider public, *On Human Nature* (1978). Wilson was an established Harvard University zoologist, an authority on the social life of ants, but his huge book on sociobiology was a bid to lay the foundations of a new science, 'the systematic study of the biological basis of all forms of social behavior, in all kinds of organisms, including man'. His ambition was staggering: he aimed to refound ethics, the humanities and the social sciences as well as human biology, all on the basis of 'truly evolutionary explanation of human behavior'. His choice of the word 'sociobiology' vividly carried the message that he thought social relations can be understood biologically, by which he meant that they are at base strategies for survival by the human animal. In Wilson's view, the application of biological knowledge is the most advanced animal strategy of all. 'Science may soon be in a position to investigate the very origin and meaning of human values, from which all ethical pronouncements and much of political practice flow.' Ultimately, however, he argued, even scientific reason comes up against our evolutionary inheritance of 'the ground rules of human behavior': 'There is a limit, perhaps closer to the practices of contemporary societies than we have had the wit to grasp, beyond which biological evolution will begin to pull cultural evolution back to itself.'[11]

Wilson's publications detailed the ways in which human behaviour, as he argued, can be predicted on the basis of natural selection theory. Sociobiologists explained the incest taboo and the practice of women marrying men of greater or at least equal wealth and status as part of a strategy to avoid the deleterious effects of inbreeding or to maximize reproductive capacity in hunter-gatherer societies. Wilson also argued that existing hunter-gatherer societies are comparable to the early evolutionary stages of mankind. He picked out four categories of behaviour – aggression, sex, altruism and religion – described them as 'elemental' and suggested how each can be analysed as part of the inherited survival strategies of a sociable animal. Like many nineteenth-century naturalists and social scientists before him, Wilson believed that knowledge of evolutionary nature, which he rephrased in modern terms as knowledge of genetic strategies, is the basis for all human science and decisions about what to do for the public good.

> The genes hold culture on a leash. The leash is very long, but inevitably values will be constrained in accordance with their effects on the human gene pool . . . Human behavior – like the deepest capacities for emotional response which drive and guide it – is the circuitous technique by which human genetic material has been and will be kept intact. Morality has no other demonstrable ultimate function.[12]

Wilson and like-minded sociobiologists made a bid to define genetic strategies as the 'ultimate' foundation of human nature, and therefore treated morality – and culture generally – as if it is only to be understood by its evolutionary function.

To sociobiology's many critics, such passages were acts of blatant disciplinary imperialism and, more profoundly, expressive of a gross reduction of human existence to one 'ultimate' biological dimension. As Wilson and other sociobiologists also wrote for the wider public, and clearly indicated that they thought their science had implications for political choices, controversy was intense. Critics associated sociobiology with the backlash against the left libertarianism of the 1960s, which had dreamed that all ways of life are possible, and with the rise of the New Right in the United States which believed in a fierce individualism as a 'natural' norm. Feminists were among Wilson's most ardent opponents as the appeal to nature featured so strongly in traditional views of gender. Conservative writers, by contrast, were happy to find support for their belief in the naturalness of heterosexuality, the family, property, pride in material reward and identification with local community. A biology of human nature that derived these values from evolutionary nature proved attractive. To other people, politically liberal or on the left, the argument denied the historically and socially constructed character and variability of human values, activity and institutions. Biologists themselves remained divided by Wilson's work and, while sociobiology and biopolitics became established as specialist fields, few scientists were as willing to generalize over such a vast area; instead, they concentrated on detailed studies of animal behaviour, population dynamics and the workings of natural selection.

Nowhere was there a closer relationship between animal research and views on human nature than in primatology, research on monkeys and the great apes. Fascination with this mirror to human nature went back beyond the eighteenth century and Buffon's and Linnaeus's preoccupation with classification. It continued in the debate about evolution – Darwin himself was an attentive visitor to the London zoo – and resurfaced in Köhler's experiments with chimpanzees in Tenerife and Robert M. Yerkes's work, also with chimps, designed to elucidate the nature of intelligence. The work differed, as ethology and comparative psychology differed, over whether priority was given to field-based study or laboratory research. Yerkes's interest, institutionalized in the Yale Laboratories of Primate Biology (opened in Florida in 1930 as the Laboratories of Comparative Psychology), encouraged intensive studies in the United States of primate learning. The main focus was on language, apparently the crucial capacity that divides people from apes, and on whether it could be learned by chimpanzees. One, Washoe, became a celebrity. She was at the centre of an intensive project at the university of Nevada in Reno begun in 1966, though extended but failed attempts to teach her language, rather than the use of signs, seemed to confirm the distance between animals and humans. Other work compared developments in young chimps and in children. Researchers did not reach agreed results and the whole topic generated diverse points of view.

Subsequently, the publicly acclaimed studies by Jane Goodall, who lived for extended periods with chimpanzees in their natural habitats in Gombe in Tanzania, questioned both the intellectual value and the ethics of laboratory studies. She was followed into the field by a number of other women, like Dian Fossey who lived and died with the Central African mountain gorilla, and they powerfully evoked different values in nature from those that were current in mechanist experimental science. After this work, it became incredible that, in 1935, Yerkes and one of his co-workers had referred to their research on chimp learning as a 'wholly naturalistic study of captive subjects'.[13] The fieldwork, which was a way of life for women who entered with empathy into the worlds of our nearest animal relatives, became a powerful model of a conservationist consciousness of nature, touched perhaps by a nostalgia for what people dreamed was humankind's own natural state.

The wider political and cultural dimensions of primate studies were examined by a US feminist historian of science, Donna Haraway. In *Primate Visions* (1989) she argued that these studies mirrored the researchers' own assumptions about human nature. Thus she compared reports about primate sexual and family life, which included descriptions of male dominance, with assumptions about gender identity and roles in contemporary society. Her work took further what had become during the 1970s the most extended critique of reference to nature in human affairs, the challenge to supposed natural gender differences. Indeed, the word 'gender' as opposed to 'sex' came into common currency to indicate that what was conventionally attributed to sex, to biology, could and should be described without any such assumption about its origin. Haraway's book was an analysis of how knowledge that appears to be determined by nature is the outcome of social relations mediated through claims about nature. Feminists, however, remained divided on the nature–culture issue: there were those who wished to assert woman's strengths because they believed she has a 'natural' closeness to nature embodied in nurturing qualities; others were suspicious of any argument couched in terms of what is 'natural' and instead sought emancipation in a woman's freedom to choose – including the choice of sexual identity.

The development of a gendered perspective, i.e., a point of view that gives priority to gender as part of the structure of an issue, influenced every aspect of the human sciences from the early 1970s – for example, descriptions of animal behaviour by ethologists, the psychoanalytically-oriented critique of gendered values in language and the history of science. In a book such as this, it became impossible to refer to human nature without asking questions about man and woman. Participants in the ensuing debate broadly divided between those who expected to find truth in the given world of nature, in reproductive biology, and those who argued that what is thought to be 'given' is constructed by human culture. In these circumstances, there was considerable interest in the biology, culture and history of sexuality, a matter of overriding political concern to feminists and also a crucial issue in knowledge of the origin and character of human differences. Some of the most challenging arguments stemmed from the work of French feminists. In the mid-1970s, the philosopher and psychoanalyst Luce Irigaray, who was then a colleague of Lacan, asked whether what we hold to be true about femininity is presupposed by the language in terms of which we speak about it. If so, she argued, it is possible to reconstruct femininity from a female vantage-point in language; this will demystify the feminine and make the feminine, not the masculine, the starting-point for speech. A huge amount of scholarship followed this line of thought because it suggested how academic subjects like literary criticism could be rethought in the light of the new consciousness of gender. As a result, as another French feminist, Hélène Cixous, observed: 'One can no longer talk about "woman" or "man" without being placed within an ideological theater where the multiplication of representations, images, reflections, myths, identifications transforms, deforms and undoes from the start all conceptualization.'[14] It was concluded that what a 'man' is and what a 'woman' is are cultural constructions; there is no vantage-point independent of these constructions on which we can stand to achieve a neutral point of view.

Such a 'theatre' of reflections was precisely what biological psychologists hoped to replace by the clear truths of natural science. Their ambition was to carry forward enlightenment, to reveal knowledge as an objective basis for a humane future. When Yerkes created a laboratory colony of chimps, his values were those of contemporary behaviourists and social scientists, his aim the prediction and control of human nature:

> It has always been a feature of our plan for the use of the chimpanzee as an experimental animal to shape it intelligently to specification instead of trying to preserve its natural characteristics. We have believed it important to convert the animal into as nearly ideal a subject for biological research as is practicable. And with this intent has been associated the hope that eventual

> success might serve as an effective demonstration of the possibility of re-creating man himself in the image of a generally acceptable ideal.[15]

This is a notable statement. Firstly, in 1943, Yerkes had no interest in the chimp as a wild animal; there was still a great distance between the assumption that nature is for human exploitation and the late twentieth-century Western concern with nature as a resource to be conserved. Secondly, he assumed that the underlying purpose of human science is a meliorist one; he drew no distinction between pure and applied work. Thirdly, the whole project required laboratory chimps and domesticated men and women to be compared. In Yerkes's world each species comes with a set of natural characteristics, and knowledge will enable us to reshape these characteristics in each species. Fourthly, Yerkes referred to a social phenomenon – 'a generally acceptable ideal' – but left it unexamined; he treated moral and political culture, in which such an ideal acquires its character, as something that is given. He certainly did not suggest that we might need knowledge of moral and political culture more than we need knowledge of chimps. He took ideals from the culture in which he lived and projected them on to the ape world, and he then used the evidence of the animal world to describe human nature. Knowledge and values about gender roles, aggression and competitiveness became locked in a circle of mirrors. But Yerkes and other researchers thought they studied nature.

The adjective 'natural' retained power in Western culture as a way to uderpin political values, in spite of a critical tradition that insisted 'we analyse relations of dominance in consciousness as well as material interests, that we see domination as derivative of theory, not of nature'.[16] The Enlightenment ideal to discover 'man as he really is', to seek knowledge by the comparison of civilization and an imagined state of nature, reappeared in transmuted form in late twentieth-century human biology. But the histories of mankind in which Vico and Herder portrayed mankind as a self-reflective creation of the human spirit also reappeared – also transmuted – in late twentieth-century human science. The debate that then occurred is discussed in the later section on the self and in the conclusion.

Some psychologists who were not biologists claimed that there is another way to enrich or possibly dissolve the discussion about the contributions of nature and culture to human nature. This was cross-cultural psychology, the empirical study of the constants in the psychological make-up of human beings. Universal constants of human nature were widely believed to exist – they were, for example, central to Freud's and Jung's ventures into anthropology – but considerable methodological ingenuity was in fact required to test for them in convincing ways. A growing body of opinion in the 1970s claimed, however, that there are, for example, constants of colour perception, mechanisms of memory and patterns in early child development. It was claimed that these are elements of a common human nature. Even so, cultural theorists argued, little is gained by reference to psychological constants in the abstract since they exist in concrete terms only as expressed in the languages of particular cultures. From this perspective, to understand a human being's psychology is to understand how a person is referred to – how the person is a focus of signification – by a language or symbol system, whether it be psychological, biological, political or religious. Critics of the comparison between animal and human nature always stressed human language use; in the human sciences great weight was placed on theories of the acquisition and use of language, such as Vygotsky's and Lacan's. If indeed there are biological constants, human scientists concluded, we are still faced by actual people whose human nature exists in the manner in which it is expressed through cultural forms.

iii *Neuroscience, Brain and Mind*

At the high tide of late nineteenth-century belief in scientific progress, many people thought that the body is the high road to knowledge of human nature. The secrets of man's nature, it was argued, must lie in the material structure, the brain, which is at one and the same time both similar and different in man and beast. The first generation of experimental psychologists felt obliged to say why their research was not subsidiary to neurophysiology, the science of nervous function, and why psychology should be an independent discipline and not just a branch of physiology. The obvious assertion was that consciousness is psychology's special province. As J. R. Angell argued in 1913: 'When she abandons the stronghold of consciousness as her peculiar institution, psychology is moderately certain to find that as an autonomous government she has ceased to exist . . .'[17] Subsequently, however, categories like behaviour, learning and cognition were used to delineate a subject matter for psychology that is physical and hence – unlike consciousness – open to natural science methods, and yet, though physical, not the province of physiology. When psychology became an autonomous science, as it did most clearly in the United States, psychologists still faced the problem of how the objects that they studied, such as visual images or behaviour, relate to the brain as studied by physiologists. As psychology and physiology both became highly specialized areas and very large disciplines, it became easy and common for scientists, on a day to day basis, to ignore problems about the relation between the fields. Some researchers, however, decried this entrenched separation in what, they argued, should be a unified science of the human organism, a science with psychologists and physiologists at work together on the study of nervous function.

The relations between psychology and physiology varied with local context. Pavlov and Watson, for example, shared an enthusiasm for conditioning as an experimental method, but Pavlov used the method to demonstrate cerebral functions and called himself a physiologist, while Watson – followed in this respect by Skinner – ignored the brain and sought only to correlate stimulus–response connections, and he called himself a psychologist. The division of labour between physiology and psychology was pragmatic, and it responded to institutional and disciplinary pressures, the training of scientists, the medical context and so forth. Scientists like Pavlov were attracted to the study of behaviour because they hoped it would lead to empirical knowledge of the correlation of nervous and psychological processes, but they did not specify how mind and brain relate. The gestalt psychologists presupposed isomorphism, a parallelism between mental and physical form in the conscious world and in the brain. Psychophysics continued to be an active area of research in the twentieth century, a branch of psychology concerned with the quantitative correlation of psychological and physiological sensory processes. In general, clarity and agreement about the relation of mind and brain was absent. Most natural scientists put to one side questions about the relation between brain and mind or consciousness and described such questions as philosophical not scientific, best left to people who like that kind of thing. This state of affairs was in part caused by confusion or disagreement about what is an empirical question and what is a conceptual question. By training and inclination, natural scientists looked for empirical answers to questions about the brain's function as the material basis of mind. This supported the search for the localization of functions, the correlation of psychological processes with particular regions of the brain – for example, visual perception with the optic lobes. But such work did not answer the question of how mind correlates with brain. Indeed, many philosophers – and some scientists – argued that this is a conceptual not an empirical matter, a matter of how we use language and claim to know about the world, not a factual matter to which there might be a concrete answer.

The subject areas that brought together the study of the brain and psychological functions expanded very rapidly in the decades after 1940, and it was one of the largest and most active areas in all of natural science. Philosophical problems about the mind–brain question provoked lively debate, at least from the 1960s when the problem of consciousness once again became prominent. The research of Karl S. Lashley (1890–1958) links the earlier period, the time of Pavlov and Watson, and modern neuroscience. Lashley initially worked with Watson and Adolf Meyer, the head of psychiatric medicine at Johns Hopkins University, and he hoped to trace conditioned reflex paths through the nervous system. He was taught the necessary surgery by S. I. Franz (1874–1933), a physiologist and psychologist who studied the question whether there are localized brain areas that correlate with learned habits. Lashley found his research project impossibly complex and perhaps theoretically confused, and he therefore turned to the study of cerebral function. He challenged the dominant belief in the localization of functions and advanced an alternative conception, which he called mass action, a theory that relates functions such as learning to the whole brain. Like Sherrington, Lashley was impressed by the integrative activity of the nervous system: 'The units of cerebral function are not simple reactions, or conditioned reflexes . . . but are modes of organization.'[18] He supported this general conclusion with experimental work that involved the destruction of successive parts of the rat's brain, followed by study of what the rat is still able to learn in a maze. As Lashley held a series of chairs in Minnesota, Chicago and Harvard (when he also worked at the Yerkes Laboratories of Primate Biology), he inspired a large number of students with this type of experimental approach to psychological questions. He also brought into focus the long-standing divergence of opinion between scientists who wanted to analyse brain function into elementary units and scientists who were committed to a more molar or holistic conception of function. Pavlov's commitment to the latter viewpoint, in what in the West was an outdated version, contributed to the separation between Pavlov and his Russian school and the mainstream in Western neurophysiology. Lashley challenged the Western mainstream and opened up broad theoretical questions about the wholeness of brain and mind.

A number of circumstances came together in the 1940s to create the feeling that the brain is an enormously exciting research topic. Ambitious scientists, large funds and institutional support transformed the scale and pace of research. Neurophysiologists, many of whom had worked with Sherrington, achieved detailed knowledge of the nerve impulse and the communication mechanism, or means of synaptic transmission, between individual nerve cells. The larger challenge was to show how events at this micro-level are organized to serve an organism's life, including its psychological functions. Investigative techniques into the brain, which included micro-electrodes, micro-surgery and neuropharmacology, promised a new level of precision in experimental work. Clinical neurology and brain surgery had passed through half a century of exploratory but systematic work, and there was the feeling that the fields had enormous potential. In 1929, the investigation by the head of neurology at the hospital in Jena, Hans Berger (1873–1941), of the brainwaves recorded by the EEG (electroencephalograph) added an important analytic tool for experimental work and clinical diagnosis. The EEG is the pattern of electrical activity in the brain made visible as waves on an oscilloscope screen or, later, by pen recorders. Research on bodily control, which brought together post-1900 knowledge of the hormonal system and nervous regulation, pointed to the significance of the mid-brain region, a complex area previously largely inaccessible to and ignored by brain research. Concrete experimental achievements, such as the report by H. W. Magoun in 1944 of an inhibitory control centre in the mid-brain, inspired further work. The war and the necessity to deal with brain damage reinforced belief in the humane value of experimental research. Work by scientists like Lashley with rats and John Fulton (1899–1960), a Yale neurophysiologist, with monkeys, created a wealth of background knowledge about the physiology of the experimental subjects.

Brain research also benefited in the 1940s and subsequently from its image as the next great frontier in natural science. The 1930s saw rapid developments in atomic physics, which made possible – in an unprecedentedly large programme of corporate science – the construction of the atom bomb. That seemed to end an era in science; some physical scientists looked elsewhere for new challenges and turned both to genetics, where they helped create molecular biology, and to brain science. There was also the belief, the significance of which is hard to assess but nevertheless real – as a flourishing genre of science fiction testifies – that the brain opens a natural-scientific door to the mystery of human existence. Some researchers even persuaded themselves that nothing more is to be known beyond the brain, that the brain is not just the next but, alongside space research, the final scientific frontier.

Whatever the precise combination of factors, brain science became one of the fastest growing areas of science during the 1950s, and the pace did not slacken thereafter. The centre of work was in the United States but, as with the sciences generally, research became increasingly international. With the relaxation of the Cold War towards the end of the 1950s, brain research was a significant area of interaction between US and Soviet scientists. Post-Pavlov research in the USSR, which deferred to Pavlov's name but was often not indebted to his theory or methods, included a good deal of brain physiology. I. S. Beritashvili (1884–1974) worked on cerebral functions – work he carried on in Tbilisi in Georgia through the Stalinist period – and Luria studied the correlation of behaviour and brain damage. At Cambridge the English physiologist E. D. Adrian (1889–1977) supported work that related knowledge of nerve properties to the psychological function of visual perception, and this sustained the precision customary in experimental physiology while it moved towards topics of interest to the psychologist. Scientists believed that knowledge results from small but rigorous steps. One implication of this was that the direction of research was often determined by technical considerations: the brain offered endless challenges to instrumentation, experimental ingenuity and the ethics of vivisection. The result was a vast field broken into technical specialities, with few scientists willing or able to provide a synthesis or to spare the energy necessary to reconsider the field's conceptual and theoretical underpinnings.

An attempt to provide a general picture, in terms relevant to psychologists, was made by a scientist in Canada, Donald O. Hebb (1904–85), in *The Organization of Behavior* (1949), a book which owed a great deal to Lashley. Hebb discussed the 'big view' and constantly referred work back to the neuronal structure of the brain. He reminded his psychological colleagues of the physiological dimension of their field and inspired them with hope that advances in knowledge of the brain might make it possible at last to answer psychological questions. Hebb worked for a while in Montreal with Wilder Penfield (1891–1976), a naturalized Canadian and innovative neurosurgeon also interested in wider questions of brain function, and then with Lashley at the Yerkes laboratories. These contacts encouraged Hebb to emphasize the brain's central organizing capacity, and this contributed to the decline of neobehaviourist learning theory. The critical direction was taken further by Karl H. Pribram (b.1919), a scientist who shared Hebb's willingness to construct general principles of brain organization. Pribram, who began life as a brain surgeon and then used his skills for research, first at Yale and then at Stanford, was willing to explore the possibility that the character of brain organization might not parallel the character of psychological organization. This implied that psychological investigation had to proceed on its own terms and not allow itself to be reduced to ideas that satisfy current knowledge of the brain. The challenge, in Pribram's view, was to explore models and metaphors of brain action, if necessary going beyond the present state of knowledge, to see how psychological events might be represented as, even if not explained by, brain structure. In the 1970s, for example, Pribram and his collaborators explored the hologram, the representation of colour in three-dimensional space, as a way to picture visual perception and memory in the brain, and these speculations seized the public imagination.

Specialist areas of neuroscience became huge research topics in their own right: the question of bilateral symmetry and of functional dominance or dependence between the left and right sides of the brain; the role of chemical substances in the midbrain in emotional control and in illnesses like schizophrenia and Parkinson's disease; the mapping of connections between the retina and the optic nerves as they enter the brain. Each topic proved more complex and taxing than originally expected. Then, during the 1970s, an unanticipated shift of public opinion against many kinds of research, a shift that created the animal rights movement, raised ethical questions that most scientists thought had been settled in the late nineteenth-century debates on vivisection. Neuroscientists tended to assume that the intrinsic value of knowledge and the major contribution of brain research to medicine legitimated experimentation within legally laid-down norms. Animal rights supporters and anti-vivisectionists questioned both points: they doubted the value of knowledge acquired by violence against nature and argued for alternative, non-vivisectionist methods in medical research. Ethical debate also began to confront issues raised by innovations in medical neuroscience, such as the propriety of animal brain tissue transplants into humans and the criteria for what constitutes brain death. All of this had great potential to disturb common views about the person and personal identity.

The investment of intelligence, experimental skills, money and institutional support in brain science caused it to flourish but also initially encouraged its isolation from philosophical questions about mind and consciousness. Nevertheless, philosophical, ethical and – for some – religious questions re-emerged whenever scientists in the area attempted to gain a public audience or to communicate their conception of human nature to academic colleagues who worked in the humanities or social sciences. Some natural scientists simply believed that all questions are factual questions, and this implied that further neuroscientific research could be expected to solve the questions about mind and brain that neuroscience itself generated. This attitude is comparable with the opinion about the problem of free will expressed by the sociobiologist Wilson: 'the paradox of determinism and free will appears not only resolvable in theory, it might even be reduced in status to an empirical problem in physics and biology'.[19] This was the voice of natural science triumphant, a voice that claimed the existence ultimately of only one meaningful discourse, one level of truth, the truth of empirical natural science. Critics denigrated it as scientism.

The mind–body problem was the philosophical parallel to the social reality of psychology's awkward relationship with physiology. Much of the time, physiologists got on with physiological research (as in Sherrington's Oxford school) and psychologists got on with psychological experiments (as in Bartlett's Cambridge school), without direct attention to each other, however much both approaches were concerned with functions of the brain and in spite of the fact that scientists slipped in and out of the use of mental and physical concepts and language. The division of labour between psychology and physiology was sometimes formulated philosophically as psychophysical parallelism, the claim that the mental and the physical constitute two separate realms and that each realm requires independent research and language. This further implies that the relation between the realms is unknowable, or at least unspecifiable in scientific terms, and that each discipline should therefore get on in its own way without wasting time on unanswerable questions. After 1945, neurophysiology productively researched perception and the emotions and provided much detail on the material dimensions of perception and emotion, but did not say what a conscious sensation or feeling is. In practice, scientists often either discussed perception and emotion simply as physical events, in effect doing physiology and not psychology, or they discussed the brain as the causal basis of perception or emotion and reiterated a dualism that left mind unexplained.

The dominant philosophical approach to the mind–brain relation during the decades of growth in brain science was the identity theory associated with the US philosopher and ex-member of the Vienna Circle Herbert Feigl and with the physicalism of the Australian philosopher J. J. C. Smart. Feigl maintained that brain events and mental events are one and the same, though they are viewed from different perspectives and discussed in different contexts. The theory was agnostic about materialism; when held informally by neuroscientists, however, it often blurred into materialism. The materialism was directly expressed in Smart's arguments, which identified consciousness with central brain states. A subtle version of the identity theory was introduced by the Harvard and then California philosopher Donald Davidson, in lapidary articles on 'Truth and Meaning' and 'Mental Events'. Davidson's writing was so brief that one reviewer described him as 'perhaps the most distinguished philosopher in history never to have written a book'.[20] What Davidson attempted was to combine two conceptions of what it is to be human, one that treats humans formally as physical systems and the other that describes ourselves in ordinary – 'mental' – language. He argued that if we specify the proper sphere of each conception, this will enable us to accept ourselves as simultaneously material and mental, causal events and rational agents. We are, he stated, single, but this singleness is describable in irreducibly distinct ways. The precise argument turned on much discussed logical points; the general direction of argument, however, which separated different but complementary modes of language, appealed widely.

There were demands for philosophers to pay attention to new neuroscientific knowledge from, amongst others, the materialist Californian philosophers Patricia Churchland and Paul Churchland. They objected to the separation of philosophy from the new knowledge generated by neuroscientists. The question of how philosophical and scientific claims relate remained a difficult and divisive issue, but materialists – as earlier in the nineteenth century – turned to science for arguments against philosophers. Modern materialists denied that the word 'mind' refers to something that can be specified independently of statements about brain processes, however much our culture has taught us to describe ourselves in terms of mental images, feelings and intentions. Materialists therefore undertook to show how it is possible to redescribe what is called mental as something physical – or, at least, to show how such redescription will become possible as factual knowledge advances. These abstract issues were tackled by the flourishing branch of philosophy called the philosophy of mind. Though philosophers and scientists had very different training and thought in different ways, the division of labour between what philosophers do and what scientists do was far from agreed. Wilson's claim that the problem of free will may be solved by physics and biology implied that science has (or will have) all the answers. Other philosophers fascinated by science, like the Churchlands, found that even when they claimed that science provides factual answers, they had to face questions about how we formulate concepts and use reason and language when we make knowledge of ourselves. It was widely accepted that natural science is not able to answer all the questions independently of philosophy. Some philosophers, including Daniel Dennett and others sympathetic to science, argued that natural science cannot in principle be autonomous and that philosophers and neuroscientists must therefore work together to understand consciousness. Many philosophers, and some scientists, in the 1990s continued to defend the view that mind–body questions are philosophical not scientific. Yet other contributors to the debate denied that there is a mind–body 'problem' at all: we have only failed to adopt the viewpoint, concepts and language of the relevant sciences. But when the language of the relevant sciences was compared with other possible languages or ways to describe people, then it appeared to most observers that there is indeed a problem. Owing to the impact of computers, discussed in the next section, the question of materialism was often reformulated as the question of consciousness: is consciousness something that a machine might have?

Not everyone accepted mind–brain identity or materialism. John Eccles (b.1903), who won a Nobel Prize for physiological research on the synapse, later joined by the Viennese-English philosopher, Karl R. Popper (1902–94), went against the dominant trend. Eccles and Popper argued for what many philosophers and scientists thought of as the 'dead' theory of mind–body interaction, the theory that the brain and the mind are distinct but causally related worlds. Eccles ventured to speculate about how mind can be said to interact with body, but Popper simply recorded ignorance. At some late stage of the process of evolution, Eccles argued, mind appeared in the universe, and this makes man truly novel. Argument in such terms connected with a century-old debate about how and when mind appeared in evolutionary history and to one influential answer known as the theory of emergence. Early evolutionists like G. H. Lewes and Lloyd Morgan described the evolution of mind as a qualitative novelty somehow produced by the evolution of a certain level of complexity in the physical system of the brain. They thought in terms of the analogy of the emergence of chemical properties, the phenomenon in which the combination of atoms into molecules results in properties ('supervenient properties') not present in the atoms alone but which appear, it was supposed, as a result of organization among atoms. As identity theories about the mind–brain relation gained ground in the 1950s and subsequent decades, theories of emergence, which accept mind–body dualism, did not seem to be needed. Nevertheless, emergence and identity theorists alike treated physical and mental languages as different levels or modes of explanation.

Another way to consider the mind–body problem came from systems theory. Though its conceptual background was in the nineteenth-century evolutionary functionalism of Herbert Spencer, systems thinking became significant after World War II as a theoretical approach to the management of large-scale undertakings like the Manhattan Project and intercontinental nuclear missile systems. Independently of the war, the Viennese scientist Ludwig von Bertalanffy (1901–72) hoped to achieve a general description for any system – the physical world, organisms or a manufacturing technology. He believed that 'general systems theory' (also the title of his book published in 1968) has the potential to unify the non-physical sciences, those of mind and brain included, through the way it specifies the properties of any system or organized complex of interacting parts. A systems approach, it was maintained, makes reference to the separate spheres of mind and brain an unnecessary abstraction; instead, the scientist is able to apply the general knowledge of systems to a particularly high level of organization.

The impact of modern analytic philosophy – philosophy directed towards the solution of problems by the analysis of what we can meaningfully say – on mind–body questions followed the argument by Gilbert Ryle (1900–76) in *The Concept of Mind* (1949). Ryle was an Oxford philosopher whose book had considerable influence and appeared to match the opinions that underlay intellectual investment in the behavioural sciences. He attacked reference to internal states of mind on linguistic and logical grounds, and he denigrated belief in what he called 'the dogma of the Ghost in the Machine', the Cartesian theory of the mind as the inhabitant of the body.[21] He suggested instead that the language of internal states, when properly understood, describes observable realities of intellect or character, not an unobservable and mysterious mind. This was a message natural scientists wanted to hear.

Ryle's argument was carried forward into the 1990s and reformulated to deal with computing and the new focus on consciousness by his student Dennett, a philosopher at Tufts

University in Massachusetts, whose enthusiasm for neuroscience as well as philosophy helped shape the discussion in the 1980s and 1990s. In *Consciousness Explained* (1991), he attacked what he called 'the Cartesian theater', belief that there is an 'I' somewhere in the brain which 'observes' the world as it passes on the stage before it. He argued that we must give up this tenaciously-held language and that we can do so because computer processing has produced a better language. He proposed to change the metaphor in terms of which we think about consciousness from the Cartesian self to the processing brain. In the terms of the new language, our knowledge of ourselves is knowledge of processing states in the brain, and all that 'distinguishes a conscious state from a non-conscious state . . . [is] the straightforward property of having a higher-order accompanying thought that is about the state in question'.[22] The exploration of parallel processing in computers suggested models of how two orders of thought might exist in the brain at the same time. To explain how such 'higher-order' thoughts originate, Dennett turned to evolutionary theory. His critics questioned both whether there is any plausible way to explain conscious awareness as the outcome of natural selection and whether his brand of materialism can indeed deal with what, in ordinary language, we know as the quality or feel of awareness (imagine, for example, a green velvet dress). This character, or phenomenology, of immediate experience remained central to debate and divided opinion in the 1990s.

Other important work on the philosophy of mind stemmed from the impact of Ludwig Wittgenstein (1889–1951). The life of this extraordinarily brilliant and proportionately anguished man was legendary; it spanned the wealth of turn-of-the-century Vienna – his sister was painted by Klimt – to service as a hospital orderly at Guy's Hospital, London, during World War II. In his later work, communicated to a small group at the university of Cambridge and then published in translation after his death as the *Philosophical Investigations* (1953), he addressed language as the source and solution of philosophical problems. He argued that language must be understood to obey the rules embedded in a way of life – it is primarily a social practice – and language cannot be said simply to refer to 'real' things. According to Wittgenstein, an emotion is not a state, as we ordinarily suppose, but a reason inscribed in a language. To understand emotions is therefore to understand how people in a particular culture use the language of the emotions. It also follows, in Wittgenstein's view, that mental and physical languages are different language games, linguistic activity according to different rules, that involve different contexts of use and meaning. There is therefore no mind–body problem, in the sense of a problem about the relation between real entities of mind and body, but only a problem about the relation between languages. There is also no such thing as a private language or a private meaning: nothing can be said about a private, 'inner' world because to say something is to join in a social, 'external' language game. Such arguments, often expressed as gnomic utterances, led to a veritable philosophical industry in the interpretation of what Wittgenstein meant and in the exploration of the implications of his statements for the philosophy of mind.

Discussion of mental states was enriched in the 1950s and 1960s by English philosophers who argued that to explain a person's action is to give a reason for it or to state the act's intention. G. E. M. Anscombe, a student of Wittgenstein, wrote on *Intention* (1959), R. S. Peters on *The Concept of Motivation* (1958) and Peter Winch on *The Idea of a Social Science and Its Relation to Philosophy* (1958); these books knocked the stuffing out of behaviourist explanation and questioned the inclusion of the human sciences in the natural sciences. The principal point was a simple one: to explain a physical event is to specify the causal law of which the particular event is an instance; to explain a human action is to specify why, in terms of reasons or intentions, someone does what they do. Physiologists talk about muscle contractions and nervous control when I stand up, but ordinary people explain my standing up as the expression of respect to someone. The argument started debate about such questions as whether reasons are causes or whether we can make sense of explanations in terms of unconscious reasons. Psychologists assimilated the argument that behaviourism is untenable since, when a behaviourist explains a person's action by reference to causes such as conditioning, this simply does not tell us what we want to know in order to understand an action. Behaviourist psychology, it was widely argued, does not explain human beings because it denies or ignores their intentionality. As we shall see, this argument was also central to controversy over computer or artificial intelligence.

Debate about language and explanation was characteristic of Anglo-Saxon philosophy. This was largely cut off from continental European thought, which was less concerned with the theory of knowledge and more concerned with the distinctiveness of what it is to be human and with the interpretation of men and women as constituted in the structure of being, language, history and society. The work of intellectuals such as Heidegger, Jürgen Habermas and Foucault, which began to be assimilated in the anglophone world in the 1960s, pointed to what still had to be done to found the human sciences rather than to problems of explanation within them. This work also used a forbiddingly abstract language, often translated only with difficulty into English. Further, many continental European writers were indifferent to the natural-scientific dimensions of human nature, and they therefore simply did not deal with questions raised by the brain sciences. All this reinforced a cultural and intellectual divide between English speakers and the French- and German-language worlds, and between philosophers who addressed questions raised by natural science and philosophers concerned with man's existential and political being. It became even more difficult for empirical psychologists or social scientists on the one side and theorists on the other to understand each other – when they bothered to try. The rapid expansion of higher education in the 1960s in Europe as well as in English-speaking countries provided institutional support for large communities of both empirically-oriented and theoretically-oriented academics and students. New disciplines and institutional support made it possible for continental theory to find a new place in Britain and North America. In these circumstances, there was a tendency for each position, whether in natural science or in philosophical theory, to become a faction, for each faction to divide into further factions, and for each faction to talk only to itself, with the end result the creation of impotent intellectual ghettos.

Debate about the character of the human sciences and the explanation of action merged, at a certain level of abstraction, with intellectual life generally. This was evident in the ferment known as 'structuralism' – though such words were used with promiscuous abandon. Piaget, for example, at one time used the word to describe his philosophy, though he shared little in common with the French-language Swiss theorist Ferdinand de Saussure (1857–1913), who was the main inspiration. Saussure's course on general linguistics, published in 1916 from student lecture notes, treated language as a sign system, that is, he detached words from any necessary relation to what they signify and studied their articulation with each other, and hence their meaning, as part of a system. His work suggested to French intellectuals in the 1950s that the human sciences must start from the fundamental rules that determine the organized structure of language, cognition and symbolic expression. The character of language, reason or expression in themselves, it was argued, reveals abstract structures that are at the foundation of the study of human action. Questions about what such structures 'are', in the sense in which an English-language natural scientist might have asked the questions, were not addressed. Rather, Claude Lévi-Strauss (b. 1908) in social anthropology, Roland Barthes in semiology (the science of signs) and Louis Althusser in Marxian philosophy took the starting-point for their respective sciences to be the abstract structures of language, reason and symbolism, in terms of which they then explained the observed activity of the human world. At one stage they were all emphatic about the scientific rigour of their enterprises.

For the most part, this francophone conception of science spoke to an audience different from the anglophone audience for psychology. There were, however, two influential points of contact, the work of the Swiss developmental psychologist Jean Piaget (1896–1980) and the US theorist of linguistics Noam Chomsky (b.1928). Both were called structuralists (though this is a label that needs much qualification) and their treatment of the internal structure of language and reason had considerable critical impact on the English-language behavioural sciences.

The description of the wider context in which this happened requires a separate section. Later, I return to the work of two other francophone writers with their intellectual roots in the structuralist debates, Foucault and Lacan.

iv *Cognitive Psychology*

Around 1970, it was common to claim that cognitive psychology was supplanting behaviourism. The new psychology investigated problem-solving, learning and memory as forms of information processing, and the new psychologists referred without embarrassment to internal psychological states. This felt like a revolution to psychologists who had been trained to study the external behaviour of rats or pigeons, but the claim that there was a cognitive take-over was an exaggeration. Firstly, psychology never was a united behaviourist discipline; non-behaviourist research like factorial analysis, gestalt psychology and Bartlett's work on internal schemata also went on. The 'revolution' overthrew regimes only within particular institutions. Secondly, it was a matter for debate whether explanatory concepts in fact changed fundamentally. It is not clear that a switch from explanation in terms of behavioural variables to explanation in terms of information-processing variables altered the basic assumptions of a causal, physical-language discourse. Scientists who explained human action in terms of intentions, in terms of the ground of being or in terms of historically constituted reasons did not see a cognitive 'revolution' but only a change of terms within a natural-scientific worldview. Cognitivists, their critics argued, like behaviourists before them, were interested in what the human machine does; they were not concerned with deeper questions about the nature of being human.

This said, cognitive psychology certainly became the heartland of late twentieth-century academic psychology. Two factors stand out as reasons for this: the many ways in which psychologists found it necessary to refer to internal psychological structures; and the development of computing technology –

something truly revolutionary, if that misused word can be applied to developments lasting several decades. The power of electronic information processing, allied to the massive investment of financial and human resources, turned cognitive science – of which cognitive psychology is only one dimension – into perhaps the most significant science of the late twentieth century.

A self-conscious revival of interest in cognitive, internal psychological activity became apparent in the 1950s. One specific influence was the anglophone rediscovery of Piaget's work on child development, though the wider dimensions of his philosophy were not always appreciated. After 1921, when Piaget arrived at the Jean-Jacques Rousseau Institute in Geneva, founded by the Swiss psychologist Claparède, he systematically studied children's development. He was interested in such questions as how and when children acquire perceptual and conceptual abilities and an understanding of the world, causal relations and the boundaries between self and other. This research was known outside French-language culture – it was, for example, a point of dialogue for Vygotsky in the early 1930s – but in the 1950s it aroused unprecedented enthusiasm among anglophone educational psychologists. Piaget argued that the child passes through a series of fixed stages, implying, for instance, that it is necessary to evaluate intelligence by reference to the sequence of ways in which the child develops her or his reasoning. The imaginative ways in which he did his research and developed his results, which built on studies of individual children, introduced a breath of fresh air into educational psychology. Piaget's research was also based on a theoretical standpoint, and he described the child's developmental stages in relation to a biologically evolved pattern, an inherited structure, which organizes experience and action as the child grows. Though his impact was slow because of the ways Piaget sometimes expressed himself in philosophical, biological and formal terms, enthusiasm for his work brought a structural conception of the mind, set in a biological framework, into a major division of English-language psychological practice.

Piaget's studies of child development were part of a large-scale programme to integrate the theory of knowledge, in the tradition of the philosopher Kant, with evolutionary biology. He hoped to square a circle and make the content of a science, evolutionary and developmental psychology, the same as the conditions of possibility for knowledge, epistemology. He called this undertaking genetic epistemology and, in later writings, explored its relation to structuralist philosophy. As he wrote in his brief autobiography, he wanted to 'be in a position to attack the problem of thought in general and to construct a psychological and biological epistemology'.[23] Whatever the outcome of his philosophical arguments, his historical impact was to legitimate reference to mental structures as explanatory categories and to familiarize psychologists with them.

The background to Piaget's own interest in intelligence was a religious one, and it was linked to the fate of liberal Protestantism at the time of World War I. As a young man during the war, Piaget hoped to integrate in his mind a social calamity – attributed by many people at the time to modern secular values – with Christian belief and with a commitment to the objectivity of science. His solution, perhaps uniquely ambitious, was to identify human reason and morality with the immanent reason of God in the world and then to study the development of reason as the objective means by which reason and morality can be known. The culmination of these arguments, and of the child studies generally, was *Le jugement moral chez l'enfant* (*The Moral Judgement of the Child*, 1932), but none of the religious inspiration appeared in print in this or, indeed, in earlier books. Rather, what Piaget's readers saw as his contribution was a psychological argument to show how the acquisition of moral belief is a stage in a child's growth and how socialization is natural and normal. It was this secularized form of his work that was influential in English-speaking countries, and it equipped educational psychologists with what were thought to be objective grounds for optimism about children's acquisition of values. Piaget persuaded teachers that education must follow the child's pattern of development; he claimed, for example, that moral learning takes place through the child's interaction with other children rather than through moral instruction.

The new field of psycholinguistics, which also accepted the existence of mental structures, had a comparable impact at more or less the same time. In this area a specific event, Chomsky's review of Skinner's book on *Verbal Behavior*, was decisive. In the late 1950s, Chomsky, a young professor at MIT, shredded behaviourist pretensions to give an adequate account of language and thus ended their pretensions to explain human beings. Skinner had long recognized language as the test case for his science and, with admirable intellectual integrity, developed his work to tackle this capacity. He apparently never finished reading the review and believed Chomsky had completely misunderstood him. It gave other psychologists pause for thought, however, to see a relatively unknown researcher in linguistics leave Skinner's claims in pieces. The concurrent rapid development of psycholinguistics as a field brought in a lot of new talent and ideas, which included structuralist concepts derived from Saussure, and all of this by-passed the issues that had seemed so important to behaviourists. In the background there was an assumption that communication among people is possible, even between people who do not share each other's language, because there are certain formal similarities in all languages. Psycholinguistics sought to relate these formal similarities in language to the structure of mind and brain, and this was a significant encouragement to pyschologists to look at central, cognitive processes.

Chomsky himself went on to elaborate what he identified as a Cartesian theory of language, a theory that presupposes the existence of universal, innate grammatical structures. The result was a concrete research programme for linguistics, to search out the grammatical universals and to trace how they underlie actual languages. This strongly stimulated the development of the field, though many researchers in linguistics with a psychological orientation soon questioned both the logic and the empirical content of Chomsky's programme, and some even wondered whether his early reputation had distorted the development of the field. Chomsky's programme – Cartesian linguistics – traced all language to a few innate mental principles, which were thought to be in effect *a priori* in the activity of reason; but it faced rival programmes based on research on how children do in fact develop their use of language in different cultures.

Piaget and Chomsky had an impact on anglophone psychology because their arguments translated into empirical research proposals that could be made part of Anglo-American science. They elaborated psychologies of mental structures rather than of observable behaviour or physiological processes, though they disagreed fundamentally about language development. The consequence was that developmental theories, like the general arguments of French structural theorists, called into question the logic and intelligibility of behaviourist and physiological explanation of human language, symbolism and action, that is, of culture in the widest sense.

The other dimension of overriding importance to cognitive psychology was computer science. The mathematical logic of the modern computer was laid down in the 1930s, when there were also some physical models of calculating machines. The war created an imperative to organize huge systems – like the Manhattan Project to build the A-bomb or the combined air and sea war in the Atlantic against submarines. The war also focused research on communication and information, for example, in connection with radar and guided weapon systems. This stimulated the development of information theory, which systematically addressed the logic of the representation, communication and storage of information. In the late 1930s, the English mathematician Alan Turing (1912–54) developed first a theory and then an actual computing machine. Later he formulated a much-debated criterion in terms of which to judge whether computers think: can a person in communication with it tell whether the respondent is machine or human? In 1943, a paper by W. S. McCulloch and W. H. Pitts, 'A Logical Calculus of the Ideas Immanent in Nervous Activity', noted the similarity between the binary character of logical inference and neural networks in the brain. This was part of the background to Hebb's work discussed earlier and was seen by later researchers to have established the logical principles of the brain as a computer. Meanwhile, large and clumsy electronic valve computers of the 1940s, a technology designed to receive, process and store large quantities of information, were quickly replaced by the more manageable machines made possible by the substitution of transistors for valves. A mathematician at MIT in the United States, Norbert Wiener (1894–1964), provided unifying principles for the new fields in terms of what he called cybernetics, the science of 'control and communication in the animal and the machine'.[24] The research attracted brilliant minds and large funds in both the US and the USSR during the period of the Cold War as the organization of weapon-delivery systems depended on the efficiency and sophistication of communication and control. Much early work was concerned with the construction of models of human activity in front of radar screens. At the same time, psychologists and social scientists began to explore information-processing techniques in relation to their own data. Lastly, the replacement of microcircuitry by the microchip made possible the mass production and mass marketing of small and cheap computers, and this was the technological dimension of a major transformation of commercial, administrative and private life. The depth of these changes was much discussed but perhaps barely fathomed at the end of the twentieth century.

Information science, visible in material form in computing technology, became what has been called a defining technology, a technology which gives a culture its dominant models for investigation and understanding and structures the way we think about the world. This 'defining' went deep in the domain of psychology. Cognitive psychology was the imagination of the computer age applied to knowledge of the mind. The word 'cognitive' connoted interest in information receiving, processing and storing functions, and in the consequent control of human capacities.

Research on the brain as the organized site of human control was well established by the 1950s. New thought in psychological terms about central processes, which in the United States required the psychologist to make a self-conscious rejection of behaviourist norms, became evident in the late 1950s, notably in J. S. Bruner, J. J. Goodnow and G. A. Austin, *A Study of Thinking* (1956) and G. A. Miller, E. Galanter and K. H. Pribram, *Plans and the Structure of Behavior* (1960). Jerome Bruner and

George Miller were eminent experimental psychologists who worked at Harvard University where they established the first cognitive science research centre. Other psychologists were quick to follow when colleagues with a reputation legitimated reference to central 'plans' in learning and activity. Outside North America, in Cambridge in England, for example, such references to central psychological processes had never disappeared, and the concept was therefore absorbed quickly. By the end of the 1960s, the language of cognition rather than behaviour had become commonplace in many areas of psychology, though most conspicuously in fields like perception and memory research.

The turn towards internal cognitive language did not just revive buried concepts from an earlier period: it created new ones in the light of distinctively twentieth-century innovations, notably statistical inference and computing. Between 1940 and 1955, inferential statistics, which constructed knowledge out of a mass of data no one piece of which is definitive, became identified with the scientific method itself in substantial areas of North American experimental psychology. By 1955, more than 80 per cent of published experimental articles in the main journals used inferential statistics to justify their conclusions. When psychologists began to think about cognition, they therefore thought of it by analogy to statistical inference, that is, they conceived of the mind as an analyser of probabilities. Cognitive psychologists were influenced by belief that people, to use Egon Brunswik's phrase coined in the 1940s, are 'intuitive statisticians', for example, when they decide what to do with a practical problem or try to remember a face. This was a clear case of the way in which the method or technology of a science sometimes becomes its substance. The same pattern was evident later in the claims by psychologists that thought is a form of computing.

The shift of attention from behaviour to cognition involved more than a shift of language and a change from the observation of peripheral events to the modelling of central processes. Information theory and computer technology, in addition to statistical methods, suggested a new way to understand people and to answer the question of the mind's relation to matter. Many people made the obvious step and compared the brain to a computer. In the 1960s, the US computer scientists Alan Newell and Herbert Simon began to explore how computers can be used to solve logical and even scientific research problems and they hoped that this work would begin to imitate how humans solve problems. One reason why Chomsky had a dramatic impact on linguistics was that what he claimed to be the natural formal grammars could be compared to the grammars used in programming. From the beginning, however, there was a deep division of opinion, and this division continued in the 1990s, between scientists who claimed that human beings *are* computers (however differently constructed from the machines) and scientists who were willing only to draw an *analogy* between computers and human beings. Within the first group there was debate about what sort of computer human beings are; within the second group there was debate about how much analogy, in what areas, there is between computers and humans.

There was nothing new about the claim that man is a machine – as earlier chapters have indicated. What was new was the description of the supposed human machinery at the level of processing, the software, not at the level of the mechanical operations, the hardware. This was why scientists were so excited. They were intellectually fascinated by the analogy, overlap or even identity of processing logics between machine and mind. Creative programming provided ways in which to try to model or reproduce, and hence to understand, explain and predict psychological activity. The result was the extremely dynamic field of artificial intelligence (AI).

The rapid growth of artificial intelligence research depended on huge financial resources integrated with national, indeed international, economies. Government and corporate decisions about the economic centrality of information technology drove the scientific and academic response to the new technology. This was evident in fields like social and economic history or human geography as well as in cognitive psychology. Political support for the new technology was accompanied by new methodologies in the human and social sciences. Funding decisions directly affected research fields: identification with the goals of behavioural science was an earlier example, investment in research on quantitative indicators, e.g., citation analysis, in order to measure academic productivity, a later one. The result, critics feared, was a concentration on methods rather than content: techniques for data analysis of a sophistication out of all proportion to the quality or significance of questions asked, concepts deployed or knowledge produced.

Computers intensified the question whether man is a machine; they even more strongly posed the converse question – are computers human? Some observers of children brought up with computers reported that the children asked questions about whether computers can be bad or whether they deliberately act in certain ways. Similarly, a fantasy genre about cyborgs explored new types of beings made from both human and computer parts. The question whether computers are or could be human was perhaps *the* question about human nature for some sections of late twentieth-century 'high-tech' culture. What this all meant for the longer term was unknown.

Meanwhile, these questions aroused passion and excitement in the field of the philosophy of mind. There was a special challenge to decide whether advanced computers might in principle have consciousness, and debate about this in many respects took over what used to be discussed as the mind–body problem. If, the argument went, computers can do what humans can do – perceive, think, evaluate, act – on what grounds can we deny them consciousness? If what people say and do is the grounds on which we attribute consciousness to them, when computers talk and act why should we not attribute consciousness to them too? What, if anything, enables us to distinguish a computer from a human being? Obviously, the physical appearance of computers was ignored for the purposes of the question, though it is not clear it should be. As a matter of fact, even in the 1990s, these questions presupposed that computers might have capacities, e.g., for language use, that they were a long way from having – though how far away was hotly debated.

Nevertheless, the matter of philosophical principle was clear. Many AI enthusiasts were sure they knew the answer: computers, in all essential respects – that is, with the exclusion of the hardware – will become human. Philosophers of mind, among whom the North American John Searle was one of the most articulate, tended to reject this, in Searle's case because he believed computers do not and cannot possess intentionality. The debate about consciousness soon became a technical debate among philosophers. Searle argued that subjective qualities cannot be explained in terms of something else, however much we might learn about the causal activity of the brain, and that subjective consciousness must therefore be regarded as a novel, 'emergent' property of being human. By contrast, another influential US theorist on the mind, Jerry Fodor, argued that intentional explanation – which refers an act back to a subjective state of mind – is in fact a species of causal explanation. Fodor conceived of states of mind in cognitive processing terms and believed (unlike Ryle earlier) that reasons, which can be computer modelled, can be causes. Computers, in his view, are capable of being rational, and minds and computers are both symbol-manipulating machines. Though Fodor was inclined to accept materialism, he conceded that 'nobody has the slightest idea how anything material could be conscious. Nobody even knows what it would be like to have the slightest idea about how anything material could be conscious.'[25] As a scientist, however, he did not much care, as he believed that the question of consciousness is marginal to the most profound and important scientific questions. Here, perhaps unwittingly, he exemplified the gulf that existed between scientists who were fascinated by the world of natural science expressed in mathematical language and lay people who wanted knowledge based on the experiential meanings and sensuous richness known to the conscious person. Independently of these disagreements, even many scientists who worked on artificial intelligence or cognitive psychology found it implausible to claim that there is much identity between computer processing, at least in then current forms, and mental processing. Their research used computer modelling to explore the unique ways in which brain processes do in fact structure and handle data, and they also hoped that theoretical ideas used in computing would turn out to be of value in psychology. In the second half of the 1980s there was excitement that a way to programme computers called parallel distributed processing made it possible to construct a plausible model of the brain as it acts consciously. The possibility also remained, however, that brain processes are very different indeed, even at the level of basic logics, from anything yet represented in computing terms. AI enthusiasts, it occasions no surprise to note, believed that only lack of time and money lie in the way of understanding.

Many ordinary people were puzzled to know what to make of the comparison between computers and humans. Sceptics and more especially humanists among them were attracted by Searle's argument that 'conscious mental states and processes have a special feature not possessed by other natural phenomena, namely subjectivity' and that this creates 'first-person existence'.[26] If some version of this argument was accepted, then it was open to science to search for knowledge of subjectivity independently of knowledge of the brain. The willingness to conceive of human life as computer life already presupposed that a particular kind of scientific research about human beings, the kind of research that led generations of psychologists to accept the leadership of natural science, has unquestionable authority. But precisely this authority continued to be questioned in the human sciences, on grounds of coherence and truth as well as on religious or ethical grounds. It is therefore necessary to turn from the vast funds and even vaster ambitions of the computer-based sciences to those psychologies that took meaning and value to lie at the foundation of human nature.

v *The Self*

The size of the psychology profession there, and the numbers of students, journals, academic and popular books, and psychological practitioners of every kind, made the United States the discipline's centre of gravity. US psychology had influence elsewhere, especially in the period after 1945, in part because of the image of rigour and objectivity that went with experimental and quantitative science. Such science looked attractive in a Europe where ideological ranting had accompanied calamity. At times, identification with North American scientific culture became almost required in professional practice. Yet, as the previous chapters have shown, psychology and the human sciences were much more diverse than this picture suggests. The concluding two sections move away from the natural-science-led and, if scale of activity is the measure, US-dominated areas of behavioural science, brain science and cognitive psychology, to forms of psychology that limited or even rejected the leadership of natural science. Even in the United States, the creation of humanistic psychology in the 1950s and 1960s, and the way clinical and educational psychology became huge areas of activity, all supported by central and not marginal figures, shows that there was more diversity than is usually acknowledged. There were also European influences in North America: psychoanalysis and ego psychology, gestalt psychology, social psychology and group dynamics, theories of children's development and education. European thought after 1945, despite its sometimes notorious abstraction and the disagreements between phenomenology, existentialism, structuralism, Marxian critical theory and a later cluster of postmodern arguments, provided many of the terms for a re-evaluation of the sciences of human nature. The issues raised brought the history of the human sciences from the past into the present.

Humanistic psychology as an institutionalized movement was principally the brain-child of Abraham H. Maslow (1908–70), in the 1950s a professor at Brandeis University in Massachusetts. He first established a network of like-minded people and then created the *Journal of Humanistic Psychology* (1961) and an association (1963). Initially represented as a 'third force' in US psychology, an alternative to behaviourism in one direction and psychoanalysis in the other, humanistic psychology then settled into an existence as yet another branch of the larger domain, with its own section in the American Psychological Association. The idea of a new psychology as a 'third force' came from Gordon Allport, who, in the 1950s, was one of the psychologists whom other psychologists most read; he spoke out against the limitations of behaviourist explanations in relation to human aspirations. Maslow and his colleagues were committed to a concrete conception of the self as the foundation for psychology.

Their intellectual problem was to formulate a coherent account of this value, the self, that translates into proposals for research, and their failure to achieve this may explain the rapid marginalization of their programme. The challenge is well illustrated by the editorial of the first issue of the *Journal of Humanistic Psychology*: the journal was

> interested in those human capacities and potentialities that have no systematic place in either positivistic or behavioristic theory or in classical psychoanalytic theory, e.g., creativity, love, self, objectivity, autonomy, identity, responsibility, psychological health, etc . . .[27]

This list consisted of topics excluded from psychology rather than the product of a systematic or coherent theory. Critics of humanistic psychology found it easy to point both to its softness on method and its conceptual vagueness. Its background was not in natural science but in therapy, counselling and existential philosophy. It grew out of a concern with what individuals make of life and how they find meaning, and precision was not of primary importance to these concerns. Humanistic psychology, however, was integrated with what ordinary Western people expected of psychology, a practical relationship to life, rather than with the commitment to objective methods claimed as the distinguishing character of science. Similar public expectations influenced the interest in psychoanalysis. The popular interest in psychology reflected an interest in the human *condition* as well as in human *nature*.

Humanistic psychology was a distinctive mixture of American values and European ideas. Maslow himself made an explicit connection between the political values of US liberal politics, the freedom, dignity and fulfilment of the individual, and the universal, essential needs and goals of the being of man. A colleague of Maslow's at Brandeis, the German émigré neurologist Kurt Goldstein (1878–1965), brought an important European influence. Goldstein had studied brain damage and disease during and after World War I in terms of the loss of integrity of the whole structure and whole personality, rather than the loss of specific functions. This holistic orientation, which interacted with Berlin gestalt theory, exemplifies a position shared by many therapists, European and American. In therapeutic work, as opposed to natural science work, it was necessary to approach loss and repair, the business of healing, in relation to the potential of a unified self. Therapeutic practice worked through the integrated physical, mental and even spiritual dimensions of the whole person, and practice presupposed a capacity for change and growth within individual people.

This orientation achieved its most influential form in the work of Carl Rogers (1902–87). He established what he called client-centred therapy, the major idea behind counselling. Rogers had a religious upbringing in which Wisconsin Protestantism stressed the individual soul's or self's relationship to the divine. His beliefs became progressively more liberal while he studied at the Union Theological Seminary (associated with Columbia University) in New York, the foremost institution of its kind, and his interests moved from theology to psychology during the 1920s. This shift of views was characteristic for people of his generation and background, and it indicates how, in individual careers, ideals became invested in psychology as a means to life and not only as a science. Rogers moved from the Seminary to Columbia Teachers College and then, in the 1930s, to work on child guidance, first in New York and then in Rochester, New York State. He began an extended conflict with the medical profession, since he had no medical training but nevertheless insisted that his work with children be recognized as therapy. When he moved to the university of Chicago and later to the university of Wisconsin, he applied his approach to mentally ill and distressed adults, and he even took the title of Professor of Psychiatry at Wisconsin. This was, to say the least, provocative to physicians. Rogers's successful opposition to the medical profession's therapeutic monopoly was very important as it opened up the social space in which non-medical therapy subsequently flourished. It also gave psychologists independent access to the huge health-care market, making it possible for medical psychology to expand until it became the largest occupational division within professional psychology.

In *Counseling and Psychotherapy* (1942) Rogers explained how he adopted a non-directive stance toward a 'client' – the term he favoured to avoid the clash with the medical term 'patient' – in order to provide an accepting context in which a person can obtain insight into her or his condition. In contrast to psychoanalysts, he focused on a person's insight as a cause for change rather than as an entry into the hidden past. His position implied belief in the presence of an innate goodness or creative power, the self, in every person, though Rogers was not interested in the articulation of philosophical or religious assumptions. The therapist's or counsellor's role, as Rogers developed it, is to provide acceptance, close to the Christian attitude of love, that enables a client to find in herself or himself insight and strength for change. He called this process actualization. His book *On Becoming a Person: A Therapist's View of Psychotherapy* (1961) reached a large audience beyond those with a psychological, psychoanalytic or medical training. More than most psychologists, Rogers gave therapy or counselling priority, that is, he consciously put theory aside in order to develop an approach through committed practice. While he did not seek to found a psychological school, his personal qualities as a leader gave him such a position, and he even acquired the status of a guru among people who wanted from psychology a way of life focused on a sensitive and loving openness to individual people.

When Rogers related his views to academic psychology, he described the presence of 'inner meaning', which he thought pivotal for a person's life, as if it is a variable open to investigation like behavioural variables. His psychology therefore analysed human life in terms of functional variables and, in this respect, was like much other American psychology. He was able to connect the academic psychology of functions with a public audience that expected psychology to help understand the 'private world of meanings'. Rogers held out the prospect of an inclusive science of the 'person who creates meaning in life' and thereby has the unity of purpose to lead a productive life.[28] This was far from the anguished reflections of European existentialism or the search for being in European phenomenology, thought that was equally concerned with the person's sense of meaning.

Rogers also wrote in accessible and concrete language, whereas the European literature on the meaning of life was often abstract.

The therapies and psychological practices that stemmed from or related to Rogers's work became legion. His contribution was made in a Western culture which more and more privatized values through the reconstruction of social values as personal feelings. Rogers – and psychotherapy generally – offered hope that this privatized world could still be humane, and he supported belief that there is a technology with which to integrate personal feelings with the feelings of others – an integration significantly known as 'interpersonal relations' rather than 'social relations'. One offshoot of Rogers's work was the encounter group movement, a translation of client-centred therapy into a group setting. Encounter groups were intended to create conditions for insight by a person into his or her relations with others and conditions in which that insight leads to change. The groups were notable for the 'leader's' refusal to lead, which forced participants back on their own creative – and often destructive – emotional engagement with others. Counselling, Rogers's direct contribution, acquired significance as a technique for the management of everyday human problems. During the 1970s and 1980s it became an occupation with many branches, it entered into all walks of life and, in the life of many churches, merged indistinguishably with pastoral care. This was one of the most obvious features of what has been called psychological society. The discourse of psychology, along with techniques of therapy and counselling, mediated late twentieth-century culture's redefinition of personal responsibility, realignment of identities of sex and gender, re-evaluation of private inner worlds at the expense of public outer lives and, indeed, generally represented order, meaning and value as qualities of the self. Psychology and the way of life became indistinguishable.

The attitude towards the person exemplified by Rogers represented one part of the background to humanistic psychology. Another part derived from European philosophy, informed by the horror of mass extermination rather than by an optimistic faith in human nature. This philosophy was rooted in the logical technicalities of phenomenology, but it also used a more accessible language with which to describe the quality of conscious being in a non-conscious world. This language referred directly to being, the human condition, and did not derive being from some supposedly more essential reality such as matter, God or self. Faced by a loss of faith in God and in moral progress, philosophers described anxiety, loss, guilt, meaninglessness and, sometimes, sheer terror. The best known of these philosophers, the Parisian intellectual Jean-Paul Sartre, who reluctantly accepted description of himself as an existentialist, combined phenomenological studies of psychological capacities like imagination, work on foundational philosophy – *L'être et le néant (Being and Nothingness,* 1943) – and plays and fiction that vividly staged his picture of harsh human truths.

For Sartre, the core of the human condition, around which everything else develops, is unconditional freedom, the irreducible and primary reality of the choices through which conscious being becomes what it is. There is, he believed, no self before reflection and action constitute a self as a freely chosen existence. It does not make sense to say, for example, that human nature is innately selfish: there is no innate self towards which a free act can be directed. As Sartre uncompromisingly stressed, we are responsible for what our being, existence, creates as the self. In this connection he formulated his much discussed account of bad faith, a notion easily translated by his readers into psychological terms and which then possessed something in common with Freud's notion of unconscious motives. By bad faith, Sartre denoted the myriad ways in which people disguise their choices, hide them from themselves and thereby abdicate responsibility for their lives. With a characteristic French male example, Sartre described how a woman, as she is seduced by a man, pretends to herself that she is desired for her company and not her physical attractiveness and thereby avoids an open choice about a sexual relationship. The woman exhibits bad faith because she does not take responsibility for what she in fact freely chooses. Logically developed, this argument reduced any causal justification for an act, and hence psychology and social science in general since they seek to determine causes, to acts of bad faith. Freedom, it seemed, gives action an ultimate meaninglessness, the state portrayed fictionally by Albert Camus (1913–60), who for a while was Sartre's friend, in *L'étranger (The Outsider*, 1942). But this state of freedom was also understood to be the source of human dignity; and in the context of the German occupation of France, followed by the Cold War and the division of the world, there was hope in the message about the irreducible freedom of being.

The work of Sartre, his life-long companion Simone de Beauvoir and Camus became known in the anglophone world in the late 1940s. In the United States, an artistic avant garde embraced meaninglessness for its anti-bourgeois message, and intellectuals and therapists stressed freedom as a restatement, albeit in stark terms, of individual dignity and independence. The Protestant theologian Paul Tillich (1886–1965), himself a socialist refugee from Germany in the 1930s, re-expressed a vision of pastoral concern in existentialist language. In his lectures published as *The Courage to Be* (1952), Tillich dealt with the anxiety he thought central to modern experience and inescapably associated with the threat of nuclear war. He asserted the possibility of an authentic encounter with man's being and hence the existence of a power to overcome individual anxiety with courage. Tillich was an extremely influential teacher at Union Theological Seminary in New York and later at Harvard University in settings where psychological therapy and pastoral concern came together. Among his students was Rollo May (1909–94), a psychoanalyst who published a series of books that linked therapy and existentialism and included the collection edited with the psychiatrists E. Angel and Henri Ellenberger, *Existence* (1958). May also co-operated with Maslow; together they wanted to turn psychology towards humanistic ends grounded on unconditional acceptance of the free person. In Britain, existentialist themes were taken up by the psychiatrist R. D. Laing (1927–89) and, filtered through his descriptions of schizophrenia as a meaningful response to intolerable family relations, became a significant element in the 1960s counter-culture.

Whatever sympathy psychologists had with humane ends, many were sceptical about the reconciliation of existentialist or humanistic psychology with a scientific discipline. May, for example, helped create a North American audience for the Danish Christian philosopher Søren Kierkegaard (1813–55), but psychologists were hard pressed to know how anything sufficiently empirical to be called science could come of Kierkegaard's anguished portrayal of life. In addition, something of the European sense of the tragedy and conflict of existence was lost as thought crossed the Atlantic, and European arguments reappeared in a more optimistic and marketable idiom in the United States. In Europe, humanistic psychology remained more tied to philosophical anthropology, a tradition of thought about human nature that characterized the individual ego as an irreducible expression of spirit – 'the directional and intentional being of man', in the words of the Leiden scholar C. A. van Peursen.[29] Literature in this vein also tended towards optimism and vagueness. Critics therefore questioned whether the humanistic psychologists possessed either the methods to make a science or the depth, or perhaps hardness, to confront humankind's tragedy. Nevertheless, humanistic psychology responded to personal needs, and it became part of the self-reflective culture characteristic of late twentieth-century society.

Ambivalence about whether humanistic psychology is a science or a philosophy of life encapsulates an unresolved issue in the human sciences. As earlier chapters have shown, psychologists and social scientists repeatedly sought knowledge as a basis for both a personal ethic and social policy. In the twentieth century, though not before, this science was partnered by a philosophy that separates factual statements and value statements, and separates the content of discovery from the context in which knowledge is applied. Scientists in the twentieth century asserted the factual content and evaluative neutrality of what they claimed to be true of human beings, and some cited Weber's contention that social science cannot address the deeper question of 'what should we do'. At the same time, many of the same scientists could not conceive of other grounds for knowledge of what to do than scientific knowledge about human nature. That is, in spite of a philosophy that separates facts and values, they valued science because, as they believed, it is the basis of right action. This was not neutrality but a firm commitment to one rather than another way of life. It was also a source of many of the theoretical problems to bedevil the human sciences. No agreement was reached that the fact–value separation is achievable either in practice or in principle. The result, exemplified by the controversy about sociobiology – a controversy in which facts and values were interwoven on all sides – was a dense debate that involved politics, ethics and philosophy, as well as the human sciences, in which it was not even possible to agree the terms for a resolution of the differences of view. Humanistic psychology at least had the openness to make the value of the autonomous self the explicit foundation of its search for knowledge. In the assertion of this value, it was close to many ordinary people's values and to their expectations that psychology should contribute to life.

A psychology that searches for knowledge of the self and despairs that such knowledge is to be had, became central to late twentieth-century Western culture. This situation was foreseen by a man who, in some accounts, is the most profound of 'psychologists'. Friedrich Nietzsche (1844–1900) did claim for psychology the primary position in the sciences, though he had a special conception of psychology in mind:

> All psychology so far has got stuck in moral prejudices and fears; it has not dared to descend into the depths . . . and the psychologist who thus 'makes a sacrifice' [of his own false morality] . . . will at least be entitled to demand in return that psychology shall be recognized again as the queen of the sciences, for whose service and preparation the other sciences exist. For psychology is now again the path to the fundamental problems.[30]

I shall explain this and then indicate the role that Nietzsche and his modern interpreters have had in the demolition – or, to use jargon, the deconstruction – of the notion of the self. Out of humanistic criticism of psychology as a natural science emerged the late twentieth-century preoccupation with the postmodern, with the fragmented, ironic and contradictory character of the human condition. All the same, to balance this view, it is necessary to remember that an attack on the notion of a coherent or unitary self was also central to mainstream scientific psychology. The argument was developed in cognitive science that there is no self, no mind or observer sitting at the centre of the brain to receive mental images or initiate movements, but that there are processing functions and that consciousness is somehow a product of the interrelation of different functions. Marvin Minsky, the leader of AI research at MIT from the late 1960s and a strong proponent of the view that mental functions can be reproduced by computing functions, also criticized ordinary notions of the self as incompatible with science. The argument, given a large audience by the philosopher Dennett, involved a radical attack on the ordinary person's sense of self and on ordinary language.

By any conventional definition of psychology as an occupation or as a science, Nietzsche was not a psychologist – a poet and moral philosopher, yes, but not a scientist. Yet Walter Kaufmann, the philosopher at Princeton University who restored Nietzsche's reputation in the English-speaking world after his name had become linked to Nazi propaganda, described him as 'the first great (depth) psychologist'.[31] Nietzsche himself, in his characteristic style that itself requires interpretation, wrote: 'that a psychologist without equal speaks from my writings, is perhaps the first insight reached by a good reader . . .'[32] Kaufmann understood this to refer to Nietzsche's penetrating capacity to go beyond a person's self-description to see hidden motives, to hear what is not said – to be 'one who has ears behind his ears', as Nietzsche wrote.[33] Freud, too, if somewhat allusively, acknowledged Nietzsche as a master, and quoted one of his well-known aphorisms: '"I have done that," says my memory. "I cannot have done that," says my pride, and remains inexorable. Eventually – memory yields.'[34] Nietzsche used an aphoristic style to force readers into their own thoughts, into an attitude of self-questioning and into the consideration of the opposite of what is being said. This raised broad questions for psychology as knowledge. What can be said about the self when what is supposed to be the self recreates itself, as in Nietzsche's example, in response to pride and not to memory? What we think we are, it seems, is formed through self-deception. It then appears to follow that there is no 'real' self but only recreations of what we persuade ourselves is the self. This was in a sense a conclusion about psychological knowledge – but it is a conclusion that threatens what is normally understood to be scientific knowledge.

If Nietzsche believed our feelings and desires are not what we say they are, what then did he think they are? Freud's answer was his theory of the unconscious. Nietzsche's answer, or rather the metaphors and rhetoric in terms of which he explored an answer, lay in his thoughts about 'the will to power'. These thoughts defy easy formulation: it was after all Nietzsche's intent to use language to destroy the banality and clichés of his time – and the capacity of later authors to repeat them. Nevertheless, when he used psychological terms, as when he said we 'strive' towards power or seek to 'overcome' self-limits, there is ambiguity about whether he was using metaphor or making claims about the psychological condition of being human. Nietzsche, who wrote mainly in the 1880s, turned to psychological discourse and against contemporary philosophical, religious, political, scientific or social discourse, for the most honest, most purifying expressions. But he was not able to control how his readers used what he wrote, especially a century later after psychological discourse had so often become banal.

The will to power was neither a will to physical domination nor a will to pleasure. To the young people at the turn of the century who proudly called themselves Nietzscheans it was part of a revolt against bourgeois values, Christian hypocrisy and academic irrelevance. They turned to Nietzsche as a philospher of life, a philosopher who identified with instincts and pointed to the vitality and beauty within special people – and it was the seduction of Nietzsche's writing to make the reader feel special. His perhaps most approachable book was entitled *Die fröhliche Wissenschaft* (*The Joyous Science*, 1882). His individualism, his admiration for strength of spirit and his scathing destruction of the pretensions of philosophical and moral systems, along with his revaluation of the power and pleasure of 'the free spirit', again appealed in the 1970s and 1980s. He was read to reinforce a personal commitment to pleasure, to find a language that locates power in every action and to support a rejection of the search for fixed foundations for knowledge. He then became an influential voice in the human sciences, an authority for a different kind of knowledge to the knowledge of natural science.

Another strand of Nietzschean interpretation, rooted in the Parisian intellectual hot-house, began to flourish in the 1960s and subsequently achieved extraordinary influence. One result was a commonplace and often superficial reference to 'the disappearance of the self'. Two French intellectuals, Jacques Lacan (1901–81) and Michel Foucault (1926–84), the latter with explicit reference to Nietzsche, drew conclusions significant to psychology and the human sciences. Their texts became part of the canon, the sacred writings, of postmodern culture in the late twentieth century.

For Lacan the starting-point was not Nietzsche but a training in psychiatry and a vehement antagonism to the direction taken by psychoanalysis in the United States during the 1930s. He believed that ego psychology enthroned an idealized and intellectually dishonest image of the self, an image which he claimed it had been Freud's achievement to destroy. He therefore attempted to impose on his psychoanalytic colleagues a return to Freud's texts, emphatically referring to it as a return to science. In 1953 his refusal to conform – the brevity of his therapeutic sessions became legendary – led the Société française de psychanalyse to split from the Société psychanalytique de Paris. He began to teach an annual series of seminars in which he lectured to audiences spellbound by his way with words. With at one time about a thousand people, including leading intellectuals, in attendance at his fortnightly classes, Lacan explored psychoanalysis as a linguistic theory that reveals the essential strangeness, unknowability and decentred character of what is most essential to us, the unconscious. His work in the 1950s drew on a structuralist approach and treated language as a symbol system that expresses the structure of the psyche; he began to incorporate the work of Saussure into his seminars. Later, especially after the political events of 1968, in which Lacan acquired the image of an anarchist hero, he emphasized more the way in which the unconscious, which speaks in language, always goes beyond what we can know or grasp and spills over in contradictions, play, seductions, irony and desires different from what our conscious, social selves pretend. What we call our body, he argued, is already structured by language and hence is other and, like the unconscious, leads the conscious ego in a dance. To match this message, Lacan's performance was theatrical, his life-style outrageous and his language inventive, playful and contradictory.

Lacan therefore explored what can be said, or not said, in language as the medium of the unconscious, the unknowable other: this is what we claim is the 'self'. For him psychoanalysis existed as the most profound form of talk and not primarily as a therapy. Disqualified as a training analyst by his peers, in 1964 Lacan founded the first of two schools, the Ecole Freudienne de Paris, during a decade when there was an extraordinary penetration of psychoanalytic ideas, which had previously been despised as an attack on reason, into French culture. In the 1970s, he attempted – and failed – to contain a feminist response to his work, a response that accused him of the mystification of the feminine as 'other'. Finally, in 1980, Lacan unilaterally dissolved his own school – a move that the newspaper *Le Monde* described as 'a surrealist farce'.[35] The legacy of all these splits and of the personal power Lacan exercised was that any statement in the French setting about analysis, language or the self was weighted with emotion. Lacan's work questioned reference to the self and at the same time it vividly voiced the unsatisfied desire that was thought inescapably a part of people's lives when they move beyond material need.

The significance of Lacan's work was subject to much debate. His arguments changed over fifty years; there were struggles over who possessed the right interpretation; and his work became known outside France piecemeal through the translation of selected essays, published as *Ecrits* (1977). In the 1930s, influenced by the surrealist celebration of unconscious forces and fascinated by the relation between images and objects, Lacan explored the unconscious self-image created by the young child's encounter with her or his mirror image. In this image, he argued, the child acquires a distinction between self and other, and as a result the sense of self, or wholeness, always remains fragile. He thus identified the self with a projected image, not with something 'real' and substantial. After the war Lacan took up in his own distinctive way a belief, derived from Hegel, in the existence of a limitless human drive to seek confirmation of being in the image of self, whether in the mirror or in the mirroring action of others. The exemplar of this condition is the condition of being in love, the fantasy of completeness achieved through the merger with the being of another. As a fantasy, though the fantasy is inescapable, love does not last. Lacan described *désir* – longing, desire, wish – as irrational, unconscious and something that precedes and always pre-empts conscious reason. His psychoanalytic activity was therefore an attempt to give his clients and his audience a language in terms of which they might address this desire. There is, he believed, no object to which desire corresponds, it is a 'hole'; but he took the phallus, understood as a symbol rather than an object, to be the most powerful signifier for this desire. Whether the privilege thus accorded to the phallus marginalized women later became the subject of heated controversy. During the 1950s, he adapted the language of structuralist linguistics to represent the unconscious as a 'language which escapes the subject in its operation and in its effects'.[36] As a consequence, he implied that play with language, which dominated his own style, is the means by which the unconscious can speak. Any claim to grasp the unconscious, 'the Other', as he expressed it, is a delusion: it is always there yet beyond reach. In a further stage of his work, in the 1960s, Lacan turned to mathematical formalisms to give his theory more rigorous or scientific expression; but this side of his work had little impact.

Lacan's work reinforced in wider intellectual culture the conviction that psychoanalysis is a discipline in the humanities and not a natural science subject, an interpretive not an explanatory science. The richness of a hermeneutic approach to Freud was also brought out in the work of other French philosophers or analysts, especially in Paul Ricoeur, *De l'interprétation: essai sur Freud* (*Freud and Philosophy: An Essay in Interpretation*, 1966) and J. Laplanche and J.-B. Pontalis's *Vocabulaire de la psychanalyse*

(1967; translated as *The Language of Psychoanalysis*). Influenced by these writers, psychoanalytic discourse spread into the intense and rebarbative world of literary, art and film criticism.

Similar emotional intensity surrounded the life – and the death in the AIDS epidemic that affected California's gay community – of Foucault. He began as an isolated and marginal intellectual; he worked in Uppsala in Sweden and then in Poland before he published *Folie et déraison: histoire de la folie à l'âge classique* (1961; abridged translation as *Madness and Civilization*), a work of richness and originality. He wrote historically and philosophically about madness in a way that questioned both the obviousness and the humanity of the modern medical view of mental disorder as illness. He traced the beginnings of the medical approach to the period around 1800. The book was also about the human sciences because, when he described concepts of madness, he made a general analysis of reason and unreason. Each version of truth about human beings, he argued, has to be understood in relation to a set of conditions for what counts as a truth claim. His history of madness and later histories set out to expose these conditions. He explored the linguistic, cognitive and social conditions that make possible different kinds of claims to truth about the human subject. In *Les mots et les choses* (1966; translated as *The Order of Things*) he argued that the human subject, as studied by the human sciences as the object of truth, became possible only early in the nineteenth century. He supported this startling and counter-intuitive claim with a discussion of radical changes in the discourse of political economy, philology and natural history, and this was complemented by earlier work on the orgins of modern clinical medicine. His extremely opaque book was a best-seller in Paris, partly because his opening was a *tour de force* – a description of the different viewing points present in Velázquez's great painting *Las Meninas*. The excitement was also partly because he showed how different areas of knowledge, as apparently unrelated as economics and natural history, could and should be understood historically in relation to deeper patterns of thought, and partly because he made the day-to-day, almost mundane administration of life – exemplified in the management of mad people, children and criminals – the centre of knowledge and power in human life. The discussion of power in the government of ordinary life was influenced by *les événements* of 1968 and it informed his later book which traced the genealogy of power, *Surveiller et punir: naissance de la prison* (1975; translated as *Discipline and Punish: The Birth of the Prison*), as well as unfinished volumes on the history of sexuality.

Foucault, as he intended, defied easy classification, though he stated clearly enough that 'the goal of my work . . . has been to create a history of the different modes by which, in our culture, human beings are made subjects'.[37] In the 1960s his readers identified him with structuralism, and his early books looked for the conditions of knowledge – what Foucault called 'the archaeology of knowledge' – in the history of *discours*, the structures, represented in language and the disciplined practices of daily life, in terms of which we know. Nevertheless, unlike the structuralist social anthropologist Lévi-Strauss, he did not wish to imply that discourse and all its expression in culture, reflects psychic structures. It was central to his argument and impact that he rendered the notion of the self a dimension of a historically located discourse, a subject with a genealogy and possibly no future. And when he wrote about the 'genealogy' of such a subject, he acknowledged a debt to Nietzsche's theory of knowledge, a theory that denies our ability to escape from the perspective of our own lives, but which enables us to trace the origins of the perspective we now have. He was also indebted to Heidegger's later philosophy, itself an intense response to Nietzsche. Foucault's work therefore encouraged thought in the human sciences to de-centre the self and direct attention to the language or conditions of knowledge in terms of which claims for truth are made about the self.

The ultimate mark of academic success in France was to become a professor at the Collège de France, and Foucault achieved this distinction in 1971 with the self-chosen title of Professor of the History of Systems of Thought. His influence, however, was not so much among historians – though he inspired significant work on crime and insanity – but in the human sciences. He suggested new ways to analyse relations between power, daily discipline, social institutions like the family, the state and the construction of what is claimed to be true about people. He argued for a new analysis of power and knowledge – initiates wrote about 'power/knowledge' to indicate the intimacy if not identity of the concepts. With this analysis he hoped to write a history of the present as a precondition for our understanding of what is currently held to be true. His work and language were in conflict with the empirical ethos of Anglo-Saxon psychology and social science, and this reinforced a sense of the separation between French theory and anglophone facts. He also stimulated important bridges, however, like Nikolas Rose's studies of 'governing the soul' or Robert Castel's work on French psychiatric legislation and on psychoanalytic power. Unsympathetic Anglo-American historians divided over whether Foucault was simply erroneous, empirically wrong about his historical claims, or whether his work was not 'history' in the sense in which the English-speaking profession understood the term.

Both Foucault and Lacan de-centred the self and made it the subject of history and language rather than the subject of a psychological science with universal pretensions. Their work flowed into a general intellectual shift, which began in the English-speaking world towards the end of the 1970s, known as the linguistic turn. The German philosopher Hans-Georg Gadamer pithily expressed the principle behind this 'turn' as the belief that we can never 'get behind the back of language'.[38] The direction of argument was taken furthest by the French philosopher Jacques Derrida (b.1930), though the linguistic turn was evident in many academic disciplines. This all supported an attempt to reanalyse human science or humanistic subjects like the mind, the self, the unconscious, the family, cultural values, history, the novel and art in terms of the self-referential character of language. The underlying claim was that language (or other systems of representation, like film images) signifies yet more language, not some outside world that can be specified apart from language. Derrida's writings sustained this argument through the method of deconstruction, a dazzling display of how every attempt to transcend language only rephrases language.

Such perfectly general arguments necessarily had implications for sciences like psychology, though for the most part the arguments interested the specialist theorists of deconstruction rather than psychologists. There was overlap, however, where psychology was made part of a wider intellectual culture, in which – during the 1980s – it became common to refer to the deconstruction of the self, the demonstration that what is conventionally referred to as 'the self' is something signified by particular self-reflective uses of language and is not a 'reality'.

Debate about these questions continued in the 1990s. Two deep ironies for psychology and the human sciences are worth reflection. Firstly, all the major intellectual and disciplinary developments in the sciences discussed in earlier sections of this chapter – sociobiology, neuropsychology and cognitive science – continued unabashed and, indeed, indifferent to French theory. The fact was that the size, autonomy and self-perpetuating character of disciplinary specialities, reinforced by the training, practical skills and technical knowledge required to be productive and have a career in any one area, made it easy for people to be narrowly focused. Narrowness was as evident among specialists in theory as among specialists in empirical science, for these same reasons. The irony, therefore, is that though the claims made for the deconstruction of human science concepts are perfectly general, the social reality was a largely unquestioned perpetuation of those concepts.

The second irony connects the earlier and later parts of this section. The climax of late twentieth-century psychological society, a society that gave unprecedented attention to the self as the ultimate value in economic, emotional and existential terms, turned into a reflection on the self as a form of words. This irony is a clue to what was described as the postmodern condition, though the 'postmodern' label referred to an ill-defined set of debates rather than a particular state. The irony masked a struggle to reformulate the identity of being human, a reformulation the outcome of which could not be known. Late twentieth-century people dwelt in thought and desire in the self, while they inhabited an intellectual culture that replaced the self by words. This was perhaps neither consistent nor sustainable. Indeed, the psychologies of the self, many of which are traceable to Rogers and to humanistic aspirations, in practice retained a deep hold with the wider public.

vi *The Human Sciences: History and Science*

The enormous endeavour in psychology and social science did not produce a synthesis at the end of the twentieth century. There is a temptation to echo the words of the French diplomat and writer Chateaubriand, at the beginning of the nineteenth century: 'The sciences are a labyrinth in which you find yourself more than ever bewildered at the very moment when you imagine that you are just at the end of it.'[39] All the same, the ideal of the unification of scientific theory about the human subject, or the unification of a particular discipline like psychology, periodically resurfaced. In 1991, for example, a new journal, *Theory & Psychology*, was founded, at least in part as a forum for psychologists who believe that theory is significant to building bridges between the diverse areas of psychology. Some psychologists, like Hearnshaw in his book *The Shaping of Modern Psychology* (1987) thought of history as a means to counter overspecialization and keep alive a search for unity. In contrast, I have not assumed the possibility of such a synthesis. At the back of my mind is the thought that the human sciences will go on being remade as long as ways of life go on being remade.

This concluding section returns to question why the debate about the identity of the human *sciences* should be a live issue at the end of the twentieth century. To recapitulate: there was a difference between continental European and English-language use of the word 'science' – in the former use it denotes systematic, rational and objective knowledge, while in the latter it denotes knowledge modelled in content and methods on the natural sciences. Most academics agreed that psychology or sociology should be 'scientific' in some sense, since to reject this view is to reject the academic goal of rational knowledge. But in the academic world there was no agreement about what the goal really is or how to achieve it. Some self-styled postmodern intellectuals regarded even the project to be rational to be impossible, though in practice they were in some sense rational in the act of communication. In the wider world, there were people who rejected the general academic goal and believed that the whole project to construct a human science is misguided because religious truths are fundamental to our nature. Even more dramatically, popular culture celebrated spontaneous feeling, rather than academic knowledge, as 'truth'.

Debate about the character of the human sciences continued to be expressed in terms that derived from the late nineteenth-century German distinction between natural-scientific knowledge and human-scientific knowledge, *Naturwissenschaften* and *Geisteswissenschaften*. The relative success of natural science explanations in the nineteenth century, and the relative decline in status of philosophy, set against a background of industrial, social and economic change, led to intense concern with the right way forward for the human sciences. The importance of systematic knowledge for human affairs had been debated at least since the seventeenth century, but it was German academics in the late nineteenth century who established the conceptual language that continued to be put to use in the twentieth-century philosophy of the social sciences. In the nineteenth century, the natural science approach to the discovery of what is given in human *nature* became so impressive that it precipitated a crisis for those who were not natural scientists but claimed knowledge of man's being.

German scholars also debated the problem of how values, the purposes for which society is organized and individuals act, relate to the way society and the individual are studied. Social theorists like Spencer or Durkheim regarded society as a natural object and studied it naturalistically like physical nature – though Durkheim, unlike Spencer, argued that sociology has its own distinct subject matter. Similarly, Darwin specifically wrote a natural history of man's origins; he studied man as a part of nature. Many German academics did not accept that either society or the individual person is a natural object in this sense. At the core of what they wanted to study were the values they themselves professed, the values that make culture, they thought, an expression of spirit and meaning and not only the outcome of material, economic and political conditions. In their debate about these questions, scholars – Weber is exemplary – recreated ultimate questions about being human as a debate about sociological and psychological knowledge.

The debate was given a twist particularly relevant to psychology by Wilhelm Dilthey, discussed earlier in connection with the development of the discipline. Unlike many academics, he was neither reactionary nor opposed to natural science. He faced up to the political and industrial transformation of German society after 1860 with support for the social sciences: 'Knowledge of the forces which prevail in society . . . has become vital for our civilization. That is why the significance of the sciences of society has been growing in comparison with natural sciences.'[40] He did not accept, however, that natural science and human science can be the same. Initially he hoped to establish a general psychology, a science of the human essence that transcends particular conditions, as the basis for social science and biography. He therefore envisaged psychology as a historical subject that studies the creation of meaning. Later, however, faced by many criticisms of this project, he turned away from psychology as the discipline that makes possible the writing of history. He concluded that a universal standpoint is unachievable in principle; he argued instead that the human sciences involve a continuous reappraisal of ourselves and our situation through study, especially historical study, 'by projecting what we have actually experienced into expression of our own and others' lives'.[41] As Dilthey understood it, such study requires an intellectual and moral act of interpretation in which the scientist enters into the meaning that life has for others – the values, intentions and feelings which constitute the world for others. He argued that the *Geisteswissenschaften*, the sciences concerned with the world as a system of human meanings, are necessarily distinct from the *Naturwissenschaften*, the sciences of objects that do not exhibit values, intentions and feelings. In the human sciences we interpret rather than explain. Dilthey, who was influenced by arguments derived from Kant, thought that interpretations of human action must refer action to its basis in the self's agency. Not to do this is to explain something other than *human* action, to refer to the causes not meanings of events. Knowledge in the human sciences therefore necessarily differs, he argued, from knowledge in the natural sciences: knowledge in the former is knowledge, in Kant's terms, of 'the intelligible world' – the world informed by human purposes; knowledge in the latter is knowledge of 'the sensible world'.

The debate subsumed the complex problem of what was later known as the fact–value distinction. The positivist philosophy accepted informally by many twentieth-century natural scientists drew a sharp distinction between factual claims about nature and evaluative claims about what it is right or wrong to do. Indeed, for many scientists, it was precisely this distinction that makes science 'science', since the distinction separates objective science from knowledge coloured by bias, prejudgement or irrational beliefs. Most psychologists and social scientists tackled the human subject in this spirit; they claimed to describe facts and to remain neutral about values. Others, however, rejected this philosophy and, in this regard the intellectual heirs of Herder, Hegel and Dilthey, thought it necessary to construct the human sciences on the basis of open discussion of the pursuit of science as a value-laden activity. These scholars conceived of the pursuit of knowledge in itself as a way of life, a mode of being human, that integrates fact and value, thought and action. The natural scientists criticized the position as it appeared to sacrifice objective neutrality. The human scientists countercharged that the natural scientists disguised their support for particular ways of life by a refusal to recognize in science anything other than 'the facts of nature'. Something of the convoluted difficulty of the issues was expressed by Weber, in his lecture in 1919 on 'Science as a Vocation', in a discussion of the academic's obligations : 'But the true teacher will guard against imposing any attitude on the student from the lectern, whether explicitly or through suggestion. "To let the facts speak for themselves" is of course the most unfair method of all.'[42] Weber opposed the belief that academic knowledge or philosophy is a source of moral or cultural leadership and he objected to politics in the classroom; at the same time, he believed that facts, unless interpreted by a theory that makes them meaningful to us, are meaningless. And any theory that creates meaning is a theory that makes evaluations.

As Dilthey's arguments about psychology indicate, there was also debate about the nature of scientific explanation; this was relevant to the philosophy of science in general and not just to the philosophy of human action. Several contributors, who included Dilthey and Weber, elaborated a distinction between *Erklärung*, explanation, and *Verstehen*, understanding, which was a common vocabulary in the debates as they were re-expressed a century later. Explanation was thought to be characteristic of the natural sciences, which seek causal laws through the observation of regularities in data. Understanding, by contrast, which was said to be the goal of the human sciences, was thought to involve research on the meaning that actions have, whether rational or irrational, for a person as a conscious social actor. How in practice this created concrete methods for the human sciences was debated at length. Weber, for example, defined sociology as the science that concerns itself 'with the interpretive understanding of social action and thereby with a causal exploration of its course and consequences'.[43] This implied that he thought meanings are causes or, at least, that meanings and causes are complementary. He therefore opposed Dilthey's attempt sharply to separate the natural and the human sciences – a separation that set causes and meanings in contrast to each other. Weber wanted to explain human actions by people's rational purposes, by subjectively intended meanings and by causes – emotions, material need, social arrangements or whatever. This search for meanings and causes is exemplified empirically in his study of Protestant sects and seventeenth-century capitalism. He thought it possible to recover meaning from the past through the examination of the whole, irreducibly social, context of human actions, and he thought that this requires the scientist to apply theories and concepts not available to people in the past. By contrast, when Dilthey discussed *Verstehen*, he denoted what a scientist does when he or she intellectually empathizes with another person in order to grasp that person's meaningful world as the reason for actions.

The neo-Kantian philosophers Wilhelm Windelband (1848–1915) and Heinrich Rickert (1863–1936), who taught in universities in the south-west of Germany, attempted to give rigorous accounts of what explanation is in the different sciences. Windelband introduced a much used distinction between explanation by general laws, called nomothetic explanation, which is characteristic of the natural sciences, and explanation at the level of particulars, called idiographic explanation, which is characteristic of history. Rickert similarly stressed the validity of non-natural science forms of explanation. He argued that particular and not general statements are proper to the explanation of human action, and it followed, he thought, that human science explanations are historical in form, that is, they refer to particular actions at a particular time.

> Methodology has to observe that . . . [natural science] treats its subject matter, nature, as devoid of value and without meaning and brings it under general concepts, whereas . . . [cultural science] represents its subject matter, culture, as meaningful and relevant to values and therefore does not content itself with the generalizing method of the natural sciences.[44]

In his view, psychology is a natural science that seeks to make general statements. It is history, Rickert argued, with its explanations by particulars, and not psychology, which is the science that grasps the meaning and value of human life. In this way he hoped to defend the scholarly pursuit of human science as both a scientific activity and an expression of the value of civilization. It followed from his argument that the researcher's own values enter into the concepts, the selection and the representation of what is studied, and hence history must belong to what Rickert called the cultural sciences, the sciences of our values.

Why describe these debates in the conclusion to a history of the human sciences? There are three reasons. The first is that this history is a history of divided opinion about the character, not just the content, of a science of men and women. Though the dominant Anglo-American view of psychology and sociology took natural science to be the model for science, there was at no time agreement about this, and there was continuous argument that some sciences are not natural sciences. The late nineteenth- and early twentieth-century debates graphically illustrate this wider view of science. The second reason is that the historical divergence of opinion about science continued into the late twentieth century. The earlier literature gave that continued debate much of its language and terms of reference. The German views were used as intellectual resources with which to support belief in the particular, historical character of knowledge in the human sciences, and belief that values are necessarily part of the content and activity of science. In the second half of the twentieth century, however, psychology, sociology, political science and other human science disciplines had become large-scale academic enterprises, each with a heavy dependence on public funds, a large involvement in social policy and administration and an investment in specialist objective methods and forms of explanation. A lot was therefore at stake in debate about the relation between values and activity in the human sciences. The previous four chapters traced the vicissitudes of scientific psychology, a complex story that culminated in this chapter with the ambitious claims made for sociobiology, brain science and cognitive science. Any debate about the coherence, sustainability and value-neutrality of these claims was an emotive matter. But there was such debate and the debate found its voice with the terms of argument elaborated earlier. Can we explain human life by natural science? And how do our values relate to knowledge about ourselves?

The third reason for outlining the earlier debates on the nature of explanation and values in the human sciences is that those debates gave due attention to history. Most twentieth-century English-speaking academics assumed that science and history are completely separate disciplines. It was a cliché of university life to oppose the arts and the sciences. This is a source of many of the difficulties in the human sciences, since the subject matter 'man' is at one and the same time natural and historical. The earlier debates suggest ways to move forward as they articulated a view of history as a human science and the human sciences as historically constituted.

In the 1980s there was some shift of opinion within psychology, as well as in the philosophy of the human sciences, away from causal, natural science explanation and towards explanation based on the understanding of meaning and language. This was one aspect of the linguistic turn mentioned earlier. There was no agreed new position but rather a range of alternative viewpoints, which ranged from the search, in the German idealist tradition, for a philosophical anthropology of 'man' to the French philosophy that deconstructed the very notion of the human subject. Within scientific psychology, especially in areas of social psychology and psychotherapy, there was an informal stress on interpretation, or hermeneutic understanding, rather than causal explanation. This pointed back to Dilthey's and Weber's descriptions of *Verstehen* and to the argument that there is a mode of value-oriented understanding of action that can create science. Some psychologists in the 1980s and 1990s were attracted by modern versions of these viewpoints because they appeared to legitimate a science of values and not only a science of events in human life.

When all this is said, however, the historical fact is that many – perhaps most – scientists gave little thought to such abstract and general issues. It was the character of the modern human as well as natural sciences to make research an empirical occupation structured by a well-defined agenda of empirical questions. Scientists did not continuously re-examine fundamentals; if they had done this, they would not have been productive on a day-to-day basis. Yet it is a conspicuous feature of the history of the human sciences that disagreement about fundamentals is often near the surface.

This situation is exemplified by the history of the split between supporters and opponents of sociobiology. Sociobiologists claimed to carry out detailed and methodologically subtle studies to show, for example, that kinship rules to regulate who reproduces with whom in East African peoples are explained by the natural selection of behaviours that favour the maximum survival of the population. Similarly, they correlated individual achievement with sibling position and related this to naturally selected strategies for reproductive success. This work was motivated by and expressed as a search for empirical truth. It also took for granted, however, the applicability of natural science to human culture, which was not an empirical claim; further, it asserted the superiority of biological explanations to other explanations for kinship rules or individual achievement – again, an evaluative and not an empirical claim. Sociobiologists maintained that they simply took seriously the facts of human evolution. Nevertheless, outside the community of sociobiologists, sociobiology was dismissed by people who certainly believed in evolution. Lévi-Strauss, for example, believed that the incest taboo is a cultural act of classification without biological import. To non-believers, the biological explanation of sibling rivalry appeared to be an elaborate projection on to nature of the culture of competitive individualism. Viewed from the outside, sociobiology exemplifies the way in which human science did not possess two of the supposed virtues of natural science, that it uniquely explains activity and that it makes value-free statements of facts. Critics supposed that this was not a failing of sociobiology alone but a failing of the philosophy that maintains there can be a value-free human science. Instead, it was argued, we must accept that human activity requires understanding in terms of culture and values. The social anthropologist Clifford Geertz (b.1926), whose writings in the 1970s and 1980s had an impact beyond the confines of his disciplinary field, wrote: 'We are in sum incomplete or unfinished animals who complete or finish ourselves through culture . . . Our ideas, our values, our acts, even our emotions, are, like our nervous system itself, cultural products – products manufactured, indeed, out of tendencies, capacities, and dispositions with which we were born, but manufactured nonetheless.'[45]

European human science scholarship largely ignored biology and studied the linguistic and symbolic representations of values as a reality, created by human beings, in its own right. From this perspective, sociobiology is itself a system of cultural representations not, or not only, a body of empirical statements. This scholarship, which influenced research in the history of science, described evolutionary theory as the form of knowledge of a culture, and thus denied it a position outside culture from which it is possible to view culture with unique objectivity. The crucial point was articulated much earlier by Rickert: 'Not only are the natural sciences an historical product of civilized man, but also "nature" itself, in a logical or formal sense, is nothing but a theoretical value of *cultural* life, a valid, i.e., objectively valuable, *conception* of reality on the part of the human intellect.'[46]

The argument suggests a significant possibility for the history of science in general and the history of the human sciences in particular. If linguistic and symbolic discourse is primary in our world, then the history of discourse, the history of the way cultures picture themselves to themselves, is central to the human sciences. In other words, from this perspective the history of the human sciences is itself a human science. This is a claim with a strange ring to many anglophone human scientists and historians of science, for whom history is a subject in the humanities and certainly not a science. In the English-speaking world, the history of science is a humanistic and liberal discipline, and for many people, especially for natural scientists, it is essentially parasitical on the natural sciences, the sciences which, as it is believed, are basic to modern notions of truth. If it is argued that human culture is primary, however, then the history of systematic understanding – the history of all the sciences – is a necessary part of science itself. In Dilthey's ambitious words: 'The totality of human nature is only to be found in history; the individual can only become conscious of it and enjoy it when he assembles the minds of the past within himself.'[47]

This is a provocative thought. Whether accepted or not, it highlights the way that the historical narrative in this book is not about something other than the narrative itself: it is not about the truths of nature or a purpose that lies outside history. The plot of the story is the story itself, the quest. The human sciences have had a dramatic life, a life lived as an attempt at reflective self-understanding and self-recreation. As this book has told the story, the journey is full of excitement, but offers few resting-places.

NOTES

CHAPTER 1

1 J. G. von Herder, *Reflections on the Philosophy of the History of Mankind*, ed F. E. Manuel (Chicago: University of Chicago Press, 1968), p. 49.

2 M. Shelley, *Frankenstein: Or, the Modern Prometheus*, ed M. Butler (reprint, Oxford: Oxford University Press, 1994), p. 97.

3 D. Hume, *A Treatise of Human Nature*, ed L. A. Selby-Bigge (Oxford: Clarendon Press, 1888), p. xix.

4 Quoted in C. N. Degler, *In Search of Human Nature: The Decline and Revival of Darwinism in American Social Thought* (New York: Oxford University Press, 1991), p. 187.

5 F. Nietzsche, *Ecce Homo*, ed W. Kaufmann (New York: Vintage Books, 1969), subtitle, p. 215.

6 K. Marx and F. Engels, *The German Ideology. Parts I & III*, ed R. Pascal (New York: International Publishers, 1963), pp. 14–15.

7 G. W. Leibniz, *New Essays on Human Understanding*, ed P. Remnant and J. Bennett (Cambridge: Cambridge University Press, 1981), III.vii.6.

8 Quoted in R. Ashcraft, *Locke's Two Treatises of Government* (London: Unwin Hyman, 1987), p. 35.

9 R. Descartes, 'Discourse on the Method', in *The Philosophical Writings of Descartes*, trans. J. Cottingham, R. Stoothoff and D. Murdoch (Cambridge: Cambridge University Press, 1985–91), Vol. 1, p. 143.

10 Quoted in I. Berlin, *Vico and Herder: Two Studies in the History of Ideas* (London: Hogarth Press, 1976), p. 197.

11 Quoted in D. Carrithers, 'The Enlightenment Science of Society', in C. Fox, R. Porter and R. Wokler (eds), *Inventing Human Science: Eighteenth-Century Domains* (Berkeley: University of California Press, 1995), p. 254.

12 Quoted in L. Goldman, 'The Origins of British "Social Science": Political Economy, Natural Science and Statistics, 1830–1835', *Historical Journal*, **26** (1983): 587–616, p. 588.

13 C. C. Gillispie, *The Edge of Objectivity: An Essay in the History of Scientific Ideas*

(Princeton: Princeton University Press, 1960).

14 R. S. Woodworth, *Contemporary Schools of Psychology*, in collaboration with M. R. Sheehan (9th edn, London: Methuen, 1965), p. 3.

15 Marx and Engels (1963), p. 29.

16 M. Foucault, *The Order of Things: An Archaeology of the Human Sciences* (London: Tavistock Publications, 1970), p. 309.

CHAPTER 2

1 Genesis, i: 26–7.

2 P. O. Kristeller, 'Humanism', in C. B. Schmitt (ed), *The Cambridge History of Renaissance Philosophy* (Cambridge: Cambridge University Press, 1988), p. 126.

3 Pico della Mirandola, 'Oration on the Dignity of Man', in E. Cassirer, P. O. Kristeller and J. H. Randall, Jr (eds) *The Renaissance Philosophy of Man* (Chicago: University of Chicago Press, 1948), p. 225.

4 Quoted in K. Park, 'The Organic Soul', in Schmitt (ed) (1988), p. 470.

5 Aristotle, *De anima*, in *The Works of Aristotle*, Vol. 3, ed W. D. Ross (Oxford: Clarendon Press, 1931), contents page.

6 Aristotle, 'On the Soul', in *The Complete Works of Aristotle. The Revised Oxford Translation*, ed J. Barnes (Princeton: Princeton University Press, 1984); the quotations are from Vol. 1, p. 661 (lines 415b 9 and 18).

7 Quoted in E. M. W. Tillyard, *The Elizabethan World Picture* (reprint, Harmondsworth: Penguin Books, 1963), p. 41.

8 Quoted in J. Kraye, 'Moral Philosophy', in Schmitt (ed) (1988), p. 312.

9 Quoted in B. P. Copenhaver, 'Astrology and Magic', in Schmitt (ed) (1988), p. 269.

10 Quoted in D. C. Lindberg, *Theories of Vision from Al-Kindi to Kepler* (Chicago: Chicago University Press, 1976), p. 203.

11 R. Burton, *The Anatomy of Melancholy*. Volume 1, ed T. C. Faulkner, N. K. Kiessling and R. L. Blair (Oxford: Clarendon Press, 1989), p. 141.

12 Shakespeare, *Twelfth Night*, II.iv.113–14.

13 Burton (1989), p. 249.

14 Shakespeare, *Julius Caesar*, II.i.67–9.

15 Quoted in H. Baker, *The Dignity of Man: Studies in the Persistence of an Idea* (Cambridge, MA: Harvard University Press, 1947), p. 237.

16 T. Browne, *Religio medici*, in *The Works of Sir Thomas Browne*, ed G. Keynes (reprint, London: Faber & Faber, 1964), Vol. 1, p. 44.

17 Quoted in S. Davies, *Renaissance Views of Man* (Manchester: Manchester University Press, 1978), p. 38.

18 Quoted in I. Maclean, *The Renaissance Notion of Woman: A Study in the Fortunes of Scholasticism and Medical Science in European Intellectual Life* (Cambridge: Cambridge University Press, 1980), p. 9.
19 Quoted in *ibid.*, p. 53.
20 Quoted in L. Schiebinger, *The Mind Has No Sex? Women in the Origins of Modern Science* (Cambridge, MA: Harvard University Press, 1989), p. 169.
21 Shakespeare, *Hamlet*, I.iii.59–61.
22 Quoted in Kraye, in Schmitt (ed) (1988), p. 341; Valla quoted Revelation, iii: 16.
23 Lucretius, *De rerum natura*, trans. W. H. D. Rouse, revised M. F. Smith (Cambridge, MA: Harvard University Press, 1975), pp. 201, 115.
24 Pico della Mirandola, in Cassirer *et al.* (eds) (1948), p. 225.
25 M. de Montaigne, *Essays*, ed J. M. Cohen (London: Penguin Books, 1958), p. 406.
26 Quoted in B. Vickers, 'Rhetoric and Poetics', in Schmitt (ed) (1988), p. 728.
27 Quoted in B. Vickers, *In Defence of Rhetoric* (Oxford: Clarendon Press, 1988), p. 274.
28 F. Bacon, 'The Two Bookes of Francis Bacon of the Proficience and Advancement of Learning Divine and Humane', Book II, in *The Works of Francis Bacon*, ed J. Spedding, R. L. Ellis and D. D. Heath (reprint, Stuttgart–Bad Cannstatt: Friedrich Frommann, Günther Holzboog, 1963), Vol. 3, p. 409.
29 Shakespeare, *Julius Caesar*, III.ii.75.
30 Quoted in L. J. Swift and S. L. Block, 'Classical Rhetoric in Vives' Psychology', *Journal of the History of the Behavioral Sciences*, **10** (1974): 74-83, p. 81.
31 Quoted in D. R. Kelley, 'The Theory of History', in Schmitt (ed) (1988), p. 749.
32 Quoted in D. R. Kelley, *Foundations of Modern Historical Scholarship: Language, Law, and History in the French Renaissance* (New York: Columbia University Press, 1970), p. 133.
33 Quoted in *ibid.*, p. 121.
34 Quoted in Kelley, in Schmitt (ed) (1988), p. 758.

CHAPTER 3

1 Quoted in R. Tuck, 'The "Modern" Theory of Natural Law', in A. Pagden (ed), *The Languages of Political Theory in Early Modern Europe* (Cambridge: Cambridge University Press, 1987), p. 113.
2 Quoted in D. R. Kelley, *The Human Measure: Social Thought in the Western Legal Tradition* (Cambridge, MA: Harvard University Press, 1990), p. 132.

3 Quoted in *ibid.*, p. 89.
4 Quoted in *ibid.*, p. 62.
5 Quoted in P. Burke, *The Italian Renaissance: Culture and Society in Italy* (2nd edn, Oxford: Polity Press, 1987), p. 188.
6 Quoted in Kelley (1990), p. 138.
7 Quoted in Tuck, in Pagden (ed) (1987), p. 111.
8 Quoted in *ibid.*, p. 113.
9 Quoted in *ibid.*, p. 112.
10 M. de Montaigne, *Essays*, ed J. M. Cohen (London: Penguin Books, 1958), p. 276.
11 *Ibid.*, p. 110.
12 *Ibid.*, p. 109.
13 Quoted in Q. Skinner, 'Political Philosophy', in C. B. Schmitt (ed), *The Cambridge History of Renaissance Philosophy* (Cambridge: Cambridge University Press, 1988), p. 408.
14 Quoted in A. Pagden, *The Fall of Natural Man: The American Indian and the Origins of Ethnology* (reprint, Cambridge: Cambridge University Press, 1986), p. 38.
15 Quoted in *ibid.*, pp. 79–80.
16 Quoted in *ibid.*, p. 153.
17 T. Hobbes, *Leviathan*, ed R. Tuck (Cambridge: Cambridge University Press, 1991), p. 28.
18 Quoted in L. T. Sarasohn, 'Motion and Morality: Pierre Gassendi, Thomas Hobbes and the Mechanical World-view', *Journal of the History of Ideas*, **46** (1985): 363–79, p. 375.
19 T. Hobbes, 'Human Nature: Or the Fundamental Elements of Policy', in *The English Works of Thomas Hobbes of Malmesbury*, ed W. Molesworth (London: John Bohn, 1839–45), Vol. 4, p. 31.
20 G. Rossini, 'The Criticism of Rhetorical Historiography and the Ideal of Scientific Method: History, Nature and Science in the Political Language of Thomas Hobbes', in Pagden (ed) (1987), p. 316.
21 Hobbes (1991), pp. 11, 35.
22 *Ibid.*, p. 88.
23 *Ibid.*, p. 110.
24 *Ibid.*, p. 9.
25 *Ibid.*, pp. 89–90.
26 S. Pufendorf, *Of the Law of Nature and Nations. Eight Books*, trans. B. Kennett (3rd edn, London: R. Sare etc., 1717), Vol.1, pp. 117, 137.
27 *Ibid.*, Vol. 2, p. 12.
28 S. Pufendorf, *On the Duty of Man and Citizen According to Natural Law*, ed J. Tully (Cambridge: Cambridge University Press, 1991), p. 116.
29 *Ibid.*, pp. 120, 123.
30 *Ibid.*, pp. 25–6.

CHAPTER 4

1 R. Descartes, 'Meditations on First Philosophy', in *The Philosophical Writings of Descartes*, trans. J. Cottingham, R. Stoothoff and D. Murdoch (Cambridge: Cambridge University Press, 1985–91), Vol. 2, p. 17.
2 T. Hobbes, *Leviathan*, ed R.

Tuck (Cambridge: Cambridge University Press, 1991), p. 61.

3 F. Bacon, 'Of the Dignity and Advancement of Learning', Book IV, in *The Works of Francis Bacon*, ed J. Spedding, R. L. Ellis and D. D. Heath (reprint, Stuttgart–Bad Cannstatt: Friedrich Frommann, Günther Holzboog, 1963), Vol. 4, p. 396.

4 R. Cudworth, *The True Intellectual System of the Universe...* (reprint, Stuttgart–Bad Cannstatt: Friedrich Frommann, 1964), p. 50.

5 R. Burton, *The Anatomy of Melancholy*. Volume 1, ed. T. C. Faulkner, N. K. Kiessling and R. L. Blair (Oxford: Clarendon Press, 1989), title page.

6 Quoted in F. Vidal, 'Psychology in the 18th Century: A View from the Encyclopaedias', *History of the Human Sciences*, **6** no. 1 (1993): 89–119, p. 91.

7 Aristotle, 'On the Soul', in *The Complete Works of Aristotle. The Revised Oxford Translation*, ed J. Barnes (Princeton: Princeton University Press, 1984), Vol. 1, 412a.

8 F. Bacon, 'The Two Bookes of Francis Bacon of the Proficience and Advancement of Learning Divine and Humane', Book I, in (1963), Vol. 3, pp. 286–7.

9 F. Bacon, 'Of the Dignity and Advancement of Learning', Book V, in (1963), Vol. 4, pp. 431, 433.

10 R. Descartes, *Discourse on the Method*, in (1985–91), Vol. 1, p. 113.

11 *Ibid.*, p. 114.

12 *Ibid.*, pp. 127, 131.

13 *Ibid.*, p. 127.

14 R. Descartes, 'Comments on a Certain Broadsheet', in (1985–91), Vol. 1, p. 303.

15 Descartes, 'Discourse on the Method', in (1985–91), Vol. 1, pp. 134, 136.

16 *Ibid.*, p. 140.

17 *Ibid.*, p. 141.

18 R. Descartes, 'The Passions of the Soul', in (1985–91), Vol. 1, p. 340.

19 R. Descartes, *Treatise of Man*, ed T. S. Hall (Cambridge, MA: Harvard University Press, 1972), p. 33.

20 Descartes, 'The Passions of the Soul', in (1985–91), Vol. 1, p. 356.

21 *Ibid.*, p. 346.

22 Extract in R. Hunter and I. Macalpine, *Three Hundred Years of Psychiatry 1535–1860: A History Presented in Selected English Texts* (London: Oxford University Press, 1963), p. 187.

23 Descartes, 'Discourse on the Method', in (1985–91), Vol. 1, pp. 111–12.

24 M. de Montaigne, *Essays*, trans. J. M. Cohen (London: Penguin Books, 1958), pp. 23, 236.

25 *Ibid.*, p. 260.

26 J. Burckhardt, *The Civilization of the Renaissance in Italy*, ed

I. Gordon (New York: New American Library, 1960), p. 225.

27 Montaigne (1958), p. 236.

28 Descartes, 'The Passions of the Soul', in (1985–91), Vol. 1, p. 348.

CHAPTER 5

1 Voltaire, *Letters on England*, trans. L. Tancock (London: Penguin Books, 1980), Letter XIII, pp. 63–4.

2 J. Locke, *An Essay Concerning Human Understanding*, ed P. H. Nidditch (Oxford: Clarendon Press, 1975), II.i.19.

3 *Ibid.*, 'The Epistle to the Reader'.

4 *Ibid.*, I.i.2.

5 *Ibid.*, I.i.6, I.vi.25, II.i.1.

6 *Ibid.*, I.ii.1, II.i.2.

7 *Ibid.*, II.i.22.

8 *Ibid.*, II.xx.14, II.xx.6.

9 *Ibid.*, II.xx.2.

10 *Ibid.*, II.xxi.47, II.xxi.51.

11 J. Locke, *The Educational Writings of John Locke*, ed J. L. Axtell (Cambridge: Cambridge University Press, 1968), p. 111.

12 P. Gay, *The Enlightenment: An Interpretation* (reprint, London: Wildwood House, 1973), Vol. 2, p. 507.

13 Locke (1968), pp. 114, 133.

14 *Ibid.*, p. 114.

15 Quoted in R. Ashcraft, *Locke's Two Treatises of Government* (London: Unwin Hyman, 1987), p. 99.

16 J. Locke, *Two Treatises of Government*, ed P. Laslett (2nd edn, Cambridge: Cambridge University Press, 1967), 2.5.49.

17 Quoted in C. Fox, *Locke and the Scriblerians: Identity and Consciousness in Early Eighteenth-Century Britain* (Berkeley: University of California Press, 1988), p. 1.

18 Locke (1975), II.xxvii.23, II.xxvii.17.

19 *Ibid.*, II.xxvii.6, II.xxvii.9.

20 Quoted in Fox (1988), p. 68.

21 Alès de Corbet, quoted in L. G. Crocker, *An Age of Crisis: Man and World in Eighteenth Century French Thought* (Baltimore: Johns Hopkins University Press, 1959), p. 229.

22 H. Aarsleff, *From Locke to Saussure: Essays on the Study of Language and Intellectual History* (Minneapolis: University of Minnesota Press, 1982), p. 24.

23 Locke (1975), III.ii.4, III.ii.2.

24 Quoted in H. Aarsleff, 'John Wilkins', in (1982), p. 240.

25 Locke (1975), III.ix.21, III.i.1, III.i.2.

26 *Ibid.*, II.ix.1.

27 *Ibid.*, II.ix.8–10.

28 G. Berkeley, *An Essay towards a New Theory of Vision*, in *The Works of George Berkeley Bishop of Cloyne*, ed A. A. Luce and T. E. Jessop, Vol. 1 (London: Nelson, 1948), para. xlv.

29 *Ibid.*, para. xxv.

30 D. Diderot, 'Letter on the Blind for the Use of Those who See', in *Diderot's Selected Writings*, ed L. G. Crocker

(New York: Macmillan, 1966), p. 27.

CHAPTER 6

1 I. Kant, *Anthropology from a Pragmatic Point of View*, trans. M. J. Gregor (The Hague: Martinus Nijhoff, 1974), p. 20.
2 Quoted in W. S. Howell, *Logic and Rhetoric in England, 1500–1700* (Princeton: Princeton University Press, 1956), p. 352.
3 B. Spinoza, *The Ethics*, in *The Philosophy of Benedict de Spinoza*, trans. R. H. M. Elwes (New York: Tudor Publishing, 1936), Part II, proposition vii.
4 *Ibid.*, Part III, definition i.
5 *Ibid.*, Part III, proposition i.
6 *Ibid.*, Part III, introduction, and proposition xxxv note.
7 *Ibid.*, Part V, proposition iii.
8 G. W. Leibniz, 'Principles of Nature and of Grace', in *The Monadology and Other Philosophical Writings*, trans. R. Latta (London: Oxford University Press, 1898), pp. 414–15.
9 *Ibid.*, p. 406.
10 G. W. Leibniz, 'Letter to Arnauld', in (1898), p. 112 note.
11 G. W. Leibniz, *New Essays on Human Understanding*, ed. P. Remnant and J. Bennett (Cambridge: Cambridge University Press, 1981), Preface, pp. 48–9, citing Romans, ii: 15, and p. 51.
12 *Ibid.*, p. 53.
13 Quoted in T. P. Saine, 'Who's Afraid of Christian Wolff?', in A. C. Kors and P. J. Korshin (eds), *Anticipations of the Enlightenment in England, France, and Germany* (Philadelphia: University of Pennsylvania Press, 1987), p. 106.
14 In R. J. Richards, 'Christian Wolff's Prolegomena to Empirical and Rational Psychology: Translation and Commentary', *Proceedings of the American Philosophical Society*, **124** (1980): 227–39, p. 231.
15 Quoted in *ibid.*, p. 229 note.
16 I. Kant, *Critique of Pure Reason*, trans. N. K. Smith (reprint, London: Macmillan, 1978), p. 68 (A24).
17 Kant (1974), p. 15.
18 *Ibid.*, p. 3.
19 *Ibid.*, pp. 183, 5.
20 *Ibid.*, pp. 147, 30, 166.
21 *Ibid.*, p. 160.
22 Quoted in B. M. Stafford, *Body Criticism: Imaging the Unseen in Enlightenment Art and Medicine* (Cambridge, MA: MIT Press, 1991), p. 100.
23 J. G. von Herder, *Reflections on the Philosophy of the History of Mankind*, ed F. E. Manuel (Chicago: University of Chicago Press, 1968), pp. 25–6.
24 Quoted in L. J. Jordanova, 'Naturalizing the Family: Literature and the Bio-medical Sciences in the Late Eighteenth Century', in L. J. Jordanova (ed), *Languages of*

Nature: Critical Essays on Science and Literature (London: Free Association Books, 1986), p. 92.

25 Quoted in G. Tytler, *Physiognomy in the European Novel: Faces and Fortunes* (Princeton: Princeton University Press, 1982), pp. 67, 62.

CHAPTER 7

1 J. O. de La Mettrie, *Man a Machine*, trans. G. C. Bussey (reprint, La Salle: Open Court, 1961), p. 85.

2 Quoted in C. Fox, 'Defining Eighteenth-Century Psychology: Some Problems and Perspectives', in C. Fox (ed), *Psychology and Literature in the Eighteenth Century* (New York: AMS Press, 1987), p. 1.

3 J. Butler, 'Upon Human Nature, Or Man Considered as a Moral Agent', Sermons I–III, in *The Works of Joseph Butler*, ed W. E. Gladstone (Oxford: Clarendon Press, 1896), Vol. 2, pp. 3–76.

4 Quoted in L. G. Crocker, *Nature and Culture: Ethical Thought in the French Enlightenment* (Baltimore: Johns Hopkins University Press, 1963), p. 480.

5 D. Hume, 'Of National Characters', in *Essays: Moral, Political, and Literary*, ed E. F. Miller (revised edn, Indianapolis: Liberty Classics, 1987), p. 198.

6 Quoted in *Oxford English Dictionary*, 'Sensibility'.

7 Quoted in J. A. Dussinger, 'Yorick and the "Eternal Foundation of Our Feelings"', in Fox (ed) (1987), p. 264.

8 L. J. Rather, *Mind and Body in Eighteenth Century Medicine: A Study Based on Jerome Gaub's De regimine mentis* (London: Wellcome Historical Medical Library, 1965), p. 64.

9 Extract in R. Hunter and I. Macalpine, *Three Hundred Years of Psychiatry 1535–1860: A History Presented in Selected English Texts* (London: Oxford University Press, 1963), p. 475.

10 Extract in *ibid.*, p. 391.

11 La Mettrie (1961), pp. 88, 122, 143–4.

12 *Ibid.*, pp. 143, 98, 128.

13 *Ibid.*, p. 95.

14 *Ibid.*, pp. 125, 117.

15 D. Diderot, 'Supplement to Bougainville's *Voyage*', in *Diderot's Selected Writings*, ed L. G. Crocker (New York: Macmillan, 1966), p. 246.

16 D. Diderot, 'D'Alembert's Dream', in (1966), pp. 189, 219.

17 D. Diderot, 'Rameau's Nephew', in (1966), p. 140.

18 Marquis de Sade, *Juliette*, ed A. Wainhouse (reprint, London: Arrow Books, 1991), p. 317.

19 Marquis de Sade, *Justine, or Good Conduct Well Chastised*, ed R. Seaver and A. Wainhouse (reprint, London: Arrow

Books, 1991), pp. 607, 603, 604, 545.

20 D. Diderot, 'Refutation of the Work of Helvétius Entitled *On Man*', in (1966), p. 284.

21 Quoted in P. Gay, *The Enlightenment: An Interpretation* (reprint, London: Wildwood House, 1973), Vol. 2, p. 560.

22 D. Diderot, 'Letter on the Blind for the Use of Those who See', in (1966), pp. 14, 17.

23 Condillac, *Logic, or the First Development of the Art of Thinking*, in *Philosophical Writings of Etienne Bonnot, abbé de Condillac*, trans. F. Philip (Hillsdale, NJ: Lawrence Erlbaum Associates, 1982), p. 410.

24 Condillac, *Essai sur l'origine des conaissances humaines: Ouvrage ou l'on reduit à un seul principe tout ce qui concerne l'entendement*, ed R. Lenoir (Paris: Armand Colin, 1924).

25 Condillac, *A Treatise on the Sensations*, in (1982), Dedication, p. 171.

26 Condillac, *An Essay on the Origin of Human Knowledge: Being a Supplement to Mr. Locke's Essay on the Human Understanding*, trans. T. Nugent (reprint, Gainsville, FA: Scholars' Facsimiles and Reprints, 1971), p.59.

27 Condillac, *A Treatise on the Sensations*, in (1982), pp. 171, 200–1.

28 Condillac (1971), p. 51.

29 Condillac, *A Treatise on the Sensations*, in (1982), p. 252.

30 Quoted in F. Vidal, 'Psychology in the 18th Century: A View from the Encyclopaedias', *History of the Human Sciences*, **6** no. 1 (1993): 89–119, p. 106.

31 F. J. Picavet, *Les idéologues. Essai sur l'histoire des idées et des théories scientifiques, philosophiques, réligieuses, etc. en France dupuis 1789* (reprint, New York: Burt Franklin, 1971), p. 345.

32 Quoted in R. L. Emerson, 'Science and Moral Philosophy in the Scottish Enlightenment', in M. A. Stewart (ed), *Studies in the Philosophy of the Scottish Enlightenment* (Oxford: Clarendon Press, 1990), p. 26.

33 Quoted in P. B. Wood, 'Science and the Pursuit of Virtue in the Aberdeen Enlightenment', in Stewart (ed) (1990), p. 137.

34 Quoted in *ibid.*, p. 132.

35 Quoted in P. B. Wood, 'The Natural History of Man in the Scottish Enlightenment', *History of Science*, **28** (1990): 89–123, p. 97.

36 Quoted in Emerson, in Stewart (ed) (1990), p. 11.

37 A. A. Cooper, Third Earl of Shaftesbury, *An Inquiry Concerning Virtue, or Merit*, ed D. Welford (Manchester: Manchester University Press, 1977), p. 59.

38 J. Butler, 'Of the Nature of Virtue', in *The Analogy of Religion Natural and Revealed*

(reprint, London: J. M. Dent & Sons, 1906), p. 264.

39 F. Hutcheson, *An Inquiry into the Original of Our Ideas of Beauty and Virtue*, in *Collected Works of Francis Hutcheson*, Vol. 1 (reprint, Hildesheim: Georg Olms, 1971), p. 164.

40 Quoted in J. Moore, 'The Two Systems of Francis Hutcheson: On the Origins of the Scottish Enlightenment', in Stewart (ed) (1990), p. 47.

41 Butler (1896), p. 64.

42 D. Hume, *A Treatise of Human Nature*, ed L. A. Selby-Bigge (Oxford: Clarendon Press, 1888), pp. xx, xxi, xix.

43 *Ibid.*, 1.I.1.

44 *Ibid.*, 1.III.6, 2.III.3.

45 D. Hume, 'A Dissertation on the Passions', in *Essays: Moral, Political, and Literary*, in *The Philosophical Works*, ed T. H. Green and T. H. Grose (reprint, Aalen: Scientia, 1964), Vol. 4, p. 166.

46 Quoted in J. P. Wright, *The Sceptical Realism of David Hume* (Manchester: Manchester University Press, 1983), p. 189.

47 Hume (1888), 3.II.2, 2.I.1, 3.I.2, 2.III.9.

48 J. Locke, *An Essay Concerning Human Understanding*, ed P. H. Nidditch (Oxford: Clarendon Press, 1975), II.xxxiii.1.

49 Hume (1888), 1.I.4.

50 J. Mill, *Analysis of the Phenomena of the Human Mind*, 2nd edn, ed J. S. Mill (reprint, New York: Augustus M. Kelley, 1967), Vol. 1, p. 78.

51 Hume (1888), 1.I.4.

52 D. Hartley, *Observations on Man, His Frame, His Duty, and His Expectations* (reprint, Hildesheim: Georg Olms, 1967), Vol. 1, pp. 114, 368, viii.

53 J. Priestley, *Hartley's Theory of the Human Mind, on the Principle of Association of Ideas; With Essays Relating to the Subject of It* (reprint, New York: AMS Press, 1973), pp. xxx, xxxii.

54 T. Reid, *Inquiry into the Human Mind on the Principles of Common Sense*, in *Philosophical Works*, ed W. Hamilton (8th edn reprint, Hildesheim: Georg Olms, 1967), Vol. 1, p. 209.

55 Quoted in G. Bryson, *Man and Society: The Scottish Inquiry of the Eighteenth Century* (reprint, New York: Augustus M. Kelley, 1968), p. 11.

56 T. Reid, *Essays on the Intellectual Powers of Man*, in (1967), Vol. 1, p. 421.

57 *Ibid.*, p. 470.

CHAPTER 8

1 J. Itard, *The Wild Boy of Aveyron*, in L. Malson, *Wolf Children* (London: NLB, 1972), p. 91.

2 D. Defoe, *The Life and Strange Surprizing Adventures of Robinson Crusoe, of York, Mariner*, ed J. D. Crowley (reprint, Oxford: Oxford University Press, 1983) pp. 165, 209.

3 Quoted in Malson (1972), p. 86.
4 Quoted in L. G. Crocker, *Nature and Culture: Ethical Thought in the French Enlightenment* (Baltimore: Johns Hopkins University Press, 1963), p. 96.
5 Quoted in M. T. Hodgen, *Early Anthropology in the Sixteenth and Seventeenth Centuries* (Philadelphia: University of Pennsylvania Press, 1964), p. 419.
6 D. Diderot, 'Supplement to Bougainville's *Voyage*', in *Diderot's Selected Writings*, ed L. G. Crocker (New York: Macmillan, 1966), pp. 248–9.
7 'The "Initial Discourse" to Buffon's *Histoire naturelle* (1749)', trans. J. Lyon, in J. Lyon and P. R. Sloan (eds), *From Natural History to the History of Nature: Readings from Buffon and His Critics* (Notre Dame, IA: University of Notre Dame Press, 1981), p. 102.
8 Quoted in J. C. Greene, *The Death of Adam: Evolution and Its Impact on Western Thought* (reprint, New York: Mentor Books, 1961), p. 182.
9 Quoted in C. Blanckaert, 'Buffon and the Natural History of Man: Writing History and the "Foundational Myth" of Anthropology', *History of the Human Sciences*, **6** no. 1 (1993): 13–50, p. 27.
10 Quoted in *ibid.*, p. 32.
11 Quoted in Greene (1961), p. 179.
12 Quoted in *ibid.*, pp. 192–4.
13 J.-J. Rousseau, *Confessions* (London: J. M. Dent & Sons, 1931), Vol. 2, p. 187 (Book XI).
14 J.-J. Rousseau, *The First and Second Discourses*, ed R. D. Masters (New York: St Martin's Press, 1964), pp. 92–3, 104.
15 *Ibid.*, pp. 114, 130, 127, 142.
16 *Ibid.*, p. 144.
17 *Ibid.*, pp. 151–2.
18 *Ibid.*, pp. 155, 179.
19 Quoted in P. Gay, *The Enlightenment: An Interpretation* (reprint, London: Wildwood House, 1973), Vol. 2, pp. 542–3.
20 J. Butler, 'Of the Nature of Virtue', in *The Analogy of Religion Natural and Revealed* (reprint, London: J. M. Dent & Sons, 1906), p. 264.
21 D. Diderot, 'Letter on the Blind for the Use of Those who See', in (1966), p. 17.
22 D. Hume, *A Treatise of Human Nature*, ed. L. A. Selby-Bigge (Oxford: Clarendon Press, 1888), 3.I.2, 3.III.3, 2.III.9.
23 A. Smith, *The Theory of Moral Sentiments*, ed D. D. Raphael and A. L. Macfie (Oxford: Clarendon Press, 1976), II.ii.2.1.
24 *Ibid.*, I.i.1.5, I.i.2.1.
25 Quoted in G. Bryson, *Man and Society: The Scottish Inquiry of the Eighteenth Century* (reprint,

New York: Augustus M. Kelley, 1968), p. 89.
26 Smith (1976), III.i.3.
27 C.-A. Helvétius, *A Treatise of Man; His Intellectual Faculties and His Education*, trans. W. Hooper (new edn, London: Albion Press, 1810), Vol. 1, pp. 124–6, 146.
28 Quoted in L. G. Crocker, *An Age of Crisis: Man and World in Eighteenth Century French Thought* (Baltimore: Johns Hopkins University Press, 1959), p. 232.
29 Quoted in Gay (1973), pp. 514–15.
30 Quoted in *ibid.*, p. 565.
31 Quoted by the editors in Montesquieu, *The Spirit of the Laws*, ed A. M. Cohler, B. C. Miller and H. S. Stone (Cambridge: Cambridge University Press, 1989), p. xi.
32 *Ibid.*, pp. 3, 9.
33 *Ibid.*, p. xliii.
34 J. Le Rond D'Alembert, *Preliminary Discourse to the Encyclopedia of Diderot*, trans. R. N. Schwab (reprint, Chicago: University of Chicago Press, 1995), pp. 35–6.
35 Montesquieu (1989), Part 6, Book 29, Chap. 14, and p. 611.
36 A. Ferguson, *An Essay on the History of Civil Society*, ed D. Forbes (Edinburgh: Edinburgh University Press, 1966), p. 206.
37 *Ibid.*, pp. 205, 5, 57, 6.
38 Quoted in R. L. Meek, *Social Science and the Ignoble Savage* (Cambridge: Cambridge University Press, 1976), p. 2.
39 Quoted in K. M. Baker, *Condorcet: From Natural Philosophy to Social Mathematics* (Chicago: University of Chicago Press, 1975), p. 197.

CHAPTER 9

1 A. Smith, *An Inquiry into the Nature and Causes of the Wealth of Nations*, ed R. H. Campbell and A. S. Skinner (Oxford: Clarendon Press, 1976), I.ii.1.
2 F. Bacon, 'Of the Dignity and Advancement of Learning', Book VIII, in *The Works of Francis Bacon*, ed. J. Spedding, R. L. Ellis and D. D. Heath (reprint, Stuttgart–Bad Cannstatt: Friedrich Frommann, Günther Holzboog, 1963), Vol. 5, p. 35.
3 Quoted in J. O. Appleby, *Economic Thought and Ideology in Seventeenth-Century England* (Princeton: Princeton University Press, 1978), p. 51.
4 Quoted in *ibid.*, p. 187.
5 Smith (1976), IV.Intro.1.
6 Quoted in Appleby (1978), pp. 56–7.
7 Quoted in R. Olson, *Science Deified & Science Defied: The Historical Significance of Science in Western Culture. Volume 2* (Berkeley: University of California Press, 1990), pp. 78, 84, 83.
8 Quoted in *ibid.*, p. 153.
9 D. Hume, 'Of the Populousness of Ancient

Nations', in *Essays: Moral, Political, and Literary*, ed E. F. Miller (revised edn, Indianapolis: Liberty Classics, 1987), p. 382.

10 A. Ferguson, *An Essay on the History of Civil Society*, ed D. Forbes (Edinburgh: Edinburgh University Press, 1966), p. 182.

11 Smith (1976), I.iv.1.

12 *Ibid.*, IV.Intro.1.

13 J. A. Schumpeter, *History of Economic Analysis*, ed E. B. Schumpeter (reprint, London: George Allen & Unwin, 1955), p. 181.

14 Smith (1976), I.xi.p.7.

15 *Ibid.*, Intro.1, Intro.3, IV.ii.9.

16 Quoted in R. Lekachman, *A History of Economic Ideas* (reprint, New York: Harper & Row, 1964), p. 153.

17 J. Bentham, *An Introduction to the Principles of Morals and Legislation*, intro. L. J. Lafleur (reprint, Darien, CT: Hafner, 1970), p. 1.

18 Helen Bevington, quoted in Lekachman (1964), p. 104.

19 Bentham (1970), pp. 64–5.

20 Quoted in P. Edwards (ed), *The Encyclopedia of Philosophy* (New York: Macmillan and Free Press, 1967), Vol. 1, p. 267.

21 T. R. Malthus, *An Essay on the Principle of Population*, ed A. Flew (Harmondsworth: Penguin Books, 1970), pp. 72, 204.

22 F. Galton, *Inquiries into Human Faculty and Its Development* (2nd edn reprint, London: J. M. Dent & Co., nd), p. 207.

23 K. Marx, 'Theories of Surplus Value. Volume II', in *Marx and Engels on Malthus: Selections from the Writings of Marx and Engels Dealing with the Theories of Thomas Robert Malthus*, ed R. L. Meek (London: Lawrence & Wishart, 1953), p. 123.

24 Quoted in E. Halévy, *The Growth of Philosophic Radicalism* (London: Faber & Faber, 1952), p. 451.

25 Quoted in H. Aarsleff, *The Study of Language in England, 1780–1860* (Princeton: Princeton University Press, 1967), p. 95.

Chapter 10

1 Quoted in C. Taylor, *Hegel* (Cambridge: Cambridge University Press, 1975), p. 35.

2 T. Carlyle, *Sartor Resartus* (reprint, London: Chapman & Hall, nd), p. 99 (Chap. 7).

3 Taylor (1975), p. 14.

4 Quoted in M. H. Abrams, *The Mirror and the Lamp: Romantic Theory and the Critical Tradition* (London: Oxford University Press, 1953), p. 21.

5 G. Vico, *The New Science of Giambattista Vico*, trans. T. G. Bergin and M. H. Fisch (reprint, Ithaca: Cornell University Press, 1984), para. 349.

6 Quoted in I. Berlin, *Vico and Herder: Two Studies in the History*

of Ideas (London: Hogarth Press, 1976), p. 15.

7 Vico (1984), paras. 330–1.

8 *Ibid.*, paras. 374, 237, 383.

9 Quoted in Berlin (1976), p. 143.

10 J. G. von Herder, *Reflections on the Philosophy of the History of Mankind*, ed F. E. Manuel (Chicago: University of Chicago Press, 1968), pp. 40, 59.

11 *Ibid.*, p. 47.

12 Quoted in Berlin (1976), p. 165.

13 P. B. Shelley, 'A Defence of Poetry', in *Selected Poetry, Prose and Letters*, ed A. S. B. Glover (London: Nonesuch Press, 1951), p. 1055.

14 Cited in H. Aarsleff, 'The Tradition of Condillac: The Problem of the Origin of Language in the Eighteenth Century and the Debate in the Berlin Academy before Herder', in *From Locke to Saussure: Essays on the Study of Language and Intellectual History* (Minneapolis: University of Minnesota Press, 1982), pp. 194–5.

15 Berlin (1976), p. 190.

16 Herder (1968), p. 82.

17 Quoted in P. Edwards (ed), *The Encyclopedia of Philosophy* (New York: Macmillan and Free Press, 1967), Vol. 7, p. 294.

18 Quoted in A. Gode-Von Aesch, *Natural Science in German Romanticism* (reprint, New York: AMS Press, 1966), p. 94.

19 *Hegel's Philosophy of Mind: Being Part Three of the Encylopaedia of the Philosophical Sciences (1830)*, trans. W. Wallace (Oxford: Clarendon Press, 1971), para. 377, p. 1.

20 Quoted in B. Willey, *Nineteenth Century Studies* (reprint, Harmondsworth: Penguin Books, 1964), pp. 13, 17–18.

21 Quoted in M. McNeil, *Under the Banner of Science: Erasmus Darwin and His Age* (Manchester: Manchester University Press, 1987), p. 79.

22 S. T. Coleridge, *Biographia Literaria Or Biographical Sketches of My Literary Life and Opinions*, ed G. Watson (London: Dent, 1956), p. 167.

23 J. S. Mill, *On Bentham and Coleridge*, intro. W. R. Leavis (reprint, New York: Harper Torchbooks, 1962), p. 40.

24 E. C. Trapp, quoted in S. Jaeger, 'The Origin of the Diary Method in Developmental Psychology', in G. Eckardt, W. G. Bringmann and L. Sprung (eds), *Contributions to a History of Developmental Psychology* (Berlin and New York: Mouton, 1985), p. 68.

CHAPTER 11

1 Extract in F. Stern, *The Varieties of History: From Voltaire to the Present* (reprint, New York: Vintage Books, 1973), p. 61.

2 Quoted in J. T. Merz, *A History*

of *European Thought in the Nineteenth Century* (reprint, New York: Dover Books, 1965), Vol. 1, p. 225 note.

3 Merz (1965), Vol. 3, p. 137.

4 W. von Humboldt, *On Language: The Diversity of Human Language-structure and Its Influence on the Mental Development of Mankind*, intro. H. Aarsleff (Cambridge: Cambridge University Press, 1988), p. 27.

5 Quoted in R. Harris and T. J. Taylor, *Landmarks in Linguistic Thought: The Western Tradition from Socrates to Saussure* (London: Routledge, 1989), pp. 141–2.

6 Quoted in H. Aarsleff, 'Wilhelm von Humboldt and the Linguistic Thought of the French *idéologues*', in *From Locke to Saussure: Essays on the Study of Language and Intellectual History* (Minneapolis: University of Minnesota Press, 1982), p. 343.

7 Humboldt (1988), pp. 49, 41.

8 Quoted in L. Taub, 'Evolutionary Ideas and "Empirical" Methods: The Analogy between Language and Species in Works by Lyell and Schleicher', *British Journal for the History of Science*, **26** (1993): 171–93, p. 187.

9 Quoted in L. Poliakov, *The Aryan Myth: A History of Racist and Nationalist Ideas in Europe* (London: Heinemann, 1974), p. 190.

10 Quoted in *ibid.*, p. 199.

11 Quoted in Taub (1993), p. 176.

12 Quoted in E. Knoll, 'The Science of Language and the Evolution of Mind: Max Müller's Quarrel with Darwinism', *Journal of the History of the Behavioral Sciences*, **22** (1986): 3–22, pp. 6, 7.

13 Extract in Stern (1973), pp. 56–7.

14 Quoted in F. Meinecke, *Historism: The Rise of a New Historical Outlook* (London: Routledge & Kegan Paul, 1972), p. xxv.

15 Quoted in B. Willey, *Nineteenth Century Studies* (reprint, Harmondsworth: Penguin Books, 1964), p. 230.

16 W. Whewell, *History of the Inductive Sciences, from the Earliest to the Present Time* (3rd edn reprint, London: Cass, 1967), Vol. 3, p. 452.

17 Quoted in G. W. Stocking, Jr, *Victorian Anthropology* (New York: Free Press, 1987), p. 74.

18 T. H. Huxley, *Man's Place in Nature and Other Essays* (reprint, London: J. M. Dent & Sons, 1906), p. 150.

19 J. G. von Herder, *Reflections on the Philosophy of the History of Mankind*, ed F. E. Manuel (Chicago: University of Chicago Press, 1968), p. 29.

20 Quoted in R. M. Young, *Mind, Brain, and Adaptation: Cerebral Localization and Its Biological Context from Gall to Ferrier*

(Oxford: Clarendon Press, 1970), p. 12.

21 J. C. Prichard, *Researches into the Physical History of Man*, ed G. W. Stocking, Jr (Chicago: University of Chicago Press, 1973), pp. 236, 233.

22 Quoted in N. L. Stepan, 'Race and Gender: The Role of Analogy in Science', *Isis*, **77** (1986): 261–77, p. 269.

23 Quoted in Stocking (1987), p. 251.

24 Quoted in *ibid.*, p. 64.

25 Quoted in W. Stanton, *The Leopard's Spots: Scientific Attitudes toward Race in America 1815–1859* (Chicago: University of Chicago Press, 1960), p. 69.

26 Quoted in C. E. Russett, *Sexual Science: The Victorian Construction of Womanhood* (Cambridge, MA: Harvard University Press, 1989), p. 54.

27 Quoted in Poliakov (1974), p. 266.

28 Quoted in D. P. Frisby, 'Georg Simmel and Social Psychology', *Journal of the History of the Behavioral Sciences*, **20** (1984): 107–27, p. 112.

29 T. Waitz, *Introduction to Anthropology*, ed J. F. Collingwood (London: Longman, Green, Longman, and Roberts, 1863), pp. 8, 12, 228.

30 Quoted in K.-P. Koepping, *Adolf Bastian and the Psychic Unity of Mankind: The Foundations of Anthropology in Nineteenth Century Germany* (St Lucia: Queensland University Press, 1983), pp. 12, 29, 48.

31 Quoted in Poliakov (1974), p. 233.

32 Quoted in G. W. Stocking, Jr, 'Paradigmatic Traditions in the History of Anthropology', in R. C. Olby *et al.* (eds), *Companion to the History of Modern Science* (London: Routledge, 1990), p. 712.

33 T. Laycock, *Mind and Brain: Or the Correlations of Consciousness and Organization* (Edinburgh: Sutherland & Knox, 1860), Vol. 1, p. 1.

34 A. Bain, *The Senses and the Intellect* (London: John W. Parker, 1855), p. v.

35 [J. Martineau], 'Cerebral Psychology: Bain', *National Review*, **10** (1860): 500-21, p. 501.

36 Quoted in F. Gregory, *Scientific Materialism in Nineteenth Century Germany* (Dordrecht: D. Reidel, 1977), pp. 89, 90.

37 Quoted in *ibid.*, p. 96.

38 I. Turgenev, *Fathers and Sons*, trans. R. Edmonds (reprint, London: Penguin Books, 1975), p. 90.

39 J. S. Mill, *A System of Logic Ratiocinative and Inductive*, in *Collected Works of John Stuart Mill*, Vol. 8 (Toronto: University of Toronto Press, 1973–4), p. 851.

40 J. S. Mill, 'Bain's Psychology', in *Essays on Philosophy and the Classics*, in *Collected Works of John Stuart Mill*, Vol. 11

(Toronto: University of Toronto Press, 1978), p. 348.
41 A. Bain, 'The Respective Spheres and Mutual Helps of Introspection and Psycho-physical Experiment in Psychology', *Mind*, new series **2** (1893): 42–53, p. 42.
42 D. Ferrier, *The Functions of the Brain* (London: Smith, Elder, 1876), p. 275.

CHAPTER 12

1 K. Marx, Preface, 'A Contribution to the Critique of Political Economy', in *Selected Writings in Sociology and Social Philosophy*, ed T. B. Bottomore and M. Rubel (2nd edn reprint, Harmondsworth: Penguin Books, 1963), p. 67.
2 Told in *Encyclopaedia Britannica*, Vol. 21 (9th edn, Edinburgh: Adam and Charles Black, 1886), p. 197.
3 J. S. Mill, *Auguste Comte and Positivism* (reprint, Ann Arbor: University of Michigan Press, 1961), p. 86.
4 A. Comte, *The Essential Comte: Selected from the Course de philosophie positive*, ed S. Andreski (London: Croom Helm, and New York: Barnes & Noble, 1974), p. 32.
5 *Ibid.*, pp. 37–8.
6 *Ibid.*, p. 24.
7 Mill (1961), p. 9.
8 Comte (1974), pp. 53, 55, 56.
9 *Ibid.*, pp. 144–5, 147.
10 *Ibid.*, p. 33.
11 Quoted in W. M. Simon, 'The "Two Cultures" in Nineteenth-Century France: Victor Cousin and Auguste Comte', *Journal of the History of Ideas*, **26** (1965): 45–58, pp. 48, 47.
12 Quoted in W. M. Simon, *European Positivism in the Nineteenth Century: An Essay in Intellectual History* (reprint, Port Washington, NY: Kennikat Press, 1972), pp. 27, 186.
13 Quoted in G. Lichtheim, *Marxism: An Historical and Critical Study* (2nd edn, New York: Frederick A. Praeger, 1965), p. 37.
14 Quoted in *ibid.*, p. 39.
15 K. Marx, 'Economic and Philosophical Manuscripts', in (1963), p. 87.
16 Quoted in E. P. Thompson, *The Making of the English Working Class* (reprint, Harmondsworth: Penguin Books, 1968), p. 395.
17 K. Marx, 'Economic and Philosophical Manuscripts', in (1963), p. 87.
18 K. Marx, Preface, 'A Contribution to a Critique of Political Economy', in (1963), p. 68.
19 L. Kolakowski, *Main Currents of Marxism: Its Origins, Growth and Dissolution* (Oxford: Oxford University Press, 1978), Vol. 1, p. 336.
20 K. Marx and F. Engels, *The German Ideology. Parts I & III*, ed R. Pascal (New York: International Publishers, 1963), pp. 22–3, 13.

21 Quoted in F. Gregory, *Scientific Materialism in Nineteenth Century Germany* (Dordrecht: D. Reidel, 1977), p. 92.
22 K. Marx, 'Economic and Philosophical Manuscripts', in (1963), p. 83.
23 F. Engels, in K. Marx and F. Engels, *Selected Works in One Volume* (London: Lawrence & Wishart, 1968), pp. 429–30.
24 K. Marx, letter to Engels, 1862, in K. Marx and F. Engels, *Collected Works*, Vol. 41 (London: Lawrence & Wishart, 1985), pp. 232, 381.
25 Quoted in R. M. Young, 'The Historiographic and Ideological Contexts of the Nineteenth-Century Debate on Man's Place in Nature', in *Darwin's Metaphor: Nature's Place in Victorian Culture* (Cambridge: Cambridge University Press, 1985), p. 196.
26 Quoted in E. Roll, *A History of Economic Thought* (4th edn, London: Faber & Faber, 1973), pp. 379, 378.
27 Quoted by the editors in Marx (1963), p. 33.

CHAPTER 13

1 C. Darwin, *The Descent of Man, and Selection in Relation to Sex* (reprint, Princeton: Princeton University Press, 1981), Vol. 1, p. 184.
2 Quoted in R. M. Young, 'The Impact of Darwin on Conventional Thought', in *Darwin's Metaphor: Nature's Place in Victorian Culture* (Cambridge: Cambridge University Press, 1985), p. 5.
3 Darwin (1981), Vol. 2, p. 405.
4 Quoted in A. Desmond and J. Moore, *Darwin* (London: Michael Joseph, 1991), p. 527.
5 T. H. Huxley, 'Time and Life: Mr. Darwin's "Origin of Species"', in *Man's Place in Nature and Other Essays* (reprint, London: J. M. Dent & Sons, 1906), p. 298.
6 Quoted in R. M. Young, 'Darwin's Metaphor: Does Nature Select?', in Young (1985), p. 124.
7 Darwin (1981), Vol. 1, p. 71.
8 T. H. Huxley, 'Darwin on the Origin of Species', in (1906), p. 336.
9 H. Spencer, *The Principles of Biology*, (revised edn, London: Williams & Norgate, 1898–9), Vol. 1, p. 530.
10 K. Pearson, *The Grammar of Science* (2nd edn reprint, London: J. M. Dent & Sons, 1937), p. 310.
11 Quoted in R. M. Young, *Mind, Brain, and Adaptation: Cerebral Localization and Its Biological Context from Gall to Ferrier* (Oxford: Clarendon Press, 1970), p. 151.
12 H. Spencer, *First Principles* (6th edn reprint, London: Williams & Norgate, 1915), p. 321.
13 W. James, 'Herbert Spencer's Autobiography', in *Memories and Studies* (London:

Longmans, Green, 1911), p. 124.
14 Quoted in Young (1970), p. 170.
15 J. S, Mill, 'Bain's Psychology', in *Essays on Philosophy and the Classics*, in *Collected Works of John Stuart Mill*, Vol. 11 (Toronto: University of Toronto Press, 1978), p. 361.
16 H. Spencer, *Principles of Psychology* (reprint, Farnborough: Gregg International, 1970), p. 606.
17 H. Spencer, 'The Filiation of Ideas', in D. Duncan (ed), *The Life and Letters of Herbert Spencer* (London: Methuen, 1908), p. 547.
18 H. Spencer, 'The Social Organism', in *The Man Versus the State: With Four Essays on Politics and Society*, ed D. MacRae (Harmondsworth: Penguin Books, 1969), p. 223.
19 H. Spencer, *The Study of Sociology* (16th edn, London: Kegan Paul, Trench, Trubner, 1892), pp. 47, 46, 21–2.
20 C. Darwin, *The Origin of Species by Means of Natural Selection* (6th edn reprint, New York: Collier Books, 1962), p. 483.
21 C. Darwin, letter to Henry Fawcett, 1861, in *The Correspondence of Charles Darwin. Volume 9: 1861* (Cambridge: Cambridge University Press, 1994), p. 269.
22 Darwin (1981), Vol. 2, p. 404.
23 *Ibid.*
24 C. Darwin, *On the Origin of Species by Means of Natural Selection*, ed J. W. Burrow (1st edn reprint, Harmondsworth: Penguin Books, 1968), p. 458.
25 Darwin (1981), Vol. 1, p. 3.
26 *Ibid.*, p. 105.
27 *Ibid.*, pp. 70, 71.
28 *Ibid.*, p. 180.
29 A. R. Wallace, 'The Origin of Human Races and the Antiquity of Man Deduced from the Theory of "Natural Selection"', *Journal of the Anthropological Society of London*, **2** (1864): clviii–clxx, p. clxviii.
30 C. Darwin, *The Expression of the Emotions in Man and Animals* (reprint, Chicago: University of Chicago Press, 1965), p. 19.
31 Quoted in J. W. Burrow, *Evolution and Society: A Study in Victorian Social Theory* (reprint, Cambridge: Cambridge University Press, 1970), p. 70.
32 Quoted in *ibid.*, p. 154.
33 Quoted in *ibid.*, p. 249.
34 Quoted in S. J. Gould, *The Mismeasure of Man* (reprint, Harmondsworth: Penguin Books, 1984), pp. 116–17.
35 Quoted in Burrow (1970), p. 254.
36 Spencer (1970), p. 374.
37 Quoted in J. M. O'Donnell, *The Origins of Behaviorism: American Psychology, 1870–1920* (New York: New York University Press, 1986), p. 168.

38 T. H. Huxley, 'On the Hypothesis that Animals Are Automata and Its History', in *Method and Results: Essays* (London: Macmillan, 1893), p. 242.
39 W. James, 'Remarks on Spencer's Definition of Mind as Correspondence', in *Collected Essays and Reviews* (London: Longmans, Green, 1920), p. 67.
40 W. James, 'Are We Automata?', in *Essays on Psychology*, intro. W. R. Woodward, *The Works of William James* (Cambridge, MA: Harvard University Press, 1983), p. 46.
41 Quoted in A. Costall, 'How Lloyd Morgan's Canon Backfired', *Journal of the History of the Behavioral Sciences*, **29** (1993): 113–22, pp. 116, 115, 117.
42 G. H. Mead, 'Social Psychology as Counterpart to Physiological Psychology', in *Selected Writings*, ed A. J. Reck (Indianapolis: Bobbs-Merrill, 1964), pp. 101–2.

CHAPTER 14

1 W. James, 'A Plea for Psychology as a "Natural Science"', in *Essays in Psychology*, intro. W. R. Woodward, *The Works of William James* (Cambridge, MA: Harvard University Press, 1983), p. 271.
2 T. Ribot, *German Psychology To-day: The Empirical School* (New York: Charles Scribner's Sons, 1886), p. 5.
3 T. Ribot, *English Psychology* (London: Henry S. King, 1873), p. 25.
4 T. Ribot, *The Diseases of Personality* (Chicago: Open Court, 1906), pp. 1–2.
5 Quoted in H. Misiak and V. M. Staudt, *Catholics in Psychology: A Historical Survey* (New York: McGraw-Hill, 1954), pp. 44, 47.
6 Quoted in E. G. Boring, *A History of Experimental Psychology* (2nd edn, New York: Appleton-Century-Crofts, 1950), p. 392.
7 E. Mach, *The Analysis of Sensations and the Relation of the Physical to the Psychical* (revised edn, New York: Dover, 1959), p. 310.
8 Quoted in M. G. Ash, 'Academic Politics in the History of Science: Experimental Psychology in Germany, 1879–1941', *Central European History*, **13** (1980): 255–86, p. 269.
9 W. Dilthey, *Descriptive Psychology and Historical Understanding* (The Hague: Martinus Nijhoff, 1977), p. 27.
10 Quoted in M. G. Ash, 'Psychology in Twentieth-Century Germany: Science and Profession', in G. Cocks and K. Jarausch (eds), *German Professions, 1800–1950* (New York: Oxford University Press, 1990), p. 295.

11 Quoted in M. M. Sokal, 'James McKeen Cattell and the Failure of Anthropometric Testing, 1890–1901', in W. R. Woodward and M. G. Ash (eds), *The Problematic Science: Psychology in Nineteenth-Century Thought* (New York: Praeger, 1982), p. 330.
12 Quoted in J. M. O'Donnell, *The Origins of Behaviorism: American Psychology, 1870–1920* (New York: New York University Press, 1986), p. 157.
13 Quoted in Sokal, in Woodward and Ash (eds) (1982), p. 332.
14 Quoted in O'Donnell (1986), p. 38.
15 Quoted in *ibid.*, p. 132.
16 Quoted in *ibid.*, p. 133.
17 Quoted in *ibid.*, p. 119.
18 Quoted in *ibid.*, p. 10.
19 E. B. Titchener, 'The Postulates of a Structural Psychology', in W. Dennis (ed), *Readings in the History of Psychology* (New York: Appleton-Century-Crofts, 1948), p. 367.
20 Quoted in O'Donnell (1986), p. 131.

CHAPTER 15

1 E. Durkheim, *The Rules of Sociological Method*, ed E. G. Catlin (reprint, New York: Free Press, 1966), pp. lvii–lviii.
2 *Ibid.*, p. 145.
3 Quoted in P. Abrams, 'The Uses of British Sociology, 1831–1981', in M. Bulmer (ed), *Essays on the History of British Sociological Research* (Cambridge: Cambridge University Press, 1985), p. 182.
4 E. Lisle, quoted in I. Hacking, 'How Numerical Sociology Began by Counting Suicides: From Medical Pathology to Social Pathology', in I. B. Cohen (ed), *The Natural Sciences and the Social Sciences: Some Critical and Historical Perspectives* (Dordrecht: Kluwer, 1994), p. 118.
5 A. Quetelet, *A Treatise on Man and the Development of His Faculties* (reprint, New York: Burt Franklin, 1968), pp. 96, 108.
6 *Ibid.*, pp. 6, 96.
7 Quoted in J. Cole, 'The Chaos of Particular Facts: Statistics, Medicine and the Social Body in Early 19th-Century France', *History of the Human Sciences*, 7 no. 3 (1994): 1–27, p. 1.
8 Quoted in *ibid.*, p. 20.
9 H. T. Buckle, *History of Civilization in England* (new edn, London: Longmans, Green, 1885), Vol. 1, p. 20.
10 Durkheim (1966), p. 110.
11 E. Durkheim, *Suicide: A Study in Sociology* (London: Routledge & Kegan Paul, 1952), pp. 258, 248.
12 E. Durkheim, *Sociology and Philosophy* (London: Cohen & West, 1953), p. 55.
13 Quoted in G. Hawthorn,

Enlightenment and Despair: A History of Sociology (Cambridge: Cambridge University Press, 1976), pp. 118–19.

14 E. Durkheim and M. Mauss, *Primitive Classification* (2nd edn, London: Cohen & West, 1969), pp. 8, 83.

15 R. Benedict, *Patterns of Culture* (reprint, London: Routledge & Kegan Paul, 1935), p. 2.

16 Quoted in L. A. Coser, *Masters of Sociological Thought: Ideas in Historical and Sociological Context* (New York: Harcourt Brace Jovanovich, 1971), p. 218.

17 M. Weber, *The Protestant Ethic and the Spirit of Capitalism* (London: Unwin, 1930), p. 181, though I use the translation given in W. J. Cahnman, [Review of A. Mitzman, *The Iron Cage: An Historical Interpretation of Max Weber*], *Journal of the History of the Behavioral Sciences*, **14** (1978): 189–91, p. 189.

18 Quoted in D. Käsler, *Max Weber: An Introduction to His Life and Work* (Oxford: Polity Press, 1988), p. 180.

19 P. Lassman and I. Velody (eds), *Max Weber's 'Science as a Vocation'* (London: Unwin Hyman, 1989), p. 18.

20 *Ibid.*, pp. 14, 18.

21 *Ibid.*, p. 30.

22 Weber (1930), p. 182, quoting Nietzsche.

23 Quoted in W. Outhwaite, *Understanding Social Life: The Method Called 'Verstehen'* (London: George Allen & Unwin, 1975), p. 47.

24 G. Simmel, 'The Metropolis and Mental Life', in *On Individuality and Social Forms: Selected Writings*, ed D. N. Levine (Chicago: University of Chicago Press, 1971), pp. 325, 338.

25 Quoted in D. Frisby, *Georg Simmel* (Chichester: Ellis Horwood, and London and New York: Tavistock, 1984), p. 48.

26 G. Simmel, *The Philosophy of Money* (London: Routledge & Kegan Paul, 1978), p. 55.

27 Quoted in T. L. Haskell, *The Emergence of Professional Social Science: The American Social Science Association and the Nineteenth-Century Crisis of Authority* (Urbana: University of Illinois Press, 1977), pp. 204–5.

28 Quoted in *ibid.*, p. 209.

29 Quoted in H. Kuklick, 'Boundary Maintenance in American Sociology: Limitations to Academic "Professionalization"', *Journal of the History of the Behavioral Sciences*, **16** (1980): 201–19, p. 205.

30 Quoted in Coser (1971), p. 520.

31 Quoted in Kuklick (1980), p. 206.

32 Quoted in Hawthorn (1976), p. 211.

33 Quoted in Kuklick (1980), p. 208.

CHAPTER 16

1 Quoted in R. Jacoby, *The Repression of Psychoanalysis: Otto Fenichel and the Political Freudians* (New York: Basic Books, 1983), p. 61.
2 G. J. Romanes, *Animal Intelligence* (3rd edn, London: Kegan Paul, Trench, 1883), p. 471.
3 Quoted in R. E. Fancher, *The Intelligence Men: Makers of the IQ Controversy* (New York: W. W. Norton, 1985), p. 95.
4 Spens Report, quoted in B. Simon, *The Politics of Educational Reform 1920–1940* (London: Lawrence & Wishart, 1974), pp. 249–50.
5 A. Binet and V. Henri, 'La psychologie individuelle', *L'année psychologique*, **2** (1895): 411–64, pp. 417, 411.
6 Quoted in Fancher (1985), p. 101.
7 Quoted in L. Zenderland, 'The Debate over Diagnosis: Henry Herbert Goddard and the Medical Acceptance of Intelligence Testing', in M. M. Sokal (ed), *Psychological Testing and American Society* (New Brunswick: Rutgers University Press, 1987), p. 63.
8 Quoted in H. L. Minton, 'Lewis M. Terman and Mental Testing: In Search of the Democratic Ideal', in Sokal (ed) (1987), p. 100.
9 Quoted in K. Danziger, *Constructing the Subject: Historical Origins of Psychological Research* (Cambridge: Cambridge University Press, 1990), pp. 146–7, 235.
10 Quoted in F. Samelson, 'World War I Intelligence Testing and the Development of Psychology', *Journal of the History of the Behavioral Sciences*, **13** (1977): 274–82, p. 276.
11 E. G. Boring, 'Intelligence as the Tests Test It', in *History, Psychology, and Science: Selected Papers*, ed R. I. Watson and D. T. Campbell (New York: John Wiley and Sons, 1963), p. 187.
12 Quoted in F. Samelson, 'Putting Psychology on the Map: Ideology and Intelligence Testing', in A. R. Buss (ed), *Psychology in Social Context* (New York: Irvington Publishers, 1979), p. 106.
13 P. Janet, *The Mental State of Hystericals*, reprint in D. N. Robinson (ed), *Significant Contributions to the History of Psychology*, Series C, Vol. 2 (Washington, DC: University Publications of America, 1977), p. 35.
14 E. G. Boring, *A History of Experimental Psychology* (2nd edn, New York: Appleton-Century-Crofts, 1950), p. 577; L. J. Cronbach, 'The Two Disciplines of Scientific Psychology', in E. R. Hilgard (ed), *American Psychology in Historical Perspective: Addresses of the Presidents of the American Psychological Association* (Washington, DC: American

Psychological Association, 1978), pp 435–58.

15 Quoted in D. S. Napoli, *Architects of Adjustment: The History of the Psychological Profession in the United States* (Port Washington, NY: Kennikat Press, 1981), p. 18.

16 J. B. Watson, 'Psychology as the Behaviorist Views It', in W. Dennis (ed), *Readings in the History of Psychology* (New York: Appleton-Century-Crofts, 1948), p. 457.

17 Quoted in Napoli (1981), p. 30.

18 Quoted in L. S. Hearnshaw, *A Short History of British Psychology, 1840–1940* (London: Methuen, 1964), p. 284.

19 C. N. Degler, *In Search of Human Nature: The Decline and Revival of Darwinism in American Social Thought* (New York: Oxford University Press, 1991), Part 2.

20 Quoted in *ibid.*, p. 134.

21 Quoted in *ibid.*, p. 135.

22 Quoted in Fancher (1985), pp. 194, 195.

CHAPTER 17

1 J. B. Watson, *Psychology from the Standpoint of a Behaviourist* (2nd edn reprint, London and Dover, NH: Frances Pinter, 1983), p. 3.

2 S. Koch, 'The Nature and Limits of Psychological Knowledge: Lessons of a Century qua "Science"', in S. Koch and D. E. Leary (eds), *A Century of Psychology as Science* (New York: McGraw-Hill, 1985), pp. 92–3.

3 Quoted in L. S. Hearnshaw, *A Short History of British Psychology, 1840–1940* (London: Methuen, 1964), p. 173.

4 F. C. Bartlett, *Thinking: An Experimental and Social Study* (London: George Allen & Unwin, 1958), pp. 132–3.

5 F. C. Bartlett, *Remembering: A Study in Experimental and Social Psychology* (reprint, Cambridge: Cambridge University Press, 1961), p. 38.

6 I. P. Pavlov, *Lectures on Conditioned Reflexes: Twenty-five Years of Objective Study of the Higher Nervous Activity (Behaviour) of Animals* (reprint, London: Lawrence & Wishart, 1963), Vol. 1, pp. 38–9.

7 J. B. Watson, 'Psychology as the Behaviorist Views It', in W. Dennis (ed), *Readings in the History of Psychology* (New York: Appleton-Century-Crofts, 1948), pp. 457, 461.

8 Quoted in F. Samelson, 'Struggle for Scientific Authority: The Reception of Watson's Behaviorism 1913–1920', *Journal of the History of the Behavioral Sciences*, **7** (1981): 399–425, p. 407.

9 Watson (1983), p. 7.

10 Watson, in Dennis (ed) (1948), pp. 461, 469, 470.

11 *Ibid.*, p. 459.

12 Watson (1983), pp. 364–5.

13 Quoted in J. M. O'Donnell, *The Origins of Behaviorism: American Psychology, 1870–1920* (New York: New York University Press, 1986), p. 202.
14 B. Harris, 'Whatever Happened to Little Albert?', *American Psychologist*, **34** (1979): 151–60, p. 158.
15 J. B. Watson, *Behaviorism* (reprint, New York: W. W. Norton, 1970), p. 104.
16 Watson (1983), p. 8.
17 Quoted from W. McDougall's public debate with Watson, in D. Bakan, 'Behaviorism and American Urbanization', *Journal of the History of the Behavioral Sciences*, **2** (1966): 5–28, p. 6.
18 J. B. Watson, 'John Broadus Watson', in C. Murchison (ed), *A History of Psychology in Autobiography*. Volume III (reprint, New York: Russell & Russell, 1961), p. 280.
19 B. D. Mackenzie, *Behaviourism and the Limits of Scientific Method* (London: Routledge & Kegan Paul, 1977), p. 1.
20 S. Koch, 'Foreword: Wundt's Creature at Age Zero – and as Centenarian: Some Aspects of the Institutionalization of the "New Psychology"', in Koch and Leary (eds) (1985), p. 23 note.
21 Quoted in G. W. Allport, 'The Historical Background of Modern Social Psychology', in G. Lindzey and E. Aronson (eds), *The Handbook of Social Psychology* (2nd edn, Reading, MA: Addison-Wesley, 1968), Vol. 1, p. 15.
22 Quoted in G. A. Kimble, 'Conditioning and Learning', in Koch and Leary (1985), p. 296.
23 Quoted in L. D. Smith, *Behaviorism and Logical Positivism: A Reassessment of the Alliance* (Stanford: Stanford University Press, 1986), p. 85.
24 E. C. Tolman, 'The Determiners of Behavior at a Choice Point', in E. R. Hilgard (ed), *American Psychology in Historical Perspective: Addresses of the Presidents of the American Psychological Association, 1892–1977* (Washington, DC: American Psychological Association, 1978), p. 364.
25 Quoted in Smith (1986), p. 119.
26 Quoted in *ibid.*, p. 135.
27 Tolman, in Hilgard (ed) (1978), p. 364.
28 Quoted in Smith (1986), p. 158.
29 Quoted in *ibid.*, pp. 358–9.
30 Quoted in *ibid.*, p. 55.
31 Koch, 'Nature and Limits of Psychological Knowledge', in Koch and Leary (eds) (1985), p. 80.
32 B. F. Skinner, *Science and Human Behavior* (New York: Macmillan, 1953), p. 64.
33 B. F. Skinner, 'Behaviorism at Fifty', in T. W. Wann (ed), *Behaviorism and Phenomenology: Contrasting*

Bases for Modern Psychology (Chicago: University of Chicago Press for William Marsh Rice University, 1964), pp. 84, 88.
34 Quoted in Smith (1986), p. 269.
35 B. F. Skinner, *Beyond Freedom and Dignity* (reprint, New York: Alfred A. Knopf, 1972), pp. 129, 136.
36 B. F. Skinner, [Review of L. D. Smith, *Behaviorism and Logical Positivism*], *Journal of the History of the Behavioral Sciences*, **23** (1987): 206–10, p. 208.
37 Quoted in J. Beloff, *Parapsychology: A Concise History* (London: Athlone Press, 1993), p. 64.
38 Quoted in S. H. Mauskopf and M. R. McVaugh, *The Elusive Science: Origins of Experimental Psychical Research* (Baltimore: Johns Hopkins University Press, 1980), p. 26.
39 Quoted in M. G. Ash, 'Gestalt Psychology: Origins in Germany and Reception in the United States', in C. E. Buxton (ed), *Points of View in the Modern History of Psychology* (Orlando: Academic Press, 1985), p. 310.
40 M. Wertheimer, 'Gestalt Theory', in W. D. Ellis (ed), *A Source Book of Gestalt Psychology* (London: Routledge & Kegan Paul, 1938), p. 2.
41 Quoted in Ash, in Buxton (ed) (1985), p. 301.
42 Quoted in H. Spiegelberg, *Phenomenology in Psychology and Psychiatry: A Historical Introduction* (Evanston: Northwestern University Press, 1972), p. 47.
43 Quoted in T. Dehue, *Changing the Rules: Psychology in The Netherlands, 1900–1985* (Cambridge: Cambridge University Press, 1995), p. 69.
44 F. J. J. Buytendijk, 'Husserl's Phenomenology and Its Significance for Contemporary Psychology', in N. Lawrence and D. O'Connor (eds), *Readings in Existential Phenomenology* (Englewood Cliffs, NJ: Prentice Hall, 1967), p. 353.
45 Quoted in Spiegelberg (1972), p. 285.
46 M. Merleau-Ponty, *Phenomenology of Perception* (London: Routledge & Kegan Paul, and New York: Humanities Press, 1962), p. xi.

CHAPTER 18

1 S. Freud, *Introductory Lectures on Psycho-analysis*, in *The Standard Edition of the Complete Psychological Works of Sigmund Freud*, ed J. Strachey (London: Hogarth Press, 1953–73), Vol. 16 [SE XVI], pp. 284–5.
2 S. Freud, *The Interpretation of Dreams*, SE V, p. 613.
3 S. Freud, letter to W. Fliess, 1900, in J. M. Masson (ed), *The Complete Letters of Sigmund Freud to Wilhelm Fliess 1887–1904* (Cambridge, MA: Belknap Press of Harvard

University Press, 1985), p. 398.

4 P. Gay, *Freud: A Life for Our Time* (New York: W. W. Norton, 1988).

5 J. Breuer, in S. Freud and J. Breuer, *Studies on Hysteria*, SE II, p. 30.

6 Freud and Breuer, 'Preliminary Communication', in *ibid.*, p. 7.

7 Quoted from Janet in H. F. Ellenberger, *The Discovery of the Unconscious: The History and Evolution of Dynamic Psychiatry* (London: Allen Lane, The Penguin Press, 1970), pp. 360–1.

8 S. Freud, *The Interpretation of Dreams*, SE IV, p. 101.

9 *Ibid.*, SE V, p. 608.

10 S. Freud, letter to W. Fliess, 1898, in Masson (ed) (1985), p. 326.

11 S. Freud, *The Interpretation of Dreams*, SE V, p. 525.

12 J. Breuer, in Freud and Breuer, *Studies on Hysteria*, SE II, p. 246.

13 S. Freud, letter to C. G. Jung, 1911, in W. McGuire (ed), *The Freud/Jung Letters: The Correspondence between Sigmund Freud and C. G. Jung* (London: Hogarth Press and Routledge & Kegan Paul, 1974), p. 469.

14 Quoted in J. Miller, 'Interpretations of Freud's Jewishness, 1924–1974', *Journal of the History of the Behavioral Sciences*, **17** (1981): 357–74, p. 365.

15 Quoted in L. E. Hoffman, 'War, Revolution, and Psychoanalysis: Freudian Thought Begins to Grapple with Social Reality', *Journal of the History of the Behavioral Sciences*, **17** (1981): 251–69, p. 251.

16 S. Freud, *Beyond the Pleasure Principle*, SE XVIII, p. 38; the words are after Schopenhauer.

17 S. Freud, in Freud and Breuer, *Studies on Hysteria*, SE II, p. 305.

18 Quoted in L. E. Hoffman, 'From Instinct to Identity: Implications of Changing Psychoanalytic Concepts of Social Life from Freud to Erikson', *Journal of the History of the Behavioral Sciences*, **18** (1982): 130–46, p. 137.

19 S. Freud, *New Introductory Lectures on Psycho-analysis*, SE XXII, p. 126.

20 S. de Beauvoir, *Nature of the Second Sex* (reprint, London: Four Square Books, 1963), p. 8.

21 S. Freud, 'The Question of Lay Analysis', SE XX, p. 212.

22 Quoted in R. M. Young, *Mental Space* (London: Process Press, 1994), p. 48.

23 Quoted in J. C. Burnham, 'From Avant-garde to Specialism: Psychoanalysis in America', *Journal of the History of the Behavioral Sciences*, **15** (1979): 128-34, p. 130.

24 Quoted in Hoffman (1982), p. 136.

25 Quoted in N. G. Hale, Jr, 'From Berggasse xix to

Central Park West: The Americanization of Psychoanalysis, 1919–1940', *Journal of the History of the Behavioral Sciences*, **14** (1978): 299–315, p. 310.

26 Freud, letter to K. Abraham, 1908, quoted in H. C. Abraham and E. L. Freud (eds), *A Psycho-analytic Dialogue: The Letters of Sigmund Freud and Karl Abraham 1907–1926* (London: Hogarth Press, 1965), p. 34.

27 C. G. Jung, *Two Essays on Analytical Psychology*, in *The Collected Works of C. G. Jung*, Vol. 7 (London: Routledge & Kegan Paul, 1953), p. 117.

28 C. G. Jung, 'Archetypes of the Collective Unconscious', in *The Archetypes and the Collective Unconscious*, in *The Collected Works of C. G. Jung*, Vol. 9, Part 1 (2nd edn, London: Routledge, 1968), p. 29.

29 Ellenberger (1970), pp. 695–6.

30 Jung (1968), p. 4.

31 *Ibid.*, p. 40.

CHAPTER 19

1 G. W. Allport, 'The Historical Background of Modern Social Psychology', in G. Lindzey and E. Aronson (eds), *The Handbook of Social Psychology*, Vol. 1 (2nd edn, Reading, MA: Addison-Wesley, 1968), p. 2.

2 *Ibid.*, p. 3.

3 Quoted in J. van Ginneken, *Crowds, Psychology, and Politics 1871–1899* (Cambridge: Cambridge University Press, 1992), p. 45.

4 Quoted in *ibid.*, pp. 130–1.

5 Quoted in *ibid.*, p. 153.

6 Quoted in *ibid.*, p. 179.

7 Quoted in *ibid.*, p. 189.

8 Quoted in *ibid.*, pp. 218, 189.

9 Quoted in Allport, in Lindzey and Aronson (eds) (1968), p. 29.

10 'William McDougall', in C. Murchison (ed), *A History of Psychology in Autobiography. Vol I* (reprint, New York: Russell & Russell, 1961), p. 207.

11 W. McDougall, *An Introduction to Social Psychology* (23rd edn, London: Methuen, 1936), p. 25.

12 W. McDougall, *The Group Mind* (Cambridge: Cambridge University Press, 1920), p. 7.

13 Quoted in Allport in Lindzey and Aronson (eds) (1968), p. 54.

14 Quoted in H. C. Greisman, 'Herd Instinct and the Foundations of Biosociology', *Journal of the History of the Behavioral Sciences*, **15** (1979): 357–69, p. 360.

15 S. Freud, *Totem and Taboo*, in *The Standard Edition of the Complete Psychological Works of Sigmund Freud*, ed J. Strachey (London: Hogarth Press, 1953–73), Vol. 13 [SE XIII], p. 157.

16 S. Freud, *The Interpretation of Dreams*, SE V, p. 548; passage added 1919.

17 S. Freud, *Group Psychology and*

the Analysis of the Ego, SE XVIII, p. 92.

18 Quoted in L. E. Hoffman, 'From Instinct to Identity: Implications of Changing Psychoanalytic Concepts of Social Life From Freud to Erikson', *Journal of the History of the Behavioral Sciences*, **18** (1982): 130–46, p. 130.

19 Quoted in K. Danziger, 'Origins and Basic Principles of Wundt's *Völkerpsychologie*', *British Journal of Social Psychology*, **22** (1983): 303–13, p. 305.

20 Quoted in L. S. Hearnshaw, *A Short History of British Psychology, 1840–1940* (London: Methuen, 1964), p. 114.

21 Quoted in Hoffman (1982), p. 132.

22 F. H. Allport, *Social Psychology* (reprint, New York: Johnson Reprint, 1975), p. 12.

23 Quoted in G. Collier, H. L. Minton and G. Reynolds, *Currents of Thought in American Social Psychology* (New York: Oxford University Press, 1991), p. 56.

24 J. Dewey, 'The Need for Social Psychology', *Psychological Review*, **24** (1917): 266–77, p. 277.

25 Allport, in Lindzey and Aronson (eds) (1968), p. 59.

26 Quoted in G. A. Cook, 'G. H. Mead's Social Behaviorism', *Journal of the History of the Behavioral Sciences*, **13** (1977): 307–16, p. 314.

27 Quoted in *ibid.*

28 Quoted in M. J. Deegan and J. S. Burger, 'George Herbert Mead and Social Reform', *Journal of the History of the Behavioral Sciences*, **14** (1978): 362–72, p. 368.

29 Quoted in *ibid.*, p. 365.

30 Quoted in J. G. Morawski, 'Organizing Knowledge and Behavior at Yale's Institute of Human Relations', *Isis*, **77** (1986): 219–42, p. 231.

31 K. J. Gergen, 'Social Psychology as History', *Journal of Personality and Social Psychology*, **26** (1973): 309–20, p. 319.

32 Z. Barbu, *Problems of Historical Psychology* (London: Routledge & Kegan Paul, 1960), p. 1.

33 Quoted in D. Joravsky, *Russian Psychology: A Critical History* (Oxford: Basil Blackwell, 1989), p. 312.

34 Quoted in *ibid.*, p. 113.

35 Quoted in *ibid.*, p. 152.

36 Quoted in A. W. Logue, 'The Origins of Behaviorism: Antecedents and Proclamation', in C. E. Buxton (ed), *Points of View in the Modern History of Psychology* (Orlando: Academic Press, 1985), p. 146.

37 Quoted in Joravsky (1989), p. 134.

38 Quoted in *ibid.*, p. 221.

39 Quoted in *ibid.*, p. 276.

40 Quoted in *ibid.*, p. 253.

41 S. Toulmin, 'The Mozart of Psychology', *New York Review*

of Books, **25** no. 14 (1978): 51–7; D. Joravsky, 'L. S. Vygotskii: The Muffled Deity of Soviet Psychology', in M. G. Ash and W. R. Woodward (eds), *Psychology in Twentieth-Century Thought and Society* (Cambridge: Cambridge University Press, 1987), pp. 189–211.

42 Quoted in Joravsky (1989), p. 260.

43 A. R. Luria, *The Making of Mind: A Personal Account of Soviet Psychology* (Cambridge, MA: Harvard University Press, 1979), p. 19.

44 Quoted in Joravsky (1989), p. 372.

45 Quoted in *ibid.*, p. 450.

CHAPTER 20

1 L. Wittgenstein, *Philosophical Investigations* (2nd edn reprint, Oxford: Basil Blackwell, 1963), p. 88e.

2 'Introduction', S. Koch and D. E. Leary (eds), *A Century of Psychology as Science* (New York: McGraw-Hill, 1985), p. 2.

3 Quoted in K. Danziger, *Constructing the Subject: Historical Origins of Psychological Research* (Cambridge: Cambridge University Press, 1990), p. 248.

4 Quoted in G. A. Cook, 'G. H. Mead's Social Behaviorism', *Journal of the History of the Behavioral Sciences*, **13** (1977): 307–16, p. 307.

5 Quoted in J. F. Ward, 'Arthur F. Bentley and the Foundations of Behavioral Science', *Journal of the History of the Behavioral Sciences*, **17** (1981): 222–31, p. 228.

6 B. Berelson (ed), *The Behavioral Sciences Today* (New York: Basic Books, 1963), p. 3.

7 Quoted in J. C. Greene, *Darwin and the Modern World View* (Baton Rouge: Louisiana State University Press, 1961), p. 101.

8 Quoted in *ibid.*, p. 112.

9 Quoted in C. N. Degler, *In Search of Human Nature: The Decline and Revival of Darwinism in American Social Thought* (New York: Oxford University Press, 1991), p. 209.

10 K. Lorenz, *On Aggression* (London: Methuen, 1966), p. x.

11 E. O. Wilson, *On Human Nature* (Cambridge, MA: Harvard University Press, 1978), pp. x, 5, 96, 80.

12 *Ibid.*, p. 167.

13 Quoted in J. G. Morawski, 'Impossible Experiments and Practical Constructions: The Social Basis of Psychologists' Work', in J. G. Morawski (ed), *The Rise of Experimentation in American Psychology* (New Haven: Yale University Press, 1988), p. 81.

14 Quoted in M. N. Evans, *Fits and Starts: A Genealogy of Hysteria in Modern France* (Ithaca: Cornell University Press, 1991), pp. 204–5.

15 Quoted in D. J. Haraway, 'The

Biological Enterprise: Sex, Mind, and Profit from Human Engineering to Sociobiology', in *Simians, Cyborgs, and Women: The Reinvention of Nature* (New York: Routledge, and London: Free Association Books, 1991), p. 47.

16 D. J. Haraway, 'The Past Is the Contested Zone: Human Nature and Theories of Production and Reproduction in Primate Behaviour Studies', in Haraway (1991), p. 23.

17 Quoted in J. M. O'Donnell, *The Origins of Behaviorism: American Psychology, 1870–1920* (New York: New York University Press, 1986), p. 204.

18 Quoted in D. Joravsky, *Russian Psychology: A Critical History* (Oxford: Basil Blackwell, 1989), p. 305.

19 Wilson (1978), p. 77.

20 C. McGinn, 'Cooling It', *London Review of Books*, 19 August 1993: 12–14, p. 12.

21 G. Ryle, *The Concept of Mind* (reprint, Harmondsworth: Penguin Books, 1963), p. 17.

22 D. C. Dennett, *Consciouness Explained* (reprint, London: Penguin Books, 1993), p. 309, in an account of a paper by D. Rosenthal.

23 Quoted in F. Vidal, 'Jean Piaget and the Liberal Protestant Tradition', in M. G. Ash and W. R. Woodward (eds), *Psychology in Twentieth-Century Thought and Society* (Cambridge: Cambridge University Press, 1987), p. 283.

24 N. Wiener, *Cybernetics: Or Control and Communication in the Animal and the Machine* (2nd edn, Cambridge, MA: MIT Press, 1965).

25 J. Fodor, 'The Big Idea: Can There Be a Science of Mind?', *Times Literary Supplement*, 3 July 1992: 5.

26 Quoted in S. P. Stich, 'Consciousness Revived: John Searle and the Critique of Cognitive Science', *Times Literary Supplement*, 5 March 1993: 5–6, p. 5.

27 A. J. Sutich, 'Introduction', *Journal of Humanistic Psychology*, **1** (1961): vii-ix, p. viii.

28 C. Rogers, 'Toward a Science of the Person', in T. W. Wann (ed), *Behaviorism and Phenomenology: Contrasting Bases for Modern Psychology* (Chicago: University of Chicago Press for William Marsh Rice University, 1964), pp. 115, 129.

29 C. A. van Peursen, *Body, Soul, Spirit: A Survey of the Body-Mind Problem* (London: Oxford University Press, 1966), p. 166.

30 F. Nietzsche, *Beyond Good and Evil: Prelude to a Philosophy of the Future* (New York: Vintage Books, 1966), Sect. 23, pp. 31–2.

31 W. Kaufmann, 'Nietzsche as the First Great (Depth)

Psychologist', in Koch and Leary (eds) (1985), pp. 911–20.

32 F. Nietzsche, *Ecce Homo* (New York: Vintage Books, 1969), p. 266.

33 F. Nietzsche, *Twilight of the Idols* (Harmondsworth: Penguin Books, 1968), Foreword, p. 21.

34 Nietzsche (1966), Part 4, Aphorism 68, p. 80.

35 Quoted in Evans (1991), p. 174.

36 Quoted in M. Sarup, *Jacques Lacan* (New York: Harvester Wheatsheaf, 1992), p. 75.

37 M. Foucault, 'Afterword: The Subject and Power', in H. L. Dreyfus and P. Rabinow, *Michel Foucault: Beyond Structuralism and Hermeneutics* (Chicago: University of Chicago Press, 1982), p. 208.

38 Quoted in D. R. Kelley, 'Horizons of Intellectual History: Retrospect, Circumspect, Prospect', *Journal of the History of Ideas*, **48** (1987): 143–69, p. 155.

39 Quoted in G. N. Izenberg, *Impossible Individuality: Romanticism, Revolution, and the Origins of Modern Selfhood, 1787–1802* (Princeton: Princeton University Press, 1992), p. 400.

40 W. Dilthey, *Introduction to the Human Sciences: An Attempt to Lay a Foundation for the Study of Society and History*, trans. R. J. Betanzos (London: Harvester Wheatsheaf, 1988), p. 77.

41 W. Dilthey, *Selected Writings*, ed H. P. Rickman (Cambridge: Cambridge University Press, 1976), p. 176.

42 P. Lassman and I. Velody (eds), *Max Weber's 'Science as a Vocation'* (London: Unwin Hyman, 1989), p. 20.

43 Quoted in D. Käsler, *Max Weber: An Introduction to His Life and Work* (Oxford: Polity Press, 1988), p. 176.

44 H. Rickert, *Science and History: A Critique of Positivist Epistemology*, ed A. Goddard (Princeton: D. Van Nostrand Company, 1962), p. xvi.

45 C. Geertz, 'The Impact of the Concept of Culture on the Concept of Man', in *The Interpretation of Cultures: Selected Essays* (New York: Basic Books, 1973), pp. 49–50.

46 Rickert (1962), p. 143.

47 Dilthey (1976), p. 120.

BIBLIOGRAPHIC ESSAY

ABBREVIATIONS

HHS *History of the Human Sciences*
JHBS *Journal of the History of the Behavioral Sciences*
JHI *Journal of the History of Ideas*

CHAPTER 1

The material for a bibliographic essay in which the subject is 'Man' is potentially uncontainable. The purpose here is a practical one, to provide access to what, with necessarily incomplete knowledge, are the best discussions of the topics in this book. By 'best', I mean texts that are readable, scholarly, historical and which say something significant. I include only: English-language sources (with a few necessary exceptions); sources relating to Western scientific culture after about 1450; and sources that are primarily historical in content and purpose. I do not normally cite reprinted primary sources, especially 'classic' works, where these can easily be found catalogued under the author's name. Like the book, the bibliography focuses on the psychological domain of human nature, but it also includes much material relevant to a wider view of the history of the human sciences.

Standard reference works are of particular value in this broad area. In addition to general encyclopaedias, the following are invaluable: P. Edwards (ed), *The Encyclopedia of Philosophy*, 8 vols (New York: Macmillan and Free Press, 1967); C. C. Gillispie (ed), *Dictionary of Scientific Biography*, 18 vols (New York: Charles Scribner's Sons, 1970–90); R. C. Olby *et al.* (eds), *Companion to the History of Modern Science* (London: Routledge, 1990); D. L. Sills (ed), *International Encyclopedia of the Social Sciences*, 17+1 vols (New York: Macmillan and Free Press, 1968, 1979); P. P. Wiener (ed), *Dictionary of the History of Ideas: Studies of Selected Pivotal Ideas,*

4+1 vols (New York: Charles Scribner's Sons, 1973–4). Brief reference works with relevant material include: W. F. Bynum, J. Browne and R. Porter (eds), *Dictionary of the History of Science* (London: Macmillan, 1981); for the post-1900 period, A. Bullock and S. Trombley (eds), *The Fontana Dictionary of Modern Thought* (London: Fontana, 1988), and A. Bullock and R. B. Woodings (eds), *The Fontana Dictionary of Modern Thinkers* (London: Fontana, 1983). A number of key figures are introduced in either the Fontana Modern Masters series (London: Fontana) or the Oxford Past Masters series (Oxford: Oxford University Press).

There are partial bibliographies for the history of psychology in: J. M. Baldwin (ed), *Dictionary of Philosophy and Psychology...*, 3 vols in 4 (1901–5; reprint, Gloucester, MA: Peter Smith, 1960); R. I. Watson, Sr (ed), *Eminent Contributors to Psychology*, 2 vols (New York: Springer, 1974–6).

The classic text, now dated, in the history of psychology is E. G. Boring, *A History of Experimental Psychology* (2nd edn, New York: Appleton-Century-Crofts, 1950). There are many one-volume histories of psychology, and textbooks abound, but it is one purpose of this book to alter their agenda. For a readable and reliable introduction: R. E. Fancher, *Pioneers of Psychology* (3rd edn, New York: W. W. Norton, 1996). For a textbook: T. H. Leahey, *History of Psychology: Main Currents in Psychological Thought* (3rd edn, Englewood Cliffs, NJ: Prentice-Hall, 1992); and for a source book, S. Diamond (ed), *The Roots of Psychology: A Sourcebook in the History of Ideas* (New York: Basic Books, 1974). Large-scale revisions have not been much attempted, with the important exceptions of P. T. Manicas, *A History and Philosophy of the Social Sciences* (Oxford and New York: Basil Blackwell, 1987), and G. Richards, *Mental Machinery: The Origins and Consequences of Psychological Ideas. Part 1: 1600–1850* (London: Athlone Press, 1992).

Two journals carry articles across the range of the history of the human sciences: *Journal of the History of the Behavioral Sciences* (f.1965); and *History of the Human Sciences* (f.1988).

The area is one of academic growth and is served by small but active specialist societies. In North America, Cheiron: The International Society for the History of the Behavioral Sciences, which publishes a *Newsletter*; the Secretary is Raymond E. Fancher, Cheiron Executive Officer, Department of Psychology, York University, North York, ON, Canada M3J 1P3. In Europe,

The European Society for the History of the Human Sciences (formerly Cheiron: The European Society for the History of the Behavioural and Social Sciences) also publishes a *Newsletter*; the Secretary is Trudy Dehue, Department of Theory and History of Psychology, University of Groningen, Grote Kruisstraat 2/1, 9712 TS Groningen, The Netherlands. In France, Société Française pour l'Histoire des Sciences de l'Homme, which publishes a *Bulletin d'information*, edited by Laurent Mucchielli, Centre Alexandre Koyré, Muséum national d'histoire naturelle, Pavillon Chevreul, 57 rue Cuvier, 75231 Paris cedex 05, France. There is a Forum for the History of the Human Sciences of the History of Science Society, which circulates a *Newsletter*; Corresponding Secretary, David Valone, 151 Spring Street, Cheshire, CT 06410, USA.

For the relations of science and religion, in a book which stresses the diversity and historical open-endedness of those relations: J. H. Brooke, *Science and Religion: Some Historical Perspectives* (Cambridge: Cambridge University Press, 1991). For a bibliography: H. Vande Kemp, in collaboration with M. H. Newton, *Psychology and Theology in Western Thought 1672–1965: A Historical and Annotated Bibliography* (White Plains, NY: Kraus International, 1984). For a clear introduction to some philosophies of human nature: R. Trigg, *Ideas of Human Nature: An Historical Introduction* (Oxford: Blackwell, 1988).

The problem of determining the subject of the history of the human sciences is discussed in: J. Christie, 'The Human Sciences: Origins and Histories', HHS, **6** no. 1 (1993): 1–12; G. Richards, 'Of What Is the History of Psychology a History?', *British Journal for the History of Science*, **20** (1987): 201–11; R. Smith, 'Does the History of Psychology Have a Subject?', HHS, **1** (1988): 147–77. Arguments for a new historiography are made in G. W. Stocking, Jr, 'On the Limits of "Presentism" and "Historicism" in the Historiography of the Behavioral Sciences', JHBS, **1** (1965): 211–18, reprinted in *Race, Culture, and Evolution: Essays in the History of Anthropology* (new edn, Chicago: University of Chicago Press, 1982), pp. 1–12; R. M. Young, 'Scholarship and the History of the Behavioural Sciences', *History of Science*, **5** (1966): 1–51. By the mid-1990s, the field had changed considerably, as this bibliography testifies; see K. Danziger, 'Does the History of Psychology Have a Future?', *Theory & Psychology*, **4** (1994): 467–84. There is an overview of historiography in E. R. Hilgard, D. E.

Leary and G. R. McGuire, 'The History of Psychology: A Survey and Critical Assessment', *Annual Review of Psychology*, **42** (1991): 79–107. The relation of history to theoretical psychology is considered in H. V. Rappard *et al.* (eds), *Annals of Theoretical Psychology, Volume 8* (New York: Plenum Press, 1993).

CHAPTER 2

Among modern Renaissance scholars, P. O. Kristeller has written work that is both accessible and influential: for the concept of the Renaissance, *Renaissance Concepts of Man and Other Essays* (New York: Harper Torchbooks, 1972); *Renaissance Thought and Its Sources*, ed M. Mooney (New York: Columbia University Press, 1979). Key sources, including Pico's 'Oration on the Dignity of Man', are translated in E. Cassirer, P. O. Kristeller and J. H. Randall (eds), *The Renaissance Philosophy of Man* (Chicago: University of Chicago Press, 1948). An invaluable survey with extensive bibliography is C. B. Schmitt *et al.* (eds), *The Cambridge History of Renaissance Philosophy* (Cambridge: Cambridge University Press, 1988). For a shorter survey of Renaissance philosophy: B. P. Copenhaver and C. B. Schmitt, *Renaissance Philosophy* (Oxford: Oxford University Press, 1992). An introduction to the Italian Renaissance is P. Burke, *The Italian Renaissance: Culture and Society in Italy* (revised edn, Oxford: Polity Press, 1987). There is an extensive review of political thought in J. H. Burns, assisted by M. Goldie, (ed), *The Cambridge History of Political Thought 1450–1700* (Cambridge: Cambridge University Press, 1991).

The study of the cultural impact of printing is summarized in E. L. Eisenstein, *The Printing Revolution in Early Modern Europe* (Cambridge: Cambridge University Press, 1983). The many roots and contexts of the Renaissance are reviewed in R. Porter and M. Teich (eds), *The Renaissance in National Context* (Cambridge: Cambridge University Press, 1992). On the university curriculum and humanist culture: A. Grafton and L. Jardine, *From Humanism to the Humanities: Education and the Liberal Arts in Fifteenth- and Sixteenth-Century Europe* (London: Duckworth, 1986); and for the persistence of this culture into the eighteenth century, L. W. B. Brockliss, *French Higher Education in the Seventeenth and Eighteenth Centuries: A Cultural History* (Oxford: Clarendon Press, 1987). For thought about what historians are more and more wary of calling the scientific revolution, D. C. Lindberg and R. S. Westman (eds), *Reappraisals of the Scientific Revolution* (Cambridge: Cambridge University Press, 1990). The transformation of man's place in the cosmos in the Renaissance was the subject of studies which, though dated as history, retain their significance in relation to debate about values: E. A. Burtt, *The Metaphysical Foundations of Modern Physical Science: A Historical and Critical Essay* (1924; 2nd edn, London: Routledge & Kegan Paul, 1932); A. N. Whitehead, *Science and the Modern World* (1926; reprint, Cambridge: Cambridge University Press, 1953). This debate about the consequences of knowledge for what it is to be human is continued in C. Taylor, 'Interpretation and the Sciences of Man', reprinted in *Philosophy and the Human Sciences: Philosophical Papers* (Cambridge: Cambridge University Press, 1985), Vol. 2, pp. 15–57.

There is a modern translation of *De anima*: 'On the Soul', in J. Barnes (ed) *The Complete Works of Aristotle. The Revised Oxford Translation* (Princeton: Princeton University Press, 1984), Vol. 1, pp. 641–92. On the soul in the Renaissance: K. Park, 'The Organic Soul', and E. Kessler, 'The Intellectual Soul' in C. B. Schmitt *et al.* (eds), (1988), pp. 464–84 and 485–534. The medieval roots are discussed in E. R. Harvey, *The Inward Wits: Psychological Theory in the Middle Ages and the Renaissance* (London: Warburg Institute, 1975); S. Kemp, *Medieval Psychology* (New York: Greenwood Press, 1990); G. Mora, 'Mind-body Concepts in the Middle Ages', JHBS, **14** (1978): 344–61, and **16** (1980): 58–72. On belief about the sensory origin of cognition, see also P. F. Cranefield, 'On the Origin of the Phrase Nihil est in intellectu quod non primus fuerit in sensu', *Journal of the History of Medicine*, **25** (1970): 77–80. For a detailed study, which covers much more than memory in the narrow modern sense: J. Coleman, *Ancient and Modern Memories: Studies in the Reconstruction of the Past* (Cambridge: Cambridge University Press, 1992). The founding text of 'the history of ideas' and still an exciting study of the chain of being is A. O. Lovejoy, *The Great Chain of Being: A Study of the History of an Idea* (1936; reprint, New York: Harper Torchbooks, 1960). For an introduction to the Neoplatonic worldview: F. A. Yates, 'The Hermetic Tradition in Renaissance Science', in C. S. Singleton (ed), *Art, Science, and History in the Renaissance* (Baltimore: Johns Hopkins University Press, 1967), pp. 255–74. For the human body as the image of the world, see L. Barkan, *Nature's Work of Art: The Human Body as Image of the World* (New Haven: Yale University Press, 1975).

E. Gombrich, *Art and Illusion: A Study in the Psychology of Pictorial Representation* (2nd edn, London: Phaidon Press, 1962) remains a fascinating study of painting and the origins of modern ways of seeing. On perspective see S. Y. Edgerton, *The Renaissance Rediscovery of Linear Perspective* (New York: Basic Books, 1975); and, from the vantage point of modern perspective theory, M. Kuboury, *The Psychology of Perspective and Renaissance Art* (Cambridge: Cambridge University Press, 1986). The claim that medieval theories of vision culminate in Kepler's discussion of the retinal image is made in D. C. Lindberg, *Theories of Vision from Al-Kindi to Kepler* (Chicago: University of Chicago Press, 1976). There is an alternative interpretation of Kepler, which also argues that his work on the pin-hole image explored by earlier artists was a step in the mechanization of man's relation to nature: S. Straker, 'The Eye Made "Other": Dürer, Kepler, and the Mechanization of Light and Vision', in L. A. Knafla, M. S. Staum and T. H. E. Travers (eds), *Science, Technology, and Culture in Historical Perspective* (Calgary: University of Calgary, 1976), pp. 7–25. Straker argues for the importance of theories of seeing to modern science in 'What Is the History of Theories of Perception the History of?', in M. J. Osler and P. L. Farber (eds), *Religion, Science, and Worldview* (Cambridge: Cambridge University Press, 1985), pp. 245–73. The significance of Kepler for Dutch art is discussed in S. Alpers, *The Art of Describing: Dutch Art in the Seventeenth Century* (London: John Murray, 1983). The argument that optics not mechanics was important for a mechanist theory of mind is made in T. Meyering, *Historical Roots of Cognitive Science: The Rise of a Cognitive Theory of Perception from Antiquity to the Nineteenth Century* (Dordrecht: Kluwer, 1989).

Almost any primary or secondary source on the Renaissance uses the language of spirits, humours and temperaments. For a readable introduction to the worldview exemplified in Shakespeare's work, see E. M. W. Tillyard, *The Elizabethan World Picture* (reprint, Harmondsworth: Penguin Books, 1963); also H. Baker, *The Dignity of Man: Studies in the Persistence of an Idea* (Cambridge, MA: Harvard University Press, 1947). The medical context is surveyed in N. G. Siraisi, *Medieval & Early Renaissance Medicine:*

An Introduction to Knowledge and Practice (Chicago: University of Chicago Press, 1990); and on the lasting tradition of Galenism, O. Temkin, *Galenism: Rise and Decline of a Medical Philosophy* (Ithaca: Cornell University Press, 1973). Illustrations of the brain and its ventricles are in E. Clarke and K. Dewhurst, *An Illustrated History of Brain Function* (Oxford: Sanford Publications, 1972), and materials for a history of brain theories are collected in E. Clarke and C. D. O'Malley, *The Human Brain: A Historical Study Illustrated by Writings from Antiquity to the Twentieth Century* (Berkeley: University of California Press, 1968). The whole question of theories of human nature and the image of the interior of the body is reassessed in J. Sawday, *The Body Emblazoned: Dissection and the Human Body in Renaissance Culture* (London and New York: Routledge, 1995); this conveys the richness of metaphorical connections of the period.

Much of the modern debate about the constitution of sexuality stems from the work of M. Foucault; see *The History of Sexuality. Volume 1: An Introduction* (New York: Pantheon, 1978). T. Laqueur, *Making Sex: Body and Gender from the Greeks to Freud* (Cambridge, MA: Harvard University Press, 1990) argues that 'sex', understood as physical difference, originated in the eighteenth century; the book references the anatomical literature. Scholarship since the 1960s has produced a large literature on woman's position. For a reliable survey of Renaissance scholarly texts: I. Maclean, *The Renaissance Notion of Woman: A Study in the Fortunes of Scholasticism and Medical Science in European Intellectual Life* (Cambridge: Cambridge University Press, 1980); and for the medieval background see J. Cadden, *Meanings of Sex Differences in the Middle Ages: Medicine, Science, and Culture* (Cambridge: Cambridge University Press, 1993). On the bodily dimensions of sexual differences, L. Schiebinger, *The Mind Has No Sex? Women in the Origins of Modern Science* (Cambridge, MA: Harvard University Press, 1989); chap. 5 discusses female personification. For the later history of imagery of the female body, L. Jordanova, *Sexual Visions: Images of Gender in Science and Medicine between the Eighteenth and Twentieth Centuries* (New York: Harvester Wheatsheaf, 1989). Discussions which indicate the historical possibilites for the inversion of the male–female hierarchy include: N. Z. Davis, 'Women on Top', in *Society and Culture in Early Modern France* (Stanford: Stanford University Press, 1975), pp. 124–51;

C. Jordan, *Renaissance Feminism: Literary Texts and Political Models* (Ithaca: Cornell University Press, 1990).

J. Kraye, 'Moral Philosophy', in Schmitt *et al.* (eds) (1988), pp. 303–86, is an extensive survey. For Padua and civic medicine: J. Bylebyl, 'The School of Padua: Humanistic Medicine in the Sixteenth Century', in C. Webster (ed), *Health, Medicine and Mortality in the Sixteenth Century* (Cambridge: Cambridge University Press, 1979), pp. 335–70. H. M. Gardiner, R. C. Metcalf and J. G. Beebe-Center, *Feeling and Emotion: A History of Theories* (reprint, Westport, CT: Greenwood Press, 1970) can to some degree be used for descriptive information on the passions. For a study of confession and attitudes to conduct: T. N. Tentler, *Sin and Confession on the Eve of the Reformation* (Princeton: Princeton University Press, 1977). On the political context of 'advice to princes': J. G. A. Pocock, *The Machiavellian Moment: Florentine Political Thought and the Atlantic Republican Tradition* (Princeton: Princeton University Press, 1975).

There is an introduction to Renaissance rhetoric in B. Vickers, 'Rhetoric and Poetics', in Schmitt *et al.* (eds) (1988), pp. 715–45; and Vickers makes an extended defence of rhetoric as a form of understanding in *In Defence of Rhetoric* (Oxford: Clarendon Press, 1988). Rhetoric is linked to the theory of the passions in M. Slawinski, 'Rhetoric and Science/Rhetoric of Science/Rhetoric as Science', in S. Pumfrey, P. L. Rossi and M. Slawinski (eds), *Science, Culture and Popular Belief in Renaissance Europe* (Manchester: Manchester University Press, 1991), pp. 71–99. The standard study of Ramus is W. J. Ong, *Ramus: Method, and the Decay of Dialogue* (Cambridge, MA: Harvard University Press, 1958); see also Grafton and Jardine (1986), chap. 7. On Vives: L. J. Swift and S. L. Block, 'Classical Rhetoric in Vives' Psychology', JHBS, **10** (1974): 74–83. The practical arts of memory are discussed in F. A. Yates, *The Art of Memory* (London: Routledge and Kegan Paul, 1966). There is a scholarly study of the transformation of logic and rhetoric in the seventeenth century in W. S. Howell, *Logic and Rhetoric in England, 1500–1700* (Princeton: Princeton University Press, 1956); and for the importance of rhetoric to the new natural philosophy of the seventeenth century: B. Shapiro, 'Early Modern Intellectual Life: Humanism, Religion and Science in Seventeenth-Century England', *History of Science*, **29** (1991): 45–71.

An introduction to historical writing is given by E. Breisach, *Historiography: Ancient, Medieval, & Modern* (Chicago: University of Chicago Press, 1983). The origin of historical scholarship in the distinction between primary and secondary sources is argued in G. Huppert, *The Idea of Perfect History: Historical Erudition and Historical Philosophy in Renaissance France* (Urbana: University of Illinois Press, 1970). See also P. Burke, *The Renaissance Sense of the Past* (London: Edward Arnold, 1969). There is a good collection of sources in D. R. Kelley (ed), *Versions of History from Antiquity to the Enlightenment* (New Haven: Yale University Press, 1991). D. R. Kelley makes an important contribution in his argument for the significance of jurisprudence to history: *Foundations of Modern Historical Scholarship: Language, Law, and History in the French Renaissance* (New York: Columbia University Press, 1970), esp. chap. 5 for Baudoin; 'The Theory of History', in Schmitt *et al.* (eds) (1988), pp. 746–61; '*Altera natura*: The Idea of Custom in Historical Perspective', in J. Henry and S. Hutton (eds), *New Perspectives on Renaissance Thought* (London: Duckworth, 1990), pp. 83–100. On Bodin, see also J. H. Franklin, *Jean Bodin and the Sixteenth-Century Revolution in the Methodology of Law and History* (New York: Columbia University Press, 1963). For the relation of the French historical school to the new natural philosophy of the seventeenth century, see S. Pumfrey, 'The History of Science and the Renaissance Science of History', in Pumfrey, Rossi and Slawinski (eds) (1991), pp. 48–70.

CHAPTER 3

The argument for the importance of jurisprudence in the origins of the human sciences is made in the work of D. R. Kelley, especially in his major study, *The Human Measure: Social Thought in the Western Legal Tradition* (Cambridge, MA: Harvard University Press, 1990), which brings together the search for order in legal, natural and moral philosophy. For earlier, much debated, studies that bring together the Christian concept of law, the rise of the absolutist state and the idea of the laws of nature: E. Zilsel, 'The Genesis of the Concept of Natural Law', *Philosophical Review*, **51** (1942): 245–79; J. Needham. 'Human Law and the Laws of Nature', in *The Grand Titration: Science and Society in East and West* (London: Allen & Unwin, 1969), pp. 299–330. The importance of the medieval theology of God's volition is stressed in J. R. Milton, 'The Origin and Development of the "Laws of Nature"', *European Journal of Sociology,* **22** (1981): 173–95, and this argument is criticized, though without sensitivity to metaphor, in J. E. Ruby, 'The Origins of Scientific "Law"', JHI, **47** (1986): 341–59. For the idea of natural law in the human sphere: L. G. Crocker, *An Age of Crisis: Man and World in Eighteenth Century French Thought* (Baltimore: Johns Hopkins University Press, 1959), chap. 1; R. Tuck, 'The "Modern" Theory of Natural Law', in A. Pagden (ed), *The Languages of Political Theory in Early-Modern Europe* (Cambridge: Cambridge University Press, 1987), pp. 99–122; R. Tuck, *Philosophy and Government 1572–1651* (Cambridge: Cambridge University Press, 1993).

The cited works by Tuck are invaluable for Grotius; for Grotius and philosophical jurisprudence, see Kelley (1990), chap. 12. He is described as pivotal for the later Enlightenment theory of natural rights in R. Tuck, *Natural Rights Theories: Their Origin and Development* (Cambridge: Cambridge University Press, 1979). The significance of the desire for political stability as a motive to pursue certainty in knowledge is argued in T. K. Rabb, *The Struggle for Stability in Early Modern Europe* (New York: Oxford University Press, 1975); and S. Toulmin, *Cosmopolis: The Hidden Agenda of Modernity* (Chicago: University of Chicago Press, 1992), which also makes suggestive links between science and social change.

The 500th anniversary in 1992 of Columbus's landing in the Caribbean islands, and a new consciousness about the status of 'the discovery' of America, produced a flood of publications. Readable discussions of Europe's relations with the wider world include: J. H. Elliott, *The Old World and the New 1492–1650* (new edn, Cambridge: Cambridge University Press, 1992); J. H. Parry, *The Age of Reconnaissance* (Berkeley: University of California Press, 1981). For the shock to Europeans as they tried to assimilate experience of the New World: A. Grafton, *New Worlds: Ancient Texts: The Power of Tradition and the Shock of Discovery* (Cambridge, MA: Belknap Press of Harvard University Press, 1992); T. Todorov, *The Conquest of America: The Question of the Other* (New York: Harper & Row, 1984); and especially A. Pagden, *European Encounters with the New World: From Renaissance to Romanticism* (New Haven: Yale University Press, 1993). On 'wildness': E. Dudley and M. E. Novak (eds), *The Wild Man Within: An Image of Western Thought from the Renaissance to Romanticism* (Pittsburgh: University of Pittsburgh Press, 1972). Europe's relations with South Africa and the East have been less studied from these points of view, but there is a discussion from both European and non-European perspectives in U. Bitterli, *Cultures in Conflict: Encounters between European and Non-European Cultures, 1492–1800* (Stanford: Stanford University Press, 1989).

There is descriptive material on the collection of what we would see as ethnographic evidence in M. T. Hodgen, *Early Anthropology in the Sixteenth and Seventeenth Centuries* (Philadelphia: University of Pennsylvania Press, 1964). For discussions of the European debate about human culture and the physical environment: C. J. Glacken, *Traces on the Rhodian Shore: Nature and Culture in Western Thought from Ancient Times to the End of the Eighteenth Century* (Berkeley: University of California Press, 1967); P. J. Bowler, *The Fontana History of the Environmental Sciences* (London: Fontana Press, 1992). Montaigne's essays exist in many editions; his Essay 31 is on 'Cannibals', but doubt is thrown on the existence of all reports of ritualized cannibalism in W. Arens, *The Man-eating Myth: Anthropology & Anthrophagy* (New York: Oxford University Press, 1979). The standard biography is D. M. Frame, *Montaigne: A Biography* (New York: Harcourt, Brace & World, 1965). For the imagery of cannibals, Caliban and Crusoe, see P. Hulme, *Colonial Encounters: Europe and the Native Caribbean, 1492–1797* (London and New York: Routledge, 1986). The theological and legal criticism of the Spanish conquests is studied in A. Pagden, *The Fall of Natural Man: The American Indian and the Origins of Ethnology* (reprint, Cambridge: Cambridge University Press, 1986), and 'Discovering the Barbarian: The Language of Spanish Thomism and the Debate over the Property Rights of the American Indian', in Pagden (ed) (1987), pp. 79–98.

There is an extensive literature on Hobbes, partly because of the power of his ideas, partly because questions about the nature of intellectual history have been debated using his work. For an introduction: R. Tuck, *Hobbes* (Oxford: Oxford University Press, 1989). Besides *Leviathan*, Hobbes's comments on human nature and morality are available in, 'De homine', trans. in B. Gert (ed), *Man and Citizen* (new edn, New York: Humanities Press, and Brighton: Harvester Press, 1978), chaps. 10–15. I accept – as some philosophers of political thought do not – that there is a close relation between Hobbes's natural philosophy and his political theory; see L. T. Sarasohn, 'Motion and Morality: Pierre Gassendi, Thomas Hobbes and the Mechanical World-view', JHI, **46** (1985): 363–79; M. Verdon, 'On the Laws of Physical and Human Nature: Hobbes's Physical and Social Cosmologies', JHI, **43** (1982): 653–63. Hobbes's theory of human nature is discussed also in: S. Goyard-Fabre, 'Right and Anthropology in Hobbes's Philosophy', in J. G. van der Bend (ed), *Thomas Hobbes: His View of Man* (Amsterdam: Rodopi, 1982), pp. 17–30; R. Olson, *Science Deified & Science Defied: The Historical Significance of Science in Western Culture*, Vol. 2 (Berkeley: University of California Press, 1990), chaps. 3 and 4; S. Shapin and S. Schaffer, *Leviathan and the Air Pump: Hobbes, Boyle, and the Experimental Life* (Princeton: Princeton University Press, 1985), chap. 3. For Hobbes and language: G. M. Ross, 'Hobbes and Descartes on the Relation beween Language and Consciousness', *Synthese*, **75** (1988): 217–29; G. Rossini, 'The Criticism of Rhetorical Historiography and the Ideal of Scientific Method: History, Nature and Science in the Political Language of Thomas Hobbes', in Pagden (ed) (1987), pp. 303–24. On the background to theories of language: D. Knox, 'Ideas on Gesture and Universal Languages, c.1550–1650', in J. Henry and S. Hutton (eds), *New Perspectives on Renaissance Thought* (London: Duckworth, 1990), pp. 101–36. The commonplace nature of the metaphor of the body politic is conveyed in: L. Barkan, *Nature's Work of Art: The Human Body as Image of the World* (New Haven: Yale University Press, 1975); D. G. Hale, *The Body Politic: A Political Metaphor in Renaissance English Literature* (The Hague: Mouton, 1971), and for the riches of the metaphor, J. Sawday, *The Body Emblazoned: Dissection and the Human Body in Renaissance Culture* (London and New York: Routledge, 1995). The question of the origins of individualism runs and runs. I cite only an introduction to the concept: S. Lukes, 'The Meanings of "Individualism"', JHI, **32** (1971): 45–66, and *Individualism* (Oxford: Basil Blackwell, 1973).

Pufendorf's work is accessible through a new edition of *On the Duty of Man and Citizen According to Natural Law*, ed J. Tully (Cambridge: Cambridge University Press, 1991), with a valuable introduction. For Pufendorf's significance for the eighteenth century, especially in connection with the opinion that sociability is a historically achieved condition: I. Hont, 'The Language of Sociability and Commerce: Samuel Pufendorf and the Theoretical Foundation of the "Four-stages Theory"', in Pagden (ed) (1987), pp. 253–76; R. Wokler, 'From *l'homme physique* to *l'homme moral* and Back: Towards a History of Enlightenment Anthropology', HHS, **6** no. 1(1993):121–38.

CHAPTER 4

Major changes occurred in natural philosophy in the seventeenth century, but these were more clear-cut in the physical sciences than in the sciences of man. Historical literature on the physical sciences often has little to say about the human sciences, except in the area of medical anatomy and physiology; see H. J. Cook, 'The New Philosophy and Medicine in Seventeenth-Century England', in D. C. Lindberg and R. S. Westman (eds), *Reappraisals of the Scientific Revolution* (Cambridge: Cambridge University Press, 1990), pp. 397–436. The argument of Burtt and of Whitehead (referenced under Chapter 2), that the physical sciences made progress by the exclusion of mental categories from nature, with negative consequences for the understanding of human nature, was influential; for a modern assessment see R. M. Young, 'Mind-body Problem', in R. C. Olby *et al.* (eds), *Companion to the History of Modern Science* (London: Routledge, 1990), pp. 702–11, and 'Persons, Organisms and . . . Primary Qualities', in J. R. Moore (ed), *History, Humanity and Evolution* (Cambridge: Cambridge University Press, 1989), pp. 375–401. The mechanical philosophers hoped to check scepticism, as discussed in R. H. Popkin, *The History of Scepticism from Erasmus to Descartes* (revised edn, Assen: Van Gorcum, 1964). There are reappraisals of the reception of the mechanical philosophy in J. H. Brooke, *Science and Religion: Some Historical Perspectives* (Cambridge: Cambridge University Press, 1991), chap. 4; K. Hutchison, 'What Happened to Occult Qualities in the Scientific Revolution?', *Isis*, **73** (1982): 233–53. The studies of B. Willey, *The Seventeenth-Century Background. Studies in the Thought of the Age in Relation to Poetry and Religion* (reprint, Harmondsworth: Penguin Books, 1962) are clear and informed.

For Pomponazzi and mortalism, see the translation, 'On the

Immortality of the Soul', in E. Cassirer, P. O. Kristeller and J. H. Randall, Jr (eds), *The Renaissance Philosophy of Man* (Chicago: Chicago University Press, 1948), pp. 257–381. The significance of the question of immortality is emphasized in E. Michael and F. S. Michael, 'Corporeal Ideas in Seventeenth-Century Psychology', JHI, **50** (1989): 31–48, and 'Two Early Modern Concepts of Mind: Reflecting Substance vs. Thinking Substance', *Journal of the History of Philosophy*, **27** (1989): 29–48. Bacon's key discussion on new knowledge of man was in two versions: *The Two Books of Francis Bacon. Of the Proficience and Advancement of Learning Divine and Humane* (1605; reprint, Bristol: Thoemmes Press, 1994), and in expanded form (published in Latin in 1623) in 'Of the Dignity and Advancement of Learning', in J. Spedding *et al.* (eds), *The Works of Francis Bacon* (1858–74; reprint, Stuttgart–Bad Cannstatt: Friedrich Frommann, Günther Holzboog, 1963), Vols. 4 and 5. On pneumatology: J. Henry, 'Medicine and Pneumatology: Henry More, Richard Baxter, and Francis Glisson's *Treatise on the Energetic Nature of Substance*', *Medical History*, **31** (1987): 15–40. For the word 'psychology', see F. Vidal, 'Psychology in the 18th Century: A View from the Encyclopaedias', HHS, **6** no. 1 (1993): 89–119. And for substantial reconsideration of the agenda for the history of psychology in the seventeenth century: G. Hatfield, 'Remaking the Science of Mind: Psychology as Natural Science', in C. Fox, R. Porter and R. Wokler (eds), *Inventing Human Science: Eighteenth-Century Domains* (Berkeley: University of California Press, 1995), pp. 184–231; G. Richards, *Mental Machinery: The Origins and Consequences of Psychological Ideas. Part 1: 1600–1850* (London: Athlone Press, 1992).

Descartes' view of the mind lies at the heart of many histories of both psychology and modernity. Modern scholarship tends to distinguish the historical figure and an emblematic figure used in moral or philosophical argument; for information about his life, see G. Rodis-Lewis, 'Descartes' Life and the Development of His Philosophy', in J. Cottingham (ed), *The Cambridge Companion to Descartes* (Cambridge: Cambridge University Press, 1992), pp. 21–57 and, in detail, S. Gaukroger, *Descartes: An Intellectual Biography* (Oxford: Clarendon Press, 1995). In a huge literature on his thought, see J. Cottingham, *Descartes* (Oxford: Basil Blackwell, 1986), chap. 5; G. Hatfield, 'Descartes' Physiology and Its Relation to His Psychology', in Cottingham (ed)

(1992), pp. 335–70. Modern translations of Descartes' works are in J. Cottingham (ed), *The Philosophical Writings of Descartes*, 3 Vols (Cambridge: Cambridge University Press, 1985–91). For Descartes and the mind–body question, see also: J. Cottingham, 'Cartesian Trialism', *Mind*, **94** (1985): 218–30, which points out how sensation and imagination, in Descartes' scheme, fall into a mixed category of mind and body; A. Gabbey, 'The Mechanical Philosophy and Its Problems: Mechanical Explanations, Impenetrability, and Perpetual Motion', in J. C. Pitt (ed), *Change and Progress in Modern Science* (Dordrecht: D. Reidel, 1985), pp. 9–84, esp. pp. 14–28; M. D. Wilson, 'Body and Mind from the Cartesian Point of View', in R. W. Rieber (ed), *Body and Mind: Past, Present, and Future* (New York: Academic Press, 1980), pp. 35–55; W. I. Watson, 'Why Isn't the Mind-Body Problem Ancient?', in P. Feyerabend and G. Maxwell (eds), *Mind, Matter, and Method* (Minneapolis: University of Minnesota Press, 1966), pp. 92–102. There is much material on theology and the soul in L. C. Rosenfield, *From Beast-machine to Man-machine: Animal Soul in French Letters from Descartes to La Mettrie* (new edn, New York: Octagon Books, 1968). On the concept of 'ideas' in Descartes and subsequently in the eighteenth century, see J. W. Yolton, *Perceptual Acquaintance from Descartes to Reid* (Oxford: Basil Blackwell, 1984); and on Descartes' usage of 'innate ideas', R. McRae, 'Innate Ideas', in R. J. Butler (ed), *Cartesian Studies* (Oxford: Basil Blackwell, 1972), pp. 35–54. Descartes is linked to the modern debate on whether man is a machine in K. Gunderson, 'Descartes, La Mettrie, Language, and Machines', in *Mentality and Machines* (2nd edn, London: Croom Helm, 1985), pp. 1–38. The argument that the mechanical philosophy disrupted organic views of nature is developed in C. Merchant, *The Death of Nature: Women, Ecology, and the Scientific Revolution* (reprint, London: Wildwood House, 1982).

Descartes' posthumous work on man is translated in T. S. Hall (ed), *Treatise of Man* (Cambridge, MA: Harvard University Press, 1972), with extensive physiological notes. For the significance of mechanical technology: J. C. Marshall, 'Minds, Machines and Metaphors', *Social Studies of Science*, **7** (1977): 475–88. G. Canguilhem, *La formation du concept de réflexe aux XVIIe et XVIIIe siècles* (2nd edn, Paris: J. Vrin, 1977), argues that the *concept* of the reflex did not develop before the second half of the eighteenth century; the relevant passage in *Treatise of Man* is on pp. 33–6. On Gassendi: M. J. Osler, 'Babtizing Epicurean Atomism: Pierre Gassendi on the Immortality of the Soul', in M. J. Osler and P. L. Farber (eds), *Religion, Science, and Worldview* (Cambridge: Cambridge University Press, 1985), pp. 163–83. The English reception of mechanical philosophy is well studied: R. G. Frank, *Harvey and the Oxford Physiologists* (Berkeley: University of California Press, 1980), and 'Thomas Willis and His Circle: Brain and Mind in Seventeenth-Century Medicine', in G. S. Rousseau (ed), *The Languages of Psyche: Mind and Body in Enlightenment Thought* (Berkeley: University of California Press, 1990), pp. 107–46; J. Henry, 'A Cambridge Platonist's Materialism: Henry More and the Concept of the Soul', *Journal of the Warburg and Courtauld Institutes*, **49** (1986): 172–95, and 'The Matter of Souls: Medical Theory and Theology in Seventeenth-Century England', in R. French and A. Wear (eds), *The Medical Revolution in the Seventeenth Century* (Cambridge: Cambridge University Press, 1989), pp. 87–113; S. Schaffer, 'Godly Men and Mechanical Philosophers: Souls and Spirits in Restoration Natural Philosophy', *Science in Context*, **1** (1987): 55–85; J. P. Wright, 'Hysteria and Mechanical Man', JHI, **41** (1980): 233–47, and 'Locke, Willis, and the Seventeenth-Century Epicurean Soul', in M. J. Osler (ed), *Atoms, Pneuma, and Tranquility: Epicurean and Stoic Themes in European Thought* (Cambridge: Cambridge University Press, 1991), pp. 239–58. There is a brief extract from Willis in R. Hunter and I. Macalpine, *Three Hundred Years of Psychiatry 1535–1860* (London: Oxford University Press, 1963), pp. 187–92, while the book as a whole is an invaluable guide to the medical literature on mind–body relations. A case of distinctive psychological interest in the activity of the early Royal Society is discussed in B. R. Singer, 'Robert Hooke on Memory, Association and Time Perception', *Notes and Records of the Royal Society*, **31** (1976): 115–31.

The question of the self and the foundations of moral action is discussed in C. Taylor, *Sources of the Self: The Making of the Modern Identity* (Cambridge: Cambridge University Press, 1989). There is a commentary on Montaigne's *Essays* in M. A. Screech, *Montaigne & Melancholy: The Wisdom of the Essays* (London: Duckworth, 1983). On the significance of the essay form for the individual empirical approach to knowledge, see J. Paradis, 'Montaigne, Boyle, and the Essay of Experience', in G. Levine (ed), *One Culture: Essays in Science and Literature* (Madison: University of Wisconsin Press, 1987), pp. 59–91. For the important historical point that 'individuality' must be studied in relation to the local character of a person's social place, N. Z. Davis, 'Boundaries and the Sense of Self in Sixteenth-Century France', in T. C. Heller, M. Soma and D. E. Wellberg (eds), *Reconstructing Individualism: Autonomy, Individuality and the Self in Western Thought* (Stanford: Stanford University Press, 1986), pp. 53–63. On the self, biography and eighteenth-century literary culture, J. O. Lyons, *The Invention of the Self: The Hinge of Consciousness in the Eighteenth Century* (Carbondale: Southern Illinois University Press, 1978), and the admired study by I. Watt, *The Rise of the Novel: Studies in Defoe, Richardson and Fielding* (London: Chatto & Windus, 1957). For discussion of 'seeing the person', as in Vermeer's painting, see S. Alpers, *The Art of Describing: Dutch Art in the Seventeenth Century* (London: John Murray, 1983), pp. 192–207; see also S. Schama, *The Embarrassment of Riches: An Interpretation of Dutch Culture in the Golden Age* (New York: Alfred A. Knopf, 1987). Josselin's diary was brought to historians' attention in A. Macfarlane, *The Family Life of Ralph Josselin, A Seventeenth-Century Clergyman: An Essay in Historical Anthropology* (Cambridge: Cambridge University Press, 1970), and A. Macfarlane (ed), *The Diary of Ralph Josselin 1616–1683* (London: Oxford University Press for the British Academy, 1976). The debate about 'the Protestant ethic' was begun by Max Weber; see the translation in *The Protestant Ethic and the Spirit of Capitalism* (London: Unwin, 1930); on the New England Puritans, see P. Miller, *The New England Mind: The Seventeenth Century* (reprint, Cambridge, MA: Harvard University Press, 1967), chap. 9. The refinement of manners is the subject of a much-cited study by N. Elias, *The Civilizing Process: The History of Manners* (German edn 1939; Oxford: Basil Blackwell, 1978); and Z. Barbu's studies of mentality are in *Problems of Historical Psychology* (London: Routledge & Kegan Paul, 1960).

CHAPTER 5

The philosophical literature on Locke, as the most influential source of British empiricism, is vast. There is a historically oriented discussion in J. W. Yolton, *John Locke and the Way of*

Ideas (reprint, Oxford: Clarendon Press, 1968). In J. W. Yolton, *Perceptual Acquaintance from Descartes to Reid* (Oxford: Basil Blackwell, 1984), there is discussion of how a lack of psychological vocabulary fostered the use, exemplified by Locke, of a spatial language for talk about awareness. Locke's *Essay* is discussed as a study in logic in J. G. Buickerood, 'The Natural History of the Understanding: Locke and the Rise of Facultative Logic in the Eighteenth Century', *History and Philosophy of Logic*, **6** (1985): 157–90; W. S. Howell, *Eighteenth-Century British Logic and Rhetoric* (Princeton: Princeton University Press, 1971). There are different views of Locke's response to the new natural philosophy, but his deepest inspiration was political and religious; see R. Ashcraft, *Revolutionary Politics & Locke's Two Treatises of Government* (Princeton: Princeton University Press, 1986); J. L. Axtell, 'Locke, Newton and the Two Cultures', in J. W. Yolton (ed), *John Locke: Problems and Perspectives* (Cambridge: Cambridge University Press, 1969), pp. 165–82; G. A. J. Rogers, 'The Empiricism of Locke and Newton', in S. C. Brown (ed), *Philosophers of the Enlightenment* (Brighton: Harvester Press, 1979), pp. 1–30. Locke's interest in what we call anthropology is discussed in G. A. J. Rogers, 'Locke, Anthropology and Models of the Mind', HHS, **6** no. 1 (1993): 73–87. For Locke and the eighteenth century: H. Aarsleff, 'The State of Nature and the Nature of Man in Locke', in Yolton (ed) (1969), pp. 99–136; C. Fox (ed), *Psychology and Literature in the Eighteenth Century* (New York: AMS Press, 1987).

Locke's work on education has been reprinted: J. L. Axtell (ed), *The Educational Writings of John Locke* (Cambridge: Cambridge University Press, 1968); *Some Thoughts Concerning Education*, ed J. W. and J. S. Yolton (Oxford: Clarendon Press, 1989); see also J. A. Passmore, 'The Malleability of Man in Eighteenth-Century Thought', in E. R. Wasserman (ed), *Aspects of the Eighteenth Century* (Baltimore: Johns Hopkins University Press, 1965), pp. 21–46; J. W. Yolton, *John Locke & Education* (New York: Random House, 1971). Locke's political thought, like that of Hobbes, has been constantly reassessed, and it is central to debate about the nature of the history of ideas; see R. Ashcraft, *Locke's Two Treatises of Government* (London: Unwin Hyman, 1987). Earlier scholarship too easily identified Locke with what Enlightenment writers made of his work; for a strongly drawn contrast, J. Dunn, 'From Applied Theology to Social Analysis: The Break between John Locke and the Scottish Enlightenment', in I. Hont and M. Ignatieff (eds), *Wealth and Virtue: The Shaping of Political Economy in the Scottish Enlightenment* (Cambridge: Cambridge University Press, 1983), pp. 119–35.

There is a discussion of the early eighteenth-century debate about identity and literary culture in C. Fox, *Locke and the Scriblerians: Identity and Consciousness in Early Eighteenth-Century Britain* (Berkeley: University of California Press, 1988); for an analysis of Locke's theological and jurisprudential, as well as philosophical, views of the attributes of man: D. P. Behan, 'Locke on Persons and Personal Identity', *Canadian Journal of Philosophy*, **9** (1979): 53–75; C. F. Goodey, 'Locke's Idiots in the Natural History of Mind', *History of Psychiatry*, **5** (1994): 215–50. For an introduction to the history of language theories: R. Harris and T. J. Taylor, *Landmarks in Linguistic Thought: The Western Tradition from Socrates to Saussure* (London: Routledge, 1989); and for more scholarly details, H. Aarsleff, 'Leibniz and Locke on Language' and 'An Outline of Language-origins Theory since the Renaissance', in *From Locke to Saussure: Essays on the Study of Language and Intellectual History* (Minneapolis: University of Minnesota Press, 1982), pp. 42–83 and 278–92, while the introduction revises Locke's significance for the history of linguistics. The seventeenth-century background is discussed in: J. Knowlson, *Universal Language Schemes in England and France 1600–1800* (Toronto: University of Toronto Press, 1975); B. J. Shapiro, *Probability and Certainty in Seventeenth-Century England: A Study of the Relationship between Natural Science, Religion, History, Law, and Literature* (Princeton: Princeton University Press, 1983). On Wilkins: Aarsleff, 'John Wilkins', in (1982), pp. 239–77; S. Clauss, 'John Wilkins' Essay Toward a Real Character: Its Place in the Seventeenth-Century Episteme', JHI, **43** (1982): 531–53, an article whose target is the discussion of language by M. Foucault, *The Order of Things: An Archaeology of the Human Sciences* (New York: Pantheon, and London: Tavistock, 1970); and for a biography, B. J. Shapiro, *John Wilkins 1614–1672: An Intellectual Biography* (Berkeley: University of California Press, 1969). G. Richards argues that there was a lack of psychological concepts precisely because mental language was constructed metaphorically from sensations derived from the physical world: 'The

Absence of Psychology in the Eighteenth Century: A Linguistic Perspective', *Studies in the History and Philosophy of Science*, **23** (1992): 195–211, and *Mental Machinery: The Origins and Consequences of Psychological Ideas. Part 1: 1600–1850* (London; Athlone Press, 1992).

M. J. Morgan, *Molyneux's Question: Vision, Touch and the Philosophy of Perception* (Cambridge: Cambridge University Press, 1977), introduces the literature of Locke, Berkeley, Cheselden, Diderot and Condillac. There is a philosophical discussion of Berkeley's work in M. Atherton, *Berkeley's Revolution in Vision* (Ithaca: Cornell University Press, 1990). For wider discussion of theories of visual perception, G. C. Hatfield and W. Epstein, 'The Sensory Core and the Medieval Foundations of Early Modern Perceptual Theory', *Isis*, **70** (1979): 363–84; and from the viewpoint of modern arguments, N. Pastore, *Selective History of Theories of Visual Perception: 1650–1950* (New York: Oxford University Press, 1971). On the moon illusion: Richards (1992), pp. 76–9. Diderot's 'Essay on the Blind for the Use of Those who See', is translated in Morgan (1977), pp. 31–58, and in L. G. Crocker (ed), *Diderot's Selected Writings* (New York: Macmillan, 1966), pp. 14–30. Locke's reference to 'association' is analysed in J. P. Wright, 'Association, Madness and the Measures of Probability in Locke and Hume', in Fox (ed) (1987), pp. 103–27.

CHAPTER 6

The literature on so-called rationalist science presents difficulties for English-language readers who are familiar with more empirical modes of argument. The study by E. Cassirer, *The Philosophy of the Enlightenment* (German edn, 1932; reprint, Boston: Beacon Press, 1955), lies at the foundation of this chapter and remains basic to study of the philosophy of knowledge of the period. The connections between philosophy and the Germanic ideal of science are discussed in F. K. Ringer, *The Decline of the German Mandarins: The German Academic Community, 1890–1933* (reprint, Hanover: Wesleyan University Press/University Press of New England, 1990), chap. 2. The importance of non-mechanistic physiology in the eighteenth century is mentioned in S. Moravia, 'The Enlightenment and the Sciences of Man', *History of Science*, **18** (1980): 247–68, and there are comments on Stahl in J. Geyer-Kordesch, 'Georg Ernst Stahl's Radical Pietist Medicine and Its Influence on the German Enlightenment', in A. Cunningham and R. French (eds), *The Medical Enlightenment of the Eighteenth Century* (Cambridge: Cambridge University Press, 1990), pp. 67–87. There is a modern translation of A. Arnauld in *The Art of Thinking: Port-Royal Logic*, intro. J. Dickoff and P. James (Indianapolis: Bobbs-Merrill, 1964); his logic is discussed in W. S. Howell, *Logic and Rhetoric in England, 1500–1700* (Princeton: Princeton University Press, 1956), and his philosophy in S. M. Nadler, *Arnauld and the Cartesian Philosophy of Ideas* (Manchester: Manchester University Press, 1989). For seventeenth- and eighteenth-century theories of language see E. F. K. Koerner and R. E. Asher, *Concise History of the Language Sciences: From the Sumerians to the Cognitivists* (Oxford and New York: Pergamon, 1995), Sects. 8 and 9. On the early modern rational and rhetorical content of French higher education, L. W. B. Brockliss, *French Higher Education in the Seventeenth and Eighteenth Centuries: A Cultural History* (Oxford: Clarendon Press, 1987). Malebranche is considered in the light of modern perception theory in E. S. Reed, 'Theory, Concept, and Experiment in the History of Psychology: The Older Tradition Behind a "Young Science"', HHS, **2** (1989): 333–56.

Spinoza and Leibniz are exceptionally difficult philosophers, precisely because of the logical structure of their metaphysics. There are introductions to the philosophy in: J. Cottingham, *The Rationalists* (Oxford: Oxford University Press, 1988); R. Scruton, *Spinoza* (Oxford: Oxford University Press, 1986). A good idea of the range of Leibniz's activity is conveyed in the biography by E. J. Aiton, *Leibniz: A Biography* (Bristol and Boston: Adam Hilger, 1985). The relation between Spinoza's account of the identity of mind and body, his metaphysics and the passions is discussed in M. Wartofsky, 'Action and Passion: Spinoza's Construction of a Scientific Psychology', in M. Grene (ed), *Spinoza: A Collection of Critical Essays* (Notre Dame: University of Notre Dame Press, 1979), pp. 329–53.

The German academies and universities in the eighteenth century are described in H. Aarsleff, 'The Berlin Academy under Frederick the Great', HHS, **2** (1989): 193–206; C. E. McClelland, *State, Society, and University in Germany 1700–1914* (Cambridge: Cambridge University Press, 1980), for the university of

Göttingen, esp. chap. 11; R. S. Turner, 'University Reformers and Professorial Scholarship in Germany 1760–1806', in L. Stone (ed), *The University in Society. Volume II. Europe, Scotland, and the United States from the 16th to the 20th Century* (Princeton: Princeton University Press, 1974), pp. 495–531. There is a systematic description of the philosophy in L. W. Beck, *Early German Philosophy: Kant and His Predecessors* (Cambridge, MA: Belknap Press of Harvard University Press, 1969). There is little in English on psychological thought, but some description in H. M. Gardiner, R. C. Metcalf and J. G. Beebe-Center, *Feeling and Emotion. A History of Theories* (reprint, Westport, CT: Greenwood Press, 1970), chap. 9; while for Wolff see R. J. Richards, 'Christian Wolff's Prolegomena to Empirical and Rational Psychology: Translation and Commentary', *Proceedings of the American Philosophical Society*, **124** (1980): 227–39, which also mentions Thomasius, and T. P. Saine, 'Who's Afraid of Christian Wolff?', in A. C. Kors and P. J. Korshin (eds), *Anticipations of the Enlightenment in England, France, and Germany* (Philadelphia: University of Pennsylvania Press, 1987), pp. 102–33. There are remarks on Tetens in I. Staeuble, 'The Relationship between Nature and Society in Early Conceptualisations of Developmental Psychology', in G. Eckardt, W. G. Bringmann and L. Sprung (eds), *Contributions to a History of Developmental Psychology* (Berlin and New York: Mouton, 1985), pp. 89–99, and in E. Scheerer, 'Pre-evolutionary Conceptions of Imitation', in *ibid.*, pp. 27–53. Reimarus is brought to attention in J. Jaynes and W. Woodward, 'In the Shadow of the Enlightenment', JHBS, **10** (1974): 3–15 and 144–59. Scholarship at the university of Göttingen is discussed in R. S. Leventhal, 'The Emergence of Philological Discourse in the German States, 1770–1810', *Isis*, **77** (1986): 243–60. Kant's work is as difficult as it is influential in modern thought, but there are succinct and relevant remarks in R. C. Solomon, *Continental Philosophy since 1750: The Rise and Fall of the Self* (Oxford: Oxford University Press, 1988), chap. 2. Kant had high ideals about enlightenment, and his famous remarks (1784) on the subject are translated in 'What is Enlightenment?', in P. Gay (ed), *The Enlightenment: A Comprehensive Anthology* (New York: Simon and Schuster, 1973), pp. 383–90, and in 'Answer to the Question: What is "Enlightening"', in S. Eliot and B. Stern (eds), *The Age of Enlightenment* (East Grinstead: Ward Lock

Educational in association with the Open University Press, 1979), Vol. 2, pp. 249–55. The political and social context of Kant's commitment is examined in detail in S. Lestitian, 'Kant and the End of the Enlightenment in Prussia', *Journal of Modern History*, **65** (1993): 57–112. His public lectures are also accessible: *Anthropology from a Pragmatic Point of View*, trans. M. J. Gregor (The Hague: Martinus Nijhoff, 1974). Kant's discussion of the relation between reason, science and psychology is studied and used to analyse relations between conceptual and empirical claims in G. Hatfield, *The Natural and the Normative: Theories of Spatial Perception from Kant to Helmholtz* (Cambridge, MA: MIT Press, 1990), and 'Empirical, Rational, and Transcendental Psychology: Psychology as Science and as Philosophy', in P. Guyer (ed), *The Cambridge Companion to Kant* (Cambridge: Cambridge University Press, 1992), pp. 200–27. An earlier study is T. Mischel, 'Kant and the Possibility of a Science of Psychology', *The Monist*, **51** (1967): 599–622. K. Arens, *Structures of Knowing: Psychologies of the Nineteenth Century* (Dordrecht: Kluwer, 1989), covers this ground and much other not readily available German-language material but is difficult to use. The relation between Kant's view of the mind and modern cognitivist theories attracts interest, but the literature is specialized; see A. Brook, *Kant and the Mind* (Cambridge: Cambridge University Press, 1994), chaps. 2–4 on the unity of the subject; P. Kitcher, *Kant's Transcendental Psychology* (New York: Oxford University Press, 1990). Post-Kant psychology is described in D. E. Leary, 'The Philosophical Development of the Conception of Psychology in Germany, 1780–1850', JHBS, **14** (1978): 113–21, and 'Immanuel Kant and the Development of Modern Psychology', in W. R. Woodward and M. G. Ash (eds), *The Problematic Science: Psychology in the Nineteenth Century* (New York: Praeger, 1982), pp. 17–42.

The German medical psychology journals are mentioned in D. B. Wiener, 'Mind and Body in the Clinic: Philippe Pinel, Alexander Crichton, Dominque Esquirol, and the Birth of Psychiatry', in G. S. Rousseau (ed), *The Languages of Psyche: Mind and Body in Enlightenment Thought* (Berkeley: University of California Press, 1990), pp. 331–90. Lavater's significance is widely noted but less studied, though see E. Shookman (ed), *The Faces of Physiognomy: Interdisciplinary Approaches to Johann Caspar Lavater*

(Columbia, SC: Camden House, 1993), and G. Tytler, *Physiognomy in the European Novel: Faces and Fortunes* (Princeton: Princeton University Press, 1982); also G. Flaherty, 'The Non-normal Sciences: Survivals of Renaissance Thought in the Eighteenth Century', in C. Fox, R. Porter and R. Wokler (eds), *Inventing Human Science: Eighteenth-Century Domains* (Berkeley: University of California Press, 1995), pp. 271–91; L. J. Jordanova, 'Naturalizing the Family: Literature and the Bio-medical Sciences in the Late Eighteenth Century', in L. J. Jordanova (ed), *Languages of Nature: Critical Essays on Science and Literature* (London: Free Association Books, 1986), pp. 86–116; B. M. Stafford, *Body Criticism: Imaging the Unseen in Enlightenment Art and Medicine* (Cambridge, MA: MIT Press, 1991), on visual representation in science; M. Shortland, 'Skin Deep: Barthes, Lavater and the Legible Body', *Economy and Society*, **14** (1985): 273–312; M. Cowling, *The Artist as Anthropologist: The Representation of Type and Character in Victorian Art* (Cambridge: Cambridge University Press, 1989). For physiognomy and Romanticism see A. Gode-Von Aesch, *Natural Science in German Romanticism* (reprint, New York: AMS Press, 1966), chap. 12; and on 'genius', S. Schaffer, 'Genius in Romantic Natural Philosophy', in A. Cunningham and N. Jardine (eds), *Romanticism and the Sciences* (Cambridge: Cambridge University Press, 1990), pp. 82–98.

CHAPTER 7

The nature and consequences of the Enlightenment (even whether there was such a historical reality) is a debate in itself; see the succinct review of historiographic issues in D. Outram, *The Enlightenment* (Cambridge: Cambridge University Press, 1995). There is much information in J. W. Yolton *et al.* (eds), *The Blackwell Companion to the Enlightenment* (Oxford and Cambridge, MA: Blackwell, 1991). I have drawn on the syntheses in L. G. Crocker, *An Age of Crisis: Man and World in Eighteenth Century French Thought* (Baltimore: Johns Hopkins University Press, 1959), and *Nature and Culture: Ethical Thought in the French Enlightenment* (Baltimore: Johns Hopkins University Press, 1963); and P. Gay, *The Enlightenment: An Interpretation*, 2 vols (reprint, London: Wildwood House, 1973), esp. Vol. 2, *The Science of Freedom*. The failure of the Enlightenment project to deliver a sustainable justification for moral action is argued, in a way that had great influence, in A. MacIntyre, *After Virtue: A Study in Moral Theory* (London: Duckworth, 1981), esp. chaps. 4–6. Approaches to this 'age of reason' often minimize Christianity and the interest in pagan religion. As a counterbalance, see C. L. Becker, *The Heavenly City of the Eighteenth-Century Philosophers* (reprint, New Haven: Yale University Press, 1976); F. E. Manuel, *The Eighteenth Century Confronts the Gods* (Cambridge, MA: Harvard University Press, 1959). The diversity of enlightenment is stressed in R. Porter and M. Teich (eds), *The Enlightenment in National Context* (Cambridge: Cambridge University Press, 1981). The political culture of the contemporary period in North America is discussed in J. G. A. Pocock, 'Enlightenment and Revolution: The Case of English-speaking North America', *Studies on Voltaire and the Eighteenth Century,* **263** (1989): 249–61. The interest in human nature is outlined in R. Smith, 'The Language of Human Nature', in C. Fox, R. Porter and R. Wokler (eds), *Inventing Human Science: Eighteenth-Century Domains* (Berkeley: University of California Press, 1995), pp. 88–111.

The culture of sensibility is described in G. J. Barker-Benfield, *The Culture of Sensibility: Sex and Society in Eighteenth-Century Britain* (Chicago: University of Chicago Press, 1992); C. Fox (ed), *Psychology and Literature in the Eighteenth Century* (New York: AMS Press, 1987); J. Mullan, *Sentiment and Sociability: The Language of Feeling in the Eighteenth Century* (Oxford: Clarendon Press, 1988); G. S. Rousseau, 'Towards a Semiotics of the Nerve: The Social History of Language in a New Key', in P. Burke and R. Porter (eds), *Language, Self, and Society: A Social History of Language* (Oxford: Polity Press, 1991), pp. 213–75. Philosophy of medicine and the soul is discussed in R. French, *Robert Whytt, the Soul, and Medicine* (London: Wellcome Institute for the History of Medicine, 1969); L. S. King, *The Philosophy of Medicine: The Early 18th Century* (Cambridge, MA: Harvard University Press, 1978); R. Porter, *Mind-Forg'd Manacles: A History of Madness in England from the Restoration to the Regency* (Cambridge, MA: Harvard University Press, 1987), and 'Medical Science and Human Science in the Enlightenment', in Fox, Porter and Wokler (eds) (1995), pp. 53–87 ; L. J. Rather, *Mind and Body in Eighteenth Century Medicine: A Study Based on Jerome Gaub's De regimine mentis* (London: Wellcome Historical Medical Library, 1965). Non-

Lockean influences on medical writings on mind are stressed in A. Suzuki, 'Anti-Lockean Enlightenment? Mind and Body in Early Eighteenth-Century English Medicine', in R. Porter (ed), *Medicine in the Enlightenment* (Amsterdam and Atlanta, GA: Rodopi, 1995), pp. 336–59. On the nervous system, see also: K. Danziger, 'Origins of the Schema of Stimulated Motion: Towards a Pre-history of Modern Psychology', *History of Science*, **21** (1983): 182–210; C. Lawrence, 'The Nervous System and Society in the Scottish Enlightenment', in B. Barnes and S. Shapin (eds), *Natural Order: Historical Studies of Scientific Culture* (Beverley Hills: Sage, 1979), pp. 19–40; J. P. Wright, 'Metaphysics and Physiology: Mind, Body and the Animal Economy in Eighteenth-Century Scotland', in M. A. Stewart (ed), *Studies in the Philosophy of the Scottish Enlightenment* (Oxford: Clarendon Press, 1990), pp. 251–302. A shift in the balance in explanations of madness with reference to body and mind is argued for in A. Suzuki, 'Dualism and the Transformation of Psychiatric Language in the Seventeenth and Eighteenth Centuries', *History of Science*, **33** (1995): 417–47. The question of the Enlightenment in the American Republic is discussed in G. Wills, *Inventing America: Jefferson's Declaration of Independence* (Garden City, NY: Doubleday, 1978). The teaching of moral philosophy in North America is described in J. G. Blight, 'Jonathan Edwards's Theory of the Mind: Its Applications and Implications', in J. Brožek (ed), *Explorations in the History of Psychology in the United States* (Lewisburg: Bucknell University Press, 1984), pp. 61–120; and R. B. Evans, 'The Origins of American Academic Psychology', in *ibid.*, pp. 17–60.

La Mettrie is translated in *Man a Machine* (reprint, La Salle: Open Court, 1961), and there is a French edition with commentary in A. Vartanian, *L'homme machine: A Study in the Origins of an Idea* (Princeton: Princeton University Press, 1960). The medical rather than Cartesian context of La Mettrie's work is argued for in K. Wellman, *La Mettrie: Medicine, Philosophy, and Enlightenment* (Durham: Duke University Press, 1992). For the controversy about the animal soul: J. C. Kassler, 'Man – A Musical Instrument: Models of the Brain and Mental Functioning before the Computer', *History of Science*, **22** (1984): 59–92; L. C. Rosenfield, *From Beast-machine to Man-machine: Animal Soul in French Letters from Descartes to La Mettrie* (new edn, New York: Octagon Books,

1968); R. M. Young, 'Animal Soul', in P. Edwards (ed), *The Encyclopedia of Philosophy* (New York: Macmillan and Free Press, 1967), Vol. 1, pp. 122–7. The debates about materialism and immaterialism are discussed in J. W. Yolton, *Thinking Matter: Materialism in Eighteenth Century Britain* (Oxford: Basil Blackwell, 1983), and *Locke and French Materialism* (Oxford: Clarendon Press, 1991); A. Vartanian, 'Trembley's Polyp, La Mettrie and Eighteenth-Century French Materialism', in P. P. Wiener and A. Noland (eds), *Roots of Scientific Thought: A Cultural Perspective* (New York: Basic Books, 1957), pp. 497–516. On mechanical models as serious science: D. M. Fryer and J. C. Marshall, 'The Motives of Jacques de Vaucanson', *Technology and Culture*, **20** (1979): 257–69.

The history of sexuality is a flourishing business; for an overview see R. Porter and M. Teich (eds), *Sexual Knowledge, Sexual Science: The History of Attitudes to Sexuality* (Cambridge: Cambridge University Press, 1994), and for the eighteenth century, R. Porter, 'Mixed Feelings: The Enlightenment and Sexuality in Eighteenth-Century Britain', in P.-G. Boucé (ed), *Sexuality in Eighteenth-Century Britain* (Manchester: Manchester University Press, and Totowa, NJ: Barnes & Noble, 1982), pp. 1–27. On the sciences and sexuality: L. J. Jordanova, 'Natural Facts: An Historical Perspective on Science and Sexuality', in C. P. MacCormack and M. Strathern (eds), *Nature, Culture and Gender* (Cambridge: Cambridge University Press, 1981), pp. 42–69. Diderot's writings are translated in L. G. Crocker (ed), *Diderot's Selected Writings* (New York: Macmillan 1966); and there are biographies in P. N. Furbank, *Diderot: A Critical Biography* (London: Secker & Warburg, 1992), and A. M. Wilson, *Diderot* (New York: Oxford University Press, 1972). There is a thoughtful discussion of Sade in D. B. Morris, 'The Marquis de Sade and the Discoveries of Pain: Literature and Medicine at the Revolution', in G. S. Rousseau (ed), *The Languages of Psyche: Mind and Body in Enlightenment Thought* (Berkeley: University of California Press, 1990), pp. 291–330.

Locke's importance to the *philosophes* is mentioned in all the literature on the Enlightenment; there is also a useful overview in M. Mandelbaum, *History, Man, & Reason: A Study in Nineteenth Century Thought* (Baltimore: Johns Hopkins University Press, 1971), chaps. 8 and 9. An accurate picture of Condillac's contribution has been slow to appear – even his date of birth has given problems; but see the essential reference work, J. Sgard (ed), *Corpus Condillac (1714–1780)* (Geneva: Editions Slatkine, 1981). There are translations in *An Essay on the Origin of Human Knowledge. Being a Supplement to Mr. Locke's Essay on the Human Understanding* (reprint, Gainsville, FA: Scholars' Facsimiles, 1971); *Philosophical Writings of Etienne Bonnot, abbé de Condillac*, trans. P. Franklin with the collaboration of H. Lane (Hillsdale, NJ: Lawrence Erlbaum Associates, 1982), which contains *A Treatise on the Sensations* and *Logic, Or the First Development of the Art of Thinking*. The significance of the *Essay* is stressed in H. Aarsleff, 'The Tradition of Condillac: The Problem of the Origin of Language in the Eighteenth Century and the Debate in the Berlin Academy before Herder', and 'Condillac's Speechless Statue', in *From Locke to Saussure: Essays on the Study of Language and Intellectual History* (Minneapolis: University of Minnesota Press, 1982), pp. 146–209 and 210–224. For Condillac and Lavoisier: W. R. Albury, 'The Order of Ideas: Condillac's Method of Analysis as a Political Instrument in the French Revolution', in J. A. Schuster and R. R. Yeo (eds), *The Politics and Rhetoric of Scientific Method: Historical Studies* (Dordrecht: D. Reidel, 1986), pp. 203–25. On psychology in the *Encyclopédie*, F. Vidal, 'Psychology in the 18th Century: A View from the Encyclopaedias', HHS, **6** no. 1 (1993): 89–119. Bonnet's anti-atheist project for psychology is discussed in L. Anderson, *Charles Bonnet and the Order of the Known* (Dordrecht: D. Reidel, 1982), and M. Grober, 'Harmony, Structure, and Force in the *Essai analytique sur les facultés de l'âme* of Charles Bonnet', JHBS, **31** (1995): 35–51. So-called sensationalist psychology is linked to animal instinct in R. J. Richards, 'Influence of Sensationalist Tradition on Early Theories of the Evolution of Behavior', JHI, **40** (1979): 85–105, and 'The Emergence of Evolutionary Biology of Behavior in the Early Nineteenth Century', *British Journal for the History of Science*, **15** (1982): 241–80. There is a discussion of psychology and the self in J. A. Perkins, *The Concept of the Self in the French Enlightenment* (Geneva: Droz, 1969).

The elevation of the *idéologues* to the central position in the foundation of the human sciences is the culmination of the massive studies in French by Georges Gusdorf; see the assessment in D. R. Kelley, 'Gusdorfiad', HHS, **3** (1990): 123–40; and also in the work in Italian by S. Moravia. See also: H. B. Acton, 'The Philosophy of Language in Revolutionary France', *Proceedings of the British Academy,* **45** (1959): 199–219; B. W, Head, 'The Origin of "idéologue" and "idéologie"', *Studies on Voltaire and the Eighteenth Century,* **183** (1980): 257–64, and *Ideology and Social Science: Destutt de Tracy and French Liberalism* (Dordrecht: Martinus Nijhoff, 1985); L. J. Jordanova, 'Earth Science and Environmental Medicine: The Synthesis of the Late Enlightenment', in L. J. Jordanova and R. S. Porter (eds), *Images of the Earth: Essays in the History of the Environmental Sciences* (Chalfont St Giles: British Society for the History of Science, 1979), pp. 119–46; M. S. Staum, 'Cabanis and the Science of Man', JHBS, **10** (1974): 135–43, and *Cabanis: Enlightenment and Medical Philosophy in the French Revolution* (Princeton: Princeton University Press, 1980); G. W. Stocking, Jr, 'French Anthropology in 1800', in *Race, Culture, and Evolution: Essays in the History of Anthropology* (reprint, Chicago: University of Chicago Press, 1982), pp. 13–41, on the Société des Observateurs de l'Homme. M. Foucault considered the years around 1800 as crucial for the intellectual conditions that made possible the human sciences, though his work is largely unassimilated into the mainstream of history of science: *The Order of Things: An Archaeology of the Human Sciences* (New York: Pantheon, and London: Tavistock, 1970).

G. Bryson, *Man and Society: The Scottish Inquiry of the Eighteenth Century* (reprint, New York: Augustus M. Kelley, 1968), is still a valuable introduction to the science of man; see also B. Willey, *The Eighteenth-Century Background: Studies on the Idea of Nature in the Thought of the Period* (reprint, Harmondsworth: Penguin Books, 1962), for moral sense theory. D. Daiches, P. Jones and J. Jones (eds), *A Hotbed of Genius: The Scottish Enlightenment 1730–1790* (Edinburgh: Edinburgh University Press, 1986), is a lively, illustrated overview. The moral sense theorists are reprinted in D. D. Raphael (ed), *British Moralists 1650–1800,* 2 vols (Oxford: Clarendon Press, 1969); T. Mautner (ed), *Francis Hutcheson: On Human Nature: Reflections on Our Common Systems of Morality: On the Social Nature of Man* (Cambridge: Cambridge University Press, 1993), which has a substantial essay by the editor. On Scottish moral philosophy: R. L. Emerson, 'Science and Moral Philosophy in the Scottish Enlightenment', in Stewart (ed) (1990), pp. 11–36; J. Moore, 'The Two Systems of Francis Hutcheson: On the

Origins of the Scottish Enlightenment', in Stewart (ed) (1990), pp. 37–60, and 'Hume and Hutcheson', in M. A. Stewart and J. P. Wright (eds), *Hume and Hume's Connections* (Edinburgh: Edinburgh University Press, 1994), pp. 22–57; P. B. Wood, 'The Natural History of Man in the Scottish Enlightenment', *History of Science*, **28** (1990): 89–123. Though the authors insist on calling Hutcheson a psychologist rather than a moral philosopher, G. P. Brooks and S. K. Aalto, 'The Rise and Fall of Moral Algebra: Francis Hutcheson and the Mathematization of Psychology', BJHS, **17** (1981): 343–56, is also useful. The culture of the Scottish Enlightenment is widely discussed; see N. Phillipson, 'The Scottish Enlightenment' in Porter and Teich (eds) (1981), pp. 19–40; P. B. Wood, 'Hume, Reid and the Science of Mind', in Stewart and Wright (eds) (1994), pp. 119–39, which gives a clear view of what sort of project Hume and Reid were engaged on and links it to natural history. Hume's status as a philosopher needs no emphasis. An important historical revision was made by J. Passmore, *Hume's Intentions* (2nd edn, London: Duckworth, 1968), and again, in a manner that makes human nature central, in J. P. Wright, *The Sceptical Realism of David Hume* (Manchester: Manchester University Press, 1983), esp. chap. 5. His connections with natural philosophy are detailed in M. Barfoot, 'Hume and the Culture of Science in the Early Eighteenth Century', in Stewart (ed) (1990), pp. 151–90. There is a non-contextual exposition of what Hume wrote about the association of ideas in J. Bricke, 'Hume's Associationist Psychology', JHBS, **10** (1974): 397–409.

An earlier emphasis on the association of ideas as the route by which scientific psychology was established is summarized in R. M. Young, 'Association of Ideas', in P. P. Wiener (ed), *Dictionary of the History of Ideas* (New York: Charles Scribner's Sons, 1973), Vol. 1, pp. 111–18. For incisive comment on 'association', K. Danziger, 'Generative Metaphor and the History of Psychological Discourse', in D. E. Leary (ed), *Metaphors in the History of Psychology* (Cambridge: Cambridge University Press, 1990), pp. 331–56. Hartley's book is reprinted (twice), and Priestley's condensed version is in *Hartley's Theory of the Human Mind, on the Principle of Association of Ideas; With Essays Relating to the Subject of It* (reprint, New York: AMS Press, 1973). On Hartley: S. H. Ford, '*Coalescence*: David Hartley's "*Great Apparatus*"', in

Fox (ed) (1987), pp. 199–223; R. Marsh, 'The Second Part of Hartley's System', JHI, **20** (1959): 264–73; B. B. Oberg, 'David Hartley and the Association of Ideas', JHI, **37** (1976): 441–54; C. U. M. Smith, 'David Hartley's Newtonian Neurophysiology', JHBS, **23** (1987): 123–36; and for the background of Hartley's book, M. E. Webb, 'A New History of Hartley's *Observations on Man*', JHBS, **24** (1988): 202–12. On Hartley and Erasmus Darwin, M. McNeil, *Under the Banner of Science: Erasmus Darwin and His Age* (Manchester: Manchester University Press, 1987), chap. 4; for an introduction to Lamarck, L. J. Jordanova, *Lamarck* (Oxford: Oxford University Press, 1984); and for Horne Tooke, H. Aarsleff, *The Study of Language in England, 1780–1860* (Princeton: Princeton University Press, 1967), and D. Rosenberg, '"A New Sort of Logick and Critick": Etymological Interpretation in Horne Tooke's *The Diversions of Purley*', in Burke and Porter (eds) (1991), pp. 300–29; while the background to theories of language is reviewed in the introduction to Aarsleff (1982), pp. 3–41. For Reid and the 'common sense' school of philosophy, see G. P. Brooks, 'The Faculty Psychology of Thomas Reid', JHBS, **12** (1976): 65–77; S. A. Grave, *The Scottish Philosophy of Common Sense* (London: Oxford University Press, 1960); and for the practical moral context, P. J. Diamond, 'Thomas Reid, Active Virtue and the Science of Man', *Studies on Voltaire and the Eighteenth Century*, **263** (1989): 525–30. For a significant example of Reid's influence in North America see E. T. Carlson, J. L. Wollock and P. S. Noel (eds), *Benjamin Rush's Lectures on the Mind* (Philadelphia: American Philosophical Society, 1981).

CHAPTER 8

It is somewhat artificial to divide the eighteenth-century science of man between chapters on natural law, Locke, natural and moral philosophy, sociability and political economy; this is equally the case for the allocation of references, and the other chapters should be consulted. There are overviews for the social sciences in D. Carrithers, 'The Enlightenment Science of Society', in C. Fox, R. Porter and R. Wokler (eds), *Inventing Human Science: Eighteenth-Century Domains* (Berkeley: University of California Press, 1995), pp. 232–70; R. Olson, *The Emergence of the Social Sciences, 1642–1792* (New York: Twayne Publishers,

1993), and *Science Deified & Science Defied: The Historical Significance of Science in Western Culture. Volume 2: From the Early Modern Age through the Early Romantic Era, ca. 1640 to ca. 1820* (Berkeley: University of California Press, 1990). The literature on wild children includes: J. Douthwaite, 'Rewriting the Savage: The Extraordinary Fictions of the "Wild Girl of Champagne", *Eighteenth-Century Studies*, **28** (1994–95): 163–92, an excellent study of the different stories told; H. Lane, *The Wild Boy of Aveyron* (London: George Allen & Unwin, 1977), which integrates Itard's work with the history of the deaf and dumb; L. Malson, *Wolf Children* (London: NLB, 1972), which includes a translation of Itard's *The Wild Boy of Aveyron;* A. Métraux, 'Victor de l'Aveyron and the Relativist-Essentialist Controversy', in G. Eckardt, W. G. Bringmann and L. Sprung (eds), *Contributions to a History of Developmental Psychology* (Berlin and New York: Mouton, 1985), pp. 101–15; and, a more popular study, R. Shattuck, *The Forbidden Experiment: The Story of the Wild Boy of Aveyron* (London: Secker & Warburg, 1980). For Séguin and the later development of theories of mental retardation in the US: J. W. Trent, Jr, *Inventing the Feeble Mind: A History of Mental Retardation in the United States* (Berkeley: University of California Press, 1994). For a modern 'psychological experiment': R. Rymer, *Genie: A Scientific Tragedy* (reprint, London: Penguin Books, 1994).

The long-term implications of the study of human types in relation to the apes are discussed in J. C. Greene, *The Death of Adam: Evolution and Its Impact on Western Thought* (reprint, New York: Mentor Books, 1961). Buffon's natural history and its background is discussed in P. Sloan, 'The Gaze of Natural History', in Fox, Porter and Wokler (eds) (1995), pp. 112–51; see also, W. F. Bynum, 'The Great Chain of Being after Forty Years: An Appraisal', *History of Science*, **13** (1975): 1–28. Buffon's status in the history of anthropology is reassessed in C. Blanckaert, 'Buffon and the Natural History of Man: Writing History and the "Foundational Myth" of Anthropology', HHS, **6** no. 1 (1993): 13–50. The relation between belief about apes, civilization and language, and the relation of nature and culture, is dicussed in a series of papers by R. Wokler: 'Tyson and Buffon on the Orang-utan', *Studies on Voltaire and the Eighteenth Century*, **155** (1976): 2301–19; 'Perfectible Apes in Decadent Cultures: Rousseau's Anthropology Revisited', *Daedalus*, (summer 1978):

107–34; 'Apes and Races in the Scottish Enlightenment: Monboddo and Kames on the Nature of Man', in P. Jones (ed), *Philosophy and Science in the Scottish Enlightenment* (Edinburgh: John Donald, 1988), pp. 145–68; 'From *l'homme physique* to *l'homme moral* and Back: Towards a History of Enlightenment Anthropology', HHS, **6** no. 1 (1993): 121–38. On views on language, see also R. Schreyer, '"Pray What Language Did Your Wild Couple Speak, when First They Met?" – Language and the Science of Man in the Scottish Enlightenment', in P. Jones (ed), *The 'Science of Man' in the Scottish Enlightenment: Hume, Reid and Their Contemporaries* (Edinburgh: Edinburgh University Press, 1989), pp. 149–77. For anatomy and questions of sexual difference in natural history: L. Schiebinger, *Nature's Body: Sexual Politics and the Making of Modern Science* (Boston: Beacon Press, and London: Pandora, 1993). For an introduction to the attitudes we call racist, though the study is unreliable in detail, see L. Poliakov, *The Aryan Myth: A History of Racist and Nationalist Ideas in Europe* (London; Heinemann, 1974). For the continuing impact of the Americas on European imagination: A. Pagden, *European Encounters with the New World: From Renaissance to Romanticism* (Hew Haven: Yale University Press, 1993); and, in relation to histories of civilization, R. L. Meek, *Social Science and the Ignoble Savage* (Cambridge: Cambridge University Press, 1976), esp. chap. 2. On attitudes to the unity of human nature: H. Vyverberg, *Human Nature, Cultural Diversity, and the French Enlightenment* (New York: Oxford University Press, 1989).

Rousseau's work is widely reprinted and discussed, especially in the history of political theory. The 'Discourse on the Origin of Inequality' is in J.-J. Rousseau, *The First and Second Discourses*, ed R. D. Masters (New York: St Martin's Press, 1964). This *Discourse* is related to the natural history of apes in F. Moran III, 'Between Primates and Primitives: Natural Man as the Missing Link in Rousseau's *Second Discourse*', JHI, **54** (1993): 37–58; and for a systematic discussion of the text, which argues for the importance of Rousseau's interest in the actual history of man, M. F. Plattner, *Rousseau's State of Nature: An Interpretation of the Discourse on Inequality* (DeKalb: Northern Illinois University Press, 1979). The standard biography is M. Cranston, *Jean-Jacques: The Early Life and Work of Jean-Jacques Rousseau* (London: Allen Lane, 1983), and *The Noble Savage: Jean-Jacques Rousseau*

1754–1762 (London: Allen Lane, The Penguin Press, 1991). For the claim that Rousseau's thought was systematically grounded on his belief in the natural goodness of man: A. M. Melzer, *The Natural Goodness of Man: On the System of Rousseau's Thought* (Chicago: University of Chicago Press, 1990). The larger question of man's perfectibility is discussed in J. Passmore, *The Perfectibility of Man* (London: Duckworth, 1970). On Pestalozzi and education, see – though rather anachronistic – R. B. Downs, *Heinrich Pestalozzi: Father of Modern Pedagogy* (Boston: Twayne Publishers, 1975).

On Scottish culture, in addition to the references listed under Chapter 7, see N. Phillipson, 'Smith as a Civic Moralist', in I. Hont and M. Ignatieff (eds), *Wealth and Virtue: The Shaping of Political Economy in the Scottish Enlightenment* (Cambridge: Cambridge University Press, 1983), pp. 179–202. Hume's theory of the moral sentiments has given rise to different interpretations; see N. Phillipson, 'Hume as Moralist: A Social Historian's Perspective', in S. C. Brown (ed), *Philosophers of the Enlightenment* (Brighton: Harvester Press, 1979), pp. 140–61. There is a study of Helvétius in D. W. Smith, *Helvétius: A Study in Persecution* (Oxford: Clarendon Press, 1965); and for his significance in North America, R. G. Weyant, 'Helvétius and Jefferson: Studies of Human Nature and Government in the Eighteenth Century', JHBS, **9** (1973): 29–41.

Montesquieu appears in many histories of sociology as a founding father, e.g., R. Aron, *Main Currents in Sociological Thought,* Vol. 1 (New York: Basic Books, 1965). D. Carrithers, 'Montesquieu's Philosophy of History', JHI, **47** (1986): 61–80, views Montesquieu as a founder of social science owing to his search for causal laws in the particularities of history. The new English translation of his main work has a valuable introduction: *The Spirit of the Laws,* ed A. M. Cohler, B. C. Miller and H. S. Stone (Cambridge: Cambridge University Press, 1989). See also D. R. Kelley, 'The Prehistory of Sociology: Montesquieu, Vico, and the Legal Tradition', JHBS, **16** (1980): 133–44. For views on the possibility of 'the science of woman': L. Jordanova, 'Sex and Gender', in Fox, Porter and Wokler (eds) (1995), pp. 152–83; S. Tomaselli, 'The Enlightenment Debate on Women', *History Workshop Journal,* no. 20 (1985): 101–24, which identifies woman's relationship to the civilizing process – woman as emblem of culture not of nature, and 'Reflections on the History of the Science of Woman', *History of Science*, **29** (1991): 185–205. The importance of such a science for the nineteenth century is argued in O. Moscucci, *The Science of Woman: Gynaecology and Gender in England, 1800–1929* (Cambridge: Cambridge University Press, 1990), chap. 1. For a more general view of gendered notions of reason, see G. Lloyd, *The Man of Reason: 'Male' and 'Female' in Western Philosophy* (2nd edn, London: Routledge, 1993).

For 'conjectural history', besides Wokler's articles cited above, see H. M. Höpfl, 'From Savage to Scotsman: Conjectural History in the Scottish Enlightenment', *Journal of British Studies*, **17** (1978): 19–40. The 'four stages theory' was introduced by R. L. Meek: 'Smith, Turgot, and the "Four Stages" Theory', *History of Political Economy*, **3** (1971): 9–27; see also Meek (1976), chap. 3. A. Ferguson's *Essay* is reprinted, *An Essay on the History of Civil Society*, ed D. Forbes (Edinburgh: Edinburgh University Press, 1966), and is examined in D. Kettler, *The Social and Political Thought of Adam Ferguson* (Columbus: Ohio State University Press, 1965). Turgot and Condorcet are examined as part of a substantial history of relations between natural and social philosophy in the eighteenth century, on which later histories build, in K. M. Baker, *Condorcet: From Natural Philosophy to Social Mathematics* (Chicago: University of Chicago Press, 1975).

CHAPTER 9

There are many textbook histories of economics but, as in the history of psychology, they are dominated by the retrospective interests of the modern academic speciality; see, e.g., E. Roll, *A History of Economic Thought* (5th edn, London: Faber & Faber, 1992). A readable introduction to economics as a search for order is P. Deane, *The State and the Economic System: An Introduction to the History of Political Economy* (Oxford: Oxford University Press, 1989); and for political economy's place in relation to the social sciences, R. Olson, *The Emergence of the Social Sciences, 1642–1792* (New York: Twayne Publishers, 1993). For more extended study, the journal *History of Political Economy* is essential. Political economy is linked to wider changes in systems of thought in M. Foucault, *The Order of Things: An Archaeology of the Human*

Sciences (New York: Random House, and London: Tavistock, 1970); for Foucault's later notion of governmentality, see *Power/ Knowledge. Selected Interviews and Other Writings 1972–1977*, ed C. Gordon (New York: Pantheon, 1981); G. Burchell, C. Gordon and P. Miller (eds), *The Foucault Effect: Studies on Governmentality with Two Lectures by and an Interview with Michel Foucault* (London: Harvester Wheatsheaf, 1991).

The book's discussion of seventeenth-century economic concepts is influenced by J. O. Appleby, *Economic Thought and Ideology in Seventeenth-Century England* (Princeton: Princeton University Press, 1978), which revised economists' assumptions about the novelty of Adam Smith's views, but critics suggest there is more change between the seventeenth century and the time of Smith than she allows. For the early period, but not including German-language work, see especially, T. H. Hutchinson, *Before Adam Smith: The Emergence of Political Economy, 1662–1776* (Oxford and New York: Basil Blackwell, 1988); also W. Letwin, *The Origins of Scientific Economics: English Economic Thought 1660–1776* (London: Methuen, 1963). Petty and Graunt are reprinted in *The Earliest Classics*, intro. P. Laslett (Farnborough: Gregg International, 1973); see also J. Mykkänen, '"To Methodize and Regulate Them": William Petty's Governmental Science of Statistics', HHS, **7** no. 3 (1994): 65–88. The development of numerical techniques and probabilistic forms of argument are discussed in I. Hacking, *The Emergence of Probability: A Philosophical Study of Early Ideas about Probability, Induction and Statistical Inference* (London: Cambridge University Press, 1975), esp. chap. 12, 'Political Arithmetic'. The unreliability of King's figures as historical information is detailed in G. S. Holmes, 'Gregory King and the Social Structure of Pre-industrial England', *Transactions of the Royal Historical Society*, 5th series **27** (1977): 41–68. On the founding and activity of the French Academy of Sciences, R. Hahn, *The Anatomy of a Scientific Institution: The Paris Academy of Sciences, 1666–1803* (Berkeley: University of California Press, 1971), chap. 1. For the cameral sciences: K. Tribe, 'Cameralism and the Science of Government', *Journal of Modern History*, **56** (1984): 263–84, and, for the later science of government, *Governing Economy: The Reformation of German Economic Discourse, 1750–1840* (Cambridge: Cambridge University Press, 1988); and for the Swedish case, K. Johannisson,

'Society in Numbers: The Debate over Quantification in 18th-Century Political Economy', in T. Frangsmyr, J. L. Heilbron and R. E. Rider (eds), *The Quantifying Spirit in the 18th Century* (Berkeley: University of California Press, 1990), pp. 343–61, and S.-E. Liedman, 'Anders Berch: A Frontrunner', in L. Engwall and E. Gunnarsson (eds), *Management Studies in an Academic Context* (Uppsala: Acta Universitatus Upsaliensis. Studia Oeconomiae Negotiorum, 35, 1994), pp. 33–44. For continental European theorists of statistics: L. Daston, *Classical Probability in the Enlightenment* (Princeton: Princeton University Press, 1988), chap. 9.

The standard study of Mandeville is M. M. Goldsmith, *Private Vices, Public Benefits: Bernard Mandeville's Social and Political Thought* (Cambridge: Cambridge University Press, 1985). There is a survey of the evolution of political economy in the Enlightenment in S. Tomaselli, 'Political Economy: The Desire and Needs of Present and Future Generations', in C. Fox, R. Porter and R. Wokler (eds), *Inventing Human Science: Eighteenth-Century Domains* (Berkeley: University of California Press, 1995), pp. 292–322. The physiocrats are studied in E. Fox-Genovese, *The Origins of Physiocracy: Economic Revolution and Social Order in Eighteenth-Century France* (Ithaca: Cornell University Press, 1976), which attributes the shift to 'economics' as a conceptual area to Quesnay's study of wealth generation; R. L. Meek, *The Economics of Physiocracy: Essays and Translations* (Cambridge, MA: Harvard University Press, 1963), which introduces primary sources, and R. L. Meek (ed), *Precursors of Adam Smith* (London: Dent, 1973). The 'civic humanist' tradition, which has been seen as part of Smith's context, is set out in J. G. A. Pocock, *The Machiavellian Moment: Florentine Political Thought and the Atlantic Republican Tradition* (Princeton: Princeton University Press, 1975). The historical relations between selfish values and commercial society are assayed in A. O. Hirschman, *The Passions and the Interests: Political Arguments for Capitalism Before its Triumph* (Princeton: Princeton University Press, 1977), which suggests that capitalist values were advocated as a way to avoid ruin brought about by the passions (reversing Weber's argument about the Protestant ethic). D. D. Raphael, *Adam Smith* (Oxford: Oxford University Press, 1985) is a succinct introduction, and J. R. Lindgren, *The Social Philosophy of Adam Smith* (The Hague: Martinus Nijhoff,

1973), a more extended study that shows the range of Smith's interests (though it still ignores jurisprudence and language theory). There is a discussion of the difficulties in the interpretation of Smith in D. Winch, 'Adam Smith's "Enduring Particular Result": A Political and Cosmopolitan Perspective', in I. Hont and M. Ignatieff (eds), *Wealth and Virtue: The Shaping of Political Economy in the Scottish Enlightenment* (Cambridge: Cambridge University Press, 1983), pp. 253–69. There is a useful discussion in A. S. Skinner, 'The Shaping of Political Economy in the Enlightenment', in H. Mizuta and C. Sugiyama (eds), *Adam Smith: International Perspectives* (London: Macmillan, and New York: St Martin's Press, 1993), pp. 113–39; see also A. S. Skinner, *A System of Social Science: Papers Relating to Adam Smith* (Oxford: Clarendon Press, 1979), and 'Adam Smith: Ethics and Self-love', in P. Jones and A. S. Skinner (eds), *Adam Smith Reviewed* (Edinburgh; Edinburgh University Press, 1992), pp. 142–67.

The classic study of psychology and utilitarian thought is E. Halévy, *The Growth of Philosophic Radicalism* (London: Faber & Faber, 1952). Beccaria's most famous essay is translated in *Alessandro Manzoni, The Column of Infamy; Prefaced by Cesare Beccaria's Of Crimes and Punishments*, intro. A. P. d'Entrèves (London: Oxford University Press, 1964), pp. 1–96. J. Dinwiddy, *Bentham* (Oxford: Oxford University Press, 1989) is a brief introduction. For psychological views of motivation in utilitarian thought: P. McReynolds, 'The Motivational Psychology of Jeremy Bentham', JHBS, **4** (1968): 230–44 and 349–64; T. Mischel, '"Emotion" and "Motivation" in the Development of English Psychology: D. Hartley, James Mill, A. Bain', JHBS, **2** (1966): 123–44. For Bentham and display in natural philosophy (including the display of his own body): S. Schaffer, 'States of Mind: Enlightenment and Natural Philosophy', in G. S. Rousseau (ed), *The Languages of Psyche: Mind and Body in Enlightenment Thought* (Berkeley: University of California Press, 1990), pp. 233–90. Wedgwood's work on factory discipline is discussed in N. McKendrick, 'Josiah Wedgwood and Factory Discipline', *Historical Journal*, **4** (1961): 30–55, and on a larger canvas in E. P. Thompson,'Time, Work-discipline, and Industrial Capitalism', *Past & Present*, no. 38 (1967): 56–97. The differences between the first and later editions of Malthus's *Essay on the*

Principle of Population need to be noted in using the various reprints. For his life and work: P. James, *Population Malthus: His Life and Times* (London: Routledge & Kegan Paul, 1979), and W. Peterson, *Malthus* (Cambridge, MA: Harvard University Press, 1979). There is discussion of the eighteenth-century background of population debates in C. J. Glacken, *Traces on the Rhodian Shore: Nature and Culture in Western Thought from Ancient Times to the End of the Eighteenth Century* (Berkeley: University of California Press, 1967), chap. 13; S. Tomaselli, 'Moral Philosophy and Population Questions in Eighteenth Century Europe', in M. S. Teitelbaum and J. M. Winter (eds), *Population and Resources in Western Intellectual Traditions,* in *Population and Development Review,* supplement to **14** (1988): 7–29. Malthus's use of 'facts' is criticized in S. Rashid, 'Malthus's *Essay on Population*: The Facts of "Super-growth" and the Rhetoric of Scientific Persuasion', JHBS, **23** (1987): 22–36. Malthus is linked to the later Darwin debates in an influential paper by R. M. Young, 'Malthus and the Evolutionists: The Common Context of Biological and Social Theory', *Past & Present,* no. 43 (1969): 109–45, reprinted in *Darwin's Metaphor: Nature's Place in Victorian Culture* (Cambridge: Cambridge University Press, 1985), pp. 23–55. For the reaction against the ahistorical psychology of utilitarian thought, exemplified by J. Mill, see J. W. Burrow, *Evolution and Society: A Study in Victorian Social Theory* (reprint, Cambridge: Cambridge University Press, 1970). There is no agreed position about Ricardo's economics; see T. Peach, *Interpreting Ricardo* (Cambridge: Cambridge University Press, 1993).

CHAPTER 10

The 'expressivist' reaction to eighteenth-century views of human nature is discussed in C. Taylor, *Hegel* (Cambridge: Cambridge University Press, 1975), chap. 1. In the literature on Romanticism, M. H. Abrams, *The Mirror and the Lamp: Romantic Theory and the Critical Tradition* (London: Oxford University Press, 1953), discusses changing metaphors of the mind, and *Natural Supernaturalism: Tradition and Revolution in Romantic Literature* (New York: W. W. Norton, 1971), argues for poetry as a secular displacement of a theological vision. See also R. Porter and M. Teich (eds), *Romanticism in National Context* (Cambridge: Cambridge University Press, 1988); C. Taylor, *Sources of the Self: The Making of the Modern Identity* (Cambridge: Cambridge University Press, 1989), Part 4. Vico studies have flourished since the late 1960s; G. Vico, *The New Science of Giambattista Vico*, ed T. G. Bergin and M. H. Fisch (reprint, Ithaca: Cornell University Press, 1984), includes a useful synopsis. There is a brief introduction in P. Burke, *Vico* (Oxford: Oxford University Press, 1985), and a meticulous analysis of the text as science in L. Pompa, *Vico: A Study of the 'New Science'* (2nd edn, Cambridge: Cambridge University Press, 1990). Vico and Herder are linked in relation to their construction of a historical consciousness in I. Berlin, *Vico and Herder: Two Studies in the History of Ideas* (London: Hogarth Press, 1976). For Vico and jurisprudence, see D. R. Kelley, *The Human Measure: Social Thought in the Western Legal Tradition* (Cambridge, MA: Harvard University Press, 1990), pp. 234–9; and for the background of theories of earth history, P. Rossi, *The Dark Abyss of Time: The History of the Earth & the History of Nations from Hooke to Vico* (Chicago: University of Chicago Press, 1984).

Herder has not been properly translated into English, but see the abridged edn, *Reflections on the Philosophy of the History of Mankind*, ed F. E. Manuel (Chicago: University of Chicago Press, 1968); and there is an exposition in G. A. Wells, *Herder and After: A Study in the Development of Sociology* (The Hague: Mouton, 1959). Relevant aspects of the background are discussed in H. Aarsleff, 'The Tradition of Condillac: The Problem of the Origin of Language in the Eighteenth Century and the Debate in the Berlin Academy before Herder', in *From Locke to Saussure: Essays on the Study of Language and Intellectual History* (Minneapolis: University of Minnesota Press, 1982), pp. 146–209; F. C. Beiser, *The Fate of Reason: German Philosophy from Kant to Fichte* (Cambridge, MA: Harvard University Press, 1987), chap. 5; K. J. Fink, 'Storm and Stress Anthropology', HHS, **6** no. 1 (1993): 51–71. The ideals of the university of Berlin are discussed in E. S. Shaffer, 'Romantic Philosophy and the Organization of Disciplines: The Founding of the Humboldt University of Berlin', in A. Cunningham and N. Jardine (eds), *Romanticism and the Sciences* (Cambridge: Cambridge University Press, 1990), pp.38–54. For historical jurisprudence in Berlin and Göttingen, Kelley (1990), chap. 13; and for the general conception of legal development, the intellectual setting for the work of Savigny, P. Stein, *Legal Evolution: The Story of an Idea* (Cambridge: Cambridge University Press, 1980).

There is a valuable assessment of the importance of freedom to human dignity in the work of Kant and Schiller in R. D. Miller, *Schiller and the Ideal of Freedom: A Study of Schiller's Philosophical Works with Chapters on Kant* (reprint, Oxford: Clarendon Press, 1970). It is possible to grasp something of the nature of Hegel's project as described in Taylor (1975), chaps. 1–5. An accessible introduction to German idealism, which emphasizes the significance of self-representation, is R. C. Solomon, *Continental Philosophy since 1750: The Rise and Fall of the Self* (Oxford: Oxford University Press,1988), and for more detail, *History and Human Nature: A Philosophical Review of European Philosophy and Culture, 1750–1850* (Brighton: Harvester Press, 1980). For the Romantic tension between the assertion of the individual self and the subsumption of the self under a higher unity, see G. N. Izenberg, *Impossible Individuality: Romanticism, Revolution, and the Origins of Modern Selfhood, 1787–1802* (Princeton: Princeton University Press, 1992). Herbart's main psychological work does not exist in English translation, while what does exist is none too reliable. D. E. Leary clarifies the nature of Herbartian psychology, though without reference to the science of the state or to education in: 'The Philosophical Development of the Conception of Psychology in Germany, 1780–1850', JHBS, **14** (1978): 113–21; 'The Historical Foundation of Herbart's Mathematization of Psychology', JHBS, **16** (1980): 150–63; 'German Idealism and the Development of Psychology in the Nineteenth Century', *Journal of the History of Philosophy*, **18** (1980): 299–317. Herbart's work and its relation to educational theory is studied in H. B. Dunkel, *Herbart and Herbartianism: An Educational Ghost Story* (Chicago: University of Chicago Press, 1970).

The classic essays (1838 and 1840) by J. S. Mill, which contrast Bentham's utilitarianism and Coleridge's idealism, retain their sharpness: *On Bentham and Coleridge* (reprint, New York: Harper & Row, 1962); the contrast was also personally significant to Mill: *Autobiography of John Stuart Mill* (new edn, New York: Columbia University Press, 1960). For a commentary: B. Willey, *Nineteenth Century Studies* (reprint, Harmondsworth: Penguin Books, 1964). On Hartley and Coleridge, R. Haven, 'Coleridge,

Hartley, and the Mystics', JHI, **20** (1959): 477–94; and for the context of Coleridge's reaction against Horne Tooke's language theory, and further references, D. Rosenberg, '"A New Sort of Logick and Critick": Etymological Interpretation in Horne Tooke's *The Diversions of Purley*', in P. Burke and R. Porter (eds), *Language, Self, and Society: A Social History of Language* (Oxford: Polity Press, 1991), pp. 300–29. Maine de Biran's work is not well known in the English-speaking world, though his diary has a reputation in French (in the definitive 1954–7 edn). The only modern full-length studies in English are P. P. Hallie, *Maine de Biran: Reformer of Empiricism, 1766–1824* (Cambridge, MA: Harvard University Press, 1959), and F. C. T. Moore, *The Psychology of Maine de Biran* (Oxford: Clarendon Press, 1970). There is a discussion of the self in relation to Cousin's influence in France in J. Goldstein, 'Foucault and the Post-Revolutionary Self: The Uses of Cousinian Psychology in Nineteenth-Century France', in J. Goldstein (ed), *Foucault and the Writing of History* (Oxford: Blackwell, 1994), pp. 99–115; see also D. S. Goldstein, '"Official Philosophies" in Modern France: The Example of Victor Cousin', *Journal of Social History*, **1** (1968): 259–79. The diary method in pedagogy is discussed in S. Jaeger, 'The Origin of the Diary Method in Developmental Psychology', in G. Eckardt, W. G. Bringmann and L. Sprung (eds), *Contributions to a History of Developmental Psychology* (Berlin and New York: Mouton, 1985), pp. 63–74.

CHAPTER 11

The eighteenth-century background to the transformation of the German university system is referenced in chapter 6. The extraordinary survey by J. T. Merz, *A History of European Thought in the Nineteenth Century*, 4 vols (1904–12; reprint, New York: Dover Books, 1964), is full of suggestive information about academic culture. The interpretation by F. K. Ringer, *The Decline of the German Mandarins: The German Academic Community, 1890–1933* (reprint, Hanover, NH: Wesleyan University Press/University Press of New England, 1990), has been influential and, despite the title, discusses reasons for the early nineteenth-century university ideals. For the origins of the research seminar and PhD degree, see W. Clarke, 'On the Dialectical Origins of the Research Seminar', *History of Science*, **27** (1989): 111–54, and 'On the Ironic Specimen of the Doctor of Philosophy', *Science in Context*, **5** (1992): 97–137. For relevant detail on North American college education: B. J. Bledstein, *The Culture of Professionalism: The Middle Class and the Development of Higher Education in America* (New York: W. W. Norton, 1976), and R. B. Evans, 'The Origins of American Academic Psychology', in J. Brožek (ed), *Explorations in the History of Psychology in the United States* (Lewisburg: Bucknell University Press, 1984), pp. 17–60. More generally, see M. Curti, *Human Nature in American Thought: A History* (Madison: University of Wisconsin Press, 1980), and R. Nisbet, *History of the Idea of Progress* (London: Heinemann, 1980). For a broad, informative survey of philosophical thought, which covers historical and evolutionary concepts, M. Mandelbaum, *History, Man, & Reason: A Study in Nineteenth-Century Thought* (Baltimore: Johns Hopkins University Press, 1971). A useful study of changes in Victorian intellectual culture, with attention to history and evolution, is T. W. Heyck, *The Transformation of Intellectual Life in Victorian England* (London: Croom Helm, 1982).

The history of theories of language is a complex domain as it covers both historical and comparative studies of languages and attempts directly to analyse language (and such huge topics as language's relation to reason). An informed reference work is E. F. K. Koerner and R. E. Asher (eds), *Concise History of the Language Sciences: From the Sumerians to the Cognitivists* (Oxford and New York: Pergamon, 1995); see also the articles by A. Ellegård, 'Study of Language', and H. M. Hoenigswald, 'Linguistics', in P. P. Wiener (ed), *Dictionary of the History of Ideas: Studies of Selected Pivotal Ideas* (New York: Charles Scribner's Sons, 1973), respectively in Vol. 2, pp. 659–73, and Vol. 3, pp. 61–73. There is a history of some topics in R. Harris and T. J. Talbot, *Landmarks in Linguistic Thought: The Western Tradition from Socrates to Saussure* (London: Routledge, 1989), and in R. H. Robins, *A Short History of Linguistics* (2nd edn, London: Longman, 1979). D. Hymes (ed), *Studies in the History of Linguistics: Traditions and Paradigms* (Bloomington: Indiana University Press, 1974), and the founding by E. F. K. Koerner of *Historiographia Linguistica: International Journal for the History of Linguistics*, also in 1974, marked a new historical seriousness in the field, in which H. Aarsleff, *From Locke to Saussure: Essays on the Study of Language and Intellec-*

tual History (Minneapolis: University of Minnesota Press, 1982) is a standard scholarly source and includes a reassessment of Humboldt's contribution. See also Aarsleff's introduction to W. von Humboldt, *On Language: The Diversity of Human Language-structure and Its Influence on the Mental Development of Mankind* (Cambridge: Cambridge University Press, 1988). For the nineteenth century: O. Amsterdamska, *Schools of Thought: The Development of Linguistics from Bopp to Saussure* (Dordrecht: D. Reidel, 1987). On Rask, see P. Diderichsen, 'The Foundation of Comparative Linguistics: Revolution or Continuation', in Hymes (ed) (1974), pp. 277–30; on Müller, E. Knoll, 'The Science of Language and the Evolution of Mind: Max Müller's Quarrel with Darwinism', JHBS, **22** (1986): 3–22, and N. C. Chaudhuri, *Scholar Extraordinary: The Life of Professor the Rt. Hon. Friedrich Max Müller, P.C.* (London: Chatto & Windus, 1974), a general biography; on Schleicher, H. Aarsleff, 'Brèal vs. Schleicher: Reorientation in Linguistics during the Latter Half of the Nineteenth Century', in (1982), pp. 293–334, and L. Taub, 'Evolutionary Ideas and "Empirical" Methods: The Analogy between Language and Species in the Works by Lyell and Schleicher', *British Journal for the History of Science*, **26** (1993): 171–93. There is an interpretation of eighteenth- and nineteenth-century theories of the origins of languages in J. H. Stam, *Inquiries into the Origin of Language: The Fate of a Question* (New York: Harper & Row, 1976). Late nineteenth-century language theory is linked to the psycholinguistics of the 1950s in A. L. Blumenthal, *Language and Psychology: Historical Aspects of Psycholinguistics* (New York: John Wiley & Sons, 1970). There are stimulating accounts of the impact of critical scholarship on the literate public in O. Chadwick, *The Secularization of the European Mind in the Nineteenth Century* (Cambridge: Cambridge University Press, 1975); B. Willey, *Nineteenth Century Studies* (reprint, Harmondsworth: Penguin Books, 1964); J. W. Burrow, 'The Uses of Philology in Victorian Britain', in R. Robson (ed), *Ideas and Institutions of Victorian Britain* (London: G. Bell & Sons, 1967), pp. 180–204. For the historian, Ranke: G. G. Iggers, *The German Conception of History: The National Tradition of Historical Thought from Herder to the Present* (Middletown, CT: Wesleyan University Press, 1968), chap. 4; G. G. Iggers and J. M. Powell (eds), *Leopold von Ranke and the Shaping of the Historical Discipline* (Syracuse: Syracuse University Press, 1990).

History and geology are linked in M. J. S. Rudwick, 'Transposed Concepts from the Human Sciences in the Early Work of Charles Lyell', in L. J. Jordanova and R. S. Porter (eds), *Images of the Earth: Essays in the History of the Environmental Sciences* (Chalfont St Giles: British Society for the History of Science, 1978), pp. 67–83, while the pictorial imagery of prehistory is examined in M. J. S. Rudwick, *Scenes from Deep Time: Early Pictorial Representations of the Prehistoric World* (Chicago: University of Chicago Press, 1992).

There is a general history of archaeology, related to the values of archaeologists, in B. G. Trigger, *A History of Archaeological Thought* (Cambridge: Cambridge University Press, 1989); see also the earlier standard history by G. Daniel, *A Hundred and Fifty Years of Archaeology* (2nd edn, London: Duckworth, 1975), and, in an illustrated version, *A Short History of Archaeology* (London: Thames & Hudson, 1981). The historiography of archaeology is reviewed in B. Trigger, 'The Coming of Age of the History of Archaeology', *Journal of Archaeological Research*, **2** (1994): 113–36. The integration of archaeology, geology, anthropology and evolutionary thought is discussed in J. W. Burrow, *Evolution and Society: A Study in Victorian Social Theory* (reprint, Cambridge: Cambridge University Press, 1970); W. F. Bynum, 'Charles Lyell's Antiquity of Man and Its Critics', *Journal of the History of Biology*, **17** (1984): 153–87; J. W. Gruber, 'Brixham Cave and the Antiquity of Man', in M. E. Spiro (ed), *Context and Meaning in Cultural Anthropology* (New York: Free Press, and London: Collier-Macmillan, 1965), pp. 373–402. See also P. J. Bowler, *The Invention of Progress: The Victorians and the Past* (Oxford and Cambridge, MA: Basil Blackwell, 1989). For T. H. Huxley's contribution to the anatomical evidence about man's relations, see M. A. di Gregorio, *T. H. Huxley's Place in Natural Science* (New Haven: Yale University Press, 1984), Part 2, and on the history of discoveries of fossil humans: P. Bowler, *Theories of Human Evolution: A Century of Debate, 1844–1944* (Baltimore: Johns Hopkins University Press, 1986, and Oxford: Basil Blackwell, 1987); J. Lyon, 'The Search for Fossil Man: Cinq personnages à la recherche du temps perdu', *Isis*, **61** (1969): 68–84. The place of archaeology and history in the separation of experts and the public is discussed in P. Levine, *The Amateur and the Professional: Antiquarians, Historians and Archaeologists in Victorian England, 1838–1886* (Cambridge: Cambridge University Press, 1986).

The major study of English-language anthropology is G. W. Stocking, Jr, *Victorian Anthropology* (New York: Free Press, 1987). See also Burrow (1970), and G. W. Stocking, Jr, *Race, Culture, and Evolution: Essays in the History of Anthropology* (reprint, Chicago: University of Chicago Press, 1982). For the US, C. M. Hinsley, Jr, *Savages and Scientists: The Smithsonian Institution and the Development of American Anthropology 1846–1910* (Washington, DC: Smithsonian Institution Press, 1981). The medical background to French anthropology is considered in E. A. Williams, *The Physical and the Moral: Anthropology, Physiology and Philosophical Medicine in France, 1750–1850* (Cambridge: Cambridge University Press, 1994). There is a review of the relation between the physical and cultural domains of anthropology in G. W. Stocking, Jr, 'Paradigmatic Traditions in the History of Anthropology', in R. C. Olby *et al.* (eds), *Companion to the History of Modern Science* (London: Routledge, 1990), pp. 712–27; and a critical review of the historiography in I. C. Jarvie, 'Recent Work in the History of Anthropology and Its Historiographic Problems', *Philosophy of the Social Sciences*, **19** (1989): 345–75. There are introductions to the emotive area of the classification of human differences in P. Bowler, *The Fontana History of the Environmental Sciences* (London: Fontana Press, 1992), chap. 10; G. Weber, 'Science and Society in Nineteenth-Century Anthropology', *History of Science*, **12** (1974): 260–83. J. C. Prichard's work is discussed at length and reprinted in G. W. Stocking, Jr (ed), *Researches into the Physical History of Man* (Chicago: University of Chicago Press, 1973). In addition to Stocking (1987), relations in Britain between ethnology and anthropology are discussed in: J. W. Burrow, 'Evolution and Anthropology in the 1860's: The Anthropological Society of London', *Victorian Studies*, **7** (1963): 137–54; R. Rainger, 'Race, Politics, and Science: The Anthropological Society of London in the 1860s', *Victorian Studies*, **22** (1978): 51–70; G. W. Stocking, Jr, 'What's in a Name? The Origins of the Royal Anthropological Institute (1837–71)', *Man*, new series **6** (1971): 369–90.

The scientific pretensions of nineteenth-century racial descriptions and craniometry are demolished from a modern perspective in S. J. Gould, *The Mismeasure of Man* (reprint, Harmondsworth: Penguin Books, 1984). The different meanings of race are traced in a historical perspective in M. Banton, *Racial*

Theories (Cambridge: Cambridge University Press, 1987). Racial theories are also discussed in: M. Olender, *The Language of Paradise: Race, Religion, and Philology in the Nineteenth Century* (Cambridge, MA: Harvard University Press, 1992); L. Poliakov, *The Aryan Myth: A History of Racist and Nationalist Ideas in Europe* (London: Heinemann, 1974); N. Stepan, *The Idea of Race in Science: Great Britain 1800–1960* (London: Macmillan 1982); and on images of monstrosity and race, S. Bann (ed), *Frankenstein, Creation and Monstrosity* (London: Reaktion Books, 1994), and H. L. Malchow, 'Frankenstein's Monster and Images of Race in Nineteenth-Century Britain', *Past & Present*, no. 139 (1993): 90–130. On the US craniologists: W. Stanton, *The Leopard's Spots: Scientific Attitudes toward Race in America 1815–1859* (Chicago: University of Chicago Press, 1960); J. S. Haller, *Outcasts from Evolution: Scientific Attitudes of Racial Inferiority, 1859–1900* (Urbana: University of Illinois Press, 1971). On Knox: E. Richards, 'The "Moral Anatomy" of Robert Knox: The Interplay between Biological and Social Thought in Victorian Scientific Naturalism', *Journal of the History of Biology*, **22** (1989): 373–436. For the development of racial typologies in France: C. Blanckaert, 'On the Origins of French Ethnology: William Edwards and the Doctrine of Race', in G. W. Stocking, Jr (ed), *Bones, Bodies, Behavior: Essays on Biological Anthropology, History of Anthropology, Volume 3* (Madison: University of Wisconsin Press, 1988), pp. 18–55; on Broca: J. Harvey, 'Evolutionism Transformed: Positivists and Materialists in the *Société d'anthropologie de Paris* from Second Empire to Third Republic', in D. Oldroyd and I. Langham (eds), *The Wider Domain of Evolutionary Thought* (Dordrecht: D. Reidel, 1983), pp. 289–310, F. Schiller, *Paul Broca: Founder of French Anthropology, Explorer of the Brain* (Berkeley: University of California Press, 1979), chaps. 8 and 9, and E. A. Williams, 'Anthropological Institutions in Nineteenth-Century France', *Isis*, **76** (1985): 331–48; on Gobineau, M. Biddiss, *Father of Racist Ideology: The Social and Political Thought of Count Gobineau* (London: Weidenfeld & Nicolson, 1970). The social history of belief about heredity, race and national hygiene in Germany is examined in P. Weindling, *Health, Race and German Politics between National Unification and Nazism, 1870–1945* (Cambridge: Cambridge University Press, 1989), and the connection between German anthropology and Nazism in, R. Proctor, 'From *Anthro-*

pologie to *Rassenkunde* in the German Anthropological Tradition', in Stocking (ed) (1988), pp. 138–79. For physical anthropology and gender: E. Fee, 'The Sexual Politics of Victorian Anthropology', in M. S. Hartman and L. W. Banner (eds), *Clio's Consciousness Raised: New Perspectives on the History of Women* (New York: Harper & Row, 1974), pp. 86–102, and 'Nineteenth-Century Craniology: The Study of the Female Skull', *Bulletin of the History of Medicine*, **53** (1979): 415–33; C. E. Russett, *Sexual Science: The Victorian Construction of Womanhood* (Cambridge, MA: Harvard University Press, 1989); N. L. Stepan, 'Race and Gender: The Role of Analogy in Science', *Isis*, **77** (1986): 262–77; and these issues are placed in a wider medical setting in L. Jordanova, *Sexual Visions: Images of Gender in Science and Medicine between the Eighteenth and Twentieth Centuries* (New York: Harvester Wheatsheaf, 1989). German ethnology, anthropology and *Völkerpsychologie* are discussed in K. Danziger, 'Origins and Basic Principles of Wundt's *Völkerpsychologie*', *British Journal of Social Psychology*, **22** (1983): 303–13; K.-P. Koepping, *Adolf Bastian and the Psychic Unity of Mankind: The Foundations of Anthropology in Nineteenth Century Germany* (St Lucia: Queensland University Press, 1983); M. Mackert, 'The Roots of Franz Boas' View of Linguistic Categories as a Window to the Human Mind', *Historiographia Linguistica*, **20** (1993): 331–51; W. D. Smith, *Politics and the Sciences of Culture in Germany 1840–1920* (New York: Oxford University Press, 1991), though this account of the political context has been challenged; J. Whitman, 'From Philology to Anthropology in Mid-Nineteenth-Century Germany', in G. W. Stocking, Jr (ed), *Functionalism Historicized: Essays on British Social Anthropology, History of Anthropology, Volume 2* (Madison: University of Wisconsin Press, 1984), pp. 214–29.

The significance of links between physiology and the human sciences in the early nineteenth century is considered in K. Figlio, 'Theories of Perception and the Physiology of Mind in the Late Eighteenth Century', *History of Science*, **12** (1975): 177–212, and 'The Metaphor of Organization: An Historiographical Perspective on the Bio-medical Sciences of the Early Nineteenth Century', *History of Science*, **14** (1976): 17–53. Idealism and materialism in German debates are discussed in G. Verwey, *Psychiatry in an Anthropological and Biomedical Context: Philosophical Presuppositions and Implications of German Psychiatry, 1820–1870*

(Dordrecht: D. Reidel, 1984), chap. 2. There is now a large literature on phrenology, which includes: R. Cooter, *The Cultural Meaning of Popular Science: Phrenology and the Organization of Consent in Ninteenth-Century Britain* (Cambridge: Cambridge University Press, 1984); D. de Giustino, *Conquest of Mind: Phrenology and Victorian Social Thought* (London: Croom Helm, and Totowa, NJ: Rowman & Littlefield, 1975); V. Hilts, 'Obeying the Laws of Hereditary Descent: Phrenological Views on Inheritance and Eugenics', JHBS, **18** (1982): 62–77; R. M. Young, *Mind, Brain, and Adaptation: Cerebral Localization and Its Biological Context from Gall to Ferrier* (Oxford: Clarendon Press, 1970). On Mesmer, medicine and politics: R. Darnton, *Mesmerism and the End of the Enlightenment in France* (reprint, New York: Schocken Books, 1970). For the general background of the science of the brain: E. Clarke and L. S. Jacyna, *Nineteenth-Century Origins of Neuroscientific Concepts* (Berkeley: University of California Press, 1987). The physiological approach to mind in Britain is discussed in K. Danziger, 'Mid-Nineteenth Century British Psychophysiology: A Neglected Chapter in the History of Psychology', in W. R. Woodward and M. G. Ash (eds), *The Problematic Science: Psychology in Nineteenth-Century Thought* (New York: Praeger, 1982), pp. 119–46; S. Jacyna, 'The Physiology of Mind, the Unity of Nature, and the Moral Order in Victorian Thought', *British Journal for the History of Science*, **14** (1981), 109–32, and 'Somatic Theories of Mind and the Interests of Medicine in Britain, 1850–1879', *Medical History*, **26** (1982): 233–58. For wider views: R. Smith, 'The Background of Physiological Psychology in Natural Philosophy', *History of Science*, **11** (1973): 75–123, and *Inhibition: History and Meaning in the Sciences of Mind and Brain* (Berkeley: University of California Press, and London: Free Association Books, 1992). The relation between empiricism and J. S. Mill's psychology is discussed (also with reference to political economy) in F. Wilson, *Psychological Analysis and the Philosophy of John Stuart Mill* (Toronto: Toronto University Press, 1990); and the limited impact of Mill's suggestions about what form psychology might take, in D. E. Leary, 'The Fate and Influence of John Stuart Mill's Proposed Science of Ethology', JHI, **43** (1982): 153–62. Bain's contribution is discussed in A. P. Greenway, 'The Incorporation of Action into Associationism: The Psychology of Alexander Bain', JHBS, **9** (1973): 42–52;

L. S. Hearnshaw, *A Short History of British Psychology, 1840–1940* (London: Methuen, 1964); Young (1970), chap. 3. For the debates about German materialism, see F. Gregory, *Scientific Materialism in Nineteenth Century Germany* (Dordrecht: D. Reidel, 1977); A. Kelly, *Descent of Darwin: The Popularization of Darwinism in Germany, 1860–1914* (Chapel Hill: University of North Carolina Press, 1981). And for the highly politicized Russian public debates on physiology: D. Joravsky, *Russian Psychology: A Critical History* (Oxford: Basil Blackwell, 1989); Smith (1992), chap. 3, for Sechenov; D. P. Todes, 'Biological Psychology and the Tsarist Censor: The Dilemma of Scientific Development', *Bulletin of the History of Medicine*, **58** (1984): 529–44; A. Vucinich, *Science in Russian Culture 1861–1917* (Stanford: Stanford University Press, 1970); and for the Russian view itself, M. G. Iaroshevskii [Yaroshevsky], 'The Logic of Scientific Development and the Scientific School: The Example of Ivan Mikhailovich Sechenov', in Woodward and Ash (eds) (1982), pp. 231–54, and *Ivan Sechenov* (Moscow: Mir, 1986). The question of the mind–body relation and responsibility is considered in R. Smith, *Trial by Medicine: Insanity and Responsibility in Victorian Trials* (Edinburgh: Edinburgh University Press, 1981), and the science of left–right differences in the brain in A. Harrington, *Medicine, Mind, and the Double Brain: A Study in Nineteenth-Century Thought* (Princeton: Princeton University Press, 1987).

CHAPTER 12

The standard study of Condorcet, the hopes and fears of 1789 and the conceptualization of social science is K. M. Baker, *Condorcet: From Natural Philosophy to Social Mathematics* (Chicago: University of Chicago Press, 1975). The origins and use of terms are clarified in K. M. Baker, 'The Early History of the Term "Social Science"', *Annals of Science*, **20** (1964): 211–26; B. W. Head. 'The Origins of "la science sociale" in France, 1770–1800', *Australian Journal of French Studies*, **19** (1982): 115–32, and 'The Origin of "idéologue" and "idéologie"', *Studies on Voltaire and the Eighteenth Century*, **183** (1980): 257–64. There is a large-scale study of Saint-Simon's confusing career and thoughts in F. E. Manuel, *The New World of Henri Saint-Simon* (Cambridge, MA: Harvard University Press, 1956), while Saint-Simon, his followers and

Comte are considered in a larger time frame in *The Prophets of Paris* (Cambridge, MA: Harvard University Press, 1962). Saint-Simon and Comte are linked to the sciences of life, and especially Bichat's physiology, in B. Haines, 'The Inter-relations between Social, Biological, and Medical Thought, 1750–1850: Saint-Simon and Comte', *British Journal for the History of Science*, **11** (1978): 19–35; J. V. Pickstone, 'Bureaucracy, Liberalism and the Body in Post-revolutionary France: Bichat's Physiology and the Paris School of Medicine', *History of Science*, **19** (1981): 115–42.

M. Pickering, *Auguste Comte: An Intellectual Biography, Volume 1* (Cambridge: Cambridge University Press, 1993) is a detailed biography that covers relations with Saint-Simon. As a 'founding father' of sociology, Comte features in all the histories of the field; among these, L. A. Coser, *Masters of Sociological Thought: Ideas in Historical and Social Context* (New York: Harcourt Brace Jovanovich, 1971) is clear and historical. For J. S. Mill's critical advocacy, see his study, *Auguste Comte and Positivism* (1865; reprint, Ann Arbor: University of Michigan Press, 1961); also, B. Willey, *Nineteenth Century Studies* (reprint, Harmondsworth: Penguin Books, 1964), chap. 7. The history of positivism as a philosophy is outlined in L. Kolakowski, *The Alienation of Reason: A History of Positivist Thought* (New York: Doubleday, 1968). On Comte's rejection of psychology as a science, especially as a response to the social prominence of Cousin, see W. M. Simon, 'The "Two Cultures" in Nineteenth-Century France: Victor Cousin and Auguste Comte', JHI, **26** (1965): 45–58; see also, F. Wilson, 'Mill and Comte on the Method of Introspection', JHBS, **27** (1991): 107–29. The fate of Comtean thought under the Second Empire is discussed in D. G. Charlton, *Positivist Thought in France during the Second Empire 1852–1870* (Oxford: Clarendon Press, 1959), and in a systematic assessment of Comte's influence, W. M. Simon, *European Positivism in the Nineteenth Century: An Essay in Intellectual History* (reprint, Port Washington, NY: Kennikat Press, 1972).

It is not hard to see why the literature on Marx and on the fate of Marxism is massive and in many languages. As well as the collected Marx-Engels works in German, Russian and English, there are many selections of the key texts. It is a form of scholarship in itself even to find one's way around the interpretations of the works and their political implications. There is a readable and reliable 'life and thought': D. McLellan, *Karl Marx: His Life and Thought* (London: Macmillan, 1973); while the texts and introduction in T. B. Bottomore and M. Rubel (eds), *Karl Marx: Selected Writings in Sociology and Social Philosophy* (2nd edn, reprint, Harmondsworth: Penguin Books, 1963), K. Marx and F. Engels, *Selected Works in One Volume* (London: Lawrence & Wishart, 1968), a selection printed in the USSR, and D. McLellan (ed), *Karl Marx: Selected Writings* (Oxford: Oxford University Press, 1977), are useful. A 'founding father', like Comte, Marx appears in general histories of sociology. Two studies by G. Lichtheim, though sometimes criticized in detail, are helpful in locating Marx in the left political movements of the nineteenth century: *Marxism: An Historical and Critical Study* (2nd edn, New York: Frederick A. Praeger, 1965), and *A Short History of Socialism* (New York: Praeger, and London: Weidenfeld & Nicolson, 1970). There is a history of Marxist thought in L. Kolakowski, *Main Currents of Marxism: Its Origins, Growth and Dissolution*, 3 vols (Oxford: Oxford University Press, 1978); this is written by a deeply disabused Polish ex-Marxist intellectual but is an invaluable history across an extended range. The influence of Marx's thought in different disciplines is reviewed in D. McLellan (ed), *Marx: The First Hundred Years* (London: Fontana, 1983); and for late twentieth-century overviews, T. Carver (ed), *The Cambridge Companion to Marx* (Cambridge: Cambridge University Press, 1991). My account of Marx is towards the idealist end of the spectrum of opinion. There is a detailed study of the Hegelian background in J. E. Toews, *Hegelianism: The Path towards Dialectical Humanism, 1805–1841* (Cambridge: Cambridge University Press, 1980). For an analysis of 'alienation', the key concept in the 1960s Western renewal of interest in the early Marx, see B. Ollman, *Alienation: Marx's Conception of Man in Capitalist Society* (2nd edn, Cambridge: Cambridge University Press, 1976), which is analytic rather than historical; for a historical overview, L. S. Feuer, 'What Is Alienation: The Career of a Concept', in M. Stein and A. Vidich (eds), *Sociology on Trial* (Englewood Cliffs, NJ: Prentice-Hall, 1963), pp. 127–47. For 'ideology', G. Lichtheim, 'The Concept of Ideology', in *The Concept of Ideology and Other Essays* (New York: Vintage Books, 1967), pp. 3–46. Marx did have a theory of human nature; the argument is summarized in

N. Geras, *Marx and Human Nature: Refutation of a Legend* (London: Verso and New Left Books, 1983). Engels's study of Manchester is examined and related to social history in S. Marcus, *Engels, Manchester, and the Working Class* (London: Weidenfeld & Nicolson, 1974).

Marx's interest in anthropology is discussed in W. H. Shaw, 'Marx and Morgan', *History and Theory,* **23** (1984): 215–28. Full references to the myth, and to the uncovery of the myth, of the 'Darwin-Marx letter', by M. A. Fay and, separately, by L. S. Feuer, are given in R. Colp, 'The Myth of the Darwin-Marx Letter', *History of Political Economy,* **14** (1982): 461–82. The intellectual relations between Darwin and Marx are reviewed in: G. Pancaldi, 'The Technology of Nature: Marx's Thoughts on Darwin', in I. B. Cohen (ed), *The Natural Sciences and the Social Sciences: Some Critical and Historical Perspectives* (Dordrecht: Kluwer, 1994), pp. 257–74; A. Taylor, 'The Significance of Darwinian Theory for Marx and Engels', *Philosophy of the Social Sciences,* **19** (1989): 409–23; and, as part of a broad Marxian programme for the history of science, R. M. Young, 'The Historiographic and Ideological Contexts of the Nineteenth-Century Debate on Man's Place in Nature', in M. Teich and R. Young (eds), *Changing Perspectives in the History of Science* (London: Heinemann, 1973), pp. 344–438, reprinted in R. M. Young, *Darwin's Metaphor: Nature's Place in Victorian Culture* (Cambridge: Cambridge University Press, 1985), pp. 164–247. On materialism, see F. Gregory, 'Scientific Versus Dialectical Materialism: A Clash of Ideologies in Nineteenth-Century German Radicalism', *Isis,* **68** (1977): 206–23; I. Mitchell, 'Marxism and German Scientific Materialism', *Annals of Science,* **35** (1978): 379–400. For Kautsky's scientific Marxism: D. Geary, *Karl Kautsky* (Manchester: Manchester University Press, 1987), chap. 7. For the search for a science of society in Russia on natural science models, A. Vucinich, *Social Thought in Tsarist Russia: The Quest for a General Science of Society, 1861–1917* (Chicago: University of Chicago Press, 1976). References to the history of political economy are given in Chapter 9. The marginalist revolution (not all economists accept there was such a thing) is described, with the argument that Jevons and Marshall were united in certain respects, in M. Schabas, 'Alfred Marshall, W. Stanley Jevons, and the Mathematization of Economics', *Isis,* **80** (1989): 60–73,

A World Ruled by Number: William Stanley Jevons and the Rise of Mathematical Economics (Princeton: Princeton University Press, 1990), and 'From Political Economy to Market Mechanics: The Jevonian Moment in the History of Economics', in Cohen (ed) (1994), pp. 235–55. The place of the history of economics in the history of science is argued for in M. Schabas, 'Breaking Away: History of Economics as History of Science', *History of Political Economy*, **24** (1992): 185–247; while the over-reliance of economics on physical models is examined in P. Mirowski, *More Heat than Light: Economics as Social Physics: Physics as Nature's Economics* (Cambridge: Cambridge University Press, 1989). A brief standard view of the contrast between classical and modern economics is H. S. Gordon, 'Alfred Marshall and the Development of Economics as a Science', in R. N. Giere and R. S. Westfall (eds), *Foundations of Scientific Method: The Nineteenth Century* (Bloomington: Indiana University Press, 1973), pp. 234–58. There is a detailed study of Marshall in J. Maloney, *Marshall, Orthodoxy and the Professionalisation of Economics* (Cambridge: Cambridge University Press, 1985), while a model for the wider development of science in relation to the interests of the state is developed in F. M. Turner, 'Public Science in Britain, 1880–1919', *Isis*, **71** (1980): 589–608.

CHAPTER 13

Though the size and activity of 'the Darwin industry' is conspicuous in the history of science, it is more concerned with the early development of Darwin's ideas than with his opinions about human evolution. Two massive biographies give due place to his long-running thoughts on 'man's place in nature': J. Browne, *Charles Darwin: Voyaging. Volume 1 of a Biography* (London: Jonathan Cape, and New York: Random House, 1995); A. Desmond and J. Moore, *Darwin* (London: Michael Joseph, 1991). P. Bowler, *Charles Darwin: The Man and His Influence* (Oxford and Cambridge, MA: Basil Blackwell, 1990) is a briefer study. There are overviews of Darwin's thoughts on man in relation to biology and the earth in P. Bowler, *Evolution: The History of an Idea* (2nd edn, Berkeley: University of California Press, 1989), and *The Fontana History of the Environmental Sciences* (London: Fontana Press, 1992); J. C. Greene, *The Death of Adam: Evolution and Its*

Impact on Western Thought (reprint, New York: Mentor Books, 1961), chap. 10. Careful consideration is given to the religious dimension and secondary literature of the Darwin debates in J. H. Brooke, *Science and Religion: Some Historical Perspectives* (Cambridge: Cambridge University Press, 1991), chaps. 7 and 8. The deeper effects of other changes on Christian faith are examined in O. Chadwick, *The Secularization of the European Mind in the Nineteenth Century* (Cambridge: Cambridge University Press, 1975). F. M. Turner, *Between Science and Religion: The Reaction to Scientific Naturalism in Late Victorian England* (New Haven: Yale University Press, 1974), introduces key evolutionists who were disturbed by the implications of the theory for 'the spiritual' and morality. Studies of the reception of evolutionary ideas include: A. Ellegård, *Darwin and the General Reader: The Reception of Darwin's Theory of Evolution in the British Periodical Press, 1859–1872* (Göteborg: Göteborgs Universitets, 1958), chap. 14; T. F. Glick (ed), *The Comparative Reception of Darwinism* (Austin: University of Texas Press, 1974); A. Kelly, *Descent of Darwin: The Popularization of Darwinism in Germany, 1860–1914* (Chapel Hill: University of North Carolina Press, 1981); G. Pancaldi, *Darwin in Italy: Science across Cultural Frontiers* (Bloomington: Indiana University Press, 1991), Part 3; C. E. Russett, *Darwin in America: The Intellectual Response 1865–1912* (San Francisco: W. H. Freeman, 1976); D. P. Todes, *Darwin without Malthus: The Struggle for Existence in Russian Evolutionary Thought* (New York: Oxford University Press, 1989); A. Vucinich, *Darwin in Russian Thought* (Berkeley: University of California Press, 1988). Darwin's relations to Malthus and to a wider debate about values is argued in papers collected in R. M. Young, *Darwin's Metaphor: Nature's Place in Victorian Culture* (Cambridge: Cambridge University Press, 1985), while Young makes a critical assessment of values derived from evolutionary theory in 'The Naturalisation of Value Systems in the Human Sciences', in *Problems in the Biological and Human Sciences,* Block VI of Open University Course, *Science and Belief: From Darwin to Einstein* (Milton Keynes: Open University Press, 1981), pp. 63–110. The earth sciences as well as evolutionary thought were related to political economy: S. Rashid, 'Political Economy and Geology in the Early Nineteenth Century: Similarities and Contrasts', *History of Political Economy,* **13** (1981): 726–44. The later connections of Darwin and political thought, in so-called social Darwinism, are systematically examined in P. Crook, *Darwinism, War and History: The Debate Over the Biology of War from the "Origin of Species" to the First World War* (Cambridge: Cambridge University Press, 1994), which stresses the range of opinion and the potential of evolutionary thought to support pacifist as well as militarist leanings. On social Darwinist ideas, see also: R. C. Bannister, *Social Darwinism: Science and Myth in American Social Thought* (Philadelphia: Temple University Press, 1979), a provocative revision of R. C. Hofstadter, *Social Darwinism in American Thought* (new edn, Boston: Beacon Press, 1955); L. L. Clark, *Social Darwinism in France* (Tuscaloosa: University of Alabama Press, 1984); G. Jones, *Social Darwinism and English Thought: The Interaction between Biological and Social Theory* (Brighton: Harvester Press, 1980); R. Weikart, 'The Origins of Social Darwinism in Germany, 1859–1895', JHI, **54** (1993): 469–88.

There are excellent studies of Spencer as an evolutionary theorist in J. W. Burrow, *Evolution and Society: A Study in Victorian Social Theory* (reprint, Cambridge: Cambridge University Press, 1970), chap. 6, and J. D. Y. Peel, *Herbert Spencer: The Evolution of a Sociologist* (London: Heinemann, 1971). His work is related to theories of mind in R. M. Young, *Mind, Brain, and Adaptation: Cerebral Localization and Its Biological Context from Gall to Ferrier* (Oxford: Clarendon Press, 1970), chap. 5, and 'The Role of Psychology in the Nineteenth-Century Evolutionary Debate', in (1985), pp. 56–78. The study by R. J. Richards, *Darwin and the Emergence of Evolutionary Theories of Mind and Behavior* (Chicago: University of Chicago Press, 1987), covers the evolutionary theorists down to James and C. Lloyd Morgan while also arguing for a naturalistic philosophy of ethics, of the kind Spencer himself initiated. Spencer's own dry account of the development of his thought is in 'The Filiation of Ideas', in D. Duncan (ed), *The Life and Letters of Herbert Spencer* (London: Methuen, 1908), pp. 533–76. There is brief but incisive comment on his crucial organic analogy in W. M. Simon, 'Herbert Spencer and the "Social Organism"', JHI, **21** (1960): 294–9. For his views on consciousness, see also: D. Blitz, *Emergent Evolution: Qualitative Novelty and the Levels of Reality* (Dordrecht: Kluwer, 1992), chap. 2; C. U. M. Smith, 'Evolution and the Problem of Mind', *Journal of the History of Biology*, **15** (1982): 55–88 and 241–62. For

Spencer's notorious phrase, 'the survival of the fittest', see D. B. Paul, 'The Selection of the "Survival of the Fittest"', *Journal of the History of Biology*, **21** (1988): 411–24. On his politics, see D. Wiltshire, *The Social and Political Thought of Herbert Spencer* (Oxford: Oxford University Press, 1978).

Darwin's *Descent of Man* has received much less attention from historians than *The Origin of Species*; the second edition (1874) differs from the first (1871) in the order of argument and has new material, but – in both cases – human evolution is discussed in chaps. 1–7 and 21. There is a survey of Darwin's thought about human evolution in J. R. Durant, 'The Ascent of Nature and Darwin's Descent of Man', in D. Kohn (ed), *The Darwinian Heritage* (Princeton: Princeton University Press, 1985), pp. 283–306; see also J. Browne, 'Darwin and the Expression of the Emotions', in *ibid.*, pp. 307–26, and R. W. Burkhardt, Jr, 'Darwin on Animal Behavior and Evolution', in *ibid.*, pp. 327–65. Darwin's importance for later psychology is described in R. Boakes, *From Darwin to Behaviourism: Psychology and the Mind of Animals* (Cambridge: Cambridge University Press, 1984). The problem that faced Darwin when he looked for evidence of mind lower down the animal scale is outlined in C. U. M. Smith, 'Charles Darwin, the Origins of Consciousness, and Panpsychism', *Journal of the History of Biology*, **11** (1978): 246–67. Darwin's early speculations were recorded in 'Metaphysical Notebooks', printed in 'Early Writings of Charles Darwin', in H. E. Gruber, *Darwin on Man: A Psychological Study of Scientific Creativity. Together with Darwin's Early and Unpublished Notebooks* (London: Wildwood House, 1974), which includes Gruber's psychological study of Darwin's thought in the 1830s. Darwin's treatment of women is discussed, with some scepticism, in E. Richards, 'Darwin and the Descent of Woman', in D. Oldroyd and I. Langham (eds), *The Wider Domain of Evolutionary Thought* (Dordrecht: D. Reidel, 1983), pp. 57–111. A. R. Wallace's papers on human evolution were reprinted (with changes) in *Natural Selection and Tropical Nature: Essays on Descriptive and Theoretical Biology* (1891; reprint, Farnborough: Gregg International, 1969), while there are substantial extracts from his views on man in C. H. Smith (ed), *Alfred Russel Wallace: An Anthology of His Shorter Writings* (Oxford: Oxford University Press, 1991), pp. 9–65; and for the relation between his life and evolutionary values, J. R.

Durant, 'Scientific Naturalism and Social Reform in the Thought of Alfred Russel Wallace', *British Journal for the History of Science*, **12** (1979): 31–58. The differences between Darwin and Wallace are systematically presented in J. S. Schwartz, 'Darwin, Wallace, and the *Descent of Man*', *Journal of the History of Biology*, **17** (1984): 271–89. The literary and cultural dimensions of imagery about human origins are considered in G. Beer, *Darwin's Plots: Evolutionary Narrative in Darwin, George Eliot and Nineteenth-Century Fiction* (reprint, London: Ark, 1985). Darwin's sometimes contradictory statements about social evolution are detailed in J. C. Greene, 'Darwin as a Social Evolutionist', *Journal of the History of Biology*, **10** (1977): 1–27. For the relation of Darwin to social psychology: R. M. Farr, 'On Reading Darwin and Discovering Social Psychology', in R. Gilmour and S. Duck (eds), *The Development of Social Psychology* (London: Academic Press, 1980), pp. 111–36.

For the literature on social evolution, see the section on anthropology and ethnology under Chapter 11 and the work on social Darwinism described above. The consolidation of the idea of primitiveness in kinship studies is discussed in A. Kuper, *The Invention of Primitive Society: Transformation of an Illusion* (London and New York: Routledge, 1988), and for an account of British anthropology in relation to a concern with 'the beast within' British society, H. Kuklick, *The Savage Within: The Social History of British Anthropology, 1885–1945* (Cambridge: Cambridge University Press, 1991). On the development of the fieldwork paradigm in anthropology, see G. W. Stocking, Jr, 'The Ethnographer's Magic: Fieldwork in British Anthropology from Tylor to Malinowski', in G. W. Stocking, Jr (ed), *Observers Observed: Essays on Ethnographic Fieldwork, History of Anthropology, Volume 1* (Madison: University of Wisconsin Press, 1983), pp. 70–120, and *After Tylor: British Social Anthropology 1888–1951* (London: Athlone Press, 1996). Use of the concept of 'survivals', and the importance of this to studies of folklore, is described in M. T. Hodgen, *The Doctrine of Survivals: A Chapter in the History of Scientific Method in the Study of Man* (London: Allenson, 1936). For North America: A. Kuper, 'The Development of Lewis Henry Morgan's Evolutionism', JHBS, **21** (1985): 3–22; M. Swetlitz 'The Minds of Beavers and the Minds of Humans: Natural Suggestion, Natural Selection, and Experiment in the Work of

Lewis Henry Morgan', in G. W. Stocking, Jr (ed), *Bones, Bodies, Behavior: Essays on Biological Anthropology, History of Anthropology, Volume 5* (Madison: University of Wisconsin Press, 1988), pp. 56–83. On the anthropology of religion: R. A. Jones, 'Robertson Smith and James Frazer on Religion: Two Traditions in British Social Anthropology', in G. W. Stocking, Jr (ed), *Functionalism Historicized: Essays on British Social Anthropology, History of Anthropology, Volume 2* (Madison: University of Wisconsin Press, 1984), pp. 31–58.

The revolutionary implication of Darwin's thought for philosophy was claimed in J. Dewey, 'The Influence of Darwin on Philosophy', in *The Influence of Darwin on Philosophy and Other Essays in Contemporary Thought* (1910; reprint, Bloomington: Indiana University Press, 1965), pp. 1–19. The development of US philosophy influenced by evolutionary thought is described in P. P. Wiener, *Evolution and the Founders of Pragmatism* (reprint, New York: Harper Torchbooks, 1965); for James and Harvard philosophy, B. Kuklick, *The Rise of American Philosophy: Cambridge, Massachusetts 1860–1930* (New Haven: Yale University Press, 1977), chaps. 9 and 10. James's life and thought has been much described; a key resource is R. B. Perry, *The Thought and Character of William James as Revealed in Unpublished Correspondence and Notes, Together with Published Writings*, 2 vols (Boston: Little, Brown, 1935); for a biography and review of his work, see G. E. Myers, *William James: His Life and Thought* (New Haven: Yale University Press, 1986); while D. W. Bjork, *William James: The Center of His Vision* (New York: Columbia University Press, 1988), argues for the integration of his apparently diverse endeavours. In the literature on James, see 'Special Issue: William James', HHS, **8** no. 1 (1985): 1–105. On his theory of conscious interests: L. Mackenzie, 'William James and the Problem of Interests', JHBS, **16** (1980): 175–85; and theory of the will, W. R. Woodward, 'William James's Psychology of Will: Its Revolutionary Impact on American Psychology', in J. Brožek (ed), *Explorations in the History of Psychology in the United States* (Lewisburg: Bucknell University Press, 1984), pp. 148–95; and the materials for a wider interpretation of James's view of consciousness, in the light of his religious sympathies, are given in E. Taylor, 'William James on Darwin: An Evolutionary Theory of Consciousness', *Annals of the New York Academy of Science,*

602 (1990): 7–33. For James and European philosophy, see B. Wilshire, *William James and Phenomenology: A Study of 'The Principles of Psychology'* (Bloomington: Indiana University Press, 1968). On Lloyd Morgan and James: R. J. Richards, 'Lloyd Morgan's Theory of Instinct: From Darwinism to Neo-Darwinism', JHBS, **13** (1977): 12–32, and Richards (1987), chaps. 8 and 9. There is a critical re-evaluation of Lloyd Morgan's canon in A. Costall, 'How Lloyd Morgan's Canon Backfired', JHBS, **29** (1993): 113–22. The general history of functional explanation in psychology is outlined in two chapters by C. E. Buxton, 'Early Sources and Basic Conceptions of Functionalism' and 'American Functionalism', in C. E. Buxton (ed), *Points of View in the Modern History of Psychology* (Orlando: Academic Press, 1985), pp. 85–111 and 113–40. J. Dewey's paper, 'The Reflex Arc Concept in Psychology', is reprinted in W. Dennis (ed), *Readings in the History of Psychology* (New York: Appleton-Century-Crofts, 1948), pp. 355–65; and G. H. Mead's, 'Social Psychology as Counterpart to Physiological Psychology', in *Selected Writings*, ed A. J. Reck (Indianapolis: Bobbs-Merrill, 1964), pp. 94–104. For one important dimension of the concept of function in social thought, see B. S. Heyl, 'The Harvard "Pareto Circle"', JHBS, **4** (1968): 316–34, and C. E. Russett, *The Concept of Equilibrium in American Social Thought* (New Haven: Yale University Press, 1966). The incorporation of a functionalist orientation in US psychology is described in Chapter 14.

CHAPTER 14

There is a brief but suggestive sketch of the contrasting origins and style of German, French and British psychology in K. Danziger, 'The Social Origins of Modern Psychology', in A. R. Buss (ed), *Psychology in Social Context* (New York: Irvington Publishers, 1979), pp. 27–45. For fine studies of individual psychologists in relation to psychology as a moral project, see C. J. Karier, *Scientists of the Mind: Intellectual Founders of Modern Psychology* (Urbana: University of Illinois Press, 1986). The establishment of French psychology is discussed in J. I. Brooks, III, 'Philosophy and Psychology at the Sorbonne, 1885–1913', JHBS, **29** (1993): 123–45; J. Carroy and R. Plas, 'The Origins of French Experimental Psychology: Experiment and Experimentalism', HHS, **9** no. 1 (1996): 73–84; J. Goldstein, 'The Advent of Psychological Modernism in France: An Alternative Narrative', in D. Ross (ed), *Modernist Impulses in the Human Sciences 1870–1930* (Baltimore: Johns Hopkins University Press, 1994), pp. 190–209. On Claparède: 'Edouard Claparède', in C. Murchison (ed), *A History of Psychology in Autobiography. Volume I* (reprint, New York: Russell & Russell, 1961), pp. 63–97. Galton is often described as pivotal in the British case (see references for Chapter 16); see also: L. Daston, 'British Responses to Psycho-physiology, 1860–1900', *Isis*, **69** (1978): 192–208; L. S. Hearnshaw, *A Short History of British Psychology, 1840–1940* (London: Methuen, 1964). The significant case of The Netherlands, where psychologists eventually became thick on the ground, is explained in: T. Dehue, 'Transforming Psychology in The Netherlands I: Why Methodology Changes', HHS, **4** (1991): 335–49, and *Changing the Rules: Psychology in The Netherlands, 1900–1985* (Cambridge: Cambridge University Press, 1995); P. J. van Strien, 'Transforming Psychology in The Netherlands II: Audiences, Alliances and the Dynamics of Change', HHS, **4** (1991): 351–69. Lombroso's position in Italy is discussed in R. A. Nye, 'Heredity or Milieu: The Foundations of Modern European Criminological Theory', *Isis*, **67** (1976): 335–55; D. Pick, 'The Faces of Anarchy: Lombroso and the Politics of Criminal Science in Post-unification Italy', *History Workshop*, no. 21 (1986): 60–86, and *Faces of Degeneration: A European Disorder, c.1848–c.1918* (Cambridge: Cambridge University Press, 1989), chap. 5. Wundt's European students are discussed in W. G. Bringmann and R. D. Tweney (eds), *Wundt Studies: A Centennial Collection* (Toronto: C. J. Hogrefe, 1980); for Lehmann and Copenhagen, see I. Nilsson, 'Alfred Lehmann and Psychology as a Physical Science', in *ibid.*, pp. 258–68. The intellectual context in Spain is sketched in H. Carpintero, 'The Introduction of Scientific Psychology in Spain 1875–1900', in W. R. Woodward and M. G. Ash (eds), *The Problematic Science: Psychology in Nineteenth-Century Thought* (New York: Praeger, 1982), pp. 255–75. The background to Russian psychology is described in D. Joravsky, *Russian Psychology: A Critical History* (Oxford: Basil Blackwell, 1989). On progressive Catholicism and the new psychology: H. Misiak and V. M. Staudt, *Catholics in Psychology: A Historical Survey* (New York: McGraw-Hill, 1954), and H. Misiak, 'Leipzig and Louvain

University in Belgium', *Psychological Research*, Wundt Centennial Issue, ed W. G. Bringmann and E. Scheerer, **42** (1980): 49–56.

The origins of experimental psychology are often traced to experimental studies of perception from the 1820s to the 1860s. E. G. Boring, *Sensation and Perception in the History of Experimental Psychology* (New York: Appleton-Century-Crofts, 1942), which includes much experimental detail only understandable by specialists, set a pattern. See also N. Pastore, *Selective History of Theories of Visual Perception: 1650–1950* (New York: Oxford University Press, 1971). Helmholtz's position is also examined in R. S. Turner, 'Hermann von Helmholtz and the Empiricist Vision', JHBS, **13** (1977): 48–58, and in a definitive study, *In the Eye's Mind: Vision and the Helmholtz-Hering Controversy* (Princeton: Princeton University Press, 1994). Relevant papers by Helmholtz are reprinted in R. Kahl (ed), *Selected Writings of Hermann von Helmholtz* (Middletown, CT: Wesleyan University Press, 1971). Moden views of Fechner are explored in J. Brožek and H. Gundlach (eds), *G. T. Fechner and Psychology, Passauer Schriften zur Psychologiegeschichte. Nr. 6* (Passau: Passavia Universitätsverlag, 1988). The centenary of Wundt's Leipzig laboratory in 1979 produced a spate of studies; see W. G. Bringmann and G. A. Ungerer, 'The Foundation of the Institute for Experimental Psychology at Leipzig University', *Psychological Research*, **42** (1980): 5–18; S. Diamond, 'Wundt before Leipzig', in R. W. Rieber (ed), *Wilhelm Wundt and the Making of a Scientific Psychology* (New York: Plenum Press, 1980), pp. 3–70. By then, reassessment of Wundt's contribution was under way: A. L. Blumenthal, 'A Reappraisal of Wilhelm Wundt', *American Psychologist*, **30** (1975): 1081–8, and 'Wilhelm Wundt: Psychology as the Propaedeutic Science', in C. E. Buxton (ed), *Points of View in the Modern History of Psychology* (Orlando: Academic Press, 1985), pp. 19–50; K. Danziger, 'Wundt's Psychological Experiment in the Light of His Philosophy of Science', *Psychological Research*, **42** (1980): 109–22, and 'Wundt's Theory of Behavior and Volition', in Rieber (ed) (1980), pp. 89–115; W. van Hoorn and T. Verhave, 'Wundt's Changing Conceptions of a General and Theoretical Psychology', in Bringmann and Tweney (eds) (1980), pp. 71–113; T. Mischel, 'Wundt and the Conceptual Foundations of Psychology', *Philosophy and Phenomenological Research*, **31** (1970): 1–26. The differentiation of modes of introspection is made in

K. Danziger, 'The History of Introspection Reconsidered', JHBS, **16** (1980): 241–62. A major re-assessment of the social construction of laboratory methodologies is undertaken in K. Danziger, *Constructing the Subject: Historical Origins of Psychological Research* (Cambridge: Cambridge University Press, 1990), and, more briefly, in 'Social Context and Investigative Practice in Early Twentieth-Century Psychology', in M. G. Ash and W. R. Woodward (eds), *Psychology in Twentieth-Century Thought and Society* (Cambridge: Cambridge University Press, 1987), pp 13–33. For the background to Wundt's *Völkerpsychologie*, see references to German ethnology in Chapter 11; also, W. Wundt, *The Language of Gestures, and Additional Essays by George Herbert Mead and Karl Bühler*, intro. A. L. Blumenthal (The Hague: Mouton, 1973), with extracts relevant to twentieth-century linguistics.

The difficult institutional position of German psychology, which persisted until after 1945, is described by M. G. Ash, 'Academic Politics in the History of Science: Experimental Psychology in Germany, 1879–1941', *Central European History*, **13** (1980): 255–86; 'Experimental Psychology in Germany before 1914: Aspects of an Academic Identity Problem', *Psychological Research*, **42** (1980): 78–86; 'Wilhelm Wundt and Oswald Külpe on the Institutional Status of Psychology: An Academic Controversy in Historical Context', in Bringmann and Tweney (eds) (1980), pp. 396–421. There is extensive detail relating to much of the history of German psychology since the late nineteenth century, as well as relevant areas of philosophy, in M. G. Ash, *Gestalt Psychology in German Culture, 1890–1967: Holism and the Quest for Objectivity* (Cambridge: Cambridge University Press, 1995). See also: M. G. Ash, 'Psychology in Twentieth-Century Germany: Science and Profession', in G. Cocks and K. Jarausch (eds), *German Professions, 1800–1950* (New York: Oxford University Press, 1990), pp. 289–307; U. Geuter, 'The Uses of History for the Shaping of a Field: Observations on German Psychology', in L. Graham, W. Lepenies and P. Weingart (eds), *Functions and Uses of Disciplinary Histories* (Dordrecht: D. Reidel, 1983), pp. 191–228; 'German Psychology during the Nazi Period', in Ash and Woodward (eds) (1987), pp. 165–87; and *The Professionalization of Psychology in Nazi Germany* (Cambridge: Cambridge University Press, 1992). The research at Würzburg is examined from a sociological perspective in M. Kusch 'Recluse, Interlocutor,

Interrogator: Natural and Social Order in Turn-of-the-century Psychological Research Schools', *Isis*, **86** (1995): 419–39. On moves away from Wundt's programme, see also: K. Danziger, 'The Positivist Repudiation of Wundt', JHBS, **15** (1979): 205–30; W. G. Bringmann and G. A. Ungerer, 'Experimental vs. Educational Psychology: Wilhelm Wundt's Letters to Ernst Meumann', *Psychological Research*, **42** (1980): 57–73. The work of the Würzburg school is described in G. Humphrey, *Thinking: An Introduction to Its Experimental Psychology* (reprint, New York: John Wiley & Sons, 1963), chaps. 2–4. Preyer and developmental psychology are discussed in G. Eckardt, W. G. Bringmann and L. Sprung (eds), *Contributions to a History of Developmental Psychology* (Berlin and New York: Mouton, 1985), especially S. Jaeger, 'Preyer and the German School Reform Movement', pp. 231–43; see also S. Jaeger, 'Origins of Child Psychology: William Preyer', in Woodward and Ash (eds) (1982), pp. 300–21. For Mach, see J. T. Blackmore, *Ernst Mach: His Work, Life, and Influence* (Berkeley: University of California Press, 1972). Brentano's 1874 text on empirical psychology was translated into English only in 1973, but there is a study of his work in A. C. Rancurello, *A Study of Franz Brentano: His Psychological Standpoint and His Significance in the History of Psychology* (New York: Academic Press, 1968); see also J. L. Sullivan, 'Franz Brentano and the Problems of Intentionality', in B. B. Wolman (ed), *Historical Roots of Contemporary Psychology* (New York: Harper & Row, 1968), pp. 248–74. Dilthey's work received attention in the anglophone world in the 1970s; for introductions, see: W. Dilthey, *Selected Writings*, ed H. P. Rickman (Cambridge: Cambridge University Press, 1976), and H. P. Rickman, *Wilhelm Dilthey: Pioneer of the Human Studies* (London: Paul Elek, 1979). The logical critique of naturalistic explanations of reason and the background to modern analytic philosophy in Frege is discussed in H. D. Sluga, *Gottlob Frege* (London and Boston: Routledge & Kegan Paul, 1980), chap. 1. For an overview of transformations in philosophy, J. A. Passmore, *A Hundred Years of Philosophy* (2nd edn, Harmondsworth: Penguin Books, 1968).

A rich account of the development of psychology as an occupation in the United States is given in J. M. O'Donnell, *The Origins of Behaviorism: American Psychology, 1870–1920* (New York: New York University Press, 1986); for a brief view of professionalization: T. M. Camfield, 'The Professionalization of American Psychology, 1870–1917', JHBS, **9** (1973): 66–75; and for comments on the rhetoric, D. E. Leary, 'Telling Likely Stories: The Rhetoric of the New Psychology, 1880–1920', JHBS, **23** (1987): 315–31. A simple story but with rich illustrations is given in J. A Popplestone and M. W. McPherson, *An Illustrated History of American Psychology* (Madison: WCB Brown & Benchmark, 1994). An information source about the profession is L. T. Benjamin *et al., A History of American Psychology in Notes and News 1883–1945: An Index to Journal Sources* (Milwood, NY: Kraus International, 1989). The continuity of moral philosophy with the new psychology is stressed in G. Richards, '"To Know Our Fellow Men to Do Them Good": American Psychology's Enduring Moral Project', HHS, **8** no. 3 (1995): 1–24, and on McCosh, see J. D. Hoeveler, Jr, *James McCosh and the Scottish Intellectual Tradition* (Princeton: Princeton University Press, 1981). There was also sometimes a positive relationship between the new psychology and religion, and a 'psychology of religion' flourished in the first decade of the new century; there is a brief description in B. Beit-Hallahni, 'Psychology of Religion 1880–1930: The Rise and Fall of a Psychological Movement', JHBS, **10** (1974): 84–90. The situation of women in the new field is discussed in L. Furomoto, 'On the Margins: Women and the Professionalization of Psychology in the United States, 1890–1940', in Ash and Woodward (eds) (1987), pp. 93–113. The divergence of US psychology from Wundt, even when he was claimed to be its founding father, is stressed in: A. L. Blumenthal, 'Wilhelm Wundt and Early American Psychology: A Clash of Two Cultures', reprinted in Rieber (ed) (1980), pp. 117–35; K. Danziger 'Wundt and the Two Traditions of Psychology', in Rieber (ed) (1980), pp. 73–87; R. D. Tweney and S. A. Yachanin, 'Titchener's Wundt', in Bringmann and Tweney (eds) (1980), pp. 380–95. Individual psychologists have in many cases been well served by historical literature. James's mixed reputation with the new psychologists is discussed in D. W. Bjork, *The Compromised Scientist: William James in the Development of American Psychology* (New York: Columbia University Press, 1983). For Cattell, see M. M. Sokal, *An Education in Psychology: James McKeen Cattell's Journal and Letters from Germany and England, 1880–1888* (Cambridge, MA: MIT Press, 1981), and 'James McKeen Cattell and American Psychology in the 1920s', in J. Brožek (ed), *Explorations in the History of Psychology in the United States* (Lewisburg: Bucknell University Press, 1984), pp. 273–323. For Hall: D. Ross, *G. Stanley Hall: The Psychologist as Prophet* (Chicago: University of Chicago Press, 1972). For Thorndike: G. Jonçich, *The Sane Positivist: A Biography of Edward L. Thorndike* (Middletown, CT: Wesleyan University Press, 1968), though subsequent historians are much less eulogistic. For Baldwin: Richards (1987), chap. 10; R. Wozniak, 'Metaphysics and Science, Reason and Reality: The Intellectual Origins of Genetic Epistemology', in J. M. Broughton and D. J. Freeman-Moir (eds), *The Cognitive-Developmental Psychology of James Mark Baldwin: Current Theory and Research in Genetic Epistemology* (Norwood, NJ: Ablex Publishing, 1982), pp. 13–45. For Titchener: R. D. Tweney, 'Programmatic Research in Experimental Psychology: E. B. Titchener's Laboratory Investigations, 1891–1927', in Ash and Woodward (eds) (1987), pp. 35–57. The history of the American Psychological Association is given in R. B. Evans, V. S. Sexton and T. C. Cadwallader (eds), *100 Years: The American Psychological Association: A Historical Perspective* (Washington, DC: The American Psychological Association, 1992).

CHAPTER 15

The history of the social sciences is considered in a broad context in P. T. Manicas, *A History and Philosophy of the Social Sciences* (Oxford and New York: Basil Blackwell, 1987), while arguments for social science are subjected to critical assessment in G. Hawthorn, *Enlightenment and Despair: A History of Sociology* (2nd edn, Cambridge: Cambridge University Press, 1987). The contest between men of letters and sociologists to interpret civilization is described in W. Lepenies, *Between Literature and Science: The Rise of Sociology* (Cambridge: Cambridge University Press, 1988). The lives and ideas of individual social scientists are described in L. A. Coser, *Masters of Sociological Thought: Ideas in Historical and Social Context* (New York: Harcourt Brace Jovanovich, 1971). There is a succinct overview of the relations between the natural and social sciences in T. M. Porter, 'Natural Science and Social Theory', in R. C. Olby *et al.* (eds), *Companion to the History of Modern Science* (London: Routledge, 1990), pp. 1024–43. M.

Foucault's studies of the conditions that made possible the human sciences in the nineteenth century are brilliant and original, though specific points are often opposed by empirical historians. New techniques of human management, and hence new possibilities for truth about 'Man', are described in *The Birth of the Clinic: An Archaeology of the Human Sciences* (New York: Vintage, and London: Tavistock, 1973); *Madness and Civilization: A History of Insanity in the Age of Reason* (New York: Pantheon, 1965, and London: Tavistock, 1967); *Discipline and Punish: The Birth of the Prison* (New York: Pantheon, and London: Allen Lane, 1977); C. Gordon (ed), *Power/Knowledge: Selected Interviews and Other Writings 1972–1977* (New York: Pantheon, and Brighton: Harvester Wheatsheaf, 1980). For the reception of Foucault's approach to madness and history in the English-speaking world, see A. Still and I. Velody (eds), *Rewriting the History of Madness: Studies in Foucault's Histoire de la folie* (London and New York: Routledge, 1992). The argument that British sociology has its roots in social concern and the social survey is made in P. Abrams, *The Origins of British Sociology 1834–1914: An Essay with Selected Papers* (Chicago: University of Chicago Press, 1968); see also 'The Uses of British Sociology, 1831–1981', in M. Bulmer (ed), *Essays on the History of British Sociological Research* (Cambridge: Cambridge University Press, 1985), pp. 181–205. There is more detail on empirical studies in M. Bulmer, 'The Development of Sociology and of Empirical Social Research in Britain', in *ibid.*, pp. 3–36, and, for the twentieth century, M. Bulmer, K. Bales and K. K. Sklar (eds), *The Social Survey in Historical Perspective 1880–1940* (Cambridge: Cambridge University Press, 1991). The dangers in taking British sociology to be a unified discipline and then looking for 'the origin' are made clear in S. Collini, 'Sociology and Idealism in Britain 1880–1920', *Archives européennes de sociologie*, **19** (1978): 3–50, and L. Goldman, 'The Origins of British "Social Science": Political Economy, Natural Science and Statistics, 1830–1835', *Historical Journal*, **26** (1983): 587–616. Collini points to the philosophical idealism current just when sociology was supposedly becoming an empirical science, while Goldman examines the intentions of the historical actors at a 'founding' moment in sociology. The character of social science as a science of reform is stressed in L. Goldman, 'A Peculiarity of the English? The Social Science Association and the Absence of Sociology in Nineteenth-Century Britian', *Past & Present*, no. 114 (1987): 133–71. For statistical societies see: D. Elesh, 'The Manchester Statistical Society: A Case Study of Discontinuity in Empirical Social Research', in A. Oberschall (ed), *The Establishment of Empirical Sociology: Studies in Continuity, Discontinuity, and Institutionalization* (New York: Harper & Row, 1972), pp. 31–72; V. L. Hilts, '*Aliis exterendum*, or, The Origins of the Statistical Society of London', *Isis*, **69** (1978): 21–43. For the LSE, see the centenary history, R. Dahrendorf, *LSE: A History of the London School of Economics and Political Science* (Oxford: Oxford University Press, 1995). There is a rich and sophisticated literature on statistics (in both earlier and later meanings), probability and determinacy, which shows how significant statistical argument was to the shaping of the human sciences in the nineteenth century, deals with the connections between natural and social science, and explains the social contexts: G. Gigerenzer *et al.*, *The Empire of Chance: How Probability Changed Science and Everyday Life* (Cambridge: Cambridge University Press, 1989); I. Hacking, *The Taming of Chance* (Cambridge: Cambridge University Press, 1990), which is especially concerned with the philosophy of indeterminacy; T. Porter, *The Rise of Statistical Thinking 1820–1900* (Princeton: Princeton University Press, 1986); and, for the mathematics, S. M. Stigler, *The History of Statistics: The Measurement of Uncertainty Before 1900* (Cambridge, MA: Belknap Press of Harvard University Press, 1986). Porter is perhaps the most accessible for Quetelet; see also: T. Porter, 'From Quetelet to Maxwell: Social Statistics and the Origins of Statistical Physics', in I. B. Cohen (ed), *The Natural Sciences and the Social Sciences: Some Critical and Historical Perspectives* (Dordrecht: Kluwer, 1994), pp.345–62; B.-P. Lécuyer, 'Probabilistic Thinking, the Natural Sciences and the Social Sciences: Changing Configurations (1800–1850)', in *ibid.*, pp. 135–52. For further details on medicine and statistics, see J. Cole, 'The Chaos of Particular Facts: Statistics, Medicine and the Social Body in Early 19th-Century France', HHS, **7** no. 3 (1994): 1–27.

Durkheim's main work is translated, and there is a detailed intellectual biography and list of his writings in S. Lukes, *Emile Durkheim: His Life and Work: A Historical and Critical Study* (London: Allen Lane, The Penguin Press, 1973). For the setting of his work: T. Clark, *Prophets and Patrons: The French University and the Emergence of the Social Sciences* (Cambridge, MA: Harvard University Press, 1973); W. P. Vogt, 'Political Conventions, Professional Advancement, and Moral Education in Durkheimian Sociology', JHBS, **27** (1991): 56–75. The background to his studies of the division of labour, of suicide and of religion are described in: C. Limoges, 'Milne-Edwards, Darwin, Durkheim and the Division of Labour: A Case Study in Reciprocal Conceptual Exchanges between the Social and the Natural Sciences', in Cohen (ed) (1994), pp. 317–43; I. Hacking, 'How Numerical Sociology Began by Counting Suicides: From Medical Pathology to Social Pathology', in Cohen (ed) (1994), pp. 101–33; R. A. Jones, 'Robertson Smith, Durkheim, and Sacrifice: An Historical Context for *The Elementary Forms of the Religious Life*', JHBS, **17** (1981): 184–205. The post-Durkheim sociologists are described in P. Besnard (ed), *The Sociological Domain: The Durkheimians and the Founding of French Sociology* (Cambridge: Cambridge University Press, and Paris: Editions de la Maison des Sciences de l'Homme, 1983). The argument for a sociology of classificatory knowledge is in E. Durkheim and M. Mauss, *Primitive Classification*, intro R. Needham (2nd edn, London: Cohen & West, 1969).

The perceived relevance of Weber's work to unresolved issues in late twentieth-century sociology, especially the debate about the nature of modernity, has created a very large literature. There is an accessible account, with a historical orientation, in D. Käsler, *Max Weber: An Introduction to His Life and Work* (Oxford: Polity Press, 1988). Weber is related to wider intellectual contexts in: H. S. Hughes, *Consciousness and Society: The Reorientation of European Social Thought 1890–1930* (reprint, London: Padadin, 1974), chap. 8; W. J. Mommsen and J. Osterhammel (eds), *Max Weber and His Contemporaries* (London: Unwin Hyman, 1987). Weber's famous 1919 lecture and the critical comments it generated are translated in P. Lassman and I. Velody (eds), *Max Weber's 'Science as a Vocation'* (London: Unwin Hyman, 1989). The culture of the academic context is described in F. K. Ringer, *The Decline of the German Mandarins: The German Academic Community, 1890–1933* (reprint, Hanover, NH: Wesleyan University Press/ University Press of New England, 1990). The context and influence of 'the Protestant ethic' is discussed in H. Lehmann and G. Roth (eds), *Weber's Protestant Ethic: Origins, Evidence, Contexts*

(Cambridge: Cambridge University Press, 1993). There is debate about the degree of Weber's influence after his death in 1920: W. J. Mommsen, 'Max Weber and Modern Social Thought', in *The Political and Social Thought of Max Weber: Collected Essays* (Oxford: Polity Press, 1989), pp. 169–96; G. Schroeter, 'Max Weber as Outsider: His Nominal Influence on German Sociology in the Twenties', JHBS, **16** (1980): 317–32. Empirical social science and the Verein für Sozialpolitik are examined in A. Oberschall, *Empirical Social Research in Germany 1848–1914* (Paris and The Hague: Mouton, 1965). The attempt to promote an autonomous social science before 1900 is described in P. Weingart, 'Biology as Social Theory: The Bifurcation of Social Biology and Sociology in Germany, circa 1900', in D. Ross (ed), *Modernist Impulses in the Human Sciences 1870–1930* (Baltimore: Johns Hopkins University Press, 1994), pp. 255–71. Relations of the city to modernity are discussed on a broad front in M. Berman, *All That Is Solid Melts into Air: The Experience of Modernity* (New York: Simon & Schuster, 1982, and London: Verso, 1983); and the question of the fragmentation of experience is considered in H. Liebersohn, *Fate and Utopia in German Sociology, 1870–1923* (Cambridge, MA: MIT Press, 1988). Simmel's essay on the city is translated as 'The Metropolis and Mental Life', in *On Individuality and Social Forms: Selected Writings*, ed D. N. Levine (Chicago: University of Chicago Press, 1971), pp. 324–39. For Simmel's interest in the psychology of social life: D. Frisby, *Sociological Impressionism: A Reassessment of Georg Simmel's Social Theory* (London: Heinemann, 1981), *Georg Simmel* (Chichester: Ellis Horwood, and London and New York: Tavistock, 1984), and 'Georg Simmel and Social Psychology', JHBS, **20** (1984): 107–27. Finally, for the work of Karl Mannheim, see D. Frisby, *The Alienated Mind: The Sociology of Knowledge in Germany 1918–33* (London: Heinemann, and Atlantic Highlands, NJ: Humanities Press, 1983), and C. Loader, *The Intellectual Development of Karl Mannheim: Culture, Politics, and Planning* (Cambridge: Cambridge University Press, 1985). I discuss further the concept of '*Verstehen*' and related ideas of explanation in Chapter 20.

The origins of sociology and the social sciences in the United States have been extensively examined in relation to the rapid change in social conditions between 1870 and 1920. The literature includes: M. O. Furner, *Advocacy & Objectivity: A Crisis in the*

Professionalization of American Social Science, 1865–1905 (Lexington: University Press of Kentucky, 1975); T. L. Haskell, *The Emergence of Professional Social Science: The American Social Science Association and the Nineteenth-Century Crisis of Authority* (Urbana: University of Illinois Press, 1977); A. Oberschall, 'The Institutionalization of American Sociology', in A. Oberschall (ed) (1972), pp. 187–251; D. Ross, 'The Development of the Social Sciences', in A. Oleson and J. Voss (eds), *The Organization of Knowledge in Modern America, 1860–1920* (Baltimore: Johns Hopkins University Press, 1979), pp. 107–38; 'American Social Science and the Idea of Progress', in T. L. Haskell (ed), *The Authority of Experts: Studies in History and Theory* (Bloomington: Indiana University Press, 1984), pp. 157–75; and *The Origins of American Social Science* (Cambridge: Cambridge University Press, 1991), which relates the social sciences to the perceived need, in the period after the Civil War, for social control compatible with liberal democracy; the arguments are summarized in 'An Historian's View of American Social Science', JHBS, **29** (1993), 99–112. The values of the Progressive Era and the new psychological and social sciences are also discussed in essays by J. C. Burnham, collected in *Paths into American Culture: Psychology, Medicine and Morals* (Philadelphia: Temple University Press, 1988), and in R. H. Wiebe, *The Search for Order 1877–1920* (New York: Hill & Wang, 1967). There are comments on the creation of a research culture in the universities in O. Zunz, 'Producers, Brokers, and Uses of Knowledge: The Institutional Matrix', in Ross (ed) (1994), pp. 290–307. Historians are more sceptical of the explanatory power of 'professionalization' than they once were: H. Kuklick, 'Boundary Maintenance in American Sociology: Limitations to Academic "Professionalization"', JHBS, **16** (1980): 201–19. The Chicago school, because of its influence, has attracted particular attention: M. Bulmer, *The Chicago School of Sociology: Institutionalization, Diversity, and the Rise of Sociological Research* (Chicago: University of Chicago Press, 1984); M. J. Deegan and J. S. Burger, 'W. I. Thomas and Social Reform: His Work and Writings', JHBS, **17** (1981): 114–25; S. J. Diner, 'Department and Discipline: The Department of Sociology at the University of Chicago, 1892–1920', *Minerva*, **13** (1975): 514–53; E. L. Faris, *Chicago Sociology 1920–1932* (reprint, Chicago: University of Chicago Press, 1970); R. K. Haerle, Jr, 'William

Isaac Thomas and the Helen Culver Fund for Race Psychology: The Beginnings of Scientific Psychology at the University of Chicago', JHBS, **27** (1981): 21–41. The scientific direction in interwar sociology is considered in R. C. Bannister, *Sociology and Scientism: The American Quest for Objectivity, 1880–1940* (Chapel Hill: University of North Carolina Press, 1987).

CHAPTER 16

The redirection of historical attention to the institutional settings and human technologies of daily life was influenced by Foucault. In addition to the sources given for Chapter 15, see M. Foucault, 'Technologies of the Self', and 'The Political Technology of Individuals', in L. H. Martin, H. Gutman and P. H. Hutton (eds), *Technologies of the Self: A Seminar with Michel Foucault* (London: Tavistock, 1988), pp. 16–49 and 145–62; G. Burchell, C. Gordon and P. Miller (eds), *The Foucault Effect: Studies on Governmentality with Two Lectures by and an Interview with Michel Foucault* (London: Harvester Wheatsheaf, 1991). This work is developed in relation to twentieth-century Anglo-American psychology in N. Rose, *The Psychological Complex: Social Regulation and the Psychology of the Individual* (London: Routledge & Kegan Paul, 1985); 'Calculable Minds and Manageable Individuals', HHS, **1** (1988): 179–200; *Governing the Soul: The Shaping of the Private Self* (London: Routledge, 1990); 'Psychology as a "Social" Science', in I. Parker and J. Shotter (eds), *Deconstructing Social Psychology* (London: Routledge, 1990), pp. 103–16; 'Engineering the Human Soul: Analyzing Psychological Expertise', *Science in Context*, **5** (1992): 351–69. Short autobiographies of famous twentieth-century psychologists are collected in C. Murchison *et al.* (eds), *A History of Psychology in Autobiography*, 8 vols. (various publishers, Vols. 1–4, 1930–52, reprinted 1961; Vol. 5, 1967; Vol. 6, 1974; Vol.7, 1980; Vol. 8, 1989).

Differential psychology, or the psychology of individual differences, has been extensively studied by historians, in large part because it offers so much scope to the examination of relations between scientific activity and social or political values. The argument for the background influence of phrenology is made in J. M. O'Donnell, *The Origins of Behaviorism: American Psychology, 1870–1920* (New York: New York University Press, 1986), which also gives an extensive description of the educational context of psychology's growth in the US. For the persistence of popular phrenology: M. B. Stern, *Heads & Headlines: The Phrenological Fowlers* (Norman: University of Oklahoma Press, 1971). A description of British work is given in L. S. Hearnshaw, *A Short History of British Psychology, 1840–1940* (London: Methuen, 1964). The general literature on Galton includes: D. W. Forrest, *Francis Galton: The Life and Work of a Victorian Genius* (London: Elek, 1974); R. E. Fancher, *The Pioneers of Psychology* (2nd edn, New York: W. W. Norton, 1990), chap. 7 , and *The Intelligence Men: Makers of the IQ Controversy* (New York: W. W. Norton, 1985), which is a clear account of the whole area; A. R. Buss, 'Galton and the Birth of Differential Psychology and Eugenics: Social, Political, and Economic Forces', JHBS, **12** (1976): 47–58. Galton is famous as the 'founding father' of eugenics, on which there is also a large historical literature, surveyed in D. J. Kevles, *In the Name of Eugenics: Genetics and the Uses of Human Heredity* (New York: Alfred A. Knopf, 1985, and Harmondsworth: Penguin Books, 1986). It is important to note that eugenics was associated with a scientific worldview, not specifically right or left politics; see, e.g., D. Paul, 'Eugenics and the Left', JHI, **45** (1984): 567–90. For the development of Galton's thought: R. Schwartz Cowan, 'Nature and Nurture: The Interplay of Biology and Politics in the Work of Francis Galton', *Studies in the History of Biology,* **1** (1977): 133–208. There is criticism of measurement and hereditarian argument from a modern perspective in S. J. Gould, *The Mismeasure of Man* (reprint, Harmondsworth: Penguin Books, 1984), chaps. 5 and 6. The links between Galton's, Pearson's and later statistical methods and social ideology are highlighted in D. A. MacKenzie, *Statistics in Britain 1865–1930: The Social Construction of Scientific Knowledge* (Edinburgh: Edinburgh University Press, 1981). Spearman's early work is examined in B. Norton, 'Charles Spearman and the General Factor in Intelligence: Genesis and Interpretation in the Light of Sociopersonal Considerations', JHBS, **15** (1979): 142–154. Burt's life and work is described in L. S. Hearnshaw, *Cyril Burt: Psychologist* (London: Hodder & Stoughton, 1979). On measurement and schooling in Britain, a topic continuously high on the political agenda: G. Sutherland, *Ability, Merit and Measurement: Mental Testing and English Education 1880–1940* (Oxford:

Clarendon Press, 1984); A. Wooldridge, *Measuring the Mind: Education and Psychology in England, c.1860–c.1990* (Cambridge: Cambridge University Press, 1994), a major study of the links between psychology and education.

For Romanes and the Darwinian interest in the concept of intelligence: R. Boakes, *From Darwin to Behaviourism: Psychology and the Mind of Animals* (Cambridge: Cambridge University Press, 1984), chap. 2; R. J. Richards, *Darwin and the Emergence of Evolutionary Theories of Mind and Behavior* (Chicago: University of Chicago Press, 1987), chap. 8. The English-language study of Binet's work is T. H. Wolf, *Alfred Binet* (Chicago: University of Chicago Press, 1973); see also Fancher (1985), chap. 2. For French psychology after Binet: W. H. Schneider, 'After Binet: French Intelligence Testing, 1900–1950', JHBS, **28** (1992): 111–32. The US development of testing has been widely discussed, and once again an introduction is Fancher (1985), while the collection M. M. Sokal (ed), *Psychological Testing and American Society* (New Brunswick: Rutgers University Press, 1987), is rich in detail – see H. L. Minton, 'Lewis M. Terman and Mental Testing: In Search of the Democratic Ideal', pp. 95–112; J. Reed, 'Robert M. Yerkes and the Mental Testing Movement', pp. 46–74; L. Zenderland, 'The Debate over Diagnosis: Henry Herbert Goddard and the Medical Acceptance of Intelligence Testing', pp. 46–74. On Goddard's studies of immigrants: S. A. Gelb, 'Henry H. Goddard and the Immigrants, 1910–1917: The Studies and the Social Context', JHBS, **22** (1986): 324–32; and for the story of the Kallikaks, J. D. Smith, *Minds Made Feeble: The Myth and Legacy of the Kallikaks* (Rockville, MD: Aspen Systems Corporation, 1985). Detail on the clinical setting is given in J. A. Popplestone and M. W. McPherson, 'Pioneer Psychology Laboratories in Clinical Settings', in J. Brožek (ed), *Explorations in the History of Psychology in the United States* (Lewisburg: Bucknell University Press, 1984), pp. 196–272. For the educational context of Terman's interests: P. D. Chapman, *Schools as Sorters: Lewis M. Terman, Applied Psychology, and the Intelligence Testing Movement, 1890–1930* (New York: New York University Press, 1988); H. L. Minton, 'Charting Life History: Lewis M. Terman's Study of the Gifted', in J. G. Morawski (ed), *The Rise of Experimentation in American Psychology* (New Haven: Yale University Press, 1988), pp. 138–62. A detailed study of the psychologists' negotiations with the army to establish their authority is J. Carson, 'Army Alpha, Army Brass, and the Search for Army Intelligence', *Isis*, **84** (1993): 278–309. And for the effects: D. J. Kevles, 'Testing the Army's Intelligence: Psychologists and the Military in World War I', *Journal of American History*, **55** (1968): 565–81; N. Pastore, 'The Army Intelligence Tests and Walter Lippmann', JHBS, **14** (1978): 316–27; F. Samelson, 'World War I Intelligence Testing and the Development of Psychology', JHBS, **13** (1977): 274–82, and 'Putting Psychology on the Map: Ideology and Intelligence Testing', in A. R. Buss (ed), *Psychology in Social Context* (New York: Irvington Publishers, 1979), pp. 103–68. On the indirect relation between psychologists and the restrictive immigration legislation: F. Samelson, 'On the Science and Politics of IQ', *Social Research*, **42** (1975): 467–92, and, more widely on race, 'From "Race Psychology" to "Studies in Prejudice": Some Observations on the Thematic Reversal in Social Psychology', JHBS, **14** (1978): 265–78. For a critical overview of the values built into testing: R. T. von Mayrhauser, 'The Practical Language of the American Intellect', HHS, **4** (1991): 371–93.

Information on William Stern is scattered, but for an introduction to his theory of personalism see, 'William Stern', in C. Murchison (ed), *A History of Psychology in Autobiography. Volume 1* (reprint, New York: Russell & Russell, 1961), pp. 335–88, and G. W. Allport, 'The Personalistic Psychology of William Stern', in B. B. Wolman (ed), *Historical Roots of Contemporary Psychology* (New York: Harper & Row, 1968), pp. 321–37. 'Personality', I think, awaits its historian, but for the social construction of terms and practices see K. Danziger, *Constructing the Subject: Historical Origins of Psychological Research* (Cambridge: Cambridge University Press, 1990), chap. 10. On the most researched personality test, the MMPI: R. D. Buchanan, 'The Development of the Minnesota Multiphasic Personality Inventory', JHBS, **30** (1994): 148–61. For psychological diagnostics in The Netherlands: T. Dehue, *Changing the Rules: Psychology in The Netherlands, 1900–1985* (Cambridge: Cambridge University Press, 1995), and P. J. van Strien, 'The Historical Practice of Theory Construction', in H. V. Rappard *et al.* (eds), *Annals of Theoretical Psychology, Volume 8* (New York: Plenum Press, 1993), pp. 149–227. For the context of the work on the authoritarian personality: F. Samelson, 'Authoritarianism from Berlin to Berkeley: On Social Psychology and History', in B. Harris, R. K. Unger and R. Stenger (eds), '50 Years of Psychology and Social Issues', *Journal of Social Issues*, **42** no. 1 (1986): 191–208, and 'The Authoritarian Character from Berlin to Berkeley and Beyond: The Odyssey of a Problem', in W. F. Stone, G. Lederer and R. Christie (eds), *Strength and Weakness: The Authoritarian Personality Today* (New York: Springer, 1993), pp. 22–43. On Fromm's relation to this work: J. Baars and P. Scheepers, 'Theoretical and Methodological Foundations of the Authoritarian Personality', JHBS, **29** (1993): 345–53.

The phrase 'occupational psychology' covers a nebulous area, especially when belief in the separation between pure and applied work is rejected, as it is in the studies cited earlier. The development of the psychology profession in relation to the social value of adjustment is discussed in D. S. Napoli, *Architects of Adjustment: The History of the Psychological Profession in the United States* (Port Washington, NY: Kennikat Press, 1981). For the early physiological study of work and fatigue: A. Rabinbach, 'The European Science of Work: The Economy of the Body at the End of the Nineteenth Century', in S. L. Kaplan and C. J. Koepp (eds), *Work in France: Representations, Meaning, Organization, and Practice* (Ithaca: Cornell University Press, 1986), pp. 475–513, and *The Human Motor: Energy, Fatigue, and the Origins of Modernity* (New York: Basic Books, 1990). Early clinical work is described in B. Richards, 'Lightner Witmer and the Project of Psychotechnology', HHS, **1** (1988): 201–19. For a biography of Taylor, see D. Nelson, *Frederick W. Taylor and the Rise of Scientific Management* (Madison: University of Wisconsin Press, 1980), and for a comparative perspective, J. A. Merkle, *Management and Ideology: The Legacy of the International Scientific Management Movement* (Berkeley: University of California Press, 1980). Taylorism also appealed to some activists in the Central Institute of Labour in the early Soviet state: S. Smith, 'Taylorism Rules OK? Bolshevism, Taylorism and the Technical Intelligentsia in the Soviet Union, 1917–41', *Radical Science Journal*, no. 13 (1983): 3–37. The effects of scientific management on labour are described in H. Braverman, *Labour and Monopoly Capital: The Degradation of Work in the Twentieth Century* (New York: Monthly Review Press, 1974). On the life and work of Münsterberg, who separated an academic and philosophical psychology from psychotechnics, see M. Hale, Jr, *Human Science and Social Order: Hugo Münsterberg* *and the Origins of Applied Psychology* (Philadelphia: Temple University Press, 1980); J. Spillmann and L. Spillmann, 'The Rise and Fall of Hugo Münsterberg', JHBS, **29** (1993): 322–38. For the 1920s: M. Sokal, 'The Origins of the Psychological Corporation', JHBS, **17** (1981): 54–67, and 'James McKeen Cattell and American Psychology in the 1920s', in Brožek (ed) (1984), pp. 273–323. Psychology is discussed in relation to the wider popularization of science in, J. C. Burnham, *How Superstition Won and Science Lost: Popularizing Science and Health in the United States* (New Brunswick: Rutgers University Press, 1987), chap. 3. The understanding of the importance of personnel selection to the history of the German psychology profession was transformed in U. Geuter, 'German Psychology during the Nazi Period', in M. G. Ash and W. R. Woodward (eds), *Psychology in Twentieth-Century Thought and Society* (Cambridge: Cambridge University Press, 1987), pp. 165–87, and *The Professionalization of Psychology in Nazi Germany* (Cambridge: Cambridge University Press, 1992); see also M. G. Ash, 'Psychology in Twentieth-Century Germany: Science and Profession', in G. Cocks and K. Jarausch (eds), *German Professions, 1800–1950* (New York: Oxford University Press, 1990), pp. 289–307. The significance of World War II to psychologists is stressed in Rose (1990), and for the US, in E. Herman, *The Romance of American Psychology: Political Culture in the Age of Experts* (Berkeley: University of California Press, 1995), chaps. 2–4. For the transformation of psychological power with the techniques exemplified by the work of the Tavistock, see P. Miller and N. Rose, 'On Therapeutic Authority: Psychoanalytical Expertise under Advanced Liberalism', HHS, **7** no. 3 (1994): 29–64.

A study of the Vienna Psychological Institute is provided by M. G. Ash, 'Psychology and Politics in Interwar Vienna: The Vienna Psychological Institute, 1922–1942', in Ash and Woodward (eds) (1987), pp. 143–64. The richness of links between psychology, education and politics is made apparent in the rushed book by S. Gardner and G. Stevens, *Red Vienna and the Golden Age of Psychology, 1918–1938* (New York: Praeger, 1992). For Peller and Montessori, R. Kramer, *Maria Montessori: A Biography* (reprint, Oxford: Basil Blackwell, 1978), pp. 285–91. For some idea of developments elsewhere in Central Europe: J. Brožek and J. Hoskovec, *Thomas Garrique Masaryk on Psychology:*

Six Facets of the Psyche (Prague: Charles University, 1995); K. Marinkovic, 'A History of Psychology in Former Yugoslavia: An Overview', JHBS, **28** (1992): 340–51; C. Pléh, 'A Hungarian Bergsonian Psychologist: Valéria Dienes', *Hungarian Studies*, **5** (1989): 141–5. The history of childcare ramifies into the history of gender, medicine and social history. On the origins of the child-study movement in the US see: D. Ross, *G. Stanley Hall: The Psychologist as Prophet* (Chicago: University of Chicago Press, 1972), especially for Hall's concern with adolescence; M. Horn, *Before It's Too Late: The Child Guidance Movement in the United States, 1922–1945* (Philadelphia: Temple University Press, 1989), a history of the Commonwealth Fund; for the clinical dimensions, written from a vantage point within the occupation: M. Levine and A. Levine, *A Social History of the Helping Services: Clinic, Court, School, and Community* (New York: Appleton-Century-Crofts, 1970), and J. M. Reisman, *A History of Clinical Psychology* (2nd edn, New York: Irvington Publishers, 1976); L. Zenderland, 'Education, Evangelism, and the Origins of Clinical Psychology: The Child-study Legacy', JHBS, **24** (1988): 152–65. F. Matthews, 'The Utopia of Human Relations: The Conflict-free Family in American Social Thought, 1930–1960', JHBS, **24** (1988): 343–62, discusses the literature on child-rearing. For Susan Isaacs, see Wooldridge (1994).

Any discussion of nature–nurture issues will be judged tendentious by one party or another, but historians bring more clarity and objectivity than many other commentators. The issues are continuous with the earlier debates over intelligence referenced above; see especially Fancher (1985); and for a bibliography, S. H. Aby and M. J. McNamara, *The IQ Debate: A Selective Guide to the Literature* (New York: Greenwood Press, 1990). For a detailed study see C. N. Degler, *In Search of Human Nature: The Decline and Revival of Darwinism in American Social Thought* (New York: Oxford University Press, 1991). A relevant examination of why scientists adopt different styles of argument is J. Harwood, 'The Race–Intelligence Controversy: A Sociological Approach', *Social Studies of Science*, **6** (1976): 369–94, and **7** (1977): 1–30. Harwood describes the political and educational background to the revived intelligence controversy of the late 1960s in 'Heredity, Environment, and the Legitimation of Social Policy', in B. Barnes and S. Shapin (eds), *Natural Order: Historical*

Studies of Scientific Culture (Beverly Hills: Sage, 1979), pp. 231–51. For hereditarian–environmentalist issues earlier in the twentieth century see also: the papers collected in J. C. Burnham, *Paths into American Culture: Psychology, Medicine, and Morals* (Philadelphia: Temple University Press, 1988); H. Cravens, *The Triumph of Evolution: American Scientists and the Heredity–Environment Controversy 1900–1941* (Philadelphia: University of Pennsylvania Press, 1978), which argues the two positions were (temporarily) reconciled in a 1930s synthesis. Galton's early use of 'nature–nurture' language is described in R. E. Fancher, 'Alphonse de Candolle, Francis Galton, and the Early History of the Nature–Nurture Controversy', JHBS, **19** (1983): 341–52. Boas's role in the stress on culture in the formation of human nature is widely acknowledged; see G. W. Stocking, Jr, 'Introduction: The Basic Assumptions of Boasian Anthropology', in G. W. Stocking, Jr (ed), *The Shaping of American Anthropology 1883–1911: A Franz Boas Reader* (New York: Basic Books, 1974), pp. 1–20, and 'Franz Boas and the Culture Concept in Historical Perspective', in *Race, Culture, and Evolution: Essays in the History of Anthropology* (reprint, Chicago: University of Chicago Press, 1982), pp. 195–233. For the 1920s background to the work of Benedict and Mead: G. W. Stocking, Jr, 'The Ethnographic Sensibility of the 1920s and the Dualism of the Anthropological Tradition', in G. W. Stocking, Jr (ed), *Romantic Motives: Essays on Anthropological Sensibility, History of Anthropology, Volume 6* (Madison: University of Wisconsin Press, 1989), pp. 208–76. The basis of Mead's work in Samoa is undermined in D. Freeman, *Margaret Mead and Samoa: The Making and Unmaking of an Anthropological Myth* (Cambridge, MA: Harvard University Press, 1983), though Freeman's viewpoint was criticized in turn. For more detail, L. Foerstel and A. Gilliam (eds), *Confronting the Margaret Mead Legacy: Scholarship, Empire, and the South Pacific* (Philadelphia: Temple University Press, 1992). On measures of masculinity and femininity: M. Lewin (ed), *In the Shadow of the Past: Psychology Portrays the Sexes. A Social and Intellectual History* (New York: Columbia University Press, 1984); J. Morawski, 'The Measurement of Masculinity and Femininity: Engendering Categorical Realities', *Journal of Personality*, **53** (1985): 196–223. A. Jensen's paper, which reanimated the IQ controversy, appeared as 'How Much Can We Boost I.Q. and Scholastic Achievement?', *Harvard Educational Review*, **39**

(1969): 1–123. The 'Burt business' is reviewed, after archival research, in the biography by Hearnshaw (1979), chap. 12, with a relevant if unpersuasive assessment of Burt's personal strengths and weaknesses in chap. 13. Hearnshaw was criticized for being too gentle on Burt's personal failings and on his profession's failure earlier to reject his findings, but he was also attacked by defenders of Burt. For a careful rejection of Hearnshaw's pro-Burt critics, see F. Samelson, 'Rescuing the Reputation of Sir Cyril Burt', JHBS, **28** (1992): 221–33. There is an account of Kamin's 'discovery' of Burt's deception in W. H. Tucker, 'Fact and Fiction in the Discovery of Sir Cyril Burt's Flaws', JHBS, **30** (1994): 335–47. The controversial claim that differences of intelligence hold the key to social problems was made again in R. J. Herrnstein and C. Murray, *The Bell Curve: Intelligence and Class Structure in American Life* (New York: Free Press, 1994).

CHAPTER 17

Sigmund Koch has maintained an extended commentary on the search for scientific psychology: S. Koch (ed), *Psychology: A Study of a Science*, 6 vols. (New York: McGraw-Hill, 1959–63); 'Foreward: Wundt's Creature at Age Zero – and as Centenarian: Some Aspects of the Institutionalization of the "New Psychology"', and 'The Nature and Limits of Psychological Knowledge: Lessons of a Century qua "Science"', in S. Koch and D. E. Leary (eds), *A Century of Psychology as Science* (New York: McGraw-Hill, 1985), pp. 7–35 and 75–97, and the volume as a whole includes reviews of the different areas of scientific psychology; for an overview see S. Toulmin and D. E. Leary, 'The Cult of Empiricism in Psychology and Beyond', pp. 594–617. For questions about the identity of scientific psychology: M. G. Ash, 'The Self-presentation of a Discipline: History of Psychology in the United States between Pedagogy and Scholarship', and U. Geuter, 'The Uses of History for the Shaping of a Field: Observations on German Psychology', in L. Graham, W. Lepenies and P. Weingart (eds), *Functions and Uses of Disciplinary Histories, Sociology of the Sciences Yearbook, Volume* 7 (Dordrecht: D. Reidel, 1983), pp. 143–89 and 191–228. The identity and representation of psychology as a field is also discussed in J. G. Morawski, 'Self-regard and Other-regard: Reflexive Practices in American Psychology, 1890–1940', *Science in Context*, **5** (1992): 281–308. The distinctiveness of British psychology is made clear in L. S. Hearnshaw, *A Short History of British Psychology, 1840–1940* (London: Methuen, 1964); and further on Bartlett, A. Costall, 'Why British Psychology Is not Social: Frederic Bartlett's Promotion of the New Academic Discipline', *Canadian Psychology/Psychologie canadienne*, **33** (1992): 633–9. For studies of the meaning and effect of quantification in experimentation: K. Danziger, *Constructing the Subject: Historical Origins of Psychological Research* (Cambridge: Cambridge University Press, 1990); G. A. Hornstein, 'Quantifying Psychological Phenomena: Debates, Dilemmas, and Implications', in J. G. Morawski (ed), *The Rise of Experimentation in American Psychology* (New Haven: Yale University Press, 1988), pp. 1–34.

There is a detailed account of the experimental basis of Sherrington's view of the nervous system in J. P. Swazey, *Reflexes and Motor Integration: Sherrington's Concept of Integrative Action* (Cambridge, MA: Harvard University Press, 1969). The background to Pavlov's work on higher nervous functions in Russia and the Soviet Union is provided in D. Joravsky, *Russian Psychology: A Critical History* (Oxford: Basil Blackwell, 1989), which is very critical of the claims made by physiologists to understand human nature. Further references to Russian psychology are given in Chapter 19. Soviet studies of Pavlov are interesting in their own right for the manner in which they relate science and values; see, e.g., P. K. Anokhin, 'Ivan P. Pavlov and Psychology', in B. B. Wolman (ed), *Historical Roots of Contemporary Psychology* (New York: Harper & Row, 1968), pp. 131–59. On Pavlov's school, G. Windholz, 'Pavlov's Position toward American Behaviorism', JHBS, **19** (1983): 394–407, and 'Pavlov and the Pavlovians in the Laboratory', JHBS, **26** (1990): 64–74. A central topic in the history of brain–mind relations, and of relations between physiology and psychology, is examined in R. Smith, *Inhibition: History and Meaning in the Sciences of Mind and Brain* (Berkeley: University of California Press, and London: Free Association Books, 1992). For debates in the German Weimar period: A. Harrington, 'Interwar "German" Psychobiology: Between Nationalism and the Irrational', *Science in Context*, **4** (1991): 429–47; *Reenchanted Science: Holism in German Culture from*

Wilhelm II to Hitler (Princeton: Princeton University Press, 1996).

Historians of behaviourism have created a rich picture of the relations between the philosophy of science, psychology and cultural and political values; see especially J. M. O'Donnell, *The Origins of Behaviorism: American Psychology, 1870–1920* (New York: New York University Press, 1986). For behaviourism and animal psychology: R. Boakes, *From Darwin to Behaviourism: Psychology and the Mind of Animals* (Cambridge: Cambridge University Press, 1984). The relationship between behaviourism and mechanistic biology is discussed in P. J. Pauly, *Controlling Life: Jacques Loeb and the Engineering Ideal in Biology* (New York: Oxford University Press, 1987). The claims by behaviourists to exemplify scientific method are examined in B. D. Mackenzie, *Behaviourism and the Limits of Scientific Method* (London: Routledge & Kegan Paul, 1977). A biography of Watson is K. W. Buckley, *Mechanical Man: John Broadus Watson and the Beginnings of Behaviorism* (New York: Guilford Press, 1989). The legacy of Watson's roots in small-town society is argued in P. G. Creelan, 'Watsonian Behaviorism and the Calvinist Conscience', JHBS, **10** (1974): 95–118, while the relation between his roots and social change is explored in D. Bakan, 'Behaviorism and American Urbanization', JHBS, **2** (1966): 5–28, and in essays by J. C. Burnham collected in *Paths into American Culture: Psychology, Medicine, and Morals* (Philadelphia: Temple University Press, 1988). Watson's contributions to advertising are described in K. W. Buckley, 'The Selling of a Psychologist: John Broadus Watson and the Application of Behavioral Techniques to Advertising', JHBS, **18** (1982): 207–21. On Watson's work as a moral project, C. J. Karier, *Scientists of the Mind: Intellectual Founders of Modern Psychology* (Urbana: University of Illinois Press, 1986), chap. 4. Claims for the existence of a behaviourist revolution are criticized in F. Samelson, 'Struggle for Scientific Authority: The Reception of Watson's Behaviorism 1913–1920', JHBS, **17** (1981): 399–425, and 'Organizing for the Kingdom of Behavior: Academic Battles and Organizational Policies in the Twenties', JHBS, **21** (1985): 33–47. Watson's relations with Meyer are detailed in R. Leys, 'Meyer, Watson, and the Dangers of Behaviorism', JHBS, **20** (1984): 128–49. The famous encounter between Watson, Rayner and 'Little Albert' is assessed in B. Harris, 'Whatever Happened to Little Albert?', *American Psychologist*, **34** (1979):

151–60. The critique of Watson's mechanistic philosophy in the 1920s, especially by William McDougall, is discussed in M. A. Boden, *Purposive Explanation in Psychology* (Cambridge, MA: Harvard University Press, 1972).

The background concern about a field dominated by applied psychology, which encouraged Boring to write his history, is examined in J. M. O'Donnell, 'The Crisis of Experimentalism in the 1920's: E. G. Boring and His Uses of History', *American Psychologist*, **34** (1979): 289–95. The study on philosophy and neobehaviourism, which revises the conventional view of the influence of logical positivism, is L. D. Smith, *Behaviorism and Logical Positivism: A Reassessment of the Alliance* (Stanford: Stanford University Press, 1986), and, more briefly, 'Psychology and Philosophy: Towards a Realignment, 1905–1935', JHBS, **17** (1981): 28–37. On operationism: T. B. Rogers, 'Operationism in Psychology: A Discussion of Contextual Antecedents and an Historical Interpretation of Its Longevity', JHBS, **25** (1989): 139–53; and on Hull at Yale, J. G. Morawski, 'Organizing Knowledge and Behavior at Yale's Institute of Human Relations', *Isis*, **77** (1986): 219–42; and for Brunswik, D. E. Leary, 'From Act Psychology to Probabilistic Functionalism: The Place of Egon Brunswik in the History of Psychology', in M. G. Ash and W. R. Woodward (eds), *Psychology in Twentieth-Century Thought and Society* (Cambridge: Cambridge University Press, 1987), pp. 115–42. For the arguments for and against Skinner's behaviourism: S. Koch, 'Psychology and Emerging Conceptions of Knowledge as Unitary', and B. F. Skinner, 'Behaviorism at Fifty', in T. W. Wann (ed), *Behaviorism and Phenomenology: Contrasting Bases for Modern Psychology* (Chicago: University of Chicago Press for William Marsh Rice University, 1964), pp. 1–45 and 79–97. There is little doubt that Skinner was often treated as emblematic of the faults of positivist science and not read; for a sympathetic assessment of his contribution, see M. N. Richelle, *B. F. Skinner: A Reappraisal* (Hove, Sussex, and Hillsdale, NJ: Lawrence Erlbaum Associates, 1993). Skinner wrote his autobiography in his own psychological terms: *Particulars of My Life* (New York: Alfred A. Knopf, 1976), *The Shaping of a Behaviorist* (New York: Alfred A. Knopf, 1979), and *A Matter of Consequences* (New York: Alfred A. Knopf, 1983). For a biography: D. W. Bjork, *B. F. Skinner: A Life* (New York: Basic Books, 1993).

Psychical research has attracted the interest of historians if not always of psychologists. For the nineteenth-century background in spiritualism and mesmerism: H. F. Ellenberger, *The Discovery of the Unconscious: The History and Evolution of Dynamic Psychiatry* (New York: Basic Books, and London: Allen Lane, The Penguin Press, 1970); A. Gauld, *The Founders of Psychical Research* (London: Routledge & Kegan Paul, 1968), and *A History of Hypnotism* (Cambridge: Cambridge University Press, 1992). For spiritualism and psychical research: J. J. Cerullo, *The Secularization of the Soul: Psychical Research in Modern Britain* (Philadelphia: Institute for the Study of Human Issues, 1982); J. Oppenheim, *The Other World: Spiritualism and Psychical Research in England, 1850–1914* (Cambridge: Cambridge University Press, 1985); A. Owen, *The Darkened Room: Women, Power and Spiritualism in Late Victorian England* (Philadelphia: University of Pennsylvania Press, 1990); J. P. Williams, 'Psychical Research and Psychiatry in Late Victorian Britain: Trance as Ecstasy or Trance as Insanity', in W. F. Bynum, R. Porter and M. Shepherd (eds), *The Anatomy of Madness: Essays in the History of Psychiatry*, Vol. 1 (London: Tavistock, 1985), pp. 233–54. For the US, R. L. Moore, *In Search of White Crows: Spiritualism, Parapsychology, and American Culture* (New York: Oxford University Press, 1977). The relations between psychical research and psychology are clearly described in J. Beloff, *Parapsychology: A Concise History* (London: Athlone Press, 1993). The work of the Rhines is examined in a detailed history: S. H. Mauskopf and M. R. McVaugh, *The Elusive Science: Origins of Experimental Psychical Research* (Baltimore: Johns Hopkins University Press, 1980). For spiritualism and psychology in France in the 1920s: F. Parot, 'Psychology Experiments: Spiritism at the Sorbonne', JHBS, **29** (1993): 22–8.

The study of gestalt psychology is placed on a definitive footing and related to German scientific culture in M. G. Ash, *Gestalt Psychology in German Culture, 1890–1967: Holism and the Quest for Objectivity* (Cambridge: Cambridge University Press, 1995). This major study is preceded by Ash's survey in 'Gestalt Psychology: Origins in Germany and Reception in the United States', in C. E. Buxton (ed), *Points of View in the Modern History of Psychology* (Orlando: Academic Press, 1985), pp. 295–344. See also 'Max Wertheimer's University Career in Germany', *Psychological Research*, **51** (1989): 52–7, and 'Gestalt Psychology in Weimar

Culture' HHS, **4** (1991): 395–415, which explores the aspirations of the gestalt psychologists to resolve questions about values. For the experimental study of perception in the nineteenth century, see references in Chapter 14. Köhler's work on Tenerife is also described in Boakes (1984), Chap. 7. For Köhler and the Nazis: M. Henle, 'One Man against the Nazis – Wolfgang Köhler', *American Psychologist*, **33** (1978): 939–44; J. Wolhwill, 'German Psychological Journals under National Socialism: A History of Contrasting Paths', JHBS, **23** (1987): 169–85. The personal side of Koffka is captured in M. Harrower, *Kurt Koffka: An Unwitting Self-portrait* (Gainseville: University Presses of Florida, 1983). On the intellectual migration: L. A. Coser, *Refugee Scholars in America: Their Impact and Their Experiences* (New Haven: Yale University Press, 1984); J. C. Jackman and C. M. Borden (eds), *The Muses Flee Hitler: Cultural Transfer and Adaptation 1930–1945* (Washington, DC: Smithsonian Institution Press, 1983), for a comparative perspective but not including psychology; M. M. Sokal, 'The Gestalt Psychologists in Behaviorist America', *American Historical Review*, **89** (1984): 1240–63.

The standard surveys of the phenomenological movement and its influence are H. Misiak and V. S. Sexton, *Phenomenological, Existential, and Humanistic Psychologies: A Historical Survey* (New York: Grune & Stratton, 1973); and, in more detail, H. Spiegelberg, *The Phenomenological Movement: A Historical Introduction*, 2 vols. (2nd edn, The Hague: Martinus Nijhoff, 1976), and *Phenomenology in Psychology and Psychiatry: A Historical Introduction* (Evanston: Northwestern University Press, 1972). Phenomenology was introduced to North American psychologists by R. B. MacLeod; see 'Phenomenology: A Challenge to Experimental Psychology', in Wann (ed) (1964), pp. 47–78. For Brentano, see the references cited for Chapter 14. For Michotte and also Catholic phenomenological psychology, H. Misiak and V. M. Staudt, *Catholics in Psychology: A Historical Survey* (New York: McGraw-Hill, 1954). For Buytendijk and the Dutch phenomenologists: T. Dehue, *Changing the Rules: Psychology in The Netherlands, 1900–1985* (Cambridge: Cambridge University Press, 1995), chap. 3; J. J. Kockelmans (ed), *Phenomenological Psychology: The Dutch School* (Dordrecht: Martinus Nijhoff, 1987).

CHAPTER 18

Psychoanalysis is the most argued about and most written about of all the psychologies. The enormous and often heated literature on Freud's life and work has all the animus of a love affair gone wrong. There is a magisterial – and readable – historical study, H. F. Ellenberger, *The Discovery of the Unconscious: The History and Evolution of Dynamic Psychiatry* (New York: Basic Books, and London: Allen Lane, The Penguin Press, 1970), with separate chapters on the main actors – Janet, Freud, Jung and Adler – as well as a picture of nineteenth-century humanistic culture. The 'magnetic' psychotherapeutic background is also stressed in A. Crabtree, *From Mesmer to Freud: Magnetic Sleep and the Roots of Psychological Healing* (New Haven: Yale University Press, 1993). On healing movements in the US see also, R. C. Fuller, *Mesmerism and the American Cure of Souls* (Philadelphia: University of Pennsylvania Press, 1982), and *Americans and the Unconscious* (New York: Oxford University Press, 1986). For widely read studies of Freud in relation to changes in the wider culture: H. S. Hughes, *Consciousness and Society: The Reorientation of European Social Thought 1890–1930* (reprint, London: Paladin, 1974); P. Rieff, *Freud: The Mind of the Moralist* (reprint, New York: Anchor Books, 1961), and *The Triumph of the Therapeutic: Uses of Faith after Freud* (reprint, Harmondsworth: Penguin Books, 1973). There are succinct introductions to Freud in R. E. Fancher, *Psychoanalytic Psychology: The Development of Freud's Thought* (New York: W. W. Norton, 1973), and 'Freud and Psychoanalysis', in R. C. Olby *et al.* (eds), *Companion to the History of Modern Science* (London: Routledge, 1990), pp. 425–41. For introductions, with references, to the debate whether psychoanalysis is a scientific psychology: S. Rosenzweig, 'Freud and Experimental Psychology: The Emergence of Idiodynamics', in S. Koch and D. E. Leary (eds), *A Century of Psychology as Science* (New York: McGraw-Hill, 1985), pp. 135–207; R. S. Steele, 'Paradigm Found: A Deconstruction of the Psychoanalytic Movement' and 'Paradigm Lost: Psychoanalysis after Freud', in C. E. Buxton (ed), *Points of View in the Modern History of Psychology* (Orlando: Academic Press, 1985), pp. 197–219 and 221–257. On Freud's language and its rendition in English, the essential study is D. G. Ornston, Jr (ed), *Translating Freud* (New Haven: Yale University

Press, 1991); also, introduction to N. Kiell, *Freud without Hindsight: Reviews of His Work (1893–1939)* (Madison, CT: International Universities Press, 1988), which also reprints reviews, and E. Timms and N. Segal (eds), *Freud in Exile: Psychoanalysis and Its Vicissitudes* (New Haven: Yale University Press, 1988), Part III. The linguistic context of Freud's work is explored in J. Forrester, *Language and the Origins of Psychoanalysis* (London: Macmillan, 1980).

In addition to Ellenberger (1970), there is a biography of Freud by P. Gay, *Freud: A Life for Our Time* (New York: W. W. Norton, 1988), with bibliographic essay. Gay's biography is considered in relation to other literature in J. E. Toews, 'Historicizing Psychoanalysis: Freud in His Time and for Our Time', *Journal of Modern History*, **63** (1991): 504–45; also, E. Young-Breuhl, 'A History of Freud Biographies', in M. S. Micale and R. Porter (eds), *Discovering the History of Psychiatry* (New York: Oxford University Press, 1994), pp. 157–73, and J. Forrester, '"A Whole Climate of Opinion": Rewriting the History of Psychoanalysis', in *ibid.*, pp. 174–90. The earlier, official biography, replaced in detail but still of interest, is E. Jones, *Sigmund Freud: Life and Work*, 3 vols. (London: Hogarth Press, 1953–7). A rich collection of pictures is in E. Freud, L. Freud and I. Grubrich-Simitis (eds), *Sigmund Freud: His Life in Pictures and Words* (London: André Deutsch, 1978). For Freud and Vienna, see the readable introduction in J. Miller (ed), *Freud, the Man, His World, His Influence* (London: Weidenfeld & Nicolson, 1972), the cultural history in C. E. Schorske, *Fin-de-siècle Vienna: Politics and Culture* (Cambridge: Cambridge University Press, 1981), chap. 4, and the detailed study in W. J. McGrath, *Freud's Discovery of Psychoanalysis: The Politics of Hysteria* (Ithaca: Cornell University Press, 1986). On the clinical and therapeutic background, besides Ellenberger (1970): A. Gauld, *A History of Hypnotism* (Cambridge: Cambridge University Press, 1992); M. S. Micale, 'Hysteria and Its Historiography: A Review of Past and Present Writings', *History of Science*, **27** (1989): 223–61, 319–51, 'Hysteria and Its Historiography: The Future Perspective', *History of Psychiatry*, **1** (1990): 33–124, and *Approaching Hysteria: Disease and Its Interpretations* (Princeton: Princeton University Press, 1995); E. Shorter, *From Paralysis to Fatigue: A History of Psychosomatic Illness in the Modern Era* (New York: Free Press, 1992). K. Lewin, *Freud's Early Psy-*

chology of the Neuroses: A Historical Perspective (Hassocks, Sussex: Harvester Press, 1978), details the early history of the theory of the neuroses. For Breuer, see A. Hirschmüller, *The Life and Work of Josef Breuer: Physiology and Psychoanalysis* (New York: New York University Press, 1989), and for an introduction to the interest in child sexuality, S. Kern, 'Freud and the Discovery of Child Sexuality', *History of Childhood Quarterly: The Journal of Psychohistory*, **1** (1973): 117–41. For the argument about the significance of Freud's 'Project for a Scientific Psychology': P. Amacher, *Freud's Neurological Education and Its Influence on Psychoanalytic Theory* (New York: International Universities Press, 1965); J. Friedman and J. Alexander, 'Psychoanalysis and Natural Science: Freud's 1895 *Project* Revisited', *International Review of Psychoanalysis*, **10** (1983): 303–18. The lasting biological dimension of Freud's work is detailed in F. J. Sulloway, *Freud, Biologist of the Mind: Beyond the Psychoanalytic Legend* (London: Burnett Books, André Deutsch, 1979). The reception of Freud's work is described in H. S. Decker, *Freud in Germany: Revolution and Reaction in Science, 1893–1907* (New York: International Universities Press, 1977), while the early analysts are the subject of P. Grosskurth, *The Secret Ring: Freud's Inner Circle and the Politics of Psychoanalysis* (London: Jonathan Cape, 1991), and P. Roazen, *Freud and His Followers* (New York: Alfred A. Knopf, 1975). Freud's many relationships with women – patients, family, disciples, friends – are discussed in L. Appignanesi and J. Forrester, *Freud's Women* (London: Virago Press, 1993).

The effect of World War I on Freud's reassessment of his theories is considered in J. van Ginneken, 'The Killing of the Father: The Background of Freud's Group Psychology', *Political Psychology*, **5** (1984): 391–414; L. E. Hoffman, 'War, Revolution, and Psychoanalysis: Freudian Thought Begins to Grapple with Social Reality', JHBS, **17** (1981): 251–69, and 'The Ideological Significance of Freud's Social Thought', in M. G. Ash and W. R. Woodward (eds), *Psychology in Twentieth-Century Thought and Society* (Cambridge: Cambridge University Press, 1987), pp. 253–69. See also P. Roazen, *Freud: Political and Social Thought* (London: Hogarth Press, 1968). For details, P. E. Stepansky, *A History of Aggression in Freud* (New York: International Universities Press, 1977), and on Freud's anthropology, E. R. Wallace, IV, *Freud and Anthropology: A History and Reappraisal* (New York:

International Universities Press, 1983). There is a brief review of the debate about the significance of Freud's Jewishness in J. Miller, 'Interpretations of Freud's Jewishness, 1924–1974', JHBS, **17** (1981): 357–74; while the practical consequences of Jewishness are detailed in relation to one of Freud's most famous cases in H. S. Decker, *Freud, Dora, and Vienna 1900* (New York: Free Press, 1991). There is a comparative survey of the Freudians in E. Kurzweil, *The Freudians: A Comparative Perspective* (New Haven: Yale University Press, 1989), while the thought of the radical Freudians is described in P. A. Robinson, *The Freudian Left: Wilhelm Reich, Geza Roheim, Herbert Marcuse* (New York: Harper & Row, 1969), and defended in R. Jacoby, *Social Amnesia: A Critique of Conformist Psychology from Adler to Laing* (Boston: Beacon Press, 1975). For Fenichel: R. Jacoby, *The Repression of Psychoanalysis: Otto Fenichel and the Political Freudians* (New York: Basic Books, 1983); B. Harris and A. Brock, 'Freudian Psychopolitics: The Rivalry of Wilhelm Reich and Otto Fenichel, 1930–1935', *Bulletin of the History of Medicine*, **66** (1992): 578–612. Many Freudians are the subject of biography: E. Young-Bruehl, *Anna Freud: A Biography* (New York: Summit Books, and London: Macmillan, 1988); S. Quinn, *A Mind of Her Own: The Life of Karen Horney* (New York: Simon & Schuster, 1987); P. Grosskurth, *Melanie Klein: Her World and Her Work* (London: Hodder & Stoughton, 1986); M. Sharaf, *Fury on Earth: A Biography of Wilhelm Reich* (reprint, London: Hutchinson, 1984); P. Roazen, *Erik H. Erikson: The Power and Limits of Vision* (New York: Free Press, 1976), and *Helene Deutsch: A Psychoanalytic Life* (New Brunswick, NJ: Transaction Books, 1985). For the reception of psychoanalysis in the US: J. C. Burnham, *Psychoanalysis and American Medicine, 1894–1918: Medicine, Science and Culture* (New York: International Universities Press, 1967), and 'From Avant-garde to Specialism: Psychoanalysis in America', JHBS, **15** (1979): 128–34; N. G. Hale, Jr, *Freud and the Americans: The Beginnings of Psychoanalysis in the United States, 1876–1917* (New York: Oxford University Press, 1971), 'From Berggasse xix to Central Park West: The Americanization of Psychoanalysis, 1919–1940', JHBS, **14** (1978): 299–315, 'Freud's Reich, the Psychiatric Establishment, and the Founding of the American Psychoanalytic Association: Professional Styles in Conflict', JHBS, **15** (1979): 135–41, and, a general interpretation, *The*

Rise and Crisis of Psychoanalysis in the United States: Freud and the Americans, 1917–1985 (New York: Oxford University Press, 1995). On psychoanalysis and the psychology profession, G. A. Hornstein, 'The Return of the Repressed: Psychology's Problematic Relations with Psychoanalysis, 1909–1960', *American Psychologist*, **47** (1992): 254–63. On the shift from European to American values in ego psychology, see H. S. Hughes, *The Sea Change: The Migration of Social Thought, 1930–1965* (New York: Harper & Row, 1975), chap. 5, and S. R. Kirschner, 'The Assenting Ego: Anglo-American Values in Contemporary Psychoanalytic Developmental Psychology', *Social Research*, **57** (1980): 821–57; while Fromm, who contributed much to this 'migration', is defended from the charge of superficiality in D. Burston, *The Legacy of Erich Fromm* (Cambridge, MA: Harvard University Press, 1991). On Freud in Britain: D. A. Rapp, 'The Reception of Freud by the British Press: General Interest and Literary Magazines, 1920–1925', JHBS, **24** (1988): 191–201, and 'The Early Discovery of Freud by the British General Educated Public, 1912–1919', *Social History of Medicine*, **3** (1990): 217–43; T. Winslow, 'Bloomsbury, Freud and the Vulgar Passions', *Social Research*, **57** (1980): 785–819. For the object-relations theorists, see E. Rayner, *The Independent Mind in British Psychoanalysis* (London: Free Association Books, 1990); also, G. Kohon, 'Note on the History of the Psychoanalytic Movement in Great Britain', in G. Kohon (ed), *The British School of Psychoanalysis: The Independent Tradition* (London: Free Association Books, 1986), pp. 24–50. The history of German psychotherapy during the Third Reich is reassessed in G. Cocks, *Psychotherapy in the Third Reich: The Göring Institute* (New York: Oxford University Press, 1985), and 'The Professionalization of Psychotherapy in Germany, 1928–1949', in G. Cocks and K. H. Jarausch (eds), *German Professions, 1800–1950* (New York: Oxford University Press, 1990), pp. 308–28. For psychoanalysis in France: E. Roudinesco, *Jacques Lacan & Co.: A History of Psychoanalysis in France, 1925–1985* (Chicago: University of Chicago Press, and London: Free Association Books, 1990), the first volume of a huge study of which the later parts are not translated. Lacan is discussed in Chapter 20, while M. N. Evans, *Fits and Starts: A Genealogy of Hysteria in Modern France* (Ithaca: Cornell University Press, 1991), is relevant, especially on questions of gender.

As well as Ellenberger (1970) on Adler and Jung, see P. E. Stepansky, *In Freud's Shadow: Adler in Context* (Hillsdale, NJ: The Analytic Press, Lawrence Erlbaum Associates, 1983). There is no full biography of Jung and the many accounts of his work are much more concerned with the ideas than their history. But on his relations with Freud, see: J. Kerr, *A Most Dangerous Method: The Story of Jung, Freud, and Sabina Spielrein* (London: Sinclair-Stevenson, 1994); P. E. Stepansky, 'The Empiricist as Rebel: Jung, Freud, and the Burdens of Discipleship', JHBS, **12** (1976): 216–39. Jung is linked to debate about modernity and psychological man in P. Homans, *Jung in Context – Modernity and the Making of Psychology* (Chicago: University of Chicago Press, 1979). The connections between Jung's thought and the *völkisch* dimensions of German-language culture are stressed in R. Noll, *The Jung Cult: Origins of a Charismatic Movement* (Princeton: Princeton University Press, 1994). The contentious matter of Jung's relations to wider political events in the 1930s are considered from different perspectives in A. Maidenbaum and S. A. Martin (eds), *Lingering Shadows: Jungians, Freudians, and Anti-Semitism* (Boston: Shambhala, 1991), and Jung's stance is subjected to thoughtful moral critiques in C. J. Karier, *Scientists of the Mind: Intellectual Founders of Modern Psychology* (Urbana: University of Illinois Press, 1986), chap. 8, a book which also has chapters on Freud and Adler, and in A. Samuels, 'National Psychology, National Socialism, and Analytical Psychology. Reflections on Jung and Anti-Semitism', *Journal of Analytical Psychology,* **37** (1992): 3–28 and 127–48. Jung's relevant writings are in *Civilization in Transition*, in *The Collected Works of C. G. Jung. Volume 10* (2nd edn, London: Routledge & Kegan Paul, 1970). On analytic psychology after Jung see A. Samuels, *Jung and the Post-Jungians* (London: Routledge, 1986), and 'The Professionalization of Carl G. Jung's Analytical Psychology Clubs', JHBS, **30** (1994): 138–47.

CHAPTER 19

There are general histories of social psychology in F. B. Karpf, *American Social Psychology: Its Origins, Development, and European Background* (1932; reprint, Dubuque, IA: Brown Reprints, 1971), which, though dated, covers a wide range, with an addendum in 'American Social Psychology – 1951', *American Journal of Sociology*, **58** (1952): 187–93; and G. Collier, H. L. Minton and G. Reynolds, *Currents of Thought in American Social Psychology* (New York: Oxford University Press, 1991). The standard text for psychologists, G. W. Allport, 'The Historical Background of Modern Social Psychology', in G. Lindzey and E. Aronson (eds), *The Handbook of Social Psychology*, Vol. 1 (2nd edn, Reading, MA: Addison-Wesley, 1968), pp. 1–80, is an interesting historical document in its own right but hardly the history it purports to be. For a critical introduction to the relations between psychological and sociological explanation, see P. F. Secord, 'Social Psychology as a Science', in J. Margolis, P. T. Manicas, R. Harré and P. F. Secord, *Psychology: Designing the Discipline* (Oxford: Basil Blackwell, 1986), pp. 128–64. A historical approach to the tendency of twentieth-century psychology to create an asocial subject is G. Jahoda, *Crossroads between Culture and Mind: Continuities and Change in Theories of Human Nature* (New York: Harvester Wheatsheaf, 1992). There is a rich historical literature on crowd psychology – for a wide perspective, see J. S. McClelland, *The Crowd and the Mob: From Plato to Canetti* (London: Unwin Hyman, 1989). For the late nineteenth century: J. van Ginneken, *Crowds, Psychology, and Politics 1871–1899* (Cambridge: Cambridge University Press, 1992); S. Barrows, *Distorting Mirrors: Visions of the Crowd in Late Nineteenth Century France* (New Haven: Yale University Press, 1981); A. Métraux, 'French Crowd Psychology: Between Theory and Ideology', in W. R. Woodward and M. G. Ash (eds), *The Problematic Science: Psychology in the Nineteenth Century* (New York: Praeger, 1982), pp. 276–99; R. A. Nye, *The Origins of Crowd Psychology: Gustave Le Bon and the Crisis of Mass Democracy in the Third Republic* (Beverly Hills: Sage, 1975), and *Crime, Politics, & Madness in Modern France: The Medical Concept of National Decline* (Princeton: Princeton University Press, 1984); D. Pick, 'The Faces of Anarchy: Lombroso and the Politics of Criminal Science in Post-unification Italy', *History Workshop*, no. 21 (1986): 60–86, and *Faces of Degeneration: A European Disorder, c.1848–c.1918* (Cambridge: Cambridge University Press, 1989). See also G. Tarde, *On Communication and Social Influence: Selected Papers*, ed T. N. Clark (Chicago: University of Chicago Press, 1969).

The evolutionary background of instinct theory is referenced in Chapter 13. In addition see R. H. Mueller, 'A Chapter in the History of the Relationship between Psychology and Sociology in America: James Mark Baldwin', JHBS, **12** (1976): 240–53. Baldwin, Ross and Hall are linked to a search for community in R. J. Wilson, *In Quest of Community: Social Philosophy in the United States, 1860–1920* (New York: John Wiley & Sons, 1968). McDougall has not attracted historians, but see: L. Krantz and D. Allen, 'The Rise and Fall of McDougall and Instinct', JHBS, **3** (1967): 326–38; H. G. McCurdy, 'William McDougall', in B. B. Wolman (ed), *Historical Roots of Contemporary Psychology* (New York: Harper & Row, 1968), pp. 111–130. On Wallas and the wider politics of social reform in Britain, R. N. Soffer, *Ethics and Society in England: The Revolution in the Social Sciences 1870–1914* (Berkeley: University of California Press, 1978). On Frazer: R. Ackerman, *J. G. Frazer: His Life and Work* (Cambridge: Cambridge University Press, 1987); R. A. Jones, 'Robertson Smith and James Frazer on Religion: Two Traditions in British Social Anthropology', in G. W. Stocking, Jr (ed), *Functionalism Historicized: Essays on British Social Anthropology, History of Anthropology, Volume 2* (Madison: University of Wisconsin Press, 1984), pp. 31–58; G. W. Stocking, Jr, *After Tylor: British Social Anthropology 1888–1951* (London: Athlone Press, 1996). For Freud's social psychology, in addition to the references in Chapter 18, see L. E. Hoffman, 'From Instinct to Identity: Implications of Changing Psychoanalytic Concepts of Social Life from Freud to Erikson', JHBS, **18** (1982): 130–46, and D. Pick, 'Freud's *Group Psychology* and the History of the Crowd', *History Workshop Journal*, no. 40 (1995): 39–61.

In general, on US social psychology, in addition to Karpf (1971) and Collier *et al.* (1991), see E. R. Hilgard, *Psychology in America: An Historical Survey* (San Diego: Harcourt Brace Jovanovich, 1987). K. Danziger, *Constructing the Subject: Historical Origins of Psychological Research* (Cambridge: Cambridge University Press, 1990), examines the field itself as a subject for historical social psychology; more briefly, see 'The Project of an Experimental Social Psychology: Historical Perspectives', *Science in Context*, **5** (1992): 309–28. See also R. M. Farr (ed), 'Special Issue: History of Social Psychology', *British Journal of Social Psychology*, **22** Part 4 (1983). On Mead, see especially G. A. Cook, *George Herbert Mead: The Making of a Social Pragmatist* (Urbana: University of

Illinois Press, 1993). See also: A. L. Blumenthal, introduction to W. Wundt, *The Language of Gestures, and Additional Essays by George Herbert Mead and Karl Bühler* (The Hague: Mouton, 1973); G. A. Cook, 'G. H. Mead's Social Behaviorism', JHBS, **13** (1977): 307–16, which emphasizes the link with functional explanation; P. Hamilton (ed), *George Herbert Mead: Critical Assessments. Volume I. Biography and Intellectual Context* (London: Routledge, 1992); D. J. Lewis and R. L. Smith, *American Sociology and Pragmatism: Mead, Chicago Sociology, and Symbolic Interaction* (Chicago: University of Chicago Press, 1980). M. J. Deegan and J. S. Burger, 'George Herbert Mead and Social Reform', JHBS, **14** (1978): 362–72, emphasize the commitment to science as social progress. For Pareto, see D. Beetham, 'Mosca, Pareto and Weber: A Historical Comparison', in W. J. Mommsen and J. Osterhammel (eds), *Max Weber and His Contemporaries* (London: Unwin Hyman, 1987), pp. 139–58, and L. A. Coser, *Masters of Sociological Thought: Ideas in Historical and Social Context* (New York: Harcourt Brace Jovanovich, 1971), pp. 386–426. G. W. Allport, *Personality: A Psychological Interpretation* (reprint, London: Constable, 1938), reviewed the meanings of 'personality' and 'trait'. On the Yale Institute: J. G. Morawski, 'Organizing Knowledge and Behavior at Yale's Institute of Human Relations', *Isis*, **77** (1986): 219–242. The Hawthorne experiments are reassessed in R. Gillespie, 'The Hawthorne Experiments and the Politics of Experimentation', in J. G. Morawski (ed), *The Rise of Experimentation in American Psychology* (New Haven: Yale University Press, 1988), pp. 114–37, and *Manufacturing Knowledge: A History of the Hawthorne Experiments* (Cambridge: Cambridge University Press, 1991); and for the wider politics of socio-psychological knowledge in industry see, L. Baritz, *The Servants of Power: A History of the Uses of Social Science in American Industry* (Middletown, CT: Wesleyan University Press, 1960). On Lewin's move to the US: M. G. Ash, 'Cultural Contexts and Scientific Change in Psychology: Kurt Lewin in Iowa', *American Psychologist*, **47** (1992): 198–207; M. van Elteren, 'Karl Korsch and Lewinian Social Psychology: Failure of a Project', HHS, **5** (1992): 33–62; A. Métraux, 'Kurt Lewin: Philosopher–Psychologist', *Science in Context*, **5** (1992): 373–84; and on the experience of refugee scholars in general, L. A. Coser, *Refugee Scholars in America: Their Impact and Their Experiences* (New Haven: Yale University Press, 1984). There is an 'in-house' history of the Research Center for Group Dynamics in A. J. Marrow, *The Practical Theorist: The Life and Work of Kurt Lewin* (New York: Basic Books, 1969), while the history of the restructuring of the self through training groups and other disciplines is discussed by N. Rose in the sources listed for Chapter 16. For anthropological responses to psychology and the culture and personality movement: E. Hatch, *Theories of Man and Culture* (New York: Columbia University Press, 1973); R. A. Jones, 'Freud and American Sociology, 1909–1949', JHBS, **10** (1974): 21–39; V. Yans-McLaughlin, 'Science, Democracy, and Ethics: Mobilizing Culture and Personality for World War II', in G. W. Stocking, Jr (ed), *Malinowski, Rivers, Benedict and Others: Essays on Culture and Personality, History of Anthropology, Volume 4* (Madison: University of Wisconsin Press, 1986), pp. 184–217. See also L. E. Hoffman, 'American Psychologists and Wartime Research on Germany, 1941–1945', *American Psychologist,* **47** (1992): 264–73. P. Buck, 'Adjusting to Military Life: The Social Sciences Go to War, 1941–1950', in M. R. Smith (ed), *Military Enterprise and Technological Change: Perspectives on the American Experience* (Cambridge, MA: MIT Press, 1985), pp. 203–52, discusses the effect of the war on research on adjustment. For an introductory study of social science responses to fascism in the interwar years, see S. P. Turner and D. Käsler (eds), *Sociology Responds to Fascism* (London and New York: Routledge, 1992). For the history of the Society for the Study of Social Issues, see B. Harris, R. Unger and R. Stagner (eds), '50 Years of Psychology and Social Issues', *Journal of Social Issues,* **42** no. 1 (1986), and especially L. J. Finison, 'The Psychological Insurgency: 1936–1945', pp. 21–33. On the ethics of Milgram's work, B. Harris, 'Key Words: A History of Debriefing in Social Psychology', in Morawski (ed) (1988), pp. 188–212. Dialectical psychology is reviewed in M. H. van Ijzendoorn and R. van der Veer, *Main Currents of Critical Psychology: Vygotskij, Holzkamp, Riegel* (New York: Irvington Publishers, 1984), the school of Holzkamp is presented in C. W. Tolman and W. Maiers (eds), *Critical Psychology: Contributions to an Historical Science of the Subject* (Cambridge: Cambridge University Press, 1991), and critical thought is deployed historically in A. R. Buss (ed), *Psychology in Social Context* (New York: Irvington Publishers, 1979). K. Danziger (1990) was a much discussed text. K. J. Gergen developed his social constructionist theory of social psychology in, 'Social Psychology as History', *Journal of Personality and Social Psychology*, **26** (1973): 309–20, *Towards Transformation in Social Knowledge* (New York: Springer, 1982), and 'The Social Constructionist Movement in Modern Psychology', *American Psychologist*, **40** (1985): 266–75.

There is a very interesting literature on Soviet psychology, though it is incomplete in its coverage and divergent in its interpretations. D. Joravsky, *Russian Psychology: A Critical History* (Oxford: Basil Blackwell, 1989), an impassioned dissection of Soviet claims to have established an objective science, has a mass of information and draws contrasts with literary understanding of human nature. A briefer history covering some areas is A. Kozulin, *Psychology in Utopia: Toward a Social History of Soviet Psychology* (Cambridge, MA: MIT Press, 1984). For the turn-of-the-century Russian background, see also: E. A. Budilova, 'On the History of Social Psychology in Russia', in L. H. Strickland (ed), *Directions in Soviet Social Psychology* (New York: Springer, 1984), pp. 11–28; A. Kozulin, 'Georgy Chelpanov and the Establishment of the Moscow Institute of Psychology', JHBS, **21** (1985): 23–32. For the debates about Marxism and science in the 1920s, see also: L. R. Graham, *Science, Philosophy, and Human Behavior in the Soviet Union* (New York: Columbia University Press, 1987); D. Joravsky, *Soviet Marxism and Natural Science 1917–1932* (London: Routledge & Kegan Paul, 1961), and 'The Construction of the Stalinist Psyche', in S. Fitzpatrick (ed), *Cultural Revolution in Russia, 1928–1931* (Bloomington: Indiana University Press, 1978), pp. 105–28; A. Vucinich, *Empire of Knowledge: The Academy of Sciences of the USSR (1917–1970)* (Berkeley: University of California Press, 1984). The standard Russian student history of psychology is M. G. Iaroshevskii [Yaroshevsky], *A History of Psychology* (Moscow: Progress Publishers, 1960). There are interesting attempts to interpret the Soviet dialectical view of psychology in R. A. Bauer, *The New Man in Soviet Psychology* (Cambridge, MA: Harvard University Press, 1952), and J. McLeish, *Soviet Psychology: History, Theory, Content* (London: Methuen, 1975). An idea of the effects of Stalinism on academic life, specifically on linguistics, is given in V. M. Alpatov, 'Marr, Marrism, and Stalinism', Recent Studies in the History of Russian Science, ed D. R. Wiener, *Russian Studies in History*, **34** (1995): 37–61. For Rubinshtein see T. R. Payne,

S. L. Rubinstejn and the Philosophical Foundations of Soviet Psychology (Dordrecht: D. Reidel, 1968). The Western literature on Vygotsky is very large but not primarily historical; but see A. Kozulin, *Vygotsky's Psychology: A Biography of Ideas* (New York: Harvester Wheatsheaf, 1990). R. van der Veer and J. Valsiner, *Understanding Vygotsky: A Quest for Synthesis* (Oxford: Blackwell, 1991), is a major study that combines history with detailed assessment of Vygotsky's work; see also A. Kozulin, 'The Concept of Activity in Soviet Psychology: Vygotsky, His Disciples and Critics', *American Psychologist,* **41** (1986): 264–74. An idea of Soviet psychology in the 1950s is given in B. Simon (ed), *Psychology in the Soviet Union* (London: Routledge & Kegan Paul, 1957).

CHAPTER 20

As the history of science merges with present science, so the distinction between work intended to contribute to science and to its history merges. Within psychology, there are attempts to survey the current US domain with a historical perspective, such as L. T. Benjamin, Jr (ed), 'Special Issue: The History of American Psychology', *American Psychologist,* **47** no. 2 (1992): 109–350, a collection with substantial historical content; E. Hearst (ed), *The First Century of Experimental Psychology* (Hillsdale, NJ: Lawrence Erlbaum Associates, 1979); E. R. Hilgard, *Psychology in America: An Historical Survey* (San Diego: Harcourt Brace Jovanovich, 1987); and S. Koch and D. E. Leary (eds), *A Century of Psychology as Science* (New York: McGraw-Hill, 1985). Presidential addresses are selected in E. R. Hilgard (ed), *American Psychology in Historical Perspective: Addresses of the Presidents of the American Psychological Association, 1892–1977* (Washington, DC: American Psychological Association, 1978). L. S. Hearnshaw, *The Shaping of Modern Psychology* (London: Routledge & Kegan Paul, 1987), attempts to use history to reassert the project of a unified, trans-historical scientific psychology, bravely opposing both over-specialization among psychologists and a relativist view of science among historians, but it is not well informed by historical writing since 1970. The activity and size of post-1945 US psychology is surveyed in A. R. Gilgen, *American Psychology since World War II: A Profile of the Discipline* (Westport, CT: Greenwood Press, 1982). The extent of psychology's involvement in social and political life is discussed in E. Herman, *The Romance of American Psychology: Political Culture in the Age of Experts* (Berkeley: University of California Press, 1995). For information on psychology as an occupation around the world: A. R. Gilgen and C. K. Gilgen, *International Handbook of Psychology* (New York: Greenwood Press, 1987); see also M. R. Rosenzweig, 'Trends in the Development and Status of Psychology: An International Perspective', *International Journal of Psychology*, **17** (1982): 117–40. There is a narrative history of the period in T. H. Leahey, *A History of Modern Psychology* (2nd edn, Engelwood Cliffs, NJ: Prentice-Hall, 1994). For material on the behavioural sciences: B. Berelson (ed), *The Behavioral Sciences Today* (New York: Basic Books, 1963), a survey of achievements, and 'Behavioral Sciences', in D. L. Sills (ed), *International Encyclopedia of the Social Sciences*, Vol. 2 (New York: Macmillan & Free Press, 1968), pp. 41–5; P. R. Senn, 'What is "Behavioral Science?" – Notes Toward a History', JHBS, **2** (1966): 105–22; D. L. Sills, 'Bernard Berelson: Behavioral Scientist', JHBS, **17** (1981): 305–11; J. F. Ward, 'Arthur F. Bentley and the Foundations of Behavioral Science', JHBS, **17** (1981): 222–31. On Lazarsfeld: A. Oberschall, 'Paul F. Lazarsfeld and the History of Empirical Social Research', JHBS, **14** (1978): 199–206.

Developments before and after 1945 in the discussion of relations between biology and society are described in C. N. Degler, *In Search of Human Nature: The Decline and Revival of Darwinism in American Social Thought* (New York: Oxford University Press, 1991), while the strength of interwar environmentalism is described in H. Cravens, *The Triumph of Evolution: American Scientists and the Heredity–Environment Controversy 1900–1941* (Philadelphia: University of Pennsylvania Press, 1978), chap. 2. For an overview of nature versus culture explanations, see P. Hirst and P. Woolley, *Social Relations and Human Attributes* (London: Tavistock, 1982). The attempt comprehensively to survey uses of the term 'culture' was made in A. L. Kroeber and C. Kluckhohn, *Culture: A Critical Review of Concepts and Definitions*, Papers of the Peabody Museum of American Archaeology and Ethnology, Harvard University, Vol. 47, no. 1 (Cambridge, MA: Peabody Museum, 1952). For Margaret Mead and her government service see, V. Yans-McLaughlin, 'Science, Democracy, and Ethics: Mobilizing Culture and Personality for World War II', in

G. W. Stocking, Jr (ed), *Malinowski, Benedict and Others: Essays on Culture and Personality, History of Anthropology, Volume 4* (Madison: University of Wisconsin Press, 1986), pp. 184–217. For the origins of ethology: J. R. Durant, 'Innate Character in Animals and Man: A Perspective on the Origins of Ethology', in C. Webster (ed), *Biology, Medicine and Society 1840–1940* (Cambridge: Cambridge University Press, 1981), pp. 157–92, and 'The Making of Ethology: The Association for the Study of Animal Behaviour, 1936–1986', *Animal Behaviour*, **34** (1986): 1601–16. D. A. Dewsbury, 'Comparative Psychology and Ethology', *American Psychologist*, **47** (1992): 208–15, argues that comparative psychology and ethology were not as separate as commonly supposed. Lorenz's papers during the early years of World War II are discussed in T. J. Kalikow, 'Konrad Lorenz's Ethological Theory, 1939–1943: "Explanations" of Human Thinking, Feeling and Behaviour', *Philosophy of the Social Sciences*, **6** (1976): 15–34, and 'Konrad Lorenz's Ethological Theory: Explanation and Ideology, 1938–1943,' *Journal of the History of Biology*, **16** (1983): 39–73, while his association with the German National Socialist Party is outlined in 'Konrad Lorenz's "Brown Past": A Reply to Alec Nisbett', JHBS, **14** (1978): 173–80, a response to the informal biography, A. Nisbett, *Konrad Lorenz* (London: J. M. Dent & Sons, 1976). For primates and language, R. A. Gardner and B. T. Gardner: 'Teaching Sign Language to a Chimpanzee', *Science*, **169** (1969): 664–72, and 'Early Signs of Language in Child and Chimpanzee', *Science*, **187** (1975): 752–3; and the readable overview in E. Linden, *Apes, Men, and Language* (reprint, Harmondsworth: Penguin Books, 1981). Aspects of the rhetoric of this debate are discussed in L. J. Prell, 'The Rhetorical Construction of Scientific Ethos', in H. W. Simons (ed), *Rhetoric in the Human Sciences* (London: Sage, 1989), pp. 48–68. D. J. Haraway's major study is *Primate Visions: Gender, Race, and Nature in the World of Modern Science* (New York: Routledge, 1990); see also *Simians, Cyborgs, and Women: The Reinvention of Nature* (New York: Routledge, and London: Free Association Books, 1991). For historical reflections on gender and psychology: M. Lewin (ed), *In the Shadow of the Past: Psychology Portrays the Sexes: A Social and Intellectual History* (New York: Columbia University Press, 1984). On the biology of aggression: J. R. Durant, 'The Beast in Man: An Historical Perspective on the Biology of Human

Aggression', in P. F. Brain and D. Benton (eds), *The Biology of Aggression* (Alphen an den Rijn: Sijthoff and Noordhoff, 1981), pp. 17–46, and 'The Science of Sentiment: The Problem of the Cerebral Localization of Emotion', in P. P. G. Bateson and P. H. Klopfer (eds), *Perspectives in Ethology*, Vol. 6 (New York: Plenum Press, 1985), pp. 1–31. E. O. Wilson, *Sociobiology: The New Synthesis* (Cambridge, MA: Belknap Press of Harvard University Press, 1975), provoked the response by a social anthropologist, M. Sahlins, *The Use and Abuse of Biology: An Anthropological Critique of Sociobiology* (reprint, London: Tavistock, 1977), who indicted sociobiology for 'scientific totemism', and a detailed refutation of both its biology and its philosophy in P. Kitcher, *Vaulting Ambition: Sociobiology and the Quest for Human Nature* (Cambridge, MA: MIT Press, 1985). A more accessible critique is S. Rose, L. J. Kamin and R. C. Lewontin, *Not in Our Genes: Biology, Ideology and Human Nature* (New York: Pantheon Books, and Harmondsworth: Penguin Books, 1984). The question of the existence of universals in human nature and of the role of culture is examined in P. Heelas and A. Lock (eds), *Indigenous Psychologies: The Anthropology of the Self* (London: Academic Press, 1981). For a balanced discussion of debates about human nature and human origins, see A. Kuper, *The Chosen Primate: Human Nature and Cultural Diversity* (Cambridge, MA: Harvard University Press, 1994).

There are no systematic histories of modern neuroscience, but see: S. Finger, *Origins of Neuroscience: A History of Exploration Into Brain Function* (New York: Oxford University Press, 1994), which catalogues earlier developments from the viewpoint of a neuroscientist; J. D. Spillane, *The Doctrine of the Nerves: Chapters in the History of Neurology* (Oxford: Oxford University Press, 1981); F. G. Worden, J. P. Swazey and G. Adelman (eds), *The Neurosciences: Paths of Discovery* (Cambridge, MA: MIT Press, 1975); F. Samson and G. Adelman (eds), *The Neurosciences: Paths of Discovery II* (Boston: Birkhauser, 1992); also Hilgard (1987), chap. 12, and K. R. Pribram and D. N. Robinson, 'Biological Contributions to the Development of Psychology', in C. E. Buxton (ed), *Points of View in the Modern History of Psychology* (Orlando: Academic Press, 1985), pp. 345–81. A presentation with historical references from a modern perspective, for a general audience, is C. Blakemore, *Mechanics of the Mind: BBC Reith Lectures 1976* (Cambridge: Cambridge University Press, 1977). The importance of Lashley is detailed in J. Orbach (ed), *Neuropsychology after Lashley: Fifty Years since the Publication of Brain Mechanisms and Intelligence* (Hillsdale, NJ: Lawrence Erlbaum Associates, 1982). There are introductions to the relevant conceptual issues in O. Flanagan, *The Science of Mind* (2nd edn, Cambridge, MA: MIT Press, 1991), and E. R. Valentine, *Conceptual Issues in Psychology* (London: George Allen & Unwin, 1982). For the debate in philosophy about the mind and dualism see the introduction in S. Priest, *Theories of the Mind* (London: Penguin Books, 1991), and, for more detail: D. Blitz, *Emergent Evolution: Qualitative Novelty and the Levels of Reality* (Dordrecht: Kluwer, 1992); H. Feigl, *The 'Mental' and the 'Physical': The Essay and a Postscript* (Minneapolis: University of Minnesota Press, 1967), with extensive bibliography; G. Ryle, *The Concept of Mind* (reprint, Harmondsworth: Penguin Books, 1963). The identification of mind states with brain states is developed philosophically in: J. J. C. Smart, *Philosophy and Scientific Realism* (London: Routledge & Kegan Paul, and New York: Humanities Press, 1963); P. Smith Churchland, *Neurophilosophy: Toward a Unified Science of the Mind-Brain* (Cambridge, MA: MIT Press, 1986); P. M. Churchland, *Matter and Consciousness: A Contemporary Introduction to the Philosophy of Mind* (2nd edn, Cambridge, MA: MIT Press, 1988). F. Crick, *The Astonishing Hypothesis: The Scientific Search for the Soul* (London: Simon & Schuster, 1994), is a scientist's declaration of faith in the power of experimental neuroscience. Wittgenstein's arguments are complex and his comments scattered; but see *Philosophical Investigations* (2nd edn reprint, Oxford: Basil Blackwell, 1963), with the line-by-line commentary in G. Hallett, *A Companion to Wittgenstein's "Philosophical Investigations"* (Ithaca: Cornell University Press, 1977). The standard biography is R. Monk, *Ludwig Wittgenstein: The Duty of Genius* (reprint, London: Vintage, 1991). Many scientific issues are briefly sketched in R. L. Gregory (ed), *The Oxford Companion to the Mind* (Oxford: Oxford University Press, 1987).

The history of artificial intelligence research and of cognitive science also largely remains to be written; the comment that exists is by participants in an area where much is debated. But see: H. Gardner, *The Mind's New Science: A History of the Cognitive Revolution* (New York: Basic Books, 1985); F. S. Kessel and W.

Bevan, 'Notes Toward a History of Cognitive Psychology', in Buxton (ed) (1985), pp. 259–94. The claim that there was a revolution, with interviews with some leading participants, is made in B. J. Baars, *The Cognitive Revolution in Psychology* (New York: Guilford Press, 1986). For aspects of the background to computing: S. J. Heims, 'Encounter of Behavioral Sciences with New Machine–Organism Analogies in the 1940's', JHBS, **11** (1975): 368–73, and *The Cybernetics Group* (Cambridge, MA: MIT Press, 1991), on the US Cybernetics Group, 1946–53; A. Hodges, *Alan Turing: The Enigma* (London: Burnett Books, 1983), a biography of the key British scientist; V. Pratt, *Thinking Machines: The Evolution of Artificial Intelligence* (Oxford: Basil Blackwell, 1987), a history from the seventeenth century to AI research. The role of statistical methodologies in directing research on cognition is stressed in G. Gigerenzer, 'From Tools to Theories: A Heuristic of Discovery in Cognitive Psychology', *Psychological Review*, **98** (1991): 254–67, 'Discovery in Cognitive Psychology: New Tools Inspire New Theories', *Science in Context*, **5** (1992): 329–50, and G. Gigerenzer and D. J. Murray, *Cognition as Intuitive Statistics* (Hillsdale, NJ: Lawrence Erlbaum Associates, 1987), chap. 1. Piaget's early life is reassessed in F. Vidal, 'Jean Piaget and the Liberal Protestant Tradition', in M. G. Ash and W. R. Woodward (eds), *Psychology in Twentieth-Century Thought and Society* (Cambridge: Cambridge University Press, 1987), pp. 271–94, and *Piaget before Piaget* (Cambridge, MA: Harvard University Press, 1994). An introduction to Piaget's theory is provided in M. A. Boden, *Piaget* (London: Fontana, 1979). For the linguistic context of Chomsky's work, see E. F. K. Koerner and R. E. Asher (eds), *Concise History of the Language Sciences: From the Sumerians to the Cognitivists* (Oxford and New York: Pergamon, 1995), sect. 10, and G. Sampson, *Schools of Linguistics: Competition and Evolution* (London: Hutchinson, 1980). An up-beat view of the development of artificial intelligence is D. Crevier, *AI: The Tumultuous History of the Search for Artificial Intelligence* (New York: Basic Books, 1993). The 'classic' riposte to the claims of the early AI researchers that computers will soon do what humans do is H. L. Dreyfus, *What Computers Can't Do: The Limits of Artificial Intelligence* (2nd edn, New York: Harper & Row, 1979). The capacity of computing models and biology to explain consciousness is argued in D. C. Dennett, *Consciousness Explained* (Boston:

Little, Brown, 1991). The assertion that they cannot is made by J. R. Searle, *The Rediscovery of the Mind* (Cambridge, MA: MIT Press, 1982), an argument for the centrality of consciousness in the human sciences, and *Intentionality: An Essay in the Philosophy of Mind* (Cambridge: Cambridge University Press, 1983), while Searle explains his views for a wider audience in *Minds, Brains and Science: The 1984 Reith Lectures* (London: British Broadcasting Corporation, 1984). For overviews of the philosophical issues and different points of view: W. Bechtel, *Philosophy of Mind: An Overview of Cognitive Science* (Hillsdale, NJ: Lawrence Erlbaum Associates, 1988); C. Blakemore and S. Greenfield (eds), *Mind-waves: Thoughts on Intelligence, Identity and Consciousness* (Oxford: Basil Blackwell, 1987); S. Guttenplan (ed), *A Companion to the Philosophy of Mind* (Oxford: Blackwell, 1994).

The history of humanistic psychology is integral to the wider history of 'psychological society' (see Chapter 16); for a history see R. J. DeCarvalho, *The Founders of Humanistic Psychology* (New York: Praeger, 1991). The history of clinical psychology in the US is told from a vantage point within the occupation in J. M. Reisman, *A History of Clinical Psychology* (2nd edn, New York: Irvington Publishers, 1976); and information on the organization of the field in the US is given in D. K. Routh, *Clinical Psychology since 1917: Science, Practice, and Organization* (New York: Plenum Press, 1994). The institutional settings of professional psychotherapy in the US are described in D. K. Freedheim, *History of Psychotherapy: A Century of Change* (Washington, DC: American Psychological Association, 1992). The background in the interwar mental hygiene movement for the successful development of clinical psychology as an occupation is discussed in J. C. Burnham, 'The Struggle between Physicians and Paramedical Personnel in American Psychiatry, 1917–41', *Journal of the History of Medicine and Allied Sciences,* **29** (1974): 93–106. On the interwar background to holism in psychology: A. Harrington, 'Interwar "German" psychobiology: Between Nationalism and the Irrational', *Science in Context,* **4** (1991): 429–47. The existentialist approach in psychological analysis has not been incorporated into existing histories. For Nietzsche, see W. Kaufmann, 'Nietzsche as the First Great (Depth) Psychologist', in Koch and Leary (eds) (1985), pp. 911–20,. The huge body of literature on Nietzsche is not primarily historical; much is concerned with

Nietzsche's aestheticism – influential studies are A. Megill, *Prophets of Extremity: Nietzsche, Heidegger, Foucault, Derrida* (Berkeley: University of California Press, 1985), and A. Nehamas, *Nietzsche: Life as Literature* (Cambridge, MA: Harvard University Press, 1986). There are historical studies of the reception of Nietzsche's work in S. F. Aschheim, *The Nietzsche Legacy in Germany 1890–1990* (Berkeley: University of California Press, 1992), and R. H. Thomas, *Nietzsche in German Politics and Society 1890–1918* (Manchester: Manchester University Press, 1983). For Jaspers's German context: H. Stierlin, 'Karl Jaspers' Psychiatry in the Light of His Basic Philosophical Position', JHBS, **10** (1974): 213–26. Saussure's work lies at the foundations of twentieth-century linguistics; for introductions: J. Culler, *Saussure* (London: Fontana/Collins, 1976); R. Harris and T. J. Taylor, *Landmarks in Linguistic Thought: The Western Tradition from Socrates to Saussure* (London: Routledge, 1989). The intellectual arguments known as structuralism are surveyed in E. Kurzweil, *The Age of Structuralism: Lévi-Strauss to Foucault* (New York: Columbia University Press, 1980). The major history of Lacan and French psychoanalysis is by E. Roudinesco, but only the first of her volumes is in English translation: *Jacques Lacan & Co.: A History of Psychoanalysis in France, 1925–1985* (Chicago: University of Chicago Press, and London: Free Association Books, 1990). In addition: M. Bowie, *Lacan* (London: Fontana, 1991); M. Evans, *Fits and Starts: A Genealogy of Hysteria in Modern France* (Ithaca: Cornell University Press, 1991), which also introduces some ideas of the French feminists; M. Sarup, *Jacques Lacan* (New York: Harvester Wheatsheaf, 1992); S. Turkle, *Psychoanalytic Politics: Jacques Lacan and Freud's French Revolution* (2nd edn, London: Free Association Books, and New York: Guilford Press, 1992). Foucault's work is referenced at the beginning of Chapters 15 and 16. In the literature on his thought, G. Gutting, *Michel Foucault's Archaeology of Scientific Reason* (Cambridge: Cambridge University Press, 1989) covers the early period, while H. L. Dreyfus and P. Rabinow, *Michel Foucault: Beyond Structuralism and Hermeneutics* (Chicago: University of Chicago Press, 1982) is valuable for the concept of 'genealogy' and includes M. Foucault, 'Afterword: The Subject and Power', pp. 208–26. Biographical details are in D. Eribon, *Michel Foucault* (Cambridge, MA: Harvard University Press, 1991, and London: Faber & Faber, 1992); they are turned into a drama for our times in J. Miller, *The Passion of Michel Foucault* (New York: Simon & Schuster, and London: HarperCollins, 1993); and made the subject of a more conventional biography in D. Macey, *The Lives of Michel Foucault* (reprint, London: Vintage, 1994). The work of the German critical 'anthropological tradition' in philosophy is discussed in A. Honneth and H. Joas, *Social Action and Human Nature* (Cambridge: Cambridge University Press, 1988). The 'linguistic turn' in historiography can be followed, with references, in W. J. Bowsma, 'From History of Ideas to History of Meaning', *Journal of Interdisciplinary History*, **12** (1981): 279–91; D. R. Kelley, 'Horizons of Intellectual History: Retrospect, Circumspect, Prospect', JHI, **48** (1987): 143–69; J. E. Toews, 'Intellectual History after the Linguistic Turn: The Autonomy of Meaning and the Irreducibility of Experience', *American Historical Review*, **92** (1987): 879–907; H. White, *The Content of the Form: Narrative Discourse and Historical Representation* (Baltimore: Johns Hopkins University Press, 1987). One major dimension of 'the linguistic turn' is a new seriousness about rhetoric: J. S. Nelson, A. Megill and D. N. McCloskey (eds), *The Rhetoric of the Human Sciences: Language and Argument in Scholarship and Public Affairs* (Madison: University of Wisconsin Press, 1987), and R. H. Roberts and J. M. M. Good (eds), *The Recovery of Rhetoric: Persuasive Discourse and Disciplinarity in the Human Sciences* (London: Bristol Classical Press, Duckworth, 1993). A second major dimension is the reinterpretation of metaphor as intrinsic to knowledge; this results in a rich reinterpretation of many aspects of twentieth-century psychology in A. J. Soyland, *Psychology as Metaphor* (London: Sage, 1994); see also D. E. Leary (ed), *Metaphors in the History of Psychology* (Cambridge: Cambridge University Press, 1990). In the reconsideration of 'the self' in the late twentieth century, R. Sennett, *The Fall of Public Man* (reprint, London: Faber & Faber, 1986), explores the political and ethical consequences of the private self's growth at the expense of the public sphere. There is a general discussion in A. Giddens, *Modernity and Self-Identity: Self and Society in the Late Modern Age* (Oxford: Polity Press, 1991); while I. Hacking, *Rewriting the Soul: Multiple Personality and the Science of Memory* (Princeton: Princeton University Press, 1995), analyses memory and the maze of multiple personality disorder as central to the identity of modern psychology as the science of the soul.

The recurrent debate about explanation and understanding in the social sciences is widely discussed, and the background is referenced in Chapters 14 and 15. There is translated primary material in W. Dilthey, *Descriptive Psychology and Historical Understanding* (The Hague: Martinus Nijhoff, 1977), and *Introduction to the Human Sciences: An Attempt to Lay a Foundation for the Study of Society and History* (Brighton: Harvester Press, 1989); H. Rickert, *Science and History: A Critique of Positivist Epistemology*, ed A. Goddard (Princeton: D. Van Nostrand Company, 1962), and *The Limits of Concept Formation in Natural Science: A Logical Introduction to the Historical Sciences (Abridged Edition)*, ed G. Oakes (Cambridge: Cambridge University Press, 1986); W. Windelband, 'History and Natural Science', *History and Theory*, **19** (1980): 165–85. For commentary: R. J. Bernstein, *The Restructuring of Social and Political Theory* (Oxford: Basil Blackwell, 1976); S. Herva, 'The Genesis of Max Weber's *Verstehende Soziologie*', *Acta Sociologica*, **31** (1988): 143–56; G. Iggers, *The German Conception of History: The National Tradition of Historical Thought From Herder to the Present* (Middletown, CT: Wesleyan University Press, 1968), chap. 6; G. Oakes, *Weber and Rickert: Concept Formation in the Cultural Sciences* (Cambridge, MA: MIT Press, 1988); W. Outhwaite, *Understanding Social Life: The Method Called 'Verstehen'* (London: George Allen & Unwin, 1975); T. E. Willey, *Back to Kant: The Revival of Kantianism in German Social and Historical Thought, 1860–1914* (Detroit: Wayne State University Press, 1978); P. Ricoeur, 'What Is a Text? Explanation and Understanding', in *Hermeneutics and the Human Sciences: Essays on Language, Action and Interpretation*, ed J. P. Thompson (Cambridge: Cambridge University Press, 1981), pp. 145–66. C. Geertz's essays are collected in *The Interpretation of Cultures: Selected Essays* (New York: Basic Books, 1973).

INDEX

The Norton History of Astronomy and Cosmology

John North

Astronomy is not only a field of scientific research capable of the most startling discoveries; it is the oldest of the exact sciences. It is fitting then that in this lucid, elegant book, Professor North devotes particular attention to both the earliest and the most recent developments.

Stressing the indispensibility of an understanding of the heavens for the elementary calendrical calculations vital in all societies, Professor North demonstrates how surveying the skies helped generate the geometrical and mathematical principles crucial to early science in the Middle East and Greece, which in turn continued to underpin advances in astronomy right through the revolutions in thinking achieved by Copernicus, Kepler and Newton.

Astronomy, North shows, has a history marked by continuity. It offers a powerful illustration of concentrated, progressive scientific endeavour and yet it has also been a complicated and many-sided enterprise, integrating metaphysical, religious and cosmological speculations with the down-to-earth practical skills needed, for example, for navigation and timekeeping. In assessing the social position of astronomers, Professor North deftly explores the tensions between these different roles.

0-00-686177-6